Essays in Physics

Essays in Physics

Thirty-two thoughtful essays on topics in undergraduate-level physics

GEOFFREY BROOKER

Department of Physics
University of Oxford

OXFORD
UNIVERSITY PRESS

Great Clarendon Street, Oxford, OX2 6DP,
United Kingdom

Oxford University Press is a department of the University of Oxford.
It furthers the University's objective of excellence in research, scholarship,
and education by publishing worldwide. Oxford is a registered trade mark of
Oxford University Press in the UK and in certain other countries

First Edition published in 2021

Impression: 1

Published in the United States of America by Oxford University Press
198 Madison Avenue, New York, NY 10016, United States of America

British Library Cataloguing in Publication Data
Data available

Library of Congress Control Number: 2021938596

ISBN 978–0–19–885724–2 (hbk.)
ISBN 978–0–19–885725–9 (pbk.)

DOI: 10.1093/oso/9780198857242.001.0001

Printed and bound in the UK by
TJ Books Limited

To: Amie, Lois, Oona, Phoebe, Alfie, Blaise, Maia, Zita and Cara.

Preface

Physics is not just a set of cut-and-dried facts. Even "standard" topics can reveal unexpected excitements or puzzles. This book contains discussions of several such topics, discussions that "sit alongside" those of standard texts. The level varies with the topic but is almost always within the four years of a Master's degree.

In each chapter, the reader should find something that is "different". It may be a more-than-usually-careful exegesis; or a cross-link between apparently unrelated areas; or even a disproof of statements commonly made but here shown to be false. Each such example illustrates the kind of intellectual agility and acuity that the student reader should be acquiring.

In illustration of the above: Chapter 1 (electromagnetic energy) polishes the student's critical faculty by examining a standard piece of bookwork in depth.

Chapter 5 (transmission lines) joins together problems from disparate areas that have a common mathematical structure. Chapters 8 (Lorentz transformation), 12 (perturbation theory), and 19 (Einstein A and B) treat standard material in a "sideways" way. Chapters 20, 21 (Boltzmann distribution, free energy) deal with topics that tend to "fall through the cracks". Chapters 15–17 (atomic physics) and 22 (magnet in field) are frankly corrective of discussions commonly given that I believe to be seriously wrong.

Inevitably, the material presented exhibits the topics that I happen to have encountered during my own teaching. Absent is any nuclear or particle physics, astrophysics, geophysics or biophysics. My exposition is unashamedly opinionated. In that circumstance, it has seemed most honest to say "I" when expressing a forthright view, rather than hiding behind the more impersonal "we".

Many of the chapters originated in worksheets that I circulated to students in advance of tutorials. These worksheets were privately circulated, and therefore could spell out which textbooks should (in my view) be used or avoided, and precisely where any given book should be treated with caution. Such forthrightness is not appropriate here. The reader will therefore encounter some passages where it is clear that a misleading discussion is being contradicted, yet with little indication as to the identity of the targets (often several). He must take my assurance that at least one target exists and is in common use by students. If he is minded to search for the source of the error and fails to find it, then he should count himself fortunate.

"Tutorials" may require explanation. At Oxford a tutor sets weekly assignments and then, in "tutorials", discusses the student's work, alone or in a small group, commonly a group of two. By contrast, evaluation of the student's progress (grades into a computer record) is done in yearly examinations that are unconnected with tutorials. The tutorial is the student's opportunity to sort out difficulties, ask searching questions of the tutor, and for the tutor to test the student's understanding by asking his own questions. Given that help and evaluation are separated, the student's best strategy in tutorials is to confess to ignorance, puzzlement or confusion, when present, not to conceal them. The essays in this book are, in many cases, this tutor's attempts at clarification or at anticipating difficulties previously found to be commonplace.

It goes without more saying that I have a great debt to students whom I have taught, though their individual contributions to my thinking cannot now be identified. Likewise, several colleagues have seen chapters in early versions and have contributed helpful and sage comments. I mention in particular Dr. C. Sukumar and Professors C.J. Foot and J.F. Gregg.

Contents

8 The Lorentz transformation   **80**

V Optics

9 Diffraction integrals and the Kirchhoff approximation   **90**

10 Diffraction at a slit   **107**

XII Electronics

29 Negative feedback   364

30 Stability of negative feedback   378

XIII Relatively advanced material

I Electromagnetism

Electromagnetic energy

Intended readership: Second year, or towards the end of an electromagnetism course that covers Maxwell's equations.

1.1   Introduction: energy density and flow

- Energy U per unit volume $= \frac{1}{2}\boldsymbol{E}\cdot\boldsymbol{D} + \frac{1}{2}\boldsymbol{H}\cdot\boldsymbol{B}$.
- Energy flow per unit area $= \boldsymbol{E}\times\boldsymbol{H}$, the Poynting vector.

This chapter concerns itself with the above two expressions. Are they trustworthy? Do they apply under all conditions, or are there restrictions? Can we give watertight derivations of them? The Poynting vector looks odd: shouldn't it be $\boldsymbol{E}\times\boldsymbol{B}/\mu_0$?

To answer these questions immediately:

- The first expression is special to fields within a medium that is linear and non-dispersive. For a general medium (non-linear or not known to be linear) we need to write instead an expression for the derivative of the energy density: $\delta U = \boldsymbol{E}\cdot\delta\boldsymbol{D} + \boldsymbol{H}\cdot\delta\boldsymbol{B}$.
- The expression given for the Poynting vector is general. The presence of $\boldsymbol{H}$ (rather than $\boldsymbol{B}$) accompanying $\boldsymbol{E}$ (rather than $\boldsymbol{D}$) is curiously asymmetric,[1] but it can be given a simple physical interpretation.
- Watertight derivations don't exist. (!) Exactly what we can prove, what loopholes remain, and how we deal with those loopholes, is a good part of what motivates this chapter.

[1] By "asymmetric" we mean that $\boldsymbol{H}$ contains within it the magnetization density $\boldsymbol{M}$, while $\boldsymbol{E}$ does not contain the electrical analogue $\boldsymbol{P}$. For the formal understanding of $\boldsymbol{M}$ and $\boldsymbol{P}$ see Chapter 2.

1.2   Electrostatic energy

Let's start by giving a standard textbook derivation of $\frac{1}{2}\boldsymbol{E}\cdot\boldsymbol{D}$, so that we know what's under discussion.[2]

Suppose we have a system of point charges in vacuum, the ith charge having charge q_i and being located at a place where the electrostatic potential it experiences (owing to all the *other* charges) is φ_i. We wish to find the energy of this system, that is, the mechanical work that must be expended in assembling it, starting from separated charges at infinity.

Bring up all the charges from infinity a little at a time, so that when the process is partly done we have charge kq_i in potential $k\varphi_i$, and all charges have the same value of k. Now bring up additional fragments of charge so that k is increased by δk. The work we have to do is

[2] The reasoning I give here is not particularly good, as we shall see. But this route is "conventional", and it's only by seeing what is wrong with it that we can understand the need for something better.

Essays in Physics: Thirty-two thoughtful essays on topics in undergraduate-level physics. Geoffrey Brooker,
Oxford University Press. © Geoffrey Brooker 2021. DOI: 10.1093/oso/9780198857242.003.0001

Work is also done in assembling the fragments of charge q_i, as fragment $q_i \, \delta k$ is brought up to kq_i (with the same i). This work is called the **self-energy** of charge q_i. It is energy associated with the *existence* of the point charge, unconnected with the environment that we put that charge into. In other words, however charge q_i is made, this energy must be supplied in the making. However, we are to think that our starting point is with all the point charges intact and with their self-energy already supplied; at this stage those charges are infinitely separated from each other. Then charge q_i must be (notionally) disassembled into fragments before reassembly into our system. The self-energy is removed during disassembly, and restored during reassembly, so it cancels out in the entire process.

$\sum_i (k\varphi_i) \, \delta(kq_i) = \sum_i \varphi_i \, q_i \, k \, \delta k$. Totalling all such work means integrating over k from 0 to 1, and we end up with[3]

$$U = \sum_i \tfrac{1}{2} \varphi_i \, q_i. \tag{1.1}$$

Now think of the system from a macroscopic point of view, where we do not enquire about the exact disposition of charges at an atomic scale of distances, but simply treat the charges as having charge density $\rho(\boldsymbol{r})$. A charge element is $\rho(\boldsymbol{r}) \, \delta V$, where δV is an element of volume at $\boldsymbol{r}$. The potential is $\varphi(\boldsymbol{r})$. Then

$$\frac{1}{2} \sum_i \varphi_i \, q_i = \frac{1}{2} \int \varphi(\boldsymbol{r}) \, \rho(\boldsymbol{r}) \, \mathrm{d}V = \frac{1}{2} \int \varphi \, \mathrm{div} \, \boldsymbol{D} \, \mathrm{d}V$$

$$= \frac{1}{2} \int \mathrm{div}(\varphi \boldsymbol{D}) \, \mathrm{d}V - \frac{1}{2} \int \nabla \varphi \cdot \boldsymbol{D} \, \mathrm{d}V$$

$$= \text{integral over surface at infinity} + \int \tfrac{1}{2} \boldsymbol{E} \cdot \boldsymbol{D} \, \mathrm{d}V. \tag{1.2}$$

The integral over a surface at infinity vanishes (why?) and we are left with an integral over volume. The integrand can be understood as giving us the energy per unit volume, so we have the expected expression for the energy density.

The magnetostatic energy $\tfrac{1}{2} \boldsymbol{H} \cdot \boldsymbol{B}$ can be obtained using methods that have a family resemblance to those given here. We need not spell out the similar reasoning that applies to it.

1.3   Electrostatic energy: explanation

The use of delta functions means that we are choosing at this stage to use the "microscopic" viewpoint of Chapter 2. Macroscopic smoothing gets built in at the next step.

In classical electromagnetism, a point charge is a problematic concept, because the self-energy in the field surrounding a point charge is infinite—show this. It's possible to escape this infinity—in classical mechanics—by treating the electron as a sphere and choosing that sphere's radius (of order 10^{-15} m) so that the surrounding $\int \tfrac{1}{2} \epsilon_0 E^2 \, \mathrm{d}V$ wholly accounts for the rest energy mc^2. But this idea is contradicted by experiment and in any case can't survive into quantum mechanics. There is a good discussion in Feynman (1964), vol. II, pp 28-3 to 28-12, as to why the electron's mass cannot be wholly electromagnetic.

The infinity of sidenote 4 need not worry us, as it disappears in the next step.

Even though the calculation given in § 1.2 is short, it contains a surprising number of non-trivial steps. In the present section I draw attention to what was going on.

We start with the first step in (1.2): $\sum_i \tfrac{1}{2} \varphi_i \, q_i = \int \tfrac{1}{2} \varphi(\boldsymbol{r}) \rho(\boldsymbol{r}) \, \mathrm{d}V$. In the sum, φ_i is the potential at charge q_i caused by all the *other* charges, while in the integral $\varphi(\boldsymbol{r})$ is the potential due to *all* charges. These expressions are therefore not the same. There is a good discussion of this in Jackson (1999), § 1.11. Suppose we think of $\rho(\boldsymbol{r})$ as a sum over δ-functions, one at the location of each point charge.[4] Then the integral can be separated into two different kinds of term: First, there are terms that represent the interaction energy of each charge with the potential from all the others; these interaction-energy terms reproduce the sum we started from. Second, there are "self-energy" terms as defined in sidenote 3. If our aim is to make (1.2) reproduce (1.1) then we ought to adjust the integral so as to exclude self-energies. But, on second thoughts, perhaps it makes better sense that self-energy should be included: if charges move, surely they carry their self-energy with them.

We'll think then of the self-energy terms as included, correctly, in the second term of (1.2), even though they are omitted from the $\sum \tfrac{1}{2} \varphi_i \, q_i$ that we started out with.[5]

We move on to the second step in (1.2). A change of viewpoint is signalled in the introduction of $\boldsymbol{D}$: we have now switched from microscopic field quantities to macroscopic, as forecast in sidenote 4.

Macroscopic fields (meaning fields that are "filtered" to vary only over macroscopic but small distances) are obtained from microscopic fields by a process of averaging (see Chapter 2). The outcome of such averaging, in the context of electric charges and dipole moments, is[6]

$$\rho_{\text{macro}} = \rho_{\text{accessible}} - \text{div}\,\boldsymbol{P}, \qquad \boldsymbol{D} = \epsilon_0 \boldsymbol{E}_{\text{macro}} + \boldsymbol{P}, \qquad \text{div}\,\boldsymbol{D} = \rho_{\text{accessible}}.$$

Once we have averaged, the averaged ρ has been smeared out so that charges are no longer point-like. Correspondingly, the $\boldsymbol{E}_{\text{macro}}$-field does not go to infinity. The self-energy part of the energy density $\frac{1}{2}\epsilon_0 \boldsymbol{E}_{\text{macro}}^2$ is finite.[7]

1.4 Electrostatic energy: critique I

We now take a harder look at the reasoning within the working of (1.2). There are several points that we can feel uncomfortable with.

First, there is a difficulty with the integral over $\varphi\rho = \varphi\,\text{div}\,\boldsymbol{D}$, by which we mean $\langle\varphi\rangle\langle\rho\rangle = \varphi_{\text{macro}}\,\rho_{\text{accessible}}$. We have one space average in the definition of φ and another in that of $\rho = \text{div}\,\boldsymbol{D}$. The integral adds a product of averages, when we should have written the average of the product. There is no obvious reason why this step should be acceptable—in any discussion of statistics we'd at once recognize it an elementary error.[8]

There is a second reason why we may be uncomfortable with our expression for electrostatic energy. We started by thinking about point charges in vacuum. These charges must have been held in place (once brought up to their final places) by non-electrical forces, otherwise they would move. Those forces did not affect the energy because no work was done against them by charges that couldn't move. But when you apply a field to a real material, there are movements of electrons (atoms acquire electric dipole moments) and sometimes of ion cores as well. We've built in such possibilities by introducing $\boldsymbol{P}$, or equivalently $\boldsymbol{D} = \epsilon_0\boldsymbol{E} + \boldsymbol{P}$. Suppose we model this situation by thinking of charges as held in place by springs. Then work is done against those springs. Has that elastic energy been included in our accountancy?[9] Incidentally, an explicit "spring" model is discussed in § 22.3.2, though in answer to a different question.

This is not the end of our difficulties. A real material does not necessarily have a linear relation between $\boldsymbol{D}$ and $\boldsymbol{E}$. A whole subject of non-linear optics exists to deal with such non-linearities induced by laser radiation. To cover all bases, and incidentally to keep a clear head, it's a good idea to remove the assumption of a linear relation between $\boldsymbol{D}$ and $\boldsymbol{E}$. This is really another way of talking about the spring mentioned in the last paragraph: the spring may have a non-linear stress–strain relation.

[6] Notice that $\rho_{\text{accessible}}$, $\boldsymbol{P}$ and $\boldsymbol{D}$ are *defined* as macroscopic-average quantities, so they need no subscript macro. Moreover, because these quantities are inherently macroscopic, their presence in any algebraic expression is a signal that a macroscopic viewpoint is being used. This is why we began this part of the discussion by saying that $\boldsymbol{D}$ "signalled" a macroscopic averaging.

[7] This outcome is not as comforting as it may sound. The average self-energy contains $\langle \boldsymbol{E}_{\text{micro}}^2 \rangle$, which is quite different from $\langle \boldsymbol{E}_{\text{micro}} \rangle^2 = \boldsymbol{E}_{\text{macro}}^2$. We have therefore shifted the origin of energy yet again, because the spikier parts of the field have been smoothed away in our averaging. Physicists agree not to worry about this. The clue comes from sidenote 4. Self-energy can be thought of as a contribution to each charged particle's mc^2, added to other unknown contributions. Changing the electromagnetic part of the self-energy is accommodated by notionally making a compensating change in the rest. This piece of intellectual gymnastics is called "renormalization".

[8] We can't sweep this difficulty under the carpet by calling it "renormalization", because the difference between the averages won't be constant if the charges move.

[9] I think it has, but I can't say it's obvious.

[10] Our difficulties here are to a good extent self-inflicted. I chose to start from expression $\sum \varphi_i \, \delta q_i$, because I was deliberately following the majority of textbooks. A better route is to consider an input of work to a system that is described macroscopically from the beginning. Such a route is taken by Landau, Lifshitz and Pitaevskiĭ, (1993), § 10. They look at only a *change* of energy, so all question of self-energy is avoided, which alone gives a great simplification.

Let us make the bold assumption that we *can* use a product of averages when proceeding from microscopic to macroscopic quantities. But we remove the assumption of linearity. Then the reasoning of (1.2) can be re-cast in the following way:

$$\delta U = \sum \varphi_i \, \delta q_i = \int \varphi(\boldsymbol{r}) \, \delta\rho(\boldsymbol{r}) \, \mathrm{d}V = \int \boldsymbol{E} \cdot \delta \boldsymbol{D} \, \mathrm{d}V. \qquad (1.3)$$

Although it has been built (here) on some dodgy assumptions,[10] I shall treat eqn (1.3) as the canonical form for electrostatic energy density U, replacing the less versatile (1.2).

1.5   Electrostatic energy: critique II

Unfortunately, we have not finished with electric-field energy, even after obtaining eqn (1.3). There is a much bigger difficulty, and it concerns the use of an electrostatic potential φ.

To obtain our starting equation (1.1), we build a charge distribution by bringing up charge elements from infinity. During the process we have charges that are moving. Moving charges constitute currents; currents set up magnetic fields; those fields are time dependent because the currents start and stop; we have non-zero $\partial \boldsymbol{B}/\partial t$; we have non-zero curl $\boldsymbol{E}$; so we don't exactly meet the condition for the existence of an electrostatic potential φ. The usual cop-out is to say: well, if we bring up the charges very slowly, we *almost* have a potential. The hope is that any corrections will go to zero if we take the limit as the building-up time goes to infinity.

Perhaps.

We might think of investigating the limit of slow building-up, but that would be a waste of time. If we succeeded, we should (at best) know for certain that we had got a correct expression for energy in a *static* field. But not many truly-static fields are very interesting. What we need is an expression for electrical energy that we can use with confidence on *time-dependent* fields. And nothing done so far in this essay is any use at all for that. We may feel an attachment for eqn (1.3), but that's all it is, not anything secure.

To make progress, we need to go back to the beginning: Maxwell's equations, with allowance for fields that are time dependent.

Similar considerations apply to the magnetic energy density, whose change is $\boldsymbol{H} \cdot \delta \boldsymbol{B}$, an expression not to be integrated because the medium may not be linear. But we must mistrust this expression and its derivation for reasons similar to those for the electric-field energy.

1.6   The Poynting-vector derivation

Consider a charge q in an electromagnetic field. The force on the charge is $\boldsymbol{F} = q(\boldsymbol{E} + \boldsymbol{v} \times \boldsymbol{B})$. The rate at which this force does work is

$$\boldsymbol{v} \cdot \boldsymbol{F} = q\{\boldsymbol{v} \cdot \boldsymbol{E} + \boldsymbol{v} \cdot (\boldsymbol{v} \times \boldsymbol{B})\} = q\{\boldsymbol{v} \cdot \boldsymbol{E} + (\boldsymbol{v} \times \boldsymbol{v}) \cdot \boldsymbol{B}\} = q\boldsymbol{v} \cdot \boldsymbol{E}.$$

The magnetic field drops out. When we have more than one charge, we have

$$\text{rate at which work is done by fields on charges} = \sum_i q_i \, \boldsymbol{v}_i \cdot \boldsymbol{E}_i. \quad (1.4)$$

When we change to a "macroscopic" viewpoint, $\sum q_i \, \boldsymbol{v}_i$ becomes $\boldsymbol{J}\,\mathrm{d}V$, and the sum is replaced by an integration[11] over volume V:

$$\text{rate at which work is done by fields on charges} = \int \boldsymbol{J} \cdot \boldsymbol{E}\,\mathrm{d}V. \quad (1.5)$$

We now use Maxwell's equations to transform expression (1.5). The rate at which work is done *by* charges *on* the field within volume V is

$$-\int \boldsymbol{J} \cdot \boldsymbol{E}\,\mathrm{d}V = -\int \left(\operatorname{curl} \boldsymbol{H} - \frac{\partial \boldsymbol{D}}{\partial t} \right) \cdot \boldsymbol{E}\,\mathrm{d}V$$

$$= \int \boldsymbol{E} \cdot \frac{\partial \boldsymbol{D}}{\partial t}\,\mathrm{d}V - \int \boldsymbol{E} \cdot \operatorname{curl} \boldsymbol{H}\,\mathrm{d}V$$

$$= \int \boldsymbol{E} \cdot \frac{\partial \boldsymbol{D}}{\partial t}\,\mathrm{d}V - \int \boldsymbol{H} \cdot \operatorname{curl} \boldsymbol{E}\,\mathrm{d}V + \int \operatorname{div}(\boldsymbol{E} \times \boldsymbol{H})\,\mathrm{d}V$$

$$= \int \left(\boldsymbol{E} \cdot \frac{\partial \boldsymbol{D}}{\partial t} + \boldsymbol{H} \cdot \frac{\partial \boldsymbol{B}}{\partial t} \right)\mathrm{d}V + \int (\boldsymbol{E} \times \boldsymbol{H}) \cdot \mathrm{d}\boldsymbol{S}. \quad (1.6)$$

The manipulation leading to eqn (1.6) is standard bookwork. What we must concentrate on is the interpretation of the terms on the right.

First, notice that the integration can be carried out over any arbitrary volume. The volume integrations do not have to be taken over the whole of space. So we have a general-purpose volume $\int \mathrm{d}V$ bounded by a surface whose element is $\mathrm{d}\boldsymbol{S}$.

Textbooks often integrate the volume integral into

$$\int \frac{\partial}{\partial t} \left(\tfrac{1}{2} \boldsymbol{E} \cdot \boldsymbol{D} + \tfrac{1}{2} \boldsymbol{H} \cdot \boldsymbol{B} \right) \mathrm{d}V,$$

but this is taking a special case: a medium that is both linear and non-dispersive.[12] We have no business taking special cases here; we are trying to find expressions that we can trust always. We leave the volume integral alone.

Most books now say "the quantities in $\boldsymbol{E} \cdot \dot{\boldsymbol{D}} + \boldsymbol{H} \cdot \dot{\boldsymbol{B}}$ are old friends, met before in electrostatics and magnetostatics". This won't do. We have indeed met these two expressions before, but as acquaintances not as trustworthy old friends.

- We have seen that $\boldsymbol{E} \cdot \dot{\boldsymbol{D}}$ has no secure derivation, even within electrostatics.
- Even if we accept that $\boldsymbol{E} \cdot \dot{\boldsymbol{D}}$ is plausible for electro*statics*, we still have nothing to justify applying it to fields that are seriously time dependent.

We need to look afresh at eqn (1.6). There are two terms on the right: a volume integral and a surface integral. Neither is securely identified from earlier work. Nor is there any hope of constructing such an identification: any attempt can only be a repetition of the working that

[11] A change of notation has been sneaked in here. In (1.4) $\boldsymbol{E}$ represents a "microscopic" field of the kind already described in §1.3. In (1.5) it has become a "macroscopic" field, a space average of the microscopic field. Now $\boldsymbol{J}$ (meaning $\boldsymbol{J}_{\text{accessible}}$) is also a macroscopic—a smoothed or averaged—quantity. So in (1.5) we have written a product of averages, an operation previously flagged as suspect. Explain why nonetheless this step is valid here.

[12] The fact that we have assumed linearity is obvious. However, the involvement of dispersion may be a surprise. In a linear-but-dispersive medium, $\boldsymbol{E} \cdot \mathrm{d}\boldsymbol{D} + \boldsymbol{H} \cdot \mathrm{d}\boldsymbol{B}$ integrates to an expression that contains the derivatives with respect to frequency $\mathrm{d}\epsilon_r/\mathrm{d}\omega$ and $\mathrm{d}\mu_r/\mathrm{d}\omega$. See, e.g., Landau, Lifshitz and Pitaevskiĭ (1993) § 80, or Jackson (1999) § 6.8.

yielded (1.6). Of course, the shape of the equation is highly suggestive, but that's not the point. The point is: what can we prove?

If we had a secure identification of *one* of the terms on the right of (1.6), we could use that equation to identify the other. But with two "new" terms, things are much more problematic.

We'd like to say that the volume term represents the (rate of change of) energy stored within the volume of interest, while the surface integral represents the rate at which energy departs through the bounding surface. Can we do this? Remember that the surface integral in (1.6) started out life as a volume term $\int \text{div}(\boldsymbol{E} \times \boldsymbol{H})\,\mathrm{d}V$. So the separation of terms into volume and surface integrals is not unique: we can "trade" between them.[13]

Our difficulty is one of those "well known" facts that is usually swept under the carpet.[14] It is a problem that has a standard solution, though one that may be a bit of a surprise. We decide that we are *going boldly to treat* $\boldsymbol{E} \cdot \dot{\boldsymbol{D}} + \boldsymbol{H} \cdot \dot{\boldsymbol{B}}$ as the (rate of increase of) energy density, and we are *going boldly to treat* $\boldsymbol{E} \times \boldsymbol{H}$ as the flux of energy per unit area. This is a confession that we have not *proved* these interpretations to be true, but also a claim that this isn't incompetence: no proof exists.

What follows? First, our chosen method of accountancy is never going to give a failure of energy conservation: eqn (1.6) ensures that. So if we were wrong, that would mean that we have somehow put energy "in the wrong place". Which raises the question: what would "the wrong place" mean, and how could we ever tell? What experiment could we do to identify with certainty the location of a chunk of energy? Once this question is asked, we see that there's no easy answer. Only in general relativity does an energy density have measurable consequences, in that it acts as a source of gravitational fields, and this doesn't help us much. We really can't tell where field energy is located.[15]

All this is well known in the trade. Oxford Physics Finals papers never ask candidates to "prove" that $\frac{1}{2}\boldsymbol{E} \cdot \boldsymbol{D}$ is energy density, because the examiners know they mustn't demand the impossible. Instead the exam question uses standard weasel words, something like "justify the use of" $\frac{1}{2}\boldsymbol{E} \cdot \boldsymbol{D}$ or of $\boldsymbol{E} \times \boldsymbol{H}$. And there are standard weasel words that the candidate is expected to use in reply: "We may regard ..." These weasel words say just what I've expressed by "going boldly to treat".

If it's right that the partitioning of energy movements between volume and surface terms isn't unique, then there must be several competing expressions for energy density and energy flow. How do we decide between them—meaning have we made the most sensible choice? I think the short answer is that we've chosen those expressions that look most sensible when tested against the tutorial examples that we can invent.[16]

Textbooks often draw attention to a detail: $\int(\boldsymbol{E} \times \boldsymbol{H}) \cdot \mathrm{d}\boldsymbol{S}$ has been shown (perhaps) to give the energy flowing outwards through the whole of the closed surface $\boldsymbol{S}$. They point out that $(\boldsymbol{E} \times \boldsymbol{H}) \cdot \mathrm{d}\boldsymbol{S}$ has not been shown to be the energy flowing through any individual surface element $\mathrm{d}\boldsymbol{S}$. This is correct. But in view of all the other (unavoidable) gaps in our reasoning, it seems a silly point to focus on.

[13] Some progress can be made by appeal to relativity, though we can hardly deal with any kind of material medium then. Jackson (1999) § 12.10 gives a good account. He shows that the "symmetrized" energy in vacuum has energy density $\frac{1}{2}\epsilon_0 \boldsymbol{E}^2 + \frac{1}{2}\boldsymbol{B}^2/\mu_0$ and flux vector $\boldsymbol{E} \times \boldsymbol{B}/\mu_0$. So our partitioning of energy between volume and surface terms is persuasive for this special case.

[14] Honourable exception: Feynman (1964) § 27.4 makes points very similar to ours.

[15] This isn't a new difficulty. In Chapter 21 we ask about the potential energy mgh of a tennis ball, mass m at height h in gravity g; where is this energy located? The only answer we can come up with is that it simply hasn't got a location.

[16] Furthermore, radio engineering uses the Poynting vector to calculate the power radiated by an aerial, equivalently the aerial's radiation resistance. Any departure from correctness in the Poynting vector would immediately show up in routine measurements.

It occasionally happens that someone publishes an objection to the Poynting vector (in particular) and suggests a rival formula. The "killer argument" for dismissing such claims is almost always that the proposed formula would conflict with long-established radio-frequency experience. See also problem 1.2.

I'll mention one paradox. Suppose we have a permanent magnet generating a static $\boldsymbol{B}$-field, and we place electrodes in that field so as to generate a static $\boldsymbol{E}$-field at right angles to $\boldsymbol{B}$. The Poynting vector says that energy is flying round the system even though all the fields are static. This sounds extremely unlikely, and in conflict with common sense. Well, given the arbitrariness of our "we may regard", it's open to us to make a new choice as to what we regard in this special case; but I don't like that. We shouldn't change the rules on grounds as slight as these. A better solution is given by Feynman (1964), § 27.5. He says: What's wrong with saying that energy is flying round the system? The energy is travelling in closed loops, it's conserved, where's the difficulty? William of Occam[17] may be unhappy, but perhaps we don't have to be.

[17] This is an allusion to "Occam's razor", an important philosophical principle (13th century) which says (in Latin) "entities should not be multiplied without necessity". In the present context, invisible energy flying round in loops could well qualify as an "entity" introduced without necessity.

1.7   Discussion

There are two asymmetries in our expressions for energy. They concern the presence of "fundamental" field vectors $\boldsymbol{E}$ and $\boldsymbol{B}$ and the "derived" partners $\boldsymbol{D} = \epsilon_0 \boldsymbol{E} + \boldsymbol{P}$ and $\boldsymbol{H} = \boldsymbol{B}/\mu_0 - \boldsymbol{M}$.

The energy density $\boldsymbol{E} \cdot \dot{\boldsymbol{D}}$ is accompanied by $\boldsymbol{H} \cdot \dot{\boldsymbol{B}}$ when we might expect the analogous quantity to be $\boldsymbol{B} \cdot \dot{\boldsymbol{H}}$. The Poynting vector $\boldsymbol{E} \times \boldsymbol{H}$ contains one "fundamental" and one "derived" field, and we might have expected a more symmetric $\boldsymbol{E} \times \boldsymbol{B}$ or $\boldsymbol{D} \times \boldsymbol{H}$ (multiplying constants apart). What can be said about either of these?

On the first: the definitions of $\boldsymbol{D}$ and $\boldsymbol{H}$ contain $\boldsymbol{P}$ and $\boldsymbol{M}$ in an asymmetrical fashion, because of the difference of signs. So any expectation that $\boldsymbol{H}$ should appear in the same way as $\boldsymbol{D}$ can't be very strong. In fact, there is an obsolete way of presenting magnetism, in which $\boldsymbol{H}$ is the fundamental field quantity giving the force on magnetic poles; and then $\boldsymbol{B} = \mu_0(\boldsymbol{H} + \boldsymbol{M})$ is the "derived" field. In this picture, the expression we expect for (rate of change of) magnetic energy is the $\boldsymbol{H} \cdot \dot{\boldsymbol{B}}$ that we actually have. The existence of this alternative viewpoint means that there is no single "what's expected", and it should weaken any unease we have.

On the Poynting vector, something useful can be done. The following idea is copied from Pershan (1963).

[18] See, e.g., Bleaney and Bleaney (2013), § 4.2.

A magnetic moment $\boldsymbol{m}$ is equivalent to a current loop[18] of area $\delta \boldsymbol{S}$ carrying current i; and $\boldsymbol{m} = i\,\delta \boldsymbol{S}$. Consider then the fields in an electromagnetic wave travelling in the $+z$-direction. Accompanying $\boldsymbol{B}$ is a magnetization $\boldsymbol{M}$ which we replace by a set of equivalent current loops. A representative current loop has current i flowing over the surfaces of volume element $\delta x\,\delta y\,\delta z$ as shown in Fig. 1.1. The $\boldsymbol{E}$-field does work on this current, negative at z where the current is flowing down, positive at $z + \delta z$ where it is flowing up. The net effect is that the current transfers energy in the $-z$-direction.

Let's make this idea quantitative. The volume element $\delta x\,\delta y\,\delta z$ has magnetic moment $M\,\delta x\,\delta y\,\delta z$, which is also $i\,\delta x\,\delta z$, so that $i = M\,\delta y$. At $z + \delta z$ the current receives energy from the field at rate $+E\,\delta x\,i$;

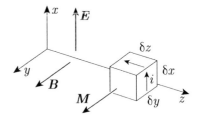

Fig. 1.1 One of the current loops accompanying magnetization $\boldsymbol{M}$.

at z the current delivers this energy at the same rate back to the field. We can think of the energy as crossing a plane in the middle of the volume element (since it starts on one side and ends up on the other): energy $-E\,\delta x\,i = -EM\,\delta x\,\delta y$ crosses the plane in the $+z$-direction in unit time. Thus there is an energy flow, per unit area per unit time, towards $+z$ of $-EM$. Given the vector directions, it is not hard to see that this is $-\boldsymbol{E} \times \boldsymbol{M}$. This flow of energy is to be added to the energy flow $\boldsymbol{E} \times (\boldsymbol{B}/\mu_0)$ that the fields would give in the absence of the magnetic medium. The total energy flow, per unit area per time, is $\boldsymbol{E} \times (\boldsymbol{B}/\mu_0 - \boldsymbol{M}) = \boldsymbol{E} \times \boldsymbol{H}$.

There is no analogous transfer of energy involving the electric dipole moment density $\boldsymbol{P}$, so in the above, $\boldsymbol{E}$ remains $\boldsymbol{E}$.

Arguments such as the present one give us confidence that we have partitioned the terms on the right of (1.6) in the way that is most sensible and insightful.[19] No one should seriously suggest that we've got it wrong. But we must still be aware that it all rests on "we may regard".

[19] You could object that the reasoning here is not wholly convincing because of the replacement of $\boldsymbol{M}$ by a claimed-equivalent current. Is this current equivalent for all purposes? I accept the objection. My intention has been to make us comfortable with the idea that there may be a $-\boldsymbol{E} \times \boldsymbol{M}$ contribution to the energy flow. I don't claim to have given a proof; we've seen that no proofs exist.

1.7.1   Parting shot I: Landau, Lifshitz and Pitaevskiĭ

In this chapter I've taken apart a "most-textbooks" treatment of energy. Don't take away an impression that the whole subject is an insecure mess. It's worth looking at Landau, Lifshitz and Pitaevskiĭ (1993). As mentioned in sidenote 10 their discussion is superior to most, though even they cannot avoid "we may regard" altogether.[20]

[20] The preface to *Electrodynamics of Continuous Media* expresses frustration: *In writing this book, we have experienced considerable difficulties, ... the customary exposition of many topics to be included does not possess the necessary physical clarity, and sometimes is actually wrong.*

1.7.2   Parting shot II: the Slepian vector

The following idea can be used only in the case where there are steady currents flowing so all fields are independent of time. Then $\boldsymbol{E} = -\nabla\varphi$ and $\operatorname{curl}\boldsymbol{H} = \boldsymbol{J}$. The Poynting vector originated in a volume term $\operatorname{div}(\boldsymbol{E} \times \boldsymbol{H})$. We can add any curl to the contents of the brackets here, since $\operatorname{div}\operatorname{curl} = 0$. Let us add $\operatorname{curl}(\varphi\boldsymbol{H})$. Nothing happens to the volume-energy terms in (1.6), but the surface term has its integrand changed to

$$\boldsymbol{E} \times \boldsymbol{H} + \operatorname{curl}(\varphi\boldsymbol{H}) = \boldsymbol{E} \times \boldsymbol{H} + \nabla\varphi \times \boldsymbol{H} + \varphi\operatorname{curl}\boldsymbol{H}$$
$$= (\boldsymbol{E} + \nabla\varphi) \times \boldsymbol{H} + \varphi\boldsymbol{J} = \varphi\boldsymbol{J}. \qquad (1.7)$$

The term $\varphi\boldsymbol{J}$ is easily interpreted. Take a length of wire, along which a current i flows with an accompanying potential drop from φ_1 to φ_2. The current density is $\boldsymbol{J}$ and the cross sectional area of the wire is $\delta\boldsymbol{S}$. According to Slepian, power $\varphi_1\boldsymbol{J}_1 \cdot \delta\boldsymbol{S}_1 = \varphi_1 i$ flows into one end of the wire, and power $\varphi_2\boldsymbol{J}_2 \cdot \delta\boldsymbol{S}_2 = \varphi_2 i$ flows out of the other end. The power left behind in the wire to be dissipated as heat is the difference $(\varphi_1 - \varphi_2)i = \text{(voltage drop)} \times \text{(current)}$. Thus eqn (1.7) predicts a result that we know to be true. The **Slepian vector**[21] $\varphi\boldsymbol{J}$ of (1.7) permits us to take the attitude that we intuitively use in low-frequency circuit theory: energy flows along the wires!

[21] Actually this is one of several possible variant energy-flow vectors that Slepian (1942) documented.

The Slepian vector is not being presented here as a serious rival to the Poynting vector, for reasons that are obvious: the assumptions that an electrostatic potential exists and that displacement current can be ignored. I mention the Slepian vector simply to rub in the fact that arbitrary choices really are being made when we interpret the terms on the right-hand side of eqn (1.6).[22]

1.8   For entertainment: a pillbox

When we prove the boundary condition that D_{normal} is continuous, we ask students to draw a "pillbox" bisected by the boundary surface. Yet

[22] *Question:* Suppose early physicists had adopted the opposite of our present convention on the sign of charge: nuclei negative, electrons positive. Then a given experimental arrangement involving currents would have J in the reverse direction to ours. Would the Slepian vector then face in the direction of $-J$?

Fig. 1.2 A pillbox, such as pills were supplied in by pharmacies sixty and more years ago. Pillboxes were made from black and white cardboard, and came in two or three different sizes. This one has had a "second use" for holding glass cover slides (for use with live specimens on a microscope stage), and the handwriting says *1″ covers for animalcule cage.*

Fig. 1.3 Here is the same pillbox opened. The contents may just be seen on a pad of cotton wool.

students have never seen the object they are asked to draw. The pillbox referred to is a familiar domestic object—of sixty-plus years ago. Today only the elderly have ever seen one.

My father was a pharmacist, and he brought these boxes home for storing small objects, such as seeds or small screws. That shown in Figs. 1.2 and 1.3 held (and holds) thin glass cover slides, used to enclose living pond life in a little cell for examination under a microscope.

The pillbox is a good object to draw when setting up the boundary condition on D or B, because its radius is fairly large compared with its thickness. There is even a white line at the join between body and lid, showing where the box should be bisected by the boundary surface. You can see why it was chosen by textbook writers—but at a time when everyone had several pillboxes at home and knew what they were being asked to draw.[23]

[23] Nowadays we might suggest drawing a tin of shoe polish, but that is becoming a rarity too.

Problems

Problem 1.1 A refinement of the Pershan model

In the discussion of the model in § 1.7, we said that equal amounts of energy $\pm E i \, \delta x$ are given to the volume element and taken from it. In fact, the energies are not quite equal, because E depends upon z. Find the power added to the volume element, and show that it is $-M_y \dot{B}_y \, \delta x \, \delta y \, \delta z$. This is in agreement with the general expression $-\boldsymbol{M} \cdot \dot{\boldsymbol{B}} \, \mathrm{d}V$ for the rate of gain of energy[24] of magnetic moment $\boldsymbol{M} \, \mathrm{d}V$ in a changing field $\boldsymbol{B}$.

The fact that this works out gives us a consistency check on the Pershan model. It gives us increased confidence in the idea that current i can be used correctly in the calculation of energies.

[24] See Chapter 22.

Problem 1.2 A tutorial example: a transmitting radio aerial

A radio engineer wishes to radiate power from a transmitting aerial. The medium surrounding the aerial is linear and loss-free so that $\boldsymbol{D} = \epsilon_r \epsilon_0 \boldsymbol{E}$ and $\boldsymbol{B} = \mu_r \mu_0 \boldsymbol{H}$.

Consider the fields within a fixed volume V bounded by surface $\boldsymbol{S}$ through which radiation is passing. We apply (1.6), but refrain from making assertions as to the interpretation of the terms on the right.

The wave radiated is periodic so its fields return to previous values after every oscillation period.[25] The expression $(\boldsymbol{E} \cdot \dot{\boldsymbol{D}} + \boldsymbol{H} \cdot \dot{\boldsymbol{B}})$ likewise returns to its previous value after a period.[26] Then when we apply (1.6) to the changes taking place over a period,

$$\int_{\text{period}} \mathrm{d}t \int (-\boldsymbol{J} \cdot \boldsymbol{E}) \, \mathrm{d}V = 0 + \int_{\text{period}} \mathrm{d}t \int (\boldsymbol{E} \times \boldsymbol{H}) \cdot \mathrm{d}\boldsymbol{S}.$$

In words: the energy given to the fields by $-\boldsymbol{J} \cdot \boldsymbol{E}$ is wholly accounted for by the flow of the Poynting vector through the bounding surface.

Argue that the reasoning has not relied on identifying $(\boldsymbol{E} \cdot \dot{\boldsymbol{D}} + \boldsymbol{H} \cdot \dot{\boldsymbol{B}})$ as a (rate of change of) stored energy. Argue that $(\boldsymbol{E} \times \boldsymbol{H})$ has not been shown to be the only correct expression for energy flow through $\boldsymbol{S}$, but that it does the business for the present, very important, case.[27]

[25] This restriction, and the loss-free condition, mean that we are dealing with a tutorial example, not a general proof. At the same time, finding the power radiated by an aerial is perhaps the most important application of the reasoning in this chapter.

[26] More precisely, $(\boldsymbol{E} \cdot \dot{\boldsymbol{D}} + \boldsymbol{H} \cdot \dot{\boldsymbol{B}})$ is the derivative with respect to time of $\frac{1}{2}(\epsilon_r \epsilon_0 E^2 + \mu_r \mu_0 H^2)$. The time integral of this, over a period, takes the same value at both limits and so yields zero. It should be easy to see that this finding results from requiring the medium to be loss-free.

[27] Argue that it further permits calculation of the **radiation resistance** (Bleaney and Bleaney 2013, p.252).

Electromagnetism: microscopic and macroscopic fields

2

Intended readership: Second year, or towards the end of an electromagnetism course that covers Maxwell's equations.

2.1 Introduction

When we learn about electromagnetic fields, we first find out about E and B, as they exist in a vacuum, and caused by point charges or currents in thin wires. And we learn how these fields may be measured from their effects on charges and currents. So far, so straightforward.

Then we learn about dielectrics and magnetic media, in which

$$D = \epsilon_0\, E + P, \qquad H = \frac{B}{\mu_0} - M.$$

Here, P is the electric-dipole-moment density within some medium, and likewise M is the magnetic-dipole-moment density. Fields D and H are introduced in order to facilitate working with these new quantities.

Later on, we try to make closer contact between P and M and the atomic structure of the medium within which they reside. Matter is made up of atoms and molecules, so "macroscopic" quantities like P and M should relate to the dipole moments possessed by individual atoms (or molecules)—"microscopic" entities. The linkage between microscopic and macroscopic is surprisingly deep, and it took several decades before a satisfactory treatment was invented. This chapter explains why there is a problem, and what the satisfactory solution is. The ideas are taken from Robinson (1973a), (1973b) and Jackson (1999).

2.2 Examples of microscopic and macroscopic fields

Suppose we have a capacitor with a slab of dielectric between the plates. The dielectric is isotropic and uniform, and it may have a linear relation between P and E which we represent as

$$D = \epsilon_r\, \epsilon_0\, E; \qquad P = (\epsilon_r - 1)\epsilon_0\, E.$$

13

Essays in Physics: Thirty-two thoughtful essays on topics in undergraduate-level physics. Geoffrey Brooker, Oxford University Press. © Geoffrey Brooker 2021. DOI: 10.1093/oso/9780198857242.003.0002

In writing equations such as these, we are thinking (at simplest) that there is a potential difference φ between the capacitor plates, and the field $\boldsymbol{E}$ is a uniform field of magnitude φ/d in a direction from the positive plate to the negative (the plates separated by d). In such a context, we are clearly dealing with a **macroscopic** $\boldsymbol{E}$-field, something measured with a voltmeter and a ruler.

Now imagine that we could turn a very powerful microscope onto the dielectric. We would find atoms, each consisting of a positively charged nucleus surrounded by a cloud of electrons. There is an electric field present, holding the atoms together. It is large. At a distance of one Bohr radius from the nucleus of a hydrogen atom, the E-field is nearly $10^{12}\,\mathrm{V\,m^{-1}}$. At the surface of the nucleus, the field is about $10^{21}\,\mathrm{V\,m^{-1}}$. Move from one side of the nucleus to the other, a distance of $10^{-15}\,\mathrm{m}$ or so, and the field is equally large but in the opposite direction. This **microscopic** electric field, with its rapid variations, is very different from the macroscopic field φ/d that we are imposing additionally.

Within the microscopic field, we can tease out a **local field**, experienced by each atom and caused by φ/d, that has the effect of inducing a dipole moment in the atom. (It is some sort of average taken over the volume of the atom.) Books on electromagnetism (e.g. Bleaney and Bleaney (2013) Chapter 10, Jackson (1999) § 4.5) discuss how the local field relates to the applied macroscopic field φ/d. The details do not concern us here. We mention the local field simply to give some feeling as to how it comes about that the atoms within a material become polarized by an applied (macroscopic) field, and so contribute to a dipole moment density and a relative permittivity.

The discussion above illustrates the considerable difference between microscopic and macroscopic fields. And it gives point to the question: how are the microscopic and macroscopic fields[1] related to each other?

The macroscopic field (and other related quantities such as charge density ρ) must be the result of space-averaging (in some way) the microscopic field, with the average defined so as to smooth away the violent variations described three paragraphs ago. This is easily said, but a number of very plausible averaging procedures turn out not to be satisfactory, for one reason or another.[2] A good discussion of these false leads is given in Robinson (1973b), Preface and Chapter 1.

2.3 An averaging procedure

We have seen the need for an averaging procedure, averaging over space, the effect of which is to smooth away the rapid variations (of charge density, or of current density, or of field) that occur on an atomic scale of distances. At the same time, the averaging must not be too fierce, as we need to retain the ability to discuss variations on "macroscopic" scales of distance.[3] Thus we envisage averaging over a "middling" scale of distances L_{av}. A discussion of that scale, and whether a suitable scale exists, is given in § 2.5.

---

[1] And the local field, but that is not the topic of the present chapter.

[2] One such false lead, the notion that the number N of charges within any interesting volume is so large that fluctuations of order $\sqrt{N}$ can be ignored, is disposed of in problem 2.1 below.

[3] We do not average over time (except in so far as time dependences differing over small distances are removed by the space averaging), as the physics of time dependence is something we wish to retain.

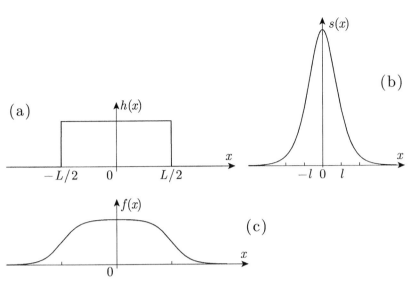

Fig. 2.1 (a) A top-hat function $h(x)$ that might be used for averaging over a distance scale L. (b) A smoothing function $s(x)$ that could be used for tapering the sharp edges of the top hat; what is drawn is $s(x) = (1/2l)\operatorname{sech}^2 x/l$. (c) An averaging function $f(x)$ that is the convolution of $h(x)$ with $s(x)$. It happens that this $f(x)$ has a resemblance to two Fermi–Dirac functions back-to-back. The graphs are drawn with $L = 6l$.

In this section, we simplify by treating everything as one-dimensional, and we think of averaging the density $\rho = \rho_{\text{micro}}(x)$ of electric charge, so that we have the additional simplification of dealing with a scalar field.

To form a macroscopic average $\rho_{\text{macro}}(X)$ of the microscopic charge density $\rho_{\text{micro}}(x)$, we form

$$\rho_{\text{macro}}(X) = \int_{-\infty}^{\infty} \rho_{\text{micro}}(x)\, f(X - x)\, \mathrm{d}x, \qquad (2.1)$$

where $f(X-x)$ is some averaging function that we use to smooth out the rapid variations with x of $\rho_{\text{micro}}(x)$. The outcome ρ_{macro} is a function of coordinate X, and varies no faster with X than the distance scale L_{av}, a scale set by the shape chosen for $f(x)$.

Figure 2.1(a) shows the simplest averaging function we might think of: a top hat $h(x)$, equal to $1/L$ over its plateau of width L (thereby giving $h(x)$ an integral of 1). The top hat is easy to think of, but it won't do because of its vertical ends. An end might fall just one side of a nucleus, or just the other side, and the charges included in the average would be different. If $h(X - x)$ replaces $f(X - x)$ in (2.1) the outcome fails to depend smoothly on the location X of the averaging function.

A better averaging function would be something like the $f(x)$ shown in Fig. 2.1(c). Here the function has a plateau having a width of about L, but it also has tapering ends, with the taper width not much smaller than L, so that it too occupies many atomic spacings.

It is not necessary to specify a precise form for the averaging function $f(x)$, because nothing about it should be critical, so long as sensible choices have been made for the widths of plateau and tapers. The important thing is: once we have decided on an averaging procedure, that procedure must be applied consistently: we average charge density ρ_{micro}, current $\boldsymbol{j}_{\text{micro}}$ and fields $\boldsymbol{E}_{\text{micro}}$, $\boldsymbol{B}_{\text{micro}}$ according to the same rule and using the same $f(x)$.

2.4   Averaging as truncation of spatial frequencies

There is a helpful complementary way of looking at (2.1). The integral in (2.1) is a convolution: of microscopic charge density $\rho_{\text{micro}}(x)$ and the averaging function $f(x)$. A convolution waves flags at us, asking us to take its Fourier transform.

Let us see why the Fourier transform is something we might want to take. Everything that was said about $\rho_{\text{micro}}(x)$ in §2.3 can be said with equal clarity about its Fourier transform. If we are to average over distances of order L_{av}, then that is equivalent to discarding space (angular) frequencies k above about $2\pi/L_{\text{av}}$. We expect that the transform $F(k)$ of $f(x)$ will be 1 for $k = 0$, and will fall towards zero for $k \sim 2\pi/L_{\text{av}}$. This behaviour may be seen in Fig. 2.2.

There is a resemblance here to the diffraction theory of image formation: Abbe theory. A microscope is understood to function by Fourier-analysing the light leaving its object, by discarding (or filtering away) the high-space-frequency components, and finally assembling an image from what remains. The microscope's resolution length (like L_{av}) is determined by the highest space frequency k_{max} that it transmits.

In this language, we have:[4]

$$\varrho_{\text{micro}}(k) = \int_{-\infty}^{\infty} \rho_{\text{micro}}(x)\, \mathrm{e}^{-ikx}\, \mathrm{d}x;$$

$$\varrho_{\text{macro}}(k) = \int_{-\infty}^{\infty} \rho_{\text{macro}}(X)\, \mathrm{e}^{-ikX}\, \mathrm{d}X;$$

$$F(k) = \int_{-\infty}^{\infty} f(x)\, \mathrm{e}^{-ikx}\, \mathrm{d}x;$$

$$\varrho_{\text{macro}}(k) = \varrho_{\text{micro}}(k) \times F(k).$$

Here the first two equations are definitions of $\varrho_{\text{micro}}(k)$ and $\varrho_{\text{macro}}(k)$; the third defines the Fourier transform $F(k)$ of $f(x)$; and the fourth equation is the Fourier transform of (2.1), showing how $F(k)$ executes its filtering. We refer to the filtering away of high-space-frequency components as **truncation**, meaning the imposing of a cut-off on the space frequencies. As is shown in Fig. 2.2 and discussed in §2.5.2 below, the cut-off is not necessarily sharp (such as a vertical step at a chosen $k = k_0$).

[4] It does not matter whether the exponentials here contain $-ikx$ or $+ikx$, so long as we make a choice and stick to it. The inverse transform contains the opposite sign, and we find it more natural for $\rho_{\text{micro}}(x)$ to be built from exponentials e^{ikx}.

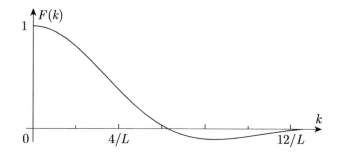

Fig. 2.2   A sample function $F(k)$ that "truncates" the Fourier transform $\varrho_{\text{micro}}(k)$ of the microscopic charge density $\rho_{\text{micro}}(x)$. The function $F(k)$ plotted here is the Fourier transform of the $f(x)$ of Fig. 2.1(c).

2.5   The length scale L_{av} for averaging

In the last two sections, we have mentioned a scale of distances L_{av} over which it is appropriate to average, and graphs have been presented of possible averaging functions. This raises the question as to what value should be chosen for L_{av}. The question can equally well be posed in relation to spatial frequencies: what value should be chosen for k_{av}, the value of k above which $F(k)$ falls off towards zero.

There are no universal answers. We can do no better than make a slight paraphrase of Robinson (1973*b*):

> The nature of the problem under discussion, rather than the physical state of the system, determines the correct choice of L_{av} or of the space-angular-frequency $k_{av} \sim 2\pi/L_{av}$. In general we find that if $L_{av} > 10^{-14}\,\mathrm{m}$ ($k_{av} < 10^{15}\,\mathrm{m}^{-1}$), we are free from the need to speculate about the internal structure of the elementary charges, while if $L_{av} > 10^{-8}\,\mathrm{m}$ ($k_{av} < 10^9\,\mathrm{m}^{-1}$) we need not discuss the internal structure of atoms. If, further, a_0 is the mean spacing of the *charges* and $k_{av}\,a_0 \ll 1$ we can always treat electricity as a fluid, while if in addition $k_{av}\,A_0 \ll 1$ where A_0 is the *interatomic* spacing we can also use Maxwell's equations in their conventional macroscopic form.

We might add the example of a "synthetic dielectric" for use at microwave frequencies. It might consist of tiny ball bearings embedded in epoxy resin, or repeated tiny conducting structures etched into a circuit board; an appropriate L_{av} might be ~ 0.1–$10\,\mathrm{mm}$.

It does not follow that a useful distance scale L_{av} (or accompanying k_{av}) always exists. If we are concerned with Bragg scattering of X-rays by atoms in a crystal, we can hardly filter away detail in the electron density without at the same time removing what we have come for.

2.5.1   A plateau is unhelpful

We refer again to the $f(x)$ in Fig. 2.1(c). Let each taper have width L_{taper} and the plateau have width $L_{plateau}$. The extreme example of the un-tapered top hat (giving a discontinuous "average") shows that L_{taper} must not be made smaller than L_{av}. But if $L_{plateau}$ is now made greater than L_{taper} it imposes an unnecessarily severe averaging, potentially discarding information about ρ_{micro} that we wish to keep.

I conclude that $f(x)$ is best chosen to have no plateau at all: a function like $s(x)$, with L_{av} replacing l, would give a very satisfactory averaging function. Indeed, as the next subsection shows, there is no need for $f(x)$ even to be everywhere positive, so use of $s(x)$ (or something like it) as the averaging function $f(x)$ is among the less surprising possibilities.

In Fig. 2.1(c) I showed a function $f(x)$ with a plateau because such a function gives an intuitive introduction to the ideas of this chapter. However, we should regard this as a pedagogical step that has largely served its purpose.[5]

[5] But a smoothed top-hat reappears as a useful model in § 2.6.

[7] The same property is well known to electronics designers, above all in the design of oscilloscopes. Here the response of the instrument must fall off at high frequencies (it cannot continue for ever), and yet the oscilloscope must display waveforms with minimal distortion; in particular, step and top-hat signals should be displayed with minimal or no overshoot. The eye will accept some rounding of a step as an unsurprising and expected artefact, but is likely to misinterpret any "ringing" as a real phenomenon. These requirements mean that the frequency response must be "rolled off" gently over a surprisingly large range of frequencies. A class of filters, known as Bessel filters, meets the requirements for oscilloscope design.

In another context, the space-frequency cut-off in an optical microscope is often rather sharp, and overshoot in the form of a "halo" is a common artefact at edges.

[8] Discussions often omit mention of the fixed charge. We are careful to include it, and to write $\rho_{\text{accessible}}$, rather than ρ_{free}, for the averaged total charge arising from fixed and free charge.

2.5.2   A property of averaging and truncation

Figures 2.1 and 2.2 display model functions for $f(x)$ and $F(k)$, drawn to be quantitatively consistent with each other. They have been chosen for mathematical simplicity:[6]

$$s(x) = \frac{1}{2l} \operatorname{sech}^2\left(\frac{x}{l}\right),$$

$$f(x) = \int_{-\infty}^{\infty} h(x')\, s(x-x')\, \mathrm{d}x' = \frac{\sinh(L/l)}{L} \frac{1}{\cosh(L/l)+\cosh(2x/l)},$$

$$F(k) = \pi \frac{l}{L} \frac{\sin(\tfrac{1}{2}kL)}{\sinh(\tfrac{1}{2}\pi kl)}.$$

Notice that $F(k)$ is negative for some values of k. This is a Fourier-transform property,[7] a consequence of the rapid fall of $f(x)$ for large $|x|$. Conversely, if we had set out to design an $F(k)$ with a rapid fall with $|k|$ then $f(x)$ might well have been negative for some values of x. While this may feel surprising, there is no necessary reason in physics or mathematics why either of $f(x)$ or $F(k)$ has to be everywhere positive.

2.6   Charge densities $-\operatorname{div} \boldsymbol{P}$ and $\boldsymbol{P}_{\text{normal}}$

We may think of $\rho_{\text{micro}}(\boldsymbol{r})$ (we are in three dimensions now) as composed of three contributions:

- "fixed" charge (nuclei and electron cores in a solid)[8]
- "free" charge (conduction electrons in a metal, or ions in an electrolyte)
- electrons that remain bound in their atoms (or molecules), but may have their wave function displaced off-centre under the influence of an electric field.

When macroscopic averaging is applied, the information that distinguishes the first two of these contributions is lost (it requires knowledge of where the charges are on a scale less than L_{av}); their averaged total is called $\rho_{\text{accessible}}$. The third contribution is the "polarization charge density" $-\operatorname{div} \boldsymbol{P}$. Thus:

$$\rho_{\text{macro}} = \rho_{\text{accessible}} - \operatorname{div} \boldsymbol{P}. \tag{2.2}$$

From here we write

$$\operatorname{div}(\epsilon_0 \boldsymbol{E}_{\text{macro}}) = \rho_{\text{accessible}} - \operatorname{div} \boldsymbol{P},$$

which we rearrange into

$$\boldsymbol{D} = \epsilon_0 \boldsymbol{E}_{\text{macro}} + \boldsymbol{P}, \qquad \operatorname{div} \boldsymbol{D} = \rho_{\text{accessible}}.$$

Associated with $\rho_{\text{accessible}}$ is a "conduction current" with density $\boldsymbol{J}$. Associated with $\boldsymbol{P}$ is "polarization current density" $\partial \boldsymbol{P}/\partial t$ and a surface charge density $\boldsymbol{P}_{\text{normal}}$ (coulombs per square metre).

The above is standard bookwork. Yet it rests on an assertion that requires derivation: eqn (2.2). To justify this, we need to demonstrate two things: that $\boldsymbol{P}$ is the dipole-moment density (in a natural sense:

the total of the dipole moments within a small volume divided by that volume), and that a non-uniform dipole-moment density yields charge densities $-\operatorname{div} \boldsymbol{P}$ and $\boldsymbol{P}_{\text{normal}}$. We give heuristic demonstrations of these properties using a model based on Robinson (1973a).[9],[10]

2.6.1   Charge density $-\operatorname{div} \boldsymbol{P}$

Figure 2.3 shows a one-dimensional array of electric dipoles. All dipoles consist of charges $\pm q$ separated by a, but the dipole moments increase in value with x (here because a increases with x). At the same time, the dipoles are all spaced by the fixed amount b. We take an average of the charge density over a distance of order L using the $f(x)$ of Fig. 2.1(c) (or similar) as averaging function. However, we do this in a slightly un-obvious way: we start off by using the top hat function $h(x)$ to "capture" charge between $x-L/2$ and $x+L/2$. This top hat is unsatisfactory for the reasons given in § 2.3: the precise locations of the ends are not something we ought to be interested in when taking a macroscopic average. We deal with this by imagining the top hat to be used not once but many times, giving its location a distribution $s(x)$, and we take an average (we might call it an ensemble average) over these locations. This means that the final averaging function $f(x)$ is the convolution of $h(x)$ with $s(x)$ —a relationship anticipated in the preparation of Fig. 2.1.

There are three ways in which the top hat can fall, relative to the dipoles. First, it can fall with both its ends between dipoles, or both ends splitting dipoles. Then the charge it "captures" is zero. Second, it may split a dipole at its left-hand end but not its right-hand end, and the probability of this happening is[11]

$$\left(\frac{a}{b}\right)_{x-L/2} \left\{ 1 - \left(\frac{a}{b}\right)_{x+L/2} \right\}.$$

If this happens the top hat captures charge $+q$, so the mean charge captured is

$$\left(\frac{(+q)a}{b}\right)_{x-L/2} \left\{ 1 - \left(\frac{a}{b}\right)_{x+L/2} \right\}.$$

Thirdly, the top hat may split a dipole at its right-hand end but not at its left-hand end. The mean charge captured is

$$\left(\frac{(-q)a}{b}\right)_{x+L/2} \left\{ 1 - \left(\frac{a}{b}\right)_{x-L/2} \right\}.$$

[9] It is common in textbooks to show mathematically that a dipole-moment density $\boldsymbol{P}$ gives the same electrostatic potential φ as a bulk charge density $-\operatorname{div} \boldsymbol{P}$ together with a surface charge density $\boldsymbol{P}_{\text{normal}}$. However, we wish to use these equivalent charges for time-dependent fields, so a demonstration restricted to a static case (φ definable by $\boldsymbol{E} = -\nabla\varphi$) is inadequate.

[10] The treatments given in §§ 2.6 and 2.7 are "teaching" models, aimed at making us feel comfortable with the macroscopic-average charges and currents—even when things are changing with time. Fully secure derivations can be given, but require a formal truncation.

[11] It is assumed that length L does not have any particular relationship with the spacing b of the dipoles.

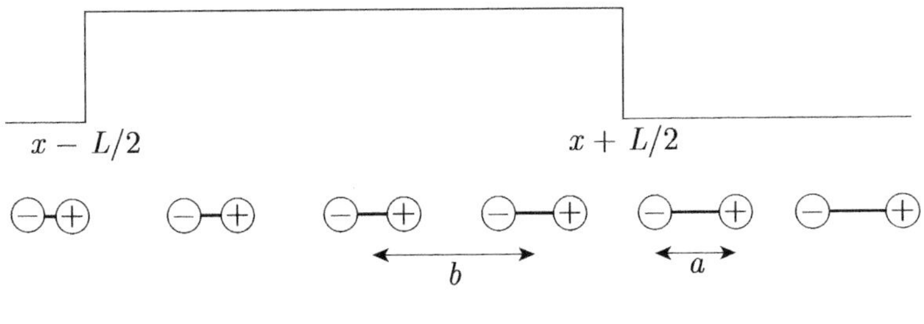

Fig. 2.3   A one-dimensional array of electric dipoles, extending in the x-direction. The dipoles all bear charges $\pm q$, but their lengths a increase with increasing x, while their spacings b remain uniform. The top hat is the $h(x)$ of Fig. 2.1(a), apart from a shift.

The reasoning is taken from Robinson (1973a) p. 61; (1973b) p. 66.

The overall average charge captured within length L is

$$\langle q \rangle = \left(\frac{qa}{b}\right)_{x-L/2} \left\{1 - \left(\frac{a}{b}\right)_{x+L/2}\right\} - \left(\frac{qa}{b}\right)_{x+L/2} \left\{1 - \left(\frac{a}{b}\right)_{x-L/2}\right\}$$

$$= -\left(\frac{qa}{b}\right)_{x+L/2} + \left(\frac{qa}{b}\right)_{x-L/2}.$$

Now $p = qa$ is the dipole moment of each dipole, and $N = 1/b$ is the number of dipoles per unit length. Thus our averaging procedure finds an average density of "bound" charge, per unit length of x-axis, of

$$\rho_{\mathrm{b}} = \frac{\langle q \rangle}{L} = \frac{-\left\{(Np)_{x+L/2} - (Np)_{x-L/2}\right\}}{L}.$$

The charge $\langle q \rangle$ varies smoothly on a macroscopic scale of distances, because it is the outcome of averaging over distance L. We assume that the same is true of (Np), so that

$$\rho_{\mathrm{b}} = -\frac{\partial}{\partial x}(Np). \tag{2.3}$$

The reader will not find it hard to see how result (2.3) can be extended into three dimensions, with the outcome that

$$\rho_{\mathrm{b}} = -\operatorname{div} \boldsymbol{P}. \tag{2.4}$$

Thus the charge density $\rho_{\mathrm{b}} = -\operatorname{div} \boldsymbol{P}$ is a "real" charge density whose presence can be demonstrated by performing the macroscopic averaging that is the topic of this chapter.

A further development is easily made. Restrict attention to those cases where the top hat captures an intact number of dipoles.[12] The total of the dipole moments within length L is $L \times (Np)$. The extension of this to three dimensions is the statement that $\boldsymbol{P}$ is the (average) dipole moment density within the medium.[13]

The top-hat model for averaging (Fig. 2.3) should be seen as a "teaching" model that exhibits an idea while pointing the way to something better.[14] An improved averaging can be done by applying $f(x)$ directly (problem 2.5), and is hardly more complicated than the reasoning given here.

One final detail needs attention. If a charge density changes with time, it must be because charge has moved. The equation of continuity requires that a time-varying charge $-\operatorname{div} \boldsymbol{P}$ should be accompanied by a current $\partial \boldsymbol{P}/\partial t$. We ought to show that this result too is a natural outcome of macroscopic averaging. Imagine that in Fig. 2.3 all positive charges move to the right with velocity v and all negative charges move to the left with velocity $-v$: each dipole moment increases with time by $dp/dt = 2qv$. Consider the probability that a randomly located plane will be crossed by a charge within time interval dt. The macroscopic-average current in the x-direction comes out to be $\partial P_x/\partial t$. You are invited to complete the steps in this reasoning in problem 2.6.

[12] Given that we have a good reason for pairing charges into the dipoles drawn on the diagram, we here "design out" the complication of having unpaired charges at one or both ends of the top hat.

[13] You are invited to make this idea formal in problem 2.4.

[14] The multiple application of a randomly shifted top hat is an awkward way of applying the smoothing function $s(x)$, and there are difficulties if the width L of the top hat happens to be an exact multiple of the inter-dipole spacing b.

2.6.2 Surface charge density P_{normal}

Next, we turn attention to the macroscopic charge in the neighbourhood of a surface, the surface bounding a polarized dielectric. Figure 2.4 shows a volume distribution of electric dipoles lying to the left of such a surface. A macroscopic average is to be taken using a procedure modelled on Fig. 2.6.1. We do not this time need to make a one-dimensional simplification, but can deal properly with three dimensions.

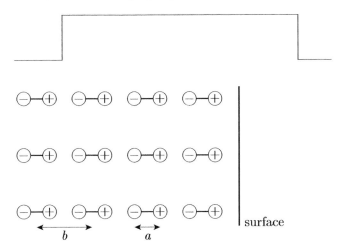

surface

Fig. 2.4 An array of electric dipole moments in the vicinity of a surface of a polarized medium. The dipole moments can all be equal in magnitude and direction. The top hat at the top of the diagram indicates how the charge distribution is sampled in the course of taking a macroscopic average.

The reasoning here is taken from Robinson (1973*b*) p. 65. That for Fig. 2.5 is from Robinson (1973*a*) p. 63.

In Fig. 2.4 the top hat represents the sampling function for a mathematical volume in the shape of a slab, covering area $A \gg b^2$ of surface, and extending a few b into the medium.

Let there be N_A dipoles per unit area of surface. For some of its possible locations, the inner surface of the slab may cut through dipoles. If it does, it captures charge $N_A A q$, and this happens with probability a/b. The mean charge captured (after repeated slightly different samplings) is (per unit area)

$$N_A A\,q\,\frac{a}{b}\,\frac{1}{A} = \frac{N_A}{b}\,(qa) = N\,(qa) = P_{\text{normal}}, \qquad (2.5)$$

since $N_A/b = N$ is the number of dipoles per unit volume. We express the result in terms of the normal component of $\boldsymbol{P}$, since any dipole-moment components parallel to the surface will not contribute.

2.7 The magnetization current curl $\boldsymbol{M}$

When a medium contains a magnetic moment density (magnetization) $\boldsymbol{M}$, a macroscopic current density $\boldsymbol{j}_{\text{m}} = \operatorname{curl} \boldsymbol{M}$ flows. We may demonstrate this using methods closely resembling those used in § 2.6.1.

A model for a magnetized medium is shown in Fig. 2.5. The magnetic moment (z-component) $m_z = IA$ of an atom is represented by a current I flowing in a loop of area A in the xy plane. A space-varying $\boldsymbol{M}$ is modelled by letting the areas A (or the currents I or both) increase

Fig. 2.5 Each atom is represented by a current I circulating round a loop of area A to yield a magnetic moment $m_z = IA$. The magnetic moments increase towards $+x$. The atoms are separated by d in the x-direction, D in the y-direction and δ in the z-direction. A strip of length L is represented (edge-on) by the bar drawn at the top of the diagram. The strip has width δ (normal to the paper), and is half above, half below, the xy plane. It is repeatedly placed into similar such positions in the xy plane but at locations random in both the x- and y-directions. On some occasions, one end falls within a current loop and intercepts current I.

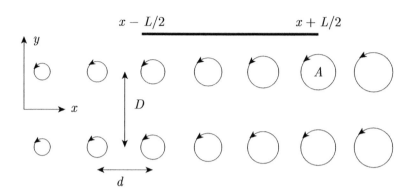

with x. The atoms are spaced d apart in the x-direction, D apart in the y-direction, and δ apart in the z-direction. The currents are sampled by imagining a strip (think of holding a card) of length L and width δ placed repeatedly (edge-on and half-way through the xy plane) into locations that vary randomly over an averaging distance l in both x- and y-directions. Thus we make a top-hat sampling in two dimensions, instead of the one dimension of § 2.6.1.

If the strip falls with both ends outside loops, or both ends inside, then it intercepts no net current. If the left-hand end falls within a loop, but the right-hand end does not, then a current crosses the strip in the $+y$-direction. This happens with probability

$$\left(\frac{A}{Dd}\right)_{x-L/2} \left\{1 - \left(\frac{A}{Dd}\right)_{x+L/2}\right\}.$$

If only the right-hand end of the strip falls within a loop, the current crossing the strip is in the $-y$-direction, and this has probability

$$\left(\frac{A}{Dd}\right)_{x+L/2} \left\{1 - \left(\frac{A}{Dd}\right)_{x-L/2}\right\}.$$

The average current in the $+y$-direction, intercepted by means of a random distribution of sampling locations, is

$$\langle I_y \rangle = -\left(\frac{IA}{Dd}\right)_{x+L/2} \left\{1 - \left(\frac{A}{Dd}\right)_{x-L/2}\right\}$$
$$+ \left(\frac{IA}{Dd}\right)_{x-L/2} \left\{1 - \left(\frac{A}{Dd}\right)_{x+L/2}\right\}$$
$$= \frac{-1}{Dd}\left\{(IA)_{x+L/2} - (IA)_{x-L/2}\right\} = \frac{-L}{Dd}\frac{\partial m_z}{\partial x}.$$

The strip has area $L\delta$, so the current per unit area crossing the strip is

$$j_y = \frac{-1}{Dd\delta}\frac{\partial m_z}{\partial x} = -\frac{\partial(Nm_z)}{\partial x}, \qquad (2.6)$$

where $N = 1/(Dd\delta)$ is the number of atoms per unit volume.

There are six diagrams that can be drawn after the fashion of Fig. 2.5, with magnetic moments increasing in each of the x, y, z-directions, and shown in planes xy, yz, zx. It is not hard to see how derivatives similar to those in (2.6) appear from these other arrangements, and that they combine into the single vector relation

$$\boldsymbol{j}_{\text{macro}} = \operatorname{curl} \boldsymbol{M}, \qquad (2.7)$$

where $\boldsymbol{M} = N\boldsymbol{m}$.

Finally, we can show that the magnetization $\boldsymbol{M}$ is accompanied by a surface current $\boldsymbol{M} \times \mathrm{d}\boldsymbol{S}$: problem 2.7.

2.8   Final comments

In the last two sections, we have used heuristic arguments to demonstrate the reasonableness of the statements that electric and magnetic dipole moments on the atoms of a material give rise to macroscopic moments $\boldsymbol{P}$ and $\boldsymbol{M}$, and that these are accompanied by macroscopic charge density ρ_{macro} and current density $\boldsymbol{j}_{\text{macro}}$ with ("accessible" contributions omitted)

$$\rho_{\text{macro}} = -\operatorname{div} \boldsymbol{P}, \qquad \boldsymbol{j}_{\text{macro}} = \operatorname{curl} \boldsymbol{M} + \frac{\partial \boldsymbol{P}}{\partial t}$$

(together with surface charge $\boldsymbol{P}_{\text{normal}}$ and surface current $\boldsymbol{M} \times \mathrm{d}\boldsymbol{S}$).

We should stress that these have been plausibility arguments, designed to make us feel comfortable with the reality of these charges and currents. Fully satisfactory calculations, from the "truncation" point of view, are given by Robinson (1973b), Chapter 8.

Problems

Problem 2.1 Macroscopic averaging is not trivial

A copper sphere of radius $a = 1\,\text{mm}$ is placed in a static electric field of magnitude $E = 1\,\text{mV\,m}^{-1}$. To maintain the conducting surface at an equipotential, electric charge moves from one hemisphere to the other. To find: how many electrons so move.

Treat this as a potential problem, to be solved by methods such as those given in Bleaney and Bleaney (2013) Chapter 2. Show that: the charge density on the surface of the sphere (using polar coordinates a, θ, ϕ) is $\sigma = 3\epsilon_0 E \cos\theta$; the charge transferred from one hemisphere to the other is $3\pi a^2 \epsilon_0 E$; and the induced electric dipole moment is $4\pi a^3 \epsilon_0 E$.

Put in numerical values to find: how many mobile electrons there are in each hemisphere, and how many cross between hemispheres.[15]

[Mass density of copper is $8933\,\text{kg\,m}^{-3}$; relative atomic mass of copper is 63.54. Take each atom as contributing one mobile electron. My answers: 1.8×10^{20}; 0.52.]

[15] This example is given by Robinson (1973b) Preface, to show that macroscopic averaging cannot be treated as trivially obvious. We might guess that N electrons in a small volume would be subject to a statistical fluctuation of order $\sqrt{N}$, which here is about 10^{10}. If this were true, electronic engineering would be impossible. The sphere is sitting in a field that is strong by the standards of radio-frequency engineering, and its induced dipole moment of order $10^{-22}\,\text{C\,m}$ is easily measurable.

Why not $\sqrt{N}$? Fluctuations of electron density create a space charge whose field suppresses the fluctuations.

[16] By "normalized" we here mean that $\int s(x)\,\mathrm{d}x = 1$; the integrand is not squared à la quantum mechanics.

Problem 2.2 Normalization of averaging functions
There is an implied assumption in the discussion of § 2.3: that if $h(x)$ and $s(x)$ are normalized[16] then so is $f(x)$.
    Show that this is the case for any convolution: if f is the convolution of h and s, and both are normalized, then f is also normalized.
    Do the integration that shows the given $s(x)$ to be normalized.

Problem 2.3 [Advanced] The Fourier transform $S(k)$ of $s(x)$
The specimen smoothing function $s(x)$ plotted in Fig. 2.1(b) is a sech^2 function. Its Fourier transform $S(k)$ is defined by

$$S(k) = \int_{-\infty}^{\infty} \frac{1}{2l} \frac{1}{\cosh^2 x/l}\, \mathrm{e}^{-\mathrm{i}kx}\, \mathrm{d}x = \frac{1}{2} \int_{-\infty}^{\infty} \frac{\mathrm{e}^{-\mathrm{i}Kz}}{\cosh^2 z}\, \mathrm{d}z,$$

where $K = kl$ and $z = x/l$.
    The function $s(x)$ is an even function of x, so its transform $S(k)$ is an even function of k. Therefore it will suffice to evaluate the integral for $\mathrm{Re}(k) > 0$. Show that this permits the path of integration to be closed by an infinitely large semicircle in the lower half of the z-plane.
    The integrand has poles on the imaginary axis at $z = -\mathrm{i}\pi(n + \tfrac{1}{2})$, where $n = 0, 1, 2, \ldots$. There is a complication because these are "double poles", and to find the residues we have to make a Laurent expansion of the integrand in the neighbourhood of each pole, and obtain the residue from the second term of the series. Show that when this is done

$$S(k) = 2\pi\mathrm{i} \sum_{n=0}^{\infty} \frac{(-\mathrm{i}K)}{2} \exp\{-K\pi(n + \tfrac{1}{2})\} = \frac{\tfrac{1}{2}\pi kl}{\sinh(\tfrac{1}{2}\pi kl)}.$$

The expression here is an even function of k, so it represents $S(k)$ for all k, not only for $\mathrm{Re}\, k > 0$.

[17] *Suggestion:* You may count dipoles directly. Alternatively, replace the positive charges by a single positive charge at the centroid of those charges. Do the same with the negative charges. Then find the dipole moment of these two.

Problem 2.4 Mean electric-dipole moment density is $\boldsymbol{P}$
Imagine the top hat of Fig. 2.3 to be so positioned that it encompasses only intact dipoles. Show formally[17] that the total dipole moment within length L is $(qa)L/b$. Extend this finding to three dimensions, and show that the space-average dipole moment density is $\boldsymbol{P}$.

Problem 2.5 Mean charge density is $-\,\mathrm{div}\,\boldsymbol{P}$
Remember that all dimensions associated with the averaging function $f(x)$ are large compared with an atomic spacing.
    One of the dipoles of Fig. 2.3 consists of a charge $-q\,\delta(x)$ together with a partner charge $+q\,\delta(x + a)$. Show that their combined charge "captured" by $f(x)$ is $qa\,f'(x) = pf'(x)$, where $p = qa$ and a prime means differentiation with respect to x. There are N dipoles per unit length, so the charge captured in δx is $Np\,f'(x)\,\delta x$. The charge captured under the whole of $f(x)$ is $N \int_{-\infty}^{\infty} pf'(x)\,\mathrm{d}x$.
    Show that if p is independent of x the total charge captured is zero.
    In the next approximation let $p = p_0 + xp'$ and show that the total charge captured is $N \int_{-\infty}^{\infty} xp'\, f'\, \mathrm{d}x$. Argue that p' can be treated as

constant over the x-range occupied by $f(x)$ so that

$$\text{total charge captured} = Np' \int_{-\infty}^{\infty} x f' \, \mathrm{d}x = -Np' = -\frac{\partial P}{\partial x},$$

since $P = Np$ is the dipole moment density.[18] We have obtained (2.3).

Make sure you are comfortable with the way in which the different rates of change with x permit the mathematical steps made in the above.

Notice that almost nothing has been said about the "shape" of the averaging function $f(x)$. Any "reasonable" function can be used and will give the same outcome. The $f(x)$ of Fig. 2.1(c) is just one of many possible averaging functions.

See if you can adapt the reasoning of this problem to the other averagings of this chapter: P_{normal}, $\text{curl}\,M$ and $M \times \mathrm{d}S$.

Problem 2.6 The "polarization current" $\partial P / \partial t$
Make formal the reasoning outlined in § 2.6.1, showing that a time-varying dipole moment density P is accompanied by a macroscopic current $\partial P / \partial t$. Do this by performing the macroscopic averaging. Check that the equation of continuity is satisfied.[19]

Problem 2.7 Surface current $M \times \mathrm{d}S$
Adapt the ideas of §§ 2.6.2 and 2.7 to show that an element $\mathrm{d}S$ of surface carries a macroscopic surface current (lying in the surface) $M \times \mathrm{d}S$.

[18] This step is an easy integration by parts. You will have to argue that xf goes to zero at both limits. Show that this must be case given that f has a shape that has permitted it to be normalized.

[19] The ideas here are taken from Robinson (1973a), p. 62.

II Mathematics

Vector theorems

3.1 Introduction

Standard vector theorems are things like

$$a \times (b \times c) = b(a \cdot c) - c(a \cdot b) \tag{3.1}$$

or Gauss's theorem

$$\int a \cdot dS = \int \operatorname{div} a \, dV. \tag{3.2}$$

Statements and proofs of these are easily found in standard texts. However, we sometimes need theorems that are hard to find, or to supply proofs, or to discover new results. An example of a "rare" theorem might be

$$\int a(b \cdot dS) = \int \Big(a \operatorname{div} b + (b \cdot \nabla)a \Big) \, dV. \tag{3.3}$$

This is easily proved by going into components, but we can be left with a feeling that there must be a more systematic way. There is, and it is the subject of this chapter.

3.2 Cartesian tensor notation

Tensor notation, for our purposes, is just a systematic way of writing down the Cartesian components of vectors. Thus a_i means the ith component of vector $a = (a_x, a_y, a_z) = (a_1, a_2, a_3)$, where the axes are now named $1, 2, 3$ and i is an integer that runs through $1, 2, 3$. To prove a relation such as (3.3) we may establish that it holds for the ith component of each side.[1]

[1] In this chapter, we use Cartesian coordinates exclusively. Tensors can be adapted to other coordinate systems, and indeed are ideal for the "curvilinear" coordinates of general relativity. But such complications do not belong here.

Exercise 3.1 Getting familiar

Let $a = (a_1, a_2, a_3)$ and $b = (b_1, b_2, b_3)$. We shall say that these vectors have components a_i and b_i where $i = 1, 2, 3$.

(1) Show that $b = \phi a$ is written tensorially as $b_i = \phi a_i$ because this encapsulates the three statements

$$b_1 = \phi a_1, \qquad b_2 = \phi a_2, \qquad b_3 = \phi a_3$$

when i is successively given the values $i = 1, 2, 3$.

(2) Show that

$$a \cdot b = a_1 b_1 + a_2 b_2 + a_3 b_3 = \sum_i a_i b_i.$$

Essays in Physics: Thirty-two thoughtful essays on topics in undergraduate-level physics. Geoffrey Brooker, Oxford University Press. © Geoffrey Brooker 2021. DOI: 10.1093/oso/9780198857242.003.0003

(3) Show that the ith component of grad ϕ is

$$\frac{\partial \phi}{\partial x_i}.$$

(4) Show that

$$\text{div } \boldsymbol{a} = \frac{\partial a_1}{\partial x_1} + \frac{\partial a_2}{\partial x_2} + \frac{\partial a_3}{\partial x_3} = \sum_i \frac{\partial a_i}{\partial x_i}.$$

(5) Show that

$$(\boldsymbol{b} \cdot \nabla)\phi = \sum_i b_i \frac{\partial}{\partial x_i}\phi,$$

and that the ith component of $(\boldsymbol{b} \cdot \nabla)\boldsymbol{a}$ is written as

$$\sum_j b_j \frac{\partial}{\partial x_j} a_i.$$

3.3   The summation convention

It is usual to omit summation signs. Then $a_i b_i$ stands for $\sum_i a_i b_i$. The rule is: if a suffix letter is repeated it is summed over.

Beginners usually find the summation convention intimidating,[2] and take fright when they see the word "tensor". There is a simple cure for this: continue writing in the summation signs! Eventually you will get tired of writing the $\sum$ signs, and that will be the time to start leaving them out.

[2] As a student I was uncomfortable: as soon as I wrote $a_i b_i$ a summation imposed itself that I might not want, and I felt I was no longer fully in control. Most often though the summation *is* wanted.

3.4   Vector products

To write $\boldsymbol{c} = \boldsymbol{a} \times \boldsymbol{b}$ in tensor notation, we need to introduce[3]

$$\varepsilon_{ijk} = \begin{cases} 1 & \text{when } i, j, k \text{ are all different and in cyclic order} \\ -1 & \text{when } i, j, k \text{ are all different and in anticyclic order} \\ 0 & \text{otherwise.} \end{cases} \tag{3.4}$$

The name for ε_{ijk} is "the totally antisymmetric tensor of the third rank". Having given this name, let's not use it again, as the name is far more frightening than the concept.

Now think about the vector product $\boldsymbol{c} = \boldsymbol{a} \times \boldsymbol{b}$. This means

$$c_1 = a_2 b_3 - a_3 b_2 \quad \text{together with cyclic permutations of } 1, 2, 3. \tag{3.5}$$

The tensor version of this is[4]

$$c_i = \sum_{j,k} \varepsilon_{ijk} a_j b_k, \tag{3.6}$$

or $\varepsilon_{ijk} a_j b_k$ if we omit the summation sign.

[3] The definition given here is correct, but there is a more fundamental version: ε_{ijk} is $+1$ if the subscripts can be brought to 123 by an even number of pairwise interchanges (and -1 if odd). Thus $312 \to 213 \to 123$ is two interchanges making $\varepsilon_{312} = \varepsilon_{123} = +1$.

Why do I say that this rule is "more fundamental"? Well, in the four dimensions of relativity there is a similar ε_{ijkl} that is non-zero only if all four of i, j, k, l are unequal; and the sign of this has to be determined by the pairwise-interchange rule.

[4] It's essential to keep careful track of the subscripts and their order of appearance.

Exercise 3.2 $a \times b$ and $\operatorname{curl} a$

(1) Take the tensor statement (3.6) and write out in full what it tells us for $i = 1, 2, 3$, thus showing that it reproduces the three statements of (3.5).

(2) Likewise show that $b = \operatorname{curl} a$ is written tensorially as

$$b_i = \sum_{j,k} \varepsilon_{ijk} \frac{\partial}{\partial x_j} a_k.$$

(3) Show that:[5]

$$a \cdot (b \times c) = (a \times b) \cdot c.$$

(4) Show that:

$$\operatorname{div}\operatorname{curl} a = -\operatorname{div}\operatorname{curl} a = 0,$$

and that

$$\operatorname{curl}(\operatorname{grad} \phi) = 0.$$

[5] The reasoning required here and for the rest of the exercise is almost trivial if you use ε_{ijk}. Even at this early stage, tensor notation is making things easy compared with other methods.

3.5   A new look at vector integral theorems

You probably encountered Gauss's theorem as your first vector-integral theorem, and regard it as the most basic such theorem. Actually, it isn't, and it's helpful to realise that it's a consequence of something more fundamental. That more fundamental statement is:

$$\int \phi \, dS = \int \nabla \phi \, dV \quad \text{or} \quad \int \phi \, dS_i = \int \frac{\partial \phi}{\partial x_i} \, dV, \qquad (3.7)$$

where ϕ is some function of position. (Although my notation suggests that ϕ is a scalar, there is no specific restriction laid on ϕ in the derivation of (3.7).)

Let's obtain Gauss's theorem by starting from (3.7). The jth component of a vector is a function of position, and can perfectly well replace ϕ in (3.7):

$$\int a_j \, dS_i = \int \frac{\partial a_j}{\partial x_i} \, dV.$$

Now take the special case where $j = i$ and where i is summed over:

$$\int \sum_i a_i \, dS_i = \int \sum_i \frac{\partial a_i}{\partial x_i} \, dV.$$

We need only disentangle the notation here: the left-hand integrand is $a \cdot dS$, and the right-hand integrand is $\operatorname{div} a \, dV$, so we have Gauss's theorem:

[Gauss] $$\int a \cdot dS = \int \operatorname{div} a \, dV. \qquad (3.8)$$

Exercise 3.3 Two Gauss-type theorems

(1) Prove (3.3) using the following suggestion. Start from (3.7) and replace ϕ with $a_i b_j$:

$$\int a_i b_j \, dS_j = \int \frac{\partial}{\partial x_j}(a_i b_j) \, dV = \int a_i \frac{\partial b_j}{\partial x_j} \, dV + \int b_j \frac{\partial a_i}{\partial x_j} \, dV.$$

Sum over j, and each term here becomes the ith component of the corresponding term of (3.3).[6]

(2) Show that:

$$\int (\text{curl}\,\boldsymbol{a}) \, dV = -\int \boldsymbol{a} \times d\boldsymbol{S}. \tag{3.9}$$

Just as Gauss's theorem is a consequence of the more fundamental (3.7), so Stokes's theorem is a consequence of

$$\int \phi \, d\boldsymbol{l} = -\int \nabla\phi \times d\boldsymbol{S}. \tag{3.10}$$

Example 3.1 Prove Stokes's theorem from (3.10)

Solution: In tensor notation, (3.10) reads (summation over j and k implied)

$$\int \phi \, dl_i = -\int \varepsilon_{ijk} \frac{\partial \phi}{\partial x_j} \, dS_k.$$

Take the special case where $\phi = a_i$:

$$\int a_i \, dl_i = -\int \varepsilon_{ijk} \frac{\partial a_i}{\partial x_j} \, dS_k = \int \varepsilon_{kji} \frac{\partial a_i}{\partial x_j} \, dS_k.$$

When i is summed over, this is the tensor equivalent of

[Stokes]$$\int \boldsymbol{a} \cdot d\boldsymbol{l} = \int \text{curl}\,\boldsymbol{a} \cdot d\boldsymbol{S}. \tag{3.11}$$

3.6   A theorem concerning ε_{ijk}

To process a product like $\boldsymbol{a} \times (\boldsymbol{b} \times \boldsymbol{c})$, we need to know how to handle two vector products juxtaposed. The statement we need is:

$$\sum_k \varepsilon_{ijk} \varepsilon_{kpq} = \delta_{ip}\delta_{jq} - \delta_{iq}\delta_{jp}, \tag{3.12}$$

where

$$\delta_{ij} = \begin{cases} 1 & \text{if } i = j \\ 0 & \text{otherwise} \end{cases} \quad \text{is the Kronecker delta.}$$

Exercise 3.4 Prove (3.12)

I suggest you start by taking the special case $i, j = 1, 2$. Then in the sum over k the only non-zero term is the one with $k = 3$. In $\varepsilon_{kpq} = \varepsilon_{3pq}$ there are two non-zero possibilities: $p, q = 1, 2$ making $\varepsilon_{3pq} = \varepsilon_{312} = +1$ and $p, q = 2, 1$ making $\varepsilon_{3pq} = \varepsilon_{321} = -1$. Since k is neither of i, j and k is neither of p, q, it follows that a non-zero outcome requires either $p = i$, $q = j$ or $p = j$, $q = i$. Look at the pattern here, and you should arrive at (3.12).

Exercise 3.5 Prove (3.1)

That is: $\boldsymbol{a} \times (\boldsymbol{b} \times \boldsymbol{c}) = \boldsymbol{b}(\boldsymbol{a} \cdot \boldsymbol{c}) - \boldsymbol{c}(\boldsymbol{a} \cdot \boldsymbol{b})$.

Example 3.2 A property needed for electromagnetic waves

Prove
$$\operatorname{curl}\operatorname{curl}\boldsymbol{a} = \operatorname{grad}\operatorname{div}\boldsymbol{a} - \nabla^2\boldsymbol{a}. \qquad (3.13)$$

Solution: The left-hand side has ith component

$$\varepsilon_{ijk}\frac{\partial}{\partial x_j}\,\varepsilon_{kpq}\frac{\partial}{\partial x_p}a_q = \left(\delta_{ip}\delta_{jq} - \delta_{iq}\delta_{jp}\right)\frac{\partial^2}{\partial x_j\,\partial x_p}a_q$$

$$= \frac{\partial}{\partial x_i}\frac{\partial a_j}{\partial x_j} - \frac{\partial^2}{\partial x_j\,\partial x_j}a_i.$$

Here a sum over j, k, p, q is implied, not written because the summation signs would be seriously cluttering. Put the last expression back into vector notation and you have the desired result.

Exercise 3.6 Now try some theorems without clues

(1) Show that:
$$\operatorname{curl}(\phi\boldsymbol{a}) = \left(\nabla\phi\right) \times \boldsymbol{a} + \phi\operatorname{curl}\boldsymbol{a}. \qquad (3.14)$$

(2) Show that:
$$\operatorname{curl}(\boldsymbol{a} \times \boldsymbol{b}) = (\boldsymbol{b} \cdot \nabla)\boldsymbol{a} + \boldsymbol{a}\operatorname{div}\boldsymbol{b} - (\boldsymbol{a} \cdot \nabla)\boldsymbol{b} - \boldsymbol{b}\operatorname{div}\boldsymbol{a}. \qquad (3.15)$$

(3) Show that:[7]
$$\operatorname{div}(\boldsymbol{a} \times \boldsymbol{b}) = \boldsymbol{b} \cdot \operatorname{curl}\boldsymbol{a} - \boldsymbol{a} \cdot \operatorname{curl}\boldsymbol{b}. \qquad (3.16)$$

[7] *Hint:* Work from right to left. Notice that this is the Poynting-theorem algebra used in obtaining (1.6).

(4) Show that:
$$\boldsymbol{a} \times \operatorname{curl}\boldsymbol{b} + \boldsymbol{b} \times \operatorname{curl}\boldsymbol{a} = \nabla(\boldsymbol{a} \cdot \boldsymbol{b}) - (\boldsymbol{a} \cdot \nabla)\boldsymbol{b} - (\boldsymbol{b} \cdot \nabla)\boldsymbol{a}. \qquad (3.17)$$

(5) Show that:
$$\int \boldsymbol{a} \times \mathrm{d}\boldsymbol{l} = \int \operatorname{div}\boldsymbol{a}\,\mathrm{d}\boldsymbol{S} - \nabla\int \boldsymbol{a} \cdot \mathrm{d}\boldsymbol{S}. \qquad (3.18)$$

Comment: Some of the theorems here look somewhat intimidating. The great thing about tensor notation is that all of the proofs are quite short; and within the working there should be no difficulty in seeing where to go next.

3.7   Some tools

Here is the three-dimensional Taylor expansion of ϕ about location r_0:

$$\phi(r) = \phi(r_0)+(x_i-x_{i0})\frac{\partial\phi}{\partial x_i}+\tfrac{1}{2}(x_i-x_{i0})(x_j-x_{j0})\frac{\partial^2\phi}{\partial x_i\,\partial x_j}+\ldots, \quad (3.19)$$

where the derivatives are evaluated at $r = r_0$, and the subscripts i and j are summed over (independently, that is, with both $i,j = 1,2$ and $i,j = 2,1$ appearing in the double sum).

In particular, to first order,[8]

$$\phi(r) = \phi(r_0) + \{(r-r_0)\cdot\nabla\}\phi; \qquad a(r) = a(r_0) + \{(r-r_0)\cdot\nabla\}a, \quad (3.20)$$

where, again, the derivatives are to be evaluated at $r = r_0$.

[8] Note that the second-order terms cannot be put into vector notation, so the only way to deal with them is by going into components (which is what tensor notation does, in an organized way).

Problems

Problem 3.1 Verifying some tools

(1) The vector $r = (x,y,z) = (x_1, x_2, x_3)$. Show that

$$\frac{\partial x_i}{\partial x_j} = \delta_{ij}, \qquad \operatorname{div} r = 3. \quad (3.21)$$

(2) Verify eqn (3.19) by differentiating twice. In the second-order term it will be helpful to take particular cases, such as $i,j = 1,1$ and $i,j = 1,2$ and $i,j = 2,1$, to check that the coefficient $\tfrac{1}{2}$ is correct.

(3) We know that $(a \times b)_i = \varepsilon_{ijk}\, a_j\, b_k$. Show from this that

$$a_i b_j - a_j b_i = \varepsilon_{ijk}(a \times b)_k. \quad (3.22)$$

(4) Likewise, show that

$$\frac{\partial a_j}{\partial x_i} - \frac{\partial a_i}{\partial x_j} = \varepsilon_{ijk}(\operatorname{curl} a)_k. \quad (3.23)$$

Problem 3.2 A vector property needed in problem 22.4
Show that

$$a \times (b\cdot\nabla c) - b\times(a\cdot\nabla c) = (b\times a)\times\operatorname{curl} c - (b\times a)\operatorname{div} c + (b\times a)\cdot\nabla c. \quad (3.24)$$

Suggestion
Start by showing that the ith component of the left-hand side is

$$\varepsilon_{ijk}(a_j\, b_p - b_j\, a_p)\frac{\partial c_k}{\partial x_p}.$$

Use (3.22) and then expand the product of two εs to get

$$\frac{\partial c_k}{\partial x_k}(a\times b)_i - \frac{\partial c_k}{\partial x_i}(a\times b)_k$$

$$= (a\times b)_i\operatorname{div} c - (a\times b)_k\left(\varepsilon_{ikp}(\operatorname{curl} c)_p + \frac{\partial c_i}{\partial x_k}\right),$$

where the last step makes use of (3.23). Each term can now be put back into vector notation, and the outcome is (3.24), as required.

III Waves

4 Longitudinal and transverse waves

Intended readership: When three-dimensional waves are discussed, and in particular when waves in plasmas are encountered.

4.1 Introduction: divergence and curl

There are many kinds of wave that can propagate in free space or in a medium. The simplest is a "scalar wave": a wave whose amplitude is a scalar. The obvious example is the Ψ describing the quantum motion of a single particle.

The next in complication is a "vector wave", where the interesting amplitude is a vector $\boldsymbol{\xi}$. Examples are sound waves in fluids (§ 5.7, amplitude $\boldsymbol{\xi}$ a displacement), and electromagnetic waves (amplitude the electric field $\boldsymbol{E}$, or perhaps the magnetic field $\boldsymbol{B}$).

More complicated waves are possible, where the amplitude is a tensor. The gravity waves of general relativity require a second-rank tensor for their description. And sound waves in solids require consideration of stress and strain, both of which are second-rank tensors.

In this chapter, we consider waves (mostly vector waves) in isotropic media.

Vector waves come in two varieties: longitudinal and transverse. In isotropic media, these have a simple mathematical definition. If the amplitude is $\boldsymbol{\xi}$,

$$\left.\begin{array}{lll} \bullet \text{ transverse:} & \operatorname{div} \boldsymbol{\xi} = 0 \\ \bullet \text{ longitudinal:} & \operatorname{curl} \boldsymbol{\xi} = 0. \end{array}\right\} \tag{4.1}$$

Given statements (4.1):

- If we encounter a new kind of wave having a vector amplitude, and we suspect it of being transverse, look at once to see if the divergence is zero.

- If we encounter a new kind of wave having a vector amplitude, and we suspect it of being longitudinal, look at once to see if the curl is zero.

The rest of this chapter gives examples to show that assertions (4.1) are in line with physical expectation, and shows how they emerge from the mathematics in each case.

Essays in Physics: Thirty-two thoughtful essays on topics in undergraduate-level physics. Geoffrey Brooker, Oxford University Press. © Geoffrey Brooker 2021. DOI: 10.1093/oso/9780198857242.003.0004

4.2 Picture drawing

Let us see why the divergence and curl should be appropriate characteristics for the two kinds of wave.

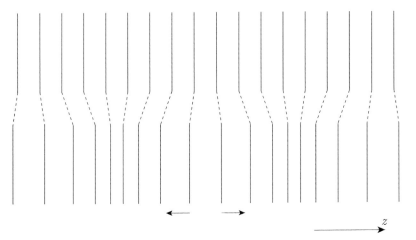

Fig. 4.1 The upper part of the figure shows planes equally spaced within some undisplaced material medium. In the lower part, a longitudinal sound wave causes some planes to get closer together, some to get further apart. The small arrows show a region where the divergence of the displacement is positive.

Figure 4.1 shows a longitudinal sound wave travelling left–right. (Because the figure is a snapshot, we cannot tell whether the wave is travelling or standing, nor if travelling which way (left or right) it is going; it does not matter.) Planes that were equispaced before the wave was imposed acquire unequal spacings, with consequent changes of density within the medium. A plane originally at z is displaced by the wave through distance $\boldsymbol{\xi}$, a function of position z and time t. Arrows show where two planes have been moved apart by the wave. Suppose we draw a closed surface composed of those two planes, joined "round the edges" out of our sight. The surface integral $\int \boldsymbol{\xi} \cdot \mathrm{d}\boldsymbol{S}$ is clearly positive, and therefore $\operatorname{div} \boldsymbol{\xi} > 0$ within the volume bounded by the closed surface. If we do the same with two planes that have been pushed together, then the surface integral is negative, and $\operatorname{div} \boldsymbol{\xi} < 0$ between those planes. Clearly, a non-zero $\operatorname{div} \boldsymbol{\xi}$ is a necessary and characteristic feature of a longitudinal wave.

In a similar picture-drawing way, we may show that a non-zero curl is a characteristic feature of a transverse wave. A transverse sound wave[1] is represented in Fig. 4.2. The wave travels in the left–right direction. (Again, it could be a standing wave, or travelling in either direction.) The lines represent planes half a wavelength apart. In one plane the displacement is up the page; in adjacent planes it is down the page. Take paths, following the arrows, in two adjacent planes, and join them as suggested by the broken lines to form a closed contour. The displacement $\boldsymbol{\xi}$ "chases its tail" along nearly all of the contour, so $\operatorname{curl} \boldsymbol{\xi}$ is non-zero on the surface of the paper between the two arrows. Clearly, a central feature of a transverse wave is that it has non-zero $\operatorname{curl} \boldsymbol{\xi}$.

We are now in a position to understand the mathematical conditions stated in (4.1). The conditions define each polarization of wave, not

[1] Transverse elastic waves (shear waves) can propagate readily in solids. Transverse waves can also propagate in a fluid, but these waves rely for their propagation on viscosity, which is a dissipative phenomenon, so they suffer very high attenuation per wavelength.

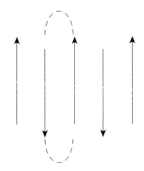

Fig. 4.2 A transverse sound wave propagates in the left–right direction. Arrows show (a snapshot of) the displacements of the medium in planes half a wavelength apart.

[2] Note that the words "longitudinal" and "transverse" can take us by surprise. The always-valid definitions are statements (4.1). By contrast, the properties $k \cdot \xi_{\mathrm{trans}} = 0$ and $k \times \xi_{\mathrm{long}} = 0$ are special cases: special to the case of a *uniform, plane* wave. An example may give substance to this warning: an electromagnetic wave travelling inside a metal waveguide constitutes a *plane* but *non-uniform* wave, and it has a component parallel to k of either E or B, even though both fields are divergence-free. The fields ξ in the waveguide remain technically "transverse" because they have zero divergence, even though $k \cdot \xi \neq 0$.

by what it is, but by what it is not. A general disturbance within a solid consists of both longitudinal and transverse waves. If we want a pure longitudinal wave, we must ensure that it contains no contribution from a transverse wave. We exclude that by requiring the characteristic feature of a transverse wave to be absent: $\mathrm{curl}\,\xi = 0$. Likewise, if we want a pure transverse wave, we must require it to be uncontaminated by any longitudinal component: $\mathrm{div}\,\xi = 0$.

We illustrate using the *special case* of a uniform plane wave. Let the wave have vector amplitude $\xi = \xi_0\,e^{i(k \cdot r - \omega t)}$. Then

$$\mathrm{div}\,\xi = i k \cdot \xi; \qquad \mathrm{curl}\,\xi = i k \times \xi.$$

Thus: if $\mathrm{div}\,\xi = 0$, the wave has ξ perpendicular to k (across the direction of k); and if $\mathrm{curl}\,\xi = 0$, the wave has ξ parallel to k (along the direction of k).[2]

4.3 Waves in a plasma

The picture-drawing arguments of the last section are persuasive, but informal. We should like to see a mathematical treatment of a vector wave leading inexorably to a separation of its motion into two independent pieces, one having zero divergence and one having zero curl. Unfortunately, sound is mathematically complicated (at heart it's a tensor wave), so we first choose a simpler physical system: a plasma.[3]

[3] I take the opportunity to discuss a plasma because the reasoning given here is hard to find elsewhere, and because most treatments of plasmas introduce special cases in a way that conceals the overall structure.

A plasma consists of positive ions and negative electrons. Each electron has been detached from some originally-neutral atom, leaving a positive ion. Ions and electrons move independently, except for the fact that each charged particle experiences electric fields from the other charges.

Electrons are more mobile than positive ions, because of their smaller mass. We shall therefore treat the positive ions as if they were stationary, providing a "background" of positive charge that makes the plasma neutral overall.

An electron (mass m_e, charge $-e$) experiences some electric field E_{local} that depends on just what the surrounding charged particles are doing. The electron has velocity v, and in the field it experiences an acceleration $\dot{v}$ given by[4]

[4] The electrons may be treated by classical mechanics because an electron's de Broglie wavelength is typically much less than the electron–electron separation. See problem 4.1.

$$m_e \dot{v} = -e E_{\mathrm{local}}. \tag{4.2}$$

We give a macroscopic description of the plasma by performing, in imagination, a space averaging after the fashion of Chapter 2. The "microscopic" charge density (a sum over delta-functions, one at each charge) is smoothed into a "macroscopic" charge density ρ; sometimes we may wish to think of ρ as composed of separate contributions ρ_+ from positive ions and ρ_- from electrons. Averaging is performed over some length scale that is large compared with the separation of the charged particles, but small compared with distances of macroscopic interest, such as the wavelength of a wave propagating in the plasma.

Let there be N electrons, on average, per unit volume. Their motion results in a macroscopic current density $\boldsymbol{j}$. Macroscopic averaging yields

$$\rho = \rho_+ + N(-e), \qquad \boldsymbol{j} = N(-e)\overline{\boldsymbol{v}}, \qquad \boldsymbol{E} = \overline{\boldsymbol{E}_{\text{local}}}. \tag{4.3}$$

Then an averaging over (4.2) gives

$$\frac{\partial \boldsymbol{j}}{\partial t} = N(-e)\overline{\dot{\boldsymbol{v}}} = N(-e)\left(\frac{-e\overline{\boldsymbol{E}_{\text{local}}}}{m_{\text{e}}}\right) = \frac{Ne^2}{m_{\text{e}}}\boldsymbol{E}. \tag{4.4}$$

In a plasma, the electrons and positive ions are not "paired up" in any obvious way to form electric dipoles. Therefore, after macroscopic averaging, we do not have a dipole-moment density $\boldsymbol{P}$ taking care of paired charges; all charges are accounted for in ρ_- and ρ_+. Given that $\boldsymbol{P}$ is not introduced, it is best not to introduce $\boldsymbol{D}$ either (even as an abbreviation for $\epsilon_0 \boldsymbol{E}$), for fear of giving misleading signals.[5] Gauss's theorem takes the form

$$\rho = \epsilon_0 \operatorname{div} \boldsymbol{E}. \tag{4.5}$$

We also have the "in vacuum" Maxwell equation

$$\boldsymbol{j} + \epsilon_0 \dot{\boldsymbol{E}} = \operatorname{curl} \frac{\boldsymbol{B}}{\mu_0}. \tag{4.6}$$

Combine the last equation with (4.4):

$$\frac{Ne^2}{m_{\text{e}}}\boldsymbol{E} + \epsilon_0 \ddot{\boldsymbol{E}} = \frac{1}{\mu_0}\operatorname{curl} \dot{\boldsymbol{B}} = \frac{1}{\mu_0}\operatorname{curl}(-\operatorname{curl} \boldsymbol{E}),$$

so that

$$\ddot{\boldsymbol{E}} + \omega_{\text{p}}^2 \boldsymbol{E} = -c^2 \operatorname{curl}\operatorname{curl} \boldsymbol{E}, \tag{4.7}$$

where

$$\omega_{\text{p}}^2 = \frac{Ne^2}{\epsilon_0 m_{\text{e}}}. \tag{4.8}$$

Equation (4.7) contains the physics governing all kinds of disturbance (within our approximations) in a plasma.

4.3.1 Longitudinal and transverse waves in a plasma

Any vector $\boldsymbol{E}$ can be expressed in the form $\operatorname{curl} \boldsymbol{A} - \nabla\phi$. The first of these terms has zero divergence while the second has zero curl. Therefore any vector $\boldsymbol{E}$ can be expressed as[6]

$$\boldsymbol{E} = \boldsymbol{E}_{\text{t}} + \boldsymbol{E}_{\text{l}} \qquad \text{where} \quad \operatorname{div} \boldsymbol{E}_{\text{t}} = 0 \quad \text{and} \quad \operatorname{curl} \boldsymbol{E}_{\text{l}} = 0. \tag{4.9}$$

Substituting this decomposition into (4.7) yields

$$\ddot{\boldsymbol{E}}_{\text{t}} + \omega_{\text{p}}^2 \boldsymbol{E}_{\text{t}} - c^2 \nabla^2 \boldsymbol{E}_{\text{t}} = -\left(\ddot{\boldsymbol{E}}_{\text{l}} + \omega_{\text{p}}^2 \boldsymbol{E}_{\text{l}}\right). \tag{4.10}$$

The divergence of the left-hand side is zero; the curl of the right-hand side is zero. Therefore both sides have zero divergence and zero curl. A function whose divergence and curl are everywhere zero can usually be taken to be zero.[7] We therefore have two independent equations of motion, one for $\boldsymbol{E}_{\text{t}}$ and one for $\boldsymbol{E}_{\text{l}}$:

$$\ddot{\boldsymbol{E}}_{\text{t}} + \omega_{\text{p}}^2 \boldsymbol{E}_{\text{t}} - c^2 \nabla^2 \boldsymbol{E}_{\text{t}} = 0 \qquad \text{with} \quad \operatorname{div} \boldsymbol{E}_{\text{t}} = 0; \tag{4.11}$$

$$\ddot{\boldsymbol{E}}_{\text{l}} + \omega_{\text{p}}^2 \boldsymbol{E}_{\text{l}} = 0 \qquad \text{with} \quad \operatorname{curl} \boldsymbol{E}_{\text{l}} = 0. \tag{4.12}$$

---

[5] Remember sidenote 6 of Chapter 1. In this chapter, we do not introduce $\boldsymbol{M}$ or $\boldsymbol{H}$ for similar reasons. All this means that there is no question of introducing an ϵ_{r} or μ_{r}. With no "medium" present, the Maxwell equation (4.6) takes its "in vacuum" form.

[6] The subscripts are chosen with obvious malice aforethought.

[7] A function $\nabla\psi$, with ψ satisfying Laplace's equation, has both divergence and curl zero. Therefore the two sides of (4.10) are not quite compelled to be zero, but we may have

$$\ddot{\boldsymbol{E}}_{\text{t}} + \omega_{\text{p}}^2 \boldsymbol{E}_{\text{t}} - c^2 \nabla^2 \boldsymbol{E}_{\text{t}} = -\nabla\psi$$

$$\ddot{\boldsymbol{E}}_{\text{l}} + \omega_{\text{p}}^2 \boldsymbol{E}_{\text{l}} = +\nabla\psi.$$

The solution to each of these consists of a complementary function and a particular integral. The complementary functions solve (4.11) and (4.12). It remains to investigate the particular solutions, containing $\nabla\psi$.

A solution of Laplace's equation is necessarily time-independent, otherwise its changes would propagate with infinite speed. Therefore the $\pm\nabla\psi$ terms are time-independent.

A particular solution can be any special-case solution that works, because the complementary function can supply the rest of a full solution. Therefore $\pm\nabla\psi$ contributes only static additions to $\boldsymbol{E}_{\text{t}}$ and $\boldsymbol{E}_{\text{l}}$. But a time-independent field has both div $\boldsymbol{E} = 0$ (electrons move to ensure charge neutrality) and curl $\boldsymbol{E} = 0$, so the separation of $\boldsymbol{E}$ into $\boldsymbol{E}_{\text{t}}$ and $\boldsymbol{E}_{\text{l}}$ ceases to be useful. We have simply a total field $\boldsymbol{E}$ obeying (4.7), which leaves ψ with no job to do.

A static field $\boldsymbol{E}$ is not ruled out: it can be imposed by electrodes and cause a static current. But this is separate from the physics of plasma waves.

Look for a solution to (4.11) as the plane wave $E_{\rm t} = E_0\,{\rm e}^{{\rm i}(\boldsymbol{k}\cdot\boldsymbol{r}-\omega t)}$. It has the dispersion relation

$$c^2 k^2 = \omega^2 - \omega_{\rm p}^2, \qquad \text{and has} \quad \boldsymbol{k}\cdot\boldsymbol{E}_{\rm t} = 0. \qquad (4.13)$$

Equation (4.11) therefore describes transverse waves, and these waves have (4.13) for their dispersion relation.

Backtracking to eqn (4.5), we see that a transverse wave has an accompanying charge density ρ of zero. The electrons move, and contribute to a current $\boldsymbol{j}$, but in such a way that charge does not accumulate.

Look for a solution to (4.12) in the form $E_{\rm l} = E_0\,{\rm e}^{{\rm i}(\boldsymbol{k}\cdot\boldsymbol{r}-\omega t)}$. It has the dispersion relation

$$\omega = \omega_{\rm p} \qquad \text{and has} \quad \boldsymbol{k}\times\boldsymbol{E} = 0. \qquad (4.14)$$

Equation (4.12) therefore describes longitudinal waves, and these waves have (4.14) for their dispersion relation. The dispersion relation (4.14) does not contain k, so k can take any reasonable value: the longitudinal wave can have any not-too-rapid dependence on position without effect on the frequency.[8] The longitudinal wave is often called a **plasma wave**.[9]

Since curl $E_{\rm l} = 0$, we also have $\dot{\boldsymbol{B}} = 0$, which implies $\boldsymbol{B} = 0$. Then (4.6) tells us that $\boldsymbol{j} + \epsilon_0\dot{\boldsymbol{E}}_{\rm l} = 0$. It may seem surprising that there can be currents without any magnetic field, but we have just shown that conduction current and displacement current cancel against each other.

Before leaving the subject of plasmas, we should mention an alternative viewpoint that is often given, though it sits a little uneasily with what we have said here. Suppose that we pair each electron with a nearby ion and treat that pair as a dipole. An electron distant $\boldsymbol{r}$ from its partner gives the pair a dipole moment $-e\boldsymbol{r}$. It also contributes $-e\dot{\boldsymbol{r}}$ to the current density $\boldsymbol{j}$. A switch of accountancy can move "conduction current" $\boldsymbol{j}$ into "polarization current" $\partial\boldsymbol{P}/\partial t$. A displacement $\boldsymbol{D} = \epsilon_0\boldsymbol{E} + \boldsymbol{P}$ is then defined in the usual way, and with it a relative permittivity $\epsilon_{\rm r} = \boldsymbol{D}/\epsilon_0\boldsymbol{E}$. In this picture, the transverse wave propagates with phase velocity $\omega/k = c/\sqrt{\epsilon_{\rm r}}$ and has[10]

$$\epsilon_{\rm r} = 1 - \frac{\omega_{\rm p}^2}{\omega^2}, \qquad (4.15)$$

which is zero at the plasma frequency $\omega_{\rm p}$. By contrast, the longitudinal plasma wave has $\boldsymbol{D} = 0$ and $\epsilon_{\rm r} = 0$ always (until spatial dispersion is included).

Finally, we observe that a plasma has supplied a physical system, not too hard to describe, within which longitudinal and transverse waves propagate as independent motions. The two kinds of wave are characterized by (4.1), and nicely illustrate the application of those definitions.

4.4   Sound waves in an isotropic solid

Sound waves have transverse and longitudinal forms, in a fashion that is very similar to what we have found for a plasma.

[8] In a higher approximation, the plasma wave has a weak dependence of ω upon k. This arises from "spatial dispersion", a correction that goes right back to the beginning of our calculation. Briefly: what an electron is doing "here and now" (at location $\boldsymbol{r}$ and time t) depends upon what field $\boldsymbol{E}$ it experienced at earlier times when it was at other places. "Earlier times" appears in the mathematics as a dependence on ω; "other places" as a dependence on k. These matters are discussed in detail in Lifshitz and Pitaevskiĭ (1995) Chapter 3.

[9] In the limit $k \to 0$, the angular frequencies ω predicted by (4.13) and (4.14) come together at $\omega_{\rm p}$. This is to be expected: for a wave with very long wavelength, the electrons "cannot tell" whether they are participating in a transverse or a longitudinal disturbance.

[10] Equation (4.15) is, of course, just a different way of expressing the dispersion relation (4.13).

The following discussion is modelled on Landau and Lifshitz (1986), Chapters 1 and 3. The setting-up calculation is rather messier than for a plasma because stress and strain are second-rank tensors. The reader uninterested in technicalities may pick up the discussion at eqn (4.21).

Let an element of matter, originally at a point $r = (x, y, z)$ in a solid, be subject to a small displacement $\boldsymbol{\xi} = (\xi_x, \xi_y, \xi_z) = (\xi_1, \xi_2, \xi_3)$. The strain tensor u_{ij} is defined as[11]

$$u_{ij} = \frac{1}{2}\left(\frac{\partial \xi_j}{\partial x_i} + \frac{\partial \xi_i}{\partial x_j}\right).$$ (4.16)

Here[12,13] i and j are suffixes indicating Cartesian components; each runs from 1 to 3.

Let a small volume V of solid be subject to a force F caused by stresses applied to its surface elements dS. The total force acting on the material within volume V is[14,15]

$$F_i = \int \sigma_{ij}\, dS_j.$$ (4.17)

This defines the stress tensor σ_{ij}.

If the total force acting on the matter within volume V is non-zero, that matter (mass density ρ) accelerates. Moreover, the surface integral of σ_{ij} can be converted to a volume integral by Gauss's theorem, so we have

$$\int \rho \ddot{\xi}_i\, dV = \int \sigma_{ij}\, dS_j = \int \frac{\partial \sigma_{ij}}{\partial x_j}\, dV,$$

and since the volume integrated over is arbitrary we have the equation of motion[16]

$$\rho \ddot{\xi}_i = \frac{\partial \sigma_{ij}}{\partial x_j}.$$ (4.18)

Hooke's law relates stress σ_{ij} to strain u_{ij}. We shall be concerned only with the simple case of an isotropic solid. Even here, two independent parameters are needed to describe the elastic properties of the solid. To start with, these will be the **bulk modulus** K and **shear modulus** μ.

The elastic moduli K and μ can conveniently be defined together by writing Hooke's law in the form:[17]

$$\sigma_{ij} = K\, u_{kk}\, \delta_{ij} + 2\mu\left(u_{ij} - \tfrac{1}{3} u_{kk}\, \delta_{ij}\right).$$ (4.19)

Here u_{kk} (a sum over k) represents a fractional change of volume,[18] and accompanies the bulk modulus K. In a similar way, $\left(u_{ij} - \tfrac{1}{3} u_{kk}\, \delta_{ij}\right)$ represents a pure shear,[19] "pure" because the last term removes changes of volume from the bracket; it is accompanied by the shear modulus μ.

By convention, the parameters that are usually chosen to describe elasticity in isotropic media, rather than the bulk and shear moduli,

[11] For an introduction to tensor notation, see Chapter 3.

[12] The two derivatives are added in (4.16), in order that a pure rotation (which would contain the difference) is not counted as a strain of interest to elasticity.

[13] A term of second order in $\boldsymbol{\xi}$ has been dropped from u_{ij}. Throughout this chapter, the strain is taken to be small, so that linearized equations can be used.

[14] The sign here can easily be a source of muddle. The force on volume V, acting through its surface, can be thought of as giving (the rate of) an *inward* flow of momentum, yet the surface element dS faces outwards. In any given system, the individual components of σ_{ij} can be obtained from (4.17) by taking special cases.

[15] As is usual in tensor analysis, we use the summation convention, in which any repeated suffix (here j) is summed over (from 1 to 3).

[16] The right-hand side can be thought of as the divergence of the stress tensor: doing the same job on the tensor as a divergence does to a vector.

[17] Equation (4.19) is the simple (!) form that applies to an isotropic solid. For a crystal, K and μ are replaced by a fourth-rank stiffness tensor.

[18] From (4.16) $u_{kk} = \partial \xi_k/\partial x_k = \mathrm{div}\,\boldsymbol{\xi}$.

[19] *Question:* Why the factor 2 in front of μ in (4.19)?

are: **Young's modulus** Y and **Poisson's ratio** σ. In terms of these,

$$\text{bulk modulus} = K = \frac{Y}{3(1-2\sigma)}, \tag{4.20a}$$

$$\text{shear modulus} = \mu = \frac{Y}{2(1+\sigma)}, \tag{4.20b}$$

$$\text{axial modulus} = A = \frac{Y(1-\sigma)}{(1+\sigma)(1-2\sigma)}. \tag{4.20c}$$

Young's modulus Y applies to a measurement such as the stretching of a wire, where the strain is the fractional elongation, and where the wire is free to contract laterally as it extends. For these same conditions, if the z-axis faces along the wire, Poisson's ratio $\sigma = -u_{xx}/u_{zz} = -u_{yy}/u_{zz}$. In practice, σ lies between 0 and $\frac{1}{2}$.

The axial modulus A relates to the strain of fractional elongation, rather as does Young's modulus, but this time the deformed body is prevented from changing lateral dimensions as it extends or contracts.

We now have enough tools to substitute for stress in the equation of motion (4.18).

$$\rho\ddot{\xi}_i = \frac{\partial\sigma_{ij}}{\partial x_j} = \frac{\partial}{\partial x_j}\left\{K\,u_{kk}\,\delta_{ij} + 2\mu\left(u_{ij} - \tfrac{1}{3}u_{kk}\,\delta_{ij}\right)\right\}$$
$$= \left(K + \frac{\mu}{3}\right)\frac{\partial}{\partial x_i}\frac{\partial\xi_k}{\partial x_k} + \mu\frac{\partial}{\partial x_k}\frac{\partial}{\partial x_k}\xi_i.$$

We may now extricate ourselves from tensor notation and write this equation of motion as a vector equation:[20]

$$\rho\ddot{\boldsymbol{\xi}} = \left(K + \frac{\mu}{3}\right)\nabla\,\text{div}\,\boldsymbol{\xi} + \mu\,\nabla^2\boldsymbol{\xi}. \tag{4.21}$$

4.4.1   Separation of longitudinal and transverse motions

The equation of motion (4.21) must now be separated into equations describing longitudinal and transverse wave motions. There will be no surprises, given what happened in § 4.3.

The displacement amplitude is written as

$$\boldsymbol{\xi} = \boldsymbol{\xi}_t + \boldsymbol{\xi}_l; \qquad \text{div}\,\boldsymbol{\xi}_t = 0, \quad \text{curl}\,\boldsymbol{\xi}_l = 0. \tag{4.22}$$

Substituting this into (4.21) we find

$$\rho\ddot{\boldsymbol{\xi}}_t - \mu\nabla^2\boldsymbol{\xi}_t = -\left(\rho\ddot{\boldsymbol{\xi}}_l - A\nabla^2\boldsymbol{\xi}_l\right). \tag{4.23}$$

Here we have used relation $A = K + 4\mu/3$, which is obtainable from (4.20).

Equation (4.23) has a property similar to that of (4.10): both sides have zero divergence and zero curl, a fact which permits us to take them to be zero.[21] We therefore have

$$\rho\ddot{\boldsymbol{\xi}}_t - \mu\nabla^2\boldsymbol{\xi}_t = 0 \qquad \text{with} \quad \text{div}\,\boldsymbol{\xi}_t = 0; \tag{4.24}$$

$$\rho\ddot{\boldsymbol{\xi}}_l - A\nabla^2\boldsymbol{\xi}_l = 0 \qquad \text{with} \quad \text{curl}\,\boldsymbol{\xi}_l = 0. \tag{4.25}$$

[20] In Chapter 5, we discuss sound in a fluid, taking the special case of a plane wave where the displacement $\boldsymbol{\xi}$ lies always in the z-direction. The discussion here can be used to supply a full three-dimensional treatment. All we need do is set $\mu = 0$ in (4.19), which then doesn't need tensor notation. The outcome is (4.21) with μ set to zero.

Note though that transverse waves in a fluid involve viscosity which has not been allowed for in setting up (4.19). Equation (4.21) is therefore restricted, for fluids, to giving us the speed of longitudinal waves.

[21] As with eqn (4.10), this is not quite true. The reasoning that validates the discussion in the text is rather different for the two cases.

Equations (4.24) and (4.25) are more correctly
$$\rho\ddot{\boldsymbol{\xi}}_t - \mu\nabla^2\boldsymbol{\xi}_t = -\nabla\psi$$
$$\rho\ddot{\boldsymbol{\xi}}_l - A\nabla^2\boldsymbol{\xi}_l = +\nabla\psi,$$
where ψ is a solution of Laplace's equation, so that $\nabla\psi$ has both zero divergence and zero curl.

A ψ satisfying Laplace's equation can only be time-independent, so the particular solutions for $\boldsymbol{\xi}_t$ and $\boldsymbol{\xi}_l$ are given by
$$-\mu\nabla^2\boldsymbol{\xi}_t = -\nabla\psi$$
$$-A\,\text{grad}\,\text{div}\,\boldsymbol{\xi}_l = +\nabla\psi.$$
The second of these integrates to
$$A\,\text{div}\,\boldsymbol{\xi}_l = -\psi + \text{constant}.$$
Thus we may identify ψ with a stress, imposed statically on the body, and giving rise to strains having both polarizations.

The outcome is that $\boldsymbol{\xi}_t$ and $\boldsymbol{\xi}_l$ (particular solutions) are not necessarily zero, nor is their sum. Static distortions of the solid may result from the imposition of static stresses, and nothing in our calculation excludes such a possibility. However, the waves that solve (4.24) and (4.25) can take place with or without such static strains, and (within our linear equations) are unaffected by them. The term $-\nabla\psi$ therefore can describe legitimate physics, but physics in which we are not currently interested.

Look for a solution to (4.24) in the form $\boldsymbol{\xi}_t = \boldsymbol{\xi}_0\, e^{i(\boldsymbol{k}\cdot\boldsymbol{r}-\omega t)}$. It has phase velocity

$$\frac{\omega}{k} = \sqrt{\frac{\mu}{\rho}} \qquad \text{and has} \quad \boldsymbol{k}\cdot\boldsymbol{\xi}_t = 0. \qquad (4.26)$$

Equation (4.24) therefore describes waves that are transverse by definition;[22] those waves are dispersion-free and have (4.26) giving their phase velocity. The strain accompanying $\boldsymbol{\xi}_t$ is a pure shear, so it makes good sense that the elastic constant appearing in (4.26) is the shear modulus.

Look for a solution to (4.25) in the form $\boldsymbol{\xi}_l = \boldsymbol{\xi}_0\, e^{i(\boldsymbol{k}\cdot\boldsymbol{r}-\omega t)}$. It has phase velocity

$$\frac{\omega}{k} = \sqrt{\frac{A}{\rho}} \qquad \text{and has} \quad \boldsymbol{k}\times\boldsymbol{\xi}_l = 0. \qquad (4.27)$$

Equation (4.25) therefore describes waves that are longitudinal by definition; those waves are dispersion-free and have (4.27) giving their phase velocity. If we have a wide plane wave, the material medium will be unable to contract or expand laterally as the wave passes, because of its inertia, so it is the axial modulus of elasticity that we expect to determine the elastic response of the medium, and that is just what (4.27) tells us.

We have confirmed that, for sound waves in a isotropic solid, independent wave motions are possible: a longitudinal wave whose displacement amplitude $\boldsymbol{\xi}_l$ obeys $\operatorname{curl}\boldsymbol{\xi}_l = 0$; and a transverse wave (strictly two, since two polarizations are possible) with displacement amplitude $\boldsymbol{\xi}_t$ obeying $\operatorname{div}\boldsymbol{\xi}_t = 0$. There is a neat conformity with definitions (4.1).

[22] The property $\boldsymbol{k}\cdot\boldsymbol{\xi}_t = 0$ in (4.26) is special to the case of a uniform plane wave; it *illustrates* the transverse property but does not define it.

Similarly, the property $\boldsymbol{k}\times\boldsymbol{\xi}_l = 0$ in (4.27) is special to the case of a uniform plane wave; it illustrates the longitudinal property but does not define it.

Problems

Problem 4.1 The validity of classical mechanics for a plasma
Investigate the order of magnitude of an electron's de Broglie wavelength, as discussed in sidenote 4.[23]

[23] It is quite hard to find situations where near-classical conditions fail to hold. Examples are the "quantum liquids" ^{3}He and ^{4}He.

Problem 4.2 Transverse and longitudinal waves in a plasma
Examine the reasoning that disposes of $-\nabla\psi$ in sidenotes 7 and 21.

Problem 4.3 The ratios of elastic constants
Prove relations (4.20).

Suggestion: Start with (4.19) and take special cases.

Stretch a wire in the z-direction so that u_{zz} is the stretching strain, while allowing the side surfaces to contract ($\sigma_{xx} = 0$). Show that

$$K = \frac{2\mu}{3}\frac{(1+\sigma)}{(1-2\sigma)}; \qquad Y = K(1-2\sigma) + \frac{4\mu}{3}(1+\sigma).$$

Stretch the wire again, this time forcing $u_{xx} = u_{yy} = 0$. Show that

$$A \equiv \frac{\sigma_{zz}}{u_{zz}} = K + \frac{4\mu}{3}.$$

Hence obtain eqns (4.20).

5   Transmission lines, the universe and everything

Intended readership: On encountering any of the problems where waves are partly reflected and transmitted at boundaries.

5.1   Introduction

Transmission-line theory was, of course, invented to take care of waves propagating along electrical transmission lines, such as coaxial cables. In particular it deals with what happens when a wave encounters a boundary where two transmission lines are joined. However, the ideas and techniques can be carried over directly to a wide variety of other waves:

all wave–boundary problems are isomorphic.

This chapter shows how the analogies are constructed.

Example: an electromagnetic wave is incident on a boundary; as a result it is partly reflected, an effect which may be either desired or undesired. We then have a problem to solve: either to enhance or to suppress the reflection. Because of the isomorphism, if you can solve one such problem, you know how to solve all others.

Transmission lines provide a natural language for discussion of wave–boundary problems: terms such as **characteristic impedance**, **impedance mismatch**, **impedance matching** encapsulate useful ideas. Only transmission-line physics seems to provide us with such terminology. It is natural, therefore, for us to use transmission lines as the paradigm with which to discuss all similar problems.

There is a similarity here to other cases where we assist understanding of a piece of physics by means of an electrical analogy:

- An LRC circuit is often used to model any kind of resonance, and the sharpness of the resonance is described by giving the **quality factor** Q.

- An electrical filter provides a natural language (**pass band**, **stop band**) for describing waves in (other) periodic structures. It can help to think of a semiconductor's band gap as a *stop band* for electrons trying to travel through the crystal. Likewise, acoustic-branch phonons in a solid propagate for frequencies up to a cut-off, so a crystal lattice is a *low-pass filter* for these vibrations.[1]

[1] And it's a band-pass filter for optical-branch phonons.

Essays in Physics: Thirty-two thoughtful essays on topics in undergraduate-level physics. Geoffrey Brooker, Oxford University Press. © Geoffrey Brooker 2021. DOI: 10.1093/oso/9780198857242.003.0005

5.2   Basics of transmission-line theory

Let a transmission line be made from two parallel conductors that are long in the z-direction.[2] There are two physical quantities V, I describing the propagation of waves along the line: the voltage difference $V(z,t)$ between the conductors and the current $I(z,t)$ flowing along one of the conductors.[3]

The equations describing an electrical transmission line, and the waves it carries, are set up in standard works such as Bleaney and Bleaney (2013). We take it that those equations are well known,[4] and proceed to show how exactly similar equations may be made to apply to other kinds of wave.

We use a complex convention so that a wave having a single frequency $\omega/2\pi$ has V and I proportional to $e^{j(\omega t \pm kz)}$. The equations governing V and I (sidenote 4) yield the following properties:

(1) For a wave travelling towards $+z$, $V/I = +Z_0$;

  for a wave travelling towards $-z$, $V/I = -Z_0$;

  quantity Z_0 is called the **characteristic impedance** of the line.

(2) The impedance Z_0 is real and positive for the case of a loss-free line.

(3) When two transmission lines are joined, V and I are continuous at the join.

(4) The power transmitted along the line is $\mathrm{Re}(V) \times \mathrm{Re}(I)$.

When discussing a wave–boundary problem in some "other" area of physics, we shall be able to identify two quantities that behave in the same way as V and I. If possible we choose the I-quantity to look like some kind of flow, and the V-quantity to look like some kind of force acting on that flow. We further choose the quantities so that properties (1) to (4) are true for them. If we can do this, then the mathematical structure of our new problem must be identical to that for joined transmission lines.

5.3   Expressions for energy flow

There are several equivalent expressions representing energy flow, and property (4) is often most easily checked in one of its other forms. Therefore we digress to obtain an expression for use later.

Let a transmission line carry a wave described by

$$V = V_0\, e^{j(\omega t - kz)},$$
$$I = I_0\, e^{j(\omega t - kz)}.$$

Here the symbols V and I stand for complex quantities, while the voltage and current are, of course, real. As always with a complex convention (outside quantum mechanics), the real physical voltage is $\mathrm{Re}(V)$ and the real physical current is $\mathrm{Re}(I)$.

[2] The conductors do not have to be identical to each other, as with two parallel wires. They could, for example, be a wire inside a tube as with a coaxial cable. Or they could be a strip and a nearby conducting plane.

[3] The "voltage difference" V at z,t is $\int -\boldsymbol{E} \cdot \mathrm{d}\boldsymbol{l}$ taken along a path "straight across" the line at z. The path must be specified because curl $\boldsymbol{E} \neq 0$. We call V the "voltage difference" because "potential difference" won't do when curl $\boldsymbol{E} \neq 0$.

The travelling-wave modes of interest are such that the currents in the two conductors are equal and opposite. To see this, imagine that the currents are unequal. We can decompose those currents into their sum (within which the currents are equal and parallel), and their difference (within which they are equal and opposite). The first of these treats the two conductors as if they were a single conductor forming a transmission line with the "outside world" as the other conductor; such a mode wastes energy and measures will be taken to suppress it. We are left with the mode described in the text: $I(z,t)$ flows along one conductor, $-I(z,t)$ along the other.

[4] The equations governing voltage difference V and current I on a transmission line are:
$$\frac{\partial V}{\partial z} = -L\,\frac{\partial I}{\partial t},$$
$$\frac{\partial I}{\partial z} = -C\,\frac{\partial V}{\partial t}.$$
Here, L is the inductance per unit length of line, and C is the capacitance per unit length. If we eliminate either V or I between these two differential equations, we obtain a wave equation for the other, with wave speed $1/\sqrt{LC}$. The characteristic impedance is $Z_0 = \sqrt{L/C}$.

The power flow towards $+z$ is N_z where

$$\begin{aligned}
N_z &= \mathrm{Re}(V) \times \mathrm{Re}(I) \\
&= \tfrac{1}{2}(V + V^*) \times \tfrac{1}{2}(I + I^*) \\
&= \tfrac{1}{4}(V^*I + VI^* + VI + V^*I^*) \\
&= \tfrac{1}{4}\left(V_0^* I_0 + V_0 I_0^* + V_0 I_0\, e^{2\mathrm{j}(\omega t - kz)} + V_0^* I_0^*\, e^{-2\mathrm{j}(\omega t - kz)}\right).
\end{aligned}$$

This gives the instantaneous value of the energy flow at time t. What concerns us more is the *average* flow of energy when the average is taken over an integer number of half-cycles of oscillation (or over a long time). Then the terms VI and V^*I^* average to zero because they contain only an oscillation at angular frequency 2ω. We are left with

$$\overline{N_z} = \frac{V^* I + V I^*}{4}. \tag{5.1}$$

5.4 Complex amplitudes

An irritating detail must be discussed at this point. This chapter is "interdisciplinary", in the sense that it discusses waves across a wide range of physics, a range that encompasses different entrenched conventions for complex amplitudes.

Two conventions are commonly used for the complex representation of waves in physics:

- amplitudes $\propto e^{-\mathrm{i}\omega t}$, e.g. in quantum mechanics where a stationary state has[5]
$$\Psi(\boldsymbol{r}, t) = \psi(\boldsymbol{r})\, e^{-\mathrm{i}Et/\hbar}$$
- amplitudes $\propto e^{\mathrm{j}\omega t}$, e.g. in electric circuit theory, including the theory of transmission lines.

There is no prospect that either of these conventions will ever oust the other from its home territory, so we have to live with both. The discussions in this chapter, and elsewhere, must be able to move smoothly from one territory to the other, without confusion.

The solution we adopt is to use the two conventions itemized above—and those two only. Hybrids ($e^{+\mathrm{i}\omega t}$ or $e^{-\mathrm{j}\omega t}$) are forbidden. It is my view that this practice ought to be universally adopted within physics.[6]

Advantages of making these choices include the following:

- Where no particular convention is entrenched (sound, mechanics ...), we may use whichever convention seems more convenient, knowing that nothing important hangs on the choice.
- The presence of i or j in a formula is a reminder of the convention within which it was derived, so you don't have to remember separately what that convention was.
- If a formula contains neither i nor j then it holds within either convention.[7]
- A switch (if desired) to the other convention is implemented simply by making the routine replacement j ↔ −i.

[5] The first convention originates in optics, where it's convenient to have $e^{\mathrm{i}kz}$ representing a wave travelling towards $+z$: $\partial/\partial z = +\mathrm{i}k$ with a + sign. The second convention arises from electric circuits where the only dependence is on time t; then $\mathrm{d}/\mathrm{d}t = \mathrm{j}\omega$, again with a + sign. The opportunity for confusion happens when these conventions "meet in the middle".

Curiously, it is teaching texts on optics that most frequently use a convention that I dis-recommend.

[6] And I cannot understand why something so sensible is not already universal.

[7] Thus eqn (5.1) holds within both conventions.

In the present chapter, we write some wave (complex) amplitudes in the form $e^{j(\omega t - kz)}$ and some as $e^{i(kz - \omega t)}$. There will be no need to comment further on the choice made.

5.5   The basic transmission-line property

In this section we give *the* transmission-line property: what happens when a line of characteristic impedance Z is terminated by a "load" impedance Z_L.

The transmission line is shown in Fig. 5.1. A voltage-current wave is launched along the line from the left (we need not enquire how). The wave encounters an impedance Z_L across the line (from one conductor to the other) at $z = 0$, as a result of which the wave is partially reflected. We calculate the amplitude of the reflected wave.

Let the incident wave have $V = A\,e^{j(\omega t - kz)}$, and the reflected wave have $V = B\,e^{j(\omega t + kz)}$. Then in total the line carries voltage and current

$$\left.\begin{aligned} V &= A\,e^{j(\omega t - kz)} + B\,e^{j(\omega t + kz)} \\ I &= \frac{A}{Z}\,e^{j(\omega t - kz)} - \frac{B}{Z}\,e^{j(\omega t + kz)}. \end{aligned}\right\} \qquad (5.2)$$

At $z = 0$ we have a boundary condition:

$$Z_L = \left(\frac{V}{I}\right)_{z=0} = Z\,\frac{A + B}{A - B},$$

which is easily rearranged into[8]

$$\frac{B}{A} = \frac{Z_L - Z}{Z_L + Z}. \qquad (5.3)$$

Expression (5.3) is what we have come for. Expressions of precisely similar shape will appear repeatedly in the remainder of this chapter. That is what it means that all one-dimensional wave–boundary problems are isomorphic.[9]

Given that the ratio B/A is known, we may now calculate the ratio V/I at any point to the left of $z = 0$; in particular we may find the **input impedance** $Z_{\mathrm{in}} = \left(V/I\right)_{z=-l}$ of the structure, as measured by equipment connected to the left-hand end of the line at $z = -l$. This is dealt with in problem 5.1.

Comment: It will be clear from (5.14) that the input impedance Z_{in} of a transmission line depends on how the line is terminated (Z_L) and on the distance l between that load and the place where the input impedance is measured. The input impedance Z_{in} of a section of line can be quite different from the characteristic impedance Z of that line.

Notice that, in the calculation above, we did not need to enquire as to the physics taking place to the right of $z = 0$: the reflection B/A depends on the impedance Z_L presented at $z = 0$ and not on any reason why Z_L takes that value. In particular, Z_L might be the impedance of some "lumped" components connected between the terminals, or it might be the input impedance of some further section of transmission

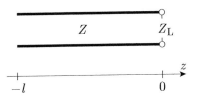

Fig. 5.1   A transmission line has characteristic impedance Z. At $z = 0$ it is "terminated" by an impedance Z_L, as a result of which waves travelling along the line are partly reflected. The effect is to give the structure an input impedance Z_{in}, as measured at $z = -l$, that depends upon Z_L and l as well as on Z.

The terminating impedance could be a "lumped" impedance such as a resistor or capacitor, or it might be another transmission line. The algebra needs no knowledge as to what kind of thing Z_L is.

Note the placing of the z-origin at the join. This simplifies the algebra by delaying the introduction of $e^{\pm jkl}$.

[8] Had we concentrated on the amplitude ratio for currents, we would have obtained expression (5.3) with reversed sign. We make a practice of displaying ratios of "voltage-type" quantities, so that the similarity with (5.3) is maintained.

[9] If we arrange things so that $Z_L = Z$ (no reflection) we say that the transmission line is terminated with a load equal to its characteristic impedance. Equivalently we say that the load impedance Z_L has been **matched** to the impedance of the line. If we had to do something to arrange $Z_L = Z$ that something was **impedance matching**. Conversely, if $Z_L \neq Z$, and in particular if Z_L/Z is far from 1, then the impedances are **mismatched**. These words are examples of the terminology referred to in § 5.1.

line. It is because we do not need to know that the circuit diagram of Fig. 5.1 is left vague at this point.

5.6 Plane electromagnetic wave

An electrical transmission line carries waves that propagate in only one dimension (i.e. along the line in one direction or the other). Then if transmission-line methods are to be carried over to radiation problems we must confine our attention to radiation in one dimension. Hence a restriction (in this chapter) to plane waves, and to normal incidence on boundaries.

Consider an electromagnetic wave (of any frequency) travelling in the z-direction, as shown in Fig. 5.2. We shall choose

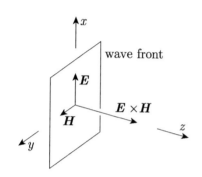

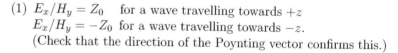

$$E_x \text{ as analogous to voltage } V$$
$$\text{and} \quad H_y \text{ as analogous to current } I.$$

We now check that the properties (1)–(4) of §5.2 are satisfied for these definitions.

Fig. 5.2 A plane electromagnetic wave travelling in the $+z$-direction.

(1) $E_x/H_y = Z_0$ for a wave travelling towards $+z$
$E_x/H_y = -Z_0$ for a wave travelling towards $-z$.
(Check that the direction of the Poynting vector confirms this.)

(2) The quantity Z_0 is the characteristic impedance of the medium within which the waves propagate. Maxwell's equations tell us that $Z_0 = \sqrt{\mu_0 \mu_\mathrm{r}/\epsilon_0 \epsilon_\mathrm{r}}$. (Here μ_r is the relative permeability of the medium, and ϵ_r is the relative permittivity.) Z_0 is real if the waves travel in a non-absorbing medium.

(3) If a travelling wave is incident normally on a plane boundary between two media, the electric and magnetic fields in the incident wave are parallel to the boundary. The tangential components of $\boldsymbol{E}$ and $\boldsymbol{H}$ are continuous at a boundary, so E_x and H_y are continuous.

(4) The power being conveyed in the $+z$-direction (per unit area) is $E_x \times H_y$, which is exactly similar to $V \times I$ (real parts taken before multiplying).

The properties enumerated in the last paragraph correspond with the list in §5.2. Therefore the quantities E_x and H_y behave in precisely the same way as the V and I on a transmission line.[10] I'll illustrate this by working out three examples. The examples are in increasing order of complexity; the first is so simple that the similarity to transmission lines does not really look useful, but the other two would be quite hard if we could not take over insights from transmission lines.

[10] Within SI definitions, we have an additional similarity. The units for E are volts per metre; those for H are amps per metre, so Z_0 has the dimensions of resistance and units of ohms. In other contexts, there is no necessity for an impedance to have the dimensions of resistance.

Example 5.1 Light reflected at an air–glass interface

Refer to Fig. 5.3. The waves travel in the $\pm z$ direction and the interface lies at $z = 0$. The electric and magnetic fields are:

$$(E_x)_\text{inc} = E_i\, e^{i(k_0 z - \omega t)}, \qquad (H_y)_\text{inc} = (E_i/Z_0)\, e^{i(k_0 z - \omega t)},$$

$$(E_x)_\text{ref} = E_r\, e^{i(-k_0 z - \omega t)}, \qquad (H_y)_\text{ref} = -(E_r/Z_0)\, e^{i(-k_0 z - \omega t)},$$

$$(E_x)_\text{trans} = E_t\, e^{i(kz - \omega t)}, \qquad (H_y)_\text{trans} = (E_t/Z)\, e^{i(kz - \omega t)}.$$

Here all three waves have the same angular frequency ω, otherwise the boundary-condition equations could not hold for all time t. Also, we have $\omega/k_0 = c$ while $\omega/k = c/\sqrt{\mu_r\, \epsilon_r}$.

The electric (E_x) and magnetic (H_y) fields at the boundary $z = 0$ are continuous:

$$E_i + E_r = E_t$$

$$\frac{E_i}{Z_0} - \frac{E_r}{Z_0} = H_i + H_r = H_t = \frac{E_t}{Z}.$$

Then

$$\frac{E_i - E_r}{Z_0} = \frac{E_i + E_r}{Z}$$

or

$$\frac{E_r}{E_i} = \frac{Z - Z_0}{Z + Z_0}. \tag{5.4}$$

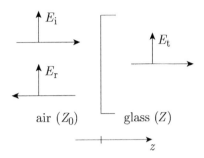

Fig. 5.3 Light is incident from air (impedance Z_0) onto a piece of glass (impedance Z), with the interface at $z = 0$. We imagine the situation so arranged that any light entering the glass does not return to the interface.

Equation (5.4) has an expected form; compare with (5.3).

Things simplify a little when our waves propagate in a non-magnetic medium so that $\mu_r = 1$. In the present example, we are thinking mainly of optical frequencies, frequencies so high that atomic magnetic moments cannot respond within a cycle. Then $Z = Z_0\sqrt{\mu_r/\epsilon_r} = Z_0/\sqrt{\epsilon_r}$ and we have

$$\frac{E_r}{E_i} = \frac{1/\sqrt{\epsilon_r} - 1}{1/\sqrt{\epsilon_r} + 1} = \frac{1 - n}{1 + n}, \tag{5.5}$$

where $n = \sqrt{\epsilon_r}$ is the refractive index of the glass.

The power per unit area carried by the incident wave is given by the z-component of the Poynting vector $N_z = \text{Re}(E_x) \times \text{Re}(H_y)$. Equation (5.1) now tells us that

$$\overline{N_z} = \frac{E^* H + E H^*}{4} = \frac{1}{4}\left(\frac{E^* E}{Z_0} + \frac{E E^*}{Z_0^*}\right)$$

$$= \frac{|E|^2}{4}\left(\frac{1}{Z_0} + \frac{1}{Z_0^*}\right) = \frac{|E|^2}{2}\, \text{Re}\left(\frac{1}{Z_0}\right),$$

which is just $|E|^2/(2Z_0)$ since Z_0 is real for air. Using this to calculate the power reflected at the boundary between air and glass, we find

$$\frac{\text{power reflected}}{\text{power incident}} = \frac{|E_r|^2/(2Z_0)}{|E_i|^2/(2Z_0)} = \left|\frac{E_r}{E_i}\right|^2 = \left|\frac{Z - Z_0}{Z + Z_0}\right|^2.$$

The (fractional) power transmitted through the boundary into the glass is evaluated from[11]

$$\frac{|E_t|^2\, \text{Re}(1/Z)}{|E_i|^2\, \text{Re}(1/Z_0)}, \qquad \text{which is } not \text{ the same as} \qquad \left|\frac{E_t}{E_i}\right|^2. \tag{5.6}$$

[11] We emphasize that the second expression in (5.6) is incorrect because it is easy to make a mistake. Elementary treatments of optics usually say "power is proportional to |amplitude|^2", because they are dealing with cases where no more than one medium is involved. This simplification must be outgrown.

A safe and unmuddling way to calculate the power transmitted is to say that it is the power that is not reflected:

$$\frac{\text{power transmitted}}{\text{power incident}} = 1 - \frac{\text{power reflected}}{\text{power incident}} = 1 - \left|\frac{Z - Z_0}{Z + Z_0}\right|^2.$$

Example 5.2 Antireflection coating ("blooming")

A layer of dielectric of impedance Z_1 is deposited on glass of impedance Z_2, as shown in Fig. 5.4. Light is incident from the air side of the structure (impedance Z_0). Such coatings[12] are used to "bloom" optical components to reduce reflections. We shall find the conditions for zero reflection.

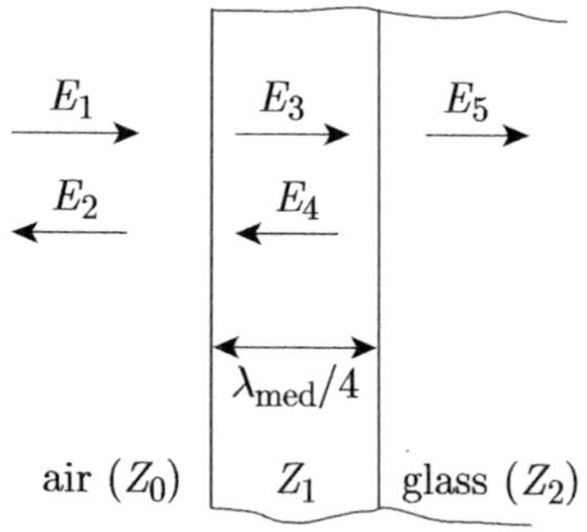

The obvious way to set this up if you don't know about transmission lines is to draw on the diagram the amplitudes of five waves E_1 to E_5 as shown. Four boundary conditions give four relations between the five amplitudes, (i.e. they give the four ratios), and these enable you to calculate the reflectance of the structure. Setting up the calculation this way is lengthy.

Quicker and more insightful: A quarter-wavelength section of transmission line acts as a transformer. If the line has characteristic impedance Z_1 and is terminated by impedance Z_2, the input impedance of the whole is

$$Z_{\text{in}} = \frac{Z_1^2}{Z_2}. \tag{5.7}$$

The quick route then is to know that this relation will be useful, and to derive it as expeditiously as possible (problem 5.2).[13]

In this example, the glass has characteristic impedance Z_2. Since it carries a wave going to the right only, the glass has input impedance Z_2. This "terminates" the quarter-wavelength section of dielectric of impedance Z_1. The input impedance of the whole structure is $Z_{\text{in}} = Z_1^2/Z_2$.

The air from which the light is incident has impedance Z_0, and it is terminated by impedance Z_{in}. Therefore the amplitude of the wave reflected off the structure is E_2 where

$$\frac{E_2}{E_1} = \frac{Z_{\text{in}} - Z_0}{Z_{\text{in}} + Z_0} = \frac{Z_1^2/Z_2 - Z_0}{Z_1^2/Z_2 + Z_0}.$$

The condition that no power be reflected is clearly[14]

$$Z_0 = Z_{\text{in}} = Z_1^2/Z_2, \qquad \text{or} \quad Z_1 = (Z_0 Z_2)^{1/2}.$$

Tidying up (and assuming optical frequencies), $Z_1 = Z_0/\sqrt{\epsilon_1} = Z_0/n_1$, and $Z_2 = Z_0/n_2$, so the condition for no reflection is

$$n_1 = \sqrt{n_2}.$$

Since we are considering lossless materials, this condition for no power to be reflected is also, of course, the condition for 100% of the incident power to be transmitted.

Example 5.3 Highly reflecting structures, e.g. laser mirrors

This example uses the same short cut as the last, but applies it to a more complicated problem. Figure 5.5 shows the structure to be discussed.

Fig. 5.4 A quarter-wave layer deposited on glass so as to make an antireflection coating. Arrows indicate the direction of travel of each wave; the electric fields are, of course, transverse.

[12] Practical coatings often have three or more layers. But the principle can be illustrated by means of the simpler arrangement discussed here.

[13] It might seem surprising that the input impedance is *inversely* proportional to the load impedance. However, an ordinary electrical transformer has this property: an impedance Z_2 is "reflected" from secondary to primary as $\omega^2 M^2/Z_2$ (Bleaney and Bleaney (2013), eqn 7.30).

[14] The fact that Z_{in} must be real explains why the only thickness considered for the coating is an integer multiple of a quarter-wavelength. It is easy to show from (5.14) that only such thicknesses give a real Z_{in}. Also, a half wavelength would give $Z_{\text{in}} = Z_2$ which is interesting but gives no help here. And a larger (odd) multiple of a quarter wavelength would give too frequency-sensitive a performance.

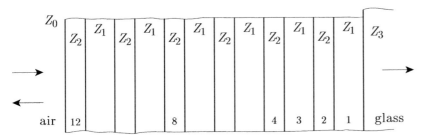

Fig. 5.5 A stack of dielectric layers of alternating refractive index is deposited onto a glass substrate to form a highly reflecting structure.

Note that the layers are numbered leftwards from the glass towards the air.

We deposit on top of a piece of polished glass a large number of layers of transparent material, of alternately high and low refractive index. All layers are a quarter wavelength thick (with the wavelength measured in the material concerned—so the actual thicknesses in metres are unequal). A multilayer coating like this is made by evaporating the two materials successively onto the glass in a high vacuum.

Light is incident on the multilayer from the left, from air whose impedance is Z_0. We wish to make the power reflected from the structure as large as possible. This means that we want to make the input impedance of the whole structure as different as possible from Z_0; either very large or very small will do.

Layer 1 is a quarter-wave transmission line (I mean layer)[15] of impedance Z_1 terminated by impedance Z_3, so its input impedance is Z_1^2/Z_3. This impedance is now what terminates layer number 2. By the same argument then, the input impedance of layer 2 must be

$$\frac{Z_2^2}{(Z_1^2/Z_3)} = \left(\frac{Z_2}{Z_1}\right)^2 Z_3.$$

We have already compared a quarter-wave transmission line to a transformer, because its input impedance is different from its load—though the two impedances are related. We may now look upon layers 1 and 2 as a composite transformer. This transformer has input impedance $(Z_2/Z_1)^2 Z_3$ when terminated by impedance Z_3 —true whatever the impedance Z_3 is, since we have not yet given Z_3 any specific value. Apply this idea to layers 3 and 4; these are a transformer identical to layers 1 and 2. Transformer $(3,4)$ is terminated by impedance $(Z_2/Z_1)^2 Z_3$, so its input impedance is

$$\left(\frac{Z_2}{Z_1}\right)^2 \times \left(\frac{Z_2}{Z_1}\right)^2 Z_3 = \left(\frac{Z_2}{Z_1}\right)^4 Z_3.$$

The rule for finding the input impedance of any layer should now be clear:[16]

the input impedance of the glass is Z_3

the input impedance of layer 2 is $(Z_2/Z_1)^2$ Z_3

the input impedance of layer 4 is $(Z_2/Z_1)^4$ Z_3

the input impedance of layer $2n$ is $(Z_2/Z_1)^{2n}$ Z_3.

Clearly, unless $(Z_2/Z_1) = 1$ (in which case the layers of dielectric are

[15] The reasoning proceeds from the glass leftwards, so the layers are numbered from right to left.

[16] Let's be in no doubt what the words mean. By "the input impedance of layer 2" is meant: the impedance E/H as measured at the left-hand surface of layer 2. The value of this impedance is determined by the properties of layer 2 and everything to the right of it.

identical), the input impedance gets farther from Z_0 as we increase the number $2n$ of layers. An n can be found that gives a desired reflectivity.

Take a numerical example. For definiteness[17] suppose that $Z_2/Z_1 < 1$. The glass substrate has refractive index 1.6. We make on it a multilayer from alternate layers of zinc sulphide ($n = 2.36$) and cryolite ($Na_3 Al F_6$, $n = 1.338$). We choose to deposit cryolite next to the glass. Then $Z_1 = Z_0/1.338$, $Z_2 = Z_0/2.36$, and $Z_3 = Z_0/1.6$. The input impedance when $2n$ layers are deposited is

$$Z_{\text{in}} = \left(\frac{1.338}{2.36}\right)^{2n} \frac{Z_0}{1.6} = \frac{(0.567)^{2n}}{1.6} Z_0.$$

Then the (power) reflectivity of the whole stack is R where

$$R = \left|\frac{Z_{\text{in}} - Z_0}{Z_{\text{in}} + Z_0}\right|^2 = 1 - \frac{4 Z_{\text{in}}/Z_0}{(1 + Z_{\text{in}}/Z_0)^2} = 1 - \frac{4(0.567)^{2n}/1.6}{\{1 + (0.567)^{2n}/1.6\}^2},$$

where the first step uses the fact that Z_{in} and Z_0 are real. Let's find out how many layers are needed to achieve a reflection coefficient of 99.8%. We require $R = 1 - 0.002$; feed this into the expression for R and solve for n. The easiest way is to do it approximately by writing[18]

$$\frac{(0.567)^{2n}}{1.6} = \frac{0.002}{4}\left(1 + \frac{(0.567)^{2n}}{1.6}\right)^2.$$

It's obvious that $(0.567)^{2n}/1.6$ is a small number, so the right-hand side is just 0.0005 to an adequate approximation, and we find $(2n) = 12.6$. Twelve layers don't quite reach the required reflectivity,[19] so we must round up to 14.

Comment: For honesty, I'll mention that dielectric mirrors usually have an odd number of layers. However, the present example is meant to teach principles, not to do a full practical design.[20]

Comment: It is feasible to make dielectric multilayers which reflect well over 99% of incident light energy. The mirrors used in (low-gain) lasers are made in this way. (For comparison, a mirror made from evaporated aluminium reflects only 93%—depending somewhat on frequency, but also on nuisances like the degree of tarnish of the metal.)

Comment: Because a dielectric mirror is made from layers a quarter-wavelength thick, the reflectivity is high for only a small range of wavelengths around the "design wavelength"; dielectric mirrors look strongly coloured.[21] Thus we trade excellent reflectivity (at the design wavelength) against a lack of versatility (small range of wavelengths).

5.7 Sound

Because this chapter is about waves travelling in one dimension, we may set up the equations for sound in a one-dimensional formulation. For convenience, we consider a fluid (liquid or gas).[22]

Figure 5.6 shows a parcel of fluid originally located (before the sound wave arrived) between z and $z + \delta z$. Its bounding surfaces move to $z + \xi$ and $z + \delta z + \xi + \delta \xi$. The density changes from ρ_0 to $\rho_0 + \delta\rho$ in such a way that a constant mass in included: $\rho_0\,\delta z = (\rho_0 + \delta\rho)(\delta z + \delta\xi)$, so that

$$\frac{\delta\rho}{\rho_0} = -\frac{\delta\xi}{\delta z + \delta\xi} \approx -\frac{\partial\xi}{\partial z},$$

since $\delta\rho/\rho_0$ is small and therefore $|\delta\xi| \ll \delta z$. Then

$$\frac{-1}{\rho_0}\frac{\partial\rho}{\partial z} = -\frac{1}{\rho_0}\frac{\partial\delta\rho}{\partial z} = \frac{\partial^2\xi}{\partial z^2} \tag{5.8}$$

(keeping small quantities of first order throughout).

Let the pressure be $P_0 + p$, where P_0 is the steady background (probably atmospheric) pressure, and $p = p(z, t)$ is the addition set up by the sound wave. The pressure difference across our element, pushing it to the right, is

$$p(z + \xi) - p(z + \delta z + \xi + \delta\xi) = -\frac{\partial p}{\partial z}(\delta z + \delta\xi) \approx -\frac{\partial p}{\partial z}\delta z.$$

The equation of motion for the matter in the element is

$$-\frac{\partial p}{\partial z}\delta z = (\rho_0\,\delta z)\frac{\partial^2\xi}{\partial t^2}, \qquad \text{so} \qquad \frac{\partial^2\xi}{\partial t^2} = \frac{-1}{\rho_0}\frac{\partial p}{\partial z}. \tag{5.9}$$

Combining (5.8) and (5.9), we find

$$\frac{\partial^2\xi}{\partial t^2} = \frac{-1}{\rho_0}\frac{\partial p}{\partial z} = \frac{\mathrm{d}p}{\mathrm{d}\rho}\left(\frac{-1}{\rho_0}\frac{\partial\rho}{\partial z}\right) = \frac{\mathrm{d}p}{\mathrm{d}\rho}\frac{\partial^2\xi}{\partial z^2}.$$

This is an equation of wave motion with wave speed v where $v^2 = \mathrm{d}p/\mathrm{d}\rho$; in this approximation, there is no dispersion and no mechanism for absorption of energy from the wave.[23]

Now that we know a wave can propagate, suppose that ξ and p are proportional to $\mathrm{e}^{\mathrm{i}(kz-\omega t)}$. Then eqn (5.9) shows that

$$\frac{p}{\xi} = \rho_0\frac{(-\omega^2)}{-\mathrm{i}k}, \qquad \text{or} \qquad \frac{p}{\dot\xi} = \rho_0\frac{\omega}{k} = \rho_0 v \times \mathrm{sign}(k). \tag{5.10}$$

In eqn (5.10), we have set up for an "impedance" discussion of waves reflected at boundaries. We wish to find two quantities that look like the V and I of a transmission line. Our first thought is probably to use p and ξ. However, these two quantities will not do, because their ratio is imaginary; we wish the characteristic impedance Z to be real. Quantities p and $\dot\xi$ do have a real ratio, and still are continuous at boundaries, so they give a better choice.

Show for yourself that conditions (1)–(4) of §5.2 are satisfied by the choice of p and $\dot\xi$ as analogous to V and I. We are now in a position to find how sound waves are reflected at boundaries, using exactly the same mathematical methods as are familiar from transmission lines.

The acoustic impedance, meaning $p/\dot\xi$ for a wave travelling towards $+z$, is $Z = \rho_0 v$.

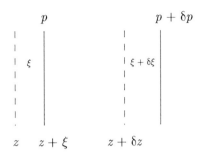

Fig. 5.6 The fluid between z and $z+\delta z$ is displaced by a sound wave passing through the medium. A surface that was at z is displaced to $z+\xi$ and similarly. The element is accelerated because the pressure p is different on the two surfaces.

[23] In more detail,

$$v^2 = \frac{\mathrm{d}p}{\mathrm{d}\rho} = \left(\frac{\partial p}{\partial\rho}\right)_S:$$

changes of density take place adiabatically, at constant entropy S.

Example 5.4 A measurement of an acoustic impedance

A plane sound wave propagates inside a slab of quartz, and repeatedly reflects normally from each end surface of the slab. Outside the quartz is a bath of liquid helium. The absorption of sound in the quartz is negligible, so the sound wave bounces back and forth inside the quartz until all of its energy has leaked out into the helium.

The sound wave has the form of a wave packet containing many periods of oscillation, so its frequency is well defined; but the length of the wave packet is small compared with the thickness of the slab. How many reflections are needed for the energy in the quartz to fall by a factor 2?

Density of quartz $= 2650 \, \mathrm{kg \, m^{-3}}$;
speed of sound in quartz $= 5720 \, \mathrm{m \, s^{-1}}$;
density of helium ($^4\mathrm{He}$) $= 145 \, \mathrm{kg \, m^{-3}}$;
speed of sound in helium $= 237 \, \mathrm{m \, s^{-1}}$.

Solution: Let the fraction of the acoustic power reflected be R where

$$R = \left| \frac{Z_{\mathrm{He}} - Z_{\mathrm{q}}}{Z_{\mathrm{He}} + Z_{\mathrm{q}}} \right|^2 ,$$

so

$$1 - R = 1 - \left| \frac{Z_{\mathrm{He}} - Z_{\mathrm{q}}}{Z_{\mathrm{He}} + Z_{\mathrm{q}}} \right|^2 = \frac{4 \, Z_{\mathrm{He}} \, Z_{\mathrm{q}}}{(Z_{\mathrm{He}} + Z_{\mathrm{q}})^2} = \frac{4 \, Z_{\mathrm{He}}/Z_{\mathrm{q}}}{(1 + Z_{\mathrm{He}}/Z_{\mathrm{q}})^2} .$$

The acoustic impedance $Z = \rho v$, so

$$\frac{Z_{\mathrm{He}}}{Z_{\mathrm{q}}} = \frac{145 \times 237}{2650 \times 5720} = 2.267 \times 10^{-3}, \qquad \text{and} \quad 1 - R = 9.027 \times 10^{-3}.$$

Let the acoustic wave packet have energy E. At each reflection it loses a fraction $-\delta E/E = 9.027 \times 10^{-3}$. We find that after n reflections $E \propto \exp(-9.027 \times 10^{-3} n)$. The energy falls to half its original value when $9.027 \times 10^{-3} n = \ln 2$, which gives $n = 77$.

Comment: This example is the basis of an experiment for investigating sound propagation in helium.[24] Its advantage is that it works even at very high frequencies ($\sim 1 \, \mathrm{GHz}$) where sound in helium is absorbed in a fraction of a millimetre,[25] and the usual "time of flight" method for speed measurement is too difficult.

[24] Keen, Matthews and Wilks, (1965).

[25] The absorption is still small per-wavelength, so it remains valid to take Z_{He} as real.

For honesty: The actual experiment used a quartz crystal, with its associated complicated elastic properties. The crystal orientation was chosen in such a way as to restore the usual simplicities. The value given for the speed of sound is that for the chosen direction of propagation.

5.8 Mechanics

A solid rod of circular cross section is twisted by applying a torque at angular frequency ω. The angle through which the rod twists is related to the applied couple (torque) Γ and to the shear modulus n of the material. The rod has radius R.

Books on mechanics show that a rod of length z twists through angle θ with

$$\Gamma = \frac{\pi R^4}{2} n \frac{\theta}{z}. \tag{5.11}$$

This applies when a static torque Γ is applied to the right-hand $(+z)$ end of the rod, so the rod twists with a constant $\partial\theta/\partial z$. Argue for yourself that in the dynamic situation we must replace eqn (5.11) with

$$\Gamma = -\frac{\pi R^4}{2}\, n\, \frac{\partial\theta}{\partial z}.$$

Here we introduce a negative sign, because the torque we wish to work with is that applied to the *left*-hand end of the rod. The equation of motion[26] for a disc of thickness δz is

$$\frac{\partial\Gamma}{\partial z}\,\delta z = -\delta z\,\rho\,\frac{\pi R^4}{2}\frac{\partial^2\theta}{\partial t^2}.$$

We may now substitute the last two equations into each other to show that there is a wave equation

$$\frac{\partial^2\theta}{\partial t^2} = \frac{n}{\rho}\frac{\partial^2\theta}{\partial z^2},\qquad \text{and that}\quad Z \equiv \frac{\Gamma}{\dot\theta} = \frac{\pi R^4}{2}\sqrt{n\rho}. \tag{5.12}$$

We have a wave equation with wave speed $\sqrt{n/\rho}$, and a pair of quantities $\Gamma, \dot\theta$ that are suitable[27] to be analogous to V and I on a transmission line, and that give Z as the characteristic impedance.

[26] A derivation can be modelled on that leading to (5.9). You may need to work out the moment of inertia of a disc of radius R and thickness δz. Problem 5.6 asks you to verify all the calculations of this section.

[27] In the second statement of (5.12), we have written the ratio of Γ and $\dot\theta$ that applies when a wave travels towards $+z$. A wave travelling towards $-z$ still has positive Z, but $\Gamma/\dot\theta = -Z$.

Example 5.5 Twisting of joined bars
Two bars are joined as shown in Fig. 5.7. The end C of the composite structure is free, and an angular displacement $\theta = \theta_0\cos\omega t$ is applied to end A. To find: the torque that must be applied at A to maintain the system in twisting oscillation. For simplicity, we take the bars to have the same radius R.

Solution
Figure 5.7 shows the two bars with their impedances and lengths.
   It is easiest to use a complex-exponential notation, so we regard $\theta_0\cos\omega t$ as the real part of $\theta_0\exp(j\omega t)$.
   Since the bars have nothing-in-particular lengths, this is an ideal example on which to demonstrate the matrices introduced in problem 5.1.

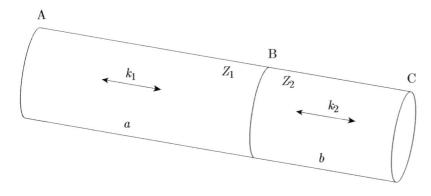

Fig. 5.7 Two bars are joined as shown. End C is free, while end A is given angular displacement $\theta = \theta_0\cos\omega t$. Matrix methods are the elegant way to find the torque required for maintaining the oscillation.

$$\begin{pmatrix} \Gamma \\ \dot{\theta} \end{pmatrix}_{\mathrm{A}} = \begin{pmatrix} \cos(k_1 a) & jZ_1 \sin(k_1 a) \\ jZ_1^{-1} \sin(k_1 a) & \cos(k_1 a) \end{pmatrix} \begin{pmatrix} \cos(k_2 b) & jZ_2 \sin(k_2 b) \\ jZ_2^{-1} \sin(k_2 b) & \cos(k_2 b) \end{pmatrix} \begin{pmatrix} \Gamma \\ \dot{\theta} \end{pmatrix}_{\mathrm{C}}.$$

At C, the free end, the torque Γ_{C} must be zero, leaving $\dot{\theta}_{\mathrm{C}}$ to take whatever value results from the imposition of Γ and $\dot{\theta}$ at A. It is now easy to multiply out the matrices, to divide Γ_{A} by $\dot{\theta}_{\mathrm{A}}$, and to find the input impedance (thereby cancelling $\dot{\theta}_{\mathrm{C}}$). Finally, after a little tidying, we arrive at

$$\Gamma = \omega Z_1 \left(\frac{Z_2 \tan(k_2 b) + Z_1 \tan(k_1 a)}{Z_2 \tan(k_1 a) \tan(k_2 b) - Z_1} \right) \theta_0 \cos(\omega t).$$

[28] The torque Γ is in phase with the given angular displacement $\theta_0 \cos(\omega t)$. Check that the energy given to the bars averages to zero over a period of oscillation. Check that this is consistent with the system having a pure imaginary input impedance.

Thus, we find the torque Γ required to maintain the twist[28] having the given $\theta_0 \cos(\omega t)$.

Notice that we have not needed to introduce wave amplitudes, similar to the A, B of (5.2), because they were eliminated, well in advance, during the setting-up of the matrices in problem 5.1.

5.9   Quantum mechanics

A particle travels in one dimension in a region of space where there is a potential energy function $V(z)$, which here will be constant or piecemeal constant with z. Its wave function satisfies the Schrödinger equation

$$i\hbar \frac{\partial \Psi}{\partial t} = -\frac{\hbar^2}{2m} \frac{\partial^2 \Psi}{\partial z^2} + V(z)\Psi = E\Psi.$$

The wave function representing a particle travelling towards $+z$ is

$$\Psi = A\,e^{i(kz - \omega t)}, \qquad \text{where} \quad \omega = E/\hbar \quad \text{and} \quad k^2 = 2m\{E - V(z)\}/\hbar^2.$$

We also have

$$-i\partial\Psi/\partial z = k\Psi.$$

Let us choose Ψ as analogous to current, and $-i\partial\Psi/\partial z$ as analogous to voltage. (Neither quantity looks like a force or a flow, so the choice of which is which is arbitrary.) With these definitions, a wave travelling along the $+z$-axis moves in a "medium" with impedance

$$Z = \frac{-i\partial\Psi/\partial z}{\Psi} = k.$$

We next check that the conditions (1)–(4) of §5.2 are satisfied by these choices. Conditions (2) and (4) need to be considered together because they involve "loss-free": what are we in danger of losing? What is conserved here is not energy but the probability current:

$$S = \frac{i\hbar}{2m}\left(\frac{\partial\Psi^*}{\partial z}\Psi - \frac{\partial\Psi}{\partial z}\Psi^* \right) = \frac{2\hbar}{m}\frac{1}{4}\left\{ \left(-i\frac{\partial\Psi}{\partial z}\right)^*\Psi + \left(-i\frac{\partial\Psi}{\partial z}\right)\Psi^* \right\}.$$

[29] We could have incorporated the coefficient $2\hbar/m$ into either "voltage" or "current", but the algebra would have been cluttered for no benefit.

Apart from multiplying constants, this expression has similar form to (5.1).[29] It is now not hard to see that the conditions of §5.2 are all satisfied by our choices of "voltage" and "current".

Finally, in the case that the particles have travelling-wave character, that is when $E > V(z)$, then k is real and the impedance is real: condition (2) is satisfied.

Example 5.6 Electrons incident on a potential step
The potential step is shown in Fig. 5.8, together with arrows indicating the directions of travel of the three waves.

The wave function has the form $e^{i(kz-\omega t)}$, where the wave vector k takes values $\pm k_1$ on the left of the step, k_2 on the right, and where

$$\frac{\hbar^2\, k_1^2}{2m} = E, \qquad \frac{\hbar^2\, k_2^2}{2m} = E - V_0.$$

The ratio

$$\frac{\text{flux of particles reflected}}{\text{flux of particles incident}} = \left|\frac{Z_2 - Z_1}{Z_2 + Z_1}\right|^2 = \left|\frac{k_2 - k_1}{k_2 + k_1}\right|^2,$$

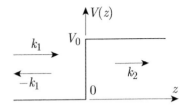

Fig. 5.8 Electrons are incident from the left on a potential step of height V_0.

since we have shown that the impedance for this problem is equal to k (that is, it is k_1 for the region to the left of the step and k_2 for the region on the right).

Physically, there are two cases here, depending on whether E is greater than or less than V_0. If the energy E of the incident particles is greater than the height V_0 of the potential step, then both k_1 and k_2 are real, and some of the particles travel off to $+\infty$, i.e., they surmount the potential step and keep on travelling to the right. In that case obviously,

$$\frac{\text{flux of particles transmitted}}{\text{flux of particles incident}} = 1 - \frac{\text{flux reflected}}{\text{flux incident}} = 1 - \left|\frac{k_2 - k_1}{k_2 + k_1}\right|^2.$$

We may, of course, use boundary conditions to obtain the wave functions on both sides of the step, and work out the probability current by brute-force quantum mechanics. The answer must be the same, because the physics of the boundary conditions has been carefully built into our impedance method.

The case where the potential step is too high for the particles to climb over is easily dealt with as well. For this case, $E - V_0 < 0$ and k_2 is imaginary. We write $k_2 = i\beta$, and then on the right of the step $\Psi \propto e^{-\beta z}$, with the sign of β determined by the fact that the particles all arrive from the left (β is therefore positive). The input impedance of the step is now $Z_{\text{in}} = k_2 = i\beta$, which is pure imaginary. An electrical component with an imaginary Z_{in} is a pure capacitance or a pure inductance. A capacitance or inductance does not absorb energy (though it can store energy, borrowed and returned within each half cycle of an oscillation). A transmission line terminated by an L or C has 100% of incident energy reflected (apart from lend and borrow within a half-cycle). Correspondingly then, we expect 100% of the particles to be reflected in the quantum mechanical case, and sure enough:

$$\frac{\text{flux of particles reflected}}{\text{flux of particles incident}} = \left|\frac{k_2 - k_1}{k_2 + k_1}\right|^2 = \left|\frac{i\beta - k_1}{i\beta + k_1}\right|^2 = 1.$$

[30] When quantum particles encounter a potential hump having $V_0 > E$, their wave function undergoes exponential decay within the hump, but a little leaks out on the far side. This phenomenon is known as "tunnelling". It is exactly analogous to the behaviour of electromagnetic waves in a short length of waveguide beyond cutoff, or to a few sections of a filter in its stop band, or to "frustrated total internal reflection" in optics. Transmission-line theory is the natural technique for handling such cases, provided that we make the adaptations appropriate to the exponential decay (imaginary impedance) within the width of the hump.

You may have seen a case analogous to this in electromagnetism: total internal reflection. Another example in electromagnetism is a waveguide beyond cutoff. Such a waveguide does not carry away any energy, though it contains fields near the input end so it does borrow and return a little energy. The input impedance of a waveguide beyond cutoff is again pure imaginary. Another example: a filter receiving signals in its stop band.[30]

A more complicated quantum-mechanical example is given, in problem format, in problem 5.7.

5.10   Final comments

This chapter has shown, by means of several tutorial examples, that all one-dimensional wave–boundary problems have the same mathematical structure. Because of this similarity, it is appropriate to use much the same language when describing any of these pieces of physics: characteristic impedance, impedance mismatch. Likewise, expressions like $(Z_2 - Z_1)/(Z_2 + Z_1)$ turn up repeatedly, and can be applied to any reflection, once the impedance quantities have been properly defined.

All this raises a question: How should we present the solution of a problem not involving actual transmission lines? Should a problem in quantum mechanics (say) be translated into a notation of V, I and Z, solved in that notation (solved, that is, as if we really did have joined transmission lines), and then translated back again? It should be clear from the models given in this chapter, especially example 5.5, that the answer is "no". In example 5.5, we worked with torque Γ and angular velocity $\dot{\theta}$, without calling those quantities V and I. We did however, use symbols Z_1, Z_2 as convenient abbreviations for quantities $Z_1 = (\pi R^4/2)\sqrt{n_1\rho_1}$ and similar, because we needed to introduce such quantities, and it was surely helpful to draw attention to the similarity to transmission-line impedances.

The power of the transmission-line analogy is that it tells us how to work things out, and it often provides us with answers in advance. This shows up in problem 5.7, where we quote results for a quarter-wave antireflecting layer, and invite verification of them.

Problems

Problem 5.1 Input impedance of a terminated transmission line
Figure 5.9 shows a section of transmission line extending from $z = z_1$ to $z = z_2$. Voltage difference $V(z_2, t)$ and current $I(z_2, t)$ are present at $z = z_2$. We write

$$V(z,t) = A\,e^{j(\omega t - kz)} + B\,e^{j(\omega t + kz)}, \quad I(z,t) = \frac{A}{Z}\,e^{j(\omega t - kz)} - \frac{B}{Z}\,e^{j(\omega t + kz)}.$$

We wish to find $V(z_1), I(z_1)$ in terms of $V(z_2), I(z_2)$. Write these voltages and currents in terms of A, B, then eliminate A, B, and show that

$$\begin{pmatrix} V \\ I \end{pmatrix}_{z=z_1} = \begin{pmatrix} \cos(kl) & \mathrm{j}\, Z \sin(kl) \\ \mathrm{j}\, Z^{-1} \sin(kl) & \cos(kl) \end{pmatrix} \begin{pmatrix} V \\ I \end{pmatrix}_{z=z_2}, \qquad (5.13)$$

in which $l = z_2 - z_1$.

Write down the square matrix for a quarter-wave line ($kl = \pi/2$) and for a half-wave line ($kl = \pi$) so that you will be able to recognize their shapes.

Voltage V and current I are continuous at boundaries, so cascaded sections of transmission line can be handled by multiplying their matrices. Check this by showing that the matrix for a half-wave line is the product of the matrices for two quarter-wave lines.

In a return to the ideas of § 5.5, let there be a "load" impedance Z_L connected across the line at z_2, while we wish to find the input impedance Z_{in} as measured at z_1. Divide upper and lower matrix elements in (5.13) to show that

$$Z_{\text{in}} = Z \left(\frac{Z_L + \mathrm{j}\, Z \tan(kl)}{Z + \mathrm{j}\, Z_L \tan(kl)} \right). \qquad (5.14)$$

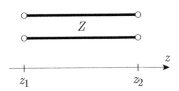

Fig. 5.9 A section of transmission line runs from $z = z_1$ to $z = z_2$. The voltage and current at z_1 are related to those at z_2. That relation is set up in problem 5.1, and is given by (5.13).

Problem 5.2 Quarter-wave layer
Apply transmission-line ideas to plane electromagnetic waves incident normally on boundaries.

Show that a quarter-wave dielectric of impedance Z_1, terminated by "something"[31] of impedance Z_2, has input impedance $Z_{\text{in}} = Z_1^2/Z_2$ in agreement with (5.7). Do this by calculation *ab initio*, and also by taking a special case in eqn (5.14).[32]

Problem 5.3 Energy conservation
In example 5.1, it was stated that a boundary between dielectrics conserves energy, because it transmits what it does not reflect. Verify this directly, using the $\frac{1}{2} |E|^2 \operatorname{Re}(1/Z)$ expressions for the energy flows.[33]

Problem 5.4 Antireflection coating
Suppose we wish to make an antireflection coating to put on glass of refractive index 1.6. The coating material should have a refractive index as close as possible to $\sqrt{1.6}$, but it also has to satisfy practical requirements: possible to evaporate; hard enough to stand abrasion; not attacked by moisture .... Such practicalities make us consider cryolite ($n = 1.338$), even though its refractive index is not quite right. Assuming that we can deposit a layer of the ideal thickness, calculate the reflectivity of the coated glass.

[My answer: 0.31%, compared with 5.3% for uncoated glass.]

Problem 5.5 Example 5.3 with zinc sulphide deposited first
How many layers are needed in example 5.3 if we use an even number of layers, depositing zinc sulphide first?

[31] The point of "something" is that it needn't be a block of dielectric of characteristic impedance Z_2; it could be any composite structure having Z_2 for its input impedance. Therefore do not obtain Z_2 by taking E/H for a wave in the second medium travelling away to the right. The calculation can be made elegant and brief if you make a point of looking for ways to make it so.

Make no mention of transmission lines in your calculation, but set up the whole problem entirely in its own language.

[32] For the *ab initio* case, look at the solution at the end of this chapter, to see how the algebra can be stripped to essentials.

[33] There are some surprising difficulties if the "first medium" absorbs (Z_0 complex), so for this problem take Z_0 to be real. But allow Z to be anything, real or complex.

Problem 5.6 Twisting of a bar
Obtain eqn (5.11) for the relation between torque Γ and twist angle θ for a twisted bar. Confirm the algebra in § 5.8.

Problem 5.7 Particles incident on a double step
The potential steps are shown in Fig. 5.10. Electrons having energy $E > V_2$ are incident from the left.

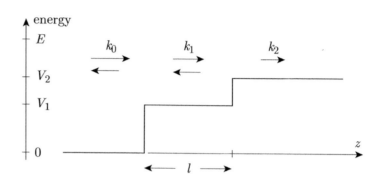

Fig. 5.10 Quantum particles having energy E are incident from the left on the two potential steps shown, steps with energy heights V_1 and $(V_2 - V_1)$. We are required to find the condition that the particles are wholly transmitted.

Show that the particles are wholly transmitted if two conditions are met:

$$k_1\, l = \pi/2, \qquad k_1 = \sqrt{k_0\, k_2}.$$

There are no prizes for noticing that this is just the physics of a quarter-wave antireflection coating again.

My solution to problem 5.2

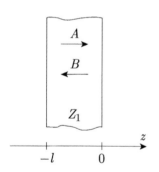

Fig. 5.11 A quarter-wave layer supporting electromagnetic waves travelling towards $\pm z$.

The quarter-wave layer is shown in Fig. 5.11. It has characteristic impedance Z_1, and its boundaries lie at $z = 0$ and $z = -l$. Waves travel within the layer with amplitudes A and B. Whatever lies to the right of $z = 0$ causes there to be impedance Z_{load} at $z = 0$; we are careful not to say anything as to how this comes about.

The fields within the quarter-wave layer take the form

$$E_x = A\, \mathrm{e}^{\mathrm{i}(kz-\omega t)} + B\, \mathrm{e}^{\mathrm{i}(-kz-\omega t)}$$

$$Z_1\, H_y = A\, \mathrm{e}^{\mathrm{i}(kz-\omega t)} - B\, \mathrm{e}^{\mathrm{i}(-kz-\omega t)}.$$

The boundary condition at $z = 0$ imposes a relation between the coefficients A and B:

$$Z_{\text{load}} = \left(\frac{E_x}{H_y}\right)_{z=0} = Z_1\left(\frac{A + B}{A - B}\right). \tag{5.15}$$

Do not waste time "undoing" this to find an expression for B/A; what we need is already on display.

We may now evaluate the fields at $z = -l$:

$$E_x = A\, \mathrm{e}^{\mathrm{i}(-kl-\omega t)} + B\, \mathrm{e}^{\mathrm{i}(kl-\omega t)} = \mathrm{i}(-A + B)\mathrm{e}^{-\mathrm{i}\omega t}$$

$$Z_1\, H_y = A\, \mathrm{e}^{\mathrm{i}(-kl-\omega t)} - B\, \mathrm{e}^{\mathrm{i}(kl-\omega t)} = -\mathrm{i}(A + B)\mathrm{e}^{-\mathrm{i}\omega t}.$$

The last step in each line uses the fact that the layer's thickness is a quarter-wavelength, so $e^{ikl} = i$, $e^{-ikl} = -i$.

The input impedance that the layer presents to whatever lies to its left is now

$$Z_{\text{in}} = \left(\frac{E_x}{H_y} \right)_{z=-l} = Z_1 \left(\frac{A - B}{A + B} \right) = Z_1 \frac{Z_1}{Z_{\text{load}}}.$$

The last step substitutes for the entire building block $\left(\frac{A-B}{A+B} \right)$ using (5.15).

Comment: Notice how much has *not* been mentioned in the above. The more we strip out inessentials, the clearer does the reasoning become, and the more inevitable does the result seem.

Counting quantum states and field modes

6.1 Introduction

[1] We can of course discuss cases where there is a real boundary affecting the physics of the particles. Two closely spaced conducting plates affect the modes between them available to the electromagnetic field, and by removing some of the modes that would otherwise exist they cause a small force pushing the plates together (Casimir effect). Electrons in a layered (heterojunction) semiconductor have quantum states that look just like the square-well states studied in a first course on quantum mechanics. These states have quantized energies that depend upon the thickness of the confining layer, a phenomenon used in fine-tuning the frequencies of optical transitions in an LED light source.

However, there is no such confinement for the discussions in this chapter.

[2] We may loosely say that we are finding the **density of states**. However, strictly, the density of states is the number of states per unit volume and per unit of energy range. The quantities are related, but not exactly the same. The density of states is discussed in § 6.6.

Suppose we have a system of quantum particles moving freely within a large volume; we need to count the quantum states available to those particles. We might be discussing the states of translational motion available to massive particles in vacuum, or the eigenfunctions available to the electromagnetic field, or the Bloch states available to electrons in a periodic lattice.

In cases such as these, we think of the particles as enclosed within a **quantization volume** (L^3 in what follows). This volume is large and arbitrary; it is unconnected with any (usually larger) real physical enclosure confining the particles.[1]

The calculation of the "number of quantum states"[2] is most often presented in textbooks on statistical mechanics. We may wish to apply a Boltzmann distribution (or Fermi–Dirac or Bose–Einstein) to some particles in order to work out thermodynamic properties of a gas composed of many such particles. However, the applications are wider than this. For example, we need the density of states (see § 6.6) when applying Fermi's golden rule to a transition (beta-decay, photon emission, . . .) with a quasi-continuum of final states.

Especially in a statistical-mechanics context, conventions have grown up as to how the calculation should proceed. Those conventions are often inappropriate, or even silly. This chapter tries to show a better way.

6.2 A "conventional" calculation

[3] We use ε to denote energy because E is used later in this chapter for electric field. Likewise, we use μ for the particle's mass because m is used as an integer eigenvalue.

Suppose we have particles of mass μ enclosed in a cubical box of side L. The Schrödinger equation for a representative particle is[3]

$$-\frac{\hbar^2}{2\mu}\nabla^2\psi + V(\boldsymbol{r})\psi = \varepsilon\,\psi, \tag{6.1}$$

where $V(\boldsymbol{r})$ is zero within the box but rises to infinity at the box walls. The effect of $V(\boldsymbol{r})$ is to impose a boundary condition on ψ, that it is zero on the walls. Then $\psi = 0$ on $x = 0$, $y = 0$, $z = 0$, $x = L$, $y = L$,

Essays in Physics: Thirty-two thoughtful essays on topics in undergraduate-level physics. Geoffrey Brooker, Oxford University Press. © Geoffrey Brooker 2021. DOI: 10.1093/oso/9780198857242.003.0006

$z = L$. The solutions for ψ have the form

$$\psi \propto \sin\left(\frac{l\pi x}{L}\right) \sin\left(\frac{m\pi y}{L}\right) \sin\left(\frac{n\pi z}{L}\right) \quad \text{with} \quad \varepsilon = \frac{\pi^2 \hbar^2}{2\mu L^2}(l^2 + m^2 + n^2),$$
(6.2)

where l, m and n are integers.

States are now counted by drawing a graphical "space" (Fig. 6.1) whose axes are l, m, n. States of nearly equal energy lie on a spherical surface whose radius in lmn space is r with $r^2 = l^2 + m^2 + n^2$. That energy is

$$\varepsilon = \frac{\pi^2 \hbar^2}{2\mu L^2}r^2, \quad \text{with differential} \quad d\varepsilon = \frac{\pi^2 \hbar^2}{2\mu L^2} 2r\, dr.$$
(6.3)

Given that the volume L^3 is arbitrarily large, the quantum states are very close together in energy, so we can say that there is on average one state per unit volume of lmn space. Then the number of states within energy interval $d\varepsilon$ is the number within a spherical shell of lmn-volume $4\pi r^2\, dr \div 8$. Division by 8 is done because l, m and n are restricted to positive values, so only the "positive octant" of lmn space contains points representing the wave functions of (6.2). Assembling the pieces, we have[4]

$$\text{number of states} = dN = (\text{volume } L^3) \times \frac{2\pi(2\mu)^{3/2}\varepsilon^{1/2}}{(2\pi\hbar)^3} d\varepsilon.$$
(6.4)

The expression in (6.4) may now be put into a variety of other forms, for example by substituting $\varepsilon = p^2/2\mu$ so that

$$\text{number of states} = dN = (\text{volume } L^3) \times \frac{4\pi p^2\, dp}{(2\pi\hbar)^3}.$$
(6.5)

Exercise 6.1 Positive integers?

Why were l, m, n restricted to positive values?[5]

If you pursue the hint, you should see that there is nothing special about positive values. But if we accept the eigenfunction with $l = 3$ we must not at the same time accept that with $l = -3$. Either will do, but not both. If we were perverse, we could accept eigenfunctions with l always positive, m always negative, and n positive when even, negative when odd. This wouldn't be wrong, just wrong-headed.

We conclude that restricting l, m and n to positive values (and the lmn space to its positive octant) is the simplest way, though not the only way, to ensure that our eigenfunctions are complete without being over-complete.

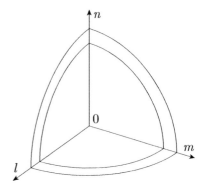

Fig. 6.1 A spherical shell within the positive octant of lmn space.

[4] Our expression counts only quantum states associated with translational motion of the particles. There may be extra factors associated with spin (for example), but such considerations are for later.

[5] *Hint:* Consider where the answer must lie. We are making a list of wave functions that solve (6.1) in order to count them. Suppose we have included in our list a wave function with $l = 3$, and we are wondering whether to add to the list another wave function with $l = -3$. Should we add it, thereby giving ourselves two wave functions to count, or should we discard it, giving ourselves only one? This kind of question always raises its head whenever we construct a set of eigenfunctions to use as an expansion set, and it is always answered in the same way.

Eigenfunctions are counted correctly if they form a complete set—they can expand anything—but they must also be only just numerous enough to be complete: no duplicates. So is the $l = -3$ wave function in excess of requirements? It will be superfluous if it can be expanded using other functions that we already have in our list, in other words if it and those functions are linearly dependent. Conversely, it will be new if linearly independent.

6.2.1 Critique of the conventional calculation

There is an unsatisfactoriness in the manipulation leading to (6.5). There is a strong hint that p is momentum—though I carefully avoided the word when introducing (6.5). But the wave functions of (6.2) represent standing waves. They are not eigenstates of momentum; we may say that they have a p but not a $\boldsymbol{p}$. Yet we almost always want to talk about states that possess momentum.[6]

Example: Suppose we are discussing the electrical resistivity of a metal. We shall wish to describe collisions where an electron of momentum $\boldsymbol{p}$ collides with another electron, or a phonon, or an impurity, and is scattered into momentum $\boldsymbol{p}'$. The point is, physics almost always requires us to work with quantum states possessing momentum. We need to start again using a more suitable set of wave functions[7] and to give a fresh derivation of (6.5).

There should also be a feeling that we may have missed a trick. The Schrödinger equation (6.1) is constructed to deal with non-relativistic massive particles: the operator $-(\hbar^2/2\mu)\nabla^2$ comes from the classical-mechanics kinetic energy $p^2/2\mu$. Our reasoning so far cannot give the "number of quantum states" for relativistic particles, or for photons, phonons, electrons in solids, ..., which have in common that $\varepsilon \neq p^2/2\mu$. But we need results that we can apply to such systems. Do we have to start all over again for every new case? Something better—more versatile—is called for.

6.3   A better approach: running waves

Given the above, we at once write a wave function $\psi_{\boldsymbol{p}}$ that is a running wave, an eigenfunction of momentum:[8]

$$\psi_{\boldsymbol{p}} \propto \exp(\mathrm{i}\boldsymbol{p} \cdot \boldsymbol{r}/\hbar) = \exp\left\{\mathrm{i}\left(l\frac{2\pi x}{L} + m\frac{2\pi y}{L} + n\frac{2\pi z}{L}\right)\right\}; \quad (6.6)$$

$$\widehat{\boldsymbol{p}}\,\psi_{\boldsymbol{p}} = -\mathrm{i}\hbar\nabla\psi_{\boldsymbol{p}} = \boldsymbol{p}\,\psi_{\boldsymbol{p}}; \qquad \boldsymbol{p} = \frac{2\pi\hbar}{L}\,(l, m, n). \quad (6.7)$$

These wave functions satisfy the need identified in § 6.2.1. (Again, l, m and n are integers.)

6.3.1   The choice of eigenfunction set

The wave functions ψ and $\psi_{\boldsymbol{p}}$ in (6.2) and (6.6) are different kinds of sinewave. When we use them as expansion functions, we are making different kinds of Fourier expansion. So let's remember what happens to Fourier expansions in one dimension. We can expand a function $f(x)$ in the interval $0 < x < L$ using

- a half-range sine series: functions $\sin(l\,\pi x/L)$
- a half-range cosine series: functions $\cos(l\,\pi x/L)$
- a full-range complex-exponential[9] series: $\exp(l\,2\pi\mathrm{i}x/L)$.

All three itemized sets of expansion functions are complete on the interval $0 < x < L$. But the use of $<$, rather than $\leqslant$, is important. Although all three function sets expand an arbitrary $f(x)$ *within* the interval 0 to L, none can handle the ends of the range as well. Figure 6.2 serves as a reminder. The sine series sums to zero at $x = 0$ and $x = L$, whether or not $f(x)$ is zero there. The cosine series has zero gradient at the ends of the range. And the complex-exponential series is discontinuous at the ends, splitting the difference between $f(0)$ and $f(L)$.

All eigenfunction sets are thus equally acceptable and equally defective.

In the three-dimensional case, the choice of eigenfunction set is usually introduced by saying that we are using "box" boundary conditions (yielding sines) or "periodic" boundary conditions (yielding complex exponentials). This is rather silly. The eigenfunction set is being chosen according to how it fails, rather than for any virtues.

It should be clear that boundary conditions are mostly an unhelpful distraction. This is well known to professional physicists, but convincing the student usually requires some delicate discussion. After all, he has only just been told the opposite.

6.3.2   Eigenfunctions of momentum

In this chapter, the particles under discussion are "free", so $\widehat{\boldsymbol{p}}$ commutes with the Hamiltonian, and the $\psi_{\boldsymbol{p}}$ of (6.6) are eigenstates of energy as well as of momentum.[10]

More generally, the $\psi_{\boldsymbol{p}}$ form a complete eigenfunction set, so they are always available even when they are not eigenfunctions of energy.

Exercise 6.2   Positive integers?

The eigenfunctions given by (6.6) contain integers l, m, n. Is there now any restriction on the range of values that these integers may take? The answer must use the same considerations as applied in sidenote 5.

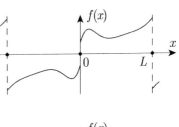
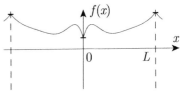
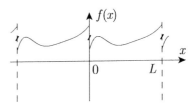

Fig. 6.2 A function $f(x)$ is Fourier-expanded using: (top) a half-range sine series; (middle) a half-range cosine series; (bottom) a full-range complex-exponential series. The function is to be represented by the series in the range $(0, L)$, but the effect of the series is to continue the function outside that range. The continuations introduce discontinuities in function or gradient (or both) at the ends of the range, and the series sum takes a half-way value there.

[10] The particles might be, e.g., phonons in a crystal, so the energy-momentum relation might be far from $\varepsilon = p^2/2\mu$.

6.4   Counting eigenstates of momentum

We set this argument up again from the beginning, choosing a method that is sympathetic to our new eigenfunctions (6.6).

We draw a graphical space, momentum space,[11] with axes p_x, p_y, p_z as shown in Fig. 6.3. A small element of momentum space is shown, having a volume in that space of $\delta p_x\, \delta p_y\, \delta p_z$. The volume element contains eigenfunctions of momentum, whose number we are to determine. From (6.7), we have

$$\delta l = \frac{L}{2\pi\hbar}\delta p_x, \qquad \delta m = \frac{L}{2\pi\hbar}\delta p_y, \qquad \delta n = \frac{L}{2\pi\hbar}\delta p_z.$$

[11] There is a reason why we switch from an lmn (integers) space to momentum space. Applications of (6.6), for example to collisions in the kinetic theory of gases, are most naturally described in momentum space. We are starting the way we mean to go on.

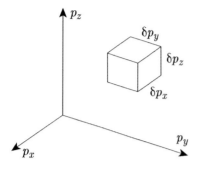

Fig. 6.3 A volume element in momentum space.

The number of momentum eigenfunctions whose momenta fall within the volume element shown in Fig. 6.3 is[12]

$$\delta N = \delta l \, \delta m \, \delta n = \left(\frac{L}{2\pi\hbar}\right)^3 \delta p_x \, \delta p_y \, \delta p_z$$

$$= \frac{(\text{volume } L^3 \text{ in real space}) \times (\text{volume in momentum space})}{(2\pi\hbar)^3}. \qquad (6.8)$$

It is often customary to work with wave vector $\boldsymbol{k} = \boldsymbol{p}/\hbar$, rather than momentum $\boldsymbol{p}$. In that case we have the corresponding expression for the number of $\boldsymbol{k}$-eigenfunctions:

$$\delta N = \frac{(\text{volume } L^3 \text{ in real space}) \times (\text{volume in } \boldsymbol{k}\text{-space})}{(2\pi)^3}. \qquad (6.9)$$

This can be obtained directly if we draw $\boldsymbol{k}$-space (axes k_x, k_y, k_z) instead of momentum space.

It should be clear that momentum $\boldsymbol{p}$ (likewise wavevector $\boldsymbol{k}$) can point in any direction, so the integers l, m, n can take both positive and negative values. But l, m, n have served their purpose now, so their signs are not interesting.

Now that Fig. 6.3 is understood, we may use any convenient volume element in momentum space or $\boldsymbol{k}$-space. In particular, if we wish to know the number of states for which the momentum lies between p and $p + \delta p$, regardless of the direction of the momentum, then we can use a spherical shell whose volume in momentum space is $4\pi p^2 \, \delta p$. We get a result identical with (6.5)—except that p now really is momentum.[13]

Comments

Quite apart from any other considerations, the reasoning that leads to eqn (6.8) is pithier than that used in § 6.2. We must be doing something right.[14]

In § 6.2.1, we criticized the use of standing-wave eigenfunctions on the ground that physics often requires us to deal with states having definite momentum. Clearly, we are now using eigenfunctions that are more appropriate to the physics we may need to describe.

Expressions (6.8–9) for the "number of momentum states" inherit the generality of the eigenfunctions (6.6). Energy has not been mentioned, so the expressions are correct even if eigenstates of momentum are not at the same time eigenstates of energy.

[12] Given that we started with waves in a cubical (or rectangular) volume of side L, it surely makes most sense if we remain consistent with that beginning and work in rectangular p_x, p_y, p_z coordinates now. The payoff is obvious: eqn (6.8) has required almost no work in its derivation. By contrast, the introduction of spherical shells in Fig. 6.1 was premature and added complication.

[13] As before (sidenote 4), our expressions (6.8) and (6.9) count only eigenstates of translational motion—of momentum. If the particles have spin, or if we're looking at "particles" such as transverse phonons having a polarization, then we must multiply by a factor to take account of the extra possibilities. We should say to ourselves that the δN expressions give the number of *momentum* states, in distinction from the number of *quantum* states.

[14] Incidentally, it is trivially easy to generalize the reasoning of § 6.4 to the case of a rectangular volume of sides L_1, L_2, L_3. An equivalent generalization is possible within the reasoning of § 6.2, but is less straightforward.

6.5   Electromagnetic waves

So far, we have dealt with particles having a scalar wave function ψ. Electromagnetic waves have vector fields $\boldsymbol{E}$ and $\boldsymbol{B}$. A different treatment is required, but not so very different.

We treat the electromagnetic field in a non-quantum way, that is, using Maxwell's equations. Maxwell's equations lead to scalar wave equations for the Cartesian components E_x, E_y, E_z of the electric field. These wave equations are obtained by eliminating field $\boldsymbol{B}$. (Alternatively, we may eliminate E in favour of B.) This means we have

$$\nabla^2 E_x = \frac{1}{c^2}\frac{\partial^2}{\partial t^2}E_x, \tag{6.10}$$

with similar equations for E_y and E_z.

Now not all solutions of (6.10) satisfy Maxwell's equations! We discarded information when eliminating unwanted field components. This is put right by reinstating the statement that, in vacuum, div $E = 0$. A travelling-wave solution of (6.10) now has the form[15]

$$E = E_0\, e^{i(k\cdot r - \omega t)} \qquad \text{with} \qquad k \cdot E = 0. \tag{6.11}$$

(Compare with statements at the end of § 4.2.) The second statement in (6.11) says, of course, that electromagnetic waves are transverse: for each k there are two possible polarizations, both having E perpendicular to k in the case of a plane wave.

Exercise 6.3 Electromagnetic travelling waves
Solve the wave equations (6.10) to obtain the solution (6.11). Check carefully that the condition div $E = 0$ has done all that is necessary to restore conformity with all of Maxwell's equations.

Show that $k \cdot E = 0$ relates together the three Cartesian components of E_0 so that only two of the three can be chosen independently. Thus each (k_x, k_y, k_z) is "used twice", having two independent field distributions associated with it. These are, of course, the two polarization possibilities, just obtained by algebra instead of by geometry.

Expression (6.11) for E already has the shape to provide an expansion set of complex-exponential functions. Within a volume L^3 these have

$$k_x = l\,\frac{2\pi}{L}, \qquad k_y = m\,\frac{2\pi}{L}, \qquad k_z = n\,\frac{2\pi}{L}. \tag{6.12}$$

A graphical space ("k-space") resembling that of Fig. 6.3 can be drawn with axes k_x, k_y, k_z. The number of k-states within a small rectangular volume of this space is given by (6.9).

Conditions (6.12) resemble those of (6.7) except that k replaces momentum p and correspondingly a factor $\hbar$ is absent.[16]

It remains only to remember that each k is associated with two polarizations, so the number of eigenfunctions we have given must be doubled:

$$\delta N_{\text{modes}} = \frac{(\text{volume in real space}) \times (\text{volume in } k\text{-space})}{(2\pi)^3}$$
$$\times\ (2 \text{ for polarizations}). \tag{6.13}$$

We say that each individual eigenfunction (having a given k and a

[15] It is perfectly possible to apply "conducting box" boundary conditions, thereby enforcing standing-wave solutions. These have the form

$$E_x = E_{0x}\cos\alpha x\,\sin\beta y\,\sin\gamma z\,e^{-i\omega t}$$
$$E_y = E_{0y}\sin\alpha x\,\cos\beta y\,\sin\gamma z\,e^{-i\omega t}$$
$$E_z = E_{0z}\sin\alpha x\,\sin\beta y\,\cos\gamma z\,e^{-i\omega t}.$$

The coefficients are linked by div $E = 0$ so that

$$\alpha E_{0x} + \beta E_{0y} + \gamma E_{0z} = 0.$$

The last equation is telling us that there are two independent solutions, not three, so that each (α, β, γ) has two eigenfunctions associated with it. Compare with exercise 6.3.

The standing-wave fields here are considerably more complicated than the travelling wave of (6.11). Moreover, the physics in the link between the coefficients is obscure. Does it represent two polarizations? It comes from div $E = 0$, so this seems likely (remember Chapter 4). But polarizations, as we would recognize them in optics, are very hard to see in the mathematics of this sidenote. So we may well wonder whether the factor 2 we have found is really to do with polarizations, or whether it's some *other* factor 2 that should be applied as well.

The penalties we suffer by choosing standing-wave eigenfunctions are very high here. And the theme of this chapter is that the penalty is self-inflicted. Wherever we are going, this is not the place to start from.

[16] Our treatment here is purely classical, based only on Maxwell's equations, hence no mention of the photon momentum p.

[17] We cannot refer to a mode as a "quantum state" because all discussion of electromagnetic waves here has been entirely non-quantum. The terminology of "modes" is important in Chapter 19.

given polarization) is a **mode**[17] of the electromagnetic field, and the δN is given a label as a reminder of this.

The result (6.13) can be put into a number of equivalent forms. If we are interested in the number of eigenfunctions within a range δk, regardless of the direction of k, then we may use a volume element in k-space in the form of a spherical shell of volume $4\pi k^2\,\delta k$. We then have

$$\delta N_{\text{modes}} = L^3 \times \frac{4\pi k^2\,\delta k}{(2\pi)^3} \times 2 = L^3 \times \frac{\omega^2\,\delta\omega}{\pi^2 c^3} = L^3 \times \frac{8\pi\nu^2\,\delta\nu}{c^3}. \quad (6.14)$$

Here $\omega = ck = 2\pi\nu$, so that ω is angular frequency and ν is frequency.

A treatment of electromagnetic waves, very similar to ours, is given by Landau and Lifshitz (1996), § 52. The thinking in the present chapter makes no claim to originality. But it remains the case that the high-grade treatments of our topic are not as frequently encountered as they should be.

6.6   The density of states

[18] But not in Chapters 24 and 28.

The expression **density of states** refers to the number of quantum states (or modes, as appropriate), per unit volume of real space, and per unit increment of energy. Very often,[18] we take no interest in the direction of momentum, so we build on (6.8) with the momentum-space volume taken as $4\pi p^2\,\mathrm{d}p$.

The density of states now becomes

$$\text{density of states} = \frac{1}{(2\pi\hbar)^3}\,4\pi p^2\,\frac{\mathrm{d}p}{\mathrm{d}\varepsilon} \times \begin{pmatrix} \text{possible factor for} \\ \text{spin or polarization} \end{pmatrix}, \quad (6.15)$$

where ε is the energy associated with momentum p.

[19] Golden Rule No. 2 requires a more careful derivation than is usually given. See problem 32.24 and eqn (32.96). Because the density of states increases as (energy)^2 for large energies we need a reason why the matrix elements fall off for large energy, otherwise a sum over final travelling-wave states is divergent.

It is this density of states that appears in "Golden Rule Number 2", giving the rate of transitions in a reaction where particles are emitted into a quasi-continuum of final states.[19]

It will be clear that evaluation of $\mathrm{d}p/\mathrm{d}\varepsilon$ requires knowledge of the energy–momentum relation for the particles under consideration. In the present general discussion we can leave this hanging, but it has to be addressed in any specific application.

Problem

Problem 6.1   cross-links
Establish links between the "number of states" of (6.8) or (6.15) and:

(1) the element of *phase space* in classical mechanics;

(2) *Liouville's theorem* (constant occupation of phase space) in classical mechanics;

(3) the conservation analogous to Liouville in quantum mechanics;

(4) the conservation of *étendue* in optics.

IV Relativity

7 Four-vectors in relativity

Intended readership: A second course in relativity, when relativistic mechanics is being applied, for example to collision reactions.

7.1 Introduction

This essay is intended to fill a gap. Textbook introductions to four-vectors often look rather complicated or obscure. Part of the trouble is that they assume you know things about ordinary vectors that you may not know; and somehow it seems to be nobody's job to tell you.

Be prepared then for a lengthy introduction on ordinary common-or-garden three-dimensional vectors, all of which *ought* to be revision, but probably isn't.

7.2 Plain ordinary three-vectors

7.2.1 Definition of a vector

A quantity a is a vector if it has all of the following properties:

(1) it has magnitude, direction, and a sense (sign) along that direction

(2) it can be represented by drawing an arrow on a fixed sheet of paper

(3) the combined physical effect[1] of applying two vectors a and b together is the same as that of applying the vector sum $(a + b)$.

[1] We may feel that "physical effect" is not precise enough for a mathematical definition. However, we shall find no difficulty in deciding how to apply this idea.

These definitions are obeyed by a displacement r, say the displacement $(r_B - r_A)$ from point A to point B. I shall regard a physical displacement as the most fundamental of all vectors.

We all know that vectors are quantities having magnitude and direction, and it may come as a surprise that I am insisting on further conditions. It will turn out, however, that all three conditions are important for the present essay.

Suppose we come across a new quantity (or thing) that has magnitude and direction: it might be angular momentum or magnetic field at the time when we first hear about it. Before we can treat this thing as a vector (as I intend using the term), we need to check that it satisfies all three conditions. The precise force of conditions (2) and (3) will be elaborated in the next subsections.

Essays in Physics: Thirty-two thoughtful essays on topics in undergraduate-level physics. Geoffrey Brooker, Oxford University Press. © Geoffrey Brooker 2021. DOI: 10.1093/oso/9780198857242.003.0007

7.2.2 Addition and axes

We define the vector sum $(a+b)$ of two vectors a and b to be the result of applying the triangle (or parallelogram) rule, as shown in Fig. 7.1. A *consequence* of this definition is that we can resolve any vector into three components referred to a set of axes (non-coplanar but not necessarily orthogonal). A further consequence is that we may add vectors by adding their components.

Comment: You may find the above irritatingly trivial, but I have a purpose in writing it. Axis systems (in relativity usually called **frames**) will play a large part in the discussions to follow, and it's helpful to be clear about where axes matter and where they don't. I have carefully defined $(a+b)$ in a way that makes no reference to any axis system, and then introduced axes afterwards. Indeed, a vector component can hardly make sense unless we have previously agreed upon a geometrical understanding of vector addition.

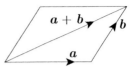

Fig. 7.1 The sum $a+b$ of vectors a and b is defined by the triangle (or parallelogram) rule.

Exercise 7.1 Vector quantities without axes
Write down definitions of the following quantities, in each case without making any mention of coordinate axes, Cartesian or otherwise:[2]

(1) $\operatorname{grad}\phi$;   (2) $a \cdot b$;   (3) $\operatorname{div} a$;   (4) $a \times b$;   (5) $\operatorname{curl} a$.

Take a and b to be vectors, and ϕ to be a scalar.

7.2.3 Genuine vectors triangle-add

Consider a particle initially at the origin. Displace it through the vector displacement r_1, and then further displace it through r_2. The position at which the particle ends up is found by completing a triangle. In other words, the effect of combining the two displacements is the same as that of making a single displacement through the vector sum $(r_1 + r_2)$. It is therefore evident that displacements satisfy condition (3). In fact, it is just this property of displacements that led us to define vector addition by the triangle rule.

Not all quantities possessing magnitude and direction have the triangle-addition property. The standard textbook counterexample is a rotation through a finite angle. We can assign a direction to the rotation by drawing a line along the axis of rotation. We can draw this line with a length proportional to the angle turned through, and we can attach an arrowhead by requiring the rotation to be clockwise when seen from the back of the arrow. The arrow represents the rotation in a way that satisfies (1) and (2). However, it is not hard to see that it fails to meet condition (3).

Two 90° rotations applied to a book are shown in Fig. 7.2. Draw your own sketch (or do the experiment) to find the result of applying the

[2] Quantities $a \times b$ and $\operatorname{curl} a$ are usually classified as **pseudovectors**, otherwise called **axial vectors** (if a and b are ordinary **polar** vectors—displacement r is a polar vector). The distinction concerns the changes of sign (or not) of vector components on transforming to an inverted (or left-handed) axis system. It helps clear thinking if we remember that definitions of $a \times b$ and $\operatorname{curl} a$ can be given with no mention of axes. Nevertheless, we have seen in Chapters 3 and 4 that components-along-axes are not always avoidable.

"Inversion" means a shift to left-handed axes, and only that, not a replacement of the real world by a mirror-image world. Thus $c = a \times b$ is the same vector whatever we do to coordinate axes. Likewise, "clockwise when seen from the back of the arrow" means what it always does.

Let $(x', y', z') = (-x, -y, -z)$ be the result of reversing all three Cartesian axes. Vectors a and b have components $a_i' = -a_i$ and $b_i' = -b_i$ along the new axes. Then in $c_i' = \varepsilon_{ijk} a_j' b_k'$ (j and k summed) there are two sign reversals on the right, giving $c_i' = c_i$: the components of c' along the new axes are the *same* as those of c along the old axes. This means that $c' = -c = -a \times b$; this happens because ε_{ijk} is (by convention) not changed.

The pseudovector $c_i = \varepsilon_{ijk} a_j b_k$ is said to be the **dual** of the second-rank tensor $a_j b_k$.

I think it would sensible to say that $c = a \times b$ defines as ordinary a vector as a, in contrast to the pseudovector c_i'; but this seems not to be the way people usually think.

In the present chapter, inversion of axes will not be considered further.

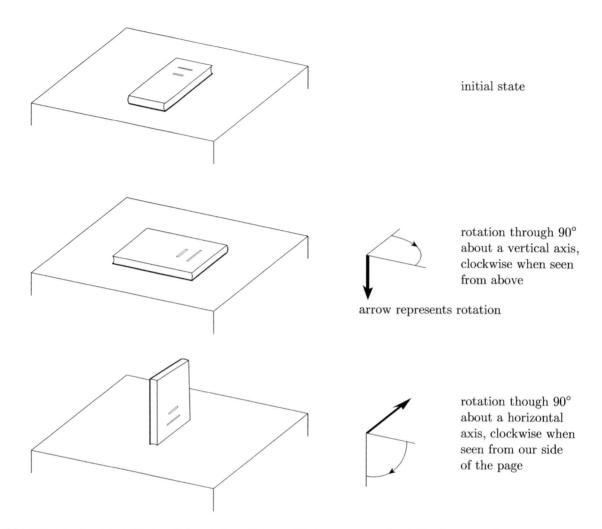

initial state

rotation through 90°
about a vertical axis,
clockwise when seen
from above

arrow represents rotation

rotation though 90°
about a horizontal
axis, clockwise when
seen from our side
of the page

Fig. 7.2 A book subjected to two finite rotations. It is possible to represent each rotation by means of an arrow as explained in the text. But rotations are not combined by triangle-adding the arrows, so a finite rotation is not a vector.

[3] The table supporting the book does not rotate, and the axes are fixed relative to the table; they are not carried round with the book.

same two rotations to the same initial state but in the opposite order.[3] You will not get to the same final state. Thus rotations are operations that do not commute: it matters which you do first. Whatever is the rule for combining rotations, it is clearly not the same as a triangle law (which gives the same result whichever operation is done first). Thus

Two finite rotations are *not* combined according to a triangle rule.

This establishes that a finite rotation fails to meet my condition (3), and so cannot be represented by a vector.[4]

[4] For future reference: an infinitesimal rotation meets condition (3) and *can* be treated as a vector—it is a *finite* rotation that does not.

Now that we have found that finite rotations do not have the property of combining according to the triangle rule, it is prudent not to use the arrow notation for them any more, lest we confuse them with true vectors; though this is a sensible precaution, not a logical necessity.

7.2.4   Rotation of (Cartesian) axes

This is the topic that is most likely to be new to you, and it is just this aspect of vectors that we need to understand for relativity.

Figure 7.3 shows a quantity with magnitude and direction, represented by the arrow O→P. To fix ideas, think of it as the position vector r of the point P relative to origin O. In Fig. 7.3(a) we use a coordinate frame $OXYZ$ of Cartesian right-handed axes. Then r is the vector whose components in this axis system are (x, y, z).

We choose to rotate our axes to the new orientation in Fig. 7.3(b). The point P is to remain the same point as before, because r is a real physical thing which is not changed merely because we happen to have turned ourselves round to look at it from a new direction. The new components (x', y', z') of r are related to the old by

$$\begin{pmatrix} x' \\ y' \\ z' \end{pmatrix} = \begin{pmatrix} l_1 & m_1 & n_1 \\ l_2 & m_2 & n_2 \\ l_3 & m_3 & n_3 \end{pmatrix} \begin{pmatrix} x \\ y \\ z \end{pmatrix}, \tag{7.1}$$

where l_1 is the cosine of the angle made by X' with X and similarly. We shall call the square matrix the **rotation matrix** T with elements T_{ij}.

In saying that "r is not changed by rotating the axes", I have re-expressed the idea contained in condition (2), where I talked about drawing r on a fixed sheet of paper.

Comment: Notice that a very restrictive mathematical condition is being placed by (7.1) on any set of three numbers (x, y, z) if they are to be the components of a vector, having its own existence independent of the coordinate frame. There is a risk attached to the use of axes and components: that the axes may "take over", meaning that the results of a calculation are true in the frame used, but only in that frame. From this point of view, the matrix rule (7.1) for transforming components is our guarantee that any chosen frame is harmless after all.[5]

The idea that axes might "take over" is being specially flagged now because it (or rather the condition for it *not* to be so) explains a lot of

[5] Of course, another way of being sure that we are not misled by the use of some special axis system is to avoid axes altogether. One reason for the usefulness of standard vector notation is that it permits us to define and manipulate vectors without using axes. You may now also see the reason for exercise 7.1.

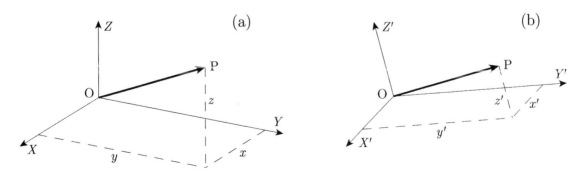

Fig. 7.3 The arrow OP represents a vector, that is, a thing having magnitude and direction (and when combined with others obeying a triangle law), and having a definite physical meaning. The vector may be resolved onto either the X, Y, Z axes or the X', Y', Z' axes. The resolved components are different, but the vector itself is the same.

what we are up to in the following. The use of axes is not avoidable in relativity, as it amounts to the use of some chosen Lorentz frame. And it becomes a serious issue that a set of physical quantities is not harmfully affected by the frame choice.

7.2.5 Definition of a scalar

Even though it seems trivial, we need a formal definition of a scalar for future use. A scalar is a quantity with the same numerical value in any frame.[6] Scalars are also called **invariants** because their values do not vary when we change to a new (rotated) set of axes. Examples of non-relativistic scalars are: the number 3; a time interval; a mass; an electric charge; a temperature.

By contrast, a component of a vector (like x or a_y) is *not* a scalar, because its numerical value depends on the frame chosen.

Exercise 7.2 Specimen vectors
Convince yourself that the following quantities are vectors:

(1) $\phi\,a$;   (2) $\operatorname{grad}\phi$;   (3) $a \times b$;   (4) $\operatorname{curl}a$,

where ϕ is a scalar and a and b are vectors. Here "convince yourself" means that you apply all three tests in turn. In those cases where you have found an axis-independent definition in exercise 7.1, you already have a reason for saying that condition (2) is satisfied.

Exercise 7.3 Specimen scalars
Convince yourself that the following quantities are scalars:

(1) $a \cdot b$;   (2) $\operatorname{div}a$,

where a and b are both known to be vectors.

7.2.6 Construction of new vector quantities (non-relativistic)

You have come across a variety of quantities in physics which are claimed to be vectors. Now that we have made more precise the conditions under which such things are in fact legitimate vectors, we ought to run through the list of purported vectors and see that they meet the conditions. I shall do this here for a selection of quantities, because I want us to see how closely the chain of reasoning is paralleled by one in relativity.

(a) *Velocity*: We have found that a displacement r is a vector. Then a small displacement δr is also a vector. Let δr be the displacement made by some body in time δt. Since $1/\delta t$ is a scalar, the quantity $\delta r/\delta t$ meets the conditions for being a vector (exercise 7.2(1)).

Proceeding to the limit $\delta t \to 0$, we obtain a new vector $\boldsymbol{v} = \mathrm{d}\boldsymbol{r}/\mathrm{d}t$: velocity is—by construction—a vector quantity.

Also, since I have included in the meaning of "vector" the idea that the physical combination of two vectors is done by the triangle rule, we have proved that velocities are combined by the triangle of velocities. This rule for combining velocities does not require independent demonstration.

(b) *Momentum*:   Momentum $\boldsymbol{p}$ is defined by $\boldsymbol{p} = m\boldsymbol{v}$, where mass m is a scalar, so it too is a vector by construction.

(c) *Force*:   A small change $\delta\boldsymbol{p}$ of momentum must again be a vector, so $\delta\boldsymbol{p}/\delta t$ is a vector. Then force, defined by $\boldsymbol{F} = \mathrm{d}\boldsymbol{p}/\mathrm{d}t$, is another vector.

In getting to this point I have, of course, quoted Newton's second law. A direct consequence of this definition is the "triangle of forces", which therefore needs no separate experimental demonstration once Newton's laws are accepted.

(d) *Electric field*:   Charge q is a scalar. Then if the force experienced by a small test charge q in an electric field is $\boldsymbol{F}$, the ratio $\boldsymbol{E} = \boldsymbol{F}/q$ defines a new vector, which we call electric field.

It should be clear that we can build chains of reasoning justifying the vector properties of all sorts of quantities. You can amuse yourself by extending my chain and seeing where you can get to.

Exercise 7.4 Infinitesimal rotation

Show that an infinitesimal rotation is a vector, and hence show that angular velocity is a vector.

Exercise 7.5 Angular momentum

Show that angular momentum is a vector.[7]

Exercise 7.6 Properties of a rotation matrix

Equation (7.1) introduced us to the 3×3 rotation matrix $\boldsymbol{T}$. Not any 3×3 matrix can be a rotation matrix. In particular, it must satisfy:

$$(\boldsymbol{T}^{-1})_{ij} = (\widetilde{\boldsymbol{T}})_{ij}, \qquad \|\boldsymbol{T}\| = 1, \tag{7.2}$$

where the tilde indicates that the transpose is taken, and $\|\boldsymbol{T}\|$ is the determinant having the same elements as the matrix (we call this the determinant of the matrix). The first of the properties in (7.2) makes $\boldsymbol{T}$ a special case of a **unitary** matrix,[8] the defining property for which is $(\boldsymbol{T}^{-1})_{ij} = (\boldsymbol{T}^{\dagger})_{ij}$.

(1) Show that the matrix $\boldsymbol{T}$ of (7.1) has an inverse that is the same as its transpose, that is, $(\boldsymbol{T}^{-1})_{ij} = (\widetilde{\boldsymbol{T}})_{ij} = T_{ji}$.

(2) Use matrix multiplication to show from (7.1) that[9,10]

$$x'^2 + y'^2 + z'^2 = x^2 + y^2 + z^2.$$

[7] *Hint*: Do not try to derive this from angular velocity. A body of irregular shape has vectors for angular velocity and angular momentum lying in different directions. The moment of inertia linking them is a second-rank tensor. The correct starting point is $\boldsymbol{r} \times \boldsymbol{p}$.

[8] The dagger indicates that the complex conjugate is taken, as well as the transpose. In quantum mechanics we should call $\boldsymbol{T}^{\dagger}$ the Hermitian conjugate of $\boldsymbol{T}$. The rotation matrix is, of course, real, but we give the definition of unitarity in its full form.

[9] *Hint*: Elegantly. There is no need to multiply out lots of matrix elements if you make proper use of part (1). Use the fact that the transpose of a matrix product is the product of the transposes in reversed order.

[10] The length of the new vector (x', y', z') is the same as that of the original vector $\boldsymbol{r} = (x, y, z)$, as it must be since only the axes have been rotated. This "conservation of length" is a consequence of the unitarity condition.

(3) Use the fact that $(T^{-1})_{ij} = (\widetilde{T})_{ij}$ to show that $\|T\| = \pm 1$. This does not quite get us to the second statement of (7.2). The ambiguity of sign is resolved by requiring T to represent a **proper rotation**, that is, axes initially right-handed remain right-handed after the operation. An "improper" rotation is exemplified by the parity operation $x \to -x$, $y \to -y$, $z \to -z$; see §11.4.

7.2.7   A last thought on the rotation of axes

I've left this until last, because I want it to be fresh when we come to four-vectors. The position vector r transforms from one axis frame to another by the matrix rule

$$
\begin{pmatrix} x' \\ y' \\ z' \end{pmatrix} = \begin{pmatrix} l_1 & m_1 & n_1 \\ l_2 & m_2 & n_2 \\ l_3 & m_3 & n_3 \end{pmatrix} \begin{pmatrix} x \\ y \\ z \end{pmatrix} = \begin{pmatrix} T_{11} & T_{12} & T_{13} \\ T_{21} & T_{22} & T_{23} \\ T_{31} & T_{32} & T_{33} \end{pmatrix} \begin{pmatrix} x \\ y \\ z \end{pmatrix}.
$$

[11] If the rotation is about one of the three Cartesian axes, it's simplest to draw a diagram. If not, exploit the simplicities of vector algebra. Write

$$r = xi + yj + zk = x'i' + y'j' + z'k'.$$

Now take a dot product with one of the unit vectors. For example,

$$x' = x\,i \cdot i' + y\,j \cdot i' + z\,k \cdot i',$$

from which we can read off $m_1 = j \cdot i'$.

[12] I have chosen to introduce four-vectors by the simplest route, that where the fourth element is (ict). This is commonly done in a first course on relativity.

Experts regard (ict) with contempt. They make their fundamental four-vector $(x^0, x^1, x^2, x^3) = (ct, x, y, z)$. Signs in the length-squared of the vector are arranged by introducing a *metric tensor* $g_{\alpha\beta}$ and putting four-vectors into contravariant and covariant forms. There are good reasons for doing this, because in general relativity we introduce gravitational fields by "bending" the metric tensor. However, this heavy mathematical apparatus gets in the way of a first understanding. So I make no apology for choosing the simpler formulation here—as something that the student will out-grow in due course.

The two conventions are held apart by making $x_4 = ($i$ct)$ the fourth component here, while $x^0 = ct$ is the zeroth component in the more advanced convention.

Because of what has been said about drawing vectors on a sheet of paper, the same rule works for *any* vector: all vectors transform in the same way. Even if we had not discovered the usefulness of vectors on other grounds, this property alone would make vectors especially important and simple.

Sometimes you can get confused over what is the correct rotation matrix (is $T_{12} = m_1$ the direction cosine of Y' relative to X or of Y relative to X'?) The cure for confusion is always this. Find by geometry[11] what equations transform (x, y, z) into (x', y', z'), and put the equations into matrix form. That gives you the matrix T_{ij}. Any other vector A is now transformed by the completely routine and unconfusing rule:

$$
\begin{pmatrix} A'_x \\ A'_y \\ A'_z \end{pmatrix} = \begin{pmatrix} \text{same matrix} \\ T \text{ as for} \\ \text{displacements} \end{pmatrix} \begin{pmatrix} A_x \\ A_y \\ A_z \end{pmatrix}.
$$

7.3   Four-vectors at last

We now have the background on three-vectors with which to appreciate four-vectors.

Consider the Lorentz transformation

$$
\left. \begin{aligned} x' &= \frac{x - vt}{\sqrt{1 - v^2/c^2}} \\[2mm] y' &= y \\ z' &= z \\[2mm] t' &= \frac{t - xv/c^2}{\sqrt{1 - v^2/c^2}} \end{aligned} \right\} \quad \text{or} \quad \left\{ \begin{aligned} x' &= \frac{1}{\sqrt{1 - v^2/c^2}}\left(x + \mathrm{i}\frac{v}{c}(\mathrm{i}ct)\right) \\[2mm] y' &= y \\ z' &= z \\[2mm] (\mathrm{i}ct') &= \frac{1}{\sqrt{1 - v^2/c^2}}\left((\mathrm{i}ct) - \mathrm{i}\frac{v}{c}x\right). \end{aligned} \right.
$$

We can put this Lorentz transformation into matrix form as[12]

$$\begin{pmatrix} x' \\ y' \\ z' \\ \mathrm{i}ct' \end{pmatrix} = \begin{pmatrix} \gamma & 0 & 0 & \mathrm{i}\beta\gamma \\ 0 & 1 & 0 & 0 \\ 0 & 0 & 1 & 0 \\ -\mathrm{i}\beta\gamma & 0 & 0 & \gamma \end{pmatrix} \begin{pmatrix} x \\ y \\ z \\ \mathrm{i}ct \end{pmatrix}, \tag{7.3}$$

where $\beta = v/c$, $\gamma = 1/\sqrt{1-v^2/c^2}$, and $\mathrm{i} = \sqrt{-1}$. This transformation bears a very close mathematical resemblance to a rotation of axes, except that now there are four "components" to each column matrix.

Exercise 7.7 The Lorentz matrix is "length preserving"

(1) Take the square matrix in equation (7.3), calling it L. Show[13] that $\mathsf{L}^{-1} = \tilde{\mathsf{L}}$.

(2) Use the result of part (1) to show that the Lorentz-transformation matrix L preserves the "length of the vector" invariant:[14]

$$x'^2 + y'^2 + z'^2 + (\mathrm{i}ct')^2 = x^2 + y^2 + z^2 + (\mathrm{i}ct)^2. \tag{7.4}$$

These two results explain why the Lorentz transformation is said to resemble a rotation of axes, generalized from three dimensions to four.

In three dimensions, we found the rule for transforming $\boldsymbol{r} = (x, y, z)$ from one set of axes to another: you multiply by the rotation matrix as in (7.1). We found a whole family of other quantities (velocity, momentum, force ...) all of which transform in the same way. We invented a noun to describe the property that all these things have in common, and called them *vectors*.

We now have a similar situation in relativity. The four quantities $(x, y, z, \mathrm{i}ct)$ can be transformed from one Lorentz frame to another by a matrix multiplication. It would be highly convenient if we could find other groups of four quantities (preferably, of course, quantities that look useful to physicists) that transform by the same matrix rule. If we find such quantities we shall call them **four-vectors**, as a way of saying that they transform in the same way as $(x, y, z, \mathrm{i}ct)$. Looking ahead for a moment, we shall find that[15]

$$(p_1, p_2, p_3, p_4) = \left(\frac{mv_x}{\sqrt{1-v^2/c^2}}, \frac{mv_y}{\sqrt{1-v^2/c^2}}, \frac{mv_z}{\sqrt{1-v^2/c^2}}, \frac{\mathrm{i}mc}{\sqrt{1-v^2/c^2}} \right)$$

is a four-vector (known as the four-momentum). Then we can tell at once what the equivalent four-vector is in any other Lorentz frame, because the matrix rule (7.3) tells us that

$$\begin{pmatrix} p_1' \\ p_2' \\ p_3' \\ p_4' \end{pmatrix} = \begin{pmatrix} \text{same} \\ \text{matrix} \\ \text{as for} \\ \text{four-position} \end{pmatrix} \begin{pmatrix} p_1 \\ p_2 \\ p_3 \\ p_4 \end{pmatrix} = \begin{pmatrix} \gamma & 0 & 0 & \mathrm{i}\beta\gamma \\ 0 & 1 & 0 & 0 \\ 0 & 0 & 1 & 0 \\ -\mathrm{i}\beta\gamma & 0 & 0 & \gamma \end{pmatrix} \begin{pmatrix} p_1 \\ p_2 \\ p_3 \\ p_4 \end{pmatrix}.$$

Here the Lorentz-transformation matrix has been "rubber-stamped" in front of the four-momentum column matrix. The ability to do Lorentz

[13] Since matrix L is complex, this does *not* make L unitary. Nevertheless, its determinant $\| L_{ij} \|$ is $+1$, and this suffices to give the "conservation of length" in (7.4).

[14] As with exercise 7.6(2), this should be proved by an easy matrix manipulation, and not by messy workings-out.
There is an incidental disadvantage of the $(\mathrm{i}ct)$ convention. When (x, y, z, t) describe a single point particle, expressions like $\mathrm{d}x^2 + \mathrm{d}y^2 + \mathrm{d}z^2 - c^2 \mathrm{d}t^2$ are negative, because the speed of a body possessing these coordinates must be less than c. The "interval" $\mathrm{d}s$ defined in sidenote 17 has a square that is positive.

[15] A vector or four-vector written out in components should properly be printed as a column matrix. We give the components here as a horizontal array simply to save space.

transformations by routine matrix multiplication is obviously desirable; we should even be prepared to go to some trouble to set up four-vectors if natural choices didn't suggest themselves.

7.3.1   Four-scalars

A four-scalar (often called an invariant) is a quantity with the same numerical value in any Lorentz frame. Examples of invariants are:

(a) the rest mass of a body (because by definition you're allowed to measure it in only one frame, the frame within which the body is at rest)

(b) an interval of "proper time", that is, the time as measured on a clock in the body's rest frame,[16] again because there's only one frame it can be measured in

(c) $x^2 + y^2 + z^2 + (\mathrm{i}ct)^2$, which was shown to be an invariant in exercise 7.7

(d) electric charge is a relativistic invariant.

A quantity like x or t is not an invariant, because it takes a different value when we change to a new Lorentz frame.

The product of a four-scalar and a four-vector is a new four-vector (compare exercise 7.2).

7.4   Construction of new four-vectors (relativistic)

I shall lead you by the hand to devising new four-vectors by a method which is almost identical to that of §7.2.6.

(a) *Four-velocity*   Let a particle have coordinates $(x, y, z, \mathrm{i}ct)$ in some Lorentz fame (i.e. the particle has space coordinates x, y, z at time t in that frame). A small change $(\delta x, \delta y, \delta z, \mathrm{i}c\,\delta t)$ in the particle's coordinates is a four-vector. We want to divide this four-vector by a scalar looking like a time interval, so as to construct a velocity. The obvious choice is δt, but this is not a four-scalar—it has different values in different Lorentz frames. Once we've seen what we can't do it's fairly obvious what would be more sensible—divide by the interval of **proper time** $\delta\tau$. Then the quantity[17]

$$\left(\frac{\delta x}{\delta\tau}, \frac{\delta y}{\delta\tau}, \frac{\delta z}{\delta\tau}, \frac{\mathrm{i}c\,\delta t}{\delta\tau} \right)$$

is a four-vector. Now the particle's ordinary three-dimensional velocity (relative to our frame) is defined just as you would expect, as $(v_x, v_y, v_z) = (\mathrm{d}x/\mathrm{d}t, \mathrm{d}y/\mathrm{d}t, \mathrm{d}z/\mathrm{d}t)$, so our new four-vector is $(v_x, v_y, v_z, \mathrm{i}c)\mathrm{d}t/\mathrm{d}\tau$. Relativistic time dilation tells us that the particle's "proper time" $\delta\tau$ relates to the time interval δt (in our frame) by $\delta\tau = \delta t\sqrt{1 - v^2/c^2}$, and so the four-velocity v becomes

$$v = \left(\frac{v_x}{\sqrt{1 - v^2/c^2}}, \frac{v_y}{\sqrt{1 - v^2/c^2}}, \frac{v_z}{\sqrt{1 - v^2/c^2}}, \frac{\mathrm{i}c}{\sqrt{1 - v^2/c^2}} \right),$$

[16] Proper frame, own frame. I find it easier to be comfortable with this sense of "proper" if I say it in French. Also, my *property* is what I *own*.

[17] There is an even better route than dividing by the interval of proper time; it's equivalent mathematically (here), but the words are better. This is to divide by the four-scalar $\delta s/c$ where the "interval" δs is defined by

$$\delta s^2 = c^2\delta t^2 - \delta x^2 - \delta y^2 - \delta z^2.$$

It is easy to show that the interval attaching to the history of a body is

$$\delta s = c\,\delta t\sqrt{1 - v^2/c^2},$$

without any need to appeal to the physics of time dilation (though it's not far away). Moreover, we shall see that this route is necessary in the definition of four-force. Incidentally, the positive square root is taken as part of the definition of δs.

where, of course, $v^2 = v_x^2 + v_y^2 + v_z^2$. Because the first three components of this vector reduce to the ordinary three-dimensional velocity when v is small,[18] our new four-vector is called the **four-velocity**. The important thing is not what we call it, but the fact that we can Lorentz-transform this quantity much more easily than any other simple quantity involving velocity.

(b) *Four-momentum*   Having found a quantity that looks rather like a velocity, we naturally want to multiply it by a scalar looking like mass, to see if we can define a momentum. The only scalar (or invariant) quantity looking like a mass is the rest mass $m = m_0$, so we define for our particle the new four-vector[19]

$$p = (p_1, p_2, p_3, p_4) = mv$$
$$= \left(\frac{mv_x}{\sqrt{1 - v^2/c^2}}, \frac{mv_y}{\sqrt{1 - v^2/c^2}}, \frac{mv_z}{\sqrt{1 - v^2/c^2}}, \frac{imc}{\sqrt{1 - v^2/c^2}} \right).$$

The first three components look like ordinary momentum when we go to the limit of small velocities. The fourth component gives (it turns out) the energy of the particle (apart from a factor c/i), so $p = (p_1, p_2, p_3, p_4)$ is known as the **energy-momentum four-vector**. Because p is a four-vector, we know how to transform it to any new Lorentz frame.

(c) *Four-acceleration and four-force*   Suppose that our particle's four-velocity $v = (v_1, v_2, v_3, v_4)$ changes by $\delta v = (\delta v_1, \delta v_2, \delta v_3, \delta v_4)$ in a small time interval. We look round for some "time" quantity to divide by, so as to build a relativistic acceleration $a = (a_1, a_2, a_3, a_4)$. We found before that δt wouldn't do, and used $\delta \tau$ instead. But $\delta \tau$ won't do here either, because the frame in which proper time is measured changes as the particle accelerates. Instead we divide by $\delta s/c$ where[20]

$$\delta s^2 = c^2 \delta t^2 - \delta x^2 - \delta y^2 - \delta z^2,$$

and $\delta x, \delta y, \delta z$ are the changes in location of the particle in time interval δt. In this way we can construct an authenticated four-acceleration a and four-force F

$$a = c \left(\frac{dv_1}{ds}, \frac{dv_2}{ds}, \frac{dv_3}{ds}, \frac{dv_4}{ds} \right);$$
$$F = ma = c \left(\frac{dp_1}{ds}, \frac{dp_2}{ds}, \frac{dp_3}{ds}, \frac{dp_4}{ds} \right).$$

We do not trouble to evaluate these expressions further.

The point we wish to make here is that a chain of definitions can be pursued, a chain that exactly parallels that of §7.2.6, in which each new quantity is known to be a four-vector (meaning: having the transformation properties of a four-vector) because of the way in which it has been built from reliable ingredients.

[18] The four-vector was, of course, constructed to make this so.

[19] I say *rest mass*, rather than just *mass*, for emphasis only. Old-fashioned accounts of relativity introduce a "relativistic mass" $m = m_0/\sqrt{1 - v^2/c^2}$. There is no place for "relativistic mass" within our discussion because it is not a four-scalar. Now that this point has been made, "mass" will always mean rest mass, and will be represented by m, not m_0.

[20] The quantity δs is set up to be a relativistic invariant, a four-scalar, even though the definition is given using coordinates and time within one Lorentz frame.

Books differ in their definitions. Sometimes the four-velocity, four-momentum and four-force are obtained by dividing by δs, rather than by $\delta s/c$.

7.5   Physics in more than one Lorentz frame

Take the definition of four-force as example. Within one Lorentz frame, four force F_i and four acceleration a_i are related by $F_i = ma_i$ (in which $i = 1, 2, 3, 4$). In another Lorentz frame,

$$F_i' = L_{ij} F_j = L_{ij}(ma_j) = m(L_{ij} a_j) = m a_i'.$$

Thus F' and a' are related by $F' = ma'$ in the primed frame, not only when set up in that frame, but also when set up in another frame and Lorentz-transformed. Four-vectors ensure that everything is consistent.

It should be clear that this idea generalizes to any physics law that can be expressed throughout in terms of four-tensors (meaning four-scalars, four-vectors, ...).[21]

[21] One way of showing that Maxwell's equations are compatible with relativity is to put those equations into four-tensor form, after which it is obvious that they hold in any Lorentz frame.

7.6   Conservation of momentum and energy

Conservation laws are intimately associated with symmetries. For the formal definition of symmetry see § 8.5.

For example, suppose we have a situation in which the physics is unaltered by translation in space through a constant vector $\boldsymbol{a}$. Then $\boldsymbol{r} \to \boldsymbol{r} + \boldsymbol{a}$ is a *symmetry operation* which we can apply and which leaves the physics looking the same. It is well known that this symmetry operation implies the conservation of linear momentum.

Here we have another symmetry: given any properly formulated physical law, you can change to a new Lorentz frame and the physics remains the same. Moreover, four vectors (and four tensors ...) are integral to the demonstration and handling of this sameness (§ 7.5). We therefore again direct our attention to four vectors, this time with a strong expectation that a conservation law is nearby.[22]

Consider some system of particles labelled by $n = 1, 2, 3, \ldots$. The nth particle has four-momentum $(p_1, p_2, p_3, p_4)_n$. We can add these to find the total four-momentum of the system in some chosen Lorentz frame:

$$P = (P_1, P_2, P_3, P_4) = \sum_n (p_1, p_2, p_3, p_4)_n. \tag{7.5}$$

[22] Textbooks often deal with energy-momentum by means of a tutorial example: an elastic collision between two identical masses. The collision is described in two different Lorentz frames, and a relativistic momentum is constructed to be "that which is found to be conserved". This is a "get you comfortable" argument that isn't easily made more rigorous.

We may now look at the system as a single composite particle. We'll assume that the constituents have no interaction with the world outside the system, even though they may interact with each other and may have relative speeds comparable with c. The composite is therefore subject to no external forces.

It is open to us to write the total four-momentum in the form[23]

$$P = (P_1, P_2, P_3, P_4)$$
$$= \left(\frac{MV_x}{\sqrt{1 - V^2/c^2}}, \frac{MV_y}{\sqrt{1 - V^2/c^2}}, \frac{MV_z}{\sqrt{1 - V^2/c^2}}, \frac{iMc}{\sqrt{1 - V^2/c^2}} \right)$$

in which $V^2 = V_x^2 + V_y^2 + V_z^2$, since we have four equations with which to find the four quantities M, V_x, V_y, V_z.

[23] For the present, this is just a mathematical form, however suggestive. No physical interpretation attaches to M or $\boldsymbol{V}$ until we have investigated further.

Given "no external forces", P must be conserved in the chosen Lorentz frame; this is where we impose the new symmetry. Moreover, P has been assembled from four-vector constituents and itself has the form of a four-vector. Therefore that conservation holds in all Lorentz frames. A further consequence is that M is a four-scalar, taking the same value in all Lorentz frames, and interpretable as the total mass of the system.[24] More importantly, the components

$$\sum_n p_{n1}, \quad \sum_n p_{n2}, \quad \sum_n p_{n3}, \quad \sum_n p_{n4}$$

are all conserved. This is just what we understand by a conservation law: the particles can change their momenta and energies, but losses balance gains overall.

Consider the four-momentum of one contributing particle, and look at it in a frame in which the velocity $v \ll c$. Then to second order in v/c the four-momentum is

$$p \approx \left(mv_x, \, mv_y, \, mv_z, \, (\mathrm{i}/c)(mc^2 + \tfrac{1}{2}mv^2) \right). \tag{7.6}$$

The fourth term shows that mc^2 has to be included in the energy that is "traded" as this particle interacts with others.

We have been able to obtain the basics of relativistic mechanics, just by insisting that the formulation has to be in terms of four-vectors.[25]

7.7 Epilogue

Here are two examples of four-vectors that have not had a mention before. Electric charge density ρ and current density j combine together into a four-vector. Likewise the scalar (φ) and vector (A) potentials (in the Lorentz gauge) combine similarly. We have four-vectors[26]

$$\left(j_x, j_y, j_z, \mathrm{i}c\rho\right) \qquad \text{and} \qquad \left(A_x, A_y, A_z, \mathrm{i}\varphi/c\right).$$

After a four-scalar and a four-vector, we come to a second-rank four-tensor, a quantity with two indices. And so on. Discussion of higher-rank tensors would take us beyond the intention of the present chapter. Once four-vectors are understood, higher-rank tensors should hold few fears. In fact, even for such familiar quantities as electric and magnetic fields, we have to move up, in complication, to a second-rank tensor.

[24] We must not over-interpret "total mass". When masses interact, the mass M of the whole is not the sum of the constituent masses because of a "mass defect" or "mass deficit". Of course, nothing said here establishes the existence of mass defect, but nothing conflicts with it either.

[25] Robinson (1995), Chapters 4,5, addresses the same questions as are discussed here, in an interestingly different way. In particular, Robinson points out that interactions often proceed via fields, and in such cases the energy and momentum possessed by a field have to be included in the accountancy of (7.5). This is a complication that I have ignored for simplicity.

[26] In the more usual presentation of relativity, these four-vectors have the (contravariant) forms

$$(c\rho, j_x, j_y, j_z)$$
$$(\varphi/c, A_x, A_y, A_z).$$

8 The Lorentz transformation

Intended readership: on a second encounter with special relativity, when the subject has ceased to seem entirely mysterious.

8.1 Introduction: relativistic postulates

When relativity is first introduced, we usually encounter postulates in something like the following form:

1. Space is uniform and isotropic (at least in the absence of gravitational fields).
2. There exist frames of reference within which a particle subject to no forces moves uniformly in a straight line; such frames are said to be **inertial**.
3. The laws of physics are identical in form in all inertial frames. In particular, nothing singles out any one inertial frame as "special".
4. The speed of light takes the same value in all inertial frames.

There is something odd about the presentation of these. Postulate (4) is already fully contained in postulate (3), so it is at most a clarification or an intensification of (3). So why is it given separately? And when we derive the Lorentz transformation we usually input (4) at once and hardly make separate use of (3). Postulate (1) may well not be mentioned at all, being taken entirely for granted.

There is a reason why things are presented in the above way, but it is rarely spelt out. Postulates (1)–(3) do not mention the speed of light, so they were in principle available to Galileo and Newton. Indeed, the *principle of relativity* implied by postulate (3) is just what Newton assumed; it is not this that distinguishes special relativity from (most) earlier theories.[1] It must be possible then to use the principle of relativity on its own, and to find the *class* of transformations that could imaginably link inertial frames. The outcome is surprising: we can *almost* obtain the Lorentz transformation from postulates (1)–(3) alone.[2] This is the programme of the present chapter.

If we follow this route, we leave all mention of light (or electromagnetism) until the very end.

I make no claim that the reasoning here is in any way superior to a more conventional derivation. Indeed, it is considerably longer: it is

[1] A good discussion of this point is given by French (1968), pp 65–6.

[2] The reasoning given here is hinted at very clearly, though not spelt out, in Rindler (2001).

Essays in Physics: Thirty-two thoughtful essays on topics in undergraduate-level physics. Geoffrey Brooker, Oxford University Press. © Geoffrey Brooker 2021. DOI: 10.1093/oso/9780198857242.003.0008

not the ideal examination answer. It's just fun to take a "sideways" approach.

8.2   Setting up

Inertial frames Σ, Σ' of Fig. 8.1 are in uniform relative motion. We choose origins O, O' in the two frames in such a way that the origins pass through each other during the motion. We define the x-axis as the path followed by O' as seen by an observer in Σ, and similarly we define the $-x'$-axis as the path followed by O as seen by an observer in Σ'.

Fig. 8.1 Frames Σ and Σ' are in uniform relative motion, with v as the velocity of O' as measured by an observer in Σ.

Axes Oy, Oz are drawn at right angles to Ox (and to each other) by an observer in frame Σ, and similarly O'y', O'z' are drawn at right angles to Ox' by an observer in Σ'.

The origins of time are chosen so that O, O' are coincident at times $t = t' = 0$.

The velocity of O' relative to O is v (as measured using rulers and clocks by an observer in Σ), so that[3]

$$x' = y' = z' = 0 \qquad \text{implies} \qquad x = vt \quad \text{and} \quad y = z = 0.$$

Frame Σ has been chosen to be an inertial frame, so the path of a free particle, moving with uniform velocity in a straight line, is given in Σ by

$$x = V_1\,t + x_0, \qquad y = V_2\,t + y_0 \qquad z = V_3\,t + z_0.$$

[3] The best presentation of this that I have seen draws frame Σ in black and frame Σ' in red. Velocity v is drawn black, because that's who measures it.

8.3   Axioms

We use axioms, based on the postulates listed in §8.1, but elaborated somewhat.

(1) In both frames, space is uniform. Uniformity means that in frame Σ there are no special values of x, y, z or t that are treated in a different way from the rest; and a similar statement applies to Σ'.

(2) In both frames, space is isotropic. Isotropy means that there are no directions singled out as special. The axiom of isotropy has consequences of two types which for clear-headedness should be kept separate:

(2a) For phenomena described entirely within one frame (say Σ), all three space directions are equivalent. For example the speed of light is the same in the x, y and z-directions.

(2b) For the transformation between the two frames, the x-axis *is* special, because we have chosen it as the direction of v. Thus we expect x to appear in the transformation equations differently from y and z. But at least y and z must appear in the transformation equations on an equal footing.

(3) We assume the **principle of relativity**, that is, nothing singles out either frame as in any way special. For example, no experiment can determine the velocity of either frame relative to a fixed "ether"; if it could then a frame at rest relative to the ether would be special. Consequences of this axiom are the following:[4]

[4] The alert reader may notice that we do not mention the Galilean transformation, even as a limiting case when $v \to 0$. Interestingly, we do not need to input this fact during the analysis that follows.

(3a) If Σ is an inertial frame then so is Σ'. Thus a free particle moving in Σ according to

$$x = V_1 t + x_0, \qquad y = V_2 t + y_0, \qquad z = V_3 t + z_0$$

is described in frame Σ' by equations of exactly similar form[5]

$$x' = V_1' t' + x_0', \qquad y' = V_2' t' + y_0', \qquad z' = V_3' t' + z_0'.$$

[5] This is stating that "free" is a frame-independent concept.

In mathematical terms, we are requiring the transformation of coordinates from Σ to Σ' to be one that maps points onto points and straight lines onto straight lines.[6]

[6] The points and straight lines are drawn in a space with four "dimensions" x, y, z, t, and the transformation maps one such space onto another. This mention of four dimensions is not really a sneaky way of introducing bits of special relativity, just a way of categorizing the mathematics we need.

(4) As mentioned in the Introduction, what follows is not a separate axiom but is contained already in (2b) and (3). But for the reasons given, I list it here because it will be used separately. Consider an "event" described in frame Σ by

$$x^2 + y^2 + z^2 - c^2 t^2 = 0,$$

where c is the speed of light. This event may be the arrival of a light pulse at (x, y, z) at time t, given that the pulse started out at the origin O at time $t = 0$. Here axiom (2a) has been used in requiring the speed of light to be the same in all directions in frame Σ. Axiom (3) now tells us that the same event must be described in frame Σ' by

$$x'^2 + y'^2 + z'^2 - c^2 t'^2 = 0.$$

Comment: In the "ideal examination answer", we state that the transformation must be linear and that $y' = y$ and $z' = z$, and from there we use axiom (4) to show that

$$(x'^2 - c^2 t'^2) = -(y'^2 + z'^2) = -(y^2 + z^2) = (x^2 - c^2 t^2).$$

[7] Invariant under those transformations where the relative velocity of the two frames lies in the x-direction. Note that $(x^2 - c^2 t^2)$ is carefully taken to be non-zero. If we too soon take the special case where it is zero, it is rather hard to deduce anything at all about the transformation.

This amounts to showing that $(x^2 - c^2 t^2)$ is an invariant.[7] From here it is straightforward algebra to obtain the Lorentz transformation.

Unfortunately, this isn't really satisfactory. It is *not* obvious that the transformation is linear. Nor, rather surprisingly, is axiom (3) on its own sufficient to establish linearity—we need axiom (1) as well. Even then, it requires justification that $y' = y$ and $z' = z$. Our aim here is to do something better, even if the reasoning is lengthened.

8.4 The "shape" of the transformation

The transformation we require is one which maps uniform motion in a straight line onto uniform motion in a straight line. A general transformation of this type is not linear, but has the perhaps unexpected form[8]

$$
\left.\begin{array}{l}
x' = \dfrac{a_1\, x + a_2\, y + a_3\, z + a_4\, t + a_5}{Ax + By + Dz + Et + F} \\[2ex]
y' = \dfrac{b_1\, x + b_2\, y + b_3\, z + b_4\, t + b_5}{Ax + By + Dz + Et + F} \\[2ex]
z' = \dfrac{c_1\, x + c_2\, y + c_3\, z + c_4\, t + c_5}{Ax + By + Dz + Et + F} \\[2ex]
t' = \dfrac{d_1\, x + d_2\, y + d_3\, z + d_4\, t + d_5}{Ax + By + Dz + Et + F}\, .
\end{array}\right\}
\tag{8.1}
$$

Notice that all the denominators are the same but all the numerators are different.[9]

Use axiom (1)
There are no special places or times in either frame. In the transformation (8.1), those values of x, y, z, t which make the denominator equal to zero are singled out as special, for they make x', y', z', t' all infinite. Worse still, if we allow $(Ax + By + Dz + Et + F)$ to pass through zero, x', y', z', t' all go to infinity and then reappear "from the other side" (for example, t' may go to $+\infty$ and then come back to finite values from $-\infty$). The only way to ensure that this kind of thing never happens for any finite values of x, y, z, t is to set $A = B = D = 0$ and $E = 0$. Thus we find, after all, that the transformation is linear.[10]

Henceforth we can absorb F into the constants in the numerator, or equivalently choose $F = 1$.

Use the requirement that the origins of the frames coincide at $t = t' = 0$
We may see at once that this makes $a_5 = b_5 = c_5 = 0$ and $d_5 = 0$.

Use the requirement that O moves along the x'-axis and O' moves along the x-axis
The first of these means that $x = y = z = 0$ forces $y' = z' = 0$ for all t. It should be obvious that $b_4 = c_4 = 0$. Knowing that $x' = y' = z' = 0$ is compatible with $y = z = 0$ while x and t are free to be non-zero, we are also able to argue that $b_1 = c_1 = 0$.

It's now convenient to summarize where our transformation equations have got to:

$$
\begin{aligned}
x' &= a_1\, x + a_2\, y + a_3\, z + a_4\, t & \text{(8.2a)} \\
y' &= + b_2\, y + b_3\, z & \text{(8.2b)} \\
z' &= + c_2\, y + c_3\, z & \text{(8.2c)} \\
t' &= d_1\, x + d_2\, y + d_3\, z + d_4\, t. & \text{(8.2d)}
\end{aligned}
$$

Of course, these equations can be put into matrix form, but that doesn't help much as yet.

[8] A proof is given by Fock (1964), Appendix A. Fock's presentation is somewhat muddled, as the equations of importance are interspersed with other equations that assume the constancy of the speed of light—just what we want to avoid at this stage. However, it is possible to weave one's way through and extract a valid argument.

[9] A three-dimensional version of this "projective transformation" can be used for deriving a generalized "lens formula" that must apply to any optical (or other) imaging system that maps points in the object onto points in the image and straight lines (rays) in the object space onto straight lines in the image space. See e.g. Joos (1934), pp 383–5; Brooker (2006), problem 13.8.

[10] You might find my reasoning pedestrian in the extreme—surely it was obvious? Well, I'm heading for a surprising result, so it's appropriate to put all steps and assumptions carefully on the line.

8.5   Symmetry

We have much more to say about symmetry in Chapter 11. Here we give a brief reminder of what symmetry means in mathematics and physics.

> A thing is symmetrical if there is something you can do to it that leaves it looking the same.

The "something you can do" is called a **symmetry operation**, and the thing is said to be "symmetric under the ... operation".

To see why symmetry might be important, let us show that eqns (8.2) are far too general. Here are two things that they would allow in their present form:

$$y' = y, \qquad x' = x + y;$$
$$y' = y, \qquad z' = 2z.$$

Both of these violate rotational symmetry about the x-axis; the first by allowing the y'-axis to be "slanted" at $45°$ in the xy plane. Clearly, we shall have to do something to exclude such silliness.

Now it's possible to go some way to dealing with symmetry "by inspection" using common sense. But if things get at all complicated we need more formal methods. So in what follows I'll invent a sequence of symmetry operations to deal with our needs.

8.6   Applying symmetries

8.6.1   90° rotation about the x-axis

We make use of axiom (2b), demanding that the transformation have rotational symmetry about the x-axis. We have already stated that Oy and Oz are perpendicular to each other and to Ox. Then a $90°$ rotation[11] about the x-axis carries y into z and z into $-y$. We likewise require that the same rotation[12] carries y' into z' and z' into $-y'$.

Applying the rotation, we arrive at

$$\left.\begin{array}{l} x' = a_1 x - a_3 y + a_2 z + a_4 t \\ y' = \qquad c_3 y - c_2 z \\ z' = \qquad -b_3 y + b_2 z \\ t' = d_1 x - d_3 y + d_2 z + d_4 t \end{array}\right\} \text{ so that } \left\{\begin{array}{l} a_3 = a_2 = 0 \\ c_2 = -b_3 \quad \text{(twice)} \\ c_3 = b_2 \quad \text{(twice)} \\ d_3 = d_2 = 0. \end{array}\right.$$

In drawing the conclusions listed, we have required that the transformation equations must "look the same" as eqns (8.2) after the rotation, and therefore we have equated coefficients.

The present symmetry requirement has guaranteed that y and z are treated on an equal footing. So we can see it as rejecting the second of the asymmetries that we met in §8.5. And the zero values of a_2, a_3 show that it has rejected the first also.

[11] The fact that $y \to z$ and $z \to -y$ imposes in the mathematics the condition that the y- and z-axes are perpendicular to the x-axis, and to each other. Draw a picture if this isn't obvious.

[12] This sneaks in a requirement that the unprimed and primed axes have the same handedness, either both right-handed or both left-handed.

8.6.2   Choice: make $O'y'$ parallel to Oy

We have as yet made no attempt at specifying the rotational orientation of the y' and z' axes around $O'x'$, in relation to the directions of Oy and Oz. We'll now require $O'y'$ to be at right angles to Oz, so that y' is independent of z; equivalently $b_3 = 0$. This further[13] makes $c_2 = 0$, so it is also the case that z' is independent of y.

8.6.3   Relative velocity of frames

Introduce the fact that O' moves with velocity v along the positive x-axis, as seen in frame Σ. Now (8.2a), with a_2 and a_3 removed, says that $x' = 0$ implies $a_1 x + a_4 t = 0$. Thus $v = x/t = -a_4/a_1$.

Let us summarize our findings so far, conveniently now in matrix form:

$$\begin{pmatrix} x' \\ y' \\ z' \\ t' \end{pmatrix} = \begin{pmatrix} a_1 & 0 & 0 & -a_1 v \\ 0 & b_2 & 0 & 0 \\ 0 & 0 & b_2 & 0 \\ d_1 & 0 & 0 & d_4 \end{pmatrix} \begin{pmatrix} x \\ y \\ z \\ t \end{pmatrix}. \tag{8.3}$$

8.6.4   180° rotations about the z-axis and z'-axis

This operation carries x into $-x$ and y into $-y$, and the equivalent in Σ'. When the operation has been done, frame Σ stands in the relation to Σ' that Σ' previously had in relation to Σ: the origin O is travelling along the (new) $+x'$ axis. Then

$$\begin{pmatrix} x \\ y \\ z \\ t \end{pmatrix} = \begin{pmatrix} a_1 & 0 & 0 & -a_1 v \\ 0 & b_2 & 0 & 0 \\ 0 & 0 & b_2 & 0 \\ d_1 & 0 & 0 & d_4 \end{pmatrix} \begin{pmatrix} x' \\ y' \\ z' \\ t' \end{pmatrix}. \tag{8.4}$$

The square matrix is the same as in (8.3), by virtue of the principle of relativity: everything depends on the *relative* motion of the two frames, and nothing else. This is another application of symmetry, the symmetry operation being the 180° rotation combined with interchange of primed and unprimed coordinates.

Note that the square matrix in (8.4) must be identical with that in (8.3). In particular, v has the same numerical value. And all the other coefficients are the same too, because they can depend on v only.

Now we rotate the axes back again. In the original coordinate axes,

$$\begin{pmatrix} -x \\ -y \\ z \\ t \end{pmatrix} = \begin{pmatrix} a_1 & 0 & 0 & -a_1 v \\ 0 & b_2 & 0 & 0 \\ 0 & 0 & b_2 & 0 \\ d_1 & 0 & 0 & d_4 \end{pmatrix} \begin{pmatrix} -x' \\ -y' \\ z' \\ t' \end{pmatrix},$$

or

$$\begin{pmatrix} x \\ y \\ z \\ t \end{pmatrix} = \begin{pmatrix} a_1 & 0 & 0 & a_1 v \\ 0 & b_2 & 0 & 0 \\ 0 & 0 & b_2 & 0 \\ -d_1 & 0 & 0 & d_4 \end{pmatrix} \begin{pmatrix} x' \\ y' \\ z' \\ t' \end{pmatrix}.$$

We have constructed the matrix relation that is inverse to (8.3).

8.6.5   There and back

It should be obvious what the next step must be. We go "there and back" and require that the combined operation is an identity.

$$
\begin{pmatrix} x \\ y \\ z \\ t \end{pmatrix} = \begin{pmatrix} a_1 & 0 & 0 & a_1\,v \\ 0 & b_2 & 0 & 0 \\ 0 & 0 & b_2 & 0 \\ -d_1 & 0 & 0 & d_4 \end{pmatrix} \begin{pmatrix} a_1 & 0 & 0 & -a_1\,v \\ 0 & b_2 & 0 & 0 \\ 0 & 0 & b_2 & 0 \\ d_1 & 0 & 0 & d_4 \end{pmatrix} \begin{pmatrix} x \\ y \\ z \\ t \end{pmatrix}.
$$

The product of square matrices here must be a unit matrix, because this matrix equation must say nothing about x, y, z, t. We find

$$a_1^2 + a_1 d_1 v = 1 \tag{8.5}$$
$$d_4^2 + a_1 d_1 v = 1 \tag{8.6}$$
$$d_1(d_4 - a_1) = 0 \tag{8.7}$$
$$a_1 v(d_4 - a_1) = 0 \tag{8.8}$$
$$b_2^2 = 1 \quad \text{(twice)}. \tag{8.9}$$

It remains only to tidy up here.

Equation (8.9) tells us that $b_2 = \pm 1$. We can choose to make $b_2 = +1$, because we now specify that the y- and y'-axes are to be parallel, rather than antiparallel. Alternatively, we can require this part of (8.3) to become an identity transformation as $v \to 0$.

The quantity $a_1 v$ cannot be allowed to be zero, so (8.8) forces $d_4 = a_1$. We can further require d_4 to be positive, otherwise t and t' would increase in opposite directions. Then a_1 is positive as well.

In eqn (8.7), the factor $(d_4 - a_1)$ is zero, which leaves d_1 unrestricted. Similarly, (8.8) has nothing more to say about $(a_1 v)$. And (8.6) has become redundant with (8.5), so (8.5) is the only equation with information still to convey.

Equation (8.5) tells us that d_1 reverses its sign when v reverses sign. Let us now write $d_1 = -a_1 v f(v)$, where the v is introduced to build in the known dependence on the sign of v. The function $f(v)$ is not known as yet, but it must be an even function of v.

[14] We already know that a_1 is positive, so we choose the positive value for the square root.

Finally, we substitute this expression for d_1 into (8.5), and we find[14] $a_1 = 1/\sqrt{1 - v^2 f(v)}$, which henceforth will be called γ.

Summarizing yet again, we now have

$$
\begin{pmatrix} x' \\ y' \\ z' \\ t' \end{pmatrix} = \begin{pmatrix} \gamma & 0 & 0 & -\gamma v \\ 0 & 1 & 0 & 0 \\ 0 & 0 & 1 & 0 \\ -\gamma v f(v) & 0 & 0 & \gamma \end{pmatrix} \begin{pmatrix} x \\ y \\ z \\ t \end{pmatrix}, \quad \text{where } \gamma = \frac{1}{\sqrt{1 - v^2 f(v)}},
\tag{8.10}
$$

and $f(v)$ is an even function of v but is otherwise unknown as yet.

8.6.6   Two transformations give a third

[15] I am indebted to Dr C.V. Sukumar for pointing out this step.

Two successive transformations of the form (8.10) must be equivalent to a single combined transformation.[15]

Consider one transformation from Σ to Σ' (relative velocity v), and then a second from Σ' to Σ'' (relative velocity v'). The combined transformation gives

$$\begin{pmatrix} x'' \\ y'' \\ z'' \\ t'' \end{pmatrix} = \begin{pmatrix} \gamma' & 0 & 0 & -\gamma' v' \\ 0 & 1 & 0 & 0 \\ 0 & 0 & 1 & 0 \\ -\gamma' v' f(v') & 0 & 0 & \gamma' \end{pmatrix} \begin{pmatrix} \gamma & 0 & 0 & -\gamma v \\ 0 & 1 & 0 & 0 \\ 0 & 0 & 1 & 0 \\ -\gamma v f(v) & 0 & 0 & \gamma \end{pmatrix} \begin{pmatrix} x \\ y \\ z \\ t \end{pmatrix}.$$

As stated, the outcome of this must be a matrix transformation of the form (8.10), with v replaced by the "third" velocity v''. Multiplying out the square matrices and equating coefficients, we obtain

$$\gamma'' = \gamma\gamma'\{1 + vv' f(v)\} = \gamma\gamma'\{1 + vv' f(v')\} \tag{8.11}$$
$$\gamma'' v'' = \gamma\gamma'(v + v') \tag{8.12}$$
$$\gamma'' v'' f(v'') = \gamma\gamma'\{v f(v) + v' f(v')\}. \tag{8.13}$$

Equation (8.11) shows that $f(v') = f(v)$, which has to be true for any values of v and v'. Then $f(v)$ can only be a constant. We shall write the constant as $1/C^2$, in which C has the dimensions of a velocity. Dividing eqns (8.11) and (8.12) now gives us the velocity addition rule

$$v'' = \frac{v + v'}{1 + v v'/C^2}, \tag{8.14}$$

while (8.13) gives no new information but provides a consistency check.
Summarizing a last time, we have finally

$$\begin{pmatrix} x' \\ y' \\ z' \\ t' \end{pmatrix} = \begin{pmatrix} \gamma & 0 & 0 & -\gamma v \\ 0 & 1 & 0 & 0 \\ 0 & 0 & 1 & 0 \\ -\gamma v/C^2 & 0 & 0 & \gamma \end{pmatrix} \begin{pmatrix} x \\ y \\ z \\ t \end{pmatrix}, \quad \text{where } \gamma = \frac{1}{\sqrt{1 - v^2/C^2}}, \tag{8.15}$$

and C is a universal constant speed which is undetermined by the reasoning so far.

Comments
As promised, the argument thus far has proceeded without our making use of the axiom involving the speed of light. So the transformation (8.15) is entirely pre-Einstein and could in principle have been derived at any time after Galileo.

What we have done then, in deriving (8.15), is to obtain a *class* of transformations, all compatible with our physical requirements, and differing only in the choice made for the constant C^2. If $1/C^2 = 0$, we are back to Galileo; if $1/C^2 = 1/c^2$ we have Lorentz.[16]

Some details are nice. The statements $y' = y$, $z' = z$ do not require any separate justification (paintbrushes in Σ' being dragged across a board in Σ?); they have come out naturally from the combination of symmetry requirements.

We have also had no need to require that the Galilean transformation is recovered for small v; we have Galileo automatically[17] in the limit $|v| \ll C$.

[16] Is there any way of determining that $f(v)$ must be positive? If not, then you could write $f(v) = -1/C^2$ and obtain another species of transformation.
    Answer: A different sign in the denominator of the velocity addition theorem (8.14) is very difficult to live with. Investigate.

[17] Well, not quite. There are two difficulties.
    First, we might have a particle moving with velocity V', not $\ll C$, in frame Σ', and transform things to frame Σ using relative velocity $v \ll C$. Thus we use a small velocity in the Lorentz transformation but discuss the physics of fast-moving particles. Then (8.14) gives a departure from the Galilean velocity addition.
    Second, even (8.15) gives trouble.
$$t' = \gamma\left(t - \frac{x}{C^2/v}\right).$$
As $v \to 0$, $\gamma \to 1$. But even so there exists a range of x for which $t' \neq t$. Admittedly, as $v \to 0$ that range is pushed out to very large x, but still it's there.
    So the Galilean transformation really requires $1/C^2 = 0$, and not just $|v| \ll C$. Acknowledgments to Prof. H. Brown for this refinement.

Notice that the velocity addition rule (8.14) makes $|v''| < C$ whenever $|v|$ and $|v'|$ are both less than C. The velocity C has the character of a limiting speed, a speed that cannot be exceeded by accelerating any particle from an initial speed $<C$. In fact, everything in (8.15) corresponds exactly to what we have in Einsteinian relativity, with the one exception that C is not yet identified with the speed of light.

8.7   Introduce the speed of light

Finally, we may impose axiom (4) onto the transformation (8.15). It's rather easy now to show that $1/C^2 = 1/c^2$. This step is so small that it is "left as an exercise for the reader".

At this point we may, if we wish, rearrange the matrices so that the columns have $(\mathrm{i}ct)$ at the bottom as the fourth element, or ct at the top as the zeroth element, but this is only tidying up.

V Optics

9 Diffraction integrals and the Kirchhoff approximation

Intended readership: On a second exposure to diffraction theory, when the Kirchhoff integral is familiar, if only for the Fraunhofer case, and the reader is ready to take a more advanced view.

9.1 Introduction

A wave (perhaps light) is incident upon a mathematical surface, which may or may not be plane. A possible arrangement is shown in Fig. 9.1, where a part of the surface is obstructed by an **obstacle** while some other part (the **aperture**) transmits. Some radiation passes through to regions downstream and thereby undergoes diffraction.

The diffracted field is obtained mathematically by integrating the fields radiated by "Huygens secondary sources": sources lying on a surface spanning the aperture and driven by the incident-field amplitude there. The integral, (9.15) below, is usually (at an undergraduate level) made plausible, rather than derived. In this chapter we improve matters by supplying a rigorous derivation, albeit of the related (9.10).

Along with investigation of the diffraction integrals, we shall address related questions:

- why do secondary sources not radiate "backwards"?
- the scalar-wave approximation: in optics, one conventionally treats the wave amplitude as a scalar, yet the electromagnetic field consists of vectors $\boldsymbol{E}$ and $\boldsymbol{B}$; how much harm is done?
- the Kirchhoff approximation:[1] simplifying assumptions are conventionally made about the field at the integration surface; how serious are they, and what might be done to improve things?
- there are alternative formulations of the diffraction integral (§§ 9.2 and 9.3), which lead to different approximate results; how do we bring order to this?

The third bullet point above mentions the "Kirchhoff approximation" (also known as St Venant's hypothesis), in which it is assumed:

- in the open areas of the integration surface, the field is the same as that of the incident wave, so "a piece has been chopped out" from

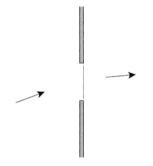

Fig. 9.1 Radiation is incident onto a mathematical surface, part of which may be obstructed. What passes through the open area undergoes diffraction. We refer to the open area as the **aperture** and the surrounding obstruction as the **obstacle**.

[1] I refer to the "Kirchhoff approximation", rather than the more usual "Kirchhoff boundary conditions", as I think it better describes what is going on.

Essays in Physics: Thirty-two thoughtful essays on topics in undergraduate-level physics. Geoffrey Brooker, Oxford University Press. © Geoffrey Brooker 2021. DOI: 10.1093/oso/9780198857242.003.0009

that incident wave (the incident-wave amplitude is multiplied by a top-hat function)

- in the opaque areas, the field amplitude on the downstream face of the obstacle is zero.

It is known from experiment that the "textbook" scalar-wave diffraction integral (9.15), incorporating a Kirchhoff approximation, gives an extremely reliable account of diffraction. We are therefore able to use such a formulation with a great deal of confidence. Nevertheless, it is somewhat unsettling that a crude treatment can be so good; nature is not usually so kind to us.[2] There is a lot to question here.

9.2   The diffraction integrals[3]

In this section, we do *not* make Kirchhoff assumptions, but work from exact equations.

Let $U(\boldsymbol{r}) = U(x, y, z)$ be a scalar amplitude ($\propto \mathrm{e}^{-\mathrm{i}\omega t}$), that represents one of the Cartesian components of $\boldsymbol{E}$ or $\boldsymbol{B}$. (It is to be understood that if more than one field component is important, then the method that follows may be applied—if only notionally—to each in turn.) From Maxwell's equations it may easily be shown that U obeys the "Helmholtz" wave equation[4],[5]

$$\nabla^2 U + k^2 U = 0, \tag{9.1}$$

in which $k = \omega/c$. In writing the homogeneous equation (9.1), we take it that all sources of radiation lie outside the region of space within which (9.1) is to be applied. Any medium present may be a vacuum, but

[2] A detailed examination of the Kirchhoff approximation is too technical for the present chapter; it is dealt with (albeit for the special case of a slit) in Chapter 10. The discussion there does something to explain why Kirchhoff is so good.

[3] We shall refer to a **Kirchhoff integral** when a Kirchhoff-approximated field has been inserted into the integrand; while a **diffraction integral** will mean a general integral that incorporates no such assumption or approximation.

[4] Equation (9.1) is obtained via

$$\mathrm{curl\,curl}\ \boldsymbol{E} = \mathrm{grad\,div}\ \boldsymbol{E} - \nabla^2 \boldsymbol{E},$$

which must be understood as applying to each Cartesian component of $\boldsymbol{E}$. This is why our forms of the diffraction integral must be applied to one or more of the *Cartesian* components of $\boldsymbol{E}$ and/or $\boldsymbol{B}$.

[5] Although we have introduced diffraction in the context of electromagnetic waves, the analysis in §§ 9.2–9.3 builds wholly on (9.1), and so could apply to any wave.

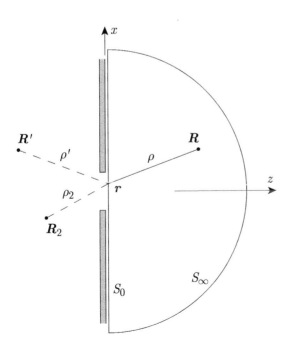

Fig. 9.2 Radiation is incident from the left onto surface S_0. Part of surface S_0 may be obstructed while the rest is open. The field amplitude diffracted to "field points" such as $\boldsymbol{R}$ is calculated, in the first instance, by means of integral (9.3), with the integration taken over area elements on the closed surface S composed of S_0 and S_∞. It is shown in the text that the integration over S_∞ drops out, and the integral reduces to (9.4), taken over S_0 only.

[6] Function G is known in the trade as a Green's function. There is no need to be intimidated by this name.
Remember that

$$\nabla^2(1/r) = -4\pi\,\delta(\boldsymbol{r}).$$

Operator $\nabla_{\boldsymbol{r}}$ differentiates with respect to $\boldsymbol{r}$ while holding $\boldsymbol{R}$ constant.

[7] Diffraction is most often discussed in contexts where a beam of light has been narrowed by passing it through some aperture such as a slit. But diffraction happens (downstream of the narrowing) whenever a beam is narrow, however the narrowing came about. Example: a Gaussian beam spreads out because it is narrow at its waist. All that is needed for diffraction spread is that the area of S_0 integrated over in (9.4) be limited in extent.

[8] See problem 9.2. For this reasoning to work, it is necessary that the area of S_0 contributing to the integrand should be finite. This could be because an aperture is finite in extent, or because non-zero fields occupy a limited width within area S_0.

[9] If radiation arrives through the hemisphere, the two terms in the left-hand integrand of (9.3) reinforce, and the result is a diffraction integral relating the field at $\boldsymbol{R}$ to secondary sources on the hemisphere. The relative signs of the two terms are determined entirely by whether the wave passing through the hemispherical surface is "incoming" or "outgoing". This statement is explored in problem 9.3.

[10] Although the similarity is not very close, it may help to remember that the Coulomb field of a dipole falls as r^{-3} at large distances, even though the field from each contributing charge falls as r^{-2}.

[11] See Fig. 9.3 for an example.

[12] For the avoidance of doubt: $U(\boldsymbol{r})$ here is the actual field amplitude (perhaps $\boldsymbol{E}$ or $\boldsymbol{B}$, not the "incident" field that would be present if the obstacle were removed.

(with revised definition of k) it could be anything LIH: linear, isotropic, homogeneous, time-independent.

We introduce a function $G(\boldsymbol{r} - \boldsymbol{R})$ with the definition and property[6]

$$G(\boldsymbol{r} - \boldsymbol{R}) \equiv \frac{e^{ik|\boldsymbol{r}-\boldsymbol{R}|}}{|\boldsymbol{r} - \boldsymbol{R}|}; \qquad \nabla_{\boldsymbol{r}}^2 G + k^2\,G = -4\pi\delta(\boldsymbol{r} - \boldsymbol{R}). \qquad (9.2)$$

It is straightforward (problem 9.1) to show from (9.1) and (9.2) that

$$\frac{1}{4\pi} \int \left(G\,\nabla_{\boldsymbol{r}} U - U\,\nabla_{\boldsymbol{r}} G \right) \cdot \mathrm{d}\boldsymbol{S} = \int U(\boldsymbol{r})\,\delta(\boldsymbol{r} - \boldsymbol{R})\,\mathrm{d}V, \qquad (9.3)$$

where S is a closed surface enclosing volume V (with $\mathrm{d}V = \mathrm{d}^3\boldsymbol{r}$) and $\mathrm{d}\boldsymbol{S}$ is a surface element with an outward-facing normal.

Figure 9.2 shows a surface over which an integration (9.3) may be taken. An obstacle/aperture is hinted at, to explain why we integrate over this particular surface.[7] The surface S consists of area S_0 spanning the aperture, and a large hemisphere S_∞. Radiation is incident from the left, and nothing arrives through the hemisphere from the right.

On the hemisphere, U, G, ∇U and ∇G all fall with distance as e^{ikr}/r, so both terms in the integrand fall as $1/r^2$ while surface elements get larger as r^2. It appears at first then that the integral over the hemisphere should tend to a finite limit. However, at large distances from the aperture both U and G behave like spherical waves radiating from the origin (or nearby),[8] so the two terms in the left-hand integrand of (9.3) cancel against each other.[9] The outcome must be that the sum of the terms acquires at least one additional power of $1/r$, making the integral over S_∞ tend to zero.[10]

The integral is therefore taken only over S_0, the obstacle/aperture part of surface S. This part of the surface may be plane (as suggested in Fig. 9.2), but may have a more general shape; S_0 might resemble a membrane stretched over the xy plane and then inflated or puckered.[11]

For a point $\boldsymbol{R}$ that lies within the closed surface S, (9.3) becomes

$$U(\boldsymbol{R}) = -\frac{1}{4\pi} \int_{S_0,\,\mathrm{d}\boldsymbol{S}_0\,+\mathrm{ve\ to\ right}} \left(G\,\nabla_{\boldsymbol{r}} U - U\,\nabla_{\boldsymbol{r}} G \right) \cdot \mathrm{d}\boldsymbol{S}_0, \qquad (9.4)$$

with $G = G(\boldsymbol{r} - \boldsymbol{R})$ given by (9.2).

Equation (9.4) shows that the field $U(\boldsymbol{R})$ at $\boldsymbol{R}$ can be evaluated by knowing the value of $U(\boldsymbol{r})$ (itself and/or its gradient) at the bounding surface S_0. This is, of course, a re-statement and validation of the Huygens principle: we may think that there are secondary sources on the surface S_0, the source at the location $\boldsymbol{r}$ of $\mathrm{d}\boldsymbol{S}_0$ having strength connected with the value of $U(\boldsymbol{r})$ there.[12] Moreover, outgoing waves do not give rise to contributions from the large hemisphere, so we are on the way to showing that Huygens sources do not radiate "backwards".

9.2.1   Huygens sources do not radiate backwards

Apply eqn (9.3) to the case where $\boldsymbol{R}$ is taken to be $\boldsymbol{R}_2$, some general point outside the closed surface S (Fig. 9.2). The contribution from the hemisphere vanishes as before, and we have

$$0 = \frac{1}{4\pi} \int_{S,\,\mathrm{d}\boldsymbol{S}\ +\text{ve outwards}} \left(G_2 \nabla_{\boldsymbol{r}} U - U \nabla_{\boldsymbol{r}} G_2\right) \cdot \mathrm{d}\boldsymbol{S}$$

$$= -\frac{1}{4\pi} \int_{S_0,\,\mathrm{d}\boldsymbol{S}_0\ +\text{ve to right}} \left(G_2 \nabla_{\boldsymbol{r}} U - U \nabla_{\boldsymbol{r}} G_2\right) \cdot \mathrm{d}\boldsymbol{S}_0, \qquad (9.5)$$

where

$$G_2 \equiv \frac{\mathrm{e}^{\mathrm{i}k|\boldsymbol{r}-\boldsymbol{R}_2|}}{|\boldsymbol{r}-\boldsymbol{R}_2|}.$$

Now apply (9.3) to a different closed surface S' (not drawn), consisting of S_0 as before but closed by a surface to the left,[13] the surface enclosing point $\boldsymbol{R}_2$. The field $U(\boldsymbol{R}_2)$ at $\boldsymbol{R}_2$ may now be evaluated by means of an integral taken over the surface S'. Our interest is in the existence, or not, of contributions to $U(\boldsymbol{R}_2)$ from Huygens sources on surface S_0. By (9.3) and (9.5) these contribute

$$U_{\text{back-diffracted}}(\boldsymbol{R}_2) = \frac{1}{4\pi} \int_{S_0,\,\mathrm{d}\boldsymbol{S}_0\ +\text{ve to right}} \left(G_2 \nabla_{\boldsymbol{r}} U - U \nabla_{\boldsymbol{r}} G_2\right) \cdot \mathrm{d}\boldsymbol{S}_0$$

$$= 0. \qquad (9.6)$$

Comment: Result (9.6) is general. It does not matter if there are obstacles or apertures lying to the left of surface S_0, though it may take a little thought to see this. Suppose there is present some obstacle, perhaps that hinted at in Fig. 9.2. The obstacle will scatter or reflect radiation. It is easiest to deal with this by shaping the surface S' so as to exclude the obstacle. Reflection from the obstacle adds to the fields entering S' and so makes its contribution to the field $U(\boldsymbol{R}_2)$. But this just adds complication: the partial field resulting from integration over S_0 is still zero. This applies even if the obstacle is close against surface S_0; nothing can happen to restrict the validity of (9.6).

Since some results to be obtained later depend upon the integration surface being plane, we note here that (9.4) and (9.6) are valid for any reasonable shape of surface S_0.

9.2.2   "Backwards" is a topological concept

Figure 9.3 shows two surfaces S_1 and S_2, each forming a closed surface when combined with the hemisphere S_∞ at infinity. These two surfaces are variants on the S_0 of Fig. 9.2. Radiation is incident from the left. Field point $\boldsymbol{R}$ lies inside surface $S_1 S_\infty$, so $U(\boldsymbol{R})$ is given by an integral of the form (9.4) taken over surface S_1. Point $\boldsymbol{R}$ lies outside surface $S_2 S_\infty$ so the field there contains a back-diffraction integral of the form (9.6) taken over S_2, an integral that gives zero. Yet a considerable portion of surfaces S_1, S_2 is common to both surfaces. In particular, a secondary source at $\mathrm{d}\boldsymbol{S}$ must make a non-zero contribution to $U(\boldsymbol{R})$ when it is thought of as part of S_1; then it must make the same contribution when considered as being a part of S_2. Integral (9.6), taken over surface S_2, contains a non-zero integrand, yet is guaranteed to evaluate to zero.

Surface S_2 may have a variety of different shapes, and incident waves may have different forms, yet integral (9.6) always evaluates to zero. It

[13] The surface S' need not be infinitely large, nor hemispherical; it must be large enough that all illuminated parts of S_0 are included in the integration over S_0, and it must enclose point $\boldsymbol{R}_2$ (and for later use $\boldsymbol{R}'$). Otherwise it may have any shape. Thus it is possible to imagine that there are sources of radiation lying even further to the left, outside surface S', while there are no sources within the surface.

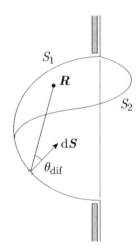

Fig. 9.3 Surfaces S_1 and S_2 are different possible variants of the surface S_0 of Fig. 9.2. Point $\boldsymbol{R}$ is downstream of S_1 but upstream of S_2. The electromagnetic field at $\boldsymbol{R}$ is given by a diffraction integral (9.4) taken over S_1. It is also given by the incident field (which here includes whatever may be scattered from any obstacle) together with a diffraction integral taken over surface S_2. "Backward" propagation from S_2 is described by (9.6) and gives zero. Surface element $\mathrm{d}\boldsymbol{S}$ is common to both integrands and must give a non-zero contribution to both. Therefore the zero outcome of (9.6) must be unconnected with anything "local" to $\mathrm{d}\boldsymbol{S}$.

[14]Here's a similar case. Put a point charge q at location $\boldsymbol{R}$ and ask about $\int \boldsymbol{D} \cdot \mathrm{d}\boldsymbol{S}$ taken over the closed surface $S_2 S_\infty$. The integrand is not zero, yet the integral always gives exactly zero.

is tempting to conclude that a zero integral must imply a zero integrand, yet we have shown that this is not the case.[14] It is also tempting to look for a reason why the integrand of (9.6) might validly be replaced by zero, finding something "local" to $\mathrm{d}\boldsymbol{S}$ that tells us when to make that replacement. For example, we might try to identify "backwards" with $\theta_{\mathrm{dif}} > \pi/2$ (notation of Fig. 9.3), yet this fails at once.

The only requirement for (9.6) to give zero is that the field point $\boldsymbol{R}$ must lie to the left of S_2, outside the closed surface $S_2 S_\infty$. This is a statement about the *topology* of $\boldsymbol{R}$ and surface S_2, and not about anything "local" to surface elements $\mathrm{d}\boldsymbol{S}$ on surface S_2.

9.2.3 Dirichlet and Neumann forms

An integral, after the fashion of (9.4), whose integrand contains U but not its gradient is said to incorporate **Dirichlet** boundary conditions. An integral containing the normal gradient of U but not U itself incorporates **Neumann** boundary conditions. We now show that (9.4) can be processed into these forms.[15]

[15]For honesty: It is shown in problem 10.3 that the boundary conditions obeyed by electromagnetic field components are in reality mixed: U is specified on some areas, $\partial U/\partial z$ over others. This complication is handled in Chapter 10.

Take a special case where the integration surface S_0 is the xy plane, and where $\boldsymbol{R}'$ is the "mirror image" of $\boldsymbol{R}$ in the xy plane (Fig. 9.2). That is, $\boldsymbol{R} = (X, Y, Z)$ and $\boldsymbol{R}' = (X, Y, -Z)$ with $Z > 0$.

Write $|\boldsymbol{r} - \boldsymbol{R}| = \rho$, $|\boldsymbol{r} - \boldsymbol{R}'| = \rho'$, with corresponding definitions of Green's functions G and G', so that

$$\rho^2 = (x - X)^2 + (y - Y)^2 + (z - Z)^2, \qquad G = \frac{e^{ik|\boldsymbol{r} - \boldsymbol{R}|}}{|\boldsymbol{r} - \boldsymbol{R}|} = \frac{e^{ik\rho}}{\rho},$$

and similarly

$$\rho'^2 = (x - X)^2 + (y - Y)^2 + (z + Z)^2, \qquad G' = \frac{e^{ik|\boldsymbol{r} - \boldsymbol{R}'|}}{|\boldsymbol{r} - \boldsymbol{R}'|} = \frac{e^{ik\rho'}}{\rho'}.$$

On the plane $z = 0$, $G' = G$ while $(\partial G'/\partial z)_{z=0} = -(\partial G/\partial z)_{z=0}$. Equations (9.4) and (9.6) may now be expressed, for $Z > 0$, as

$$U(\boldsymbol{R}) = \frac{-1}{4\pi} \int \frac{e^{ik\rho}}{\rho} \left(\frac{\partial U(\boldsymbol{r})}{\partial z} \right)_{z=0} \mathrm{d}x\,\mathrm{d}y + \frac{1}{4\pi} \int \left(\frac{\partial}{\partial z} \frac{e^{ik\rho}}{\rho} \right)_{z=0} U(\boldsymbol{r})\,\mathrm{d}x\,\mathrm{d}y, \quad (9.7\mathrm{a})$$

$$0 = \frac{1}{4\pi} \int \frac{e^{ik\rho}}{\rho} \left(\frac{\partial U(\boldsymbol{r})}{\partial z} \right)_{z=0} \mathrm{d}x\,\mathrm{d}y + \frac{1}{4\pi} \int \left(\frac{\partial}{\partial z} \frac{e^{ik\rho}}{\rho} \right)_{z=0} U(\boldsymbol{r})\,\mathrm{d}x\,\mathrm{d}y. \quad (9.7\mathrm{b})$$

These equations may be added and subtracted to yield (again for $Z > 0$)

$$\text{Dirichlet:} \quad U(\boldsymbol{R}) = \frac{1}{2\pi} \int \left(\frac{\partial}{\partial z} \frac{e^{ik\rho}}{\rho} \right)_{z=0} U(x, y, 0)\,\mathrm{d}x\,\mathrm{d}y, \quad (9.8\mathrm{a})$$

$$\text{Neumann:} \quad U(\boldsymbol{R}) = \frac{-1}{2\pi} \int \frac{e^{ik\rho}}{\rho} \left(\frac{\partial U(x, y, z)}{\partial z} \right)_{z=0} \mathrm{d}x\,\mathrm{d}y. \quad (9.8\mathrm{b})$$

[16]Advanced books give a more elegant derivation of (9.8) by defining cleverer Green's functions (problem 9.4). Here we use no more sophistication than is necessary.

Thus we may evaluate the (downstream) diffracted field by integrating over the bounding surface *either* the amplitude U *or* its normal derivative $\partial U/\partial z$; it is not necessary to know both.[16]

Dirichlet–Neumann integrals are usually available only in the case where the surface being integrated over is plane (or otherwise extremely simple);[17] any more general surface requires that we return to (9.4) and use an integrand containing both U and $\nabla_r U$.

Comment: As with (9.4), eqns (9.8) are exact: there may be an obstacle present (or not); the quantities U, $\partial U/\partial z$ in the integrand represent the actual field, however it may be affected by the electromagnetic properties of any obstacle. The only restrictions on (9.8) are that the integration surface S_0 is the xy plane, and that nothing arrives from $z = +\infty$.

The Dirichlet integral (9.8a) may be tidied by evaluating $\partial G/\partial z$. Let the line from surface element dS_0 to field point R make an angle θ_{dif} with the forward normal to the surface at dS_0, as shown in Fig. 9.4. At the surface S_0, with dS_0 positive to the right,[18]

$$\nabla_r G \cdot dS_0 = \left\{ \frac{\partial}{\partial z} \left(\frac{e^{ik\rho}}{\rho} \right) \right\}_{z=0} dS_0 = \frac{d}{d\rho} \left(\frac{e^{ik\rho}}{\rho} \right) (- \cos \theta_{\text{dif}}) \, dS_0$$

$$= -ik \frac{e^{ik\rho}}{\rho} \left(1 - \frac{1}{ik\rho} \right) \cos \theta_{\text{dif}} \, dS_0. \qquad (9.9)$$

We now have from (9.8a) the exact diffraction integral (Dirichlet form)

$$U(R) = \frac{-ik}{2\pi} \int_{xy \text{ plane}} \frac{e^{ik\rho}}{\rho} \left(1 - \frac{1}{ik\rho} \right) \cos \theta_{\text{dif}} \, U(x, y, 0) \, dx \, dy, \qquad (9.10)$$

in which $\rho = |R - r|$ and $r = (x, y, 0)$.

9.2.4   Diffraction in the Fraunhofer limit

A particular case of (9.10) is that where the radiation is diffracted to a very large distance ρ. Then we have a Fraunhofer case and can write

$$\text{Fraunhofer}: \quad \rho = R - x \sin \theta - y \sin \phi. \qquad (9.11)$$

Here $\pi/2 - \theta$ and $\pi/2 - \phi$ are the angles made by the line to R with the x- and y-axes, while $R = |R|$. The bracket $(1 - 1/ik\rho)$ reduces to 1. The large distance $\rho \approx R$ in the denominator can be taken outside the integral, as can $\cos \theta_{\text{dif}}$. We obtain the Fourier transform[19]

$$\frac{U_{\text{far}}(R)}{\cos \theta_{\text{dif}}} = \frac{-ik}{2\pi} \frac{e^{ikR}}{R} \int_{xy \text{ plane}} U(x, y, 0) \, e^{-ikx \sin \theta - iky \sin \phi} \, dx \, dy. \qquad (9.12)$$

Similarly, the Neumann form (9.8b) yields another Fourier transform:

$$U_{\text{far}}(R) = \frac{-e^{ikR}}{2\pi R} \int_{xy \text{ plane}} \left(\frac{\partial U}{\partial z} \right)_{z=0} e^{-ikx \sin \theta - iky \sin \phi} \, dx \, dy. \qquad (9.13)$$

Comment: We draw particular attention to the fact that all equations given so far in this chapter are unapproximated. Even the Fourier-transform statements are exact[20]—albeit in the large-R limit.

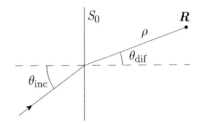

Fig. 9.4 Radiation is incident on surface S_0, arriving at angle θ_{inc} to that surface's normal. Diffracted radiation is collected at angle θ_{dif} to the normal.

[18] For this step the surface of integration need not be plane. We use a notation that would permit (9.9) to be used for simplifying (9.4).

[19] One condition is being assumed here. The expression for ρ in (9.11) relies on $|x|$ and $|y|$ remaining $\ll R$ for all places where the integrand is non-zero. (Indeed this condition was noted as necessary as early as sidenote 8.) The point is not trivial, since particular attention will be paid, in problem 9.6 and Chapter 10, to the case of a long slit, where (9.11) does not apply because ρ depends on y as y^2. There is yet another complication in the case of a long slit: for one polarization choice the fields spread widely in the direction *across* the slit, into the shadow of the slit jaws (§ 10.7).

[20] A Fourier-transform property is usually derived for students via the Kirchhoff approximation, so it is noteworthy that both relations (9.12) and (9.13) are in fact exact.

9.3   The Kirchhoff approximation

9.3.1   The Kirchhoff integral

For the first time, we now introduce simplifying assumptions and approximations—assumptions that may or may not be valid in real cases.

Let a wave be incident on a surface S_0, which will be our integration surface. This surface may be, but need not be, plane, so we are removing the restriction to a plane surface used in §§ 9.2.3 and 9.2.4. The incident wave arrives wholly from the left.

We introduce the "textbook" Kirchhoff approximation:

- the field amplitude $U(r)$ is zero at locations r on the downstream surface of any opaque obstacle
- in the aperture[21] (the open area), $U(r) = U_{inc}(r)$ the amplitude of the incident wave, and the same is true of $\nabla_r U$.

We further assume that the incoming wave has wavefronts that are not too strongly curved, so they are "locally plane". We have a simplification similar to that used in finding $\nabla_r G$: the gradient $\nabla_r U$ is close to $ik_{inc} U$ in magnitude and direction. Then, with θ_{inc} defined in Fig. 9.4, $\nabla_r U \cdot dS_0 \approx ikU \cos\theta_{inc}\, dS_0$. With these simplifications, eqns (9.4) and (9.6) become[22]

$$U_{\text{diffracted}}(R) \approx \frac{-ik}{2\pi} \int_{S_0,\, dS_0 \text{ +ve to right}} \frac{e^{ik\rho}}{\rho} \left(\frac{\cos\theta_{inc} + \cos\theta_{dif}}{2} \right) U_{inc}(r)\, dS_0, \qquad (9.14a)$$

$$0 = -U_{\text{back-diffracted}}(R_2) \approx \frac{-ik}{2\pi} \int_{S_0,\, dS_0 \text{ +ve to right}} \frac{e^{ik\rho_2}}{\rho_2} \left(\frac{\cos\theta_{inc} + \cos\theta_{2dif}}{2} \right) U_{inc}(r)\, dS_0. \qquad (9.14b)$$

In these equations, the angle between the diffracted ray and the forward normal is θ_{dif} for R and θ_{2dif} for R_2. The likelihood is that $\cos\theta_{2dif}$ is negative, though we can invent geometries where it is not always so (example in Fig. 9.3).

The integrals in (9.14) have the same form, so they may be combined into a single statement about the field $U(R)$ that holds whether R lies upstream or downstream from surface S_0:

$$U(R) \approx \frac{-ik}{2\pi} \int_{S_0,\, dS_0 \text{ +ve to right}} \frac{e^{ik\rho}}{\rho} \left(\frac{\cos\theta_{inc} + \cos\theta_{dif}}{2} \right) U_{inc}(r)\, dS_0, \qquad (9.15)$$

where $\rho = |R - r|$.

Equation (9.15) gives us (approximately) the field amplitude diffracted to location R. It is the canonical form of the Kirchhoff integral.[23]

9.3.2   The obliquity factor

Within (9.15) we see the

obliquity factor $\equiv \frac{1}{2}\left(\cos\theta_{inc} + \cos\theta_{dif}\right)$. $\qquad (9.16)$

This form for the obliquity factor[24] is needed when we use Kirchhoff approximations and the integration surface is not given to be plane.

[21] For simplicity, this condition is stated on the assumption that the open area is just that, open. It is easy to adapt the condition to cater for an aperture that is partly absorbing or phase-shifting, or for that matter is originating a reflected wave rather than transmitting.

[22] Given the crudity of the Kirchhoff approximation, there is no point in retaining the $-1/ik\rho$ in (9.9).

[23] Despite superficial appearance, (9.15) is not a Dirichlet integral: ∇U is incorporated, not eliminated. We are reminded that, for an integration surface that's not known to be plane—and that's the case here—Dirichlet and Neumann forms are hard to obtain because they rely on first finding an appropriate G. If (9.15) *were* a Dirichlet integral, it would be "too good".

[24] We shall reserve "obliquity factor" for a function of angles appearing in a Kirchhoff (i.e. approximated) integral. Functions of angle appearing in an exact diffraction integral will be called "angle factor" or similar.

For a plane integration surface, we may see the obliquity factor in (9.15) as a half-and-half mixture of the $\cos\theta_{\text{dif}}$ of (9.10) or (9.17a), and the $\cos\theta_{\text{inc}}$ of (9.17b). We are being told that no single obliquity factor accompanies the Kirchhoff approximation.[25]

9.3.3 Zero back-diffraction

In the exactly-backward direction where $\cos\theta_{\text{dif}} = -\cos\theta_{\text{inc}}$, the obliquity factor is zero. Textbooks often state that this zero value of the obliquity factor gives (or in history gave) respectability to Huygens secondary waves, by showing that there is no radiation into the backward direction. We take the view that the respectability is real, but this is the wrong place to look for it.[26]

Equation (9.6) has told us that back-diffraction is exactly zero. We need to know this before we can obtain the approximate (9.15). It is therefore silly to call upon the obliquity factor to dispose—approximately—of back-diffraction, already known to be absent exactly.

9.3.4 Kirchhoff integrals (plural)

We return to the special case of a plane integration surface. We can obtain Dirichlet and Neumann forms of the Kirchhoff integral by substituting approximated expressions for $\nabla_r G$ and $\nabla_r U$ into (9.8):

$$\text{Dirichlet:} \quad U(\boldsymbol{R}) \approx \frac{-\mathrm{i}k}{2\pi} \int U_{\text{inc}}(x,y,0) \frac{\mathrm{e}^{\mathrm{i}k\rho}}{\rho} \cos\theta_{\text{dif}}\, \mathrm{d}x\,\mathrm{d}y, \quad (9.17a)$$

$$\text{Neumann:} \quad U(\boldsymbol{R}) \approx \frac{-\mathrm{i}k}{2\pi} \int U_{\text{inc}}(x,y,0) \frac{\mathrm{e}^{\mathrm{i}k\rho}}{\rho} \cos\theta_{\text{inc}}\, \mathrm{d}x\,\mathrm{d}y. \quad (9.17b)$$

The first of these two equations looks like a rewrite of (9.10) (apart from omission of the small $-1/\mathrm{i}k\rho$), but now $U = U_{\text{inc}}$ is a piece cut from the incident wave, rather than the actual field as modified by the obstacle defining an aperture. The second equation makes a similar replacement, but in the gradient $\partial U/\partial z$. Both equations have been Kirchhoff-approximated.[27]

The equations that we started from, (9.8), are exact, so both must yield the correct diffracted field if supplied with a correct integrand. The same is not true of the "Kirchhoff" versions (9.17). We ask:

- Given the two rival approximate expressions (9.17)—containing different obliquity factors—is there any reason why one should be more reliable than the other?
- We may see the canonical Kirchhoff integral (9.15) as "splitting the difference" between (9.17a) and (9.17b); indeed a whole family of linear mixtures could be built similarly. Does any advantage accrue to any of these "intermediate" formulations?
- In particular, we might hope to justify the canonical Kirchhoff integral (9.15) if there were some reason to believe that (9.17a) and (9.17b) make errors in opposite directions?

The answer seems to be "no" to all of these.

[25] And Kirchhoff does not tell us which of the possibilities to use!

If we ask "which obliquity factor is right?" we miss the point that we are working within an approximation. The existence of rival obliquity factors is a signal: that they all lie within the accuracy of the Kirchhoff approximation(s), and they may perhaps help us to see what the limit on the accuracy is.

If we want better, we must return to the exact expressions obtained earlier in this chapter and somehow find U in the aperture—which may well mean facing up to doing a full solution of Maxwell's equations subject to boundary conditions imposed by the obstacle. Just such a full solution is discussed in Chapter 10.

If we are using exact expressions, then there is, of course, no uncertainty as to the angular function that should be used in the diffraction integral: $\cos\theta_{\text{dif}}$ in (9.10); 1 in (9.13).

[26] Section 9.2.2 has shown that no factor "local" to a surface element $\mathrm{d}\boldsymbol{S}$ can possibly account for the absence of back-diffraction. Therefore it is wrong in principle to look to the obliquity factor to do this for us.

[27] We say "Kirchhoff" to indicate that a field component, or its gradient, is replaced by "a piece cut from the incident wave", and not to indicate the specific form (9.15). We have already seen that the obliquity factor is subject to some choice—"choice" meaning ambiguity.

9.3.5 Contradictory assumptions

It is often pointed out that it is in principle wrong to specify both U and $\partial U/\partial z$ on the xy plane, since this over-constrains the wave field. If $U(x, y, z{=}0)$ is known, then (9.10) gives $U(\mathbf{R})$—exactly—for all downstream points $\mathbf{R}$, even those very close to the xy plane. Thus $(\partial U/\partial z)_{z=0}$ is determined by knowledge of $U_{z=0}$ itself. If we fix both U and $\partial U/\partial z$ at $z = 0$ we necessarily input contradictory information.[28]

Putting it starkly: If Kirchhoff's $U(x, y, 0)$ was correct, then his $\partial U/\partial z$ would necessarily be wrong, and vice versa.

It is my view that, while this is true, not too much should be read into it. In (9.17a), we should not think that $U(x, y, z{=}0) = U_{\text{inc}}(x, y, z{=}0)$ is a truth-statement about what the field is doing; rather, we are inputting an approximation that we hope may be helpful. It is entirely legitimate to try a different intended-to-be-helpful approximation when evaluating (9.17b). More important than the "internal" contradiction is the fact that the two Kirchhoff inputs both conflict with Maxwell's equations.

9.4 Diffraction integrals and electromagnetism

9.4.1 Scalar and vector amplitudes

Each Cartesian component of $\mathbf{E}$ or $\mathbf{B}$ in the diffracted field is a $U(\mathbf{R})$ "sourced" by the corresponding field component $U(\mathbf{r})$ at the aperture, the two related by a diffraction integral. Diffraction integrals may validly be applied to more than one Cartesian component, even to all six taken in turn—if we know the appropriate $U(x, y, 0)$ or $(\partial U/\partial z)_{z=0}$ for each. But we don't know them. We have to approximate (or guess); and this is where we need Kirchhoff—and where inconsistencies lurk.

Just as U and $\partial U/\partial z$ should not both be assigned chosen values, so there are dangers in fixing an inappropriate Cartesian field component in the Kirchhoff manner. Figure 9.5 shows one such risk. Let radiation arrive in the z-direction: $\theta_{\text{inc}} = 0$. Then $E_z^{\text{inc}} = 0$ in the incident wave,[29] and Kirchhoff (though not Maxwell) says that $E_z(x, y, 0)$, being cut from the incident wave, is zero over the entire xy plane. A diffraction integral then says that E_z^{dif} must be zero everywhere in the diffracted field. This is clearly false, since radiation diffracted into the direction θ_{dif} (Fig. 9.5) necessarily has a non-zero E_z^{dif}. Working back, there must be a non-zero $E_z(x, y, 0)$ in the integration plane, even though there is none in the far-upstream incident wave.[30]

To handle diffraction within a Kirchhoff approximation, there is usually a sensible way to make progress. We choose a field component that is believed to dominate the problem in hand, perhaps E_x or E_y, and we use this as $U(x, y, 0)$ in a Kirchhoff integral. Other field components, such as E_z, may perhaps be dismissed as "not very interesting", but we do not go further and assume them to be zero: we hope that they will somehow look after themselves—an idea elaborated below.

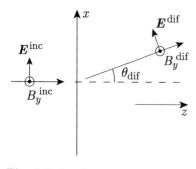

Fig. 9.5 An electromagnetic wave is incident normally onto the xy plane. Field component E_z^{inc} is zero. On Kirchhoff assumptions, $E_z(x, y, 0) = 0$ also. The implications for $\mathbf{E}^{\text{dif}}$ are discussed in the text.

[28] Equations (10.1a) show, in a special case, that $\partial U/\partial z$ is related to B_x if $U = E_y$, and is related to E_x if $U = B_y$. Thus the conclusions of this subsection may be stated equivalently by saying that we may not specify both $\mathbf{E}$ and $\mathbf{B}$ simultaneously in the plane of the aperture.

[29] Henceforth, in this chapter and in Chapter 10, we write adjectives "incident", "diffracted" etc. as superscripts to $\mathbf{E}$ and $\mathbf{B}$ so that there is room to indicate a Cartesian component by a subscript.

[30] It is really not surprising that the electromagnetic disturbance set up by the obstacle should deform the field into having $E_z \neq 0$, even when the incident wave arrives normally. In the case of diffraction from a conducting slit, a non-zero E_z or B_z is generated by currents and charges induced in the slit jaws. Field B_z is shown in Fig. 10.3.

In the procedure promulgated here, we may be said to use a **scalar-wave approximation**. In its most naïve and unsatisfactory form, a scalar-wave approximation can mean that we work out what a truly scalar wave would do, and then hope that an electromagnetic wave will be not much different. But there are more respectable possibilities. Here we take "scalar-wave approximation" to mean the use of U to represent a single dominant (Cartesian) field component. And we intend to "think Maxwell" in anchoring U firmly to that field component.

Suppose we know $E_x(x, y, 0)$ and $E_y(x, y, 0)$ in the xy plane. This suffices to give knowledge also of $\partial E_z/\partial z$, $\partial B_x/\partial z$, $\partial B_y/\partial z$ and B_z in that plane (problem 9.5). From here, all six field components of the diffracted field can be found using either Dirichlet or Neumann.[31] Suppose we find E_x^{dif} and E_y^{dif} via Dirichlet integrals (9.10). The other four field components may notionally be evaluated via further diffraction integrals, but a better idea is to think of them as found by applying Maxwell's equations to E_x^{dif} and E_y^{dif} (a procedure that guarantees Maxwell respectability even if E_x or E_y has been approximated). In any event, four of the six field components can indeed "look after themselves".

Jackson (1999), his eqn (10.101), gives a neat version of the above:[32]

$$E(R) = \frac{-1}{2\pi} \text{curl}_R \int \frac{e^{ik\rho}}{\rho} E(r) \times dS. \qquad (9.18)$$

Here dS is a surface element at location $r = (x, y, z)$, $E(r)$ is the field at r, and $\rho = |R - r|$. In the integrand, $E(r) \times dS$ projects $E(r)$ onto the integration surface, thereby removing the normal component (E_z if the surface is the xy plane)—consistent with findings in problem 9.5. Once $E(R)$ is known, $B(R)$ may be found from it using Maxwell's equations, and the whole diffracted field is guaranteed to be Maxwell-respectable. Of course, evaluation of the integral (9.18) requires that we have an $E(r)$ to put into the integrand, and for that we shall probably need to make a Kirchhoff-type approximation. If we do this, then we are using the Smythe–Kirchhoff formulation of diffraction theory.[33]

If in (9.18) we replace the actual E_x and E_y at the aperture by Kirchhoff approximations to them, we are likely to end up with "the right solution to the wrong problem", but at least it will be a Maxwell-respectable solution.

[31] Similarly, knowledge of B_x and B_y suffices to determine all six field components.

[32] Here curl_R differentiates with respect to the coordinates of the field-point location R.

In eqn (9.18) the surface over which the field is integrated does not have to be plane, which is why dS has not been replaced by $dx\,dy$.

[33] That is, the statement (9.18) is due to Smythe, and we make a Kirchhoff replacement for the field $E(r)$ in the integrand.

9.4.2   The special case of a slit

In Chapter 10 we give detailed attention to one special case of diffraction: that where a wave passes through a long uniform slit. We set up for that discussion here by obtaining some statements about diffraction at such a slit. The reasoning is presented, in problem form, in problem 9.6.

An electromagnetic plane wave falls normally on the slit shown in Fig. 9.6. Everything is independent of the y-coordinate (along the slit), but otherwise the field $E(x)$ or $B(x)$ at the slit may have any reasonable shape (modified from E^{inc}, B^{inc} by currents and charges induced in the slit jaws). In problem 9.6 (8) and (9), it is shown that, in the far field,

$$U_{\text{along}}(\boldsymbol{R}) = \cos\theta \times \frac{\mathrm{e}^{\mathrm{i}(kR-\pi/4)}}{\sqrt{\lambda R}} \int U_{\text{along}}(x)\exp(-\mathrm{i}kx\sin\theta)\,\mathrm{d}x, \quad (9.19\mathrm{a})$$

$$U_{\text{trans}}(\boldsymbol{R}) = 1 \times \frac{\mathrm{e}^{\mathrm{i}(kR-\pi/4)}}{\sqrt{\lambda R}} \int U_{\text{across}}(x)\exp(-\mathrm{i}kx\sin\theta)\,\mathrm{d}x. \quad (9.19\mathrm{b})$$

Here U_{along} is a field component, E_y or B_y, that lies parallel to the length of the slit. Field $U_{\text{across}}(x)$ is a component, $E_x(x)$ or $B_x(x)$, that faces across the slit. And $U_{\text{trans}}(\boldsymbol{R})$ is the magnitude of a diffracted field lying in the xz plane which, for reasons in Fig. 9.5, is perpendicular to the outgoing direction of travel and therefore does not face exactly in the x-direction. The angle θ of deviation is defined in Fig. 9.6.

Equations (9.19) have not been approximated, except by the taking of a far-field limit.

9.4.3 Contradictory assumptions, again

It is, of course, possible to introduce a Kirchhoff approximation into one or both of eqns (9.19). If both, the outcomes are (9.27) and (9.28). One of these contains obliquity factor $\cos\theta$, while the other contains obliquity factor 1. Taken together, (9.27) and (9.28) yield $cB_y^{\text{dif}} = E^{\text{dif}} \times \cos\theta$, while Maxwell insists that $cB_y^{\text{dif}} = E^{\text{dif}} \times 1$.

It follows that Kirchhoff replacements must not be made for $B_y(x,y,0)$ and $E_x(x,y,0)$ simultaneously.[34] This prohibition is related to the finding in § 9.3.5 governing U and $\partial U/\partial z$.

Notice that the assumption leading to this contradiction looked harmless enough: $E_x(x) = cB_y(x)$ in the plane of the slit, which is equivalent to saying that the incident wave arrives normally and is "locally plane". Yet even this apparently weak assumption is shown to be inconsistent.

We may ask: Given that only one of $\boldsymbol{E}$ or $\boldsymbol{B}$ can consistently be given a Kirchhoff replacement in the plane of an aperture, does the choice matter, and which is better? Some answers are given, for normal incidence on a slit, in Chapter 10. Kirchhoff agrees best with exact calculation if the obliquity factor is set equal to 1. Consistently, it is the field component facing across the slit from one jaw to the other, whether $\boldsymbol{E}^{\text{inc}}$ or $\boldsymbol{B}^{\text{inc}}$, that is most closely similar to a top-hat function cut from the incident wave.[35]

9.5 The limits on Kirchhoff

The Kirchhoff approximation is intended to give us a good idea of how radiation is diffracted by apertures that are fairly large compared with a wavelength. It does this with impressive success. A Fourier-optics reason why we can get away with Kirchhoff is given in § 10.8.

If we are to conduct a stringent test, we must look at apertures that are not large compared with a wavelength.[36]

When an aperture is small compared with the wavelength, the usual "interference" effects in the angular distribution have dropped out. As

[34] This is part of the reason why we stated in § 9.3.5 that the Kirchhoff replacements for U and its derivative conflict with Maxwell's equations.

[35] As judged by an "eyeball" inspection of Figs. 10.3 and 10.6. This finding holds for a single slit with normal incidence. It is not known whether it continues to hold for more complicated apertures or for non-normal incidence. It would therefore be dangerous to generalize.

[36] Or we look at large angles to the forward direction. This accounts for our paying what may seem an inordinate attention to obliquity/angle factors. In most experimental configurations, diffraction is observed so close to forwards that any obliquity factor is 1, and asking for more precision is pointless.

example, the Kirchhoff–Fraunhofer expression $\sin(\frac{1}{2}ka\sin\theta)/(\frac{1}{2}ka\sin\theta)$ for a slit of width a becomes 1 when $(ka)\ll 1$. What remains is the angle factor: maybe the $\cos\theta_{\mathrm{dif}}$ of (9.12) or maybe the 1 of (9.13).

It makes sense then to concentrate attention on the total energy transmitted.[37] We may quantify this by giving a cross section σ or a transmission coefficient T.

In the case of a slit, we need the cross section[38] per unit length of slit, which means that the cross section is an effective width, rather than an effective area. A slit of width a having $\boldsymbol{E}^{\mathrm{inc}}$ along it has cross section $\sigma_{\boldsymbol{E}\text{ along slit}}$. A similar slit having $\boldsymbol{B}^{\mathrm{inc}}$ along it has cross section $\sigma_{\boldsymbol{B}\text{ along slit}}$. It is found (Bouwkamp 1954) that for normal incidence and $ka\ll 1$,

$$T_{\boldsymbol{E}\text{ along slit}}=\frac{\sigma_{\boldsymbol{E}\text{ along slit}}}{a}=\frac{\pi^2}{256}(ka)^3;\tag{9.20a}$$

$$T_{\boldsymbol{B}\text{ along slit}}=\frac{\sigma_{\boldsymbol{B}\text{ along slit}}}{a}=\frac{2}{ka}\frac{1}{1+(4/\pi^2)\{\ln(ka/8)+\gamma\}^2}.\tag{9.20b}$$

Here γ is the Euler constant ≈ 0.5772. These expressions are very different, so the diffraction is strongly polarization dependent for small ka. We cannot expect Kirchhoff to deal with polarization dependence, so Kirchhoff approximations are completely out of their validity range when $ka\ll 1$.

For a slit at normal incidence, Kirchhoff gives a cross section[39] of $\sigma_{\mathrm{Kirchhoff}}=a(ka)/2$, independent of polarization. This gives the correct energy transmission, within about 10%, for $a\gtrsim 0.4\lambda$ and for both polarization possibilities, so a slit can be surprisingly small before this aspect of Kirchhoff breaks down.

As with a narrow slit, there is a failure of Kirchhoff when light is diffracted by a small circular aperture (or scattered from a small disc). For normal incidence,[40]

$$\frac{\sigma}{\pi a^2}=\frac{64}{27\pi^2}(ka)^4;\qquad \frac{\sigma_{\mathrm{Smythe-Kirchhoff}}}{\pi a^2}=\frac{(ka)^2}{3}.\tag{9.21}$$

Here a is the radius of the aperture and the limit $ka\ll 1$ is taken. The very different dependences on (ka) tell their own story. Smythe–Kirchhoff makes a very bad guess at the field in the aperture,[41] even though it creates a Maxwell-respectable diffracted field from that: a clear case of "the right solution to the wrong problem".

9.6 Final comments

In §9.2, we have obtained a number of exact mathematical statements of the Huygens principle. These statements apply to each Cartesian component of $\boldsymbol{E}$ and of $\boldsymbol{B}$, so they apply within a fully vector understanding of the electromagnetic field. And they relate the diffracted field to the field at an aperture, however much that field may be perturbed by the proximity of the obstacle delimiting that aperture.

[37] The transmission T is the power transmitted, divided by the power incident on the geometrical area of the aperture.

[38] In the case of obstacles that are plane, perfectly conducting and of negligible thickness, there is an electromagnetic **Babinet principle** that applies to **complementary apertures**: problem 10.3(7); also Born & Wolf (1999) §§11.2–11.3. A particular application of this concept relates the fields diffracted by a slit to those scattered by a complementary strip: two for the price of one.

Equation (9.20a) gives the forward scattering for a strip with $\boldsymbol{B}^{\mathrm{inc}}$ along the strip, and (9.20b) gives the forward scattering for a strip with $\boldsymbol{E}^{\mathrm{inc}}$ along it.

[39] This expression for $\sigma_{\mathrm{Kirchhoff}}$ is obtained by patching in an obliquity factor of 1. This is expected if it is known, or believed, that it is the "across" fields at the slit that are best mimicked by a Kirchhoff top-hat (eqns 9.19). That is, in fact, the case. However, if we are needing to appeal to Kirchhoff it will be because we do not have that knowledge.

[40] A derivation of the first of these formulae may be found in Landau, Lifshitz and Pitaevskiĭ (1993), §95 problem 2. The Smythe–Kirchhoff expression is derived in Jackson (1999) §10.9.

The Smythe–Kirchhoff calculation evaluates expression (9.18), and gives workable algebra for any size of (ka). There is no explicit use of any obliquity factor, because all dependence on angles is taken care of in the calculation. However in (9.21) we quote only the result for normal incidence and $ka\ll 1$.

[41] It should not be a surprise that Smythe–Kirchhoff is so wrong in this case. The $\boldsymbol{E}$-field in the open area of a small hole in a conductor must surely be reduced (as compared with the field in the incident wave) by the proximity of the conductor.

Within these exact statements, we have a proof that secondary sources do not radiate "backwards", a proof that is similarly robust whatever influence the obstacle may have on the field at the aperture. (An obliquity factor takes no part in this reasoning.)

We even have Fourier-transform relationships (9.12) or (9.13) relating the far field (Fraunhofer regime) to the field at the aperture, again obtained without approximation.

Even exact statements are of little practical use if they require us to integrate an amplitude $U(x, y, 0)$ at an aperture when that amplitude is unknown. (Except for a very few cases including a single slit) we need the Kirchhoff approximation, or something very like it, to work out anything concrete. There is no doubt that the Kirchhoff approximation is extremely reliable: experiment says so.[42] Less confidence is justified if we push things too far and ask for high quantitative accuracy. In particular, the conventional obliquity factor $\frac{1}{2}\left(\cos\theta_{\text{inc}} + \cos\theta_{\text{dif}}\right)$ is unsatisfactory on every occasion where a critical test can be made, and seems to have nothing to recommend it. Sometimes there is rivalry between different possible obliquity factors, and it seems best to take this as an indication of where accurate predictions cease.

[42] Provided that the aperture is not too small on the scale of a wavelength.

Problems

Problem 9.1 Derivation of eqn (9.3)
Supply the mathematics that obtains (9.3) from (9.1) and (9.2).

Problem 9.2 Zero contribution from the hemisphere of Fig. 9.2
In §9.2, it is stated that the contribution to $(G\,\nabla_r U - U\,\nabla_r G)\cdot d\mathbf{S}$ from the large hemisphere S_∞ tends to zero as the radius of the hemisphere tends to infinity. This comes about because U, G, $\nabla_r U$ and $\nabla_r G$ behave like spherical waves radiating from almost the same centre. It is claimed that this puts at least one factor r^{-1} into the integrand, and thereby makes the integral tend to zero.

Take U as a spherical wave centred on the origin and G as a spherical wave centred on a point at $\mathbf{a}$. Show that $(G\,\nabla_r U - U\,\nabla_r G)\cdot d\mathbf{S}$ is of order a^2/r^2 multiplied by a function of angles. The integral therefore behaves like r^{-2} as the radius r of the hemisphere tends to infinity.

Problem 9.3 Outgoing waves do not contribute to the surface integral in (9.3)
Look at the surface integral in (9.3) in another way, using as model a one-dimensional case.

Let $U = f(x)\,e^{-i\omega t}$ be a solution of the scalar wave equation, in one dimension, with wave speed c. Then
$$f(x) = A\,e^{ikx} + B\,e^{-ikx} \qquad \text{with} \quad k = \omega/c.$$
Consider
$$\left(\frac{\partial}{\partial x} + ik\right)U = \left(2ik\,A\,e^{ikx} + 0\times e^{-ikx}\right)e^{-i\omega t}.$$

Thus the linear operator $(\partial/\partial x + \mathrm{i}k)$, acting on U, picks out the portion $\mathrm{e}^{\mathrm{i}(kx - \omega t)}$ travelling towards $+x$. And similarly, $(\partial/\partial x - \mathrm{i}k)$ picks out the piece travelling towards $-x$.

Guided by this model, show that $(G \nabla_r U - U \nabla_r G)$ in (9.3) has the effect of picking out of U that part that is travelling inwards through the surface S_∞, while giving zero coefficient to that part which is travelling outwards.[43]

[43] There is a similarity here to an operation encountered in the quantization of waves. See §31.4 and problem 31.3, where a similar idea is put to use, albeit with t and x interchanged.

Problem 9.4 Alternative Green's functions
There is a more elegant way of handling the reasoning between eqns (9.4) and (9.8).[44] Define new Green's functions

$$G_\mathrm{D} = \frac{1}{2} \left(\frac{\mathrm{e}^{\mathrm{i}k\rho}}{\rho} - \frac{\mathrm{e}^{\mathrm{i}k\rho'}}{\rho'} \right), \qquad G_\mathrm{N} = \frac{1}{2} \left(\frac{\mathrm{e}^{\mathrm{i}k\rho}}{\rho} + \frac{\mathrm{e}^{\mathrm{i}k\rho'}}{\rho'} \right),$$

[44] Jackson (1999) §10.5 presents very similar reasoning, and indeed is well worth looking at alongside much of the present chapter.

with ρ and ρ' defined in Fig. 9.2. Show that these functions are as acceptable as the original G, in that they obey the differential equation (9.1) and they tend to zero correctly on the large hemisphere S_∞. Therefore either may be used[45] (with some consequential changes in (9.2)–(9.4)) in place of the original G. Show that on the xy plane $G_\mathrm{D} = 0$ and $\partial G_\mathrm{N}/\partial z = 0$, and that consequently

[45] The original G remains helpful for the reasoning that disposes of the "backward" wave.

$$U(\boldsymbol{R}) = \frac{1}{2\pi} \int \frac{\partial G_\mathrm{D}}{\partial z} U \, \mathrm{d}x \, \mathrm{d}y, \qquad U(\boldsymbol{R}) = \frac{-1}{2\pi} \int G_\mathrm{N} \frac{\partial U}{\partial z} \, \mathrm{d}x \, \mathrm{d}y.$$

By starting with G_D and G_N, we avoid the need to add and subtract eqns (9.7a) and (9.7b), but instead we get to the Dirichlet and Neumann forms (9.8) "in one go".

Notice that the construction of G_D and G_N has a loose similarity to the method of images that we sometimes find helpful for solving potential problems in electrostatics. We added a "charge" at $\boldsymbol{R}'$, partnering the "charge" at $\boldsymbol{R}$, so as to give a zero "potential" on the xy plane (Dirichlet) or to give a zero "potential gradient" on the xy plane (Neumann). This suggests that further developments may be possible. Show that the field $U(\boldsymbol{R})$ within the "quarter-space" bounded by the xy and xz planes can be obtained as a Dirichlet integral of U over the two plane boundaries, once we have invented a suitable G for the purpose. Find that G. Find a G that similarly gives a Neumann integral over the same two planes.[46]

[46] If you're really ingenious, you may be able to apply a Dirichlet condition over one surface and a Neumann condition over the other. Experience with "image" problems is definitely a help here.

Problem 9.5 Obtaining all six field components from just two
Suppose that $E_x(x, y, 0)$ and $E_y(x, y, 0)$ are known in the xy plane. Show from Maxwell's equations that we have enough information to determine the values in the xy plane of $\partial E_z/\partial z$, $\partial B_x/\partial z$ and $\partial B_y/\partial z$ and B_z. Show that these may be put into Dirichlet or Neumann integrals (as appropriate) to determine all six field components in the diffracted field.

Problem 9.6 Fraunhofer diffraction at a long slit
We build on exact equation (9.10). A Kirchhoff approximation will be used eventually, but there are advantages in delaying that step for as long as possible.

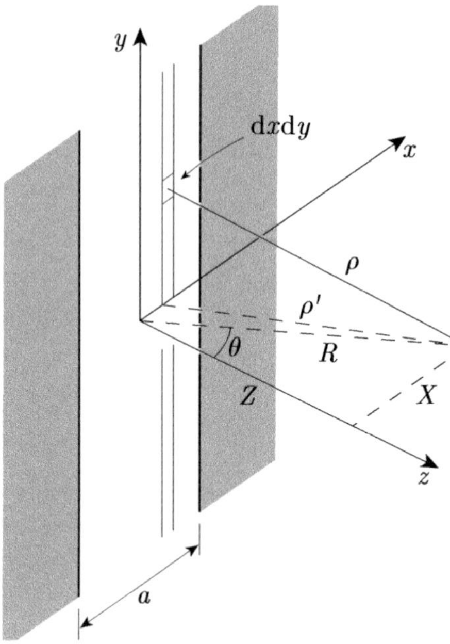

Fig. 9.6 A slit having thin opaque jaws is illuminated by a wave arriving from negative z. Incidence does not need to be normal, but the incident wave's $\mathbf{k}$-vector(s) must have no y-component so that everything is independent of y. We are to evaluate the field that is diffracted downstream. Point $\mathbf{R}$ is a representative "field point".

For diffraction from the surface element $\mathrm{d}x\,\mathrm{d}y$, the angle θ_{dif} of diffraction is the angle (not marked to avoid clutter) between the line marked ρ and the z-axis. Integration over y is dealt with in problem 9.6(3), where it is shown that for all interesting y angle θ_{dif} is negligibly different from the angle between ρ' and the z-axis. Problem 9.6(7) deals with integration over x, and it is argued that for all interesting x this angle is itself negligibly different from the deviation angle θ.

[47] Because of the way that Fresnel zones are defined, each zone makes a contribution to the diffraction integral that is π out of phase with those of its neighbours. We may get an informal idea of what is going on in the integration by thinking of it as the addition of the effects of the zones. The zones make diminishing contributions as y is increased, because the zone widths decrease, and this is why the integration over y converges. The formal mathematics is expressed in terms of things called Fresnel integrals.

To find that 20 zones make the integration over y "mostly complete", I required the integral over y of part (4) to be within 10% of its final value.

Figure 9.6 shows a long uniform slit in the xy plane with its length parallel to the y-axis. A wave arrives from $-z$ and undergoes diffraction after passing through the slit. The incident-wave fields are uniform in the y-direction; then all resulting fields are independent of y (so $\partial/\partial y = 0$). The fields at the slit may have any reasonable dependence on x.

(1) Show that Maxwell's equations couple together field components E_y with B_x and B_z, and separately B_y with E_x and E_z, cf. (10.1). Thus there are two polarization cases, one where $\mathbf{E}$ lies parallel to the length of the slit, and one where $\mathbf{B}$ is parallel to the slit length.

(2) We shall use eqn (9.10) to find the field diffracted to a representative field point $\mathbf{R} = (X, 0, Z)$ from area elements $\mathrm{d}x\,\mathrm{d}y$. The distance ρ from $\mathrm{d}x\,\mathrm{d}y$ to $\mathbf{R}$ is given by

$$\rho^2 = (X - x)^2 + y^2 + Z^2 = \rho'^2 + y^2.$$

Because we are going to proceed to a large-R limit, we can approximate $(1 - 1/\mathrm{i}k\rho)$ to 1 in (9.10).

(3) We integrate first over y along the strip drawn of width $\mathrm{d}x$; for this step X, Z and x are held constant. To understand what is going on we can borrow ideas from textbook discussions of Fresnel diffraction: we divide the strip of width $\mathrm{d}x$ into Fresnel half-period zones, each zone boundary defined by incrementing ρ by $\lambda/2$. Thus the first zone's limits are defined by $\rho - \rho' = \lambda/2$, giving $y^2 = \lambda\rho' + \lambda^2/4 \approx \lambda\rho'$. The angle subtended by each half of this first zone at $\mathbf{R}$ is $|y|/\rho' = \sqrt{\lambda/\rho'}$, which becomes small as ρ' becomes large. The 20th zone (at which point the integration over zones is mostly complete)[47] has outer extremities at $y = \pm\sqrt{20\lambda\rho'}$, at angles $\pm\sqrt{20\lambda/\rho'}$; this shows a similar limiting

behaviour with only the coefficient changed. Therefore the entire family of active Fresnel zones occupies a small (vertical) angle range when seen "looking backwards" from $\boldsymbol{R}$. The angle of diffraction θ_{dif} is approximately independent of y and can be taken outside the integration over y. It has become the angle (not marked in Fig. 9.6 to avoid clutter) between the line marked ρ' and the z-direction.

Given that the interesting values of y are much less than ρ', show that we may make the expansion $\rho \approx \rho' + y^2/2\rho'$.

(4) Now build these ideas into (9.10).[48]

$$U(\boldsymbol{R}) = \frac{-\mathrm{i}k}{2\pi} \int \frac{\exp(\mathrm{i}k\rho)}{\rho} \cos\theta_{\text{dif}} \, U(x) \, \mathrm{d}x \, \mathrm{d}y$$

$$\approx \frac{-\mathrm{i}k}{2\pi} \int \mathrm{d}x \, U(x) \cos\theta_{\text{dif}} \frac{\exp(\mathrm{i}k\rho')}{\rho'} \int_{-\infty}^{\infty} \mathrm{d}y \, \exp(\mathrm{i}ky^2/2\rho').$$

In this, we know that $U(x, y, 0)$ is independent of y, so it has been written as a function of x only; we leave its dependence on x unspecified, so no Kirchhoff or Fraunhofer approximation is yet being made.

(5) We have said that the integration over y is mostly done when 20 zones have been summed over. Show that then $ky^2/2\rho' = 20\pi$. In part (3) we expanded ρ in powers of y, stopping at y^2. Show that the next term is $-y^4/8\rho'^3$, contributing a factor $\exp\left(-\mathrm{i}ky^4/8\rho'^3\right)$ which has reached $\exp\left(-\mathrm{i}100\pi\lambda/\rho'\right)$. Argue that this is negligibly different from 1 for sufficiently large ρ', so we were right to stop the expansion at y^2.

(6) The real and imaginary parts of $\int_{-\infty}^{\infty} \exp\left(\mathrm{i}\pi t^2/2\right)\mathrm{d}t$ are both 1. You may take this as a given property of Fresnel integrals, but with a little ingenuity you may be able to adapt the evaluation of a Gaussian integral. Either way, show that

$$U(\boldsymbol{R}) = \mathrm{e}^{-\mathrm{i}\pi/4} \int \frac{U(x)}{\sqrt{\lambda\rho'}} \cos\theta_{\text{dif}} \exp(\mathrm{i}k\rho')\mathrm{d}x. \qquad (9.22)$$

(7) So far, we have integrated along the length of the strip drawn in Fig. 9.6, a strip of width $\mathrm{d}x$. Integration over y has been possible because all fields are independent of y. The only approximations required have been those that put $\boldsymbol{R}$ at a large distance downstream from the slit.

Next integrate over x. For the first time, take it that the strip's width $a \ll R$ so that

$$\rho' \approx R - x\sin\theta. \qquad (9.23)$$

This gives an integral over x with an exponent that is linear in x; for the purpose of integrating over x we have Fraunhofer mathematics,[49,50] even though Fresnel mathematics applied to the integration over y. Argue that also $\cos\theta_{\text{dif}} \approx \cos\theta$ (Fig. 9.6) which can be taken outside the integral. Show that

$$U(\boldsymbol{R}) = \frac{1}{\sqrt{\lambda R}} \, \mathrm{e}^{\mathrm{i}(kR - \pi/4)} \cos\theta \int U(x) \exp(-\mathrm{i}kx\sin\theta) \, \mathrm{d}x. \qquad (9.24)$$

(8) The notation U, for wave amplitude, used so far may suggest a naïve scalar-wave approximation. The situation is better than that. We may use U to represent E_x or E_y or B_x or B_y, and for each component the scalar additions leading to (9.24) remain valid.[51]

[48] The Fresnel half-period zones have served their purpose now, and we revert to an algebraic evaluation.

[49] This step is permitted because the range of integration over x is much less than the observation distance R. However, for one polarization, the fields extend into the shadow of the slit, covering an x-range somewhat larger than the slit width (Fig. 10.5). This point therefore needs an eye kept on it.

[50] Remember that we are not yet using a Kirchhoff assumption as to the behaviour of $U(x)$, though we have assumed that everything is independent of y. The actual $U^{\text{inc}}(x)$ could have a phase dependent upon x, depending on the geometry of the incident wave. For example, if what is incident is a plane wave arriving at an angle, the phase of $U^{\text{inc}}(x)$ is a linear function of x. A cylindrical incident wave would have $U^{\text{inc}}(x)$ possessing a quadratic dependence of its phase upon x. Within optics, a plane incident wave (combined with observation at a large distance) would be said to give Fraunhofer diffraction, while a cylindrical incident wave would give Fresnel diffraction.

We shall depart from the usual nomenclature here. As we prefer to use words, eqn (9.24) exhibits a Fraunhofer case—and a Fourier transform as in §9.2.4—because the exponent on display is linear in x, regardless of phase factors that may be hiding within $U(x)$.

[51] Application of (9.24) to E_z or B_z is not ruled out in principle, but we have seen that these field components are best left to "look after themselves" after other components have been dealt with.

Suppose that the incident wave has $\boldsymbol{B}^{\text{inc}}$ parallel to the y-axis, as suggested in Fig. 9.5 (though incidence need not now be normal). From part (1) above, this means that the diffracted $B_y(\boldsymbol{R})$ relates to $B_y(x)$; this is without making any Kirchhoff approximation. Argue that

$$B_y^{\text{dif}}(\boldsymbol{R}) = \cos\theta \times \frac{\mathrm{e}^{\mathrm{i}(kR - \pi/4)}}{\sqrt{\lambda R}} \int B_y(x)\exp(-\mathrm{i}kx\sin\theta)\,\mathrm{d}x \qquad (9.25)$$

in the far field, since the θ_{dif} of Fig. 9.5 can be replaced by θ.

(9) Refer again to Fig. 9.5, and apply (9.24) to E_x. Show that

$$E_x^{\text{dif}}(\boldsymbol{R}) = \cos\theta \times \frac{\mathrm{e}^{\mathrm{i}(kR - \pi/4)}}{\sqrt{\lambda R}} \int E_x(x)\exp(-\mathrm{i}kx\sin\theta)\,\mathrm{d}x.$$

Now, in the far field, the diffracted wave is nearly plane, so $\boldsymbol{E}^{\text{dif}}(\boldsymbol{R})$ makes an angle θ to the x-axis. Argue that $E_x^{\text{dif}}(\boldsymbol{R})$ is a fraction $\cos\theta$ of the whole $\boldsymbol{E}^{\text{dif}}(\boldsymbol{R})$ (Fig. 9.5). Thus $\cos\theta$ cancels, and we have

$$E^{\text{dif}}(\boldsymbol{R}) = 1 \times \frac{\mathrm{e}^{\mathrm{i}(kR - \pi/4)}}{\sqrt{\lambda R}} \int E_x(x)\exp(-\mathrm{i}kx\sin\theta)\,\mathrm{d}x. \qquad (9.26)$$

(10) Construct a similar analysis for the case where $\boldsymbol{E}^{\text{inc}}$ faces in the y-direction. Check with the summary given in eqns (9.19).[52]

(11) Obtain (9.26) by another route. Use (10.1a) and (9.8b) to show

$$E^{\text{dif}}(\boldsymbol{R}) = cB_y(\boldsymbol{R}) = \frac{-1}{2\pi}\int \frac{\mathrm{e}^{\mathrm{i}k\rho}}{\rho}\left(\frac{\partial cB_y}{\partial z}\right)_{z=0}\,\mathrm{d}x\,\mathrm{d}y$$

$$= \frac{-\mathrm{i}k}{2\pi}\int \frac{\mathrm{e}^{\mathrm{i}k\rho}}{\rho}E_x(x)\,\mathrm{d}x\,\mathrm{d}y,$$

which again shows an angle factor of 1.

(12) For the first time, introduce a Kirchhoff assumption about the value of $U(x)$ in the plane of the slit. Take the special case of a plane incident wave arriving normally. Show that, if $U(x) = B_y(x) = B_y^{\text{inc}}$ within the open area $|x| < a/2$, and zero on the opaque area, then

$$\frac{B_y^{\text{dif}}}{B_y^{\text{inc}}} = a\,\frac{\mathrm{e}^{\mathrm{i}(kR - \pi/4)}}{\sqrt{\lambda R}}\,\frac{\sin\left(\frac{1}{2}ka\sin\theta\right)}{\frac{1}{2}ka\sin\theta} \times (\text{obliquity factor } \cos\theta). \qquad (9.27)$$

(13) In a similar way, assume instead that $U(x) = E_x(x) = E_x^{\text{inc}}$ within the open area and is zero on the opaque area, and show that

$$\frac{E^{\text{dif}}}{E_x^{\text{inc}}} = \frac{cB_y^{\text{dif}}}{E_x^{\text{inc}}} = a\,\frac{\mathrm{e}^{\mathrm{i}(kR - \pi/4)}}{\sqrt{\lambda R}}\,\frac{\sin\left(\frac{1}{2}ka\sin\theta\right)}{\frac{1}{2}ka\sin\theta} \times (\text{obliquity factor } 1).$$

$$\qquad (9.28)$$

Comment: Equations (9.27) and (9.28) demonstrate once again that $B_y(x)$ and $E_x(x)$ cannot be simultaneously be taken to be "pieces chopped out of the incident wave", because the result is two contradictory values for B_y^{dif}.

(14) Investigate a case of non-normal incidence where the incident wave has $\boldsymbol{k}_{\text{inc}} = (k\sin\theta_{\text{inc}}, 0, k\cos\theta_{\text{inc}})$ and has its $\boldsymbol{B}$ parallel to the length of the slit. Show that, within a Kirchhoff approximation, B_y^{dif} possesses an obliquity factor $\cos\theta$, but that E^{dif} has obliquity factor $\cos\theta_{\text{inc}}$.

[52] *Comment:* Equations (9.24)–(9.26) are exact (in the far-field limit). The same applies to the field components requested in part (10). Because no approximation has been made, these field components are correct solutions of the scalar wave equation (9.1). Other field components (E_x and E_z accompanying B_y, or B_x and B_z accompanying E_y) can now be found via Maxwell's curl equations. Built this way, the diffracted field is fully Maxwell-respectable. This is what we meant in § 9.4.1 when we said that a dominant field component could be calculated, with other components "looking after themselves".

Diffraction at a slit

10

Intended readership: in an advanced course discussing diffraction, in optics or electromagnetism or quantum mechanics, where it's appropriate to question fundamentals.

10.1   Introduction

In the case of diffraction, there is a conventional wisdom[1] as to how we ought to proceed: we use a Kirchhoff approximation (§ 9.1) in a diffraction integral. The results of such calculations agree excellently with experiment: we must be doing something right. And yet . . . .

The discussions in Chapter 9 have yielded the following results and non-results of relevance here:

- There are exact statements of the diffraction integral, in particular (9.10), which is general except that integration must be taken over a plane boundary surface (the surface where there are secondary sources). If a field amplitude $U(x, y, 0)$ in that (xy) plane is known, the diffracted amplitude $U(\mathbf{R})$ at downstream locations $\mathbf{R}$ can be calculated from the integral, without approximation.

- Usually $U(x, y, 0)$ is not known, and to make progress we have to make a guess at it. The usual guess is the Kirchhoff approximation: in an open area of the xy plane we take $U(x, y, 0)$ to be a piece cut from the incident wave; in regions where the xy plane is obstructed by an opaque screen $U(x, y, 0)$ is zero.

- A Kirchhoff approximation can be applied to either $\mathbf{E}$ or $\mathbf{B}$, but not to both at once (§§ 9.3.5 and 9.4.3, and problem 9.6 parts (12) and (13)). And if to at most one, then which?

- No calculation from Maxwell's equations has been offered to justify a belief that either $\mathbf{E}$ or $\mathbf{B}$ (or any hybrid such as $\sqrt{EB}$) is well approximated by a Kirchhoff chopped-out piece of field. We have no theoretical underpinning to accompany the experimental observation that the results are good.[2] I have long thought that this isn't good enough.

Unfortunately, a full solution of Maxwell's equations cannot be given except for one or two very special cases—that is why the Kirchhoff approximation is needed. But some progress can be made. We concentrate here on Fraunhofer diffraction at a single slit, because this is the case of diffraction first encountered by every student.

[1] I have already disagreed with conventional wisdom in §§ 9.3.3 and 9.4.2 by dismissing the customary obliquity factor $\frac{1}{2}(\cos\theta_{\mathrm{inc}} + \cos\theta_{\mathrm{dif}})$ as having anything to do with removing back-diffraction, and indeed as having anything at all to recommend it.

[2] As mentioned in § 9.4.1, the Smythe–Kirchhoff integral (9.18) is an improvement that is guaranteed to yield diffracted fields obeying Maxwell's equations. However, it is still necessary to input a guess at the behaviour of the aperture-plane $\mathbf{E}(x, y, 0)$. If we guess wrong (as is likely), we shall end up with "the right solution to the wrong problem".

We have encountered a spectacular failure of Smythe–Kirchhoff in § 9.5. Radiation normally incident by diffracted by a small circular aperture of radius a. The Smythe–Kirchhoff dependence on $(ka)^2$ should correctly be $(ka)^4$.

Essays in Physics: Thirty-two thoughtful essays on topics in undergraduate-level physics. Geoffrey Brooker,
Oxford University Press. © Geoffrey Brooker 2021. DOI: 10.1093/oso/9780198857242.003.0010

10.2   Fields diffracted at an ideally conducting slit

A plane electromagnetic wave travels in the $+z$-direction until it encounters a slit in the xy plane. The open area of the slit extends from $x = -a/2$ to $x = a/2$. The slit is long in the y-direction; everything is independent of y. Some of the radiation penetrates the slit opening and is diffracted into the "downstream half-space" ($z > 0$). Such an arrangement, for two possible polarization cases, is drawn in Figs. 10.1 and 10.2.

To apply Maxwell's equations, we must assign some electromagnetic properties to the jaws of the slit. For simplicity we take the jaws to be very thin in comparison with a wavelength. To make thin jaws opaque we must give them a very high electrical conductivity σ so that the skin depth $\delta = \sqrt{2/(\mu_r \mu_0 \sigma \omega)}$ is even smaller than the jaw thickness. Such properties make each jaw into an almost perfect reflector.[3]

For simplicity also, the surrounding medium is a vacuum, so it has $\epsilon_r = \mu_r = 1$. Every field component varies with time as $e^{-i\omega t}$.

The Maxwell curl equations now give two sets of equations:[4]

$$\dot{B}_x = \frac{\partial E_y}{\partial z}, \qquad\qquad \mu_0 J_x + \frac{\dot{E}_x}{c^2} = -\frac{\partial B_y}{\partial z}, \qquad (10.1a)$$

$$\dot{B}_z = -\frac{\partial E_y}{\partial x}, \qquad\qquad \frac{\dot{E}_z}{c^2} = \frac{\partial B_y}{\partial x}, \qquad (10.1b)$$

$$\mu_0 J_y + \frac{\dot{E}_y}{c^2} = \frac{\partial B_x}{\partial z} - \frac{\partial B_z}{\partial x}, \qquad\qquad \dot{B}_y = \frac{\partial E_z}{\partial x} - \frac{\partial E_x}{\partial z}. \qquad (10.1c)$$

Here all derivatives with respect to y have been set to zero. Eliminating between these equations, we find that every field component U satisfies the "scalar wave equation" (away from the conductors so $\boldsymbol{J} = 0$):

$$\nabla^2 U + k^2 U = 0, \qquad \text{where} \quad k = \omega/c. \qquad (10.2)$$

This, of course, just repeats (9.1), though for a special case.

Aside on terminology

For any configuration of fields and pieces of ideal conductor, charges and currents induced in the conductors give rise to "scattered" fields, which add to the incident field[5] to give the total field. As example, Fig. 10.2 shows lines of $\boldsymbol{B}^{\text{scatt}}$ circulating round induced currents. Then

$$\boldsymbol{E}^{\text{total}} = \boldsymbol{E}^{\text{inc}} + \boldsymbol{E}^{\text{scatt}}; \qquad \boldsymbol{B}^{\text{total}} = \boldsymbol{B}^{\text{inc}} + \boldsymbol{B}^{\text{scatt}}.$$

Equations (10.1) are linear, so they apply separately to all the incident, scattered, and total (diffracted) fields.

On the left of (10.1), there are three field components: E_y, B_x and B_z; on the right are the other three: B_y, E_x and E_z. Field components appearing on the left are absent from the right, and vice versa. This means that there are two independent polarization cases.

Before we think further about the field equations, it will be helpful—perhaps surprisingly so—to obtain some more qualitative information.

[3] For calculation purposes, this model obstacle may be further idealized: to an infinitely conducting sheet of negligible (or zero) thickness. This idealization is conventional: it is used in the whole of Born and Wolf (1999) Chapter 11.

Our first thought might be that the obstacle should be "black": absorbing what's incident. In order not to reflect, the obstacle would then have to have a characteristic impedance Z not too far from that of free space, which in turn would enforce a smallish conductivity and a skin depth of many wavelengths. To model the obstacle we would need to think about the shape of its edges and take that shape into account when calculating the fields.

However, if we were to try to handle a slit with partially conducting jaws of non-negligible thickness, we would still need solutions for the idealized case, to act as a standard of comparison, so that we could see what is made different by a changed conductivity or a changed jaw thickness.

[4] The two divergence equations add nothing to what is written. In particular, div $\boldsymbol{D} = \rho$ tells us about ρ without adding any constraint on the behaviour of $\boldsymbol{D}$.

[5] For the avoidance of doubt: the "incident" field extends over the whole of space; it does not stop at the slit. Then, for example, if the total $\boldsymbol{E}$-field is zero somewhere, perhaps in the shadow of a slit jaw, it is because $\boldsymbol{E}^{\text{scatt}} = -\boldsymbol{E}^{\text{inc}}$ there.

10.3   Field lines

We are discussing classical (non-quantum) optics, where E and B behave in the same way as do classical electromagnetic fields in any other frequency range. A microwave engineer would at once demand to know "what are the field lines doing?". We too should ask that question.

In Fig. 10.1, a slit in the plane $z = 0$ is illuminated normally by a plane wave whose E^{inc} faces across the slit. This field must induce charges on the slit jaws, the charges having travelled from elsewhere within each jaw. In turn, the charges give rise to "scattered" fields E^{scatt}. At a time when the charges have the signs indicated, the field lines of E^{scatt} fan out from the left-hand jaw and "fan in" to the right-hand jaw. At a small distance $r \ll \lambda$ from a jaw edge, the scattered field strength $\left| E^{\text{scatt}} \right| \propto r^{-1/2}$ (problem 10.1), that is, it goes to infinity as $r \to 0$.

The field lines in Fig. 10.1 of E^{scatt} have mirror symmetry about the xy plane. Then we can say:

- On both faces of a jaw, $E_x^{\text{total}} = 0$; in particular the downstream face of a jaw is "dark" so far as E_x^{total} is concerned.
- But E_x^{total} has an $r^{-1/2}$ infinity (not at all Kirchhoff) in the open area at each jaw edge.
- Field $E_z^{\text{total}} = 0$ in the open area. For reasons in Chapter 9 sidenote 30, E_z^{total} cannot be zero everywhere, so it must be non-zero somehow on the downstream faces of the jaws.
- In the xy plane, open area, $B_y^{\text{scatt}} = 0$ (eqn 10.1c), so $B_y^{\text{total}} = B_y^{\text{inc}}$.
- At this stage we have no information as to the behaviour of B_y^{total} on the downstream faces of the jaws.

We turn to the polarization case of Fig. 10.2. The incident wave has E^{inc} facing along the slit. This field sets up currents J_y in the slit jaws, out of the paper at times when $J_y > 0$. In turn, these currents set up a B^{scatt} that loops round the jaws as shown. At small distances $r \ll \lambda$ from a jaw edge, and in the open area, the scattered field strength $\left| B^{\text{scatt}} \right| \propto r^{-1/2}$ (problem 10.1).

The field lines for B^{scatt} have mirror antisymmetry about the xy plane. It follows that:

- In the exact slit plane, open area, $B_x^{\text{scatt}} = 0$ so that $B_x^{\text{total}} = B_x^{\text{inc}}$.
- In the slit plane, open area, B_z^{scatt} goes to infinity at each jaw edge as $r^{-1/2}$. And it does so with opposite signs at the two edges.
- In the open area and close to a jaw edge (eqn 10.1b), $E_y^{\text{scatt}} \propto r^{1/2}$: field E_y^{total} rises from zero (of course) with infinite gradient. Not quite the step-function rise that Kirchhoff assumes, but "having a good try".
- On the upstream face of a jaw, and well away from the open area, $|x| \gg a/2$, field B_x^{scatt} reinforces B_x^{inc} giving a doubling, in line with experience elsewhere with reflection from an ideal conductor.
- On the downstream face of a jaw, and well away from the open area, $|x| \gg a/2$, field B_x^{scatt} cancels B_x^{inc} giving a zero total.

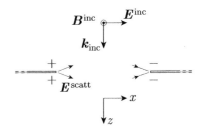

Fig. 10.1 A slit having highly conducting jaws lies in the plane $z = 0$. It is long in the y-direction (normal to the paper) and extends from $-a/2$ to $a/2$ in the x-direction. A plane electromagnetic wave is incident normally and has its E^{inc} field in the x-direction.

Charges are induced on the slit jaws. At times when the charges have the signs shown, field lines of E^{scatt} fan out from the left-hand jaw and "fan in" to the right-hand jaw.

Do not be tempted to "join across" the field lines in the middle of the diagram: the slit is envisaged to be several wavelengths across, so there is plenty of room between the jaws for the field to oscillate.

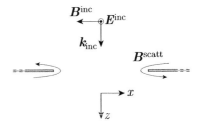

Fig. 10.2 The layout is similar to that of Fig. 10.1, except that E^{inc} now lies in the y-direction, parallel to the slit length. Currents J_y are induced, running along the jaws. In turn these set up a scattered B-field. At times when $J_y > 0$ (out of the paper), the lines of B^{scatt} loop round the jaws as shown. (Of course, everything is reversed in sign half a period later.)

[6] The reader should annotate each of the eleven bullet points as to whether it is in line, or not, with a Kirchhoff approximation.

Either now, or after seeing Figs. 10.3–10.6, you may be able to add a couple more bullet points.

- But in the region drawn, where the field lines curve, that cancellation cannot be total, so B_x^{total} "leaks round the edge" to some extent into the shadow regions.[6]

If we want to know more, then the only recourse is to make a full solution of eqns (10.1). Only then can we know how much harm is done by replacing $\boldsymbol{E}^{\text{total}}(z{=}0{+})$ or $\boldsymbol{B}^{\text{total}}(z{=}0{+})$ with a Kirchhoff approximation to it, and how much harm this does to the predicted far-field diffraction pattern. Nevertheless, we have already gained a great deal of useful information from little more than picture drawing. A mindset borrowed from microwave physics has served us very well indeed.

10.4   Solving the field equations (10.1)

The algebra involved in solving eqns (10.1) is somewhat technical. I shall say just enough to show that there are no surprises.

We solve a scalar wave equation for E_y (or B_y depending on the polarization case chosen), after which the x and z components of $\boldsymbol{B}$ or $\boldsymbol{E}$ are found using the curl equations of (10.1).[7]

[7] We see again the idea of § 9.4.1 that a single field component, here E_y or B_y, can be calculated, after which the other components will "look after themselves" in a Maxwell-compliant way.

We use a coordinate system (here elliptic-hyperbolic cylinders) within which ∇^2 has a form that permits a separation of variables. With a boundary condition this results in differential equations having eigenvalues and eigenfunctions. The field E_y or B_y is expressed as a sum over eigenfunctions, and boundary conditions (again) yield the expansion coefficients. No surprises.

This is not the place to give more detailed information, and we refer the reader to the original paper (Brooker 2008).

10.5   Fields in the plane of the slit: $\boldsymbol{E}^{\text{inc}}$ along slit

We wish to compare the $\boldsymbol{E}$ and $\boldsymbol{B}$ fields as calculated exactly and as predicted by Kirchhoff. For this purpose, a fairly limited range of slit widths needs to be considered. For a very wide slit, Kirchhoff must give good answers, because "edge effects" from the slit jaws then occupy a small fraction of the slit's area.[8] For a *very* narrow slit, exact expressions reduce to (9.20). Here we display results for an illustrative intermediate case,[9] where the slit width $a = 5.5\lambda$.

[8] These "edge effects" can manifest themselves as Fresnel diffraction patterns. For a wide slit, the two edges can be treated independently as in § 10.6. Also, the mathematical methods of the present chapter can be applied directly to the case of a wide slit ($a \gg \lambda$), but we do not pursue this case here.

[9] The graphs given in the present chapter apply to values of a/λ different from those in the paper cited, where we chose $a/\lambda = 5$. They have therefore not appeared in print before.

[10] Further discussion of this cylindrical-wave interpretation is given in § 10.6.

Figure 10.3 shows the computed fields in the plane of the slit when the incident wave has its $\boldsymbol{E}^{\text{inc}}$ parallel to the length of the slit. We know from eqns (10.1) that the three field components plotted are the only ones to be non-zero for this polarization choice.

The middle panel of Fig. 10.3 shows $B_z^{\text{total}} = B_z^{\text{scatt}}$ in the plane of the slit. At small distances $r = a/2 - |x| \ll \lambda$ from a jaw edge, B_z goes to infinity as $r^{-1/2}$, and does so with opposite signs at the two jaw edges. We may see B_z^{scatt} as made up from two roughly-cylindrical waves, one radiating from each slit jaw.[10]   The cylindrical waves originate from

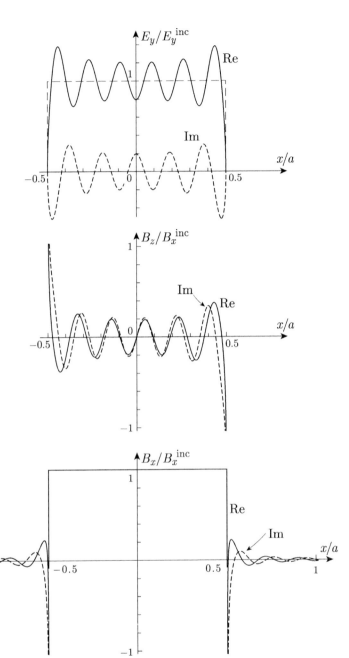

Fig. 10.3 A slit of width a is illuminated normally by an electromagnetic wave having its $\boldsymbol{E}^{\mathrm{inc}}$ parallel to the length of the slit (the y-direction). The three non-zero electromagnetic field components (total fields) are plotted in the plane of the slit. The fields are complex, meaning that they have phase shifts relative to the incident wave. In each case, the real part is shown by a continuous curve, and the imaginary part by a broken curve. The slit width $a = 5.5\lambda$.

Field $B_x = B_x^{\mathrm{total}}$ is equal to the incident B_x^{inc} in the open area of the slit, but also spills just a little into the "shadow" of the slit jaws. By contrast, field components $E_y = E_y^{\mathrm{total}}$ and $B_z = B_z^{\mathrm{total}}$ do not spill at all into the shadow, and are non-zero in the open area of the slit only.

The oscillations of B_z may be seen as the superposition of two "cylindrical" waves which are scattered from the slit jaws and then undergo multiple reflection between the jaws. The oscillations in E_y are interference between three waves: the two cylindrical waves and the incident wave.

A Kirchhoff assumption would make all fields zero in the shadow of the slit jaws, which is almost true except for the spillage of B_x. It would also make E_y and B_x into top hats within the open area (while B_z would be zero). This is exactly the case for B_x. The Kirchhoff assumption for E_y is shown as a broken-line top hat superimposed on the plot of E_y; we can see E_y^{total} "trying" to conform to the top hat, though being forced to oscillate as well.

Note that $cB_x^{\mathrm{inc}} = -E_y^{\mathrm{inc}}$. This sign difference has been allowed for in the normalizing of the magnetic fields, dividing by B_x^{inc} rather than E_y^{inc}.

currents in the slit jaws, induced by the E_y-field of the incident wave. Each cylindrical wave propagates (we concentrate on the xy plane) until it encounters the other jaw, where further travel is prevented by the conductor, and it must be reflected. There is therefore a kind of standing wave between the jaws.

The top panel of Fig. 10.3 shows field E_y^{total} in the plane of the slit. A coarsely broken line shows the Kirchhoff top-hat assumption.[11] We

---

[11] There are no adjustable constants, so a comparison of absolute magnitudes is displayed.

[12] In confirmation: The oscillations of B_z and E_y have wavelengths very close to the free-space wavelength λ. We plot for $a = 5.5\lambda$, so it is to be expected that roughly 5 periods of oscillation will be seen, and that is the case.

may see E_y as "trying" to follow the top hat. At the same time, the oscillation is the superposition (interference) of three waves: the incident wave and the two cylindrical waves.[12] The cylindrical-wave constituents account for a rise of E_y as $r^{1/2}$ close to each jaw edge.

Field component B_x^{total} (Fig. 10.3 bottom panel) is exactly equal to B_x^{inc} in the open area of the slit, but it also leaks a little into the shadow region just downstream of each slit jaw (it is plotted for $z = 0+$).

The field components conform to the bullet points of § 10.3 in all cases. At the same time, Fig. 10.3 shows that the field components E_y^{total} and B_x^{total} in the plane of the slit resemble quite closely those assumed in a Kirchhoff approximation.

10.5.1 Fields diffracted to large distances: the Fraunhofer limit

[13] For the polarization case of Fig. 10.4, the actual field E_y^{dif} is zero in the exact-sideways directions ($\theta = \pm\pi/2$). This is the case for all values of the slit width, as E_y must be zero on the surfaces of the conducting slit jaws. Notice that this boundary condition is not obeyed by the Kirchhoff expression. Indeed, I have chosen to display the case $a/\lambda = 5.5$ (a non-integer) in order to exhibit this feature. (If a/λ is an integer, the Kirchhoff prediction is zero at $\sin\theta = \pm 1$, thereby giving an appearance of agreement with exact calculation that is merely the consequence of choosing this a/λ.)

The far field, where there is Fraunhofer diffraction, is the region where experiment has told us that Kirchhoff works well. Let's check. We continue to look at the polarization case of § 10.5.

Figure 10.4 shows the field E_y^{dif} ($= E_y^{\text{total}}$) diffracted to the far field, as a function of $\sin\theta$, covering the range from $\theta = 0$ (forwards) to $\theta = \pm\pi/2$ (sideways); the deviation angle θ is defined in Fig. 9.6.[13]

In Fig. 10.4, E_y^{dif} is plotted rather than its square, because a plot of intensity would not show detail where the deviation angle θ is large. And the modulus (meaning the absolute value of the complex quantity) is plotted because the exact field has phase shifts (fairly small, so not detailed here) even when the factor $e^{i(kR-\pi/4)}$ of (9.19) is divided away.

Fig. 10.4 The far-field amplitude $|E_y^{\text{dif}}|$ for the case of a slit of width $a = 5.5\lambda$, illuminated normally by a field $\boldsymbol{E}^{\text{inc}}$ parallel to the slit's length. Solid line: exact calculation. Broken line: outcome of a Kirchhoff approximation using an obliquity factor of 1. The exact graph is normalized to 1 at $\sin\theta = 0$; the Kirchhoff graph is divided by the same normalizing constant, so an absolute comparison of magnitudes is displayed.

The Kirchhoff expression is real, so it has a constant phase apart from reversals of π at the zeros. The exact expression is not quite real, and its imaginary part is responsible for "filling in" the minima.

The Kirchhoff prediction gives a non-zero amplitude at $\theta = \pi/2$, because the obliquity factor has been set to 1. Other choices of obliquity factor are, of course, possible, but give less good agreement overall.

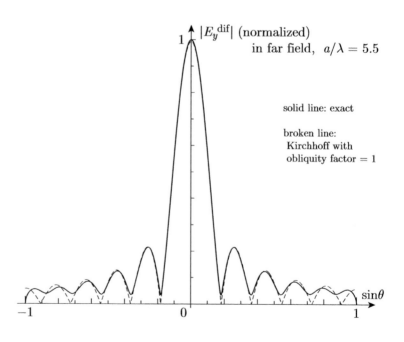

The exact curve of Fig. 10.4 shows minima that do not quite fall to zero. But those minima look much smaller when the diffracted intensity, proportional to $\left|E_y^{\mathrm{dif}}\right|^2$, is plotted.

The Kirchhoff (broken) curve is plotted using an obliquity factor of 1. A factor $\cos\theta$, suggested by (9.19a),[14] might seem more justifiable, and would make the Kirchhoff curve fall to zero at $\theta = \pi/2$. But such an obliquity factor makes the Kirchhoff curve underestimate the energy diffracted to large angles, an outcome that (I think) is worse overall.

The curves plotted in Fig. 10.4 have been normalized so that the exact curve takes the value 1 at its peak. The Kirchhoff curve is presented on the same scale, so it is not exactly 1 at its peak, though the difference is barely visible in the figure. A comparison of absolute magnitudes is on display.

The curves of Fig. 10.4 are remarkably close to each other: Kirchhoff gives a very reliable prediction. And the agreement is similarly good for other choices of slit width (other than very narrow slits). We have a justification from Maxwell's equations for the success of Kirchhoff, the justification called for in the introduction to this chapter.[15]

10.6 The fields diffracted by a conducting half-plane

In the last section, we explained the wiggly fields E_y and B_z in the slit plane in terms of "roughly-cylindrical" waves radiated from the slit jaws. A simpler physical system showing similar behaviour is that where we have just one "jaw" present: a single conducting half-plane illuminated by a plane electromagnetic wave.[16]

Let the conductor lie in the xy plane at $x < 0$, so its edge lies along the y-axis. The incident wave travels in the z-direction and is polarized with its $\boldsymbol{E}^{\mathrm{inc}}$ in the y-direction.[17] The field lines for $\boldsymbol{B}^{\mathrm{scatt}}$ resemble those on the left side of Fig. 10.2, and have

$$B_z^{\mathrm{total}} = B_z^{\mathrm{scatt}} \propto \frac{\mathrm{e}^{\mathrm{i}kx}}{\sqrt{kx}}$$

in the xy plane for $x > 0$. We confirm that a cylindrical wave radiates away from a conductor's sharp edge.[18] Also, since $B_z^{\mathrm{total}} \propto x^{-1/2}$, we confirm, via (10.1b), that $E_y^{\mathrm{total}} \propto x^{1/2}$.

10.7 The case where B^{inc} faces along the slit

Figure 10.5 shows scattered "roughly-cylindrical" waves radiating from the sharp edges of the slit jaws. These now propagate away from the slit opening into the shadow of each jaw. This behaviour is displayed for B_y^{total}; E_z participates similarly but is not shown.[19] There is a serious conflict between this and the Kirchhoff prediction that there will

[14] Remember problem 9.6(3) and 9.6(7). It is there shown that, in the case of diffraction at a slit, the far-field diffraction angle θ_{dif} is closely approximated by the angle of deviation θ.

[15] The agreement of Fig. 10.4 is not as convincing as it may seem. For the Kirchhoff curves I have patched in an obliquity factor of 1. Such an obliquity factor is to be expected if it is known, or believed, that in the slit plane it is the "across-the-slit" field component (here B_x) that most closely resembles a Kirchhoff "piece cut from the incident wave" (eqns 9.19). We could not have known this in advance of computing the field components of Fig. 10.3 (and making a judgement as to which field component is "better"). An obliquity factor of 1, asserted without justification, makes Kirchhoff look better than perhaps he should.

[16] The solutions of Maxwell's equations for this system are given by Born & Wolf (1999) Chapter 11.

[17] The coordinates have been chosen to resemble those for the case of the slit, except that the half-plane has its edge at $x = 0$, not at $x = -a/2$.

[18] The cylindrical wave is only part of the scattered field. There are also contributions to B_x^{scatt} and E_y^{scatt} that together are responsible for a reflected wave and a shadow. These field shapes merge into each other. We are able to isolate the cylindrical wave only by agreeing to look at a special case: the field component B_z in the xy plane.

[19] Nor do I plot the far-field amplitude for this polarization. The case I wish to make in this chapter is already well illustrated by Figs. 10.3 and 10.4.

Field E_z is zero in the open area. In the "shadow" of the slit jaws ($z = 0+$, $|x| > a/2$), E_z oscillates much as does B_y; this is expected since a cylindrical wave travelling towards $\pm x$ has $E_z \approx \mp c B_y$.

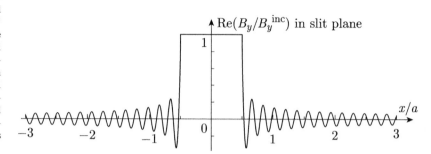

Fig. 10.5 An ideally conducting slit having width $a = 5.5\lambda$ is illuminated normally by a plane wave whose $\boldsymbol{B}^{\mathrm{inc}}$ lies parallel to the slit's length. The $\boldsymbol{B}$-field in the slit's open area is equal to $\boldsymbol{B}^{\mathrm{inc}}$, but there is additionally a cylindrical wave propagating away from each jaw edge, giving non-zero B_y and E_z in the "shadow" region where Kirchhoff predicts darkness. Only the real part of the B_y-field is plotted; the imaginary part has similar wiggles but is zero in the open area.

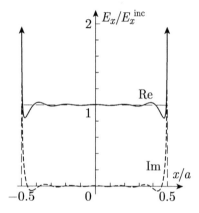

Fig. 10.6 An ideally conducting slit having width $a = 5.5\lambda$ is illuminated normally by a plane wave whose $\boldsymbol{B}^{\mathrm{inc}}$ lies parallel to the slit's length. In the plane of the slit, the electric-field component E_x is non-zero in the open area only. Both real and imaginary parts go to infinity as $(a/2 - |x|)^{-1/2}$ close to the sharp edges of the slit jaws. Apart from these rather narrow infinities, the field closely resembles a Kirchhoff top-hat.

[20] There is a complication in the Fourier expansion. A Fourier component $e^{i\beta x}$ extends over x from $-\infty$ to ∞. The Fraunhofer condition (9.23), used in obtaining (10.3), does not hold for all $e^{i\beta x}$, only for a part of it in the vicinity of the origin. We ought to go back to (9.22) and Fourier-expand the $U(x)$ there. Fortunately, we can derive (10.3) afresh from that starting point, using the fact that $kR \gg 1$, and without making explicit use of the Fraunhofer condition (9.23) (problem 10.4). It is therefore acceptable to apply a Fourier expansion to the $U(x)$ in (10.3).

be "dark" in the shadow regions. Nevertheless, this part of B_y makes little contribution to the far-field diffraction pattern, for which Kirchhoff gives good results (if anything slightly better for this polarization). For explanation see the next section.

Field E_x goes to infinity as $(a/2 - |x|)^{-1/2}$ close to each jaw edge (Fig. 10.6). This behaviour exactly corresponds to the similar square-root rise of B_z discussed in § 10.6, and again in problem 10.1. The infinite rise of E_x is quite narrow, so that over most of slit's open area E_x^{total} resembles a Kirchhoff top-hat. This justifies a Kirchhoff approximation, and because an "across" field component is involved we again find that the Kirchhoff-friendly choice (eqns 9.19) of obliquity factor is 1.

10.8 Why can we get away with Kirchhoff?

The Kirchhoff approximation to the field (components) in the slit plane turns out to represent the actual field rather closely. We ask whether there is a simple explanation.

A helpful insight lies in Fourier optics. We rewrite (9.24):[20]

$$U_{\mathrm{dif}}(\boldsymbol{R}) = \frac{1}{\sqrt{\lambda R}}\, e^{i(kR - \pi/4)} \cos\theta \int U(x)\, \exp(-ikx\sin\theta)\, \mathrm{d}x, \qquad (10.3)$$

where $\boldsymbol{R} = (R, \theta)$ in the cylindrical coordinates of Fig. 9.6; we are in the far field where R is large. The integral in this is the Fourier transform $\tilde{U}(\beta)$ of $U(x)$ with "transform variable" $\beta = (k\sin\theta)$.

$$\tilde{U}(\beta) = \int U(x)\, e^{-i\beta x}\, \mathrm{d}x; \qquad U(x) = \int_{-\infty}^{\infty} \tilde{U}(\beta)\, e^{i\beta x}\, \frac{\mathrm{d}\beta}{2\pi}.$$

We refer to $\beta/2\pi$ as the **spatial frequency** of the sinewave $e^{i\beta x}$.

Equation (10.3) tells us that

> the process of diffraction to the far field creates a display of $\tilde{U}(\beta)$, with each β sent to its own θ according to $\sin\theta = \beta/k$.

Now, in (10.3), $|\sin\theta| \leqslant 1$, so $|\beta| \leqslant k$; the entire far-field diffraction pattern is assembled from $\tilde{U}(\beta)$s within the range $|\beta| \leqslant k$. Diffraction to the far field acts as a low-pass filter, with space frequencies $|\beta/2\pi| \leqslant 1/\lambda$ accepted for propagation, while space frequencies $|\beta/2\pi| > 1/\lambda$ are rejected. If we measure the diffraction pattern, we can infer what is going on in the plane of the slit—but for spatial frequencies $|\beta/2\pi| < 1/\lambda$ only.[21,22]

Where the Kirchhoff approximation predicts a correct $U_{\mathrm{dif}}(R,\theta)$, it must be getting that spatial frequency right; where there is a discrepancy Kirchhoff is making a guess that is less good. Kirchhoff gets the low spatial frequencies mostly right—and what he may have to say about very high spatial frequencies ($> 1/\lambda$) does not matter.

These ideas help to answer this section's question. The top graph of Fig. 10.3 shows that $E_y(x)$ wiggles rather than being flat-topped. Those wiggles have wavelength close to the free-space wavelength λ. They contribute mainly to a sinewave component whose wavelength is λ, and diffracts (mainly) towards $\sin\theta \approx \pm 1$. It is easy to see "by eye" that the difference between the actual E_y and the Kirchhoff top-hat can have little effect on the diffraction pattern: the spatial frequencies are too high.[23]

We may apply similar reasoning to the near-cylindrical waves exhibited in Fig. 10.5. These waves have period close to the wavelength of the radiation. The Fourier components that they create most strongly must have period d close to λ. This in turn means that they contribute most to diffraction at angle $\theta = \pm\pi/2$ to the forward direction—which is where we came in because that is what they are known to be doing.

In Fig. 10.5, Kirchhoff gets the field "behind" the slit spectacularly wrong, yet because the space frequencies are high the diffraction pattern (not shown but a bit better-fitting than in Fig. 10.4) is nearly right.[24]

10.9   Diffraction in quantum mechanics

It is common in the teaching of physics to show experimental evidence that electrons (or other quantum particles), passing through a slit, exhibit a diffraction pattern resembling that familiar in optics. We ought to explain why. Fortunately, the solution to one problem supplies us with the solution to the other.

Let a ψ-wave be incident on the xy plane, part-covered by thin impervious obstacles (that might, for example, be the jaws of a slit). Away from the obstacles, the ψ-wave travels freely in a region where the potential-energy function $V(\boldsymbol{r}) = 0$, so the time-independent Schrödinger equation gives[25]

$$\nabla^2\psi + k^2\psi = 0, \qquad \text{where} \quad \frac{\hbar^2 k^2}{2m} = E,$$

a Helmholtz wave equation identical in form to (10.2).

We can model each obstacle as imposing a strong ($\gg E$) potential V_0 on the particles. Within the thickness the wave function decays as

$$\psi \propto e^{-\kappa z}, \qquad \text{where} \quad \frac{\hbar^2\kappa^2}{2m} = V_0 - E.$$

In order for the obstacle to be impervious to particles, we must make κ so large that the wave function decays in a distance much less than the thickness, which in turn must be much less than the (external) de Broglie

[21] An amplitude $e^{i\beta x}$ with $|\beta| > k$ radiates an evanescent wave, decaying exponentially with downstream distance z, and therefore makes no contribution to the far field; see problem 10.2(2).

There are no prizes for recognizing that $e^{i\beta x}$ is periodic with period $d = |2\pi/\beta|$, and that $\sin\theta = \pm\lambda/d$. This is, of course, the equation describing the action of a diffraction grating.

[22] This is the reason why all forms of microscopy have limitations on their resolution. We have given a potted version of Abbe theory.

[23] In fact, nothing does get diffracted to $\theta = \pm\pi/2$, for this polarization, for two reasons. One reason is the angle factor $\cos\theta$ in (9.19a), which is zero in the exactly-sideways direction. But a more down-to-earth reason is that E_y must be zero on the downstream faces of the conducting jaws.

The observant reader may question why the $\cos\theta$ factor does not similarly kill off the "sideways" radiation of Fig. 10.5. As we make θ approach $\pm\pi/2$ the $\cos\theta$ factor does go to zero, but the rest of the integral in (9.19a) diverges, yielding a finite outcome when the limit is taken. Diverges? The integrand behaves as $1/\sqrt{|x| - a/2}$, coming from the two cylindrical waves.

[24] Nearly right if we again patch in an obliquity factor of 1.

[25] A non-relativistic approximation is being taken. Each particle has mass m and kinetic energy E.

wavelength $2\pi/k$. Then, just as we made the electrical conductivity g[...]
to infinity before, we must let $V_0 \to \infty$ now; when this condition i[...]
imposed $\psi^{\text{total}} = 0$ on the obstacle.

The "impervious" boundary condition now says that[26]

$$\psi^{\text{total}} = \psi^{\text{inc}} + \psi^{\text{scatt}} \quad \text{with} \quad \psi^{\text{scatt}} = -\psi^{\text{inc}} \quad \text{on obstacle.}$$

This states that ψ^{scatt} is "driven" by a source $-\psi^{\text{inc}}(z{=}0)$ occupying[...]
part of the xy plane. The boundary condition makes no mention o[...]
what happens at any z other than $z = 0$; likewise it makes no mention[...]
of how the source of ψ^{scatt} happened to be there. Therefore the scattered
wave is symmetrical about the source plane:[27] ψ^{scatt} is an even function
of z. Then $\partial\psi^{\text{scatt}}/\partial z$ is an odd function of z. Over the open part of the
xy plane, $\partial\psi^{\text{scatt}}/\partial z$ must be continuous, and is therefore zero.

Putting the above together, we have boundary conditions

$$\psi^{\text{scatt}} = -\psi^{\text{inc}} \text{ on obstacle}; \qquad \frac{\partial\psi^{\text{scatt}}}{\partial z} = 0 \text{ in open area}$$

or

$$\psi^{\text{total}} = 0 \text{ on obstacle}; \qquad \frac{\partial\psi^{\text{total}}}{\partial z} = \frac{\partial\psi^{\text{inc}}}{\partial z} \text{ in open area.} \quad (10.4$$

Compare (10.4) with (10.5), the boundary condition governing E_y^{total}[...]
For the same layout of impervious/conducting obstacles, the boundary
conditions are identical. The solutions for ψ^{total} and E_y^{total} must b[...]
identical likewise, if the waves arrive in the same way. This equality
holds whether or not we have a hammered-out solution for either field
(i.e. it is not confined to any special case such as a slit).

[26] The most fundamental condition
for the obstacle to be impervious is
that the probability current through it
should be zero. This does not quite
force ψ^{total} to be zero within the thickness. However, I have been unable to
think of physical properties, even idealized, to assign to the obstacle, that
would permit ψ to be non-zero within
the thickness yet prohibit penetration
of the downstream surface. The condition given in the text seems to be the
most realistic that we can invent.

[27] This step holds whether or not the
incidence is normal. Think about it.

Problems

[28] Help in Landau, Lifshitz & Pitaevskiĭ
(1993) § 3 problem 3. The maths is easy
if you look up ∇^2 in cylindrical coordinates and do a separation of variables.
The ϑ boundary condition is that φ is
an even function of ϑ and is zero at
$\vartheta = \pm\pi$. The simplest solution to this
is $\cos\left(\frac{1}{2}\vartheta\right)$. This gives the separation
constant, via which you can find the
function of r.

[29] Help in Landau, Lifshitz & Pitaevskiĭ
(1993) § 30 second footnote. Again, a
separation of variables is straightforward, and all you need is to decide the
function of ϑ that makes A_y behave appropriately. It is also possible to solve
the problem using a magnetic scalar
potential.

Problem 10.1 Fields near a thin conductor's edge
Consider a half-plane as in § 10.6 (conductor occupying $x < 0$ in the xy[...]
plane, with its edge on the y-axis). Use cylindrical coordinates $(r, \vartheta, y$[...]
having $r = 0$ on the conductor's edge and having $\vartheta = 0$ in the xy
plane with $x > 0$. When E_x is incident, lines of $\boldsymbol{E}$ fan out from the
conductor's sharp edge, as in Fig. 10.1. Over distances $r \ll \lambda$, the
E-field takes the same shape as it would in electrostatics, and can be
obtained from a potential $\varphi(r, \vartheta)$ obeying Laplace's equation. Show[28]
that $\varphi \propto r^{1/2} \cos(\frac{1}{2}\vartheta)$. Show that $|\boldsymbol{E}| \propto r^{-1/2}$, going to infinity as $r \to 0$,
cf. Fig. 10.6.

Consider next the case where E_y is incident. Currents are induced in
the conductor. Show that the vector potential $\boldsymbol{A}$ has a single component
A_y which is zero on the conductor and behaves[29] as $A_y \propto r^{1/2} \cos(\frac{1}{2}\vartheta)$.
Show that $|\boldsymbol{B}| \propto r^{-1/2}$ with lines as sketched in Fig. 10.2. Show that
the lines of $\boldsymbol{B}$ for this case are perpendicular to the lines of $\boldsymbol{E}$ for the
former polarization. Show that $E_y \propto r^{1/2}$ and $B_z \propto r^{-1/2}$ close to the
conductor. Thus E_y has an infinite gradient but not a step discontinuity
whilst B_z has an infinity. See these behaviours in Fig. 10.3.

Problem 10.2 Introduction to Fourier optics

Refer to eqn (10.3). Let a field component $U(x)$ in the plane of a slit consist of a single sinewave $U(x) = \mathrm{e}^{\mathrm{i}\beta x}$. Equation (10.3) tells us that the amplitude diffracted[30] to distance R and angle θ is

$$U_{\mathrm{dif}}(R,\theta) \propto \int U(x)\,\mathrm{e}^{-\mathrm{i}(k\sin\theta)x}\,\mathrm{d}x = \int_{-\infty}^{\infty} \mathrm{e}^{\mathrm{i}(\beta - k\sin\theta)x}\,\mathrm{d}x$$

$$= 2\pi\,\delta(\beta - k\sin\theta).$$

The given $U(x) = \mathrm{e}^{\mathrm{i}\beta x}$ sends radiation into the single direction θ satisfying $\sin\theta = \beta/k$.

(1) Show that this finding can be put into the form $\sin\theta = \lambda/d$, where d is the period of the sinewave $\mathrm{e}^{\mathrm{i}\beta x}$. Identify this with the equation governing the behaviour of light at a diffraction grating, and explain why such a similarity is to be expected. Allow for the possibility that β may have either sign.

(2) Use the Helmholtz wave equation (10.2) to show that a field equal to $\mathrm{e}^{\mathrm{i}\beta x}$ in the xy plane, and having $|\beta| > k$, must decay exponentially with z, so it cannot contribute to the far-field diffraction pattern.

(3) Argue that if an actual $U(x)$ is represented as a Fourier sum

$$U(x) = \int \widetilde{U}(\beta)\,\mathrm{e}^{\mathrm{i}\beta x}\,\frac{\mathrm{d}\beta}{2\pi},$$

information about $\widetilde{U}(\beta)$ is to be found by looking at what is diffracted to direction θ satisfying $\sin\theta = \beta/k$. Moreover, the diffraction pattern contains information about $\widetilde{U}(\beta)$ for $|\beta| < k$ only: all Fourier components outside this range are lost.

Problem 10.3 Symmetries of the scattered field components

When radiation falls on a conductor, charges and currents are induced in the conductor. We can divide the resulting "total" field into "incident" and "scattered" contributions:

$$\boldsymbol{E}^{\mathrm{total}} = \boldsymbol{E}^{\mathrm{inc}} + \boldsymbol{E}^{\mathrm{scatt}}; \qquad \boldsymbol{B}^{\mathrm{total}} = \boldsymbol{B}^{\mathrm{inc}} + \boldsymbol{B}^{\mathrm{scatt}}.$$

It is the currents that are responsible for generating a scattered field via the integral that relates vector potential $\boldsymbol{A}^{\mathrm{scatt}}(\boldsymbol{R})$ at field point $\boldsymbol{R}$ to current elements $\boldsymbol{J}(\boldsymbol{r})\,\mathrm{d}V$ at locations $\boldsymbol{r}$ in the scatterer.[31]

(1) For us, the conduction currents flow in a thin conductor in the xy plane. Argue that A_x^{scatt} and A_y^{scatt} must be even functions of z. Use this to argue that E_x^{scatt} and E_y^{scatt} are likewise even functions of z. Use div $\boldsymbol{E} = 0$ to show that E_z^{scatt} must be an odd function of z. Use the curl equations, with $\partial/\partial y$ included, not the special cases of (10.1), to show that B_x^{scatt} and B_y^{scatt} are odd functions of z while B_z^{scatt} is even.[32]

(2) The equation cited in sidenote 31 applies in the Lorentz gauge, when there is also present a scalar potential φ contributing to $\boldsymbol{E}$. Investigate.[33]

(3) Check that the field lines drawn in Fig. 10.2 are consistent with the symmetries of B_x^{scatt} and B_z^{scatt} found here.

(4) Show that in open areas $B_x^{\mathrm{scatt}} = 0$ and therefore $B_x^{\mathrm{total}} = B_x^{\mathrm{inc}}$. Compare with Fig. 10.3.

[30] See sidenote 20. A justification has to be given for the validity of integrating x over such a large range. The point is investigated in problem 10.4.

[31] See, e.g., Bleaney & Bleaney (2013) eqn (8.66):

$$\boldsymbol{A} = \frac{\mu}{4\pi} \int \frac{[\boldsymbol{J}]\,\mathrm{d}\tau}{r}.$$

[32] Notice that nothing has needed to be said here about the geometry of the incident wave. In particular that wave need not be plane or incident normally for these symmetries to hold.

The setting-up of symmetries in this problem may look tedious, but it is an essential step in determining the form of the solutions from which this chapter's graphs were obtained.

[33] We are modelling a conductor as having a thickness of many skin depths, so the upstream and downstream surfaces carry charges and currents that are established independently (they cannot communicate through the conductor's thickness). Show that the charges and currents do in fact have to be equal (for the same x, y). Show that the scalar potential (if there is one) makes a contribution to $\boldsymbol{E}$ having the same symmetry with z as that from $\boldsymbol{A}$. Be prepared to draw pictures after the fashion of Figs. 10.1 and 10.2.

(5) Using (10.1) show that the boundary condition for the E_y polarization case can be written

$$E_y^{\text{total}} = 0 \text{ on conductors;} \qquad \frac{\partial E_y^{\text{total}}}{\partial z} = \frac{\partial E_y^{\text{inc}}}{\partial z} \text{ in open areas.} \quad (10.5)$$

(6) For the other polarization, show by similar reasoning that

$$B_y^{\text{scatt}} = 0 \text{ in open areas;} \qquad \frac{\partial B_y^{\text{scatt}}}{\partial z} = -\frac{\partial B_y^{\text{inc}}}{\partial z} \text{ on conductors.}$$

$$(10.6)$$

(7) From the similarity of (10.5) and (10.6), set up an equivalence[34] between the fields diffracted downstream from complementary screens: the E_y^{total} resulting from diffraction at a slit has the same form as B_y^{scatt} scattered by the complementary strip. (For this, the incident waves for the two cases must have the same geometry, except for the different polarizations.)

(8) The Babinet equivalence obtained in part (7) is special to a case where everything is independent of y. Make a proof that holds for any shape of plane obstacle/aperture.[35]

Problem 10.4 More Fourier optics

In sidenote 20 it is envisaged that we diffract light whose amplitude is the one-dimensional $U(x)$. To make contact with Fourier mathematics, we extended the x-range to $(-\infty, \infty)$, even though the Fraunhofer condition (9.11) cannot hold for large $|x|$. Show[36] that this step is valid in the limit $R \to \infty$.

[34] We have said already that such an equivalence is called the Babinet principle. In textbooks, an approximate equivalence is usually obtained from a Kirchhoff-approximated diffraction integral, but what is discussed here is an exact electromagnetic version.

[35] Help in Born and Wolf (1999) § 11.3. Also in Jackson (1999) § 10.8. The Babinet equivalence between a slit and a strip was used as a simplifying step in the calculations leading to Fig. 10.3.

[36] You may find the idea of Fresnel zones helpful.

VI Quantum mechanics

11 Symmetry in quantum mechanics

> SHOPKEEPER (to inexperienced assistant who has just made a rather shapeless display of goods): *Haven't you heard of symmetry?*
>
> ASSISTANT (startled): *What, you mean the place they bury people?*

Intended readership: Around the time of introducing degenerate perturbation theory.

This chapter holds an idea that is extremely helpful, particularly with atomic physics, to make sense of coupling schemes. And we meet symmetry again in connection with Bloch-function waves in condensed-matter physics .... So it may well be a chapter to keep coming back to.

11.1 Introduction: commuting operators

Reminder of a standard theorem:

- If two operators $\widehat{A}$ and $\widehat{B}$ commute, then they have a complete set of eigenfunctions in common.

Remember what this means. If we find an eigenfunction of $\widehat{A}$, accompanying a non-degenerate eigenvalue A (of $\widehat{A}$), then it *must* be an eigenfunction of $\widehat{B}$ as well. If we find two or more degenerate eigenfunctions[1] of $\widehat{A}$, they are not automatically eigenfunctions of $\widehat{B}$, but linear combinations can be found which *are* eigenfunctions of $\widehat{B}$. In all cases then,

- when operators $\widehat{A}$ and $\widehat{B}$ commute, the eigenfunctions of $\widehat{A}$ are eigenfunctions of $\widehat{B}$, or can be chosen so to be.

"Or can be chosen so to be" is permission/encouragement to take linear combinations of $\widehat{A}$ eigenfunctions if the A states are degenerate.

Two problems at the end of this chapter may help to bring this idea to life. Problem 11.1 takes a three-dimensional harmonic oscillator and finds eigenstates of (orbital) angular momentum. Problem 11.2 considers a hydrogen atom in an electric field, and shows how an awareness of this section's theorem can give a structure and a simplification to a potentially tedious calculation. Work these problems now, as the discussion that follows assumes you are familiar with them.

[1] I mean: two or more eigenfunctions of $\widehat{A}$ all having the same eigenvalue A. It is the common eigenvalue A that is said to be degenerate. "Degenerate eigenfunctions" is not correct jargon, but avoidance is clumsy.

Essays in Physics: Thirty-two thoughtful essays on topics in undergraduate-level physics. Geoffrey Brooker, Oxford University Press. © Geoffrey Brooker 2021. DOI: 10.1093/oso/9780198857242.003.0011

11.2   Commuting operators and the problems

Problem 11.1 discusses a three-dimensional harmonic oscillator. The Hamiltonian $\widehat{H} = \frac{1}{2}m\omega^2 r^2$ is spherically symmetrical, so it commutes with orbital-angular-momentum operators $\widehat{l^2}$ and $\widehat{l}_z$. So these operators ($\widehat{H}$ and the $\widehat{l}$s) have complete sets of eigenfunctions in common. Problem 11.1 shows how eigenfunctions common to both $\widehat{H}$ and $\widehat{l}_z$ can be found in an easy case. The four lowest-energy states suffice to show the kind of thing that can happen. In two cases we are handed the correct functions on a plate; in two cases we have to take linear combinations. The theorem on commuting operators guarantees that suitable linear combinations will be available.

11.2.1   Solve the easier problem

The theorem on commuting operators can be used in a more creative way. Suppose $\widehat{A}$ and $\widehat{B}$ commute, and we wish to find the eigenfunctions of $\widehat{A}$. Since the operators commute, the eigenfunctions of $\widehat{A}$ are the same as those of $\widehat{B}$ (or can be chosen to be). If $\widehat{A}$ is complicated and $\widehat{B}$ is simple, we may save a lot of pain by solving $\widehat{B}$ instead. Devious!

We can bring this idea to life by thinking about the linear Stark effect in hydrogen, as in problem 11.2. Spin and the spin–orbit interaction are ignored for now. Look at the Hamiltonian in (11.7). All three terms commute with $\widehat{l}_z$: the first two because they are spherically symmetrical, so they commute with every kind of angular momentum; and the third because an electric field $\mathcal{E}$ along the z-direction does not exert any torque about the z-axis.[2] Therefore $\widehat{H}$ commutes with $\widehat{l}_z$: $\left[\widehat{l}_z, \widehat{H}\right] = 0$. We want eigenfunctions of $\widehat{H}$; find eigenfunctions of $\widehat{l}_z$.

There are four wave functions available having $n = 2$. The wave functions are already eigenfunctions of $\widehat{l}_z$, but we remember that linear combinations may be required. However, *we can linearly combine only wave functions having the same l_z.* We can list at once[3]

$$\left.\begin{array}{l} \psi_{211} \\[4pt] \psi_{21-1} \\[4pt] (\text{coeff } a_3)\,\psi_{210} + (\text{coeff } a_4)\,\psi_{200} \quad (\text{two combinations}). \end{array}\right\} \quad (11.1)$$

The first two of these cannot be linearly combined with anything, so they are eigenfunctions of $\widehat{H}$ by inspection. We do not need to work this out by evaluating a great collection of matrix elements, as in problem 11.2 parts (3) and (4), all of which turn out to be zero. The remaining two wave functions can be combined, but from there on the problem involves 2 simultaneous equations (for a_3, a_4), not 4. We've achieved a great simplification, exploited in problem 11.2.

[2] You can, of course, verify these statements by brute-force algebra.

[3] The three subscripts give:

- the principal quantum number n, here $n = 2$;
- the orbital-angular-momentum quantum number l, here 1 or 0;
- the z-component of orbital angular momentum ($\div\hbar$).

In case it isn't obvious, let me spell out the reasoning in more detail. We might think of forming a linear combination of, say, ψ_{211} with ψ_{210}. This combination would no longer be an eigenstate of $\widehat{l}_z$, being a mix of $l_z = 1$ with $l_z = 0$. Measurements of l_z would yield 1 sometimes and 0 sometimes; an advantage would have been thrown away.

By contrast, the two linear combinations printed are made from building blocks both of which have $l_z = 0$. In spite of our taking the linear combinations, we still have eigenstates of l_z. The outcome of measurement is $l_z = 0$ with probability $|a_3|^2$ plus $l_z = 0$ with probability $|a_4|^2$, that is, $l_z = 0$ always.

11.2.2 Prefabricating wave functions

Let me change the emphasis a little. Imagine we're going to solve problem 11.2 afresh, this time with the full benefit of hindsight. To find: the consequences of applying the perturbation $\widehat{H'} = e\mathcal{E}_z\,\widehat{z}$. We've realised that the solutions are eigenfunctions of $\widehat{l}_z$ as well as of $\widehat{H'}$, and that $\widehat{l}_z$ is easier to handle. So we start by building eigenfunctions of $\widehat{l}_z$ as in (11.1). We may do this *before* attempting to solve for the energies[4] by applying perturbation $\widehat{H'}$. That is, we *prefabricate* wave functions that are eigenfunctions of the simpler operator, here $\widehat{l}_z$. This greatly simplifies the perturbation-theory calculation that follows. There may be some coefficients still to be found, or we may be lucky and find wave functions in full from the start (in (11.1) both happen).

Prefabrication can be used in all kinds of problem. But the biggest payoff usually comes when there is degeneracy, just because the cost of *not* exploiting prefabrication is then so high. We have seen in problem 11.2, parts (2)–(4), how degenerate perturbation theory can be messy if done by brute force.

Let's spell out how the formalism of degenerate perturbation theory can (in fortunate cases) be bypassed. Imagine that we have somehow obtained the correct linear-combination wave functions for the $n = 2$ levels of hydrogen:[5]

$$\psi_+ = \frac{\psi_{210} + \psi_{200}}{\sqrt{2}}, \qquad \psi_- = \frac{\psi_{210} - \psi_{200}}{\sqrt{2}}.$$

The energy shifts work out to be

$$\int \psi_+^*(e\mathcal{E}_z\,z)\,\psi_+\,\mathrm{d}V = -3e\,a_0\,\mathcal{E}_z, \qquad \int \psi_-^*(e\mathcal{E}_z\,z)\psi_-\,\mathrm{d}V = +3e\,a_0\,\mathcal{E}_z.$$

The shapes of these integrals display an important property: the calculation of energies proceeds exactly as it would if there were no degeneracy. When prefabrication is available and exploited, the saving of labour can be very great.[6]

The situation with ψ_{211} and ψ_{21-1} is even simpler, because linear combinations are not needed at all.

11.3 Finding commuting operators

The theorem quoted at the beginning of this chapter does not help unless we have insights that can point us towards operators that commute. Part of the answer is the symmetry that is the name of the chapter. Symmetry is made formal in § 11.4, but we give one more introductory discussion first.

[4] Though we do need to know what $\widehat{H'}$ is coming up in order to see what simpler operator might be exploited to provide a set of helpful eigenfunctions.

[5] In this particular case, there does not seem to be a short-cut route to these wave functions, so my example is rather artificial. Teaching examples often yield wave functions where the expansion coefficients are equal, or equal apart from a sign, and that is the case here. However, there is no rule that this has to happen, so the wave functions written have nothing obvious about them.

[6] I confess again that this example is over-optimistic because we do not have an "eyeball" route to ψ_+ and ψ_-. However, even without it, we have reduced the problem from solving four simultaneous equations to solving two.

Example 11.1 Linear momentum

Consider a particle moving in one dimension x. We suppose that the potential $V(x)$ is constant, so the Hamiltonian $\widehat{H}$ is independent of x.

This means that $\partial \widehat{H} / \partial x = 0$. Then

$$\left[\frac{\partial}{\partial x}, \widehat{H} \right] \psi = \frac{\partial}{\partial x} \left(\widehat{H} \psi \right) - \widehat{H} \frac{\partial \psi}{\partial x} = \frac{\partial \widehat{H}}{\partial x} \psi = 0 \psi$$

for any ψ, so $\partial / \partial x$ commutes with $\widehat{H}$. Equivalently we have

$$\widehat{H}(x + \delta x) = \widehat{H}(x) \qquad \text{and} \qquad \left[\widehat{p_x}, \widehat{H} \right] = 0, \qquad (11.2)$$

since $\widehat{p_x} = -i\hbar \partial / \partial x$ is the operator for x-component momentum.[7]

It is shown in quantum-mechanics textbooks that *mean values obey classical equations*. Two equations exemplifying this are:[8]

$$\frac{\mathrm{d}\langle p_x \rangle}{\mathrm{d}t} = -\left\langle \frac{\partial V}{\partial x} \right\rangle = -\left\langle \frac{\partial \widehat{H}}{\partial x} \right\rangle$$

$$= \frac{1}{i\hbar} \left\langle \left[\widehat{p_x}, \widehat{H} \right] \right\rangle. \qquad (11.3)$$

The first of these statements says that momentum changes if there is a gradient of potential energy V, and conversely momentum p_x is conserved if the region is field-free; the second says that the conservation of $\langle p_x \rangle$ happens when $\widehat{p_x}$ commutes with the Hamiltonian.

Equations (11.2) and (11.3) establish an intimate connection between:

- a symmetry: the operation $x \to x + \delta x$ leaves $\widehat{H}$ looking the same[9]
- a zero commutator $\left[\widehat{p_x}, \widehat{H} \right] = 0$
- a conservation law inherited from classical mechanics: in zero field $\langle p_x \rangle$ is independent of time
- eigenstates of zero-field $\widehat{H}$ are eigenstates of $\widehat{p_x}$, or can be so chosen.

Your favourite book on quantum mechanics will tell you that (11.3) is a special case of a more general statement. For any B:

$$\frac{\mathrm{d}\langle B \rangle}{\mathrm{d}t} = \frac{1}{i\hbar} \left\langle \left[\widehat{B}, \widehat{H} \right] \right\rangle.$$

Any quantity B that is conserved classically has an operator $\widehat{B}$ that commutes with the Hamiltonian: $\left[\widehat{B}, \widehat{H} \right] = 0$. Eigenstates of $\widehat{H}$ are also eigenstates of $\widehat{B}$, or can be chosen to be.

Look back to the Stark effect in hydrogen. The electric field $\mathcal{E}_z$ permits angular momentum l_z to be "conserved classically", and this gives us a new way of understanding why we had $\left[\widehat{l_z}, \widehat{H} \right] = 0$. I made use of this informally when I said that "$\mathcal{E}$ does not exert any torque about the z-axis".

(At least some) operators that commute with the Hamiltonian are not so very hard to find.

11.4 Symmetry

The word **symmetry** appears in this section heading and in the heading to the whole chapter, yet the relevance of symmetry may so far be only

---

[7] In case the first statement of (11.2) is not obvious, write a Taylor series

$$\widehat{H}(x + \delta x) = \widehat{H}(x) + \delta x \frac{\partial}{\partial x} \widehat{H} + \dots$$

If $\partial \widehat{H} / \partial x = 0$ then $\widehat{H}(x + \delta x) = \widehat{H}(x)$: $\widehat{H}$ is unaffected by a shift through δx.

[8] In classical mechanics, Newton's laws can be cast into the form of Hamilton's equations:

$$\frac{\partial p_x}{\partial t} = -\frac{\partial H}{\partial x}, \qquad \frac{\partial x}{\partial t} = \frac{\partial H}{\partial p_x}.$$

A quantum analogue of the first of these is on display in the text. The second Hamilton equation can be compared with the statements in elementary quantum mechanics that the velocity of a particle is the group velocity of its ψ-wave: $\partial H / \partial p_x$ is close to $\partial \omega / \partial k_x$.

The second formulation in the text, involving the commutator, also has a classical analogue, here the Poisson-brackets equation of motion.

There is no need to be conversant with these topics in classical mechanics in order to understand the present chapter.

[9] The form of words here looks ahead to § 11.4. It has also been encountered in § 8.5.

dimly apparent. To show why symmetry is helpful, I need the formal definition:[10]

> *A thing is symmetrical if there is something you can do to it that leaves it looking the same.*

This is a beautiful definition: short; clear; precise; using only everyday words; giving exactly the required emphasis. The emphasis is on the "something you can do": a **symmetry operation**. The system is said to be "symmetric under the ... operation".

Whenever we state that there is a symmetry, we should[11] specify at the same time the operation that "leaves the system looking the same".

11.4.1   Examples of symmetries

1. A blank sheet of paper has left–right symmetry: the operation of mirror-imaging it about a plane bisecting the paper from top to bottom leaves it looking the same.
2. A particle moving in a field-free region has $\partial \widehat{H}/\partial x = 0$. This implies $\widehat{H}(x+\delta x) = \widehat{H}(x)$. That is, $\widehat{H}(x)$ has symmetry under the operation $x \to x + \delta x$; and the commutator $\left[\widehat{p_x}, \widehat{H}\right] = 0$.
3. A hydrogen 2p wave function has odd parity, that is, it has symmetry under the combined operation

$$x \to -x; \quad y \to -y, \quad z \to -z, \quad \text{multiply by } -1.$$

We may define a parity operator $\widehat{P}$ that reverses the signs of x, y, z all at once. In zero external field (but only then) the Hamiltonian is unaffected if we operate on it with $\widehat{P}$ so we have the commutator $\left[\widehat{P}, \widehat{H}\right] = 0$. Eigenstates of $\widehat{H}$ are eigenstates of $\widehat{P}$, or can be chosen to be. The eigenvalues P of $\widehat{P}$ are $+1$ and -1; for a hydrogen atom in zero field $P = (-1)^l$.

Look again at problem 11.2, and you'll see that

- when taking linear combinations, you mixed states having the same l_z only, those with $m_l = 0$
- you mixed states having different parity (2s and 2p), so the final states are not eigenstates of parity.
- the mixing of states having different parity took place because the perturbing Hamiltonian $e\mathcal{E}_z z$ is an odd-parity operator.

4. A hydrogen atom, field-free or in an electric field $\mathcal{E}_z$, has $\widehat{H}$ independent of azimuth angle ϕ. Then $\widehat{H}$ has symmetry under the operation $\phi \to \phi + \delta\phi$; and the commutator $\left[\widehat{l_z}, \widehat{H}\right] = 0$.
5. A hydrogen atom in no external field has a Hamiltonian independent of the polar angles θ, ϕ, so it has conservation of its *total* angular momentum (orbit and spin taken together) $\boldsymbol{j} = \boldsymbol{l} + \boldsymbol{s}$, even when there is a spin–orbit interaction linking orbital angular momentum $\boldsymbol{l}$ to spin $\boldsymbol{s}$. There is still symmetry under the operation $\phi \to \phi + \delta\phi$ so quantum states are eigenstates of $\widehat{j_z}$, or can be chosen to be. And the commutator $\left[\widehat{j_z}, \widehat{H}\right] = 0$.

6. Each wave function of (11.1) is symmetric under one of the following symmetry operations (which?):

   - rotate about the z-axis through angle ϕ_0; multiply by $e^{-i\phi_0}$
   - rotate about the z-axis through angle ϕ_0; multiply by $e^{+i\phi_0}$
   - rotate about the z-axis through angle ϕ_0.

7. The Hamiltonian for electrons in a helium atom (electrons labelled $1, 2$ with locations $\boldsymbol{r}_1, \boldsymbol{r}_2$) is

$$\widehat{H} = -\frac{\hbar^2}{2m_e}\nabla_1^2 - \frac{\hbar^2}{2m_e}\nabla_2^2 - \frac{Ze^2}{4\pi\epsilon_0\, r_1} - \frac{Ze^2}{4\pi\epsilon_0\, r_2} + \frac{e^2}{4\pi\epsilon_0\, r_{12}} \quad (11.4)$$

(plus terms tiny compared with these). This has a number of symmetries, but I'll point out one for now: it is symmetric under the operation

$$\widehat{I}_r : \quad (\text{location } \boldsymbol{r}_1) \longleftrightarrow (\text{location } \boldsymbol{r}_2). \quad (11.5)$$

8. Several symmetries, and symmetry operations, have been encountered in Chapter 8. A Lorentz transformation must leave the basic physics of a system unaltered, so it is a symmetry operation.

Comment: You might feel: this has taken us a long way from the shopkeeper's symmetry: $\partial\widehat{H}/\partial x = 0$ is not something I'd have thought of as a symmetry before now. Well, yes, that's right. But that is why we introduced the formal definition of symmetry. $\widehat{H}(x + \delta x) = \widehat{H}(x)$ is exactly in line with the definition: there is something $(x \to x + \delta x)$ you can do to $\widehat{H}$ that leaves it looking the same.

11.4.2 Example: spin–orbit interaction in hydrogen

Think about the 2p level of a hydrogen atom, including spin but at first with spin–orbit interaction neglected. Let the electron states for the 2p level be represented by functions $|m_l, m_s\rangle$, where m_l and m_s are the quantum numbers for the z-components of orbital angular momentum $\boldsymbol{l}$ and spin angular momentum $\boldsymbol{s}$. For the 2p level there are three possible values for m_l and two for m_s, making six wave functions in all.

Now apply the spin–orbit interaction, previously omitted, as a perturbation. The perturbing Hamiltonian $\widehat{H}'$ contains $\hat{l}\cdot\hat{s}$. Example 5 above shows that $\widehat{H}'$ commutes with $\widehat{j}_z$, so we know that we are going to construct wave functions that are eigenstates of $\widehat{j}_z$. The sensible thing is to *prefabricate* these linear combinations, to make subsequent work simple. The combinations are:

$$j_z = \tfrac{3}{2}\hbar : \quad |1, \tfrac{1}{2}\rangle$$
$$j_z = \tfrac{1}{2}\hbar : \quad a|0, \tfrac{1}{2}\rangle + b|1, -\tfrac{1}{2}\rangle \quad (\text{two possibilities})$$
$$j_z = -\tfrac{1}{2}\hbar : \quad \alpha|0, -\tfrac{1}{2}\rangle + \beta|-1, \tfrac{1}{2}\rangle \quad (\text{two possibilities})$$
$$j_z = -\tfrac{3}{2}\hbar : \quad |-1, -\tfrac{1}{2}\rangle.$$

What has happened here is very similar to what we found with the Stark effect. Two wave functions have not been mixed into linear combinations at all, and they give their part of the answer at once. And for the rest there are two separate linear combinations of two wave functions only. Instead of having 6 simultaneous equations for coefficients, we have at worst 2: solving a quadratic.

Comment: The case of spin–orbit interaction is a little unusual, in that there exists a short cut for finding the energies. We write

$$l \cdot s = \frac{j^2 - l^2 - s^2}{2}$$

and use the fact that the states sought are eigenstates of all three operators on the right. If we wish to go on from here to obtain the six new wave functions (the values of a, b and α, β), then the symmetry property has given us a very useful start, by showing that the problem can be solved in pieces, and no part is worse than solving a quadratic.

11.5   Symmetries of helium

It's hard to work out the wave functions and energies for helium (Chapter 15), so we ought to exploit symmetry to the full: we need all the help we can get. Fortunately, the Hamiltonian (11.4) for helium has a very rich set of symmetries.

1. The parity operator $\widehat{P}$ for helium reverses the signs of all six of $x_1, y_1, z_1, x_2, y_2, z_2$. It is not hard to see that $\widehat{P}$ commutes with $\widehat{H}$.

2. We've already seen in (11.5) that $\widehat{H}$ commutes with $\widehat{I}_r$, the operator that interchanges r_1 and r_2. At that stage, spin had not been mentioned, so $\widehat{I}_r$ should be understood as interchanging space coordinates only. Now invent also a spin interchange operator $\widehat{I}_s$ which interchanges the labels attached to the electron spins (leaving the space coordinates alone). A complete interchange of labels on the two electrons is performed by $\widehat{I} = \widehat{I}_r \widehat{I}_s = \widehat{I}_s \widehat{I}_r$. We may easily show that $\widehat{H}$ commutes with all three of $\widehat{I}_r$, $\widehat{I}_s$ and $\widehat{I}_r \widehat{I}_s$, and that they commute with each other.

3. Argue for yourself that $\widehat{I}_r^{\,2} \psi = \psi$, and this is identically true where ψ is any two-particle wave function. Hence the eigenvalues of $\widehat{I}_r$ are ± 1. A similar property applies to $\widehat{I}_s$ and $\widehat{I} = \widehat{I}_r \widehat{I}_s$.

4. The two electrons in helium are identical particles.[12] Their space coordinates and spin *must* appear in $\widehat{H}$ in the same way for each. This means that any imaginable $\widehat{H}$ for two identical particles must satisfy

$$\widehat{I}\,\widehat{H} = \widehat{H}, \qquad \text{or equivalently} \qquad [\widehat{I}, \widehat{H}] = 0.$$

Identical particles *always* have wave functions that are eigenstates of $\widehat{I}$; bosons have eigenvalue $+1$; fermions have eigenvalue -1. Electrons are fermions, so for them the only permitted eigenvalue of $\widehat{I}$ is -1. A multi-electron wave function is *antisymmetric* under interchange of all of one electron's labels with those of any other.

5. Think about the spin of the electrons in helium. Spin is not mentioned in the Hamiltonian (11.4) at all, so $\widehat{H}$ commutes with every spin operator you can invent.[13] With malice aforethought I'll suggest

[12] In Chapter 14 we explore what would happen if the electrons were not identical. This is done in order to show that several phenomena are *not* caused by the indistinguishability of the electrons, even though it is often claimed that they are. However, we know that electrons *are* identical, and it's appropriate to use that fact here.

[13] In helium there are tiny effects from spin, so tiny that they don't concern us while are still trying to get the electrostatics right. This means in particular that we ignore spin–orbit interaction.

we concentrate on

$$\left[\widehat{\boldsymbol{S}^2}, \widehat{H}\right] = 0; \qquad \left[\widehat{S_z}, \widehat{H}\right] = 0, \qquad \text{where} \quad \boldsymbol{S} = \boldsymbol{s}_1 + \boldsymbol{s}_2.$$

Why do I select these operators from among all possibilities? It's not hard to see: $\widehat{\boldsymbol{S}^2}$ and $\widehat{S_z}$ commute with $\widehat{I_s}$. Any operators like $\widehat{s_{1x}}$, that relate to only one electron, do not commute with $\widehat{I_s}$, so they fail to exploit/respect one of the symmetries known to exist for helium.

6. Figure 11.1 shows the Coulomb-repulsive forces acting on two electrons in an atom (treated classically). Each force separately exerts a torque about the nucleus, so neither electron can have conservation of its orbital angular momentum. Yet the two electrons taken together have their totalled orbital angular momentum conserved classically. What is conserved classically is quantized: its operator commutes with the Hamiltonian. This is the case with $\boldsymbol{L} = \boldsymbol{l}_1 + \boldsymbol{l}_2$; therefore in particular[14]

$$\left[\widehat{\boldsymbol{L}^2}, \widehat{H}\right] = 0; \qquad \left[\widehat{L_z}, \widehat{H}\right] = 0, \qquad \text{where} \quad \boldsymbol{L} = \boldsymbol{l}_1 + \boldsymbol{l}_2.$$

Now draw threads together. We want to solve the Schrödinger equation for a lightish[15] atom, perhaps helium; its Hamiltonian is given in (11.4), or an obvious extension. The term(s) involving e^2/r_{12} makes the mathematics difficult. So we'll do everything we can to exploit tricks.

Prefabricate (in imagination if not in reality) wave functions that possess all the known symmetry properties. Only when these wave functions are ready shall we attempt to feed them into a Schrödinger equation. The wave functions we prefabricate will be eigenstates of[16]

parity	eigenvalues $+1$, -1
$\widehat{\boldsymbol{L}^2}, \widehat{L_z}$	eigenvalues $L(L+1)\hbar^2$, $m_L\hbar$
$\widehat{\boldsymbol{S}^2}, \widehat{S_z}$	eigenvalues $S(S+1)\hbar^2$, $m_S\hbar$
$\widehat{I_r}\widehat{I_s}$	eigenvalue must be -1 for electrons
$\widehat{I_r}, \widehat{I_s}$	eigenvalues $+1, -1$ or $-1, +1$ to conform to symmetry under $\widehat{I_r}\widehat{I_s}$.

Not only do all these operators commute with $\widehat{H}$, they have been constructed so that they also commute with each other. Therefore the wave functions sought can be chosen to be simultaneously eigenfunctions of *all* these operators.

There are so many constraints here that we have a really useful ability to "target" our attention onto the right kind of wave function. Not only that, but even before doing any work on the Schrödinger equation, we know most of the quantum numbers that will be possessed by the solution!

Comment: Depending on what you have studied already you may recognize $e^2/(4\pi\epsilon_0 r_{12})$ as an **interaction**, and the constructing of the wave functions as **coupling**:[17] eigenstates of $\widehat{I_r}$ as space-symmetric and space-

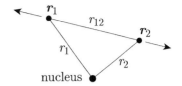

Fig. 11.1 The arrows show the repulsive forces acting on each electron because of the presence of the other. These forces exert torques about the nucleus, so neither electron has conservation of its orbital angular momentum $\boldsymbol{l}$. Nevertheless, the two forces are equal and opposite, so the torque they exert on the entire atom is zero.

[14] Helium is a very special case, because all of its bound states have one electron in a 1s state, contributing nothing to the overall orbital angular momentum. The total orbital angular momentum $\boldsymbol{L}$ is just that of the "other" electron. But the list we have built applies to more atoms than helium, and in other atoms item 6 does have something to say.

[15] The point of "lightish" is that relativistic effects, in particular spin–orbit interaction, can be ignored beside electrostatics. Such effects become more important as the atomic number Z is increased.

[16] Atoms are ingenious in circumventing our expectations. A good example is neon, which we might expect to conform to LS coupling, given that its Z is only 10. However, the excited states of neon have one outer electron that is hardly noticed by the others. The "others" form a $^2\mathrm{P}_{1/2}$ or $^2\mathrm{P}_{3/2}$ "core", and the outer electron is loosely coupled to that core. Overall, the neon atom has neither L nor S as a "good quantum number", but has a classification scheme of its own. The same applies to all noble gases other than helium.

[17] When we prefabricate wave functions, we are executing the *coupling* before, but in anticipation of, applying the *interaction*. I have written this sentence to show that there is a distinction between *interaction* and *coupling*, and to show how the words are correctly used within that distinction.

antisymmetric; eigenstates of $\widehat{I}_s$ as spin-symmetric and spin-antisymmetric; eigenfunctions of the L and S operators as conforming to LS coupling.

Comment: The whole object of applying symmetry is to produce a correct set of wave functions. If these still have degeneracy (and they do) then the choice is not unique. We aim here to find the simplest set, which means here that they are eigenstates of $\widehat{L}_z$ and $\widehat{S}_z$.

At a later stage, we might well introduce the spin–orbit interaction. In preparation for that, we would take the degenerate $|m_L, m_S\rangle$ wave functions and linearly combine them (prefabrication again) into eigenstates of $\widehat{J}^2$ and $\widehat{J}_z$. But it would be silly to anticipate this complication at the stage when we are still trying to get the electrostatics right. This example shows that we may need to be judicious in choosing symmetries when several are available.

Comment: There is a possible confusion because I have mentioned symmetries in two slightly different contexts. The fundamental symmetry of a physical problem is the symmetry whereby the *Hamiltonian* commutes with some other operator, such as parity. A *consequence* may be the possession of a symmetry property by the resulting wave function. If these two kinds of symmetry are getting at all muddled, then you should reread until things are straight.[18]

Further applications of symmetry appear in later chapters of this book, in particular in dealing with atomic structure and in the construction of Bloch functions for electrons in solids (problem 25.3).

[18] An example to point the distinction. A symmetry of the Hamiltonian is $\widehat{I}_r\widehat{H} = (+1)\widehat{H}$. This requires that the wave function ψ obey $\widehat{I}_r\psi = (\pm 1)\psi$. The symmetry of the wave function follows from that of the Hamiltonian, but the rule obeyed by ψ is not the same as the rule obeyed by $\widehat{H}$.

Problems

Problem 11.1 Three-dimensional harmonic oscillator

Consider a particle moving in three dimensions and subject to a potential $V = \frac{1}{2} m\omega^2(x^2 + y^2 + z^2)$.

(1) Write $\psi = X(x)\,Y(y)\,Z(z)$ and aim to solve the (time-independent) Schrödinger equation in separated form.

(2) Show that each of X, Y, Z is a solution for the harmonic oscillator in one dimension.

(3) Hence show that the energy eigenvalues are $E = \hbar\omega\left(\frac{3}{2} + n_1 + n_2 + n_3\right)$ where the n_i are integers $\geqslant 0$ associated with the x, y, z motions.

(4) What are the possible (n_1, n_2, n_3) values that yield energy $\frac{3}{2}\hbar\omega$? Hence work out the degeneracy.[19] Do the same for the energy $\frac{5}{2}\hbar\omega$.

(5) The potential can be written in the form $V(r) = \frac{1}{2}m\omega^2 r^2$, dependent on r only (not on θ or ϕ). It should therefore be possible to solve the problem by writing $\psi = R(r)\,Y_{lm}(\theta, \phi)$. Our particle should have (orbital) angular-momentum quantum numbers[20] l, m_l. From the degeneracies found in part (4), guess values for l appropriate to states having the two energies $\frac{3}{2}\hbar\omega$ and $\frac{5}{2}\hbar\omega$.

[19] The degeneracy is the number of independent quantum states sharing this energy.

[20] Function $Y_{lm}(\theta, \phi)$ is a spherical harmonic.

I recommend you don't actually attempt to re-solve the Schrödinger equation in r, θ, ϕ coordinates, as the algebra is quite sticky. And you don't need to do it: you won't find anything new because our solutions in terms of x, y, z constitute a complete set.

(6) Look up (one-dimensional) harmonic-oscillator wave functions, and write down the wave functions belonging to the states identified in part (4). Convert your wave functions to functions of r, θ, ϕ, and verify the values that you guessed for the angular-momentum quantum numbers l, m_l in part (5).[21]

(7) What are the parities of the states you found in part (4)?[22] Check whether your values of parity agree with $(-1)^l$, with l taking the values found in part (6).[23]

(8) Work out the degeneracies of the states with energy $\frac{7}{2}\hbar\omega$, $\frac{9}{2}\hbar\omega$. Work out also their parities. Use these two pieces of information to make informed guesses of the angular momenta associated with these two energies.

Comment: The methods used in this problem can be applied fluently once you have seen the idea. They have an application to the shell model of nuclear structure. See Enge (1966), Fig. 6–5.

Problem 11.2 The linear Stark effect in hydrogen

Consider a hydrogen atom (ignoring spin), placed in a uniform electric field $\mathcal{E}_z$ along the z-axis. This field causes an additional energy in the Hamiltonian, which to a good approximation is[24]

$$\widehat{H'} = -(\text{dipole moment})\mathcal{E}_z = -(-ez)\mathcal{E}_z = e\mathcal{E}_z\, z, \qquad (11.6)$$

where z is the z-component of the electron's position $\boldsymbol{r}$ relative to an origin at the nucleus.

We write hydrogen-atom wave functions in forms like $|2\,1\,0\rangle$, where the quantum numbers in the ket mean $n = 2$, $l = 1$ and $m_l = 0$.

(1) Show that, to first order in the field strength $\mathcal{E}_z$, the ground-state energy is unchanged[25] by the perturbation, i.e. that $\langle 1\,0\,0|z|1\,0\,0\rangle = 0$.

(2) In the unperturbed problem, the first excited energy, having $n = 2$, is four-fold degenerate: the three 2p states $\psi_{211} = |2\,1\,1\rangle$, $\psi_{210} = |2\,1\,0\rangle$, $\psi_{21-1} = |2\,1\,-1\rangle$ and the single 2s state $\psi_{200} = |2\,0\,0\rangle$ all have the same energy; call that energy E_2. Consider this four-fold degenerate level. The equation we want to solve is the Schrödinger equation, including the perturbation:

$$\left(-\frac{\hbar^2}{2m_e}\nabla^2 - \frac{e^2}{4\pi\epsilon_0\, r} + e\mathcal{E}_z\, z\right)\psi = E\,\psi. \qquad (11.7)$$

We shall try to find a solution that is correct to first order in $\mathcal{E}_z$ by writing ψ as a *linear combination of the degenerate wave functions*:[26]

$$\psi = a_1\,\psi_{211} + a_2\,\psi_{21-1} + a_3\,\psi_{210} + a_4\,\psi_{200}. \qquad (11.8)$$

This will give the lowest-order correction to the energy: other states (non-degenerate with this one, but possibly degenerate with each other) will mix within higher orders only.

We know that the unperturbed E_2 wave functions satisfy

$$\widehat{H_0}\,\psi_{2lm} = \left(-\frac{\hbar^2}{2m_e}\nabla^2 - \frac{e^2}{4\pi\epsilon_0\, r}\right)\psi_{2lm} = E_2\,\psi_{2lm}. \qquad (11.9)$$

[21] You may have to take linear combinations.

[22] Use the wave functions of part (6) and work out what happens to X, Y, Z when you reverse the signs of x, y, z.

[23] The parity exists and is $(-1)^l$ for any *radial* potential. In fact, the potential doesn't even have to be radial: all it needs is to possess a centre of symmetry so that $V(\boldsymbol{r}) = V(-\boldsymbol{r})$; this is enough to permit the parity operator to commute with the Hamiltonian. If you first encountered parity in connection with hydrogen, recognize now that it exists more generally.

[24] The electron has charge $-e$ and mass m_e.

[25] Here and later, we do not want the integrations to be burdensome, so look for the quickest valid route. Suggestion for this integral: parity.

[26] This procedure means that we are following the standard no-tricks route into degenerate perturbation theory.

130 Symmetry in quantum mechanics

Substitute (11.8) into (11.7) and use (11.9) to show that

$$e\mathcal{E}_z\, z\Big(a_1\,\psi_{211} + a_2\,\psi_{21-1} + a_3\,\psi_{210} + a_4\,\psi_{200}\Big)$$
$$= \big(E - E_2\big)\Big(a_1\,\psi_{211} + a_2\,\psi_{21-1} + a_3\,\psi_{210} + a_4\,\psi_{200}\Big). \qquad (11.10)$$

(3) It should be obvious what to do with eqn (11.10): multiply it by things like ψ_{211}^* and integrate over all space. We are going to need some matrix elements, and it's tidy to work them out first. Show that[27]

$$\langle 211|z|211\rangle = \langle 211|z|21-1\rangle = \langle 211|z|210\rangle = \langle 211|z|200\rangle = 0.$$

Now multiply (11.10) by ψ_{211}^* and integrate. Show that one eigenstate, to this order in $\mathcal{E}_z$, is still ψ_{211} and has energy $E = E_2$. Similarly, multiply (11.10) by ψ_{21-1}^* and show that another eigenstate is ψ_{21-1}, again with energy E_2.

(4) For the next step we need the remaining matrix elements. Show that

$$\langle 210|z|211\rangle = \langle 210|z|210\rangle = \langle 210|z|21-1\rangle = 0,$$

but that

$$\langle 210|z|200\rangle = -3\,a_0,$$

where a_0 is the Bohr radius.[28] Knowing these building blocks, multiply (11.10) by ψ_{210}^* and integrate. You should find

$$-3\,a_0\,e\mathcal{E}_z\,a_4 = (E - E_2)\,a_3.$$

Also multiply (11.10) by ψ_{200}^* and integrate to obtain

$$-3\,a_0\,e\mathcal{E}_z\,a_3 = (E - E_2)\,a_4.$$

(5) Solve the last pair of equations to show that the remaining two energy eigenvalues are (to this order in $\mathcal{E}_z$) given by $E_\pm = E_2 \pm 3\,a_0\,e\mathcal{E}_z$. Thus the original four degenerate states have become a pair at the unshifted energy E_2, together with two others having the separated energies E_+ and E_-.

(6) What are the eigenfunctions associated with the energies E_+, E_-? Are they eigenfunctions of: (a) $\widehat{l}^2$? (b) $\widehat{l}_z$? Discuss.

[27] Suggestions: parity and the integral over ϕ.

[28] Look up the wave functions, and use

$$\int_0^\infty r^n\,e^{-r/a}\,dr = a^{n+1}\,n!,$$

remembering that $z = r\cos\theta$. The sign of $\langle 210|z|200\rangle$ may depend on whose wave functions you look up.

Successive approximation; perturbation theory in quantum mechanics

Intended readership: towards the end of a first course on quantum mechanics.

The three successive-approximation exercises could usefully be worked before the introduction of perturbation theory.

The presentation of perturbation theory here is a little unusual, and is best seen shortly after a first encounter with a standard-textbook treatment.

12.1   Introduction

There is a presentation of perturbation theory that seems to have become the standard way of doing things. In it, we consider the effect of applying a Hamiltonian

$$\widehat{H} = \widehat{H}_0 + \lambda \widehat{H}', \tag{12.1}$$

in which $\widehat{H}_0$ is the Hamiltonian of the unperturbed system and $\widehat{H}'$ is an added "perturbing" small piece that has the effect of slightly changing energies and wave functions. The coefficient λ is used as a label for sorting contributions into different orders, and is set to 1 once this has been done.

The logic of this "standard route" is faultless. Nevertheless I have never been happy with it. Symbol λ is not needed for sorting terms of different order, as we can do exactly the same thing by counting powers of $\widehat{H}'$. Moreover, by "automating" the process of picking out terms, we risk giving ourselves licence to ignore signals that are trying to tell us to take a more intelligent approach.

Perturbation theory is at heart a successive-approximation procedure: an iteration. In computer language, it's a DO loop. It helps to recognize this structure—surely familiar to every physicist these days when computers are part of the furniture.

It's a good idea to walk before running. My experience with teaching this material is that students have often not encountered successive approximation in easier contexts, so it's hardly surprising if they find perturbation theory a bit of a shock. This chapter starts with three introductory exercises that are intended to supply the deficiency.

Essays in Physics: Thirty-two thoughtful essays on topics in undergraduate-level physics. Geoffrey Brooker, Oxford University Press. © Geoffrey Brooker 2021. DOI: 10.1093/oso/9780198857242.003.0012

12.2 Solution of equations by successive approximation

$$x = 0.1 + x^3$$

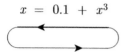

Fig. 12.1 The information flow in the solving of eqn (12.2). A rough x fed into the right-hand side is used (heavy line) to calculate a new x. We carry this x back (thin line) to the right and do it again.

[1] If you've done any computer programming at all, you'll recognize what's happening here as a DO loop. Just how that loop is programmed depends on the computer language you're familiar with (it might be called a FOR loop or a WHILE loop), but the following (in Mathematica language) should be comprehensible:

```
Do [ x = 0.1 + x^3, { n,10 } ]
```

This uses n to count the number of iterations, starting at n = 1 and continuing until n = 10. On each passage round the loop, the command x = 0.1+x^3 is executed, thereby replacing x with an improved value.

Every iterative procedure—every successive-approximation recipe—has the structure of a DO loop: this cross reference to a familiar procedure may help us to see where to go next and why. By contrast, the "λ method" of (12.1) tends to conceal the similarity to a DO loop.

[2] Suppose you start with $x_0 = 0.1$. Then $x_1 = 0.101$. In the next step you need to calculate x_1^3. Because x_1 is close to a power of 10, you can write $x_1^3 = 10^{-3}(1 + 0.01)^3$ and do a binomial expansion. Given that x_1 is good to the third decimal place, and things improve by about 2 significant figures at each iterative step, the best you can hope for is 5-figure accuracy in x_2. This can be achieved by keeping only two figures in the bracket, and for that mental arithmetic is more than adequate. When finding x_3, you can play the same game again, though you now need to retain more figures; mental arithmetic is still up to the job—just.

Exercise 12.1 $x - x^3 = 0.1$: a longish exercise, with a surprising amount in it

Consider the equation $x - x^3 = 0.1$. We wish to find a solution for x close to $x = 0.1$, correct to 7 significant figures, and without using a computer or programmable calculator (an ordinary calculator is permitted except where stated). I'll lead you through a procedure that may not be obvious at first.

Rewrite the equation in the form

$$x = 0.1 + x^3. \tag{12.2}$$

If x is about 0.1, x^3 is about 0.001, so it contributes only about 1% to the right-hand side. Equation (12.2) is *almost* an explicit equation for x. With this knowledge, we can invent a routine for solving the equation as follows (Fig. 12.1): Make a guess at x, say $x = x_0$, and then work out $x_1 = 0.1 + x_0^3$. The x_1 that results from this is still not a solution to (12.2), but it ought to be closer to it than was x_0. Continuing the process,[1] we may **iterate** (i.e. perform repeated similar operations) as follows:

$$x_1 = 0.1 + x_0^3$$
$$x_2 = 0.1 + x_1^3$$
$$x_3 = 0.1 + x_2^3$$
$$\vdots \tag{12.3}$$

The numbers x_n that we produce by this method should converge towards the solution of (12.2), so that the solution is $\lim_{n\to\infty} x_n$.

(1) Before undertaking to calculate x by method (12.3), convince yourself that there is nothing critical about the initial guess x_0 at the solution. Do this by feeding in a few trial values of x_0 and seeing what values of x_1 you get. You should find that there is a range of values of x_0, all of which yield an x_1 that is nearer to the right answer. Find *roughly* the extent of this range. (Of course, the farther x_0 is from the right answer, the more steps in the iteration would be needed to achieve a given accuracy if you started from there.)

(2) Choose a suitable starting value for x_0, and solve eqn (12.2) to 7 significant figures. There is no need to use a calculator here, as you can find a pencil-and-paper procedure that isn't unduly laborious.[2] Nevertheless I'll assume you prefer to use a calculator.

(3) How much improvement did each iterative step give? Find values for things like $(x_2 - x_1)/(x_1 - x_0)$, just by subtracting and dividing with the numbers you obtained. Is the ratio (new change)/(old change) roughly the same for all steps in the iteration?

(4) Use eqns (12.3) to find a tidy expression for $(x_2 - x_1)/(x_1 - x_0)$. Find its numerical value. Does it agree with what you found by "experimental arithmetic" in part (3)?

(5) Sketch the graph of

$$y(x) = x - x^3 - 0.1$$

versus x. Find out what goes on—in terms of the geometry of the graph—when you make a step in the iteration. You should begin to understand why and how the whole process works.

(6) Use your newly found insight to consider again what is the range of values (let me call it the **capture range** of the method) within which we must place x_0 if the iterative method is to converge to the required solution. Your new findings should agree (within claimed accuracy) with those in part (1), but should be more definite.[3]

(7) Over what range of initial guesses (i.e. within what capture range) would it be safe to use Newton's approximation?[4] You should find that my iteration has a larger capture range than Newton's.

Comment: Capture range is an important consideration when we solve equations on a computer. In a big calculation, you cannot afford to watch the behaviour of the numbers as the iteration proceeds, so you have no means of knowing (in "real time") whether all is going according to plan. It is far more important to use a method that is guaranteed to get to the right answer—eventually—than to minimize the number of steps performed.

(8) Consider the region of the graph close to the solution (the solution near $x = 0.1$). Consider the geometrical interpretation of my iteration, and of Newton's. You should see a new reason why $(x_2 - x_1)/(x_1 - x_0)$ took the value found in parts (3) and (4). My method approaches the solution slower than Newton's: a price we pay for the larger capture range.[5]

(9) Rearrange eqn (12.2) into a form[6] suitable for finding the solution near $x = 1$. Find the solution to 3 significant figures.

(10) The methods you have been discovering work just as well for doing algebra as for doing arithmetic. Replace eqn (12.2) by

$$x = \varepsilon + x^3,$$

where ε is "small", and the solution sought for x is not very different from ε. Use a method[7] exactly mimicking that of part (2) to solve for x as a power series in ε, keeping terms up to ε^7.

Solution to exercise 12.1

(1) The rough capture range is from -1 to 1.

(2) The obvious starting point is $x_0 = 0.1$. Punching numbers into a calculator quickly gives x_1; hold this number in memory to be used as the start for a second calculation, and so on. We reach the required number of significant figures with $x_4 = 0.101\,031\,26$.

[3] Parts (6) to (8) can be answered quickly by using a judicious mixture of algebra and geometry on the graph of part (5). They are propaganda for looking at a problem, even a computational problem, in all available ways.

[4] The Newton capture range is *not* delimited by the turning values of the graph. Why not? Working out the precise range is messy, and isn't required. But you should be able to explain how it could be found.

[5] Part (8) should trigger inventiveness: couldn't we hybridize Brooker and Newton by inventing the improved iteration

$x_1 = x_0 - y(x_0)/(\text{the fixed value } 0.97)$.

Yes, it's a good idea. Indeed, I often do things like this if I have to solve an implicit equation numerically. The "economic choice" is to improve the gradient (in the above from 1 to 0.97) just once; any second improvement costs more in labour to produce than it saves in subsequent iterative steps.

[6] There is no single right answer to this, but some are much more sensible than others.

[7] Be observant as to how many powers of ε should be retained at each step. This is really good experience for understanding what to keep at each stage of an iteration.

(3) I'll assume that you solved for x using a calculator, that you wrote down the outcome of each step to as many figures as your calculator gives, and that each x_n was used unaltered as input for the next step. You should find that the ratio of successive changes is about $1/30$. This pattern may not have shown up too clearly at the beginning if your starting value x_0 was too far from 0.1; and it will have gone to pieces when the changes became so small as to be dominated by rounding errors.

(4) The tidy expression is

$$\frac{x_2 - x_1}{x_1 - x_0} = \frac{(0.1 + x_1^3) - (0.1 + x_0^3)}{x_1 - x_0} = \frac{x_1^3 - x_0^3}{x_1 - x_0} = x_1^2 + x_1 x_0 + x_0^2$$

$$\approx (0.1)^2 + (0.1)(0.1) + (0.1)^2 = 0.03.$$

It should be obvious that the numerical result here is not critically dependent on its inputs,[8] so it will hold so long as x_0 is reasonably close to 0.1. The behaviour found in part (3) is accounted for.

(5) We can write $(x_1 - x_0) = -y(x_0)$. This has a simple geometrical interpretation. At x_0 draw a line from the x-axis to the curve; its length is $y(x_0)$. We find x_1 by measuring this same length, leftwards (if $y > 0$) from x_0. Equivalently: we move vertically from x-axis to curve; then we move back to the axis along a line of slope $+1$. This is shown in Fig. 12.2.

[8] This is in contrast to almost any other algebraic expression that we might construct from the same starting point. If we feed numbers into an expression that's at all complicated, or subtract nearly equal quantities, we can have no feel for how the result might change with changed inputs.

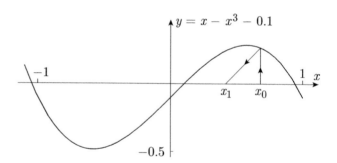

Fig. 12.2 A graph of the function $y(x) = x - x^3 - 0.1$. Also shown is a possible location for our iteration's start x_0 and the way in which the iteration proceeds from there to fix the location of x_1. Make your own addition to the diagram to show what happens if x_0 is located so as to make $y(x_0)$ negative.

(6) It's now at once obvious that the capture range of the iteration is delimited by the "outer" roots, one just left of $+1$, the other just left of -1.

(7) In Newton's approximation we move from the x-axis to the curve, then back to the axis along a tangent to the curve. If we start close to a turning value, the tangent meets the axis so far away that we're lost. We need successive steps to spiral in towards the root (meaning for now the root near $x = 0.1$). So the capture range is delimited by the regime where the process spirals neither in nor out, tracing a closed path. That path is shown in Fig. 12.3. Iteration (12.3) has a bigger capture range than Newton: the x_0 shown in Fig. 12.2 is acceptable for our iteration, while it would cause Newton to go in the wrong direction. The actual ratio of capture ranges is shown in Fig. 12.3.[9]

[9] The capture range for Newton is uncomfortably small. We've learned something that is well known: Newton's method is notoriously precarious when applied to a graph with an S-bend.

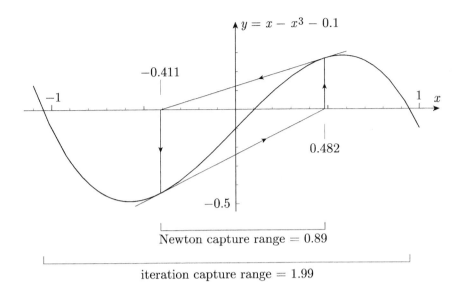

Fig. 12.3 The capture range of Newton's approximation is determined by finding the condition for his iteration to trace repeatedly a closed path. Here we show that figure, and compare the resulting capture range with that of our iterative procedure. The ratio of capture ranges is 2.23.

(8) The gradient of the graph is $y' = 1 - 3x^2$. Close to $x = 0.1$ this takes the value $1 - 3(0.1)^2 = 0.97$. Taking our cue from Newton, we can see that the best way to approach the root is to go (from the curve back to the axis) along a line of gradient 0.97. We have been using a line of slope 1: too steep by 3%, resulting in an undershoot, in which each movement is 3% too small and must be corrected in the next step. This explains why each change was found in part (3) to be 1/30 of the previous change. It also explains the modified iteration suggested in sidenote 5.

(9) I give a menu of ideas, some good, some poor; no claim to complete-ness.

(a) $x = (x - 0.1)^{1/3}$. This is perhaps the first thing we'd think of, since it involves taking x from the other term containing it in eqn (12.2). It's not a very good idea, because evaluation of the cube root is labour-intensive, even inside a computer, compared with the other ideas in our list.

(b) Opportunism. Reshape eqn (12.2) as $x(1 - x)(1 + x) = 0.1$, and extract $(1 - x)$ to get

$$x = 1 - \frac{0.1}{x(1 + x)}.$$

A rough x fed into the right-hand side of this yields a better x: the basis for an iteration that works in the same way as (12.3).

(c) Newton. This is sensible now, since the capture range extends from the right-hand turning value at $x = 1/\sqrt{3}$ all to the way to $+\infty$.

(d) We can copy the logic of part (5), returning to the x-axis using lines whose slope is $y' = 1 - 3x^2 \approx 1 - 3(1)^2 = -2$. This means $x_1 = x_0 - y(x_0)/(-2)$. The capture range has a left-hand limit at the root near $x = 0.1$. There is a right-hand limit as well, because a first step must not yield an x_1 that is too far to the left.

(e) I've left this idea until last, for emphasis, because it's the most forward-looking. One thing that made the iteration (12.3) user-friendly was the presence in the equation of a small quantity. If we have a small quantity available, we badly miss a trick if we don't exploit it. We have a small quantity available here because x is known to be close to 1: the small quantity is $z = x - 1$. So substitute this into our original equation to obtain

$$z = -\frac{0.1 + 3z^2 + z^3}{2}. \qquad (12.4)$$

This equation has been shaped up ready for an iteration.[10] Feed a rough $z = z_0$ into the expression on the right, and the result will be a z_1 that is closer to the required root.

(10) If this iteration is to be kept tame we must know what powers of ε must be kept and what discarded at each step. I'll write out what results from doing things unintelligently.

$$x_0 = 0$$
$$x_1 = \underline{\varepsilon}$$
$$x_2 = \varepsilon + \underline{\varepsilon^3}$$
$$x_3 = \varepsilon + \varepsilon^3 + \underline{3\varepsilon^5} + 3\varepsilon^7 + \varepsilon^9$$
$$x_4 = \varepsilon + \varepsilon^3 + 3\varepsilon^5 + \underline{12\varepsilon^7} + 28\varepsilon^9 + \cdots$$
$$x_5 = \varepsilon + \varepsilon^3 + 3\varepsilon^5 + 12\varepsilon^7 + \underline{55\varepsilon^9} + \cdots . \qquad (12.5)$$

Underlining highlights the left-most term that has changed from the previous line.[11] Notice that underlined terms do not change in subsequent iterations, so they identify where each power of ε is got right. There is a clear pattern: we gain one power of ε^2 in each iterative step.

Any term to the right of an underlined term makes no contribution to the reliable terms in any subsequent step; it is a complete waste of space, and of effort in the evaluation of it.[12] Therefore, in each step

we should keep just enough terms to give us the
new expression as far as its underlined term.

Keeping any more terms than these is worse than just a waste of time: surplus terms clutter the working and make mistakes more likely. Because the underlined terms follow a strict pattern, we know where they will lie and therefore what terms must be kept.

In any iteration, one of our first tasks is to identify this kind of pattern. It is a pretty good rule of thumb that you gain one order (it's usually clear what that means)—don't expect more—with each iterative step.

---

[10] "Shaped" means that we've put on the right the things that are known and things that are small—meaning small compared with the z on the left. We have been doing this "shaping" as far back as (12.2). You can already see there a right-hand side that contains "the known and the smaller".

[11] It's instructive to follow the progress of our original numerical iteration in part (2), underlining each decimal digit as it acquires its final value, and looking at how the underlines behave.

[12] The above list was prepared "unintelligently" because we now see that useless terms are included in the expressions for x_3 and x_4. It may be hard to force yourself to discard a term of order ε^7 in x_3, but you should. You won't get the ε^7 term right in this iterative step whatever you do. Don't worry, it'll come right in the next step, when you find x_4.

12.2.1 Recipe for iteration

Experience with successive-approximation methods leads us to formulate a recipe, already suggested in sidenote 10:

> Put the quantity you want on the left of an equation containing it. On the right, put "the known and the smaller".[13]

An iteration method is available if this "shaping" can be done.

Exercise 12.2 Simultaneous equations
Solve the following simultaneous equations to 5 significant figures, given that x and y are both expected to lie near 0.1:

$$\left.\begin{array}{l} x = 0.1 + y^3 \\ y = 0.1 - x^3. \end{array}\right\} \qquad (12.6)$$

It's intended that you bring to this the insights gained in exercise 12.1.

Solution to exercise 12.2
Don't substitute one equation into the other, so as to eliminate either x or y; it doesn't help.

It's tempting to put two rough values into the two right-hand sides to get two new values for x and y, and then repeat. I'll write this out symbolically:

step 1: $\quad x_1 = 0.1 + y_0^3 \qquad y_1 = 0.1 - x_0^3$

step 2: $\quad y_2 = 0.1 - x_1^3 \qquad x_2 = 0.1 + y_1^3$

step 3: $\quad x_3 = 0.1 + y_2^3 \qquad y_3 = 0.1 - x_2^3$

$\quad\quad\quad\cdots \qquad\qquad\quad \cdots \qquad\qquad\quad \cdots\,.$

Notice that we have two "chains" here: one set of numbers propagates down the left-hand column, and another set of numbers goes down the right-hand column, while the two sets are not linked. This means we are doing two lots of work when one would do. Delete one of the columns, and the survivor gives all we need.[14]

The way information is used can be represented by the "figure-8" shape shown in Fig. 12.4. We start somewhere, let's say with a value for y (which we might call y_0), and we use the top equation to obtain from it a value for x. We carry that x into the lower equation and use it to get a better y. We carry that y to the top equation and use it to get a better x. And so on in a figure-8 loop until we have the accuracy we require.

The numerical solutions don't matter much beside the understanding that we've gained. But for the record: $x = 0.100\,969$, $y = 0.989\,706$.

Comment: Look ahead to the description of time-independent perturbation theory in § 12.3. We see a pair of equations, printed boxed, one giving a value for the energy eigenvalue and the other giving wave-function coefficients. We solve between these using a figure-8 information flow that parallels what we have found in exercise 12.2. If you're comfortable

[13] It's possible to find pathological cases where this isn't quite the right thing to do. What's important about the terms on the right is not that they're small, but that they're insensitive. If the quantity sought is x, then the whole right-hand side must be little changed by changes of x, whether or not individual terms on the right are small compared with x.

[14] Another way of saying this: In the evaluation of y_1, we used x_0 when we already had the better value x_1. So we were using poor information (about x) when we already had better. Always use the best information you have.

$$x = 0.1 + y^3$$

$$y = 0.1 - x^3$$

Fig. 12.4 The information flow used in our iterative solution of eqns (12.6).

with exercise 12.2 (and above all if you invented the correct procedure for yourself), then perturbation theory will have no fears for you.

Exercise 12.3 Pendulum swinging with a finite amplitude
This is a problem in classical mechanics which can be tackled by successive approximation. You fully understand the physics, so can concentrate on handling the approximations.

The equation of motion for a pendulum is of the form

$$\ddot{\theta} + \omega_0^2 \sin\theta = 0, \tag{12.7}$$

where θ is the angle made by the pendulum with the downward vertical. Usually we approximate by setting $\sin\theta \approx \theta$, and that gives simple harmonic motion. But if we want to do accurate work using a pendulum as a timing device, we may well want to know how large we can afford to make the amplitude, and how constant we have to hold it. Therefore we need to know the effect of a finite amplitude on the period. A rough idea of the finite-amplitude correction will be quite sufficient, so we'll be content with a first-order approximation.

In the crudest approximation, when we set $\sin\theta \approx \theta$, the equation of motion reduces to

$$\ddot{\theta} + \omega_0^2 \theta \approx 0. \tag{12.8}$$

We know that the solution to this looks like the sine or cosine of $(\omega_0 t)$. Let's require the angular displacement to be zero at time $t = 0$, so we have a "zeroth" approximation for the angular displacement as

$$\theta = A \sin\omega_0 t. \tag{12.9}$$

In the next approximation, we make a series expansion of $\sin\theta$ in powers of θ, and keep one term more than we retained before. The equation of motion (12.7) becomes in the second approximation

$$\ddot{\theta} + \omega_0^2 \theta \approx \tfrac{1}{6} \omega_0^2 \theta^3.$$

The term on the right is a small correction—after all, we were satisfied to ignore it altogether up until now.[15] Since the right-hand side is small, it should suffice if we substitute into it a rough value for θ, just as in exercise 12.1 we were satisfied to substitute a rough value for x into the x^3 term. The equation to be solved is therefore

$$\ddot{\theta} + \omega_0^2 \theta \approx \tfrac{1}{6} \omega_0^2 A^3 (\sin\omega_0 t)^3. \tag{12.10}$$

(1) Obtain the solution to eqn (12.10). Require your solution to be zero for $t = 0$. Require also that your solution be the same as (12.9) except for small corrections.[16]

(2) The angular displacement of the pendulum is zero at time $t = 0$. Find an equation for the time at which the displacement is next zero. Solve this equation to find the period of oscillation, as corrected for the finite amplitude A of the pendulum's movement.[17]

The finite-amplitude correction is given (calculated "properly") in e.g. Newman and Searle (1957), p. 26. Do you agree?

[15] Notice our obedience to the recipe: *put the known and the smaller on the right.* Here the θ that we want is on the left, and the small correction—the θ^3 term—is on the right.

[16] You may need to make use of the identity $\sin 3x = 3\sin x - 4\sin^3 x$.

[17] *Hint:* You will have to be prepared to make approximations. I expect you to be worried about what you are and are not allowed to do. In deciding what terms to keep, bear in mind that quantities of order A^5 were discarded on the right-hand side of (12.10).

Solution to exercise 12.3

(1) The solution to eqn (12.10) is

$$\theta = A\sin(\omega_0 t) - \frac{A^3}{16}(\omega_0 t)\cos(\omega_0 t) + \frac{A^3}{192}\sin(3\omega_0 t). \qquad (12.11)$$

(2) There is an unusual feature to expression (12.11). The second term grows as $(\omega_0 t)$, so for large t this term is no longer small. On second thoughts perhaps we shouldn't be surprised. The oscillation described by (12.7) has lower frequency than that of (12.8),[18] so graphs of the two functions must slowly draw apart with increasing t. What we learn from this is that solution (12.11) is valid for small $(\omega_0 t)$ but must not be used thoughtlessly for large $(\omega_0 t)$. It should be safe to use it to find the first or second occasions on which θ passes through zero after $t = 0$.

We have to solve

$$\sin(\omega_0 t) - \frac{A^2}{16}(\omega_0 t)\cos(\omega_0 t) + \frac{A^2}{192}\sin(3\omega_0 t) = 0. \qquad (12.12)$$

The presence of $(\omega_0 t)$ multiplying the cosine in the second term means that no analytic solution can be possible. We must do something approximate.[19]

Let's eliminate a possibility. The student may try to expand the sines and cosines in power series, hoping to obtain an algebraic equation for $(\omega_0 t)$. This isn't sensible. The smallest value of $(\omega_0 t)$ that we hope to find is about π, so we're expanding something like

$$\cos\pi = 1 - \frac{\pi^2}{2!} + \frac{\pi^4}{4!} - \frac{\pi^6}{6!} + \frac{\pi^8}{8!} - \cdots$$
$$= 1 - 4.93 + 4.06 - 1.34 + 0.24 - \cdots.$$

For $(\omega_0 t) \sim \pi$ the expansions must be carried so far that the equation we end up with will be too complicated to be useful.[20]

It's often useful to do something wrong like this, because it may give a pointer to what ought to be done instead. What we did wrong was to use an expansion with too big an expansion variable. Expansions are useful when we have a small quantity. Remember the comment made in the solution to exercise 12.1 part (9)(e): if we've got a small quantity, we're foolish if we don't exploit it. We do have a small quantity here. If we're looking for a solution for $(\omega_0 t)$ near π then $\varepsilon = (\omega_0 t - \pi)$ is small. An expansion in powers of ε *would* be sensible.

Let's state this another way. Equation (12.12) has many solutions: one at 0, one near π, one near 2π, and so on.

> If we're to get the mathematics to give us a chosen
> one of these solutions, we've got to give it some help.

By working with ε and requiring ε to be small, we tell the mathematics to focus in on the solution near π.

Substituting $(\omega_0 t) = \pi + \varepsilon$ into (12.12), we obtain

$$\sin\varepsilon = \frac{A^2}{16}(\pi + \varepsilon)\cos\varepsilon - \frac{A^2}{192}\sin(3\varepsilon). \qquad (12.13)$$

[18] We know this because $|\sin\theta| < |\theta|$: the actual restoring force is weaker than that for the harmonic oscillator.

[19] You can invent a graphical method of solution, modelled on the procedure in Fig. 12.2. But the teaching points I want to make are best exhibited by pursuing an algebraic route.

[20] This is notwithstanding the fact that the expansion is convergent for all values of $(\omega_0 t)$. The mathematical knowledge that the series converges provides comfort where we should have none.

Expanding the trigonometric functions and keeping only the leading terms (in ε and in A^2, which we rapidly discover are of the same order of magnitude as each other), we obtain

$$\varepsilon = \frac{A^2\pi}{16} \qquad \text{oscillation period} = 2t = \frac{2\pi}{\omega_0}\left(1 + \frac{A^2}{16}\right). \qquad (12.14)$$

This is the standard expression for the finite-amplitude correction to a pendulum's oscillation period.

Comment on brutality: We may well feel that what we did in getting (12.14) was brutal: discarding far too much. We may wish to expand the trigonometrical functions, keeping more terms. Something like:[21]

$$\varepsilon = \left(\frac{\varepsilon^3}{6} - \cdots\right) + \frac{A^2}{16}(\pi+\varepsilon)\left(1 - \frac{\varepsilon^2}{2} + \cdots\right) - \frac{A^2}{192}\left(3\varepsilon - \frac{(3\varepsilon)^3}{6} + \cdots\right)$$

We can substitute rough expressions for ε into the right-hand side to obtain a better approximation, and keep on doing it until we are satisfied that we have gone to high enough powers of A^2. As a solution of (12.13) this is fine, but as a solution to our physical problem it is **WRONG**.

Remember that (12.10) carried the expansion of $\sin\theta$ to order A^3 beside terms in A, that is, to order (leading term) $\times A^2$. Terms of order (leading term) $\times A^4$ were discarded. If we extract terms of order A^4 from (12.13), we shall get only some of them, because others were discarded long ago. And it's a rule:

never keep only some terms of a given order;

those you've rejected might well be larger than those retained and of opposite sign, so you can't even know that you've got the sign right. It's all or none.[22] And since it can't be all, it has to be none. The "brutality" in (12.14) was in fact *compulsory*.

Comment on what this chapter has done so far: The three exercises have introduced several teaching points in connection with successive approximation methods, in particular:

- the "shaping" of equations so that known terms and small terms are placed on the right (and in descending order of size if that can be done tidily)[23]
- an alertness to the order of magnitude of different terms, so you know how to do the "shaping"
- the feeding of rough values into small terms on the right, as a valid successive-approximation step
- the advice: exploit a small quantity if you have one
- the rule: keep all terms of a given order or none
- the rule: be aware of the order of accuracy that a given iterative step can give, and work to that accuracy and no higher; if you want more accuracy, don't fuss now, go round the loop again.
- an opportunity: you can be brutal to the coefficient of a small term.[24]

[21] Here I expand $\sin\varepsilon$ on the left of (12.13) but transfer all terms after the first over to the right-hand side. As always, we "shape" our equations so that the known and the smaller are put on the right.

[22] We are re-stating a conclusion already encountered in the keeping of powers of ε in eqns (12.5).

[23] I did this as a matter of habit when writing eqn (12.4).

[24] This is implicit in the discarding of powers of ε in (12.5). It has application in Chapters 27 and 32.

12.3   Perturbation theory: time independent, non-degenerate

The presentation given here is a little different from that in most books. It is meant to offer a complementary approach; I'm not claiming any unique merit. My intention is to give a clearer idea of the logical steps. Display of those steps can be done with the simplest case of perturbation theory, so most of the discussion is of the time independent non-degenerate case.

The Schrödinger equation to be solved is

$$(\widehat{H}_0 + \widehat{H}')\psi = E\,\psi, \tag{12.15}$$

where $\widehat{H}'$ is regarded as a "small" perturbation on top of the larger $\widehat{H}_0$. Operator $\widehat{H}_0$ has (unperturbed) eigenfunctions ψ_n with energy eigenvalues E_n, where n labels the state.

The state we are perturbing has wave function ψ_0, which is the eigenfunction of $\widehat{H}_0$ having energy eigenvalue E_0. The wave function ψ_0 is to be understood as "the initial state". There is no implication that it is the ground state.[25]

We'll write the perturbed wave function as

$$\psi = a_0\,\psi_0 + \sum_{n\neq 0}^{\infty} a_n\,\psi_n. \tag{12.16}$$

Here a_0 is a number rather close to 1 (since in the unperturbed problem $a_0 = 1$ exactly); and all the other a_n are much less than 1 (correctly $|a_n| \ll 1$). In (12.16) we pick out the term $a_0\,\psi_0$ for special treatment, rather than including it in the sum, just because it is distinguished by being so much larger than the other terms.

Substituting the wave function (12.16) into the Schrödinger equation (12.15) we have

$$\left(\widehat{H}_0 + \widehat{H}'\right)\left\{a_0\,\psi_0 + \sum_{n\neq 0}^{\infty} a_n\,\psi_n\right\} = E\left\{a_0\,\psi_0 + \sum_{n\neq 0}^{\infty} a_n\,\psi_n\right\}. \tag{12.17}$$

Since the wave functions that we are using to expand ψ are all eigenfunctions of $\widehat{H}_0$, we have of course

$$\widehat{H}_0\,\psi_0 = E_0\,\psi_0; \qquad \widehat{H}_0\,\psi_n = E_n\,\psi_n.$$

Using these relations in (12.17) we now have

$$a_0\,E_0\,\psi_0 + \sum_{n\neq 0}^{\infty} a_n\,E_n\,\psi_n + a_0\widehat{H}'\,\psi_0 + \sum_{n\neq 0}^{\infty} a_n\widehat{H}'\,\psi_n = E\left\{a_0\psi_0 + \sum_{n\neq 0}^{\infty} a_n\psi_n\right\}. \tag{12.18}$$

At this stage, we break up (12.18) into separate pieces. First multiply (12.18) by ψ_0^* and integrate. We get

$$a_0\,E_0 + a_0\,H'_{00} + \sum_{n\neq 0}^{\infty} a_n\,H'_{0n} = E\,a_0,$$

which we rearrange into the more meaningful shape

[25] Textbook treatments usually perturb a state that is overtly general, with wave function written as ψ_i or something similar. This is fine when you fully understand what is going on. My fear is that a beginner will get lost in the subscripts if this more general notation is used. When the present section is understood, the reader can kick away the ladder—he's encouraged to do so—by which he has climbed, and be happy writing his own version with ψ_i as the initial state.

One equation for energy E.

$$E = E_0 + H'_{00} + \sum_{n \neq 0}^{\infty} \left(\frac{a_n}{a_0} \right) H'_{0n}. \tag{12.19}$$

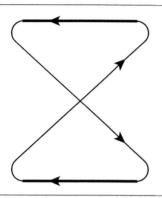

Here the matrix elements of $\widehat{H}'$ are defined as always by

$$H'_{mn} \equiv \int \psi_m^* \, \widehat{H}' \, \psi_n \, \mathrm{d}V. \tag{12.20}$$

Now go back to (12.18), multiply it by ψ_m^* and integrate. Since we've already dealt with what happens when we multiply by ψ_0^*, we'll restrict m to $m \neq 0$.

$$a_m E_m + a_0 H'_{m0} + \sum_{n \neq 0}^{\infty} a_n H'_{mn} = E \, a_m.$$

We again perform a rearrangement into a more insightful shape:

A *set* of equations, one for each $m \neq 0$, giving (a_1/a_0), (a_2/a_0), (a_3/a_0), ...

$$\left(\frac{a_m}{a_0} \right) = \frac{H'_{m0} + \sum_{n \neq 0}^{\infty} \left(\dfrac{a_n}{a_0} \right) H'_{mn}}{E - E_m} \qquad \text{for } m \neq 0. \tag{12.21}$$

It is important to notice what we have done in deriving eqns (12.19) and (12.21). We have not approximated at any stage, so both equations are exactly true. For some simple problems it may be possible to solve the boxed equations exactly, in which case there is no need to call upon the mathematical apparatus of perturbation theory at all. However, it usually happens that we are not able to solve (12.19) and (12.21) exactly, and it is then that perturbation theory has to be used to give us an approximate answer. We may think of "perturbation theory" as a long name for a method for solving (12.19) and (12.21) by successive approximation.

Equations (12.19) and (12.21) could have been written in any number of equivalent forms if we were going to do only algebra with them. But what we are going to do is closer to the kind of thing we did in exercises 12.1 to 12.3. The precise way that I've shaped (12.19) and (12.21) makes it as easy as possible to solve by successive approximation. That is, what we want to find is on the left, while the right-hand side contains the known and the smaller—as we'll see.

The procedure we shall use mimics that of Fig. 12.4: following a figure-8 pattern. We put something rough into the right-hand side of (12.19) to obtain a first shot at E. We carry this E into (12.21) to obtain a first shot at the wave-function coefficient (a_m/a_0). We carry this back into (12.19) to get a better E. We carry this E back into (12.21) to get a better wave function. And so on.[26]

We need to be able to "eyeball" the orders of magnitude of the various terms in (12.19) and (12.21). We start with (12.19). The three terms on the right have been printed in descending order of smallness. First,

[26] As we shall see, there are some differences of detail between what I've said here and what actually happens, but the principle is correct.

E_0 is the energy of the state before the perturbation is applied. It's there even if the perturbation is removed, so we'll say this has zeroth order of smallness. The next term H'_{00} is of first order because it is obtained (eqn (12.20)) by a linear operation from $\widehat{H}'$: if you doubled $\widehat{H}'$ you'd double H'_{00}. The changes of wave function, quantified by the coefficient ratios (a_n/a_0), are also small, because they become zero if $\widehat{H}'$ is removed. Then all terms in the sum $\sum (a_n/a_0)H'_{0n}$ are of the second order of small quantities. I'll assume that the *sum* of all these second-order quantities is still of second order. Then we may label the orders of magnitude of the terms in (12.19) as follows:

each quantity arrowed is first order, but (a_n/a_0) contains corrections of higher order

$$
E = E_0 \; + \; H'_{00} \; + \; \sum_{n \neq 0}^{\infty} \left(\frac{a_n}{a_0} \right) H'_{0n} \, . \tag{12.22}
$$

zeroth order   first order   second order of small quantities, but containing corrections of higher order

Make a similar analysis for yourself on the terms within (12.21).

12.3.1   Systematic solution of eqns (12.19) and (12.21)

Start with (12.19). The right-hand side contains known terms of zeroth and first order. Experience[27] tells us that we shouldn't worry yet about the second-order terms: we gain only one order at a time as we move through an iteration, and must not keep more. So

$$
E = E_0 + H'_{00}. \qquad \text{[first order]} \tag{12.23}
$$

Experience with exercise 12.2 tells us what to do next. Having got as much as we can from one of our two equations, we switch attention to the other, carrying our new knowledge with us. Once again, we keep only the largest of the terms on the right for now, knowing that smaller terms can't be got right until next time round. In the numerator we keep H'_{m0}, and the denominator is accurate enough[28] if we write it as $E_0 - E_m$. Then[29]

$$
\left(\frac{a_m}{a_0} \right) = \frac{H'_{m0}}{E_0 - E_m}. \qquad \text{[first order]} \tag{12.24}
$$

Experience with exercise 12.2 tells us what to do next. Having got as much as we can from the second of our two equations, we switch attention back to the first, carrying our new knowledge with us. Knowing (a_n/a_0) to first order, we can evaluate the sum on the right, correct to second order, and that will give us E to second order:[30]

$$
E = E_0 + H'_{00} + \sum_{n \neq 0}^{\infty} H'_{0n} \frac{H'_{n0}}{E_0 - E_n}. \qquad \text{[second order]} \tag{12.25}
$$

[27] In particular with exercise 12.1 part (10).

[28] It's a slight surprise that the knowledge we have of E is better than is needed at this stage in the processing of (12.21).

[29] Equation (12.24) shows us that a_m/a_0 is indeed of first order. The right-hand side has numerator of first order on a zeroth-order denominator. I assumed that a_m/a_0 is of first order (it might have been smaller) when making the annotated version of (12.19) in (12.22).

[30] This chapter started by looking at $\widehat{H} = \widehat{H}_0 + \lambda \widehat{H}'$, accompanied by a recipe that orders of smallness were to be identified by picking out powers of λ. It should be obvious that if we reinsert λ into (12.25) H'_{00} will be multiplied by λ, while all terms in the sum will contain λ^2. Application of the recipe tells us that the H'_{00} term is of first order, while the terms in the sum are of second order. However, we can see directly in (12.25) that H'_{00} contains $\widehat{H}'$ in first order, while $H'_{0n}H'_{n0}$ contains $\widehat{H}'$-quantities squared, giving second order: we do not need the λs to guide us to these insights.

[31] If this procedure is followed, we gain one order, one power of $\widehat{H}'$ or of λ, for each circuit round the loop of eqns (12.19–21). There is an obvious similarity to the gaining of one power of ε^2 for each step in (12.5).

There is however one way in which λ is helpful to us after all. The expressions we obtain by iterating, within perturbation theory, can be seen as building a power series in λ.

As with any other power series, there is a question as to whether the series is convergent. When using perturbation theory, we are usually content to assume that its series is convergent, or at least that it is an asymptotic series within its range of usefulness. And in this chapter we take convergence as a "for later" possible complication.

However, I should mention that there are cases in physics which require a "non-perturbative" treatment because perturbation theory fails in some way. The most notorious such case is superconductivity. See e.g. Bardeen, Cooper and Schrieffer (1957) eqn (2.43): the energy has the form $e^{-1/V}$, where V is a dimensionless quantity derived from the perturbing Hamiltonian. This expression has no Maclaurin series, as every derivative at $V = 0$ is zero. No power series, no possibility of applying perturbation theory.

It's open to us to continue this process, carrying E, and a rough (a_n/a_0), into (12.21) to get a better set of wave-function coefficients and so on. We almost never go so far because the algebra gets messy.[3]

12.3.2   Normalization of the wave function

Equation (12.16) can be rewritten as

$$\psi = a_0 \left\{ \psi_0 + \sum_{n \neq 0}^{\infty} \left(\frac{a_n}{a_0} \right) \psi_n \right\}. \tag{12.26}$$

Equations (12.19) and (12.21) contain the coefficients in the expansion of ψ only in the form (a_n/a_0). Thus the only thing we can hope to calculate from (12.19) and (12.21) is the *ratios* of the coefficients; the coefficient a_0 is left undetermined. We have to obtain a_0 separately by requiring the wave function to be normalized:

$$1 = |a_0|^2 \left\{ 1 + \sum_{n \neq 0}^{\infty} \left| \frac{a_n}{a_0} \right|^2 \right\}. \tag{12.27}$$

Show this.

The failure of (12.19) and (12.21) to give us a_0 should make sense. The Schrödinger equation is *homogeneous*, which implies that the solution always comes out with an arbitrary coefficient in front. You find the coefficient afterwards (after finding the *form* of the solution) by normalizing—if you so choose. Normalizing is something *you do*, not something forced on you by the differential equation. There would be something wrong if perturbation theory (which is just a special way of solving a Schrödinger equation after all) seemed to tell you the value of the normalizing constant.

Comment: It's time to take stock: Why have I have written out this presentation of perturbation theory?

First (and least important), I have chosen to find the change to energy E_0 and wave function ψ_0, rather than dealing with an *i*th state. This is entirely a get-you-started measure, to be discarded as soon as you no longer need the simplification.

Second (and most important), the figure-8. Once we have understood from exercise 12.2 that this is the right way to process a pair of simultaneous equations, we understand the "structure" of the iteration: we always know where to go next. This is the case even though we probably go only $1\frac{1}{2}$ times round the DO loop.

Third (quirky?), I have departed from convention by keeping coefficient a_0 in the equations. "Usual practice" sets this coefficient to 1 at the beginning, thereby building an un-normalized wave function; this is put right at the end by dealing with normalization as an afterthought. By shaping the coefficients into building blocks (a_n/a_0), I try to make it clear from the beginning that what we are finding is ratios; nothing strange is going on with the coefficients.

Fourth, the hope is that by now you're secure with spotting the orders of magnitude of terms by eye, without the need for a mechanical prompt, such as the λ mentioned at the beginning of the chapter. There is an advantage in having this maturity. Look at eqn (12.21) again. The numerator H'_{m0} is first-order small, so the whole right-hand side is. But wait: what if the denominator happens by accident to be small too? There is a danger signal here, which we must notice at need. By outgrowing an "automated" method, we've a better chance of being alert to such signals.

12.4   Perturbation theory: time dependent, non-degenerate

We give a treatment here of time-dependent perturbation theory, adopting the same spirit as that in §12.3. The differences from textbook convention are, however, relatively small.

We have to solve the time-dependent Schrödinger equation:

$$i\hbar \frac{\partial \Psi}{\partial t} = (\widehat{H}_0 + \widehat{H}')\Psi, \tag{12.28}$$

where $\widehat{H}'$ is a small perturbation to the Hamiltonian, and in general $\widehat{H}'$ (though—usually—not $\widehat{H}_0$) may be a function of time. Before we solve the perturbed problem, we need to know the solutions to the unperturbed problem. These are the eigenfunctions Ψ_n defined by

$$i\hbar \dot{\Psi}_n = \widehat{H}_0 \Psi_n,$$

and which have the property[32]

$$\Psi_n(\boldsymbol{r}, t) = \psi_n(\boldsymbol{r}) e^{-iE_n t/\hbar}. \tag{12.29}$$

As previously, we look for a solution to the perturbed problem (12.28) in the form of an expansion in the unperturbed eigenfunctions:

$$\Psi = \sum_n a_n \Psi_n = a_0 \Psi_0 + \sum_{n \neq 0}^{\infty} a_n \Psi_n. \tag{12.30}$$

Also as previously, we shall be perturbing the zeroth state, and so we must have $a_0 \approx 1$, while all the other a_n are small. We build this into the algebra by displaying the a_0 term separately. The fact that the problem involves a dependence on time will appear in the expansion (12.30), because the coefficients can be dependent on time.

Substituting (12.30) into (12.28), we have

$$i\hbar \left\{ \dot{a}_0 \Psi_0 + \sum_{n \neq 0}^{\infty} \dot{a}_n \Psi_n + a_0 \dot{\Psi}_0 + \sum_{n \neq 0}^{\infty} a_n \dot{\Psi}_n \right\} = (\widehat{H}_0 + \widehat{H}') \left\{ a_0 \Psi_0 + \sum_{n \neq 0}^{\infty} a_n \Psi_n \right\}.$$

Subtracting off the unperturbed Schrödinger equations for the eigenfunctions yields

$$i\hbar \left\{ \dot{a}_0 \Psi_0 + \sum_{n \neq 0}^{\infty} \dot{a}_n \Psi_n \right\} = \widehat{H}' \left\{ a_0 \Psi_0 + \sum_{n \neq 0}^{\infty} a_n \Psi_n \right\}. \tag{12.31}$$

[32] We use the customary convention that ψ represents a wave function independent of time, while Ψ signals that a time dependence is present.

In this equation, we must pick out $\dot{a}_0$, $\dot{a}_n$ separately in order to find differential equations for these coefficients. There should be no difficulty in seeing what to do. First multiply through by Ψ_0^* and integrate:

$$i\hbar\,\dot{a}_0 \int \Psi_0^* \,\Psi_0 \,\mathrm{d}V = a_0 \int \Psi_0^* \,\widehat{H}' \,\Psi_0 \,\mathrm{d}V + \sum_{n\neq 0}^{\infty} a_n \int \Psi_0^* \,\widehat{H}' \,\Psi_n \,\mathrm{d}V.$$

Now comes a trap! We want to express these integrals in terms of the matrix elements of $\widehat{H}'$; but the matrix elements are defined using the time-independent wave functions ψ_n, rather than the Ψ_n. We use (12.29) and find[33]

$$\int \Psi_0^* \,\Psi_0 \,\mathrm{d}V = 1; \qquad \int \Psi_0^* \,\widehat{H}' \,\Psi_0 \,\mathrm{d}V = H_{00}';$$

$$\int \Psi_0^* \,\widehat{H}' \,\Psi_n \,\mathrm{d}V = H_{0n}' \,\exp\{i(E_0 - E_n)t/\hbar\}.$$

Using these in the last displayed equation, we obtain

[33] Matrix elements are defined according to the pattern

$$H_{0n}' = \int \psi_0^* \,\widehat{H}' \,\psi_n \,\mathrm{d}V$$

and similarly.

| One equation for a_0. | $$i\hbar\,\dot{a}_0 = a_0 H_{00}' + \sum_{n\neq 0}^{\infty} a_n H_{0n}' \,\exp\{i(E_0 - E_n)t/\hbar\}. \qquad (12.32)$$ |

To get the other coefficients, multiply (12.31) through by Ψ_m^* and integrate (taking it that $m \neq 0$). Show for yourself that the result is

---

| A *set* of equations one for each $m \neq 0$, giving a_1, a_2, ... | $$i\hbar\,\dot{a}_m = a_0 H_{m0}' \,\exp\{i(E_m - E_0)t/\hbar\} + \sum_{n\neq 0}^{\infty} a_n H_{mn}' \,\exp\{i(E_m - E_n)t/\hbar\}.$$ <div align="right">(12.33)</div> |

Equations (12.32) and (12.33) correspond to (12.19) and (12.21) in the time independent case, and we should be able to solve them in a similar way. As a first step, we must identify the sizes of the various terms, so that we can see what approximations are permissible. In (12.32) we know that H_{00}' and H_{0n}' are first-order small, and that a_n should also be small, except that a_0 is not far from 1. Therefore we have

$$i\hbar\,\dot{a}_0 = a_0 H_{00}' + \sum_{n\neq 0}^{\infty} a_n H_{0n}' \,\exp\{i(E_0 - E_n)t/\hbar\}.$$

first order

each factor is first order, so entire sum is second order

Do a similar analysis of (12.33) for yourself.

The problem remaining now is to solve (12.32) and (12.33) by a successive approximation method. As before, this is a matter of dodging back and forth between (12.32) and (12.33)—a figure-8 procedure—picking up an extra order of accuracy each time round. Keeping only first-order terms in (12.32) we have

$$i\hbar\,\dot{a}_0 = a_0 H_{00}'. \qquad \text{[first order]} \qquad (12.34)$$

Keeping first-order terms on the right of (12.33) we find

$$i\hbar\,\dot a_m = a_0\,H'_{m0}\exp\{i(E_m - E_0)t/\hbar\}. \qquad \text{[first order]} \qquad (12.35)$$

We cannot write out explicit expressions for a_0 and a_m because we have not been told how H'_{00} and H'_{m0} vary with time, and so we cannot integrate. But it is enough for this general analysis to know that for any given problem the coefficients a_0 and a_m can be found by integrating (12.34) and (12.35) over time. It is also easy now to see that we may substitute the first-order expressions for a_0 and a_m into (12.32) and (12.33), after which we may obtain expressions correct to second order.[34] A further substitution could lead to expressions correct to third order, and so on.

[34] It's *very* unlikely that you will need to find an expression to second order, let alone anything higher. The point here is to understand the nature of the figure-8 DO loop that you have just started, so that there is no mystery to the process.

And, to address a difficulty I remember having as a student: you always know where to go next.

12.4.1   A standard example

An example often described in books is a perturbation that is zero for $t < 0$, and which is "turned on" to a constant final value for $t > 0$. Then for $t > 0$, the perturbing operator $\widehat{H}'$ is independent of time. In (12.34) and (12.35), the matrix elements H'_{00} and H'_{m0} are now independent of time (for $t > 0$), so we may integrate at once:[35]

$$a_0 = \exp(-i\,H'_{00}\,t/\hbar) \qquad \text{[first order]} \qquad (12.36a)$$

$$a_m = \frac{H'_{m0}}{E_0 - E_m}\Big\{\exp\{i(E_m - E_0)t/\hbar\} - 1\Big\}. \qquad \text{[first order]} \qquad (12.36b)$$

[35] *Question*: Why the -1 on the right of (12.36b)?

It is now possible (but unlikely to be useful) to find a_0 and a_m to second order by substituting these expressions back into (12.32) and (12.33).

12.4.2   Normalization in time-dependent perturbation theory

There is a difference between the present calculation and that in §12.3. In §12.3, eqn (12.19) told us the energy of the perturbed state, and we were left with no equation to use for finding the coefficient a_0. We could calculate only the *ratios* of the coefficients. Then we normalized (by our choice) the wave function at the end of the calculation.

By contrast, in the time-dependent case, (12.32) tells us the value of a_0, and we have no equation for the energy. There doesn't seem to be an opportunity to normalize the wave function, because all coefficients have been determined already. Is the wave function already normalized; and what about the energy?

Answers at once: The wave function *is* already normalized (or rather it retains whatever normalization it started with); and there is *no* equation for the energy.

Let's notice that in a problem where the wave function varies with time, the system (electron or whatever) may (indeed should) not have a well defined energy. If the wave function changes significantly over a time Δt, the energy must be uncertain by $\Delta E \sim \hbar/\Delta t$. Thus we should not expect to get a clean equation for energy coming out in the time-dependent case. This means that (12.32) is "left over" for determining something else—which happens to be a_0. Notice too that we haven't lost *all* information about energy: combine (12.29) and (12.36a) and examine the result; you should find a resemblance to (12.23).

Problems

Problem 12.1   A check on normalization
Use the time dependent Schrödinger equation to show that[36]

$$\int_{\text{all space}} |\Psi|^2 \, dV = \text{constant},$$

meaning constant with respect to time.

[36] *Hint:* Adapt the textbook calculation that derives the probability current.

Problem 12.2   Controlling the convergence of an iteration
Consider some equation that we want to solve, and which we have put into the form

$$x = f(x).$$

We are to solve this after the manner of (12.3):

$$x_1 = f(x_0)$$
$$x_2 = f(x_1)$$
$$\vdots$$

In this problem, we assume that ratios of changes, like $(x_2-x_1)/(x_1-x_0)$, do not change much as the iteration proceeds, just as they took values close to 0.03 in exercise 12.1(3).[37] Set $r_n = (x_n-x_{n-1})/(x_{n-1}-x_{n-2})$ to be the ratio of changes in the nth step and the previous step. If $r_n < 1$, the iteration undershoots, and could usefully be speeded up, after the fashion of sidenote 5. If r_n is negative, the iteration overshoots, and would be improved if the change $(x_n - x_{n-1})$ were reduced in size.

[37] This assumption means that a graph of $f(x)$ is close to a straight line in the region we are exploring: a Newton approximation would work well if we chose to use it.

(1) Show that the iteration is improved by setting

$$(x_n - x_{n-1})_{\text{better}} = \frac{(x_n - x_{n-1})_{\text{found}}}{1 - r_n}. \tag{12.37}$$

(2) Show that what we have just done is to obtain the change (x_n-x_{n-1}) that Newton would have given us in the solution of

$$y = x - f(x) = 0.$$

Comment: If the derivative dy/dx can be calculated, then the best thing to do here is to use it in a Newton approximation. But it may happen that the function we are handling is too complicated for that to be practical. In that case, an empirical adjustment along the lines of (12.37) gives a sensible way of speeding up the iterative process.

(3) Recipe (12.37) can even be used to "rescue" a divergent iteration, one where $|r_n| > 1$: successive steps get bigger, not smaller. Show that (12.37) gives us a best estimate of the next step that we should use, even in this case.

VII Atomic physics

These diminutive gentlemen are called atoms.

Flann O'Brien, *The Third Policeman*

Intended readership: Shortly after seeing a lecture or textbook presentation of hydrogen quantum mechanics, when a second look supplying refinements can be appreciated.

13.1 Introduction

The hydrogen atom is usually the first system that the student sees for which quantum mechanics makes predictions that can be directly compared with experiment. To my surprise, almost every textbook treatment perpetrates an error (or at least infelicities) somewhere; one really would expect such basic material to have "settled down" by now. This chapter tries to give a treatment that is securely correct. At the same time, opportunity is taken to give material on radial wave functions that I consider helpful, and which is absent from the usual presentations.

It is unfortunate that the quantum-mechanical calculations for hydrogen are somewhat lengthy. I try to keep the physics presentation uncluttered by putting a rather large proportion of the mathematics into end-of-chapter problems.[1]

[1] This chapter is concerned only with "gross structure". We ignore relativistic corrections, including the existence of spin (possessed by both electron and nucleus).

13.2 Classical mechanics of orbital motion

Consider a body of mass m orbiting round a fixed centre[2] under the influence[3] of an inverse-square-law force $-\alpha/r^2$. The body can be taken as point-like, so we refer to it as a "particle".

Angular momentum $\boldsymbol{L} = \boldsymbol{r} \times \boldsymbol{p}$ is conserved. Position vector $\boldsymbol{r}$ is always at right angles to $\boldsymbol{L}$, so the motion lies in a plane[4] perpendicular to $\boldsymbol{L}$. Put the z-axis along the direction of $\boldsymbol{L}$, and describe the motion using (r, θ) coordinates in the xy plane. Then

$$L = mr^2\dot{\theta}, \tag{13.1}$$

and the equation governing radial motion is

$$m(\ddot{r} - r\dot{\theta}^2) = -\alpha/r^2. \tag{13.2}$$

These equations integrate (problem 13.1) to give

$$\tfrac{1}{2}m\dot{r}^2 + \frac{L^2}{2mr^2} - \frac{\alpha}{r} = E, \tag{13.3}$$

[2] Mention of a fixed centre means that centre-of-mass motion is excluded for now. Centre-of-mass motion is formally separated away in problems 13.2 and 13.3.

[3] When mass m orbits in the gravitational field surrounding a mass M, the force is $-GMm/r^2$ where G is the gravitational constant. When mass m carries charge $-e$ and orbits in the electric field surrounding a charge Ze, the force is $-Ze^2/(4\pi\epsilon_0\, r^2)$. Our classical-mechanics discussion covers all such cases because we leave the physical meaning of α unspecified at this stage.

[4] To show this, find $\boldsymbol{r} \cdot \boldsymbol{L}$. Alternatively, spot that $\boldsymbol{L} = \boldsymbol{r} \times \boldsymbol{p}$ defines $\boldsymbol{L}$ to be always perpendicular to $\boldsymbol{r}$.

Essays in Physics: Thirty-two thoughtful essays on topics in undergraduate-level physics. Geoffrey Brooker, Oxford University Press. © Geoffrey Brooker 2021. DOI: 10.1093/oso/9780198857242.003.0013

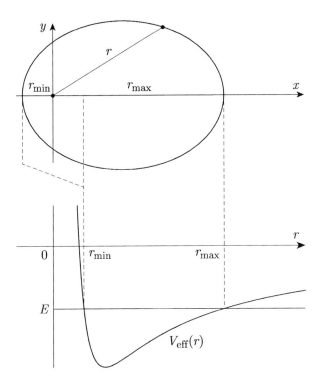

Fig. 13.1 (Top) A classical elliptical orbit.
(Bottom) The effective radial potential $V_{\text{eff}}(r)$ is given by (13.4). It includes both the actual potential energy $V(r) = -\alpha/r$ of attraction to the centre and the "centrifugal potential" from angular motion. The total energy is E, negative for a bound state. Where E crosses $V_{\text{eff}}(r)$ we have the limits of classical motion, at radii r_{min} and r_{max}. The two parts of the diagram taken together show how these $r_{\text{min}}, r_{\text{max}}$ relate to the least and greatest radii of the ellipse.
Although the motion envisaged is purely classical, the orbit shape and V_{eff} have been plotted as if the energy E and angular momentum L could take the values for quantum numbers $n = 9$, $l = 6$. The proportions of the ellipse are worked out at the end of problem 13.6.

in which E is a constant of integration that is identified with the total energy (kinetic plus potential) of the particle.

We recast (13.3) into the form

$$\tfrac{1}{2}m\dot{r}^2 = E - \left(\frac{L^2}{2mr^2} - \frac{\alpha}{r}\right) = E - V_{\text{eff}}(r),$$

where

$$V_{\text{eff}}(r) = \frac{L^2}{2mr^2} - \frac{\alpha}{r} \qquad (13.4)$$

is an **effective potential** for radial motion.

The implications of $V_{\text{eff}}(r)$ may be seen from the lower part of Fig. 13.1. The kinetic energy consists of a radial component $\tfrac{1}{2}m\dot{r}^2$ and an angular component $\tfrac{1}{2}mr^2\dot{\theta}^2 = L^2/2mr^2$. As the particle moves towards smaller r, conservation of angular momentum enforces an increase of the angular kinetic energy, leaving less energy available for radial motion. At radius r_{min} nothing is left for radial motion, so r cannot decrease further. Within $V_{\text{eff}}(r)$, the term $L^2/2mr^2$ gives energy that is unavailable for radial motion, and it's well described as the **centrifugal barrier**.

We are looking for a bound state, implying an outer limit r_{max} on r. From Fig. 13.1 this requires $E < 0$.

The two limiting radii, r_{min} and r_{max}, are the **limits of classical motion**. If we were discussing a planet orbiting the Sun, r_{min} and r_{max} would be called perihelion and aphelion. These limits continue to have significance in the quantum treatment below.

Given $E < 0$ we recast (13.3) again as

$$\tfrac{1}{2} m \dot{r}^2 = \frac{(-E)}{r^2} \left(-r^2 + \frac{\alpha}{(-E)} r - \frac{L^2}{2m(-E)} \right). \qquad (13.5$$

For the right-hand side to be positive, r must lie between the roots of the quadratic expression, that is between $r_{\min}$ and $r_{\max}$ where[5]

$$r_{\min} = \frac{\alpha - \sqrt{\alpha^2 - 2L^2(-E)/m}}{2(-E)}, \qquad r_{\max} = \frac{\alpha + \sqrt{\alpha^2 - 2L^2(-E)/m}}{2(-E)}$$

$$(13.6$$

13.3   The reduced mass and the relative coordinate

Two particles with masses m_1 and m_2 attract each other with an inverse square law of force $-\alpha/r^2$, where r is the separation of the masses. Work problem 13.2 to show that in classical mechanics the centre-of-mass motion can be separated away, while the equation of relative motion is

$$\mu \ddot{\boldsymbol{r}} = \frac{-\alpha \boldsymbol{r}}{r^3}. \qquad (13.7$$

Here

$$\mu = \frac{m_1 m_2}{m_1 + m_2}; \qquad \boldsymbol{r} = \boldsymbol{r}_1 - \boldsymbol{r}_2, \qquad (13.8$$

in which μ is called the **reduced mass** while $\boldsymbol{r}$ gives the separation of the two particles: the **relative coordinate**.

Comment: Notice that two new quantities have been introduced in (13.8): the reduced mass μ and the particle separation $\boldsymbol{r}$. These go together as a "package": use one and you must use the other as well. It's important to remember this when applying hydrogen-like equations to other systems. See § 13.6 below for just such a trap.

An exactly similar pair, μ and $\boldsymbol{r}$, describes a quantum-mechanical system of two attracting particles. Work problem 13.3 to obtain the time-independent Schrödinger equation for the "relative" motion of the two particles:[6]

$$\frac{-\hbar^2}{2\mu} \nabla_{\boldsymbol{r}}^2 \psi(\boldsymbol{r}) + V(r)\,\psi(\boldsymbol{r}) = E\,\psi(\boldsymbol{r}). \qquad (13.9$$

Here μ and $\boldsymbol{r}$ have the same definitions as in (13.8). Also, E is the energy of "relative" motion of the system, i.e. the kinetic energy of centre-of-mass motion has been subtracted away.

13.4   Radial and angular motion

We proceed to solve the Schrödinger equation (13.9) for "relative" motion.

Given a partial differential equation, there is an obvious line of attack. Using spherical polar coordinates (r, θ, ϕ), effect a (partial) separation of variables by writing[7]

[5] Of course, the quantity under the square root must be positive. Show that it is if E lies above the minimum of the V_{eff} curve.

The classical orbit traced out by the particle is an ellipse. However, hardly any of the discussion in the present chapter needs to make use of this knowledge.

[6] Here $V(r) = -\alpha/r = -Ze^2/(4\pi\epsilon_0\, r)$ is the attractive Coulomb potential energy function. The centrifugal potential will appear later; for the moment it is hiding in $\nabla_{\boldsymbol{r}}^2 \psi$.

For hydrogen itself we have $Z = 1$, but we retain Z so that the analysis holds equally for other hydrogen-like atoms.

[7] The location $\boldsymbol{R}$ of the centre of mass, used in problems 13.2 and 13.3, will not be mentioned again, so there should be no confusion with the new $R(r)$.

$$\psi(\boldsymbol{r}) = R(r)\,Y(\theta,\phi). \tag{13.10}$$

The manipulation required is rehearsed in problem 13.4, with outcome:

$$\frac{\mathrm{d}^2 R}{\mathrm{d}r^2} + \frac{2}{r}\frac{\mathrm{d}R}{\mathrm{d}r} - \frac{l(l+1)}{r^2}R + \frac{2\mu}{\hbar^2}\Big\{E - V(r)\Big\}R = 0; \tag{13.11}$$

$$-\left\{\frac{1}{\sin\theta}\frac{\partial}{\partial\theta}\left(\sin\theta\,\frac{\partial Y}{\partial\theta}\right) + \frac{1}{\sin^2\theta}\frac{\partial^2 Y}{\partial\phi^2}\right\} = l(l+1)Y. \tag{13.12}$$

The quantity $l(l+1)$ is the separation constant, expressed in this form because it turns out later that l is a positive integer.

The operator on the left of (13.12) can be recognized to be the operator $\widehat{l^2}$ for the square of orbital angular momentum (divided by $\hbar^2$); the eigenvalues of this operator are $l(l+1)$ with $l \geqslant 0$ an integer.[8] This is all we need to know about $Y(\theta,\phi)$ for the remainder of the present chapter. In case the quantum mechanics of angular momentum is not completely familiar,[9] we rehearse some of the mathematics in problem 13.5.

[8] It will be no surprise that a dependence of $Y(\theta,\phi)$ on angles θ, ϕ implies an angular motion (just as a wave function dependent upon x describes some kind of motion in the x-direction). Angular momentum was to be expected.

[9] Even those readers who are familiar with the mathematics of angular momentum may find something new in problem 13.5.

13.5   The radial wave function

We return to the equation (13.11) that specifies the radial part $R(r)$ of the wave function, and put it back into the shape of a Schrödinger equation

$$\frac{-\hbar^2}{2\mu}\left(\frac{\mathrm{d}^2 R}{\mathrm{d}r^2} + \frac{2}{r}\frac{\mathrm{d}R}{\mathrm{d}r}\right) + \left(\frac{\hbar^2 l(l+1)}{2\mu r^2} + V(r)\right)R = E\,R. \tag{13.13}$$

Obtaining the possible energies and wave functions for hydrogen requires that we solve (13.13). The mathematics is somewhat technical, and is rehearsed in problem 13.8. Here we concentrate on the physical behaviour that is exhibited by the radial wave function.

Equation (13.13) may be brought to a more familiar shape if we define $\chi = r\,R(r)$:

$$\frac{-\hbar^2}{2\mu}\frac{\mathrm{d}^2\chi}{\mathrm{d}r^2} + \left(\frac{\hbar^2\,l(l+1)}{2\mu r^2} + V(r)\right)\chi = E\,\chi. \tag{13.14}$$

This is the Schrödinger equation for a particle having mass μ and wave function χ, moving in a one-dimensional effective potential

$$V_{\mathrm{eff}}(r) = \frac{\hbar^2 l(l+1)}{2\mu r^2} + V(r) = \frac{\hbar^2\,l(l+1)}{2\mu r^2} - \frac{Ze^2}{4\pi\epsilon_0 r}. \tag{13.15}$$

The interpretations of Fig. 13.1 carry over to the present case. In (13.15), the "centrifugal potential" corresponds to the classical $L^2/2mr^2$, in which L^2 has been given its quantum value[10] of $\hbar^2 l(l+1)$. The Coulomb potential $V(r)$ is given its specific value $-Ze^2/4\pi\epsilon_0 r$. And $V_{\mathrm{eff}}(r)$ is the sum of these. In Fig. 13.2(a) there are shown "limits of classical motion" r_{min} and r_{max} where E crosses the curve of $V_{\mathrm{eff}}(r)$. Between these limiting radii would lie a classical elliptical orbit, if it could have the same energy E and angular momentum L.

We apply to (13.14) a well known property of one-dimensional wave functions. Where $E - V_{\mathrm{eff}} > 0$, the sign of $\mathrm{d}^2\chi/\mathrm{d}r^2$ is opposite to that

[10] A difference from (13.4) is that r is now the separation of the bodies, rather than the distance from a fixed centre. This is dealt with by the replacement of m by the reduced mass μ; no other change is required to the form of $V_{\mathrm{eff}}(r)$.

Fig. 13.2 (a) The effective potential $V_{\text{eff}}(r)$ and the bound-state energy E for hydrogen in a quantum state having $n = 9$, $l = 6$. The effective potential is given by (13.15); it is the same as that plotted in Fig. 13.1. Where E crosses $V_{\text{eff}}(r)$ there are the limits $r_{\min}$, $r_{\max}$ of classical motion, given by (13.6) or (13.16).
(b) The function $\chi(r) = rR(r)$, in which $R(r)$ is the radial part of the wave function as in (13.10). Equation (13.14) shows that χ is an oscillating function of r for $r_{\min} < r < r_{\max}$, a non-oscillating function outside that range, and has an inflexion at each of the limits of classical motion.
(c) The radial probability density $|\chi|^2 = r^2 |R(r)|^2$ (solid curve) and the "classical radial probability" (broken curve). The classical probability is calculated by taking a time average of the motion in the elliptical orbit of Fig. 13.1. It is zero outside the classical limits, and between those limits rises to infinity as each limit is approached (the orbiting particle lingers close to each limit because $\dot{r}$ is small there).

The quantum probability $r^2 |R(r)|^2$ should resemble the classical probability in the limit of large quantum numbers. The quantum numbers n, l are not large here, so the approach to the classical limit is not well displayed. For a more convincing case, see Fig. 13.3(a).

The classical and quantum radial probabilities are given by (13.30) and (13.38). They are normalized in the same way, so their magnitudes are directly comparable.

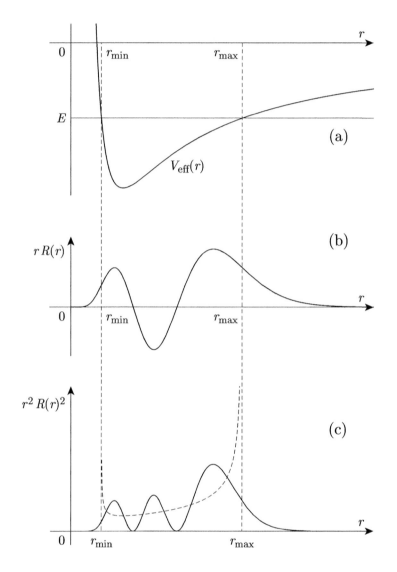

[11] Note that it is $\chi = rR(r)$ that has inflexions at $r_{\min}, r_{\max}$, as required by (13.14). Other functions, such as $R(r)$ or $r^2R(r)^2$, have inflexions nearby, but not exactly at those radii.

[12] The ideas pursued here are taken from Kleppner *et al.* (1981). An atom having large n is often called a *Rydberg atom*.

of χ, which makes χ an oscillating function of r; this happens when r lies between the limits of classical motion. Where $E - V_{\text{eff}} < 0$, $\mathrm{d}^2\chi/\mathrm{d}r^2$ has the same sign as χ, which makes χ a non-oscillating function of r. Where $E = V_{\text{eff}}$, $\mathrm{d}^2\chi/\mathrm{d}r^2 = 0$ and a graph of χ versus r has an inflexion. Just such a behaviour may be seen in Fig. 13.2(b).[11]

In one dimension, the quantum probability for finding a particle in δx is given by $|\psi(x)|^2 \, \delta x$. The corresponding classical probability can be calculated by finding the dwell time for a classical particle in each δx. It is customary for these distributions to be compared for a harmonic oscillator, which we do here in Fig. 13.3(b). Our comparison is made for $n = 24$ because the predictions of quantum mechanics approach those of classical mechanics when quantum numbers are large.[12]

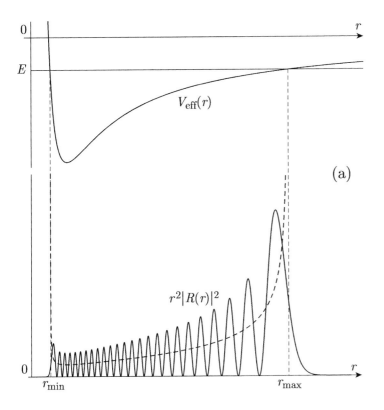

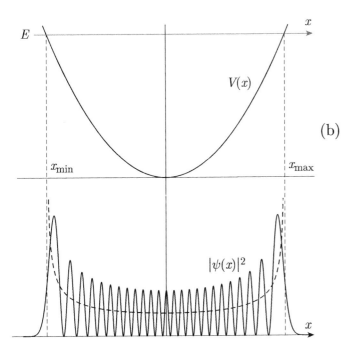

Fig. 13.3 (a) The effective potential $V_{\mathrm{eff}}(r)$ and the energy E (negative) for a hydrogen bound state, plotted for $n = 50$, $l = 25$.

The lower panel shows $|\chi|^2 = r^2 |R(r)|^2$ for this quantum state (solid line), and the radial probability for a classical elliptical orbit having the same energy and angular momentum (broken line). The quantum and classical radial probabilities are normalized in the same way, so their relative magnitudes are directly comparable.

The classical probability rises towards infinity at the classical limits because $\dot{r} = 0$ there.

A plot of χ versus r (not shown) has inflexions at $r_{\min}, r_{\max}$. The $|\chi|^2$ curve drawn has inflexions, but these do not lie precisely at $r_{\min}, r_{\max}$.

(b) The potential $V(x)$ for a harmonic oscillator having energy E; the limits of classical motion lie at $x_{\min}, x_{\max}$.

The lower panel shows the squared wave function for a harmonic oscillator having quantum number $n = 24$, a value chosen so that $\psi(x)$ has the same number of zeros as does the hydrogen $R(r)$. Again, a broken line shows the classical probability distribution for an oscillator having the same energy.

For the harmonic oscillator, the classical probability approaches infinity at the classical limits because the oscillator is stationary ($\dot{x} = 0$) there. Near the classical limits, the wave function "tries" to reproduce this steep rise towards infinity. Near the middle of the well, the classical probability looks like a low-resolution view of the quantum probability.

Harmonic-oscillator wave functions are often drawn accompanied by the limits $x_{\min}, x_{\max}$ of classical motion, as here, as these limits give a good indication as to the x-range within which the oscillator lies. I argue that the classical limits $r_{\min}, r_{\max}$ similarly give the best indication of the r-extent for hydrogen-atom wave functions, and that those limits are given by (13.16).

[13] Given Fig. 13.3(a), it seems obvious (to me anyway) that hydrogen wave functions—of every energy—are best presented along with the classical limits r_{min}, r_{max} for comparison. Yet many textbook presentations of hydrogen omit the classical limits entirely (a missed opportunity), or over-approximate them to $r_{min} = 0$, $r_{max} = 2n^2(a_0/Z)$, effectively ignoring the centrifugal potential. Confirm that in Fig. 13.3(a), $r_{min} = 349(a_0/Z)$, $r_{max} = 4651(a_0/Z)$, while $r_{max} \approx 2n^2(a_0/Z)$ yields $r_{max} \approx 5000(a_0/Z)$.

The behaviour of highly excited hydrogen is shown in Fig. 13.3(a) for a bound state having $n = 50$, $l = 25$ (problems 13.6 and 13.8). Quantum and classical probabilities are displayed; the limiting radii are:

$$\left. \begin{aligned} r_{min} &= (a_0/Z)\, n\{n - \sqrt{n^2 - l(l+1)}\} \\ r_{max} &= (a_0/Z)\, n\{n + \sqrt{n^2 - l(l+1)}\} \end{aligned} \right\}. \tag{13.16}$$

I have prepared Fig. 13.3 to emphasize the family resemblance between its (a) (hydrogen) and (b) (harmonic oscillator): each looks like a "bent" version of the other. It seems uncontroversial that the classical limits x_{min}, x_{max} for the oscillator give a good, if rough, indication of the oscillator's x-extent, and do so for all n. Similarly, I argue from Fig. 13.3(a) that the classical limits r_{min}, r_{max} give the best indication of the r-range for hydrogen[13] and do so for all n. For a more extreme

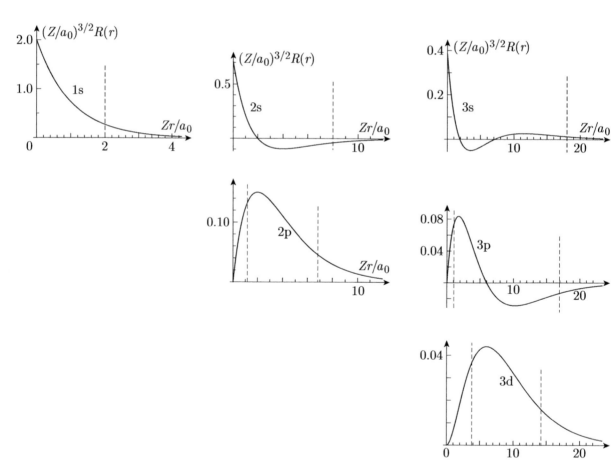

Fig. 13.4 Hydrogen radial wave functions for $n = 1, 2, 3$. The limits r_{min}, r_{max} of classical motion are indicated by broken vertical lines. The wave functions plotted are $R(r)$, not $r\,R(r)$, so inflexions are not located at the classical limits. Note that the classical limits depend upon l as well as on n. For these low-quantum-number states (farthest from classical behaviour), the electron's probability distribution extends well beyond the classical limits, both "outwardly" and (for $l \neq 0$) "inwardly".

The p wave functions start out from the origin "linearly", proportional to r; on the scale we have had to use here, these rises are so steep that they might be misinterpreted as vertical. The 3d wave function exits from the origin as r^2.

counterexample to other possibilities, see the "nearly circular orbit" of problem 13.7.

Radial wave functions for hydrogen are displayed in Fig. 13.4 for $n = 1, 2, 3$. The limits on classical motion $r_{\min}, r_{\max}$ are calculated from (13.16) and are indicated by broken vertical lines.

13.6   Positronium

Positronium is an "atom" consisting of an electron and a positron forming a bound state. The quantum numbers and eigenfunctions are similar to those for hydrogen. The difference lies in the mass of the positron, equal to that of the electron, replacing the relatively large mass of the nucleus in hydrogen.

Equation (13.8) tells us the reduced mass, from which we obtain

$$\mu_{\text{hydrogen}} = \frac{m_{\text{electron}} m_{\text{nucleus}}}{m_{\text{electron}} + m_{\text{nucleus}}} \approx m_{\text{electron}},$$

$$\mu_{\text{positron}} = \frac{m_{\text{electron}} m_{\text{positron}}}{m_{\text{electron}} + m_{\text{positron}}} = \frac{m_{\text{electron}}}{2}.$$

The dimensions of any radial wave function are expressed in terms of the Bohr radius, which is

$$a_{\text{hydrogen}} = \frac{\hbar^2 (4\pi\epsilon_0)}{\mu_{\text{hydrogen}} e^2},$$

$$a_{\text{positronium}} = \frac{\hbar^2 (4\pi\epsilon_0)}{\mu_{\text{positronium}} e^2} \approx 2 a_{\text{hydrogen}}.$$

It appears, then, that the wave functions for positronium extend over doubled distances, relative to those for hydrogen.

This conclusion is wrong; we were warned in § 13.3.[14] Remember the definition of r: it is the distance between the two particles, in both cases. For hydrogen, $r = a_{\text{hyd}}$ gives roughly the orbit radius for the electron, because the centre of mass is close to the nucleus. For positronium, r is the right-across distance from positron to electron, *double* the distance to the centre of mass. Figure 13.5 gives a cartoon of this, showing how we might draw the rough size of each orbit.[15]

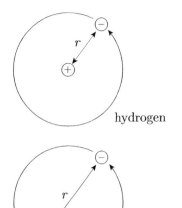

Fig. 13.5 In hydrogen, the electron orbits round the nucleus, which is close to the centre of mass of the atom (upper sketch). In the corresponding state of positronium, the electron orbits at twice the distance from the positron, which puts it at the same distance from the centre of mass.

[14] Our reasoning exactly parallels that given by Cohen-Tannoudji *et al.* (1977), pp 808–9.

[15] For the hydrogen and positron ground states, Fig. 13.5 might show the e^{-1} points of the $e^{-r/a}$ wave functions. Of course, the arrows indicating rotational motion would be inappropriate for such 1s states—hence "cartoon".

Problems

Problem 13.1   Classical orbit and its energy

Use (13.1) to eliminate $\dot{\theta}$ from (13.2) and so obtain a differential equation for r. Multiply through by $\dot{r}$ and integrate term by term to obtain (13.3).

Backtrack to show that

$$\frac{L^2}{2mr^2} = \tfrac{1}{2} m (r\dot{\theta})^2,$$

so the first two terms in (13.3) give the two contributions to the kinetic energy. Hence identify E with the total energy of the orbiting particle.[16]

[16] Check that in § 13.2 the origin was placed at the source—fixed—of the inverse square force. That choice of origin excluded the possibility of centre-of-mass motion.

Problem 13.2 Classical orbit: reduced mass

A classical-mechanics system consists of two particles 1 and 2, having masses m_1 and m_2, located at r_1, r_2, and interacting via an attractive inverse-square-law force $-\alpha/(\text{separation})^2$.

Convince yourself that the equations of motion of the two masses can be written in the form:

$$m_1 \ddot{r}_1 = \frac{-\alpha(r_1 - r_2)}{|r_1 - r_2|^3}, \qquad m_2 \ddot{r}_2 = \frac{-\alpha(r_2 - r_1)}{|r_2 - r_1|^3}. \qquad (13.17)$$

(1) Make a change of variables from r_1, r_2 to R, r where R is the location of the centre of mass and r is the location of particle 1 relative to particle 2:

$$R = \frac{m_1 r_1 + m_2 r_2}{m_1 + m_2}, \qquad r = r_1 - r_2. \qquad (13.18)$$

Show that the equations of motion (13.17) take the form

$$m_1 \left(\ddot{R} + \frac{m_2}{m_1 + m_2} \ddot{r} \right) = \frac{-\alpha r}{r^3}, \qquad m_2 \left(\ddot{R} - \frac{m_1}{m_1 + m_2} \ddot{r} \right) = \frac{+\alpha r}{r^3},$$

where $r = |r|$.

(2) Add the last two equations to show that the centre of mass moves with constant velocity. Show that this was foreseeable from (13.17).

(3) Subtract the last two equations to show that

$$\mu \ddot{r} = \frac{-\alpha r}{r^3}, \qquad \text{where} \quad \mu = \frac{m_1 m_2}{m_1 + m_2}.$$

Here μ is the reduced mass, while r is the distance from mass 2 to mass 1. These equations confirm the similar statements (13.7) and (13.8) in § 13.3.

(4) Re-work this problem with a general potential $V(|r_1 - r_2|)$ replacing the Coulomb potential, and show that the conclusions continue to hold.[17]

Problem 13.3 Schrödinger equation: reduced mass

Confirm that the non-relativistic time-dependent Schrödinger equation for particles having masses m_1 and m_2 in potential $V(|r_1 - r_2|)$ is

$$i\hbar \frac{\partial \Psi}{\partial t} = \frac{-\hbar^2}{2m_1} \nabla_1^2 \Psi + \frac{-\hbar^2}{2m_2} \nabla_2^2 \Psi + V \Psi.$$

Here the particles have locations r_1 and r_2, and ∇_1^2 differentiates with respect to (x_1, y_1, z_1) and similarly.

We are to express[18] the Schrödinger equation in terms of the centre-of-mass location $R = (X, Y, Z)$ and the relative location $r = (x, y, z)$. Following (13.18) write

$$X = \frac{m_1 x_1 + m_2 x_2}{m_1 + m_2}, \qquad x = x_1 - x_2, \qquad M = m_1 + m_2.$$

Using these, show that

$$\frac{\partial \Psi}{\partial x_1} = \left(\frac{m_1}{m_1 + m_2} \frac{\partial}{\partial X} + \frac{\partial}{\partial x} \right) \Psi.$$

[17] Thus the conclusions of this problem hold for the movements of the two nuclei in a diatomic molecule, as well as for a hydrogen atom.

[18] In the classical-mechanics problem, we were able to show that the overall linear momentum is conserved, and therefore that the centre of mass moves with constant velocity. This was done before transforming coordinates to R and r, thus motivating that transformation. We have no similar pointer here in the mathematics. However, the precedent from classical mechanics—and our physics knowledge—strongly suggest that the corresponding transformation should be tried.

With this start, find $\partial^2 \Psi / \partial x_1^2$. Follow a similar route to $\partial^2 \Psi / \partial x_2^2$, and then show

$$\frac{1}{m_1} \frac{\partial^2 \Psi}{\partial x_1^2} + \frac{1}{m_2} \frac{\partial^2 \Psi}{\partial x_2^2} = \frac{1}{(m_1 + m_2)} \frac{\partial^2 \Psi}{\partial X^2} + \frac{1}{\mu} \frac{\partial^2 \Psi}{\partial x^2}.$$

Similar expressions result after cyclic replacements of x with y and z. Add these to obtain

$$i\hbar \frac{\partial \Psi}{\partial t} = \frac{-\hbar^2}{2M} \nabla_{\boldsymbol{R}}^2 \Psi + \frac{-\hbar^2}{2\mu} \nabla_{\boldsymbol{r}}^2 \Psi + V(r) \Psi, \qquad (13.19)$$

where $\nabla_{\boldsymbol{R}}^2$ differentiates with respect to X, Y, Z and $\nabla_{\boldsymbol{r}}^2$ differentiates with respect to x, y, z.

   Equation (13.19) is a partial differential equation for Ψ, friendlier than what we started with because V depends upon r and not upon R. There is only one trick that we know for dealing with such an equation: separate the variables! Write

$$\Psi(\boldsymbol{R}, \boldsymbol{r}, t) = S(\boldsymbol{R}) \times \psi(\boldsymbol{r}) \times T(t),$$

and substitute into (13.19) to obtain

$$\left(\frac{-\hbar^2}{2M} \frac{\nabla_{\boldsymbol{R}}^2 S}{S} \right) + \left(\frac{-\hbar^2}{2\mu} \frac{\nabla_{\boldsymbol{r}}^2 \psi}{\psi} + V(r) \right) = i\hbar \frac{\dot{T}}{T}.$$

This permits a double separation of variables:

$$\frac{-\hbar^2}{2M} \nabla_{\boldsymbol{R}}^2 S = (E_{\text{total}} - E)S, \qquad i\hbar \frac{dT}{dt} = E_{\text{total}} T, \qquad (13.20\text{a})$$

$$\frac{-\hbar^2}{2\mu} \nabla_{\boldsymbol{r}}^2 \psi + V(r)\psi = E\psi, \qquad (13.20\text{b})$$

in which E_{total} and E are two separation constants.

   Show that S and T, taken together, describe uniform motion of the total mass M. The energy E belongs to the "internal" motion of the electron in the field of the nucleus, which leaves $E_{\text{total}} - E$ for translational motion of the atom as a whole. See if you agree with this identification of the energies.

   The Schrödinger equation (13.20b) contains the reduced mass μ and the separation $\boldsymbol{r}$ of the two particles. Although the algebra looks very different, we have recovered (13.9), a Schrödinger equation that has the same "structure" as applies to the classical orbit of problem 13.2.

Problem 13.4 Separation of radial and angular motion
Show that (13.9), expressed in spherical polar coordinates r, θ, ϕ, reads

$$\frac{-\hbar^2}{2\mu} \left\{ \frac{1}{r^2} \frac{\partial}{\partial r} \left(r^2 \frac{\partial \psi}{\partial r} \right) + \frac{1}{r^2 \sin \theta} \frac{\partial}{\partial \theta} \left(\sin \theta \frac{\partial \psi}{\partial \theta} \right) + \frac{1}{r^2 \sin^2 \theta} \frac{\partial^2 \psi}{\partial \phi^2} \right\} + V(r)\,\psi = E\,\psi. \quad (13.21)$$

In this, make a partial separation of variables by writing

$$\psi(r, \theta, \phi) = R(r)\, Y(\theta, \phi), \qquad (13.22)$$

as in (13.10), and show that (13.21) becomes

$$
\frac{1}{R}\frac{d}{dr}\left(r^2\frac{dR}{dr}\right) - \frac{2\mu}{\hbar^2}r^2\,V(r) + \frac{2\mu}{\hbar^2}r^2\,E
$$

$$
= \frac{-1}{Y}\left\{\frac{1}{\sin\theta}\frac{\partial}{\partial\theta}\left(\sin\theta\frac{\partial Y}{\partial\theta}\right) + \frac{1}{\sin^2\theta}\frac{\partial^2 Y}{\partial\phi^2}\right\}
$$

$$
= \text{(separation constant)} = l(l+1). \tag{13.23}
$$

Here the variables are separated because the left-hand side does not depend on θ or ϕ, while the right-hand side does not depend on r. Therefore both sides are constant. We have, with a great deal of malice aforethought,[19] written the separation constant as $l(l+1)$.

Write out the two outcomes of (13.23) and confirm that they give eqns (13.11) and (13.12).

Problem 13.5 *Orbital angular momentum*

Aim for a further separation of variables by writing

$$
Y(\theta,\phi) = \Theta(\theta) \times \Phi(\phi),
$$

and recast eqn (13.12) into the form

$$
\frac{\sin\theta}{\Theta}\frac{d}{d\theta}\left(\sin\theta\frac{d\Theta}{d\theta}\right) + l(l+1)\sin^2\theta
$$

$$
= -\frac{1}{\Phi}\frac{d^2\Phi}{d\phi^2} = \text{(separation constant)} = m^2.
$$

Show that the second part of this yields

$$
\Phi \propto e^{im\phi}. \tag{13.24}
$$

Argue that m must be real, otherwise Φ grows indefinitely with either increase or decrease of ϕ. Further, if we can assume that Φ is a single-valued function of angle ϕ, then[20]

$$
m = \text{integer (positive, negative or zero).} \tag{13.25}
$$

Now turn attention to the equation giving Θ as a function of θ:

$$
\frac{1}{\sin\theta}\frac{d}{d\theta}\left(\sin\theta\frac{d\Theta}{d\theta}\right) + l(l+1)\Theta - \frac{m^2}{\sin^2\theta}\Theta = 0.
$$

This equation can be put into a standard form by making the substitution $\nu = \cos\theta$. Show that this results in[21]

$$
(1-\nu^2)\frac{d^2\Theta}{d\nu^2} - 2\nu\frac{d\Theta}{d\nu} + l(l+1)\Theta - \frac{m^2}{1-\nu^2}\Theta = 0. \tag{13.26}
$$

This is the "associated" Legendre equation, whose properties can be looked up in works such as Abramowitz and Stegun (1965), Chapter 8. The solution[22] is a "well behaved" function of θ provided that $l - |m|$ is an integer $\geqslant 0$. Since m is already being taken to be an integer, this makes l an integer as well:

$$
l = \text{integer}, \qquad l \geqslant |m|.
$$

[19] It is an open secret that l will turn out to be a positive integer. But this is not being assumed at the present stage.

[20] This is not trivially obvious. We might, for example, take seriously the possibility that $m = \pm\frac{1}{2}$, so that Φ is double-valued, joining onto itself after two rotations round the z-axis. Such possibilities can be excluded, but not by any simple argument. We duck the question here, as the concern of this chapter is with less esoteric matters.

[21] It is more usual to use μ to represent $\cos\theta$, but we have already used μ to represent the reduced mass.

[22] One way to solve (13.26): make the substitution $\Theta = (\sin\theta)^{|m|}G(\nu)$ and obtain the "ultraspherical" (Gegenbauer) differential equation for G:

$$
(1-\nu^2)G'' - 2(|m|+1)\nu G'
$$
$$
+ \left\{l(l+1) - |m|\,(|m|+1)\right\}G = 0
$$

(Abramowitz and Stegun 1965, relation 22.6.5). This equation may be solved by the series method. There are two solutions (one even in ν, one odd), both well behaved at $\nu = 0$ (the expansion is not about a singular point). Both solutions blow up, and cause Θ to blow up, at $\nu = \pm 1$ unless a series is terminated. One series can be terminated by a suitable choice of l, while the other (which does not terminate for the same l) must be rejected. The condition for termination—the eigenvalue condition—is $l - |m| = \text{(integer)} \geqslant 0$.

There can be problems (not hydrogen) that yield (13.24) and (13.26) but for which m is non-integral. The eigenvalue condition on $l - |m|$ is as above. The wording used in the text, admittedly unusual in a hydrogen context, is therefore correct.

Textbook solutions of the associated Legendre equation usually follow a different route from that given here, noticing that the ultraspherical equation is equivalent to the mth derivative of the Legendre equation. I outline the "ultraspherical" route here because it is what one would do if one did not know m to be an integer.

The solution for Θ is now an associated Legendre polynomial $\mathrm{P}_l^m(\cos\theta)$. Combining this with (13.24), we have

$$Y = Y_{l,m}(\theta,\phi) = (\text{constant})\,\mathrm{P}_l^{|m|}(\cos\theta)\,e^{im\phi}. \qquad (13.27)$$

Your favourite quantum-mechanics textbook should show you that $Y_{l,m}(\theta,\phi)$ represents a state where the electron is in an eigenstate of $\hat{l}_z$ with eigenvalue $m\hbar$, and also it's in an eigenstate of $\hat{l}^2$ (angular momentum squared) with eigenvalue $l(l+1)\hbar^2$.

Problem 13.6 The radial probability for a classical elliptical orbit
Consider a "half-orbit" as r increases from $r_{\min}$ to $r_{\max}$. The time that the particle spends between r and $r+dr$ is dt where $\dot{r} = dr/dt$, so $dt = dr/\dot{r}$. The fraction of its time that the particle spends within range dr is

$$P_{\mathrm{cl}}(r)\,dr = \frac{1}{T_{1/2}}\frac{dr}{\dot{r}}, \qquad (13.28)$$

where $T_{1/2}$ is the time required for tracing out the "half-orbit":

$$T_{1/2} = \int_{r_{\min}}^{r_{\max}} \frac{dr}{\dot{r}}.$$

Substitute for $\dot{r}$ from (13.5) to put the integral for $T_{1/2}$ into the form

$$T_{1/2} = \sqrt{\frac{\mu}{2(-E)}} \int_{r_{\min}}^{r_{\max}} \frac{r\,dr}{\sqrt{-r^2 + \dfrac{\alpha}{(-E)}r - \dfrac{L^2}{2\mu(-E)}}}.$$

Complete the square in the denominator to create the shape

$$\left(\frac{\alpha^2}{4E^2} + \frac{L^2}{2\mu E}\right) - \left(r + \frac{\alpha}{2E}\right)^2 = a^2 - r'^2,$$

where a and r' are defined by this statement.

Show[23] that $a^2 > 0$, so that a is real (and it will be taken to be positive).

Show that

$$r_{\min} + \frac{\alpha}{2E} = -a, \qquad r_{\max} + \frac{\alpha}{2E} = +a,$$

so that, when the variable of integration is changed from r to r', the limits of integration become $-a$ and a. Show that

$$T_{1/2} = \sqrt{\frac{\mu}{2(-E)}} \int_{-a}^{a} \frac{(r' - \alpha/2E)}{\sqrt{a^2 - r'^2}}\,dr' = \pi\alpha\sqrt{\frac{\mu}{8(-E)^3}}.$$

Backtrack to (13.28) to show that the classical radial probability is $P_{\mathrm{cl}}(r)$ where

$$P_{\mathrm{cl}}(r)\,dr = \frac{2(-E)}{\pi\alpha}\frac{r\,dr}{\sqrt{-r^2 + \dfrac{\alpha}{(-E)}r - \dfrac{L^2}{2\mu(-E)}}}. \qquad (13.29)$$

One final transformation is required for computing a graph such as that of Fig. 13.2(c). We continue to assume classical mechanics to hold,

[23] *Hint:* Find the minimum of $V_{\mathrm{eff}}(r)$ and use the knowledge that E lies above that minimum. Look also at the square root in (13.6).

but we insert into (13.29) the energy and angular momentum of an (n, l) quantum state. Thus

$$\alpha = \frac{Ze^2}{4\pi\epsilon_0}, \qquad E = -\frac{\mu e^4}{2\hbar^2(4\pi\epsilon_0)^2}\frac{Z^2}{n^2}, \qquad L^2 = l(l+1)\hbar^2,$$

$$a_0 = \frac{\hbar^2(4\pi\epsilon_0)}{\mu e^2}, \qquad \rho = \frac{Z}{n}\frac{r}{a_0}.$$

Show that with these substitutions,

$$P_{cl}(\rho)\,d\rho = \frac{\rho\,d\rho}{\pi n\sqrt{-\rho^2 + 2n\rho - l(l+1)}}. \tag{13.30}$$

Show that, in terms of the scaled radius ρ, the limits of classical motion occur at radii

$$\rho_{min} = n - \sqrt{n^2 - l(l+1)}, \qquad \rho_{max} = n + \sqrt{n^2 - l(l+1)}. \tag{13.31}$$

These are the zeros of the denominator in (13.30); show that they result in (13.16).

Figure 13.1 shows a classical elliptical orbit, calculated by giving the quantum eigenvalues to energy and angular momentum. From the quadratic denominator of (13.30), show that,[24] in ρ units, the ellipse has semi-major axis n and semi-minor axis $\sqrt{l(l+1)}$. Show that, for the case of Fig. 13.1, the ellipse semi-axes (in metres) are $81a_0/Z$ and $58.3a_0/Z$, while the greatest and least orbit radii are $137.2a_0/Z$ and $24.8a_0/Z$.

Problem 13.7 An atom having $n - l - 1 = 0$
The radial wave function χ is determined by Q, a solution of (13.34). The coefficient of Q in (13.34) is now zero. The differential equation is of second order so must have two solutions. Show that one solution is $Q = $ constant, and the other behaves as $e^{+\rho}$ for large ρ, and so has to be discarded. Show that therefore $\chi \propto \rho^{l+1}e^{-\rho}$, $R \propto \rho^{n-1}e^{-\rho}$. Show that this function has a simple peak with its maximum at $(a_0/Z)n(n-1)$.
Show from (13.16) that the classical-limit radii are $(a_0/Z)n(n \pm \sqrt{n})$:

$$\frac{\text{separation of classical-limit radii}}{\text{radius at peak}} = \frac{(a_0/Z)2n\sqrt{n}}{(a_0/Z)n(n-1)} = \frac{2\sqrt{n}}{n-1}.$$

For a given n, setting $l = n-1$ gives the smallest possible $(r_{max} - r_{min})$: we have the nearest that quantum mechanics can give to a classical circular orbit.[25]

To display near-classical behaviour, we always look at the limit of large quantum numbers, so sketch the wave function on a to-scale radial scale for $n = 30$ (other values at your choice).

In sidenote 13 it is mentioned that the outer limit of a hydrogen wave function is sometimes given to be a radius of $(a_0/Z)2n^2$. Include this in your diagram, and check whether it's a sensible indication for the size of the atom.[26]

Problem 13.8 The radial wave function: technicalities
We are to find the eigenfunction solutions to eqn (13.14).

---

[24] *Hint:* You may assume that the orbit is an ellipse. Show that the major axis is $r_{min} + r_{max}$ and that the semi-minor axis is $\sqrt{r_{min}r_{max}}$. Then use the properties of the quadratic expression in (13.30). If you find yourself hammering out the product of the expressions in (13.31) you've missed a trick.

[25] For another view of the "near-circular orbit", see Woodgate (1980) problem 2.3.

[26] My values for $n = 30$:

$$r_{min} = (a_0/Z) \times 736,$$
$$r_{max} = (a_0/Z) \times 1064,$$
$$(a_0/Z)2n^2 = (a_0/Z) \times 1800.$$

So the claimed outer limit is an overestimate by nearly a factor 2.

It helps first to clear out constants by expressing things in terms of a scaled radius ρ, and a scaled energy, so write

$$\rho^2 = \frac{2\mu(-E)}{\hbar^2}r^2; \qquad -E = \frac{\mu e^4}{2\hbar^2(4\pi\epsilon_0)^2}\frac{Z^2}{n^2},$$

in which $1/n^2$ is a new quantity related to the energy.[27] Show that these lead to

$$\rho = \frac{Z}{n}\frac{r}{a_0} \qquad \text{where} \qquad a_0 = \frac{\hbar^2(4\pi\epsilon_0)}{\mu e^2}.$$

Show that (13.14) becomes

$$\frac{\mathrm{d}^2\chi}{\mathrm{d}\rho^2} + \left(\frac{-l(l+1)}{\rho^2} + \frac{2n}{\rho} - 1\right)\chi = 0. \qquad (13.32)$$

We do not yet know that n is an integer, so it appears simply as a constant that may in due course be subject to an eigenvalue condition.

Find familiar properties in the coefficient of χ in (13.32). It is a quadratic form with roots $\rho = n \pm \sqrt{n^2 - l(l+1)}$, which agrees with (13.31). From the sign of $(1/\chi)\mathrm{d}^2\chi/\mathrm{d}\rho^2$, show that χ is an oscillating function of ρ when ρ lies between the roots, non-oscillating when outside them. Identify the roots with the limits of classical motion.

Differential equation (13.32) is accompanied by boundary conditions. It may help to look back at Fig. 13.2(b). We must require that $\chi = rR$ falls when r (equivalently ρ) lies outside the limits of classical motion, that is, as $\rho \to 0$ and as $\rho \to \infty$.

Look first at the limit $\rho \to \infty$. We can ignore all but the largest term in the coefficient of χ in (13.32), and we have approximately

$$\frac{\mathrm{d}^2\chi}{\mathrm{d}\rho^2} \approx \chi.$$

The outcome can be seen at once to be $\chi \propto e^{\pm\rho}$. The possibility $e^{+\rho}$ is unsuitable, so we reject it and accept $e^{-\rho}$. The solution for χ can now be expected to have the form

$$\chi = e^{-\rho} \times \text{(slower-varying function of }\rho).$$

Think about the limit $\rho \to 0$. To see this limiting behaviour we can try a series solution

$$\chi = e^{-\rho}\rho^s \sum_{i=0}^{\infty} a_i \rho^i$$

in (13.32). The indicial equation for s yields $s = -l$ or $s = l+1$. For small ρ we have $\chi \sim \rho^{-l}$ or $\chi \sim \rho^{l+1}$ giving $R \sim \rho^{-l-1}$ or $R \sim \rho^l$. For $l \geqslant 1$, $R \sim \rho^{-l-1}$ gives a divergent normalizing integral so it must be rejected.[28]

We build in the above findings by looking for a solution to (13.32) in the form

$$\chi(\rho) = e^{-\rho}\rho^{l+1}Q(\rho), \qquad (13.33)$$

and we start again from here, rather than pursuing further the series solution begun above.

[27] We are looking to find bound states, so E will be negative, making n^2 real. Moreover, we want ρ to have the same sign as r, so n is positive.

[28] The case $l = 0$ needs a little more care because $R \sim \rho^{-l-1} = \rho^{-1}$ is normalizable. However this R does not solve the Schrödinger equation (13.9) even though it solves (13.32); remember that $\nabla^2(1/r) = -4\pi\delta(r)$. (Ask yourself how such a non-solution could have turned up.) Therefore $\chi(\rho)$ behaves as ρ^{l+1} as $\rho \to 0$, for all l including $l = 0$.

[29] If we use a trial series for $Q(\varrho)$, we find that the series diverges as e^{ϱ} unless it is terminated by setting $(n - l - 1)$ to an integer (or zero), and this results in the quoted Laguerre polynomial. There must still be a second solution (the differential equation for Q is of second order, so it has two independent solutions). That second solution can be shown to be badly behaved, so it must be rejected.

[30] There are many different notations for Laguerre polynomials. We use that of Abramowitz and Stegun (1965), relation 22.6.15.

[31] This is not the conventional way to normalize (why not?), but it's the same as is implied by (13.30). We are making P_{qm} and P_{cl} directly comparable.

[32] Abramowitz and Stegun (1965), relation 22.7.30.

[33] Abramowitz and Stegun (1965), relation 22.2.12.

Substitute (13.33) into (13.32), using $\varrho = 2\rho$, and obtain

$$\varrho \frac{\mathrm{d}^2 Q}{\mathrm{d}\varrho^2} + \left\{ (2l+1) + 1 - \varrho \right\} \frac{\mathrm{d}Q}{\mathrm{d}\varrho} + (n - l - 1)Q = 0. \tag{13.34}$$

Equation (13.34) is a form of Laguerre's equation. For it to have a well behaved solution, it is necessary that[29]

$$n = \text{integer} \geqslant l + 1, \tag{13.35}$$

and then the solution is a Laguerre polynomial:[30]

$$Q(\varrho) \propto \mathrm{L}^{(2l+1)}_{n-l-1}(\varrho).$$

Backtracking to (13.33), the radial wave function is now

$$R(\rho) = (\text{constant } B)\, \mathrm{e}^{-\rho}\, \rho^l\, \mathrm{L}^{(2l+1)}_{n-l-1}(2\rho). \tag{13.36}$$

In order to plot graphs like Fig. 13.2(c) and Fig. 13.3(a), we need to know the normalization constant for $R(\rho)$. We choose to normalize[31] by

$$1 = \int_0^\infty \rho^2\, |R(\rho)|^2\, \mathrm{d}\rho = |B|^2 \int_0^\infty \mathrm{e}^{-2\rho}\, \rho^{2l+2}\, \left\{ \mathrm{L}^{(2l+1)}_{n-l-1}(2\rho) \right\}^2 \mathrm{d}\rho.$$

This makes $P_{qm}(\rho)\, \mathrm{d}\rho = \rho^2\, |R(\rho)|^2\, \mathrm{d}\rho$ correspond to the $P_{cl}(\rho)\, \mathrm{d}\rho$ of (13.30). The integral here is not the normalization integral for Laguerre polynomials, which contains ρ^{2l+1} rather than ρ^{2l+2}. However,[32]

$$\mathrm{L}^{(2l+1)}_{n-l-1}(x) = \mathrm{L}^{(2l+2)}_{n-l-1}(x) - \mathrm{L}^{(2l+2)}_{n-l-2}(x),$$

and when this is used it results in two normalization integrals and a cross-term that vanishes by orthogonality. Show that[33]

$$R(\rho) = 2^{l+1} \left(\frac{(n-l-1)!}{n(n+l)!} \right)^{1/2} \mathrm{e}^{-\rho}\, \rho^l\, \mathrm{L}^{(2l+1)}_{n-l-1}(2\rho). \tag{13.37}$$

The radial probability, for comparison with the $P_{cl}(\rho)$ of (13.30), is

$$P_{qm}(\rho) = \frac{(n-l-1)!}{n(n+l)!}\, \mathrm{e}^{-2\rho}\, (2\rho)^{2l+2}\, \left\{ \mathrm{L}^{(2l+1)}_{n-l-1}(2\rho) \right\}^2. \tag{13.38}$$

As mentioned in the figure captions, we need $P_{qm}(\rho)$ and $P_{cl}(\rho)$ so that an absolute comparison of probabilities can be displayed in Fig. 13.3(a)

Identical particles and the helium atom

Would it surprise you to be told ... that the Atomic Theory is at work in this parish?

Flann O'Brien, *The Third Policeman*

Intended readership: At the point in a quantum-mechanics course when bosons and fermions are introduced. Again in a course on atomic physics when the identity of electrons is seen to control the quantum states of atoms such as helium.

This chapter and the next two give accounts of topics in atomic physics that suffer from misleading treatments in several standard textbooks. There are correct accounts in advanced books, but the beginner must use "soft" introductions, often in "compendium" books, that are surprisingly unreliable.

14.1   Introduction: harmless labelling

All electrons are identical: that is, each electron is identical to all others. Similar statements can be made about all other fundamental particles (and larger groupings, such as atoms of the same isotope). We often say *indistinguishable* as a synonym for *identical*.[1] Special quantum mechanical rules apply when we have two or more identical particles together.

To write a Schrödinger equation, and a wave function, for N particles, you must **label** the particles with space coordinates r_1, r_2, ..., r_N, and likewise label their spins. We must make sure that the labelling is **harmless**, so that it does not subvert the particles' indistinguishability.[2]

You might think that we ought to set up an N-particle wave function without introducing any labels at all, rather as vector notation avoids the need for mentioning components along coordinate axes. Unfortunately we can't think of a way of avoiding labels. We resign ourselves to using labels, and then engineer things to make sure the labelling is harmless.

Once the particles are labelled, only two kinds of wave function satisfy the indistinguishability condition: either the N-particle wave function is **symmetric** (no change) under every pairwise interchange of labels (such as $1 \leftrightarrow 2$); or the N-particle wave function is **antisymmetric** (reversal of sign) under every pairwise interchange of labels. Interchange means that we exchange 1 and 2 in both the space coordinates and the spin—all at once, as in the operation $\widehat{I} = (\widehat{I}_r \widehat{I}_s)$ of § 11.5.

[1] There is a difficulty with the English word *identity*, which can mean both unique-ness (as in identity card) and the exact opposite: same-ness (as in identical). So *identity* can be ambiguous and we frequently use *indistinguishability* instead.

[2] Compare with the situation in relativity. Any given physical system may be described in more than one Lorentz frame, and each frame's description is different from others'. However, the underlying physics must be the same, and a requirement is laid on the Lorentz frames, and the descriptions given within them, that makes this so. If we choose, as we must, to describe the physics within some one chosen frame, then that choice of frame must be made *harmless*. Four-vectors and four-tensors are introduced as the mathematical vehicle that ensures the choice of frame is harmless. A four-vector's components are different in different frames, but in such a way as to describe the exact-same physics.

165

Essays in Physics: Thirty-two thoughtful essays on topics in undergraduate-level physics. Geoffrey Brooker, Oxford University Press. © Geoffrey Brooker 2021. DOI: 10.1093/oso/9780198857242.003.0014

A collection of N bosons (integer spin) has a wave function that is symmetric under every pairwise interchange of labels. A collection of N fermions (half-integer spin) has the antisymmetric possibility.

There will be no surprises in the above. But care is needed with what is done next.

14.2   The Slater determinant for fermions

The particles we deal with most often are electrons, which are fermions. It is therefore appropriate to spend much of our time thinking about fermions.

A standard mathematical way of constructing a wave function obeying the fermion rules is to build a **Slater determinant**.

We imagine, to start with, that we have obtained wave functions for the N particles by a process in which those particles are dealt with one at a time. That is, no cognizance is yet taken of the fact that the particles must be indistinguishable: the particles are labelled as if particle 1 was distinguishable from particle 2. The particles have spin—necessarily, since we are talking about fermions, which can't have zero spin.

A representative one of these single-particle wave functions may be written after the fashion $\psi_a(\boldsymbol{r}_1)|a\rangle_1$. Here ψ_a is the (space) wave function for a quantum state whose quantum numbers are indicated[3] by a; the particle occupying this state is number 1 with space coordinates $\boldsymbol{r}_1 = (x_1, y_1, z_1)$. Likewise, the particle has a spin state indicated by the ket $|a\rangle$, again with a subscript 1 to indicate that it is particle-labelled-1 that has this spin state.[4] The Slater determinant is assembled using these building blocks.

The general form of a Slater determinant can now be written:

$$\psi_{N\text{-fermion}} = \frac{1}{\sqrt{N!}} \begin{vmatrix} \psi_a(\boldsymbol{r}_1)|a\rangle_1 & \psi_a(\boldsymbol{r}_2)|a\rangle_2 & \cdots & \psi_a(\boldsymbol{r}_N)|a\rangle_N \\ \psi_b(\boldsymbol{r}_1)|b\rangle_1 & \psi_b(\boldsymbol{r}_2)|b\rangle_2 & \cdots & \psi_b(\boldsymbol{r}_N)|b\rangle_N \\ \cdots & \cdots & \cdots & \cdots \\ \psi_z(\boldsymbol{r}_1)|z\rangle_1 & \psi_z(\boldsymbol{r}_2)|z\rangle_2 & \cdots & \psi_z(\boldsymbol{r}_N)|z\rangle_N \end{vmatrix}.$$
(14.1)

Here we have labelled the quantum states from a to z and the particles from 1 to N. Each row contains a single quantum state (space and spin), while each column contains a single label.[5]

The Slater determinant has a form that ensures the labelling of the fermions is harmless, because any interchange of labels, such as $1 \leftrightarrow 2$, interchanges two columns of the determinant, which does no more than change the sign of $\psi_{N\text{-fermion}}$. At the same time, the sign change is just what is required by the rules for fermions. It is this building-in of correct behaviour that motivates our constructing the Slater determinant.

Notice particularly that each element in the Slater determinant must contain both a space and a spin "wave function". Only in that way do we get a complete determinant that is antisymmetric under interchange of *all* labels attached to the particles.

[3]More correctly: the particle has a full set of four quantum numbers; the space wave function has three of these quantum numbers, and the spin has the fourth. We use a to indicate the full set of four, it being understood that each constituent takes what it needs from the four.

[4]With a given choice of z-axis, the spin states for a spin-½ particle might be written as $|\uparrow\rangle$ and $|\downarrow\rangle$, for example, or even as $\uparrow$ and $\downarrow$, with the arrow indicating the sign of quantum number m_s. In other contexts, we might write the spin-½ state as a two-element column matrix, available to be operated on by Pauli spin matrices.

Some fermions have spin greater than ½. In such cases a more elaborate notation is needed for specifying the z-component of spin, but nothing needs to be changed in (14.1).

[5]Of course, the rows and columns could equally well be laid out the other way round, because the value of a determinant is not changed by mirror-imaging its elements across the main diagonal. We have had to make an arbitrary choice.

We use ψ_N to represent the state of the entire system, so ψ_N encompasses the spin matrices as well as algebraic functions of the space coordinates.

For simplicity, the wave functions are all written as if they were independent of time. Think about what should be done if the physics is time dependent.

Comment 1: Physicists often say that a wave function changes sign (fermions) or not (bosons) under interchange of *particles*. As a shorthand this is acceptable.[6] But there isn't any way that you could interchange the particles themselves when they're identical—it doesn't make sense. So I prefer to be more correct (or pedantic) and insist that it is the *labels* that are being interchanged.

Comment 2: It is common for textbooks to present the Slater determinant in a more concise form than ours. Usually $\psi_a(\boldsymbol{r}_1)|a\rangle_1$ is further abbreviated to $\psi_a(\boldsymbol{r}_1)$ or to $\psi_a(1)$. The statement is made that this is the "total" wave function, thereby conveying that spin is included. The author is correct—but the student reading the account may well not be. Unfortunately, the abbreviated notation can easily lead us into some tempting errors: exercise 14.1(2).

Comment 3 on bosons: Although this section is about fermions, we may usefully draw a contrast here with bosons. What do bosons have in place of a Slater determinant? The answer is easily given: you write down the same set of permuted products as you get from multiplying out a Slater determinant ($N!$ of them), but you put a plus sign in front of every term, instead of letting the determinant rules insert minus signs. Try writing down such an expression for two particles, then for three, and it should be obvious that the rule stated here yields something that is, as required, symmetric under every pairwise interchange of labels.[7]

The Slater determinant is to be thought of as a building block, just as Schrödinger eigenfunctions are often used as an expansion set. The "wave function" for an atom (for example) may well be a linear combination of several Slater determinants. But the discipline remains: such a linear combination must be possible, otherwise the symmetry requirement is violated.

[6] The present author takes particular exception to statements like "in a solid, you could never tell if two atoms changed places", because this statement could only make sense if the particles had an underlying individuality that quantum mechanics denies.

[7] A many-boson wave function is subject to conditions fully as strict as those applicable to fermions. This may come as a surprise. When we learn statistical mechanics, we obtain the Fermi–Dirac distribution by requiring fermions to obey the Pauli exclusion principle; we obtain the Bose–Einstein distribution by not doing so. It is easy to get a sloppy idea that bosons are unconstrained, or less constrained than fermions, perhaps nearer-to-classical, because they aren't given orders by Pauli.

Exercise 14.1 The Slater determinant

(1) Consider the Slater determinant of (14.1). Show that if b is the same as a the determinant is zero. So two fermions cannot both be in state a. Show that this property applies equally to any sum of Slater determinants, and therefore to *any* possible fermion wave function:

Pauli exclusion principle: *two fermions cannot occupy the same quantum state.*

(2) In the same determinant, set $\boldsymbol{r}_1 = \boldsymbol{r}_2$, so that two fermions are at the same location in space. Show that this does not make two columns equal and therefore does *not* force the determinant to be zero. The first two columns could be equal only if it was possible to set $|a\rangle_1 = |a\rangle_2$, $|b\rangle_1 = |b\rangle_2$ and so on down the column, which is equivalent to setting $1 = 2$; but this is a nonsense because the labels attached to different particles cannot be equated.[8]

[8] This explains why it is not a good idea to consolidate the spin kets like $|a\rangle_1$ in with the space wave function $\psi(\boldsymbol{r}_1)$. If this consolidation is done, it is not at all obvious that equating two columns is a disallowed operation. And, inevitably, you can find instances where people, including the writers of respectable textbooks, have been misled.

14.3   Wave functions for two particles

Let us illustrate by first taking the case of two identical fermions having spin $\frac{1}{2}$. In the spirit of Slater determinants, let the fermions occupy single-particle space states ψ_a, ψ_b and spin states $\uparrow, \downarrow$. A *possible* way of building more useful two-body wave functions is to take linear combinations of two Slater determinants, resulting in[9]

$$\psi_s = \frac{1}{2}\left\{\psi_a(\boldsymbol{r}_1)\,\psi_b(\boldsymbol{r}_2) + \psi_a(\boldsymbol{r}_2)\,\psi_b(\boldsymbol{r}_1)\right\} \times \left\{\uparrow_1\downarrow_2 - \uparrow_2\downarrow_1\right\}, \quad (14.2)$$

$$\psi_t = \frac{1}{2}\left\{\psi_a(\boldsymbol{r}_1)\,\psi_b(\boldsymbol{r}_2) - \psi_a(\boldsymbol{r}_2)\,\psi_b(\boldsymbol{r}_1)\right\} \times \left\{\uparrow_1\downarrow_2 + \uparrow_2\downarrow_1\right\}. \quad (14.3)$$

These wave functions ψ_s, ψ_t have been constructed so that they factorize into space and spin pieces, giving total spins $S = 0, 1$ respectively, hence the subscripts s(inglet) and t(riplet).[10]

Now consider how things are different for a pair of bosons. To make everything needful show up, we need the bosons to have non-zero spin, which means that the smallest spin available is 1; we can't use the $\uparrow\downarrow$ notation now. Let the space wave functions be $\psi_c(\boldsymbol{r}_1)$ and $\psi_d(\boldsymbol{r}_1)$ for particle 1, with spin states $s_c(1)$, $s_d(1)$; similarly for particle 2. Then states corresponding to those above can be written in the forms:

$$\psi_- = \frac{1}{2}\left\{\psi_c(\boldsymbol{r}_1)\,\psi_d(\boldsymbol{r}_2) - \psi_c(\boldsymbol{r}_2)\,\psi_d(\boldsymbol{r}_1)\right\} \times \left\{s_c(1)\,s_d(2) - s_c(2)\,s_d(1)\right\}$$
$$(14.4)$$

$$\psi_+ = \frac{1}{2}\left\{\psi_c(\boldsymbol{r}_1)\,\psi_d(\boldsymbol{r}_2) + \psi_c(\boldsymbol{r}_2)\,\psi_d(\boldsymbol{r}_1)\right\} \times \left\{s_c(1)\,s_d(2) + s_c(2)\,s_d(1)\right\}$$
$$(14.5)$$

These wave functions have correct boson symmetry, as can be seen by inspection. If we prepare symmetrized building blocks as suggested above (multiplying out as with a Slater determinant but forcing all signs positive), then each of our two-boson wave functions is the sum of two such building blocks; see problem 14.1.

14.4   Consequences for fermions and bosons

A consequence of the Slater determinant is the Pauli exclusion principle obtained in exercise 14.1(1): No two fermions may occupy the same quantum state.

It is my view that the Pauli exclusion principle is given more prominence than (with today's knowledge) it deserves. We should be focusing on the antisymmetry rule that underlies Pauli.

The Pauli principle has consequences for reactions involving fermions. A collision (or other reaction) cannot put a fermion into a state that is already occupied. Expressions for reaction rates must include factors to build in this prohibition.

There is also a tendency for bosons to gather together in the same quantum state. That's why lasers work: photons emitted by stimulated emission go into the exact same quantum state as is occupied by the stimulating photon (§ 19.8). Reaction rates contain factors that build in

[9] Wave functions of the forms (14.2–3) are frequently written when discussing the two electrons in a helium atom, in which case a and b might be 1s and 2p, for example. We do so ourselves in (15.2–3). Nevertheless it should be mentioned at once that helium is not well described by (14.2–3). Of course, an expansion in terms of Slater determinants must be possible in principle; but many terms are needed, such that the expansion is not arithmetic-friendly. See sidenote 14.

[10] In conformity with § 14.1, I have written ψ_s and ψ_t so that the second term in each bracket is a labels-exchanged version of the first. The same applies to (14.4) and (14.5).

Problem 14.1 confirms that each of (14.2), (14.3) is indeed the sum of two Slater determinants.

The factors $\frac{1}{2}$ in expressions (14.2)–(14.5) ensure that the wave functions given are normalized if the constituent wave functions are normalized.

this tendency, a sort of inverse of the fermion prohibition.[11] However, there is no *obvious* way of obtaining this behaviour from wave functions such as those in (14.4) or (14.5), so discussion of this boson property belongs elsewhere.

Some textbooks sloganize the above statements about reaction rates: "fermions avoid each other's quantum states; bosons tend to huddle together in the same quantum state". This is correct. Unfortunately, an unwarranted extension is also to be found in print: "fermions keep out of each other's space; bosons huddle together in space".[12] Debunking this is easy, but we must make sure it is done. Particles avoid each other *in space* if their *space* wave function is antisymmetric (under interchange of the particles' labels); this is the case in eqns (14.3) and (14.4), so it's an option available equally to bosons and fermions. Likewise, bosons and fermions *may* be found together *in space* if their *space* wave function is symmetric; this is the case in eqns (14.2) and (14.5), so again both bosons and fermions can do it.[13] In fact, the first equation in each displayed pair is a counterexample to the statements that we're attacking.

14.5 Symmetric and antisymmetric space wave functions

One thing can be proved quickly. Let $\psi_{\mathrm{anti}}(\boldsymbol{r}_1, \boldsymbol{r}_2)$ be a two-particle *space* wave function that is antisymmetric under $1 \leftrightarrow 2$. Then

$$\psi_{\mathrm{anti}}(\boldsymbol{r}_1, \boldsymbol{r}_2) = -\psi_{\mathrm{anti}}(\boldsymbol{r}_2, \boldsymbol{r}_1)$$

and hence

$$\psi_{\mathrm{anti}}(\boldsymbol{r}, \boldsymbol{r}) = 0. \tag{14.6}$$

The two particles can never be found together at the same point in space.

> There is **no "inverse"** to (14.6): *no general rule that "space-symmetric means together".*

Discussions of space anti/symmetry frequently make use of (14.2–5) as examples, so it important to remember that these are special-case wave functions in the following ways:[14]

- the wave function factorizes into a product of a space part and a spin part;
- the space part has the shape $2^{-1/2}\{\psi_a(\boldsymbol{r}_1)\psi_b(\boldsymbol{r}_2) \pm \psi_a(\boldsymbol{r}_2)\psi_b(\boldsymbol{r}_1)\}$.

Neither characteristic is compulsory (a Slater determinant has neither property!), though both forms are found quite often.[15]

The space-symmetric wave functions (14.2) and (14.5) have the property that the particles have an increased probability of being found together (as compared with a notional simple-product wave function); see exercise 14.2. But these are special-case wave functions, however much they are highlighted in textbooks. In Chapter 17 we exhibit several space-symmetric wave functions that are counterexamples to any idea that "space symmetric means together"; an extreme counterexample[16] is in (16.5).

[11] This may be seen clearly in eqn (19.22) for the case of photon emission.

[12] As elsewhere, I do not identify the source of this misconception.

"Explanations" of this non-effect are also supplied: "exchange forces", said to be "a purely quantum effect": i.e. don't try to understand—it's magic.

We have mentioned that the Slater determinant is often presented in an abbreviated notation that can conceal the presence of spin in each element. My suspicion is that the notion "fermions keep out of each other's space" must be a misreading of this notation: the confusion of space-antisymmetric with overall-antisymmetric that is corrected here in exercise 14.1(2).

[13] The only exception is the case of spin-zero bosons, where the lack of a spin means that the space wave function does have to be symmetric.

[14] The quantum states of helium conform to the first of these but not the second: space and spin factorize in the wave function (until states are mixed by the tiny spin–orbit interaction), but the space wave function is not well described by (14.2) or (14.3); see Chapter 15.

[15] Factorization, of the wave function into a space part and a spin part, is not available, save in a few special cases, when we have more than two particles present. See problem 17.5.

[16] The wave function of (16.5) is built from a linear combination of four Slater determinants, rather than the two of (14.2–3).

Exercise 14.2 Space-symmetric wave functions

(1) Compare the wave functions:

$$\text{simple product:} \quad \psi_a(\boldsymbol{r}_1)\,\psi_b(\boldsymbol{r}_2) \tag{14.7}$$

$$\text{space symmetric:} \quad \frac{1}{\sqrt{2}}\Big\{\psi_a(\boldsymbol{r}_1)\,\psi_b(\boldsymbol{r}_2) + \psi_a(\boldsymbol{r}_2)\,\psi_b(\boldsymbol{r}_1)\Big\}. \tag{14.8}$$

Put $\boldsymbol{r}_1 = \boldsymbol{r}_2$ in the second of these, and show that the probability of finding the two particles together is doubled, relative to that for the simple-product wave function (14.7). It is this finding, common but not universal, that is too often generalized into a statement that particles in a space-symmetric wave function have an increased tendency to be found together.

(2) The lowest-energy state of helium has both electrons in 1s wave functions. Try to form a "symmetrized" space wave function after the fashion of (14.8), and you'll find the normalization is wrong. The only possibility is the simple product $\psi_{1s}(\boldsymbol{r}_1)\,\psi_{1s}(\boldsymbol{r}_2)$. Of course: this is symmetric already so it doesn't need to be symmetrized. This wave function is space-symmetric, yet there is no doubling of the probability for finding the particles together.

(3) Look at the 3p 4p ^{1}P wave function of (16.5). The electron labelled 1 has coordinates (r_1, θ_1, ϕ_1) while the electron labelled 2 has coordinates (r_2, θ_2, ϕ_2). Show that the space part of the wave function is symmetric, meaning (as always) symmetric under interchange of labels 1 and 2. Show that the electrons are never found together: the wave function is zero[17] when $r_1 = r_2$. This is our extreme counterexample to the idea that "space-symmetric means close together".

[17] Indeed the electrons are "more apart" than is suggested by a zero at $\boldsymbol{r}_1 = \boldsymbol{r}_2$; see § 16.3.1.

14.6   Helium and the (anti)symmetry of its electrons

We present here a not-quite-conventional treatment of helium. A "conventional" discussion makes much of the indistinguishability of the two electrons, from the start requiring the two-electron wave function to be properly antisymmetrized. In an advanced calculation, I would do the same.[18] But for a beginner the conventional route can increase confusion, because it conceals what does and what does not depend on indistinguishability and (anti)symmetry.

[18] Just such antisymmetrized wave functions are assumed in Chapters 15 to 17.

14.6.1   Helium with non-identical electrons

Let's see how far we can get with working out the energy-level diagram for helium, pretending (for now) that the electrons are non-identical;

spin is an irrelevance at this stage and is ignored. I can justify this curious approach in three ways:

- This is in fact the way that the algebra most naturally sets itself up, starting with (14.9) in which spin is not mentioned.
- This kind of calculation was first done in 1927, when the Schrödinger equation was known, but people hadn't properly understood electron spin, and didn't know about antisymmetric wave functions for fermions.[19] So, in understanding how helium helped people *then* to appreciate the indistinguishability of the electrons, it helps *us now* to know which results depend on indistinguishability and which don't.
- There is a nasty tendency to "explain" things as caused by electron indistinguishability when they are not. As a result, students can't make sense of the physics and end up believing in magic. This won't do.

We start from the two-electron Schrödinger equation[20]

$$\left(\frac{-\hbar^2}{2m_e}\nabla_1^2 + \frac{-\hbar^2}{2m_e}\nabla_2^2 - \frac{Ze^2}{4\pi\epsilon_0\, r_1} - \frac{Ze^2}{4\pi\epsilon_0\, r_2} + \frac{e^2}{4\pi\epsilon_0\, r_{12}} \right)\psi = E\,\psi. \quad (14.9)$$

Here $\psi = \psi(\boldsymbol{r}_1, \boldsymbol{r}_2)$ is the two-electron wave function, a function of the six coordinates $(x_1, y_1, z_1, x_2, y_2, z_2)$; ∇_1 differentiates with respect to (x_1, y_1, z_1) and similarly; the nuclear charge number is Z (which for helium is 2); and r_{12} is the separation of the two electrons. The electrons have been labelled 1 and 2, and we are treating them as if they were distinguishable.[21]

Let's spell this out. The Hamiltonian on the left of (14.9) gives each electron charge $-e$ and mass m_e. That is all. The electrons are very similar to each other, but have not been made *identical*.[22] The labelling represents (or could do so) a real difference between the particles.

For teaching purposes, it is common first to omit the electron–electron interaction $e^2/(4\pi\epsilon_0\, r_{12})$, then put it in as a perturbation. But I make a point of *not* doing this here.

The Hamiltonian is the five-term operator on the left of (14.9). It is unaltered if we interchange the labels 1 and 2. We may see this interchange as being done by the operator $\widehat{I}_r$ introduced in (11.5) and §11.5 item 3. The interchange operator commutes with the Hamiltonian, so the two operators have a complete set of eigenfunctions in common. The eigenfunction solutions of (14.9) are therefore space-symmetric or space-antisymmetric (or can be chosen to be):

$$\left.\begin{array}{l} \psi_{\text{symm}}(\boldsymbol{r}_1, \boldsymbol{r}_2) = +\psi_{\text{symm}}(\boldsymbol{r}_2, \boldsymbol{r}_1) \\[4pt] \psi_{\text{anti}}(\boldsymbol{r}_1, \boldsymbol{r}_2) = -\psi_{\text{anti}}(\boldsymbol{r}_2, \boldsymbol{r}_1). \end{array}\right\} \quad (14.10)$$

This conclusion is general, and does not rely on the use of perturbation theory, or any other approximation method.[23]

Wave functions (14.2–3) are consistent with the symmetries of (14.10). Nevertheless these forms do not well describe helium, so we do not make use of them in the discussion of helium that follows.

[19] The Slater determinant dates from 1929 (Slater 1929).

[20] This Schrödinger equation ignores the finite mass of the helium nucleus. It is also non-relativistic and ignores some tiny effects associated with electron spin, such as spin–orbit interaction. Corrections are four orders of magnitude smaller than what is retained, and are far too small to affect the reasoning here.

[21] This step—treating the electrons temporarily as distinguishable—is in line with what we do in preparation for building a Slater determinant.

[22] There are cases, K^0 mesons, where particles have the same mass, the same charge, and almost the same interactions, yet are not identical. Nature doesn't seem to have supplied us with two nearly-identical kinds of electron. But so far as we know there isn't any rule that prevented it from doing so. We can therefore imagine a helium atom containing one electron of one kind and one of another, genuinely non-identical.

[23] Textbooks often set out to calculate the energies of helium by means of degenerate perturbation theory, as is outlined in §15.3. Indeed, the helium atom makes a good example for illustrating the method. Unfortunately, some misleading results emerge for reasons discussed in Chapter 15. This is why I focus on the fact that symmetric–antisymmetric is an exact property of (14.9). As a colleague said to me apropos of helium: if you've got an exact result you'd be a fool not to use it.

[24] An example of such a procedure is given in problem 15.5.

[25] The name comes from a certain approximation within which the energy difference is given by the **exchange integral**. The relevant approximation is not well satisfied by helium, but the name "exchange energy" persists. For the exchange integral, see § 15.3 and in particular eqn (15.5).

[26] As mentioned above, in *helium*, the space-antisymmetric possibility has the lower energy, within any given configuration. It is tempting to find something simple about space-(anti)symmetric that accounts for the sign of this energy separation. Almost every textbook attempts such a simple explanation based on Exercise14.2(1). The correct explanation (Chapter 15) is less simple and quite different. However, this energy difference is not the main business of the present chapter.

The calculation of the actual space wave functions for helium must rely on numerical methods, because (14.9) is not algebra-friendly.[24] Here we may imagine that such calculations have been done, to as much accuracy as is needed. We do not need to write out explicit expressions for the wave functions or energies.

It is to be expected that the space-symmetric and space-antisymmetric quantum states will have different energies, and they do. In helium (but not always in other atoms) it is a space-antisymmetric state that has the lower energy (within any one configuration such as 1s 2p).

The energy difference between space-symmetric and space-antisymmetric states is often called an **exchange energy**, for technical reasons associated with its mathematical form.[25] Here we emphasize that nothing has been added to the physics contained in the five terms on the left of (14.9): the electron kinetic energies and Coulomb energies of attraction and repulsion. The energy difference is fully accounted for by the interplay of these five terms, nothing else.

Let's pause to think about what has happened. We started from the Schrödinger equation (14.9). We have come up with two wave functions and two energies. One (space) wave function is symmetric under interchange of labels 1 and 2, the other antisymmetric. All this with non-identical particles. So if you have picked up any of the following ideas (from reading or mis-reading books), unlearn them now:

✗ The symmetry and antisymmetry of the space wave functions is forced on the problem because the particles are identical.

✗ The exchange energy arises because of the identity of the particles which keep exchanging places in the atom because you can't tell which is which.

✗ Electrons in space wave function ψ_{symm} can be found together (true for helium) and those in ψ_{anti} are never found together (always true for two particles). They are pushed into these relative positions by a mysterious thing called an exchange force, which originates with the identity of the electrons.

As the Duke of Wellington said when accosted by a stranger with "Mr Smith I believe?": "Sir, if you believe that you will believe anything". No, the wave functions and energies are entirely the consequence of our starting with the Schrödinger equation (14.9), nothing more.[26] In more detail: the space-symmetry and space-antisymmetry have come about because the Hamiltonian in (14.9) is symmetric under $1 \leftrightarrow 2$.

14.6.2   The helium energy-level diagram, with non-identical electrons

The fact that we already have symmetric and antisymmetric *space* wave functions is fortunate because it will make things easy when eventually we think about identical electrons with spin. But that time is not yet.

We are now in a position to draw an energy-level diagram for helium, one that we pretend we have calculated: Fig. 14.1.

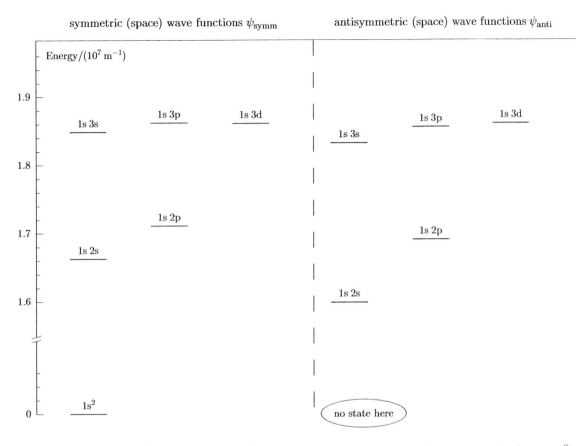

Fig. 14.1 The energies for a helium atom, calculated for non-identical spinless electrons. Each configuration (except $1s^2$) gives rise to two different energies, according as the space wave function is symmetric or antisymmetric under interchange of labels attached to the electrons.

The diagram brings to life what we have said about the eigenstates having the properties of ψ_{symm} and ψ_{anti}. For example, each panel contains a state labelled $1s\,2p$. These have different energies, with (as it happens) ψ_{anti} lying lower.

Notice also that you can't make an antisymmetric space wave function for $1s^2$ because everything you attempt gives zero. There is no wave function with energy at the bottom of the diagram on the right.[27]

It is not difficult to work out one of the selection rules for electric-dipole transitions connecting our energy levels. We find (easily) that transitions are forbidden between the (space-) symmetric and (space-) antisymmetric quantum states. Experiment agrees: there are no transitions linking the two sides of the energy-level diagram.[28]

What has this pretend-calculation achieved, and where has it failed?

Success: Spectacular. We have the whole energy-level diagram for helium, except for very fine detail, with the energies in the right places, and no second $1s^2$ level. Also we have the right selection rules, and therefore the right spectrum (again except for fine detail).

[27] This absence is sometimes attributed to electron indistinguishability and the Pauli principle. I disagree, because it has happened with non-identical electrons.

[28] Except for some very weak "intercombination lines".

Failure: Electrons *do* have spin. The quantum states we have (pretend-) calculated are misdescribed because spin is ignored.

So we can't stop here. We still do not make the electrons indistinguishable, but we now take into account the existence of electron spin.

14.6.3   The helium energy-level diagram, with spin

In this second step, we keep the electrons non-identical, but we add to our previous picture the knowledge that each electron has spin $\tfrac{1}{2}$.

With obvious foresight, we'll "prefabricate" (in the sense of § 11.2.2) the spin states of the two electrons to give combinations possessing a total S of 0 or 1. There is no energy involving spin,[29] so we make *no change* in the Schrödinger equation (14.9). In consequence, none of the energy levels moves from the positions we assigned in Fig. 14.1. What

[29] There are three small interactions involving the spin's magnetic moment, of which the spin–orbit interaction is the most familiar. See e.g. Woodgate (1980) § 7.3. These are similar in order of magnitude to each other, but are four orders of magnitude ($\sim \alpha_{fs}^2$) smaller than the electrostatic energies under discussion. Therefore, to the accuracy needed here, there is *no* energy directly involving spin.

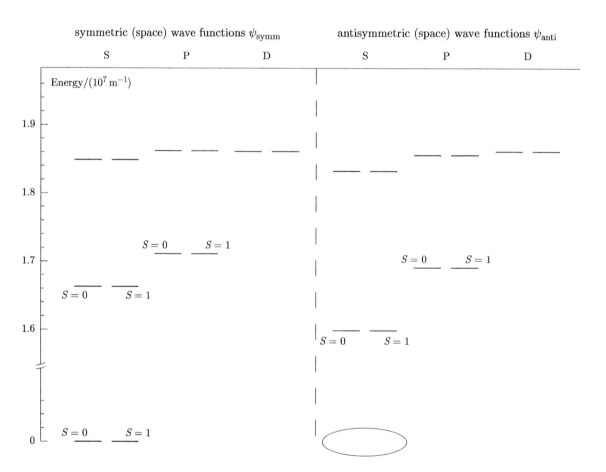

Fig. 14.2 The energies for a helium atom, calculated for electrons that are still non-identical but have spin $\tfrac{1}{2}$. In preparation for a later stage, the spins are combined into states with $S = 0$ and $S = 1$. To indicate this, each energy level is drawn doubled, the left-hand having $S = 0$, the right-hand having $S = 1$. The levels near the top of the diagram follow the same pattern, but are not labelled to avoid clutter.

happens is that, where before there was one energy level, there are now two (coincident) energy levels, one with $S = 0$ and one with $S = 1$.

This brings us to Fig. 14.2. The energy-level diagram is drawn again, with each level doubled up. Each left-hand member of a pair has $S = 0$, while the right-hand member has $S = 1$. They have the same energy because *the spin is not mentioned in the Hamiltonian*.[30]

At risk of being boring, I want to rub that in: the differences of energy between the two panels of Fig. 14.2 started out life as caused by electrostatic effects; so they are *still* due to electrostatics.

14.6.4   Helium: only totally antisymmetric wave functions are permitted

Now we proceed to the final step in the argument. Electrons, being identical fermions, must have an overall (space and spin) wave function, like the ψ_N of (14.1), such that if you interchange labels 1 and 2—throughout—the wave function changes sign. I've built that information into Fig. 14.3. We are allowed only $S = 0$ (spin part of the wave function antisymmetric) on the left-hand side of the diagram, so we cross out all the $S = 1$ levels there. On the right, we are allowed only $S = 1$ (spin part of the wave function symmetric), so we cross out levels having $S = 0$. The indistinguishability of the electrons has been put into the problem by abolishing wave functions having the wrong symmetry. We may now label the surviving quantum states as singlets (meaning $S = 0$) on the left, and triplets (meaning $S = 1$) on the right. This has been done in the column headings (^{1}S, ^{3}S etc.) of Fig. 14.3.

As ever, let's look round and see what has happened. Crossing out some quantum states as disallowed has not changed in any way the states that are allowed to remain. All old insights attached to them remain in force. Take an example. Compare $1s\,2p\,^1P$ with $1s\,2p\,^3P$. These have different energies, with ^3P lying lower, because electrostatics affects the two space wave functions differently. Spin never had anything to do with this. Spin has entered the problem only via the back door, because the symmetry rule has forced us to put $S = 1$ with only one kind of space wave function (ψ_{anti}), and $S = 0$ with the other (ψ_{symm}). The difference of energy is *associated with* spin, but is not *caused by* spin.[31] Please read this over again until you really get it right. All your instincts (aided by some books) may well be trying to make you get it wrong.[32]

14.6.5   "Exchange interaction"?

The energy differences between the two panels of Fig. 14.3 come from the electrostatic energies in (14.9) and the way in which the space wave functions ψ_{symm} and ψ_{anti} form themselves as eigenfunctions for that Hamiltonian. It happens that for ψ_{anti} the electrons are able to snuggle down into a lower-energy arrangement than is the case for ψ_{symm}; see Chapter 15.

As mentioned above, each singlet–triplet energy difference is often

[30] When something new is introduced, we usually have to regroup previous findings by taking linear combinations of wave functions. This happens in § 17.2 where we form $|m_L, m_S\rangle$ wave functions from $|m_l, m_s\rangle$ functions. We are fortunate here that $S = 0$ and $S = 1$ can be attached to already-found space wave functions without a need to build linear combinations This is what we meant by "make it easy" in § 14.6.2.

[31] That energy difference was on display in Fig. 14.1, long before we said anything about the electrons being identical.

[32] A book that gets this right is Foot (2005). His problem 3.4 asks us to imagine a helium atom in which one of the electrons is an exotic particle with spin $\frac{3}{2}$, but is otherwise just like an ordinary electron. Such an atom would have our Fig. 14.2 for its energy-level diagram, except that the spin S would take three values: $0, 1, 2$.

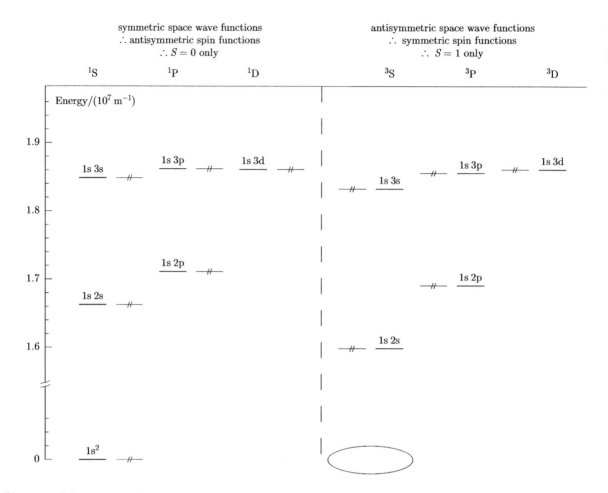

Fig. 14.3 Helium as it really is. At this stage, we input the fact that the two electrons of helium are identical and must have an overall wave function that is antisymmetric under interchange of labels attached to the electrons. To meet this condition, we have to delete all $S = 1$ states on the left side of the diagram, and all $S = 0$ on the right.

called an *exchange energy*, although the exchange integral, from which it takes its name, does not well account for the energy difference in helium.

In solid-state magnetism, "parallel spins" and "antiparallel spins" likewise differ in energy; in that context people say that the energy difference is caused by an *exchange interaction*. There is a difference of language here that requires explanation.

Consider, as example, the energies of helium's terms $1s\,2p\,^1\mathrm{P}$ and $1s\,2p\,^3\mathrm{P}$. We can introduce dimensionless spin vectors $\boldsymbol{\sigma}_1, \boldsymbol{\sigma}_2$ for the two electrons such that $\boldsymbol{\sigma}_1 \cdot \boldsymbol{\sigma}_2 = -\frac{1}{4}$ in (14.2) and $+\frac{1}{4}$ in (14.3). Then

$$E = \tfrac{1}{2}(E_{\mathrm{symm}} + E_{\mathrm{anti}}) - 2(E_{\mathrm{symm}} - E_{\mathrm{anti}})\boldsymbol{\sigma}_1 \cdot \boldsymbol{\sigma}_2$$

encapsulates the singlet and triplet energies in a single formula. It should be clear that what we have just done is physically empty: it adds nothing to our understanding of E_{symm} and E_{anti}—and indeed sends a confusing signal that $(E_{\text{symm}} - E_{\text{anti}})$ might be something to do with a spin–spin interaction when we have been at pains to insist that it is not.

Nevertheless, in solid-state magnetism, it is customary—and highly convenient—to introduce an "effective Hamiltonian" $-2\mathcal{J}\boldsymbol{\sigma}_1 \cdot \boldsymbol{\sigma}_2$, called the "exchange interaction", for handling energy differences that affect spin. Yet those energy differences, and the value of $\mathcal{J}$, are (*still*) of electrostatic origin; they only *look* as though they are caused by inter-action of the spins. This is the significance of "effective" in "effective Hamiltonian".

14.6.6   "Exchange force"?

Is there an "exchange force" accompanying "exchange energy"? It would not occur to me to think of such a thing, but unfortunately this idea can be found in some textbooks: a "force" that pushes electrons into being apart (space antisymmetric) or close together (space symmetric). Suppose that we could turn off the Coulomb repulsion between the elect-rons in (14.9); this is notionally done in § 15.3 as an initial step when that repulsion is to be put in later as a perturbation. Then the space-symmetric and space-antisymmetric states (of the same configuration) have *equal energy*.[33] This kind of exchange force is there to account for an energy difference of zero; in other words, it has no job to do.

[33] This equality is often referred to as *exchange degeneracy.*

14.6.7   Helium as a case of *LS* coupling

We pick up the idea of "prefabrication" introduced in Chapter 11 and in particular in § 11.2.2. We have performed two different prefabrication steps in the present chapter:[34] one in imposing space-symmetric and space-antisymmetric structure on the wave functions (in advance of any attempt at calculating those wave functions or their energies); and one in combining spin states to yield $S = 0$ and $S = 1$.

After this, the spin states are eigenstates of $\widehat{\boldsymbol{S}^2}$ with eigenvalues $S(S + 1)\hbar^2$. This is (part of) what **LS coupling** means: the *formation* of wave functions possessing quantum numbers L and S.

Because helium has one electron in a 1s wave function,[35] the orbital angular momentum is wholly that contributed by the other electron, and there is no operation of linear-combination that forms eigenstates of $\widehat{\boldsymbol{L}^2}$. To see *LS* coupling doing everything that it can do, we need to look at Fig. 16.3. But helium is always regarded as exemplifying *LS* coupling because of the way that the spin states are combined.

Note that *coupling* is being used in its correct sense here: the form-ation of linear-combination wave functions (in this case combining spin m_s states to form eigenstates of $\widehat{\boldsymbol{S}^2}$). This combining is necessitated by the physics, but it also something *we may do* in anticipation (prefabric-ation), to simplify the process of finding the quantum states.

[34] In a more advanced treatment, we would combine these two steps into one. There is no point in wasting our time calculating energies and wave functions that are destined to be rejected as be-longing to the wrong symmetry.

[35] It is possible to excite helium into high-energy states, such as 2s 2p, in which neither electron is in a 1s wave function. So much energy has to be given in forming such states that the energies lie above the ionization energy of neutral helium. There is a good rea-son why these highly excited states are usually ignored. However, this does mean that helium exhibits a somewhat limited repertoire of quantum states.

Comment
Because nothing is done in the formation of its $\boldsymbol{L}$, helium is a rather stunted example of *LS* coupling. For understanding *LS* coupling prop-

erly it is best to start afresh as we do in § 16.2.

Given the above, it is unfortunate that helium seems often to be taken as a model for understanding other atoms. In particular, we show in Chapter 15 that the singlet–triplet energy differences in helium are a really complicated business that does not lend itself to any simple interpretation.

By contrast, larger atoms do lend themselves to relatively simple discussion. Contrast helium with carbon, still quite a "small" atom. In helium, $e^2/4\pi\epsilon_0 r_{12}$ is one energy contribution beside 4 others of comparable magnitude, which makes it unsuitable for a perturbation treatment. In carbon ($Z = 6$), there are 6 kinetic energies, 6 potential energies of the electrons in the field of the nucleus, and 15 electron–electron repulsion energies, of which 14 can be combined into a benign central field: the repulsion $e^2/4\pi\epsilon_0 r_{12}$ between the two outermost electrons is one energy contribution beside 26 others. That one contribution is small enough by comparison that it *can* be treated as a perturbation.

In helium, triplet states lie below singlets (within a common configuration). There is no authority here for believing that other atoms will have triplets below singlets—let alone all triplets below all singlets—within a single configuration. It seems likely that an over-reliance on the precedent of helium may be part of the reason why the (wrong) diagrams of Figs. 16.1 and 16.2 are so commonly found in print.

Problems

Problem 14.1 Wave functions built from Slater determinants
Show that the wave function of eqn (14.2) is a linear combination of two Slater determinants. Do the same with (14.3).

Show that the wave functions of (14.4) and (14.5) can be written as linear combinations of two building-block boson wave functions (constructed like a multiplied-out Slater determinant but with all terms positive).

Problem 14.2 Wave functions having shapes (14.2–3) are not compulsory
Consider the function $(x-y)^3$, which is antisymmetric under interchange of x and y. Multiply out the bracket and show that

$$(x - y)^3 = \begin{vmatrix} x^3 & y^3 \\ 1 & 1 \end{vmatrix} - 3 \begin{vmatrix} x^2 & y^2 \\ x & y \end{vmatrix}.$$

Try to combine these two determinants into a single one of the form

$$\begin{vmatrix} f(x) & f(y) \\ g(x) & g(y) \end{vmatrix},$$

and I think you'll fail.

Helium: energies for singlets and triplets

... spend the whole night proving a bit of it with rulers and cosines and similar other instruments ....

Flann O'Brien, *The Third Policeman*

Intended readership: In a second course on atomic physics, when helium is to be understood in detail, and there is a need for explanation of the singlet–triplet energy difference.

15.1 Introduction: triplets below singlets?

In Chapter 14 it is shown that the quantum states of helium are classified into singlets ($S = 0$) and triplets ($S = 1$), the singlets having space-symmetric wave functions and the triplets having space-antisymmetric wave functions. It was stated there that the singlets and triplets differ in energy, but no attempt was made at calculating the difference or even at accounting for its sign.

In the present chapter we address the energy difference: the reason for it, and its sign. For a given configuration of helium, such as 1s 2s or 1s 2p, the triplet lies lower in energy than the singlet. Why?

15.1.1 Rival explanations

The focus in this chapter will be on some of the 1s np, 1s ns configurations of helium, because numerical values for them are available in the literature. Only a few configurations are described. However, even these are sufficient to challenge and correct descriptions commonly given.

There are two explanations that are current for the singlet–triplet energy difference. Both are based on correct facts:

1. The textbook explanation. Fact:

   When two electrons occupy a space-antisymmetric wave function $\psi_{\text{anti}}(\boldsymbol{r}_1, \boldsymbol{r}_2)$ they are never found together at the same location: $\psi_{\text{anti}}(\boldsymbol{r}, \boldsymbol{r}) = 0$: eqn (14.6).

   Interpretation:

   (a) In the space-antisymmetric wave function, the electron–electron repulsion energy $\langle e^2/4\pi\epsilon_0 r_{12}\rangle$ is the smaller (as compared with the space-symmetric) because $\langle r_{12}\rangle$ is expected to be the greater;[1] this makes the triplet state lie lower in energy than the singlet.

[1] Of course this step is suspect, since a greater $\langle r_{12}\rangle$ does not enforce a smaller $\langle r_{12}^{-1}\rangle$. But a lot more is wrong, as we show below.

Essays in Physics: Thirty-two thoughtful essays on topics in undergraduate-level physics. Geoffrey Brooker, Oxford University Press. © Geoffrey Brooker 2021. DOI: 10.1093/oso/9780198857242.003.0015

[2] In atomic physics, we refer to 1s 2p and its wave functions as a *configuration* and 1s 2p ^{1}P, 1s 2p ^{3}P as *terms*. To avoid confusion with "term" in its mathematical sense, in this chapter we italicize *term* when using it to refer to some L, S parts of a *configuration*.

[3] When wave functions are worked out accurately, they do not take the forms (14.2–3), with the consequence that the electron–electron repulsion energy does not separate into direct and exchange pieces.

It can be tempting to feel that all space-symmetric and space-antisymmetric wave functions (for two particles) must be expressible in the forms (14.2–3), repeated here without the spin factors.

$$\psi = \frac{1}{\sqrt{2}} \times$$
$$\left\{ \psi_a(\boldsymbol{r}_1)\,\psi_b(\boldsymbol{r}_2) \pm \psi_a(\boldsymbol{r}_2)\,\psi_b(\boldsymbol{r}_1) \right\}.$$

This is not so: we have said in § 14.5 that these are very special shapes. In Problem 17.3 we give six eigenfunctions for a pp configuration, and none has precisely the shape of (14.2) or (14.3). As a further counterexample, consider $e^{-r_{12}}(r_1 \pm r_2)$, which has the required exchange symmetries but again won't regroup into the forms given above.

[4] More precisely, eqn (15.1) treats the helium nucleus as infinitely massive, as well as omitting relativistic contributions. When corrections are made, it is the mass that matters most.

[5] That is to say: we have outgrown the idea of pretend-non-identical electrons that was used in Chapter 14, and are exploiting to the full the knowledge that electrons are identical fermions.

[6] This is within the approximation contained in (15.1). There exist spin-dependent energies, but they are smaller than those considered by a factor of order α_{fs}^2. We therefore omit mention for now of the spin parts of the wave functions (and with them the splitting of ^{3}P$_0$, ^{3}P$_1$, ^{3}P$_2$), and mention spin only indirectly when using "singlet" and "triplet" as convenient labels.

(b) The singlet–triplet energy difference is accounted for primarily by the exchange-integral part of $\langle r_{12}^{-1} \rangle$.

2. The triplet-orbitals-more-compact explanation. Facts:
The space-antisymmetric (triplet) wave function (both 1s and np pieces in the case of a 1s np configuration, similarly both 1s, ns for 1s ns) is more compact than the space-symmetric (singlet). In the triplet *terms*,[2] the more-compact wave function means that the electrons are closer to each other (than in the singlet), and the electron–electron repulsion is actually the *greater*.

Interpretation:

(a) It is the Coulomb attraction between nucleus and electrons that dominates the energy differences. The more compact wave functions of the triplet *terms* profit from a greater closeness of the electrons to the nuclear charge, more than they are penalized by the greater closeness of the electrons to each other, and by a greater kinetic energy; it is the balance of all these contributions that makes a triplet *term* lie lower than its companion singlet.

This essay shows that interpretation 1(a) is definitely wrong, while 1(b) has meaning only in an approximation that is too crude to be helpful.[3] Interpretation 2(a) is correct, though it re-describes what happens more than offering an explanation.

15.2   Setting the scene

As in Chapter 14, we start by writing the non-relativistic two-electron Schrödinger equation (14.9), repeated here:[4]

$$\left(\frac{-\hbar^2}{2m_e}\nabla_1^2 + \frac{-\hbar^2}{2m_e}\nabla_2^2 - \frac{Ze^2}{4\pi\epsilon_0\, r_1} - \frac{Ze^2}{4\pi\epsilon_0\, r_2} + \frac{e^2}{4\pi\epsilon_0\, r_{12}} \right) \psi = E\,\psi. \quad (15.1)$$

We are now at an intellectual level where we know that electrons are identical fermions. The wave function for two electrons is antisymmetric under interchange of labels 1 and 2 attached to the electrons, the interchange being made in both space and spin parts of the wave function.[5] We know, from the symmetry of the Hamiltonian in (15.1) under interchange $\boldsymbol{r}_1 \leftrightarrow \boldsymbol{r}_2$, that the space part of the wave function is either space-symmetric or space-antisymmetric (or can be chosen to be). Overall antisymmetry then requires space-symmetric to go with $S = 0$ (singlet) and space-antisymmetric to go with $S = 1$ (triplet).

Reminder: The Hamiltonian in (15.1) makes no mention of spin, so the energies we seek contain no contribution arising *directly* from spin. The energies and energy differences arise wholly from Coulomb forces acting on the space part of the wave function.[6]

It is the case that, when (15.1) is solved for the space part of the two-electron wave function, the space-antisymmetric solution does lie lower in energy than the space-symmetric—for helium, at least for low-energy configurations.

15.3 The "textbook" explanation

We first give the usual "textbook" explanation, to show what it is we shall be arguing against in that case.

Solve (15.1) in two stages, first ignoring electron–electron repulsion, then putting it in as a perturbation.

When the electron–electron repulsion is neglected, each electron moves in a spherically symmetric potential, and has a quantized orbital angular momentum. We can describe the electrons by means of a **configuration**, giving for each the principal quantum number n and the orbital angular-momentum quantum number l according to the usual conventions. Thus we may refer (e.g.) to a $1s^2$ configuration or a $1s\,2p$ configuration.

For definiteness, consider a $1s\,2p$ configuration. The building-block wave functions ψ_{1s}, ψ_{2p} are hydrogen-like, except that they are both appropriate to nuclear-charge number Z (2 for helium). Then electron–electron repulsion is imposed using perturbation theory, resulting in[7]

$$\psi_s = \frac{1}{\sqrt{2}}\Big\{\psi_{1s}(\boldsymbol{r}_1)\,\psi_{2p}(\boldsymbol{r}_2) + \psi_{1s}(\boldsymbol{r}_2)\,\psi_{2p}(\boldsymbol{r}_1)\Big\}, \qquad \Delta E_s = J + K, \quad (15.2)$$

$$\psi_t = \frac{1}{\sqrt{2}}\Big\{\psi_{1s}(\boldsymbol{r}_1)\,\psi_{2p}(\boldsymbol{r}_2) - \psi_{1s}(\boldsymbol{r}_2)\,\psi_{2p}(\boldsymbol{r}_1)\Big\}, \qquad \Delta E_t = J - K, \quad (15.3)$$

in which

$$J = \int \psi_{1s}(\boldsymbol{r}_1)^*\,\psi_{2p}(\boldsymbol{r}_2)^*\,\frac{e^2}{4\pi\epsilon_0\,r_{12}}\,\psi_{1s}(\boldsymbol{r}_1)\,\psi_{2p}(\boldsymbol{r}_2)\,\mathrm{d}^3\boldsymbol{r}_1\,\mathrm{d}^3\boldsymbol{r}_2, \quad (15.4)$$

$$K = \int \psi_{1s}(\boldsymbol{r}_1)^*\,\psi_{2p}(\boldsymbol{r}_2)^*\,\frac{e^2}{4\pi\epsilon_0\,r_{12}}\,\psi_{1s}(\boldsymbol{r}_2)\,\psi_{2p}(\boldsymbol{r}_1)\,\mathrm{d}^3\boldsymbol{r}_1\,\mathrm{d}^3\boldsymbol{r}_2. \quad (15.5)$$

The integrals J and K are called the **direct** and **exchange** integrals, respectively.

A (textbook) explanation (interpretation 1(a)) of the energy difference $\Delta_{st}E = E_s - E_t = 2K$ now follows. The triplet state has a space-antisymmetric wave function, so the electrons are never close together; the singlet state has a space-symmetric wave function, for which the electrons can be close together, and indeed have an enhanced probability of being found together (in comparison with a simple-product wave function), as is shown[8] in exercise 14.2. A zero wave function at $r_{12} = 0$ (space-antisymmetric ψ_t) must mean that the two-electron wave function "piles up" elsewhere, and "elsewhere" can only be at larger r_{12}. Conversely, the space-symmetric wave function ψ_s has no need to pile up at larger separations. The Coulomb repulsion of the electrons should raise the energy more for the "together" state than for the "apart" state, so the triplet should lie lower in energy.

The exchange integral supplies the energy difference (interpretation 1(b)) and confirms that the sign has come out as expected—at least for those quantum states for which K can be shown to be positive.

The above is the account most frequently given, and it's almost a shame that it isn't right.[9]

[7] This calculation is spelt out clearly in Woodgate (1980) § 5.1; Foot (2005) § 3.2—who are both fully aware of the limitations of what they are presenting. It has notionally been simplified here because we know in advance that the final (space) eigenfunctions must be symmetric and antisymmetric under interchange of labels 1 and 2.

It is unfortunate that J is used both for the direct integral and for the total angular momentum quantum number. No confusion should arise in the present chapter because further mention will not be made of the total angular momentum.

[8] Shown, that is, for a special-case wave function having the form (15.2). As counterexample, eqn (17.18) shows a space-symmetric wave function within which the electrons do *not* pile up at zero separation. In fact, the electrons are "more apart" in that ^{1}P *term* than in any other, for the given 3p 4p configuration.

[9] The "textbook" discussion contains a rarely-flagged danger signal. When J and K are worked out for the 1s 2p ^{1}P state of helium (let alone higher states), the calculation gives an energy above that of He$^+$ 1s, thereby predicting a state that is unbound; the 2p electron can wander off. See problem 15.4. An error of this magnitude should cast doubt on the entire picture.

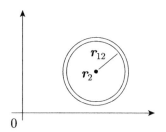

Fig. 15.1 The repulsion energy of electrons located at r_1 and r_2 is given by (15.6). The integral is evaluated by first holding r_2 constant and integrating over r_1. For this purpose, we draw spherical shells of radius r_{12} and thickness dr_{12} surrounding r_2. The energy is an integral over $|\psi(r_1, r_2)|^2 \, r_{12}$, so it pays no attention to the value of $\psi(r_1, r_2)$ at $r_1 = r_2$.

---

[10] Atoms know of numerous cases where the space-symmetric possibility has the lower energy; see e.g. the "alternation" referred to by Kuhn (1969) p. 267; Condon and Shortley (1951) p. 200.

[11] The largeness of $e^2/4\pi\epsilon_0 \, r_{12}$ has two consequences. The computations that led to Table 15.1 had to build on wave functions whose space parts are not of the forms (14.2–3). Instead, the wave functions were from the start functions of r_{1s}, r_{np} and $r_{12} = |r_{np} - r_{1s}|$. This structure precludes the repulsion energy from being decomposed into direct and exchange pieces.

Another consequence shows up in Table 15.1. The 1s and np wave functions in each singlet state are significantly different from those in the corresponding triplet state. Were we to attempt an evaluation of J and K, those energies would likewise be different in the singlet and triplet states: there can be no $J \pm K$ that serves both singlet and triplet with a common J, K.

It is possible to refine § 15.3 by putting different ψ_{1s}, ψ_{np} into (15.2) and (15.3), and adjusting them separately by a variational method. Problem 15.5 attempts this using just two adjustable parameters. Unfortunately, the resulting 1s wave functions show differences of compactness in the wrong direction. There seems to be no simple "friendly amendment" that will rescue § 15.3.

15.3.1   A refutation

We can show quickly that interpretation 1(a) will not do. Consider the Coulomb repulsion energy of two electrons held together in an atom, not necessarily helium.

Let the two-electron wave function be $\psi(r_1, r_2)$. We need not specify whether this is space-symmetric or space-antisymmetric: the reasoning will be "two for the price of one".

The electron–electron repulsion energy is

$$E_r = \int dV_2 \int dV_1 \, \psi(r_1, r_2)^* \, \frac{e^2}{4\pi\epsilon_0 \, r_{12}} \, \psi(r_1, r_2)$$
$$= \frac{e^2}{4\pi\epsilon_0} \int dV_2 \int dV_1 \, |\psi(r_1, r_2)|^2 \, \frac{1}{r_{12}}.$$

Usually we might think of dividing volume V_1 into spherical shells centred on the nucleus, but now instead we divide it into spherical shells centred on r_2, as indicated in Fig. 15.1. Then

$$E_r = \frac{e^2}{4\pi\epsilon_0} \int dV_2 \int d\Omega_1 \int dr_{12} \, r_{12}^2 \, |\psi|^2 \, \frac{1}{r_{12}}$$
$$= \frac{e^2}{4\pi\epsilon_0} \int dV_2 \int d\Omega_1 \int dr_{12} \, r_{12} \, |\psi|^2. \qquad (15.6)$$

Because we are now using an awkward coordinate system, $|\psi|^2$ is a messy function of the angles integrated over in $\int d\Omega_1$. But fortunately all we need to notice is that the volume element has made the integration over r_{12} contain r_{12} as a weight factor in the *numerator*.

The repulsion energy of (15.6) pays zero attention to the value of ψ at $r_{12} = 0$. In fact, the weight factor r_{12} gives most attention to the value of $\psi(r_1, r_2)$ at largish values of r_{12}, where nothing said here can give an idea of the relative $|\psi(r_1, r_2)|^2$ values of the space-symmetric and space-antisymmetric possibilities.[10] Interpretation 1(a) is based on a false premise.

What then of the direct and exchange integrals? Isn't the algebra leading to energies $J \pm K$ convincing? Unfortunately not, because the $e^2/4\pi\epsilon_0 \, r_{12}$ perturbation is too large.[11]

15.4   Energies from the literature

The information we need can be extracted from a series of papers published in the 1950s to 1970s, when computers became available that could handle massively complicated approximation methods. Two papers are relevant here: Schiff *et al.* (1965) and Accad *et al.* (1971). Some findings from these papers are given in Table 15.1. The story told by the four configurations is very similar, and we shall therefore concentrate discussion mainly on 1s 2p.

The largest contribution to the singlet–triplet energy difference comes from the 1s electron, which for the triplet state lies deeper in the nuclear potential energy by 7736 cm^{-1}. A smaller contribution in the same

Table 15.1 The energies contributing to some $1s\,np$ *terms* of helium. The total energy in the last column is that relative to a $^4\mathrm{He}^{++}$ ion (plus two remote electrons at rest); it is obtained from current spectroscopic values for $^4\mathrm{He}$ and $^4\mathrm{He}^+$. The kinetic energy in the first column is obtained from the total energy via the virial theorem. It is that for the two electrons combined (and also from motion of the nucleus); we are unable to separate it into $1s, np$ contributions. The np potential energy (in the field of the nucleus) is obtained from the values of $\langle 1/r_1 \rangle$ tabulated in Schiff *et al.* (1965) and Accad *et al.* (1971); likewise the electron–electron repulsion energies are obtained from their $\langle 1/r_{12}\rangle$ values. The $1s$ potential energy is obtained by difference. Energies here are all in units of the Rydberg $R = R_\infty hc$, except that energy differences have been converted to cm^{-1}.

The figures in the references are given to 9 or 10 significant digits, so they take account of relativity, spin, the finite nuclear mass, and even QED. Here I have rounded to five decimal places. Even so, the precision goes just beyond that of the Schrödinger equation (15.1) (which is non-relativistic and also takes no account of the finite nuclear mass).

	KE/R	1s PE/R	np PE/R	repulsion/R	total/R
$1s\,2p\,^1\mathrm{P}$	4.24728	−7.86142	−1.12318	0.49005	−4.24728
$^3\mathrm{P}$	4.26594	−7.93192	−1.13324	0.53328	−4.26594
$(^1\mathrm{P} - {}^3\mathrm{P})/\mathrm{cm}^{-1}$	−2048	7736	1105	−4744	2048
$1s\,3p\,^1\mathrm{P}$	4.10991	−7.97013	−0.46888	0.21919	−4.10991
$^3\mathrm{P}$	4.11578	−7.99319	−0.47023	0.23186	−4.11578
$(^1\mathrm{P} - {}^3\mathrm{P})/\mathrm{cm}^{-1}$	−645	2531	149	−1390	645
$1s\,4p\,^1\mathrm{P}$	4.06176	−7.98950	0.25775	0.12372	−4.06176
$^3\mathrm{P}$	4.06427	−7.99949	−0.25807	0.12901	−4.06427
$(^1\mathrm{P} - {}^3\mathrm{P})/\mathrm{cm}^{-1}$	−276	1096	35	−580	276
$1s\,5p\,^1\mathrm{P}$	4.03944	−7.99506	−0.16318	0.07936	−4.03944
$^3\mathrm{P}$	4.04073	−8.00022	−0.16329	0.08205	−4.04073
$(^1\mathrm{P} - {}^3\mathrm{P})/\mathrm{cm}^{-1}$	−142	567	12	−295	142

direction comes from the 2p electron (1105 cm^{-1}). These energy differences are partly offset by a greater kinetic energy (for the triplet state), a consequence of a "more curved" wave function (−2048 cm^{-1}). They are also partly offset by a greater electron–electron repulsion energy (for the triplet state), as the electrons are closer to each other (−4744 cm^{-1}).[12]

The story is the same for all four of the configurations detailed here.

Table 15.1 thus validates statement 2 of § 15.1.1 and its interpretation 2(a). The triplet states have *more compact* wave functions (both 1s and np) than the singlets, in such a way as to lie lower in energy.

For each energy eigenstate, the atom's wave function must take a shape that yields the best compromise between the five energy contributions in (15.1). Table 15.1 shows that the singlet and triplet states involve different compromises.[13] However, there seems to be no simple way of seeing *why* things work out as they do.

In the hope of gaining better understanding, we have attempted a tutorial calculation of the 1s 2p energy, using the variational method and including only two adjustable parameters (problem 15.5). Some features of the physics are quite well reproduced by the calculation. However, the calculation predicts a *less* compact 1s orbital for the triplet state, thereby

[12] Notice that this gives the lie to the textbook statement that the triplet state has the *smaller* electron–electron repulsion energy, consequent upon a zero wave function at zero separation. Our statement in the "triplet more compact" explanation is correct: the triplet state has the *larger* electron–electron repulsion energy.

[13] In the calculations leading to Table 15.1, $e^2/4\pi\epsilon_0 r_{12}$ is too large to be approximated, e.g. by a central field common to both *terms* of a 1s np configuration. The two *terms* have to be dealt with separately from the beginning. This contrasts with the use of a central field in Chapter 16, where a single central field gives a first shot at an entire configuration of 36 or 60 wave functions.

failing to reproduce what is perhaps the most conspicuous finding in Table 15.1. This attempt at "seeing why" has been worth trying, but the result is disappointing.

15.4.1   Technicalities of Table 15.1

The two papers cited give their numerical values in "atomic units": distances are given in units of a_0, the Bohr radius; energies are in units of $2R$. The electrons' coordinates are r_2 standing for r_{1s}, r_1 for r_{np} (both with origin at the nucleus) and $r_3 = r_{12}$. Atomic units make good sense when the Hamiltonian $\widehat{H}$ is written as

$$\frac{\widehat{H}}{2R} = -\frac{a_0^2}{2}\left(\nabla_1^2 + \nabla_2^2\right) - Z\left(\frac{a_0}{r_1} + \frac{a_0}{r_2}\right) + \frac{a_0}{r_{12}}. \tag{15.7}$$

In (15.7), the potential energy for the np electron (in the field of the nucleus), and the electron–electron repulsion energy, are obtained from $-\langle a_0/r_{2p}\rangle$ and from $\langle a_0/r_{12}\rangle$. This explains two of the five columns in Table 15.1.

Unfortunately some adjustments have to be made. For the $1s\,np$ states of helium, the np electron has an orbital that is large (roughly a factor 7 even for $n = 2$) compared with the charge distribution of the 1s electron. This means that the np electron is almost fully screened from one of the two proton charges.[14] Its wave function should resemble that of an np electron in a hydrogen atom having nuclear charge $Z = 1$. For such a hydrogenic wave function $\langle a_0/r\rangle = 1/n^2$, so we expect $\langle a_0/r_{np}\rangle \approx 1/n^2$. Similarly, for the electron–electron repulsion, we can treat the 1s electron roughly as a point charge at the nucleus, and so $\langle a_0/r_{12}\rangle \approx \langle a_0/r_{np}\rangle$ should also be close to $1/n^2$. For $\langle a_0/r_{12}\rangle$, the tabulated values are close enough to $1/n^2$ to look sensible. The same is not the case for $\langle a_0/r_1\rangle$; instead of being close to $1/n^2$ they are all close to 1, so we have to understand that $\langle a_0/r_1\rangle$ stands for $n^2\langle a_0/r_{np}\rangle$. Table 15.1 has been constructed within these understandings.[15]

I have obtained the total energy in all cases from spectroscopic data in Kramida *et al.* (2021).[16]

To make further progress, we invoke the virial theorem.[17] In the present context, where all forces are electrical with a $1/r$ potential, the virial theorem says that

$$\langle \text{total kinetic energy}\rangle = -\langle\text{total energy}\rangle. \tag{15.8}$$

This accounts for the contents of the first column of Table 15.1, in which the values of total energy have been copied with a change of sign.

Given the total kinetic energy of the two electrons (plus that of the nucleus, if the desired accuracy requires it to be taken into account), we can find the potential energy of the 1s electron in the field of the nucleus. This explains where the second column of Table 15.1 comes from. The numerical value is different for the different atomic *terms* because there is some screening by the np electron.

15.4.2   Other configurations

Accad *et al.* (1971) give computed results[18] for helium configurations $1s\,ns\,^1S$, ^3S and for $1s\,np\,^1P$, ^3P for $n = 2$-5. And they do the same for ions in the isoelectronic sequence from He to Ne^{8+}. The results validate wider application of "triplet more compact", though the evidence is a bit mixed for ^1P, ^3P in O^{6+} to Ne^{8+}. There is sufficient here to show that

---

[14]We repeat reasoning given in problem 15.2.

[15]I have found no mention of n^2 in the published papers. There are unexplained factors $\frac{1}{2}$ in the mean values $\langle r_1\rangle$ and $\langle r_1^2\rangle$ as well.

[16]In principle, the total energy of each atomic state should be obtainable from the tables in the cited papers. However, the explanations are not entirely straightforward, and my attempts at disentanglement do not achieve numerical consistency.

[17]See e.g. Schiff (1968) p. 180.

The virial theorem has a relativistic form, given by Landau and Lifshitz (1996), § 34. In it, the "total energy" on the right of (15.8) has to be understood as excluding rest energy.

A test can be made by looking at the $1s^2$ ground state of helium. Its wave functions and energies have been calculated by Pekeris (1959). Here the total energy, and all potential energies, are obtainable, so the kinetic energy can be found by difference. It agrees, to the 8th significant digit, with the total energy (apart from the sign difference, of course). That is to say, the virial theorem is confirmed to hold in a regime where relativistic effects are highly significant.

[18]For the P *terms*, tabulated values of $\langle a_0/r_1\rangle$ and $\langle a_0/r_{12}\rangle$ relate directly to electron potential energies as shown in (15.7). Values of $\langle a_0/r_{12}\rangle$ are also available for helium 2S, 3S *terms* (Liverts and Barnea 2011), and show the same behaviour: the triplet lies lower and yet has the greater electron–electron repulsion energy.

A loose end was left in sidenote 1. Values of $\langle r_{12}\rangle$ are tabulated for the quantum states investigated in this chapter. In all cases for which I have information, a larger $\langle r_{12}\rangle$ is accompanied by a smaller $\langle r_{12}^{-1}\rangle$, so the association flagged in sidenote 1 as untrustworthy did not in fact mislead. However it remains that the accompanying "expectation" is false.

"triplet more compact" is not just a quirk of helium, but extends to ss and sp configurations in at least some other two-electron atoms/ions.[19]

15.5   Conclusions

The "textbook" interpretation 1(a) is doubly wrong:

- Whether $\psi(\boldsymbol{r}_1, \boldsymbol{r}_2)$ is or is not zero at $\boldsymbol{r}_1 = \boldsymbol{r}_2$ has no bearing on the electron–electron repulsion energy, because the repulsion pays attention only to "largish" values of the separation $r_{12} = |\boldsymbol{r}_1 - \boldsymbol{r}_2|$.
- For many configurations, the electron–electron repulsion energy is *greater* for the triplet state than for the singlet, so has no part in explaining why the triplet lies lower. This difference in the repulsion energy is swamped by other energy differences acting in the opposite direction.

The "textbook" interpretation 1(b) is unhelpful:

- The wave functions of Schiff *et al.* and Accad *et al.* take forms showing that the electron–electron repulsion energy cannot be separated neatly into direct and exchange parts.[20]
- An exchange integral K, worked out from (15.5) using $Z = 2$ hydrogenic wave functions, is positive for the configurations 1s ns, 1s np for $n = 2$–5. It has the right sign to account for the singlet–triplet total-energy differences, but the wrong sign to account for the differences in electron–electron *repulsion* energies.

It is a fact that helium 1s, ns and np wave functions (at least for $2 \leqslant n \leqslant 5$) are more compact in the triplet *terms* than in the singlet. This results in a deeper potential energy in the field of the nucleus,[21] and more than compensates for a greater kinetic energy and a greater repulsion energy. In this, we restate and endorse interpretation 2(a).

I remain uncomfortable: I am unable to explain *why* a space-antisymmetric two-particle wave function has the opportunity to snuggle down to a lower energy than does a space-symmetric subject to the same Hamiltonian.[22] What insight is missing?[23]

[19] Of course, more work on helium has been done since 1971, taking advantage of faster computers and improved numerical methods. See Drake (1996), who lists energies for helium up to $n = 10$, $l = 7$. Unfortunately, Drake gives total energies only, so we are unable to extract information relevant to "compactness". Importantly though, there is no suggestion that the conclusions we have drawn from Accad *et al.* (1971) require amendment.

I am indebted to Prof. C.J. Foot for drawing my attention to this reference.

[20] This prohibition applies to the states of helium, for which a high numerical accuracy is being sought, and where the computational physics is distinctly unfriendly. There is no suggestion here that direct and exchange energies are inadmissible in other contexts.

In particular, direct and exchange energies are just what we need in Chapter 16.

[21] Historians of science may find it interesting that this correct explanation was given by Slater as early as 1928 (Slater 1928).

[22] This has been established only for ss and sp configurations in helium-like atoms and ions. It can't be emphasized too often that two-electron configurations do *not* always have triplets below the corresponding singlets. See Fig. 16.3 for a counterexample.

[23] Our one attempt (problem 15.5) at setting up a tutorial calculation has proved too naïve to be useful.

Problems

Problem 15.1 The direct integral of (15.4)
Use a procedure from elementary electrostatics to evaluate the direct integral of (15.4).

Take a "point charge" $-e\,|\psi_{2p}(\boldsymbol{r}_2)|^2\,\mathrm{d}V_2$ at location $\boldsymbol{r}_2$ relative to the nucleus. This is embedded in a spherically symmetrical charge distribution $-e\,|\psi_{1s}(r_1)|^2$.

Divide the charge of ψ_{1s} into two: a part where $r_1 < r_2$ and a part where $r_1 > r_2$. For the first part, show that the potential experienced by ψ_{2p} is

$$\frac{1}{4\pi\epsilon_0} \frac{1}{r_2} \int_0^{r_2} \mathrm{d}r_1\, r_1^2 \int \mathrm{d}\Omega_1 (-e)\, |\psi_{1s}(r_1)|^2 ,$$

(draw a Gaussian sphere of radius r_2 and replace the charge enclosed by a point charge at the centre). For the second part, divide the outer part of ψ_{1s} into thin spherical shells. Find the potential at r_2 caused by such a shell of radius r_1. Total the potentials at r_2 caused by all such shells to give

$$\int_{r_2}^{\infty} dr_1\, r_1^2 \int d\Omega_1 \left(\frac{-e}{4\pi\epsilon_0} \frac{1}{r_1} \right) |\psi_{1s}(r_1)|^2 .$$

Combine these contributions to the potential at r_2, and show that the energy of electron 2p in the field of electron 1s is

$$\frac{e^2}{4\pi\epsilon_0} \int dV_2\, |\psi_{2p}(r_2)|^2 \int dV_1\, |\psi_{1s}(r_1)|^2\, \frac{1}{r_>},$$

in which $r_>$ is the greater of r_1, r_2. This same form is obtained again in problem 15.3(8).

Problem 15.2   Rough estimates of energies
Make rough estimates of helium energies using the following picture.

In a 1s np configuration, the np electron sees an effective nuclear charge of $1e$ because it is screened from one proton charge by the 1s electron; so its wave function resembles that for an np electron in hydrogen ($Z = 1$). Conversely, the 1s electron is hardly screened from any of the nuclear charge by penetration of the np electron's charge, so its wave function resembles that of a 1s electron in hydrogenic He$^+$ ($Z = 2$). In this approximation show that:[24]

1. The np electron's orbit radius $\sim n^2 a_0$; the 1s electron's orbit radius $\sim a_0/2$; so the ratio of these is at least 8, confirming the reasonableness of the approximation.[25]

2. The 1s electron has potential energy $-8R$ in the field of the bare nucleus.[26]

3. The np electron has potential energy $-2R/n^2$ in the field of the nucleus as screened by the 1s electron.

4. The kinetic energies sum to $R(4 + 1/n^2)$.

5. The total energy of the atom from all causes is $-R(4 + 1/n^2)$.

Compare these rough estimates with the computed values in Table 15.1.

Problem 15.3   The energies for hydrogen-like wave functions
In this problem, we assume that the electrons in helium take the hydrogen-like forms[27]

$$\psi_{1s} = \left(\frac{\alpha}{a_0}\right)^{3/2} \frac{1}{\sqrt{\pi}} e^{-\alpha r/a_0}; \qquad \psi_{2p} = \left(\frac{\beta}{a_0}\right)^{5/2} \frac{1}{\sqrt{\pi}} r\, e^{-\beta r/a_0} \cos\theta.$$

$$(15.9)$$

Here α and β will be given values depending on what is being discussed.

(1) Show that if the 1s electron were moving independently round a nucleus of charge Ze it would have $\alpha = Z$. Show that on the same assumptions the 2p electron would have $\beta = Z/2$.

(2) Check the normalization of the wave functions given in eqn (15.9).

---

[24] In this problem, as in Table 15.1, $R = R_\infty hc$ is the Rydberg energy. The Bohr radius is a_0.

[25] More carefully: The mean radius $\langle r \rangle$ for an n, l electron bound to a nuclear charge of Ze is given by

$$\frac{\langle r \rangle}{a_0/Z} = \tfrac{1}{2}\{3n^2 - l(l+1)\}$$

(Landau & Lifshitz (1991), eqn (36.16)). This gives

$$\frac{\langle r_{np} \rangle}{\langle r_{1s} \rangle} = 2n^2 - \frac{4}{3},$$

which, even for 2p, is 6.7.

There is a hint here that, if followed up, may explain the odd behaviour of neon noted in Chapter 11 sidenote 16.

[26] *Hint:* You know the bound-state energy, and the virial theorem helps from there.

[27] We have chosen a 2p wave function for simplicity. Also we have chosen the one having $m_l = 0$ because this wave function does not depend upon azimuth ϕ; this slightly simplifies the algebra to come. The outcome cannot depend upon this choice of m_l because the 2p electron is accompanied by the spherically symmetrical 1s.

3) Show that ψ_{1s} and ψ_{2p} are orthogonal to each other because of their different angular parts. Show that therefore they can be combined into space-symmetric and space-antisymmetric whole-atom space wave functions (spin omitted)

$$\psi_\pm = \frac{1}{\sqrt{2}}\left\{\psi_{1s}(\boldsymbol{r}_1)\,\psi_{2p}(\boldsymbol{r}_2) \pm \psi_{1s}(\boldsymbol{r}_2)\,\psi_{2p}(\boldsymbol{r}_1)\right\} \qquad (15.10)$$

with $1/\sqrt{2}$ as the correct normalization.

4) The kinetic-energy operator (normalized to $2R$) is given in (15.7). Using the wave function (15.10), show that the expectation value of the kinetic energy takes the form[28]

$$\frac{\langle\mathrm{KE}\rangle}{R} = -a_0^2\left(\int dV_1\,\psi_{1s}^*(\boldsymbol{r}_1)\,\nabla_1^2\,\psi_{1s}(\boldsymbol{r}_1) + \int dV_2\,\psi_{2p}^*(\boldsymbol{r}_2)\,\nabla_2^2\,\psi_{2p}(\boldsymbol{r}_2)\right)$$

$$= \alpha^2 + \beta^2. \qquad (15.11)$$

5) The operator for the potential energy of the electrons in the field of the nucleus (normalized to $2R$) is given in (15.7). Using the wave function (15.10), show that the expectation value of the potential energy takes the form

$$\frac{\langle\mathrm{PE}\rangle}{R} = -2Za_0\left(\int dV_1\,\psi_{1s}^*(\boldsymbol{r}_1)\,\frac{1}{r_1}\,\psi_{1s}(\boldsymbol{r}_1) + \int dV_2\,\psi_{2p}^*(\boldsymbol{r}_2)\,\frac{1}{r_2}\,\psi_{2p}(\boldsymbol{r}_2)\right)$$

$$= -Z(2\alpha + \beta). \qquad (15.12)$$

In (15.11) and (15.12), the 1s and 2p electrons make individual contributions to the energies, easily identified by the presence in those energies of α or β.

6) Now deal with the energy of electron–electron repulsion. The calculation is lengthy, but is intimidating more than difficult.
   The repulsion energy E_r is given by

$$\frac{E_r}{R} = \int dV_1\,dV_2\,\psi_\pm^*\left(\frac{2a_0}{r_{12}}\right)\psi_\pm.$$

Substitute into this $\psi_\pm$ from (15.10), and multiply out the two curly brackets. Show that terms are equal in pairs so that the integral reduces to

$$\frac{E_r}{R} = 2a_0\int dV_1\,dV_2\,\psi_{1s}^*(\boldsymbol{r}_1)\,\psi_{2p}^*(\boldsymbol{r}_2)\,\frac{1}{r_{12}}$$

$$\times \left\{\psi_{1s}(\boldsymbol{r}_1)\,\psi_{2p}(\boldsymbol{r}_2) \pm \psi_{1s}(\boldsymbol{r}_2)\,\psi_{2p}(\boldsymbol{r}_1)\right\}. \qquad (15.13)$$

What we have done reproduces the mathematics that leads from (15.2–3) to (15.4) and (15.5). The first integral in (15.13) is recognizable as the "direct" integral J/R, and the second is the "exchange" integral K/R.

7) Next, we look at $1/r_{12}$:

$$\frac{1}{r_{12}} = \frac{1}{\sqrt{r_1^2 + r_2^2 - 2r_1r_2\cos\Theta}},$$

in which Θ is the angle between vectors $\boldsymbol{r}_1$ and $\boldsymbol{r}_2$. This expression is the generating function for Legendre polynomials, but the expansion must be expressed in terms of either r_1/r_2 or r_2/r_1, whichever is less than [?]. Write $r_<$ and $r_>$ for the lesser and the greater of r_1, r_2, and then

$$\frac{1}{r_{12}} = \frac{1}{r_>} \sum_{n=0}^{\infty} \left(\frac{r_<}{r_>}\right)^n P_n(\cos\Theta).$$

[29] See e.g. Jackson (1999), § 3.6. Courage! It simplifies.

The angle Θ looks inconvenient, but fortunately we have the "addition theorem" for spherical harmonics:[29]

$$P_n(\cos\Theta) = \sum_{m=-n}^{m=n} \frac{(n-|m|)!}{(n+|m|)!} P_n^{|m|}(\cos\theta_1) P_n^{|m|}(\cos\theta_2) e^{im(\phi_1-\phi_2)}.$$

Return to (15.13) and spell out the details of the integration over volume V_1. We may use the fact that ψ_{1s} and ψ_{2p} are real.

$$\frac{E_r}{R} = 2a_0 \Bigg\{ \int dV_2\, \psi_{2p}(\mathbf{r}_2)^2 \int dV_1\, \frac{1}{r_{12}} \psi_{1s}(\mathbf{r}_1)^2$$

$$\pm \int dV_2\, \psi_{1s}(\mathbf{r}_2)\, \psi_{2p}(\mathbf{r}_2) \int dV_1\, \frac{1}{r_{12}} \psi_{1s}(\mathbf{r}_1)\, \psi_{2p}(\mathbf{r}_1) \Bigg\}$$

$$= 2a_0 \Bigg\{ \int dV_2\, \psi_{2p}(\mathbf{r}_2)^2 \int_0^\infty dr_1\, r_1^2\, \psi_{1s}(r_1)^2 \int_0^\pi d\theta_1 \sin\theta_1$$

$$\pm \int dV_2\, \psi_{1s}(r_2)\, \psi_{2p}(\mathbf{r}_2) \int_0^\infty dr_1\, r_1^2\, \psi_{1s}(r_1) \int_0^\pi d\theta_1 \sin\theta_1\, \psi_{2p}(\mathbf{r}_1)$$

$$\times \int_0^{2\pi} d\phi_1\, \frac{1}{r_{12}}.$$

Here we make use of the fact that $\psi_{1s}(\mathbf{r})$ depends upon the magnitude r of $\mathbf{r}$ and not upon its direction (θ, ϕ). Likewise, neither $\psi_{1s}(\mathbf{r})$ nor $\psi_{2p}(\mathbf{r})$ depends upon ϕ, so integration over ϕ_1 involves $1/r_{12}$ only. Show that the integral over ϕ_1 introduces a friendly simplification:

$$\int_0^{2\pi} d\phi_1\, \frac{1}{r_{12}} \text{ contains } \int_0^{2\pi} d\phi_1\, e^{im\phi_1} = 2\pi\, \delta_{m0}.$$

The sum over m in $P_n(\cos\Theta)$ reduces to the single term with $m = 0$.

$$\int_0^{2\pi} d\phi_1\, \frac{1}{r_{12}} = \frac{2\pi}{r_>} \sum_{n=0}^{\infty} \left(\frac{r_<}{r_>}\right)^n P_n(\cos\theta_1)\, P_n(\cos\theta_2).$$

(8) It's tidiest now to work separately on the direct and exchange integrals. The direct integral J becomes

$$\frac{J}{R} = 2a_0 \int dV_2\, \psi_{2p}(\mathbf{r}_2)^2 \int_0^\infty dr_1\, r_1^2\, \psi_{1s}(r_1)^2$$

$$\times \frac{2\pi}{r_>} \sum_{n=0}^{\infty} \left(\frac{r_<}{r_>}\right)^n P_n(\cos\theta_2) \int_0^\pi d\theta_1 \sin\theta_1\, P_n(\cos\theta_1).$$

Recognize that the integral over θ_1 is an orthogonality integral for P_n and P_0, so it evaluates to $\frac{2}{2n+1}\delta_{n0} = 2\delta_{n0}$. Now it is the sum over n that reduces to a single term.

$$\frac{J}{R} = 8\pi a_0 \int dV_2\, \psi_{2p}(\mathbf{r}_2)^2 \int_0^\infty dr_1\, r_1^2\, \frac{1}{r_>} \psi_{1s}(r_1)^2.$$

We now need to separate functions of radius from functions of angle, so we write

$$\psi_{1s}(\mathbf{r}) = R_{1s}(r) \times \frac{1}{\sqrt{4\pi}}; \qquad \psi_{2p}(\mathbf{r}) = R_{2p}(r) \times \sqrt{\frac{3}{4\pi}} \cos\theta,$$

so that $R_{1s}(r)$ and $R_{2p}(r)$ are separately normalized. Then when we write out the integration over V_2 in full we have

$$\frac{J}{R} = 8\pi a_0 \frac{3}{4\pi} \frac{1}{4\pi} \int_0^\infty dr_2\, r_2^2\, R_{2p}(r_2)^2 \int_0^\pi d\theta_2\, \sin\theta_2 \cos^2\theta_2 \int_0^{2\pi} d\phi_2$$

$$\times \int_0^\infty dr_1\, r_1^2\, \frac{1}{r_>}\, R_{1s}(r_1)^2.$$

The integration over θ_2 yields $2/3$ and that over ϕ_2 yields 2π, so this simplifies to

$$\frac{J}{R} = 2a_0 \int_0^\infty dr_2\, r_2^2\, R_{2p}(r_2)^2 \int_0^\infty dr_1\, r_1^2\, \frac{1}{r_>}\, R_{1s}(r_1)^2. \qquad (15.14)$$

The integral over r_1 is a function of r_2 because of the presence in it of $r_>$.

(9) Given the steps spelt out rather fully in part (8), work out the exchange integral K in a similar way, and show that

$$\frac{K}{R} = \frac{2a_0}{3} \int_0^\infty dr_2\, r_2^2\, R_{1s}(r_2)\, R_{2p}(r_2) \int_0^\infty dr_1\, r_1^2\, \frac{r_<}{r_>^2}\, R_{1s}(r_1)\, R_{2p}(r_1).$$

$$(15.15)$$

(10) In (15.14), $r_>$ is handled by writing

$$\int_0^\infty dr_1\, r_1^2\, \frac{1}{r_>}\, R_{1s}(r_1)^2 = \int_0^{r_2} dr_1\, r_1^2\, \frac{1}{r_2}\, R_{1s}(r_1)^2 + \int_{r_2}^\infty dr_1\, r_1^2\, \frac{1}{r_1}\, R_{1s}(r_1)^2,$$

and the r_1 integral in (15.15) can have its range split in a similar way. Be prepared to simplify the algebra by making substitutions such as $x = 2\alpha r_1/a_0$ or $y = (\alpha + \beta)r_1/a_0$. Show that:

$$\frac{J}{R} = \beta - \frac{\beta^5(3\alpha + \beta)}{(\alpha + \beta)^5}; \qquad \frac{K}{R} = \frac{56}{3} \frac{\alpha^3 \beta^5}{(\alpha + \beta)^7}.$$

Summarizing, the expectation value $\langle E \rangle$ of the helium atom's energy (relative to $He^{++} + 2e$) is given by

$$\frac{\langle E \rangle}{R} = (\alpha^2 + \beta^2) - Z(2\alpha + \beta) + \left(\beta - \frac{\beta^5(3\alpha + \beta)}{(\alpha + \beta)^5} \right) \pm \frac{56}{3} \frac{\alpha^3 \beta^5}{(\alpha + \beta)^7}. \quad (15.16)$$

In (15.16), the terms have been grouped so as to display (in order):
the kinetic energies of the two electrons
the potential energies of the two electrons in the field of the nucleus
the electron–electron repulsion energy, direct integral
the electron–electron repulsion energy, $\pm$ exchange integral.

Problem 15.4 The "textbook" energy of helium $1s\, 2p\,{}^1P$
Suppose that the two electrons in helium move independently round a nucleus of charge $2e$, with the electron–electron repulsion neglected. Then the electrons' wave functions are given by (15.9) with $\alpha = Z = 2$ and $\beta = Z/2 = 1$; and the atom's energy is given by the first two terms

[30] To be fair to the writers of textbooks, they usually make it clear that their calculation has a pedagogical purpose, introducing the direct and exchange integrals, rather than aiming to yield reliable numerical values. Nevertheless it is useful to be reminded how unsatisfactory is this calculation.

Any atom other than helium would be treated differently (Chapter 16), putting as much as possible of the electron–electron repulsion into a **central field**, and solving the Schrödinger equation with that replacement for the actual repulsion. Only after that is perturbation theory applied for dealing with the left-over, the **residual electrostatic interaction**. It is tempting to think that this might be a better procedure for helium also. Unfortunately it isn't. The central-field and residual-electrostatic energies are similar in size, so the residual-electrostatic energy is still not a small perturbation. Helium really is a numerically-unfriendly case—not to be used as a model for understanding other atoms.

[31] In an anthropomorphic way, we may think of the ground-state wave function as "snuggling down" into the shape that minimizes its energy.

A hydrogen atom takes the shape that it does (Fig. 13.4) because of a trade-off between the two terms in its Hamiltonian: the kinetic energy and the potential energy. If we guess a wave function that is too compact, the ⟨kinetic energy⟩ will be too big, in a way that swamps the advantage we gain by putting the electron close to the nucleus. If we guess a wave function that is too puffed-out, the ⟨kinetic energy⟩ will be small but there is a greater penalty in raised ⟨potential energy⟩. The correct wave function can be thought of (and indeed obtained) as the outcome of a trial-and-error process in which we find the best trade-off between the potential and kinetic energies.

Helium is in principle similar, except that the trade-off is between the five energies given in (15.1).

[32] You will need to do the minimization numerically, using a suitable program such as Mathematica.

in (15.16). A "textbook" calculation next applies the electron–electron repulsion using first-order perturbation theory, that is, leaving the wave functions unaltered but using them to evaluate the repulsion energy. The result is the total energy given in (15.16).

Feed the given values of α and β into (15.16) and show that the total energy of the helium atom works out to be

$$\frac{E}{R} = -4 + \frac{7}{3^8}(-27 \pm 64).$$

It is most insightful if we shift the origin of energy to that of a He$^+$ 1s ion (plus one distant electron at rest), which is an energy of $-4R$. Relative to this origin, the predicted energy is given by

$$\frac{E}{R} = \frac{7}{3^8}(-27 \pm 64),$$

which is positive for the singlet *term* (upper sign). If *term* 1s 2p ^{1}P had this predicted energy the atom could become ionized: the 2p electron would be free to wander off with positive kinetic energy—and real helium is not like that. The "textbook" attempt at finding the helium energy by first-order perturbation theory is pretty hopeless.[30]

Problem 15.5 An attempted variational calculation
The **variational principle** tells us that the eigenstates of a Hamiltonian are stationary with respect to variations of the wave function. If the correct eigenfunction is ψ, a small change $\delta\psi$ from ψ results in a change $\delta\langle E \rangle$ of the (expectation-value) energy $\langle E \rangle$ that is of second order in $\delta\psi$. In particular, the ground-state wave function[31] is that which gives the lowest possible $\langle E \rangle$. An excited state also has a minimum $\langle E \rangle$ provided that its trial wave function is constrained to be orthogonal to all wave functions of lower energy.

Apply (15.16) to helium 1s 2p and minimize $\langle E \rangle$ by adjusting α and β. Do this separately for the ^{1}P and ^{3}P *terms*.[32] I think you will find that the 1s wave function is *less* compact in the triplet *term* than in the singlet.

The use of only two variational parameters (α and β) in this problem is a failure: two parameters are simply not enough. This experience may help to explain why the authors responsible for Table 15.1 used as many as 1000 adjustable parameters in their attempts at getting 9–10-figure accuracy. Whatever you want to do, helium really does seem to be obstructive.

LS coupling; Hund's rules

16

Did you never study atomics when you were a lad?
Flann O'Brien, *The Third Policeman*

Intended readership: At the time when *LS* coupling is first encountered, to steer the student through some very misleading treatments.

16.1   Introduction: common errors

This chapter deals with two topics that, in principle, have nothing to do with each other. What they have in common is: first, that they are more often than not presented incorrectly; second, part of that incorrectness consists in making a link between them where there is none.

Diagrams similar to Figs. 16.1 and 16.2 purport to show energies of an *LS*-coupled atom. Most books' accounts include one diagram or the other more-or-less as drawn here.[1] Both diagrams are seriously wrong: in their shape, and in the impression they give of the underlying physics.

Later in this chapter (§ 16.4), we discuss Hund's rules. Many textbook statements of the rules are either wrong or incomplete, and we give the correct version here. We also give counterexamples from the behaviour of real atoms to show why those rules must be stated as we give them.

Returning to *LS* coupling: it is hard to see how Figs. 16.1 and 16.2 came into being unless the originator (and many imitators), misunderstood Hund's rules, and applied those misunderstood rules to draw the splittings shown. This is the link that we have said should not exist.

[1] On a quick survey of commonly used textbooks, I find one of these bad diagrams in seven out of ten; one book has both. As usual, I do not identify those books, but they are easily found. Three honourable exceptions who do get it all right: Corney (1977) p. 84; Foot (2005) p. 82; and Woodgate (1980) p. 119. However, these books are quite advanced, and the unfortunate student, consulting less demanding sources, is very likely to encounter one of the wrong treatments in their first-time study.

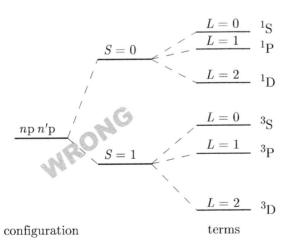

configuration   terms

Fig. 16.1 An $npn'p$ configuration is shown split (by the residual electrostatic interaction) into *LS*-coupled *terms*. We omit here the smaller splitting of the *LS terms* into *J* levels because this is caused by the spin–orbit interaction, not by the residual electrostatic interaction. It is assumed that $n' \neq n$, so that all six *terms* are permitted to exist by the antisymmetrization rules. For a p^2 configuration, in which the electrons are "equivalent" ($n' = n$), terms ^{1}P, ^{3}S, ^{3}D are absent.

Drawn in this way, the diagram is seriously misleading, for reasons given in §§ 16.2–3.

Essays in Physics: Thirty-two thoughtful essays on topics in undergraduate-level physics. Geoffrey Brooker, Oxford University Press. © Geoffrey Brooker 2021. DOI: 10.1093/oso/9780198857242.003.0016

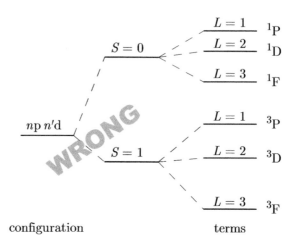

Fig. 16.2 An $np\,n'$d configuration is shown split into *LS*-coupled *terms*. Again we omit the smaller splitting caused by the spin–orbit interaction. Here n' may be the same as n or different, as no *terms* are excluded in either case. As with Fig. 16.1, this diagram gives an entirely wrong impression of how atoms behave.

[2] Those same "explanations" can also be found in books that do not give either diagram.

[3] "All the *other* electrons" is not quite right.

One might expect that a 2p 3p configuration would require one $S_{2p}(r)$ for the 2p electron, taking into account the fact that it feels the Coulomb repulsion from the charge density of a 3p electron, while the 3p electron feels a different repulsion $S_{3p}(r)$ from the more compact charge density of its 2p companion. This is what is done in the Hartree "self-consistent field".

However, the central field uses a single potential $S(r)$ acting on all the electrons in the atom. Only then can the calculated wave functions form an orthogonal set; and only with such an orthogonal set can we build a proper Slater determinant. The energies and wave functions obtained are necessarily imperfect—each electron contributes to a part of the $S(r)$ that it itself feels!— but they are corrected in the residual-electrostatic step.

All this is for one given configuration. To investigate another configuration we start all over again with a new $S(r)$.

Figures 16.1–2 are the more pernicious because they are accompanied by physical "explanations" as to why atoms behave in this way—when they don't.[2] More debunking is done in this chapter than in any other.

16.2 *LS* coupling: what it is

We start with a brief explanation of the context. It is helpful to outline the physics in the form of a pretend-solution of the Schrödinger equation for the electrons within a many-electron atom. The equation is first solved using a rough approximation, and then corrections are made to accommodate refinements. We do not do a real calculation, which would involve very heavy computation. Nevertheless, the introduction of each refinement in the pretend-calculation corresponds neatly to the appearance of a new physical behaviour.

The outline procedure is as follows:

- First, imagine solving the Schrödinger equation, for *each* electron, using a **central-field** Hamiltonian. The potential-energy function $U(r_i)$ acting on the ith electron is the "central field"

$$U(r_i) = \frac{-Ze^2}{4\pi\epsilon_0 r_i} + S(r_i). \qquad (16.1)$$

Here the first term is the attraction of the nucleus (charge Ze), while $S(r_i)$ is a spherically symmetrical potential caused by an averaged charge density (perhaps provisional) from all the other electrons.[3] Wave functions are obtained from the potential. If necessary, a revised charge density is obtained from the wave functions, and the process is repeated .... Such a successive-approximation procedure should hold no mystery after what we have seen in Chapter 12. Because the potential $U(r)$ is spherically symmetrical, each electron has a well defined orbital angular momentum, and we assign it quantum numbers n, l, together (for simplicity) with m_l, m_s. The energy depends on the n, l but not on m_l or m_s. The wave functions for a given n, l constitute a **configuration**. For example, one of the excited configurations of carbon is $1s^2\, 2s^2\, 2p\, 3p$, otherwise (core) $2p\, 3p$ or simply $2p\, 3p$. Because there are 9 values of m_l and 4 values of m_s, this configuration is a set of 36 wave functions, all of which have, in this approximation, the same energy (degeneracy). Likewise a configuration $1s^2\, 2s^2\, 2p\, 3d$ comprises 60 same-energy wave functions.[4]

Whole-atom wave functions, properly antisymmetrized, can now be constructed as Slater determinants. This operation introduces spin, but degeneracy, 36-fold or 60-fold, remains.

- In the central-field stage, we are able to deal only roughly with the Coulomb repulsion of one electron on another. As much of this repulsion as possible has been included in the central-field $U(r)$, but there is a left-over—a residue—the **residual electrostatic interaction**.[5] We imagine that we now improve our energies and wave functions by including this correction, perhaps by using it as the $\widehat{H}'$ in perturbation theory. Each configuration splits (in energy) into **terms**, where each *term* possesses new quantum numbers L and S (and, for simplicity, m_L, m_S). The accompanying wave functions are formed by taking linear combinations of the Slater determinants,[6] and the whole process is called ***LS* coupling**. In a $2p\, 3p$ configuration, the 36 central-field wave functions yield six *terms*. (Clearly a good deal of degeneracy must remain.) Those six *terms* are the six shown in Fig. 16.1—but although the diagram correctly shows the *terms* and their quantum numbers L, S the energy order is all wrong and there are other faults.

- Once we have sorted out the first two steps in this list, we shall have dealt fully with the Coulomb forces within the atom: the attraction of each electron towards the nucleus, and the repulsion between that electron and all the others. The next step is to include usually-smaller effects, in particular the spin–orbit interaction. This splits each *term* into **levels**, each level[7] being characterized by all the previous quantum numbers (the n, l quantum numbers identifying the configuration, and the L, S of the *term*) together now with J (and m_J): more linear combinations.

We need not say much about this **fine structure** here, as our beef is with the second step in this list. However, I must mention the **interval rule** because I need to refer to it later. In spin–orbit coupling, the J-levels belonging to a given *term* have a simple relation between their spacings: the energy difference ("gap") between two levels is proportional to the larger of the Js of the two levels either side of the gap.[8]

The hierarchy outlined above holds provided that each step (except the first) in the list introduces only small corrections to what went before. The residual electrostatic interaction is assumed not (significantly) to form linear combinations of wave functions taken from *different* configurations.[9] And the spin–orbit interaction is assumed not (significantly) to form linear combinations of wave functions taken from *different* terms. When this hierarchy holds, its second step yields *LS* coupling.

[4] The total energy of the atom is not the sum of the one-electron energy eigenvalues, because there is some double-counting of the electron–electron repulsion. In principle this is corrected at the residual-electrostatic stage. Details are not of concern in our pretend-calculation.

[5] Consider an atom containing N electrons. The residual electrostatic interaction introduces the actual electron–electron repulsion energy while removing the previous approximation to it:

$$\widehat{H}'_{\text{re}} = \sum_{i=1}^{N}\left\{ \sum_{j>i}^{N} \frac{e^2}{4\pi\epsilon_0\, r_{ij}} - S(r_i)\right\}.$$

See e.g. Foot (2005) eqn (5.1).

Contrast with §15.3 where we started with hydrogenic wave functions for nuclear charge $Z = 2$, and used the whole of $e^2/4\pi\epsilon_0\, r_{12}$ as perturbation. That was a highly inaccurate step intended to make things simple for a beginner. Here, the starting wave functions have been (notionally) calculated from a central field; they and their energies start out fairly close to the "right answer" and require relatively little correction by $\widehat{H}'_{\text{re}}$.

[6] We are making a useful verbal distinction, previously encountered in §11.5. *LS* coupling is a consequence of the residual electrostatic *interaction*.

Some exemplar linear combinations are built in Chapter 17.

[7] A *level* may be split in energy into its m_J states by applying a magnetic field (Zeeman effect).

This nomenclature—configuration-term-level-state—is standard in atomic physics. Reversing: a *state* is a single wave function, with all quantum numbers specified. A *level* is a group of states, differing only in their m_Js. A *term* is a group of *levels* differing only in their Js (and m_Js).

[8] The wording is careful here. The larger of the two Js. Not the J of the level with the higher energy.

[9] Such a mixing of configurations is rather common, a fact usually brushed under the carpet.

[11] It is no accident that the 3p 4p configuration of silicon is a favourite example in books on atomic physics. Not many other cases of *LS* coupling are so free from complication, conforming to the hierarchy with no irregularity—above all configuration mixing—needing to be explained away.

[12] By "pure", we mean that there is no "configuration mixing": application of the residual electrostatic interaction (plus any smaller interactions) does not cause the wave functions of the resulting *terms* to contain contributions taken from other configurations. This is easy to say, and sounds plausible. Yet there are remarkably few atomic configurations that conform accurately to equations such as (16.2), just because complications such as configuration mixing are rather common.

Comment for honesty

Real calculations sometimes use different central fields $U(r)$ for different *terms*, even *terms* within a single configuration.[10] It is then unrealistic to attribute all energy shifts within such a configuration to a single $\widehat{H}'_{\text{re}}$.

The 3p 4p configuration of silicon (Fig. 16.3) is an unusually simple[11] exemplar of *LS* coupling, in that all six *terms* can be constructed from a common set of 36 building-block (central-field) wave functions acted on by a single $U(r)$ and a single residual-electrostatic $\widehat{H}'_{\text{re}}$.

16.2.1   Critique of figures 16.1 and 16.2

We are now in a position to see how Figs. 16.1 and 16.2 fit (or fail to fit) into the picture.

In each diagram, we start at the left with the energy of a chosen configuration. This energy is what we imagine we have calculated using a central-field approximation. Although a single energy is shown, it represents a bunch (36 or 60) of quantum states that, in this approximation, have the same energy as each other.

We have said that the next stage in a pretend-calculation of atomic energy levels is to introduce the residual electrostatic interaction, which splits the *configuration* into *terms*, each *term* being identified by quantum numbers L and S: *LS* coupling. Figures 16.1 and 16.2 purport to show the nature of this splitting.

Let me now spell out what is wrong with Figs. 16.1 and 16.2.

- Two-stage splitting: The configuration is shown as first splitting into singlets and triplets ($S = 0$ and $S = 1$); then each of these is further split into the levels identified additionally by L. This is wrong. The splitting is a one-stage process (caused by a single $\widehat{H}'_{\text{re}}$), not two-stage, and should be drawn that way.

- All singlets are shown located above all triplets, and the Ls are in descending order within each S. This ordering is not how atoms behave. It is also quite different from that of "pure" *LS* coupling (eqns 16.2–3).[12]

- Diagrams in the style of Figs. 16.1–2 often suggest (whether or not the accompanying text makes it explicit) that the L terms (within each S) have separations obeying an interval rule: the gap between two *L-terms* (having the same S) is drawn proportional to the larger of the L-values either side of the gap. There is such a thing as an interval rule, governing the separation of J-levels split by the spin–orbit interaction, but there is *no* interval rule governing the energy spacings of the *L-terms*.

If Figs. 16.1 and 16.2 were correct, it would surely be easy to find plentiful examples in the periodic table of atoms or ions having their energies ordered as shown. I have not made an exhaustive search, but I have fossicked extensively in
https://physics.nist.gov/PhysRefData/ASD/levels_form.html
a website of the US National Institute of Science and Technology, and

I have failed to find a single example that looks like either of Figs. 16.1 or 16.2.[13]

Correct energy-level diagrams are not hard to find; sometimes in the same books as give the wrong versions!

Wrong diagrams are supported by various wrong "explanations".

- It is sometimes said that one interaction separates $S = 0$ from $S = 1$, and then a different and weaker interaction[14] separates the Ls.

- Sometimes "exchange forces" explain everything.[16]

- Sometimes the reasoning[17] of §15.3 is invoked to show that singlets are raised in energy more than triplets by $e^2/4\pi\epsilon_0 r_{12}$ because $\psi_{\text{triplet}}(\boldsymbol{r}_1, \boldsymbol{r}_2) = 0$ when $\boldsymbol{r}_1 = \boldsymbol{r}_2$.

- Several books argue that the Ls lie in descending order within each S because electrons like to keep themselves apart by spacing themselves round the periphery of the atom and rotating together.[18]

- One book seeks to account for a (non-existent remember) interval-rule separation of the Ls by proposing an "effective Hamiltonian" $(a_1 \boldsymbol{s}_1 \cdot \boldsymbol{s}_2 + a_2 \boldsymbol{l}_1 \cdot \boldsymbol{l}_2)$ that yields it.[19]

- Sometimes it is claimed that the energy order in each of Figs. 16.1 and 16.2 is necessitated by Hund's Rules.[20]

Not only must we reject the seriously wrong energy-level diagrams of Figs. 16.1–2, we must reject alongside them the textbook "explanations" that have been invented to account for them.

16.3   The correct description of LS coupling

Two correct energy-level diagrams are given in Figs. 16.3 and 16.4. They display experimentally measured energies, showing what the atoms in question really do.

Figure 16.3 shows the $(1s^2\, 2s^2\, 2p^6\, 3s^2)\, 3p\, 4p$ configuration of neutral silicon, split by the residual electrostatic interaction into LS-coupled terms. The splitting is shown as the one-stage process that it is, not as a first splitting into singlets and triplets, followed by a second splitting according to L.

Wave functions for the states shown in Fig. 15.3 are worked out in the problems to Chapter 17. Once those wave functions are seen, the correct energy order of the states (very different from that in Fig. 16.1) is near-obvious.

The correct structure (eqns 16.2–3) of the residual electrostatic interaction and its consequences has been known for a very long time. It is given by Condon and Shortley (1951) pp 199–200; my 1964 copy is a reprint, with only small amendments, of a book first published in 1935.

[13] This should not be a surprise. An atom's energies depart from (16.2–3) when things are being complicated by configuration mixing. It would take an extraordinary conspiracy for the mixing to pummel the energies into the order of Fig. 16.1.

[14] Not true. The only interaction here is the residual electrostatic interaction.+

In what may be the index case,[15] the author introduces a "spin-spin correlation energy" (undefined) to separate singlets from triplets, and a weaker "residual electrostatic energy" (also undefined by him) to arrange the Ls in descending order. Although Hund is not mentioned, it is hard to see this fabrication as other than a creative misapplication of Hund's Rules.

[15] The "index case" is a term from epidemiology for the first case in an outbreak. Ours is no later than 1959. After 60+ years it is high time for a robust refutation of the whole package.

[16] "A purely quantum effect", aka magic. We have disposed of exchange force in §14.6.6.

[17] This reasoning is shown to be irrelevant in §15.3.1. Also, Fig. 16.3 and eqn (16.5) show that the electrons in some singlets can be "more apart" than in corresponding triplets.

[18] If this held, it would show up best in states having $m_L = L$, for which the z-component of total orbital angular momentum is greatest. For ^{3}D and ^{1}D the wave functions (eqns 17.14, 17) contain $\sin\theta_1\, e^{i\phi_1} \sin\theta_2\, e^{i\phi_2}$: no correlation, so the "explanation" won't do.

In fact, the wave functions that come closest to "opposite sides" contain

$(\sin\theta_1\, e^{i\phi_1} \cos\theta_2 - \sin\theta_2\, e^{i\phi_2} \cos\theta_1)$;

this factor appears in the wave functions for ^{3}P and ^{1}P, and goes part-way to explaining why ^{1}P lies lowest.

[19] Science fiction.

[20] See §16.4. Hund tells us which S, L *term* lies lowest in an atom's *ground configuration*. He can have nothing to say about Figs. 16.1–2, which describe (or claim to) excited configurations with two incomplete sub-shells.

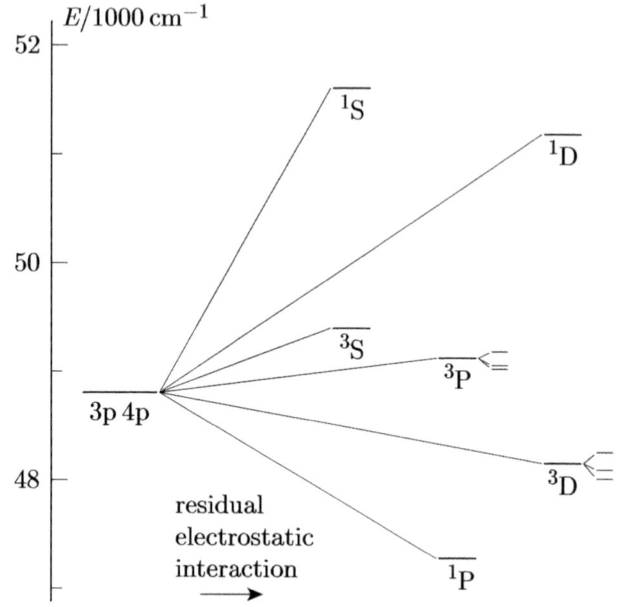

Fig. 16.3 A example of *LS* coupling as it really is: the 3p 4p configuration of neutral silicon. The 3p 4p configuration is split by the residual electrostatic interaction, in a single step, into the *LS* terms. The spin–orbit splitting of the ^{3}P and ^{3}D terms is shown, merely to give reassurance that it is small enough not to disrupt things. Note how very different is the energy order from that in Fig. 16.1: in particular the lowest level is not even a triplet. Diagram modelled after Woodgate (1980) p. 119; energy levels from Kramida *et al.* (2021).

Note a feature that further contradicts Fig. 16.1. In "pure" *LS* coupling there is in general an alternation in the energy order of singlets and triplets with increasing *L*. See e.g. Kuhn (1969) p. 267; Condon & Shortley (1951) p. 200. Here ^{3}S lies below ^{1}S, then ^{1}P below ^{3}P, then ^{3}D below ^{1}D.

For a pp configuration (meaning $np\,n'$p with $n' \neq n$) that is purely *LS* coupled, Condon and Shortley (p. 199) give the following:

$$E(^1\text{S}) = F_0 + 10F_2 + G_0 + 10G_2, \quad E(^3\text{S}) = F_0 + 10F_2 - G_0 - 10G_2,$$
$$E(^1\text{P}) = F_0 - 5F_2 - G_0 + 5G_2, \quad E(^3\text{P}) = F_0 - 5F_2 + G_0 - 5G_2,$$
$$E(^1\text{D}) = F_0 + F_2 + G_0 + G_2, \quad E(^3\text{D}) = F_0 + F_2 - G_0 - G_2. \tag{16.2}$$

In these, the energies are given relative to the location of the uncorrected (i.e. central-field) configuration. The quantities F_0, F_2, G_0, G_2 are matrix elements of $\widehat{H}'_{\text{re}}$ taken between the $np\,n'$p building-block wave functions (actually they are integrals over radii r_1, r_2; angular integrations have been done and have yielded the numerical coefficients). Integrals F_0 and F_2 are direct integrals; G_0 and G_2 are exchange integrals. The four integrals are in principle calculable if we know enough about the atomic wave functions, but usually we don't. Therefore the integrals are often treated as adjustable parameters, fitted to experimental data.[21] Woodgate (1980) (pp 117–18) gives a detailed analysis for the case of silicon 3p 4p, part-repeated here in exercise 16.1.

[21] There are six energies and only four parameters. Therefore there are proportions that the energy-level diagram must have if the parameters are to fit. See exercise 16.1.

[22] The averaging is a device for simplifying things by removing the exchange integrals.

[23] The residual-electrostatic $\widehat{H}'_{\text{re}}$ is given in sidenote 5. When it is applied, the central-field terms $S(r_1)$, $S(r_2)$ give the same outcome for all linear-combination wave functions, so they contribute only to F_0. The other three integrals contain $e^2/(4\pi\epsilon_0\, r_{12})$ only. This is always positive, so the direct integral F_2 is positive also. Integrals G_0 and G_2 tend to be positive, though this is not required. The sign of F_0 depends upon where the central-field step left us.

Exercise 16.1 The energy order of *LS* terms for pp configurations

(1) In eqns (16.2), average[22] the energies for each *L* after the fashion $D_{\text{m}} = \frac{1}{2}\{E(^1\text{D}) + E(^3\text{D})\}$. Show that these averages have sequence P–D–S. If Fig. 16.1 were correct the averages would have order D–P–S. This is perhaps the simplest way of confirming my statement in § 16.2.1 that the energy order of Fig. 16.1 violates "pure" *LS* coupling.[23]

(2) Use eqns (16.2) to show that

$$s_1 \equiv \frac{E(^1S) - E(^1D)}{E(^1D) - E(^3P)} = \frac{9F_1 + 9G_2}{6F_2 + 6G_2} = \frac{3}{2};$$

$$s_2 \equiv \frac{E(^3S) - E(^3D)}{E(^3D) - E(^1P)} = \frac{9F_2 - 9G_2}{6F_2 - 6G_2} = \frac{3}{2}.$$

A further test ratio, though not independent of s_1, s_2, is

$$s_3 \equiv \frac{S_m - D_m}{D_m - P_m} = \frac{9F_2}{6F_2} = \frac{3}{2}.$$

For Si: 3p 4p, these ratios are found by measurement to be 1.34, 1.41 and 1.37. See Woodgate (1980), p. 118.[24]

Exercise 16.2 The energy order of *LS* terms for pd configurations

(1) The energies of an *LS*-coupled pd configuration are given by:

$$E(^1P) = F_0 + 7F_2 + \ \ G_1 + 63G_3, \quad E(^3P) = F_0 + 7F_2 - \ \ G_1 - 63G_3,$$

$$E(^1D) = F_0 - 7F_2 - 3G_1 + 21G_3, \quad E(^3D) = F_0 - 7F_2 + 3G_1 - 21G_3,$$

$$E(^1F) = F_0 + 2F_2 + 6G_1 + \ \ 3G_3, \quad E(^3F) = F_0 + 2F_2 - 6G_1 - \ \ 3G_3.$$

$$(16.3)$$

Here, F_0 and F_2 are again direct integrals, while G_1 and G_3 are exchange integrals (all different from before). Form average energies as in exercise 16.1(1). Show that these averages have energy order D–F–P thereby invalidating Fig. 16.2. For a real atom having a pd configuration, see Fig. 16.4.

(2) Take the energies of (16.3) and show that

$$s_4 \equiv \frac{P_m - F_m}{F_m - D_m} = \frac{5}{9}.$$

16.3.1   The terms 3p 4p ^{1}P and 3p 4p ^{3}P of silicon

Triplets often lie below (same-configuration) singlets; they do in helium.[25]

Why then does ^{1}P lie, not only below ^{3}P but at the very bottom?

If this is a surprise, we have picked up false expectations. To recapitulate: A space-antisymmetric wave function $\psi_a(\mathbf{r}_1, \mathbf{r}_2)$ is necessarily zero at $\mathbf{r}_1 = \mathbf{r}_2$. But §15.3.1 has shown that the repulsion energy $\langle e^2/4\pi\epsilon_0 r_{12} \rangle$ pays no attention to ψ_a at $\mathbf{r}_1 = \mathbf{r}_2$; if we focus attention on $\mathbf{r}_1 = \mathbf{r}_2$ we are looking in the wrong place for an explanation of the energy. Also, we have said in §14.5 that a space-symmetric $\psi_s(\mathbf{r}_1, \mathbf{r}_2)$ does not necessarily have a "doubled" $|\psi_s(\mathbf{r}_1, \mathbf{r}_2)|^2$ at $\mathbf{r}_1 = \mathbf{r}_2$ (not that that would have any relevance to the energy anyway). There are no reasons here, either way, as to the energy order of ψ_a and ψ_s.

Wave functions for representative 3p and 4p states are[26]

$$\left| 3p,\, m_l{=}1,\, m_s{=}{+}\tfrac{1}{2} \right\rangle = R_{3p}(r)\,(-\sin\theta)\,e^{i\phi} \uparrow,$$

$$\left| 4p,\, m_l{=}0,\, m_s{=}{-}\tfrac{1}{2} \right\rangle = R_{4p}(r)\,\cos\theta \downarrow,$$

[24] The reader may be surprised to see energies of silicon used to illustrate *LS* coupling. Isn't this a rather obscure atom to choose? The obvious choice would be the first case in the periodic table having an $np\,n'p$ configuration: carbon 2p 3p. This configuration has its energies in almost the same order as those in Fig. 16.3: ^{1}P lies at the bottom, and the triplets form a bunch in the middle—very far from the "bunch at the bottom" of Fig. 16.1. However, the test ratios s_1–s_3 are less convincing for carbon 2p 3p than for silicon 3p 4p: 1.10, 1.18 and 1.14.

Incidentally, these energies in carbon are drawn incorrectly in some textbooks, even recent ones. The error seems to date back to Herzberg p. 143 (1944 but info of 1937), whose diagram was clearly provisional. The values of L, S were soon corrected, already in the monumental catalogue of atomic energies by Moore (1949) p. 22. The carbon energies are of some interest because (in right or wrong forms) they are easily-found counterexamples to Fig. 16.1 and 16.2, even though sometimes printed in the same books.

[25] My suspicion is that the placing of all triplets below all singlets in Figs. 16.1 and 16.2 arises, in part but only in part, from an unwarranted generalization of this property of helium. (This and wrong-Hund can easily reinforce each other into a slogan "triplets lie lower".) But Chapter 15 has shown that helium is a particularly awkward case, not to be used as a precedent.

[26] The radial functions $R_{3p}(r)$ and $R_{4p}(r)$ are to be thought of as the outcomes of a central-field calculation. Fortunately we do not need to know anything about them.

The negative sign in the first wave function arises from a convention as to how the spherical-harmonic part of the wave function should be defined. It is, of course, not responsible for the relative signs in (16.4) and (16.5).

and similarly. The simplest wave functions for 3p 4p ^{3}P and 3p 4p ^{1}P are (omitting normalization constants):[27]

$$\left|^3\mathrm{P},\, m_L{=}1, m_S{=}1\right\rangle = \uparrow_1 \uparrow_2 \left\{ R_{3\mathrm{p}}(r_1)\,R_{4\mathrm{p}}(r_2) + R_{3\mathrm{p}}(r_2)\,R_{4\mathrm{p}}(r_1) \right\}$$
$$\times \left(\sin\theta_1\, e^{i\phi_1} \cos\theta_2 - \sin\theta_2\, e^{i\phi_2} \cos\theta_1 \right); \quad (16.4)$$

$$\left|^1\mathrm{P},\, m_L{=}1, m_S{=}0\right\rangle = \left(\uparrow_1 \downarrow_2 - \uparrow_2 \downarrow_1\right)$$
$$\times \left\{ R_{3\mathrm{p}}(r_1)\,R_{4\mathrm{p}}(r_2) - R_{3\mathrm{p}}(r_2)\,R_{4\mathrm{p}}(r_1) \right\}$$
$$\times \left(\sin\theta_1\, e^{i\phi_1} \cos\theta_2 - \sin\theta_2\, e^{i\phi_2} \cos\theta_1 \right). \quad (16.5)$$

The triplet state has a symmetric spin part and an antisymmetric space part. That is, the space part reverses its sign if we interchange r_1, θ_1, ϕ_1 with r_2, θ_2, ϕ_2. The singlet state has the opposite symmetries; the ground rule of overall antisymmetry is satisfied.[28]

Look first at the triplet wave function of (16.4). It is zero whenever $\theta_1 = \theta_2$ and $\phi_1 = \phi_2$: the electrons are prohibited from ever lying on a common radial line out from the nucleus. This separation is opposed by the radial part { } whose square is doubled (yes) when $r_1 = r_2$. Contrast this with the singlet wave function of (16.5). This is again zero whenever the electrons lie on a common radial line, but it is also zero when $r_1 = r_2$: the electrons are prohibited from being at the same distance from the nucleus, regardless of the directions they lie in. These electrons are apart for two reasons, they really are "more apart" than those in the triplet state.[29]

Wave function (16.5) provides a powerful counterexample to any idea that space-antisymmetric implies "more apart" than space symmetric.

Fig. 16.4 The energies of the 3d 4p configuration of vanadium 3+, given here to confront the textbook diagram of Fig. 16.2. The configuration chosen may look even more obscure than that of Fig. 16.3, but it happens that V^{3+} 3d 4p is unusually free from configuration mixing: the worst-case level has only a 7% admixture from elsewhere. This same configuration is used as example by Corney (1977) p. 84.

Note similarities to the energy order of Fig. 16.3: the lowest term is a singlet, while the triplets again all lie in the middle.

Also, there is the expected alternation: ^{3}P lies below ^{1}P, then ^{1}D lies below ^{3}D, then ^{3}F lies below ^{1}F.

The ratio s_4 of exercise 16.2 is 0.487, not too far from the 0.555 that is expected for a case of "pure" *LS* coupling.

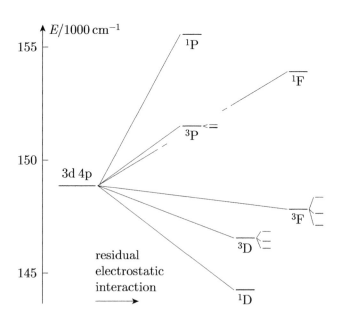

16.4 Hund's rules

We move on to the second of this chapter's topics. Every student remembers the rules.[30] In an atom's or ion's ground configuration:

1. to find the lowest-energy *term*, take the largest possible S;
2. if any choice remains, take the largest possible L.

But what's far more important than the rules themselves is the knowledge of how and when to apply them, the precise conditions under which they hold:

- the energy sought is that of the very-lowest-energy *term* (meaning L and S) in the ground configuration of the atom or ion
- nothing is said, even about that, if the ground configuration has more than one incomplete (i.e. partly filled) sub-shell[31]
- nothing is said about the energy order of the *terms* above that very-bottom *term*, even of *terms* within the same configuration
- and nothing at all is said about the energy order within excited configurations.

Most textbooks do little to spell out these restrictions.[32] Indeed, my dark suspicion is that Figs. 16.1 and 16.2 must originate from a massive misunderstanding of Hund: "maximize S" (misapplied rule 1) puts all triplets below all singlets; then "maximize L" (misapplied rule 2) puts the Ls in descending order within the triplets and within the singlets; possibly also "apply an interval rule" (particularly shameful) spaces the Ls; and all this is done on an excited configuration with two part-filled sub-shells.[33] This is the non-link, mentioned in § 16.1, between the two topics of this chapter.

Students often fail to take on board the itemized restrictions above. In Figs. 16.1 and 16.2 they find every tempting misunderstanding confirmed. Not a good idea.

There does not seem to be a straightforward way of showing that Hund's rules follow from theory; they should be taken as just rules that happen to be reliable—when correctly applied.

16.4.1 Extensions to Hund's strictly-stated rules?

I've stated that Hund's rules apply only to the very lowest *term* (quantum state with specified L and S) within the very lowest *configuration* (specified ns and ls) for an atom or ion. Given the propensity of textbooks to offer us "extensions" to those rules (permitting their application to the ordering of energies or to excited configurations or both), we must ask whether I have been too strict. Does experiment permit any such extensions? Answer: no. I'll back this up by presenting counterexamples.

The energy levels of atoms and ions are tabulated. I have used the NIST website (see bibliography under Kramida *et al.*).

The requirement that an atom's overall wave function be antisymmetric under interchange of labels on any pair of electrons (loosely, the

[30] Well, actually, often not. I've worded the rules carefully here. It only takes a slight variation in the wording to yield a statement that's wrong.

[31] Paraphrasing: Hund is always restricted to dealing with "equivalent" electrons within a ground configuration having a single incomplete sub-shell.

[32] My search of 17 textbooks has revealed 14 that give incorrect or incomplete statements of Hund's rules. Even authors commanding the greatest respect do not always say clearly enough that Hund's rules are restricted to the ground configuration only—and further restricted within that. The honourable exceptions are Foot (2005) pp 81–2 and Woodgate (1980) p. 124.

[33] At least two of the nine texts alluded to in sidenotes 1 and 2 make this link explicit: they say that their (wrong) energy-level diagrams have been constructed to be in conformity with their (wrong) Hund. It is not only students who have been tempted.

[34] Standard nomenclature: A **shell** is all the quantum states with a given n. A **subshell** is all the quantum states with a given n and l. Thus the $n = 2$ *shell* consists of two *subshells* 2s and 2p.

[35] The reasoning that shows ^{3}P, ^{1}D and ^{1}S to be the only allowed *terms* for a p^2 configuration is given in § 17.3. Likewise, the *terms* for a p^3 configuration are dealt with in problem 17.5.

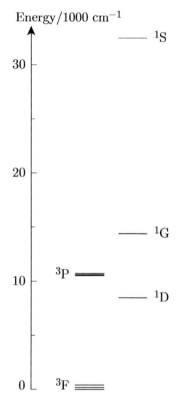

Fig. 16.5 The terms of the ground configuration 3d^2 of the ion Ti^{2+}.

Given two electrons outside closed shells, each term must be either a triplet or a singlet. To find the lowest-energy term, apply Hund's rules. Rule 1: the lowest term has the largest available S, so it is a triplet. Rule 2: a choice remains, between ^{3}F and ^{3}P; the lowest term has the largest L: ^{3}F.

Hund has now finished. We must refrain from hoping he might do more. In particular, we must not think that ^{3}P comes next in energy, as a sort of second-best. As for the singlets, Hund is silent.

Pauli principle) restricts the LS values that are allowed to exist. Here are the possibilities (for a single part-filled subshell—"equivalent" electrons—accompanying other filled shells or subshells):[34]

p^2 and p^4: ^{3}P ^{1}D ^{1}S

p^3: ^{4}S ^{2}D ^{2}P

d^2 and d^8: ^{3}F ^{3}P ^{1}G ^{1}D ^{1}S

d^3 and d^7: ^{4}F ^{4}P ^{2}H ^{2}G ^{2}F ^{2}D (twice) ^{2}P

d^4 and d^6: ^{5}D ^{3}H ^{3}G ^{3}F (twice) ^{3}D ^{3}P (twice) ^{1}I ^{1}G (twice) ^{1}F ^{1}D (twice) ^{1}S (twice)

d^5: ^{6}S and many others

f^2 and f^{12}: ^{3}H ^{3}F ^{3}P ^{1}I ^{1}G ^{1}D ^{1}S

f^3 and f^{11}: ^{4}I ^{4}G ^{4}F ^{4}D ^{4}S and many doublets

f^4 and f^{10}: ^{5}I ^{5}G ^{5}F ^{5}D ^{5}S and many others

f^5 and f^9: ^{6}H ^{6}F ^{6}P and many others

f^6 and f^8: ^{7}F and many others

f^7: ^{8}S and many others.

The *terms* are listed in the sequence that *would* be the energy order if the more daring books were right.

Ground configurations with one part-filled subshell

So what does experiment have to say? For atoms and ions whose ground configuration contains one part-filled subshell, Hund (properly applied) gives the correct bottom *term* in the 325 cases (neutral atoms and ions) I've been able to check, with no exceptions.

With this reassurance that we're doing something right, we go on to look at the order of energies above the bottom.

Atoms and ions with a part-filled subshell of p-electrons
The energy order within the ground configuration is:[35]

p^2: ^{3}P – ^{1}D – ^{1}S   54 instances, no exceptions
p^3: ^{4}S – ^{2}D – ^{2}P   56 instances, no exceptions
p^4: ^{3}P – ^{1}D – ^{1}S   55 instances, no exceptions.

These findings are compatible with almost any statement, right or wrong, that claims to assert Hund's rules. But, for the avoidance of doubt, Hund rule 1, correctly applied, places ^{3}P, ^{4}S and ^{3}P at the bottom. No choice remains, so rule 2 is not activated. Hund says nothing more. In particular, the D terms come next, but

Hund did not order that; he does not award consolation prizes.

Atoms and ions with a part-filled subshell of d-electrons
The crunch comes when we look at atoms with ground configurations containing a part-filled d- or f-subshell. Hund always gets the lowest

term right, which is all he aims to do. What happens above that lowest *term* follows no simple pattern. Examples follow.

Atoms and ions with ground configuration d^2 or d^8: The NIST website gives complete lists of energies for Ti^{2+} $3d^2$, V^{3+} $3d^2$ and Ni^{2+} $3p^6$ $3d^8$. For all three we have energy order (Fig. 16.5):

$$Ti^{2+}\ 3d^2:\ {}^3F - {}^1D - {}^3P - {}^1G - {}^1S.$$

We do *not* have all the singlets above all the triplets, and the singlets do *not* have their Ls in descending order.

Atoms and ions with ground configuration d^3 or d^7: The only complete d^3 example I can find is:

$$V^{2+}\ 3d^3:\ {}^4F - {}^4P - {}^2G - {}^2P - {}^2D - {}^2H - {}^2F - {}^2D.$$

For d^7 we have complete listings for Co^{2+}, Ni^{3+}, Cu^{4+} and Zn^{5+}, all of which have the order:

$$Co^{2+}\ 3d^7:\ {}^4F - {}^4P - {}^2G - {}^2P - {}^2H - {}^2D - {}^2F - {}^2D.$$

In these d^3 and d^7 cases, all the quartets do happen to lie at the bottom, but the doublets do not have their Ls is descending order.

Atoms and ions with configuration d^4 or d^6: The most complete example I can find is V^+:

$$V^+\ 3d^4:\ {}^5D - {}^3P - {}^3H - {}^3F - {}^3G - {}^1G - {}^3D - {}^1I - {}^1S - {}^1D - {}^1F -$$
$$ {}^3F - {}^3P - {}^1G - {}^1D -$$

with a ${}^1S$ *term* unaccounted for but expected to lie higher than the *terms* listed. What we have is enough to show that: the triplets do not lie below all the singlets; and there is no simple order of the Ls within the triplets or within the singlets.

Atoms and ions with configuration d^5: An almost-complete example is:

$$Mo^+\ 4d^5:\ {}^6S - {}^4G - {}^4P - {}^4D - {}^2D - {}^2I - {}^4F - {}^2F - {}^2G - {}^2F - {}^2H -$$
$$ {}^2S - {}^2D - {}^2G -$$

with ${}^2P$ and a third ${}^2D$ unaccounted for. Neither quartets nor doublets have their Ls in descending order.

Atoms and ions with a part-filled subshell of f-electrons
Ions Pr^{2+} and Nd^{3+} have ground configuration $4f^3$. Of these, Pr^{2+} is more fully documented, and has *terms* in the energy order (Fig. 16.6):

$$Pr^{2+}\ 4f^3:\ {}^4I - {}^4F - {}^2H - {}^4S - {}^2G - {}^4G - {}^2K - {}^2D - {}^2P - {}^4D - {}^2I -$$
$$\phantom{Pr^{2+}\ 4f^3:\ } {}^2L - {}^2D - {}^2H - {}^2F - {}^2G - {}^2F.$$

The lowest *term* here is a quartet (Hund rule 1). Of the quartets, ${}^4I$ lies lowest (Hund rule 2). After this Hund has finished. Above ${}^4I$, the quartets do not have their Ls in descending order. We have now falsified any thought that "for the largest S the Ls are in descending

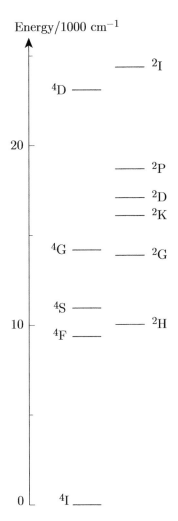

Energy/1000 cm^{-1}

Fig. 16.6 The ground configuration $4f^3$ of Pr^{2+}. Only the lowest-lying level of each *term* is shown (the levels interlace in a complicated way), and the six highest doublets are omitted. Other configurations (having many energies in the range drawn) are also omitted. Hund places ${}^4I$ at the bottom, and having done that has finished. Note that even the quartets do not have their Ls in descending order.

order". Showing this has required that we have at least three Ls for the greatest S, something that doesn't happen with p or d configurations.

Excited configurations

As soon as we look at excited configurations, we find that atoms are even more seriously ignorant of how textbooks would tell them to behave.

Start with p^2 and p^3 excited configurations, for direct comparison with the comparable ground configurations already described. We have doubly ionized silicon[36] and triply ionized sulphur:

Si^{2+} (excited configuration) $3p^2$: $^1\text{D} - {}^3\text{P} - {}^1\text{S}$

S^{3+} (excited configuration) $3p^3$: $^2\text{D} - {}^4\text{S} - {}^2\text{P}$.

The triplet and quartet do not lie lowest. We can't extend Hund's rules to excited configurations, even for a case (which we have here) of equivalent electrons in a single part-filled subshell.

For more complicated configurations, with two or more part-filled subshells, we need look no further than Si: 3p 4p, Fig. 16.3. This has the triplets bunched in the middle, not at the bottom, while the lowest term is the singlet ^1P.

Finally, I give the energy order for three pd (excited) configurations, thereby confronting the diagram of Fig. 16.2:[37]

C $2s^2\, 2p\, 3d$: $^1\text{D} - {}^3\text{F} - {}^3\text{D} - {}^1\text{F} - {}^1\text{P} - {}^3\text{P}$

Si $3s^2\, 3p\, 3d$: $^1\text{D} - {}^3\text{F} - {}^3\text{P} - {}^1\text{F} - {}^1\text{P} - {}^3\text{D}$

$\text{V}^{3+}\, 4p\, 3d$: $^1\text{D} - {}^3\text{D} - {}^3\text{F} - {}^3\text{P} - {}^1\text{F} - {}^1\text{P}$.

Ground configurations having more than one part-filled subshell

I have found 39 cases of atoms and ions whose ground configurations have more than one part-filled subshell. Most are Hund-like. But I have found two counterexamples:

Ce $4f\, 5d\, 6s^2$: $^1\text{G} - {}^3\text{F} - {}^3\text{H} - {}^3\text{G} - {}^1\text{D} -$

$\text{Ce}^+\, 4f(^2\text{F})\, 5d^2(^3\text{F})$: $^4\text{H} - {}^4\text{I} -$

Both of these are listed incompletely, but we're only concerned with the lowest energies. Neutral cerium does not have the largest S at the bottom (nor the largest L). The Ce^+ ion[38] has the largest S at the bottom but not accompanied by the largest available L.

Summary

Hund's Rules are stated correctly in §16.4. Any proposed widening of their application is proved inadmissible.

---

[36] There must be something a bit odd about these two configurations. For a $3p^2$ configuration, free from configuration mixing, the energies of the three terms are given, according to the *LS*-coupling integrals, by

$$E(^1\text{S}) = F_0 + 10F_2$$
$$E(^3\text{P}) = F_0 - 5F_2$$
$$E(^1\text{D}) = F_0 + F_2.$$

These expressions do not permit ^3P to have an energy between the two singlets. However, the ^1S term is strongly configuration-mixed with $3s\, 4s\, {}^1\text{S}$, while ^1D is even more strongly mixed (33%) with $3s\, 3d\, {}^1\text{D}$. Likewise, the $3p^3\, {}^2\text{D}$ and ^2P are strongly configuration-mixed (40% for ^2D).

It is interesting that Si^{2+} has ^1S mixed with ^1S, so that the L and S quantum numbers remain good in spite of the configuration mixing. The same applies to ^1D. In this way, LS coupling can remain a good description of the atomic states, even though the energies of those states depart from formulae such as (16.2). Of course: L^2 and S^2 commute with the Hamiltonian.

[37] The terms for carbon 2p 3d do not obey the "alternation" rule mentioned in the caption to Fig. 16.3. Configuration mixing is never far away.

[38] The notation $4f(^2\text{F})\, 5d^2(^3\text{F})$ indicates "parentage": the two 5d electrons are LS-coupled to form a ^3F *term*, as if the 4f electron was not there. The 4f electron joins on afterwards, interacting only weakly with the two d-electrons.

Terms in LS coupling

Atomics is a very intricate theorem ....
> Flann O'Brien, *The Third Policeman*

Intended readership: In an advanced course on atomic physics.

17.1   Introduction: building wave functions

The topic of this chapter counts as belonging to advanced atomic physics. We present it here because: it is not particularly difficult to understand; and it is available in few places elsewhere.[1]

To find the energy levels and wave functions of a light-to-middle-sized atom, we proceed (in reality or in imagination) according to the procedure given in § 16.2. First, we calculate wave functions for electrons in a *central field*. Then we apply the *residual electrostatic interaction*, which causes the wave functions to be formed into linear combinations that are eigenstates of $\widehat{L^2}$ and $\widehat{S^2}$. At the end of this, we have wave functions and energies that are said to conform to LS coupling.[2]

In an elementary treatment, we find the possible values of L and S by applying vector-model rules. But this leaves questions:

- How is a proper quantum-mechanical justification to be given for the creation of those values of L and S?
- When the electrons are "equivalent", some combinations of L and S are disallowed; for example a p^2 configuration has ^1P missing; how do we understand which combinations are allowed?
- Wave functions that are eigenstates of $\widehat{L^2}$, $\widehat{L_z}$, $\widehat{S^2}$ and $\widehat{S_z}$ are linear combinations of m_l, m_s wave functions; which wave functions participate in the required linear combinations?
- How could we proceed to find the coefficients in those linear combinations?

These are the questions that this chapter aims to answer.

17.2   Example: pp configuration

An example of a pp configuration (meaning $np\, n'p$ with $n' \neq n$), and its splitting into LS-coupled terms, is shown in Fig. 16.3; it makes a good example for us to discuss in detail. For definiteness, we shall say that we are dealing with that case: a $3p\, 4p$ configuration.

---

[1] Tables 17.1 and 17.2 are given by Condon and Shortley (1951). A p^3 configuration is discussed in a similar way by Budker, Kimball and DeMille (2008).

[2] It is assumed that interactions not included, in particular the spin–orbit interaction, are smaller than those that have been dealt with. This is usually, but not always, the case in light atoms.

Essays in Physics: Thirty-two thoughtful essays on topics in undergraduate-level physics. Geoffrey Brooker, Oxford University Press. © Geoffrey Brooker 2021. DOI: 10.1093/oso/9780198857242.003.0017

[3] Potential $U(r)$ contains as much as possible of the electron–electron Coulomb repulsion. The LS-coupling regime that we are investigating results from improving matters beyond the central-field approximation by adding in the residual electrostatic interaction. However, as is always sensible, we are anticipating the residual-electrostatic stage by finding out how to prefabricate the linear-combination wave functions that will result.

[4] The 3p and 4p electrons must be constituents of an atom containing many more electrons. If the atom is neutral silicon there are twelve others. Therefore we ought to be using a 14×14 Slater determinant. However, there are theorems in atomic physics that permit us to ignore the twelve electrons in the "core" (except for their rôle in determining the form of the central-field $U(r)$), and work with a determinant involving only the two electrons in part-filled subshells.

[5] Since the electrons are indistinguishable, and are labelled only so that the Slater determinant can be written—at all—we ought not to talk about "the 4p electron". A more correct statement would be something like "a full specification for the 4p building block is that it has $m_l = 0$ and $m_s = -\frac{1}{2}$". But correct words are clumsy.

We start by pretend-solving the Schrödinger equation for each electron in a central field whose potential-energy function $U(r)$ is spherically symmetrical.[3] The resulting wave functions are eigenfunctions of orbital angular momentum, that is of $\widehat{l}^2$, and (unless we make a perverse choice) of $\widehat{l}_z$ (lower-case letters because these apply to each electron). They are also eigenstates of $\widehat{s}^2$ (necessarily) and (again most simply) of $\widehat{s}_z$. In the case of a p electron there are six quantum states, all degenerate, identified by their values of m_l (three) and m_s (two). These wave functions are the building blocks from which everything else will be constructed.

We are going to deal with indistinguishable electrons "properly", so the single-particle wave functions described above are at once to be assembled into Slater determinants. Given six wave functions for each electron, there are 36 combinations; we need a succinct notation. Let us write:[4]

$$(1^+0^-) = \frac{1}{\sqrt{2!}} \begin{Vmatrix} 3\text{p}, \, m_{l1}=1, \, m_{s1}=+\frac{1}{2} \rangle & |3\text{p}, \, m_{l2}=1, \, m_{s2}=+\frac{1}{2} \rangle \\ 4\text{p}, \, m_{l1}=0, \, m_{s1}=-\frac{1}{2} \rangle & |4\text{p}, \, m_{l2}=0, \, m_{s2}=-\frac{1}{2} \rangle \end{Vmatrix}. \tag{17.1}$$

This is "definition by example", but the intention should be clear. We have two electrons, labelled 1 and 2, within the determinant. These electrons occupy two quantum states, one 3p with (in the example) $m_l = 1$ and $m_s = +\frac{1}{2}$; the other 4p with $m_l = 0$ and $m_s = -\frac{1}{2}$. In (1^+0^-), the first entry 1^+ gives the value (1) of m_l for the 3p state, and the superscript gives the sign (+) of m_s for that state. The second entry 0^- likewise gives $m_l = 0$ for the 4p electron[5] and the sign of its m_s.

We shall find out how the 36 Slater determinants are assembled into LS-coupled wave functions. Let us again explain by example. Suppose we wish to construct a wave function with $m_L = 1$, $m_S = 1$. Then we might have

$$|m_L=1, \, m_S=1\rangle = (\text{coefficient } a)\,(1^+0^+) + (\text{coefficient } b)\,(0^+1^+). \tag{17.2}$$

The constituents on the right have been chosen so that the m_l-values for the 3p and 4p electrons add up to the required total of 1, and the m_s-values likewise add up to 1. If a measurement were made of the total z-component of orbital angular momentum, then we should get 1 with probability $|a|^2$ from the first constituent and 1 with probability $|b|^2$ from the second, meaning 1 always. This is the only acceptable outcome compatible with $m_L = 1$, a state for which measurement of L_z must always yield 1. Similar reasoning applies to the z-component of spin. Furthermore, it is easy to see that none of the other 34 Slater determinants can be added to the right-hand side of (17.2), as none has the required $\{m_l(3\text{p}) + m_l(4\text{p})\} = 1$ and $\{m_s(3\text{p}) + m_s(4\text{p})\} = 1$.

Once the above is understood, it is clear what we should do next. Table 17.1 shows which Slater determinants can contribute to each m_L and m_S. To build a wave function having $m_L = 1$ and $m_S = 1$, we look at the row having $m_L = 1$ and the column having $m_S = 1$. Where these intersect we see the two Slater determinants of (17.2) listed: (1^+0^+) and

Table 17.1 The Slater determinants for a pp configuration (perhaps 3p 4p) that contribute to LS-coupled states having each combination of m_L and m_S. Tables 17.1 and 17.2 are taken from Condon and Shortley (1951).

pp		m_S		
		1	0	-1
m_L	2	$(1^+ 1^+)$	$(1^+ 1^-)\,(1^- 1^+)$	$(1^- 1^-)$
	1	$(1^+ 0^+)\,(0^+ 1^+)$	$(1^+ 0^-)\,(1^- 0^+)\,(0^+ 1^-)\,(0^- 1^+)$	$(1^- 0^-)\,(0^- 1^-)$
	0	$(1^+ -1^+)\,(0^+ 0^+)$ $(-1^+ 1^+)$	$(1^+ -1^-)\,(0^+ 0^-)\,(-1^+ 1^-)$ $(1^- -1^+)\,(0^- 0^+)\,(-1^- 1^+)$	$(1^- -1^-)\,(0^- 0^-)$ $(-1^- 1^-)$
	-1	$(0^+ -1^+)\,(-1^+ 0^+)$	$(0^+ -1^-)\,(0^- -1^+)\,(-1^+ 0^-)\,(-1^- 0^+)$	$(0^- -1^-)\,(-1^- 0^-)$
	-2	$(-1^+ -1^+)$	$(-1^+ -1^-)\,(-1^- -1^+)$	$(-1^- -1^-)$

$(0^+ 1^+)$. These are to be added, with coefficients not yet determined, to yield a required $|m_L, m_S\rangle$ wave function.

Table 17.1 can be used straightforwardly to show us which values of L and S can exist for the given 3p 4p configuration. All the information we need resides in the top left-hand quarter of the table: the rows having $m_L \geqslant 0$ and the columns having $m_S \geqslant 0$.

Start at the top left-hand corner: $m_L = 2$, $m_S = 1$. We have a single Slater determinant $(1^+ 1^+)$. Given $m_L = 2$, this must contribute to an L of at least 2. But a larger L, say 3, would have to have wave functions with $m_L > 2$ and there are none. Therefore $L = 2$. Similarly $S = 1$. We have one of the wave functions for ^{3}D. And we're fortunate, because the wave function is found completely; no linear combination needs to be made.[6]

This establishes that one of the LS *terms*[7] formed by 3p 4p is ^{3}D.

Now ^{3}D necessarily has five values of m_L and three values of m_S, in all 15 combinations, so 15 of the 36 linear combinations that we can build must be assigned to ^{3}D. In every box of the table, one linear combination[8] belongs to ^{3}D.

Look at the table entry for $m_L = 1$, $m_S = 1$. It contains two Slater determinants. Two functions can form exactly two independent linear combinations. One of these has been commandeered by ^{3}D. One remains, to be allocated to some other *term*. By a repetition of the reasoning used previously, this linear combination must build a wave function for ^{3}P: $L = 1$ since $m_L = 1$ and no larger m_L is available with this m_S; and $S = 1$ because $m_S = 1$ and no larger m_S is available. Moreover, ^{3}P must be capable of existing "all ways up" so it accounts for a further 9 linear-combination wave functions.

Proceeding further is now easy. For $m_L = 0$, $m_S = 1$ we have three Slater determinants in the table. Two linear combinations of these are accounted for, allocated to ^{3}D and ^{3}P. The one remaining must be a part of ^{3}S. At the top of the table, $m_L = 2$, $m_S = 0$, one linear

[6] Such an un-mixed wave function is sometimes said to constitute a "pure state".

[7] We are using the terminology introduced in Chapter 16 sidenote 7. A *term* belongs to a specified *configuration* (given ns and ls) and has additionally specified values of L and S.

We are also continuing the convention of Chapter 15 sidenote 2 by italicizing *term* when it means an L, S state, to distinguish it from "term" in the mathematical sense.

[8] Yes: linear combination. We cannot assign any single Slater determinant to ^{3}D except in the four extreme corners of the table. For example, the linear combination for $m_L = 1, m_S = 1$ is given below, in (17.5).

combination is unaccounted for; it must give ^{1}D. Below this, we have a box containing four Slater determinants; three combinations contribute to ^{3}D, ^{3}P and ^{1}D; the fourth must give ^{1}P. Finally, at the centre of the table there are six Slater determinants. Five linear combinations of these are required by ^{3}D, ^{3}P, ^{3}S, ^{1}D and ^{1}P. What remains, a single linear combination, must be what we need to make ^{1}S.

Summarizing: the LS combinations that can be formed by the 3p 4p configuration are ^{3}D, ^{3}P, ^{3}S, ^{1}D, ^{1}P and ^{1}S. A calculation in the margin[9] confirms that 36 Slater determinants have yielded 36 linear combinations, so nothing remains unaccounted for.

The singlet and triplet *terms* just listed are those that we could have obtained by elementary methods, using vector-model diagrams. There are no surprises. That is because we have chosen to discuss a rather straightforward case; other cases yield results that could not be foreseen if we had only elementary methods available. One such is the p^2 configuration dealt with in the next section, and another is d^3, dealt with in problem 17.1.

The reasoning we have given in this section amounts to no more than counting entries in the boxes of Table 17.1. To obtain fully the wave functions of the various *terms*, we would have to find the expansion coefficients. This can be an untidy business; nevertheless, we pursue it in problems 17.2 and 17.3. In particular, we obtain a wave function for the ^{1}P *term*; this is the *term* used as an important counterexample in Chapter 16 and eqn (16.5).

17.3   Example: a p^2 configuration

The case of "equivalent" electrons, electrons in the same subshell, is significantly different from the case dealt with in § 17.2. The number of wave functions is reduced from 36 to 15, and in consequence some of the LS combinations are disallowed. There is no way to deduce which are the allowed combinations merely by drawing vector diagrams, so the methods of this chapter are the only way forward.[10]

To work out what happens, we build a new table after the fashion of Table 17.1. This time, we must exclude those possibilities, like $(1^+ 1^+)$, in which the two electrons are put into the same quantum state. Also, there is now no difference between the first and second entries in a bracket.[11] Thus we write down only one of $(1^+ 0^+)$, $(0^+ 1^+)$, and similarly. The outcome is given in Table 17.2.

The LS assignments for a p^2 configuration can be reasoned from Table 17.2 in the same way as for the pp case from Table 17.1.

Starting at the top left-hand corner of Table 17.2, we have no wave function in the corner box, with $m_L = 2$, $m_S = 1$. This excludes the possibility of there being a ^{3}D *term*. Next, we work down the left-hand column to $m_L = 1$, $m_S = 1$. There is a wave function, $(1^+ 0^+)$, which must have $L = 1$, $S = 1$, giving a ^{3}P *term*. This accounts for 9 Slater determinants (or linear combinations of them) in the 9 central boxes of

[9] Consistency check:

^{3}D	15
^{3}P	9
^{3}S	3
^{1}D	5
^{1}P	3
^{1}S	1
	36

[10] We must end up with 15 wave functions. Even this requirement does not suffice to determine what happens. We can get a total of 15 with ^{3}D or with ^{1}D + ^{3}P + ^{1}S or with ^{3}P + ^{1}P + ^{3}S.

[11] Remember that in Table 17.1 the first entry of the $(a\,b)$ notation gave the quantum numbers for a 3p state, and the second entry for a 4p state, so $(1^+ 0^+)$ was different from $(0^+ 1^+)$. Now both entries relate to the same np (perhaps 3p).

Table 17.2 The Slater determinants for a p^2 configuration that contribute to LS-coupled states having each combination of m_L and m_S.

p^2		m_S		
		1	0	−1
m_L	2		$(1^+\,1^-)$	
	1	$(1^+\,0^+)$	$(1^+\,0^-)\;(1^-\,0^+)$	$(1^-\,0^-)$
	0	$(1^+\,-1^+)$	$(1^+\,-1^-)\;(1^-\,-1^+)\;(0^+\,0^-)$	$(1^-\,-1^-)$
	−1	$(0^+\,-1^+)$	$(0^+\,-1^-)\;(0^-\,-1^+)$	$(0^-\,-1^-)$
	−2		$(-1^+\,-1^-)$	

the table. At top centre, $m_L = 2$, $m_S = 0$, Slater determinant $(1^+\,1^-)$ must belong to $L = 2$, $S = 0$, a ^{1}D *term*. A ^{1}D *term* has five states, with $m_L = 2, 1, 0, -1, -2$, so this accounts for a further five wave functions. Only the box with $m_L = 0$, $m_S = 0$ has more entries in it than can be accounted for by the *terms* already identified, so we have just one further linear combination; this must be ^{1}S.

Summarizing: the *terms* allowed for a p^2 configuration are: ^{3}P, ^{1}D and ^{1}S.

Aside on Hund's rules and non-rules:

We are not concerned here with the energy order of the *terms* revealed by Tables 17.1 and 17.2. But since Hund's rules are frequently misapplied, I issue a reminder that the pp configuration of Table 17.1 cannot be a ground configuration—and has two part-filled subshells. Hund has nothing to say, as is confirmed by a glance at Fig. 16.3.

An atom *may* have a p^2 configuration as its ground configuration (example carbon). In such a case (and only then), Hund tells us that ^{3}P is the *term* with the lowest energy. However he has nothing to say as to the energy order of ^{1}D and ^{1}S; remember: no consolation prizes.

17.4   Further developments

The methods developed in the last two sections can be applied to other configurations. As an example, problem 17.1 asks you to show that a d^3 configuration has two different ^{2}D terms.

Tables 17.1 and 17.2 show us which Slater determinants are formed into linear combinations for building each possible m_L, m_S combination. Things are not as complicated as we might have feared. For example, the ^{1}S wave function of a p^2 configuration is a linear combination of no more than three Slater determinants. An obvious next question to ask is: how do we find the coefficients in these linear combinations? We show one way in which this can be done in problem 17.2.[12]

[12] You might think that we could insist that the linear combination be an eigenstate of $\widehat{L}^2$ and $\widehat{S}^2$. This works, but the manipulation is a bit messy. The route followed in problem 17.2 is actually simpler.

Problems

Problem 17.1 A surprise

Show that for a d^3 configuration[13] there are two *terms* ^2D.

Problem 17.2 Wave functions for $3p\,4p$, ^3P and ^1P

To obtain the Slater-determinant constituents of the ^3P and ^1P states given in (16.4) and (16.5), we first define the lowering operators for orbital and spin angular momentum. We have[14]

$$\widehat{l_x} = -i\hbar\left(y\frac{\partial}{\partial z} - z\frac{\partial}{\partial y}\right); \qquad \widehat{l_y} = -i\hbar\left(z\frac{\partial}{\partial x} - x\frac{\partial}{\partial z}\right).$$

The lowering operator is

$$\widehat{l_-} = \frac{\widehat{l_x} - i\widehat{l_y}}{\hbar} = e^{-i\phi}\left(-\frac{\partial}{\partial\theta} + i\cot\theta\frac{\partial}{\partial\phi}\right).$$

The effect of the lowering operator on an $|l, m_l\rangle$ wave function is:

$$\widehat{l_-}\,|l, m_l\rangle = \sqrt{l(l+1) - m_l(m_l - 1)}\,|l, m_l-1\rangle. \qquad (17.3)$$

The square root is obtained by standard angular-momentum algebra that we do not reproduce here.[15]

Spin likewise has a lowering operator. The eigenfunctions for $m_s = \frac{1}{2}$ and $m_s = -\frac{1}{2}$ are $\binom{1}{0}$ and $\binom{0}{1}$. The associated operators are:

$$\widehat{s_x} = \tfrac{1}{2}\hbar\begin{pmatrix} 0 & 1 \\ 1 & 0 \end{pmatrix}; \quad \widehat{s_y} = \tfrac{1}{2}\hbar\begin{pmatrix} 0 & -i \\ i & 0 \end{pmatrix}; \quad \widehat{s_-} = \frac{\widehat{s_x} - i\widehat{s_y}}{\hbar} = \begin{pmatrix} 0 & 0 \\ 1 & 0 \end{pmatrix}.$$

The effect of $\widehat{s_-}$ on a spin state $|s, m_s\rangle$ has the same form as (17.3).

We are now ready to begin the calculation.

(1) A representative Slater determinant may be written after the fashion of (17.1) as:

$$\left(a^+ b^-\right) = \frac{1}{\sqrt{2!}}\begin{vmatrix} |3p, m_{l1}=a\rangle \uparrow_1 & |3p, m_{l2}=a\rangle \uparrow_2 \\ |4p, m_{l1}=b\rangle \downarrow_1 & |4p, m_{l2}=b\rangle \downarrow_2 \end{vmatrix}.$$

Show that

$$\left(\widehat{l_{1-}} + \widehat{l_{2-}}\right)\left(a^+ b^-\right) = \sqrt{l(l+1) - a(a-1)}\left(a{-}1^+ b^-\right)$$
$$+ \sqrt{l(l+1) - b(b-1)}\left(a^+ b{-}1^-\right). \qquad (17.4)$$

(2) Show that a property similar to (17.4) applies to the corresponding lowering operators for spin.

(3) Look at a two-electron wave function ψ from a different point of view. We can define operators

$$\widehat{\boldsymbol{L}} = \widehat{l_1} + \widehat{l_2}, \qquad \widehat{\boldsymbol{S}} = \widehat{s_1} + \widehat{s_2},$$

and require that ψ is an eigenstate of $\widehat{\boldsymbol{L}}^2, \widehat{L_z}, \widehat{s}^2, \widehat{s_z}$. Then ψ is probably a linear combination of Slater determinants such as $(a^+ b^-)$. Angular-momentum algebra tells us that we can define lowering operators

$$\widehat{L_-} = \frac{\widehat{L_x} - i\widehat{L_y}}{\hbar}, \qquad \widehat{S_-} = \frac{\widehat{S_x} - i\widehat{S_y}}{\hbar},$$

and that these lower m_L, m_S after the fashion of (17.3).

(4) Now apply these tools to the wave functions listed in Table 17.1.[16] Wave function (1^+1^+) has the m_l, m_s values that we can read off from the contents of the bracket. But it is also known to be the LS-coupled wave function $|m_L{=}2, m_S{=}1\rangle$. We shall lower m_L by operating on $|m_L{=}2, m_S{=}1\rangle$ with $\widehat{L_-}$, and also use operator $(\widehat{l_{1-}} + \widehat{l_{2-}})$ on (1^+1^+). The results of these operations are:

$$\widehat{L_-}\,|^3\mathrm{D}, m_L{=}2, m_S{=}1\rangle = \sqrt{2(2+1) - 2(2-1)} \times$$
$$|^3\mathrm{D}, m_L{=}1, m_S{=}1\rangle;$$
$$(\widehat{l_{1-}} + \widehat{l_{2-}})(1^+1^+) = \sqrt{1(1+1) - 1(1-1)}\,(0^+1^+)$$
$$+ \sqrt{1(1+1) - 1(1-1)}\,(1^+0^+).$$

Tidying up, we find

$$|^3\mathrm{D}, m_L{=}1, m_S{=}1\rangle = \frac{1}{\sqrt{2}}\Big\{(1^+0^+) + (0^+1^+)\Big\}. \qquad (17.5)$$

Check the steps undertaken here.

(5) Look in Table 17.1 at the box containing the Slater determinants contributing to terms with $m_L = 1, m_S = 1$. There are two entries in the box: the two building blocks appearing in (17.5). Two building blocks can give two linear combinations, and we know already that the other combination belongs to $^3\mathrm{P}$. Argue that this combination must be orthogonal to the one already evaluated, so

$$|^3\mathrm{P}, m_L{=}1, m_S{=}1\rangle = \frac{1}{\sqrt{2}}\Big\{(1^+0^+) - (0^+1^+)\Big\}. \qquad (17.6)$$

(6) Return to the original wave function (1^+1^+) and this time reduce the z-component of spin. Show that

$$|^3\mathrm{D}, m_L{=}2, m_S{=}0\rangle = \frac{1}{\sqrt{2}}\Big\{(1^+1^-) + (1^-1^+)\Big\}. \qquad (17.7)$$

(7) Now continue after the same fashion, to obtain the following sequence of statements:[17]

$$|^1\mathrm{D}, m_L{=}2, m_S{=}0\rangle = \frac{1}{\sqrt{2}}\Big\{(1^+1^-) - (1^-1^+)\Big\}. \qquad (17.8)$$

$$|^3\mathrm{D}, m_L{=}1, m_S{=}0\rangle = \frac{1}{2}\Big\{(1^+0^-) + (1^-0^+) + (0^+1^-) + (0^-1^+)\Big\}. \qquad (17.9)$$

$$|^1\mathrm{D}, m_L{=}1, m_S{=}0\rangle = \frac{1}{2}\Big\{(1^+0^-) - (1^-0^+) + (0^+1^-) - (0^-1^+)\Big\}. \qquad (17.10)$$

$$|^3\mathrm{P}, m_L{=}1, m_S{=}0\rangle = \frac{1}{2}\Big\{(1^+0^-) + (1^-0^+) - (0^+1^-) - (0^-1^+)\Big\}. \qquad (17.11)$$

$$|^1\mathrm{P}, m_L{=}1, m_S{=}0\rangle = \frac{1}{2}\Big\{(1^+0^-) - (1^-0^+) - (0^+1^-) + (0^-1^+)\Big\}. \qquad (17.12)$$

[16] The method we use here is taken from Woodgate (1980), Chapter 7. Woodgate works on a p^2 configuration, so his calculation is less complicated than ours.

[17] *Suggestions*
Obtain (17.8) by requiring the $^1\mathrm{D}$ wave function to be orthogonal to the $^3\mathrm{D}$ wave function of (17.7).
Obtain (17.9) by lowering m_L on (17.7) or by lowering m_S on (17.5).
Obtain (17.10) by lowering m_L on (17.8).
Obtain (17.11) by lowering m_S on (17.6).
Finally, obtain (17.12) by requiring the wave function to be orthogonal to all three expressions in (17.9), (17.10) and (17.11).

Problem 17.3 The rest of configuration 3p 4p

Obtain the wave functions of (17.13–18). For ^{1}S and ^{3}S this first requires more work along the lines in problem 17.2. Then write out in full the Slater determinants, as in (17.1).[18]

[18] We have chosen to display the wave functions that have the largest m_L and m_S. One reason for this is to provide support for the statements in Chapter 16 sidenote 18.

The wave functions of (17.13–18) are listed in the energy order of Fig. 16.3.

$$\left|\,^1\text{S},\, m_L{=}0, m_S{=}0\right\rangle = (\uparrow_1 \downarrow_2 - \uparrow_2 \downarrow_1)$$
$$\times \left\{ R_{3p}(r_1)\, R_{4p}(r_2) + R_{3p}(r_2)\, R_{4p}(r_1) \right\}$$
$$\times \cos \Omega \qquad (17.13)$$

$$\left|\,^1\text{D},\, m_L{=}2, m_S{=}0\right\rangle = (\uparrow_1 \downarrow_2 - \uparrow_2 \downarrow_1)$$
$$\times \left\{ R_{3p}(r_1)\, R_{4p}(r_2) + R_{3p}(r_2)\, R_{4p}(r_1) \right\}$$
$$\times \sin \theta_1 \sin \theta_2\, e^{i\phi_1}\, e^{i\phi_2} \qquad (17.14)$$

$$\left|\,^3\text{S},\, m_L{=}0, m_S{=}1\right\rangle = \uparrow_1 \uparrow_2$$
$$\times \left\{ R_{3p}(r_1)\, R_{4p}(r_2) - R_{3p}(r_2)\, R_{4p}(r_1) \right\}$$
$$\times \cos \Omega \qquad (17.15)$$

$$\left|\,^3\text{P},\, m_L{=}1, m_S{=}1\right\rangle = \uparrow_1 \uparrow_2$$
$$\times \left\{ R_{3p}(r_1)\, R_{4p}(r_2) + R_{3p}(r_2)\, R_{4p}(r_1) \right\}$$
$$\times \left(\sin \theta_1\, e^{i\phi_1}\, \cos \theta_2 - \sin \theta_2\, e^{i\phi_2}\, \cos \theta_1 \right) \quad (17.16)$$

$$\left|\,^3\text{D},\, m_L{=}2, m_S{=}1\right\rangle = \uparrow_1 \uparrow_2$$
$$\times \left\{ R_{3p}(r_1)\, R_{4p}(r_2) - R_{3p}(r_2)\, R_{4p}(r_1) \right\}$$
$$\times \sin \theta_1 \sin \theta_2\, e^{i\phi_1}\, e^{i\phi_2} \qquad (17.17)$$

$$\left|\,^1\text{P},\, m_L{=}1, m_S{=}0\right\rangle = (\uparrow_1 \downarrow_2 - \uparrow_2 \downarrow_1)$$
$$\times \left\{ R_{3p}(r_1)\, R_{4p}(r_2) - R_{3p}(r_2)\, R_{4p}(r_1) \right\}$$
$$\times \left(\sin \theta_1\, e^{i\phi_1}\, \cos \theta_2 - \sin \theta_2\, e^{i\phi_2}\, \cos \theta_1 \right), \quad (17.18)$$

where $\cos \Omega = \cos \theta_1 \cos \theta_2 + \sin \theta_1 \sin \theta_2 \cos(\phi_1 - \phi_2)$; Ω is the angle between the directions of vectors $\boldsymbol{r}_1$ and $\boldsymbol{r}_2$. Normalization constants have been omitted.

Comment: None of the wave functions here has the shape of (15.2) or (15.3). Dismiss any idea that those shapes are "the way atoms have to be". Dismiss any idea that "space-symmetric means close together".

Comment: Expressions $\left\{ R_{3p}(r_1) R_{4p}(r_2) \pm R_{3p}(r_2) R_{4p}(r_1) \right\}$ contain the scalar radii r_1, r_2, not the vector locations $\boldsymbol{r}_1, \boldsymbol{r}_2$. Thus the probability of finding the two electrons at the same distance $r_1 = r_2$ from the origin is zero or doubled, by comparison with $R_{3p}(r_1) R_{4p}(r_2)$, regardless of the directions $(\theta_1, \phi_1; \theta_2, \phi_2)$ in which they lie.

Make your own interpretation along these lines of the functions of angle: $\left(\sin \theta_1\, e^{i\phi_1} \cos \theta_2 - \sin \theta_2\, e^{i\phi_2} \sin \theta_1 \right)$ prohibits electron 2 from lying along the same radial line as electron 1.[19]

In particular, wave functions (17.13–18) exhibit "closeness" or "apartness" that is not concerned with behaviour at points $r_1 = r_2$ (as seemed important in (15.2) and (15.3) but was discredited in § 15.3.1).

[19] The ^{1}P wave function is remarkable for having both kinds of "strong" separation. If electron 1 is found at distance r_1 from the nucleus, then electron 2 cannot lie anywhere on a sphere having that same radius. And if electron 1 is found in direction θ_1, ϕ_1 out from the nucleus, then electron 2 cannot lie anywhere on that same radial line. It is easy to see why ^{1}P has the lowest energy. We are repeating the discussion of § 16.3.1.

Problem 17.4 The energy order for the 3p 4p *terms*

We cannot easily calculate the energies of the six terms catalogued in problem 17.3, because we do not know the radial functions $R_{3p}(r)$ and $R_{4p}(r)$. The best we can do is work out the integrals over angles and leave integrals over radii as unevaluated parameters; this is the procedure that results in eqns (16.2).

Nevertheless, we can get a remarkably good insight into the physics by looking at the shapes of the wave functions, and interpreting those shapes as was done in § 16.3.1 and in the last comment. The energy differences come from the Coulomb repulsion $e^2/4\pi\epsilon_0 r_{12}$ of the electrons, so the energy should be greatest for wave functions where the electrons are closest together, and least when the electrons are most widely separated. Let us "score" the wave functions for the "togetherness" that they give to the electrons.

A simple product wave function, like $(\sin\theta_1 \, e^{i\phi_1})(\sin\theta_2 \, e^{i\phi_2})$, scores zero, because the electrons' probability distributions are uncorrelated in their angular dependence.

A "sum" wave function, like $\{R_{3p}(r_1) R_{4p}(r_2)+R_{3p}(r_2) R_{4p}(r_1)\}$, (when properly normalized) gives a doubled probability when $r_1 = r_2$ (relative to a simple-product function) so give it a "togetherness" score of 2.

A "difference" wave function, with a negative sign in its { } or ( ) bracket, is the opposite of a "sum" wave function, so give it a score of -2.

I suggest we give a score of 3 when a wave function contains $\cos\Omega$; think about why this could be sensible.

With the suggested scoring system, the scores for the six wave functions catalogued in problem 17.3 are (in order): $5, 2, 1, 0, -2, -4$. These scores give the correct energy order for the six *terms*![20]

[20] You can try other scoring systems. They may not get all the energies in the right order, but (if they're reasonable) they must get the extremes right, and just vary a bit in the middle. A "togetherness score" is a perfectly sensible way of extracting a *rough* idea of what each wave function is telling us.

You can't sensibly adjust the scoring system to reproduce the proportions of Fig. 16.3; you'd be fitting data, not making predictions. If you want to calculate energies, you have to do just that, working out integrals over wave functions. Back to eqns (16.2).

The observant reader will notice that my "togetherness scores" cannot be reconciled with the F_0, F_2, G_0, G_2 of (16.2). It would be amazing if such a crude scoring—close to being irresponsible—could withstand a quantitative scrutiny.

Problem 17.5 A p^3 configuration

Construct a table resembling that of Table 17.2 for the case of a p^3 configuration. Show that there are 20 "wave functions" (meaning space and spin) in all. Show that a quadrant of the table containing seven Slater determinants is sufficient for identifying all the *terms* that can exist. Show that the permitted *terms* are ^4S, ^2D and ^2P. Check that these *terms* have the required total of 20 states.

Show that the "wave function" $\big|\,^4S, m_L=0, m_S=\tfrac{3}{2}\big\rangle$ is the product of a spin function and a space function. Show that the spin part is symmetric under every interchange of electron labels, while the space part is antisymmetric under every such interchange.

Obtain the "wave function" for $\big|\,^2D, m_L=2, m_S=\tfrac{1}{2}\big\rangle$. Show that the spins within it cannot be assembled to make a spin function for the three electrons that is antisymmetric under every interchange of electron labels. Show that the same is true of the space part. Thus a factorization into (spin)×(space) is not general once we have more than two electrons.

Show that $\big|\,^2D, m_L=2, m_S=\tfrac{1}{2}\big\rangle$ is an eigenstate of $\widehat{S^2}$ with eigenvalue $\tfrac{1}{2}(\tfrac{1}{2} + 1)\hbar^2$, even though there is no factorization of the wave function into space and spin factors.

18 | The Zeeman effect

[1] I am indebted to the late Dr G. Smith for much of the information upon which this chapter is based.

They are lively as twenty leprechauns doing a jig on top of a tombstone.

Flann O'Brien, *The Third Policeman*

Intended readership: after a first course on the Zeeman effect, when the student is ready to see the Zeeman effect applied diagnostically to the assignment of atomic quantum numbers.[1]

18.1 Introduction: use for diagnosis

The Zeeman effect is the splitting of spectral *lines* into *components* separated in frequency, when an emitting or absorbing atom makes its radiative transition in an applied magnetic field. The effect is understood as arising from an energy splitting of the quantum *states* comprising each *level*, with a state of magnetic quantum number m_J shifted[2] in energy by $g_J \mu_B B m_J$; each level has its own value of g_J.

[2] Here μ_B is the Bohr magneton, B is the magnetic field, and g_J is the Landé g-factor.

In this chapter, we are concerned wholly with the "weak-field" Zeeman effect.

This chapter does not give an account of the basics of the Zeeman effect, which are well described in books on atomic physics. Rather, we discuss two particular sub-topics:

- the assignment of atomic quantum numbers, as revealed by measurements of a Zeeman-effect spectrum;
- the relative intensities of the Zeeman components, and the sum rules that restrict the intensity ratios.

These sub-topics are related, because the intensity ratios are helpful in the assignment of quantum numbers.

It is commonly stated in textbooks that the Zeeman effect has no place in the diagnosis of atomic structure: the assignment of quantum numbers to the energy levels found from spectroscopic measurement. Rather (it is said), the Zeeman effect can do no more than confirm assignments already made on other evidence. I disagree. It is very often possible to make a primary identification of quantum numbers on the basis of a Zeeman measurement.[3] This possibility is illustrated below in §18.5, by handling a particular example—and that example is not specially selected for any rare helpfulness.

[3] We are concerned here with the 'anomalous' Zeeman effect. If we see the 'normal' effect, we learn that both levels involved in the transition have $S = 0$, but nothing is learned about J or L. Fortunately, the 'anomalous' effect is more informative and far more commonly encountered.

18.2 Experimental layout

This section sets up definitions of directions and axes that will be used consistently throughout the present chapter.

Essays in Physics: Thirty-two thoughtful essays on topics in undergraduate-level physics. Geoffrey Brooker, Oxford University Press. © Geoffrey Brooker 2021. DOI: 10.1093/oso/9780198857242.003.0018

An apparatus for observation of the Zeeman effect is outlined in Fig. 18.1. The $\boldsymbol{B}$-field is applied in the z-direction to atoms that are excited in a gas-discharge tube. Emitted radiation may be collected in the x-direction, an arrangement that is most convenient when the magnet has iron pole tips as here. It is also possible to collect radiation in the z-direction; this can be done by using a pierced pole tip, as hinted at here (with a rather limited solid angle for light gathering), or along the axis of a powerful solenoid. What is observed in the two cases is different, but either way it carries sufficient information for our purposes.[4]

Spectroscopic equipment is omitted from Fig. 18.1. In a student laboratory, it is likely to include a Fabry–Perot étalon, together with the means for selecting a single spectral line for analysis: either an interference filter or a low-resolution spectrometer.

Figure 18.1 shows the apparatus in a somewhat uncommon orientation. Usually the magnet's yoke rests in its most natural position (on the base of its ☐ shape), and the discharge tube, being vertical, is easily imaged optically onto the entrance slit of a spectrometer. I have chosen

[4] For simplicity, the description in this chapter is always of radiation emitted by excited atoms—an emission spectrum. It is, of course, equally possible and practical to work with an absorption spectrum.

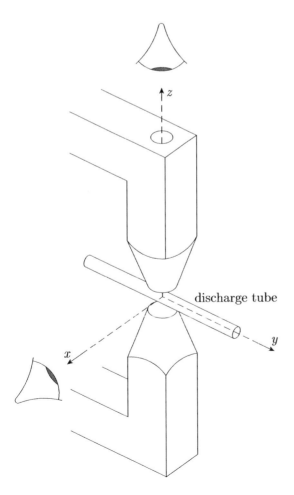

discharge tube

Fig. 18.1 A possible layout for a measurement of the Zeeman effect. Radiation may be collected for observation and measurement in the x-direction (transverse to the $\boldsymbol{B}$-field) or in the z-direction (along the $\boldsymbol{B}$-field).

A more usual configuration of the apparatus would have the $\boldsymbol{B}$-field horizontal and the discharge tube vertical, but we have set up axes here in an orientation that is uniform with that in Fig. 18.2.

the unusual layout so that the xyz axes are in the same orientation here as in Fig. 18.2.

18.3   Selection rules and polarizations

The selection rules are derived in textbooks on atomic physics, and are quoted only here:

$$\Delta m_J = \pm 1 \qquad \sigma \text{ transitions,}$$
$$\text{or} \quad \Delta m_J = 0 \qquad \pi \text{ transitions}$$
$$\text{but } m_J = 0 \text{ does not connect with } m_J = 0 \text{ when } \Delta J = 0.$$

For diagnostic purposes, it is important that the radiation has characteristic angular distributions and polarizations that are different for the σ and π transitions. Figure 18.2 shows two of the three possible motions of the atom's electric dipole moment during a transition.

On the left, the atom has an electric dipole moment that rotates in the xy plane. The emitting atom undergoes a transition in which its z-component of angular momentum changes by $-\hbar$: its m_J quantum number changes by -1. (This may be seen from the sense of the rotation, which is giving $+\hbar$ of z-component angular momentum to the outgoing radiation.) The electromagnetic field radiated carries information about the behaviour of the dipole moment that originated it. In particular, radiation travelling towards $\pm z$ is circularly polarized, with the rotation in the same sense as that of the originating dipole. For a transition having $\Delta m_J = +1$ the sense of rotation is reversed. The $\Delta m_J = \pm 1$ transitions are often called σ transitions.[5] Radiation travelling in the x-direction is linearly polarized with its $\boldsymbol{E}$ in the y-direction, because the atomic motion is "edge-on" when seen from there.

On the right of Fig. 18.2, the atom undergoes a transition in which $\Delta m_J = 0$; such transitions are called π transitions. Here the atom's electric dipole moment oscillates along the z-axis. Radiation travelling

[5] In the trade, the two signs of Δm_J are distinguished by calling the transitions σ^+ and σ^-. A σ^+ transition is one in which the m_J of the upper state is greater (by 1) than that of the lower state (this wording applies to both emission and absorption); see Corney (1977) p. 123. Notice that the signs of the m_J are determined by the direction chosen for the z-axis, and are reversed if that axis is reversed. Figure 18.2 (left) shows the atomic dipole moment during a σ^+ emission.

The σ^+ and σ^- transitions can be separated experimentally, by means of circular polarizers, if observation is made of radiation travelling in the z-direction. That separation is not possible for radiation collected in the x-direction because the circular motion is seen "edge-on". However, in simple cases, such as those in Figs. 18.3 and 18.4, the σ^+ and σ^- transitions form two well separated families, and it is then easy to allocate the transitions to each category.

Fig. 18.2 The heavy arrows centred on the origin show the motion of the atom's electric dipole moment. On the left, circular motion in the xy plane happens during a σ^+ transition ($\Delta m_J = -1$ for emission). Observed along the z-axis, the radiation is circularly polarized; observed along the x-axis, the radiation emitted is linearly polarized in the y-direction. For a σ^- transition the sense of the circular motion is reversed. On the right, electron motion along the z-axis happens during a transition having $\Delta m_J = 0$: a π transition. Observed along the x-axis, the radiation is polarized in the z-direction; nothing is radiated into the z-direction.

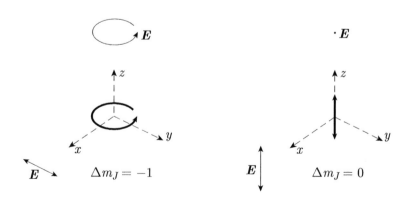

in the x-direction is linearly polarized with its $\boldsymbol{E}$ in the z-direction. Nothing is observed to travel in the $\pm z$ direction, because an electric dipole "does not radiate off its end".

Very simple diagnostic rules emerge from Fig. 18.2. For radiation travelling in the x-direction, the (linear) polarization tells us at once whether we are looking at a σ or π transition: all we need to do is rotate a Polaroid, and we can pick out transitions[6] having $\Delta m_J = \pm 1$ or $\Delta m_J = 0$. Different information is obtained if we collect radiation travelling along the z-direction: we do not see the π transitions, but we can use optical equipment to determine the handedness of the circular polarization, so we can identify which spectral components come from $\Delta m_J = +1$ and which from $\Delta m_J = -1$.

18.4   Intensity patterns

The relative intensities of Zeeman components can be worked out from atomic matrix elements (see problem 18.1). For the present section, we do not need such detailed information. Instead, we need three statements:

- When the level having the larger J has the larger g_J, the σ components are "shaded in".

- When the level having the larger J has the smaller g_J, the σ components are "shaded out".

- When the two levels have equal Js ($\Delta J = 0$), the σ components with $\Delta m_J = \pm 1$ form two separately symmetrical groups.

These three cases are illustrated by the examples shown in Fig. 18.3.[7]

In Fig. 18.3, the left-hand panel shows a case where the level having the larger J has the larger g_J. The σ components have intensities that diminish from the outside of the pattern towards the middle; this is what we mean by "shaded in". The middle panel shows transitions involving the same values of J and m_J as before, but the transitions appear in a different frequency order. Now the level having the larger J has the smaller g_J, and the σ components are weakest towards the outside of the pattern; they are "shaded out". Finally, the right-hand panel shows a case where both levels have the same value of J. The σ components are now not shaded either in or out, but have a symmetry within the $\Delta m_J = +1$ group and within the $\Delta m_J = -1$ group. There is also a different shape to the intensities of the π components, and this too can contribute to the diagnosis of atomic quantum numbers.[8]

For use in the next section, we note a property that can be readily seen from the displays in Fig. 18.3: the spacing of adjacent π components (and likewise that of adjacent σ components within each of the two groups) is $|g_J - g_{J'}|\mu_B B$ in energy units, where g_J and $g_{J'}$ are the g-factors of the two levels.

[6] We should pause to savour the moment. An operation as simple as rotating a polarizer gives access to an atomic property as intimate as a selection rule.

[7] In Fig. 18.3, all six levels have g_J positive. The magnetic field faces along the $+z$-direction, so the atomic states have m_J greatest towards the top of each diagram where the energy is greatest (not so labelled to avoid clutter). All this means that the σ^+ transitions lie to the right in each spectrum.

In Fig. 18.3, the energy-level splittings are all drawn to the same vertical scale. Likewise, the frequency splittings are all drawn to a common horizontal scale. However, in energy units, the horizontal splittings are magnified, relative to the vertical, by a factor 2.

[8] The transitions displayed in Fig. 18.3 have been chosen to have a friendly property: each group of σ components lies wholly "outside" the π components, rather than interlacing them (or, indeed, interlacing the other σs). These choices have been made in order to have un-messy diagrams. Real Zeeman spectra are not always so tidy; see problem 18.6.

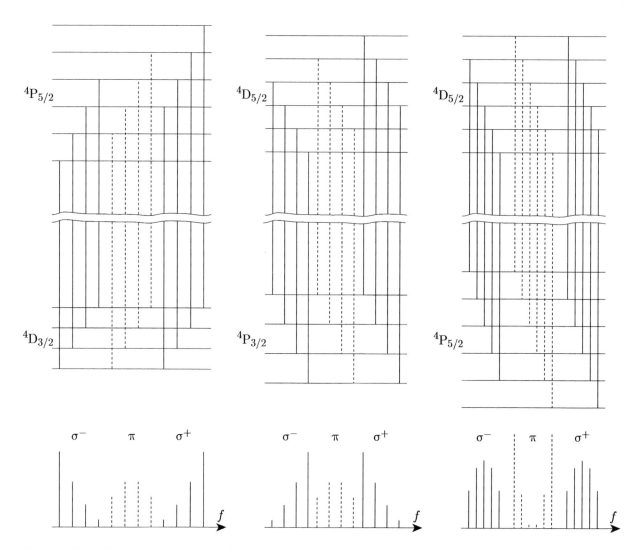

Fig. 18.3 Three Zeeman patterns showing how the σ components may be "shaded in" (left), "shaded out" (centre), or form two symmetrical groups (right). The σ components have their transitions drawn with unbroken lines, while the π components are drawn with broken lines. The spectra at the bottom (f is frequency) show the components with the relative intensity indicated by the length of the vertical line.

The intensities drawn indicate the power radiated by the atoms into all directions; they are proportional to the Einstein A coefficients for the respective transitions. Because of the shape of the angular distributions, the relative intensities observed in a direction perpendicular to B (in the x-direction in Fig. 18.1) are different: the π components there have doubled intensity relative to the σ components (see § 18.6).

18.5   An example of Zeeman-effect analysis

A simulation of an experimental recording is shown in Fig. 18.4. The graph represents the output from a pressure-scanned Fabry–Perot étalon. In such an apparatus, the pressure of air between the étalon plates is raised linearly with time, so that the optical path between the plates rises linearly with time. As a result, each frequency is presented in turn

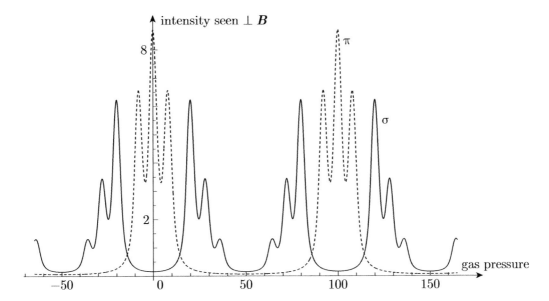

Fig. 18.4 A simulated recording of a Zeeman pattern, as displayed with a pressure-scanned Fabry–Perot. Observation is in the x-direction of Fig. 18.1, light being collected at right angles to the $\boldsymbol{B}$-field. The horizontal scale is marked so that there are 100 units between the repetitions of the same pattern in adjacent interference orders. In the absence of a $\boldsymbol{B}$-field, the pattern has peaks at locations 0, 100 (and multiples) only. The $\boldsymbol{B}$-field spreads each of these into a display of the Zeeman components, one such display lying between locations -50 and 50.

The intensity scale is marked in the units of eqn (18.7). The actual peak heights do not correspond exactly to those units because each peak sits on the wings of its neighbours.

The $\boldsymbol{B}$-field is $1\,\mathrm{T}$, and the étalon plates are spaced by $2.57\,\mathrm{mm}$.

σ and π transitions are displayed separately. The σ transitions are recorded via a polarizer set to transmit E_y; the π transitions via a polarizer transmitting E_z.

This recording shows that the σ transitions are "shaded out", but is otherwise not ideal for quantitative measurement of the Zeeman splittings. For that purpose, the settings chosen for Fig. 18.5 are preferable.

to a detector, but the pattern repeats when the same frequencies arrive in the next interference order—and the next .... A frequency scale can be attached to the peaks *within one single order* by measuring the spacing of those peaks in relation to the inter-order peak spacing.[9]

We now show how diagnostic information can be extracted from the recording of Fig. 18.4. It can be assumed that the atom concerned has, when neutral, two electrons outside closed shells, and that the discharge excitation is likely to be giving us radiative decays in neutral atoms, rather than in ions (which would require more violent excitation). *LS* coupling can be assumed.

With two electrons, the possible values of spin quantum number S are 0 and 1. If S were 0 we would have $g_J = 1$ for both levels, resulting in a "normal" Zeeman effect. What we have is clearly not a "normal" effect, so S must take the only remaining possibility: $S = 1$. Moreover, there is a selection rule $\Delta S = 0$, so $S = 1$ must apply to both upper and lower levels involved in the transition.

Figure 18.4 shows three π components, including a strong component at the centre of the pattern. This component, by symmetry, connects

[9]The étalon fringes obey the relation $2nd\cos\theta = p\lambda$, where n is the refractive index of the gas between the plates (a linear function of gas pressure), d is the spacing between the plates, θ is the angle made by light with the normal to the plates, λ is the wavelength of the light, and p is the order of interference (a large integer, here nearly 10^4). When $d\cos\theta$ and p are held constant, λ is a linear function of gas pressure, and therefore frequency increases to the left in Figs. 18.4 and 18.5.

We emphasize that a frequency scale applies *within a single order only*: the two tallest peaks in Fig. 18.4 belong, of course, to the *same* frequency; they have the same λ but different values of p.

states both of which have $m_J = 0$. Now transitions from $m_J = 0$ to $m_J = 0$ do not happen when $\Delta J = 0$, so here $\Delta J \neq 0$. The selection rule $\Delta J = 0, \pm 1$ means that the two levels have J-values that differ by 1. We shall arbitrarily take the upper level as having the larger J, for convenience of discussion.[10]

Figure 18.4 shows that there are three π components.[11] This means that both levels have at least three m_J-values, implying $J \geqslant 1$. At the same time, there are only three π components, so one of the two levels must be limiting that number to three: the smaller J is 1. That is: the J-values are $J = 2$ and $J = 1$.

It is consistent that the recording seems to display three σ components on each side of the pattern. However, we could be just uncertain as to whether there is an additional weak σ component out in the wings of each pattern.[12] This is why the argument identifying the Js has concentrated on the π components. However, what is very clear is that the σ components are "shaded out", so the level having the larger J (that is, $J = 2$) has the smaller g_J. Call the g-values of the two levels g_2 and g_1 (for $J = 2$ and $J = 1$ respectively). Then $g_2 - g_1 < 0$.

Knowing the S and J quantum numbers for both levels, we have a rather limited set of possibilities for the L quantum numbers, possibilities that are listed in Table 18.1. Our remaining task is to identify (if possible) a single one of these possibilities that is consistent with experiment. To make that identification, we need to use the data in a more quantitative way.

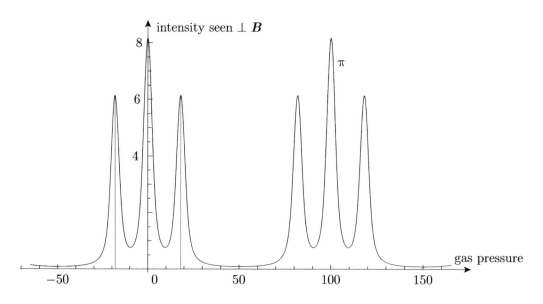

Fig. 18.5 This (simulated) recording displays the π components only. The B-field is the same as for Fig. 18.4, but a new étalon has been used, with a larger plate separation, so that the π-peaks appear more spread out.

The B-field is 1 T, and the étalon plate spacing is 5.84 mm. Together with the recording, this information suffices for the determination of $(g_2 - g_1)$, and that in turn identifies which are the correct choices from the quantum states listed in Table 18.1.

It is hard to obtain information of sufficient accuracy from a recording such as that of Fig. 18.4, because adjacent peaks lie rather close together. One cure for this is to display the π components only, using a wider-spaced étalon to spread the π-peaks more widely across the interference order.[13] Such a recording (simulated) is shown in Fig. 18.5.

Fabry–Perot theory tells us that the frequency separation between two peaks (such as that between two adjacent π components in Fig. 18.5) is given by

$$\frac{\text{separation of those peaks}}{\text{separation of pattern repetitions}} \times \text{free spectral range of étalon.}$$

Here the peak separation is about 18 display units, and the pattern repetitions happen 100 display units apart. The étalon has plate spacing $d = 5.84\,\text{mm}$, which implies a free spectral range $c/(2d) = 25.7\,\text{GHz}$. Then

$$\pi \text{ separation} = |g_2 - g_1|\,\mu_\text{B} B/h = \frac{18}{100}\,25.7\,\text{GHz.}$$

For a B-field of $1\,\text{T}$, $\mu_\text{B} B/h = 14.0\,\text{GHz}$, and so

$$|g_2 - g_1| = \frac{18}{100}\,\frac{25.7}{14.0} = 0.33.$$

The shaded-out σ pattern has told us the sign. Thus we find

$$g_2 - g_1 = -0.33. \tag{18.1}$$

The g-difference of (18.1) is to be compared with those listed in Table 18.1. We hope that it will agree with one of those differences, while excluding all others. For this, it does not suffice to find a closest match: we must *definitely exclude* all others, as lying outside the possible error of measurement.

This throws emphasis onto experimental error. What we need here is a *greatest possible error*, rather than the more usual statistical estimate involving standard deviations. Given the recording of Fig. 18.5, we might reasonably put the error at 5% or less. However, we must remember that Fig. 18.5 is a simulation (albeit as realistic as it can be made—see § 18.5.1 below), and a real recording will be afflicted by detector noise and environmental disturbances. We must also think of uncertainty and non-uniformity in the B-field .... To be realistic, we should be very cautious with the "greatest possible error"; I'll be pessimistic and raise it to 20%. Then our experimental finding is

$$g_2 - g_1 = -(0.33 \pm 20\%), \quad \text{so between } -0.396 \text{ and } -0.264. \tag{18.2}$$

[13] In energy units, the splitting of the π components in Fig. 18.5 is the same as in Fig. 18.4, because the **B**-field has the unchanged value of 1 tesla. It is the free spectral range that has been reduced, so as to be only a little greater than the frequency range to be recorded.

Table 18.1 The values of $(g_2 - g_1)$ for transitions between the atomic levels given at the top ($J = 2$) and side ($J = 1$) of the table.

	$^3\text{P}_2,\ g_J = 3/2$	$^3\text{D}_2,\ g_J = 7/6$	$^3\text{F}_2,\ g_J = 2/3$
$^3\text{S}_1,\ g_J = 2$	$-1/2$		
$^3\text{P}_1,\ g_J = 3/2$	0	$-1/3$	
$^3\text{D}_1,\ g_J = 1/2$	1	$2/3$	$1/6$

Table 18.1 lists the three possible upper levels and the three possible lower levels. For each transition between these permitted by the selection rule $\Delta L = 0, \pm 1$, the table gives the value of $(g_2 - g_1)$. Only two candidates have the correct sign for $g_2 - g_1$, so the possible transitions are $^3P_2 \to {}^3S_1$ with $g_2 - g_1 = -1/2$ and $^3D_2 \to {}^3P_1$ with $g_2 - g_1 = -1/3$. The greatest possible experimental error eliminates the first of these, so we assign the transition to

$$\text{the transition is} \quad {}^3D_2 \to {}^3P_1. \tag{18.3}$$

Finally, we ask whether the experimental data permit us to decide which of the two levels lies higher in energy. You are asked to think about this in problem 18.2. (The answer is no: you need additional spectroscopic information.) Even so, we have come a long way by now restricting the possibilities, so it is likely that the jig-saw of energy levels will fit together in only one way.

In the analysis leading to (18.3), the reasoning has been pursued "opportunistically", applying the known rules as seems best for the Zeeman pattern given. A different pattern might well be dealt with by following a different route. It is possible to systematize the rules for analysis, but the decision tree becomes rather complicated. The opportunistic route is to be preferred. One alternative procedure is explored in problem 18.5.

18.5.1   Aside on the preparation of Figs. 18.4 and 18.5

The Oxford Physics Practical Course has an experiment in which the student measures an anomalous Zeeman effect in the spectrum of mercury, and can identify the quantum numbers by using reasoning similar to that here. The example analysed in the present chapter has been chosen so that it does not give the game away for that experiment, or for similar experiments likely to be offered in other physics departments.

I have modelled the recordings of Figs. 18.4 and 18.5 on a spectral line at 552 nm in the spectrum of barium. The element chosen had to lie some way along the periodic table in order to have a decent separation of the ^3D term into its J-levels. Even now, the ^3D—^3P manifold has another spectral line only 56 cm^{-1} away from the line under investigation, so the experimental equipment would need to incorporate a fairly good spectrometer to isolate the required spectral line from others nearby.

Within the above, I have tried to make the simulated recordings as realistic as possible, and to assume equipment no better than what might be found in a teaching laboratory. The spectral lines are broadened by three mechanisms:

- The Doppler broadening of the line. I assumed that the atoms in the gas discharge have a kinetic-energy temperature of 400 K, and have the mass number, 137, of barium.
- The limited finesse of the étalon. I have assumed a finesse of 30 which is affordable on a teaching-laboratory budget.

- The finite diameter of the pinhole, at the focus of a fringe-forming lens, transmitting energy to a photoelectric detector. This pinhole must not be too large, or it degrades resolution; and it must not be too small, or too little energy reaches the detector. The line profile contributed by the pinhole alone is a top-hat (see problem 18.4). I have assumed a pinhole whose top-hat, on its own, would make a single frequency appear to have a width 1/30 of the free spectral range, i.e. giving a "pinhole finesse" of 30.

The overall line profile is the convolution of the profiles contributed by these three broadening mechanisms.[14] Mathematica was instructed to perform that convolution, repeating for each Zeeman component in turn, and then repeating the whole thing for different interference orders.

[14] The convolution of the étalon pattern with the pinhole's top-hat can be done analytically, see problem 18.4. Then only one remaining convolution needs to be done numerically by the computer program.

18.5.2 Comparison of observations made across and along the B-field

Figure 18.5 is obtained by using a linear polarizer to select the π components alone, and using an étalon whose plate spacing is such that it spreads that pattern well across the free spectral range.

We could equally well record the Zeeman pattern radiated into the z-direction (along B), using optical equipment to select only one of the two σ polarizations.[15] The three σ components (and there are three in each of σ^+ and σ^- here) can be spread out by use of the thicker étalon, and the $(g_2 - g_1)$ difference can be measured, just as it was above for the π components. Therefore the opportunity for extracting information is just as good for this case. Indeed, by concentrating on the σ components, we can obtain the shaded in/out/not information, and the g-difference information, from a single recording.

[15] It's a good idea to select only one circular polarization at a time. Some Zeeman patterns have the two sets of σ components interlaced, and so difficult to disentangle from each other; see problem 18.6. We can, of course, record both circular polarizations, in separate experiments, but this is unlikely to be necessary.

With the ambitious facilities of a dedicated research laboratory, the Zeeman effect can be an extremely powerful diagnostic tool. A large spectrograph, with a powerful superconducting magnet, can photograph an entire spectrum, with the Zeeman pattern of every line recorded, on a single photograph (Smith and Tomkins 1976). There is now no need to make a laborious Fabry–Perot examination of each spectral line in turn, with fresh equipment needed to deal with each new frequency.

In the experiments of Smith and Tomkins, a superconducting solenoid supplied the B-field of 5 T, and an absorption spectrum was measured with radiation travelling along the axis of the solenoid. Absorption transitions from the $4f^7 6s^2\, {}^8S_{7/2}$ ground level of europium were photographed in the wavelength range 210–720 nm. Sufficient optical resolution for recording of the Zeeman splittings was provided by a 9-metre concave-grating spectrograph.[16] To appreciate the advantage of this, we should know that the rare earth elements have such complicated energy-level structures that they defy less informative methods of investigation.[17]

[16] Nine metres is the diameter of the Rowland circle.

[17] Such a complexity is manifest in the energy-level diagram of praseodymium in Fig. 16.6. The figure caption mentions a severe culling of the levels that I had to impose in order to de-clutter the diagram for display.

18.6   Use of sum rules for determining relative intensities

The relative intensities of the Zeeman components can be worked out straightforwardly by using the matrix elements tabulated in problem 18.1. But in simple cases there is a more elementary method, exploiting sum rules, that gives the same results. Even when the sum rules do not suffice to determine the relative intensities completely, those rules provide a valuable consistency check.

Before we demonstrate the use of the sum rules, we need some preliminary discussion. The Zeeman splitting of a spectral line is a very small fraction of that line's unperturbed frequency. For the transition analysed in the last section, we have a total splitting (between the outermost σ components) of $3\mu_{\rm B}B$, which translates into frequency as $42\,{\rm GHz}$ for a field of $1\,{\rm T}$. The spectral line has wavelength $\lambda = 552\,{\rm nm}$, so its unperturbed frequency is $5.4 \times 10^{14}\,{\rm Hz}$, implying that the splitting is less than a fraction 10^{-4} of the original frequency.

It now follows that the atomic wave functions are negligibly altered by application of the $\boldsymbol{B}$-field. Therefore the transition matrix elements take the same values as they had in zero field. Also, the transition rates[18] are proportional to $\omega^3 \times |{\rm matrix\ element}|^2$, and we can treat ω^3 as a constant for all Zeeman components within a single spectral line.

[18] See eqn (19.16).

The transition probabilities must be unaltered by a reversal of the z-axis (leaving the $\boldsymbol{B}$-field unchanged), which leaves the Zeeman pattern unaltered but reverses the signs of all the m_Js. The matrix elements that used to apply to the right-hand side of the pattern now describe the left-hand side. This makes the Zeeman pattern symmetrical about its centre.

In zero $\boldsymbol{B}$-field, there is no preferred direction in space, and so an isotropic population of atoms among the upper m_J-states decays in such a way as not to create a preferred direction for the population after the atoms have all decayed to the lower level. As a corollary to this: the radiation emitted is unpolarized if $\boldsymbol{B} = 0$, and has unchanged relative intensities when $\boldsymbol{B} \neq 0$.

We may now apply these ideas to the transitions shown in Fig. 18.6. All upper states decay at the same rate, so

$$c = b + d = 2a + e. \tag{18.4}$$

If all five upper states start off equally populated, the rates of filling of the three lower states must be equal:

$$a + c + d = 2b + e. \tag{18.5}$$

For the next step, refer to Fig. 18.2. In a σ transition, energy is radiated by dipole moments p_x and p_y, equal amounts from each. For observation into the x-direction, only the radiation from p_y is received. By contrast, a π transition has energy radiated by one dipole-moment component p_z. Therefore, for equal decay rates, a σ transition radiates only half the intensity into the x-direction, compared with a π transition.

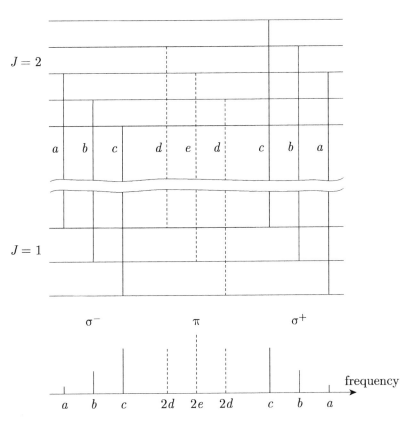

$J = 2$

| a | b | c | d | e | d | c | b | a |

$J = 1$

σ^- $\qquad$ π $\qquad$ σ^+

frequency

a $\quad$ b $\quad$ c $\quad$ 2d $\quad$ 2e $\quad$ 2d $\quad$ c $\quad$ b $\quad$ a

Fig. 18.6 The relative intensities of the transitions between a $J = 2$ level and a $J = 1$ level. Transitions drawn with solid lines are σ components; those with broken lines are πs. In the top panel, the intensities a to e are the rates of decay of the upper states along the "channels" drawn; they are proportional to the Einstein A coefficients for the transitions. The spectrum at the bottom shows intensities as observed along the x-direction in Fig. 18.1 (obviously with a multiplying coefficient throughout). For this direction of observation, the π components have doubled intensity relative to the σs, hence the factors 2.

Equivalently, a π transition gives a double intensity into the x-direction. This is indicated in Fig. 18.6, where the π components are given intensities, for observation transverse to B, of $2d, 2e, 2d$, although the upper-state decay rates are d, e, d. Now in the absence of a preferred direction in space (or equivalently, for equal initial populations of the upper states), the radiation travelling towards $+x$ must have the same E_y^2 as E_z^2, so we have

$$2(a + b + c) = 2d + 2e + 2d. \qquad (18.6)$$

We now have four equations with which to find four ratios of intensities. It is easy to solve the equations and to find the relative intensities for transverse observation:

$$a:b:c:\ 2d:2e:2d:\ c:b:a = 1:3:6:\ 6:8:6:\ 6:3:1 \quad (18.7)$$

These proportions have been used in drawing the spectrum in Fig. 18.6.

We have illustrated the reasoning here by applying the three rules to one particular spectral "line". Similarly the spectral analysis of §18.5 was explained by application to a specific example (which might be the same one). In both cases, there should be no difficulty seeing how the rules can be applied to any other Zeeman spectrum.

The sum rules give sufficient equations only in simple cases: for integer J the larger J must be no greater than 2; while for half-integer J at least

one of the Js must be $\frac{1}{2}$. You are asked to verify this in problem 18.3
But even in those cases where the sum rules do not suffice on their own
to determine all the intensity ratios, they provide valuable consistency
checks that can correct (for example) careless looking-up of matrix elements.[19]

[19] A full set of relative intensities is always available from eqns (18.8).

Problems

Problem 18.1   The shading rules

The electric-dipole moment operators for transitions with $\Delta m_J = 0, \pm 1$
are[20]

$$p_z = (-e)z, \qquad p_{\pm} = (-e)\left(\mp \frac{(x \pm iy)}{\sqrt{2}}\right).$$

Then for transitions from $|J, m\rangle$ to $|J', m'\rangle$, the matrix elements are:[21]

$$
\left.
\begin{aligned}
\text{Case } J' = J + 1: \quad & \left|\langle J', m{+}1 | p_+ | J, m\rangle\right|^2 = B\tfrac{1}{2}(J + m + 1)(J + m + 2) \\
& \left|\langle J', m | p_z | J, m\rangle\right|^2 = B\left\{(J + 1)^2 - m^2\right\} \\
& \left|\langle J', m{-}1 | p_- | J, m\rangle\right|^2 = B\tfrac{1}{2}(J - m + 1)(J - m + 2) \\
\text{case } J' = J: \quad & \left|\langle J', m{+}1 | p_+ | J, m\rangle\right|^2 = A\tfrac{1}{2}(J - m)(J + m + 1) \\
& \left|\langle J', m | p_z | J, m\rangle\right|^2 = A\,m^2 \\
& \left|\langle J', m{-}1 | p_- | J, m\rangle\right|^2 = A\tfrac{1}{2}(J + m)(J - m + 1). \\
\text{case } J' = J - 1: \quad & \left|\langle J', m{+}1 | p_+ | J, m\rangle\right|^2 = C\tfrac{1}{2}(J - m)(J - m - 1) \\
& \left|\langle J', m | p_z | J, m\rangle\right|^2 = C\,(J^2 - m^2) \\
& \left|\langle J', m{-}1 | p_- | J, m\rangle\right|^2 = C\tfrac{1}{2}(J + m)(J + m - 1).
\end{aligned}
\right\} \quad (18.8)
$$

[20] The $\mp$ sign in $p_{\pm}$ may look strange.
It comes from the convention, men-
tioned in sidenote 26 of Chapter 16, as
to a sign in the definition of the spherical harmonics $Y_{lm}(\theta, \phi)$.

Here A, B, C are integrals taken over the radial parts of the wave functions; they are of no interest when we are considering only the ratios of
transition strengths within a single line.[22]

Use these expressions to calculate the relative decay rates for the transitions drawn in Fig. 18.3. Show that the heights of the lines there drawn
in the spectrum are presented in the correct proportions, and that these
are:

[21] The last four of these are easily obtained from the first four.

for ${}^4P_{5/2} \to {}^4D_{3/2}$: $10:6:3:1:$ $4:6:6:4:$ $1:3:6:10$
for ${}^4D_{5/2} \to {}^4P_{3/2}$: $1:3:6:10:$ $4:6:6:4:$ $10:6:3:1$
for ${}^4D_{5/2} \to {}^4P_{5/2}$:
$\quad 10:16:18:16:10:$ $25:9:1:1:9:25:$ $10:16:18:16:10.$

[22] The matrix elements of (18.8) are
taken from Corney (1977), table 5.1.

Confirm:

(1) the g_J-values for the levels shown in Fig. 18.3 are:[23]
${}^4P_{5/2}$: 8/5; ${}^4D_{3/2}$: 6/5; ${}^4D_{5/2}$: 48/35; ${}^4P_{3/2}$: 26/15;

[23] You may assume throughout that LS
coupling holds, and that the Landé
g_J-factor is given to adequate accuracy
by

$$g_J = \frac{3}{2} + \frac{S(S+1) - L(L+1)}{2J(J+1)}.$$

(2) the differences $g_{\text{upper}} - g_{\text{lower}}$ for the three plots of Fig. 18.3 are
$+2/5, -38/105, -8/35$;

(3) the σ components are shaded in when the level having the larger J
has the larger g_J;

(4) the σ components are shaded out when the level having the larger J has the smaller g_J;

(5) the σ components form two separately symmetrical groups whenever $\Delta J = 0$.

Problem 18.2 Can we tell the energy order of the levels?
The analysis of § 18.5 identifies the L, S, J quantum numbers of both levels involved in the transitions. For simplicity, I have drawn energy-level diagrams with the larger J uppermost. Draw a diagram, after the same fashion, with the smaller J uppermost. Pay particular attention to the signs of the circular polarizations of σ radiation. Show that everything about the observed spectrum is unchanged. Therefore the Zeeman effect cannot tell us which of the levels has the higher energy.

Problem 18.3 The condition for sum rules to suffice
You will probably have to sketch out several special cases, along the lines of Figs. 18.3 and 18.6, to see how things go.

Show that, as claimed in the text: the sum rules suffice for finding the intensity ratios provided that: for integer J neither level has $J > 2$; while for half-integer J at least one level must have $J = \frac{1}{2}$.

Problem 18.4 The pinhole for centre-spot scanning
The intensity pattern given by a Fabry–Perot (when illuminated by a single frequency) can be represented as an infinite sum over Lorentzians (Koppelmann 1969, eqn (1.13)). This is the convenient way to think in the present problem.

To display a Fabry–Perot pattern, we need a fringe-forming lens to map the directions of arrival of light onto locations in the lens's focal plane. For pressure scanning, we put a pinhole at the centre of the fringe pattern, and allow what is transmitted through the pinhole to fall on a photoelectric detector. Show that the effect of the pinhole (on its own) on the range of frequencies transmitted has the shape of a top hat. Find the pinhole radius[24] that gives transmission over frequency range $\Delta\nu$.

When a single frequency is input to the Fabry–Perot, what is recorded by the detector is a convolution of the pinhole's top hat with the Fabry–Perot ("Airy") function. Since the Airy function is the sum of many Lorentzians, we can find out how things go by convolving a top hat with a single Lorentzian. Do this. The convolution of a top hat with a single Lorentzian results in a peak that is broader than the original Lorentzian. Show that this peak's width (defined as the full width at half maximum height, FWHM) is given by[25]

$$\left(\text{FWHM}_{\text{convolution}}\right)^2 = \left(\text{FWHM}_{\text{Lorentzian}}\right)^2 + \left(\text{FWHM}_{\text{pinhole}}\right)^2.$$

Comment: A similar "addition in quadrature" applies to the convolution of two Gaussians. However, there is *no* rule that addition in quadrature applies generally. Indeed the convolution of two Lorentzians has a width that is the simple sum of the contributing widths. So the quadrature addition here seems to be something of an accident, albeit a

[24] *Answer*: Let the pinhole's radius subtend angle θ at the fringe-forming lens. Then

$$\theta^2 = \frac{2\lambda}{c}\,\Delta\nu.$$

[25] Care may be needed with the range of angles assigned to an arctangent.

fortunate one. It is the reason why the simulation of § 18.5.1 was able to use a "pinhole finesse" equal to the "étalon finesse" with no more than a factor $\sqrt{2}$ degradation in the combined resolution.

Problem 18.5 Alternative analysis
The analysis of § 18.5 concentrates on measuring $(g_2 - g_1)$ from either the π components only or from one group of σ components only. Figure 18.4 shows that other possibilities exist: for example working with the spacing between the two strongest σ components (one taken from each of the two circular polarizations).

For $J_2 \neq J_1$, the frequency spacing between the two strongest σ components (one σ^+ and one σ^-) is $2|J_2 g_2 - J_1 g_1|\mu_B B/h$. Show that this is the case, and that it holds whether the Zeeman pattern is shaded in or shaded out.

The spectrum of Fig. 18.4 was (notionally) obtained using an étalon having its plates separated by 2.57 mm. Show that the free spectral range is 58.3 GHz. The σ peaks are separated in Fig. 18.4 by 40 display units; show that their frequency separation is 23.3 GHz. Show that therefore $|4g_2 - 2g_1| = 1.67$.

Make a table of $4g_2 - 2g_1$ along the lines of Table 18.1. Show that the possible transitions giving a shaded-out spectrum are:
$^3P_2$—$^3S_1$, giving $4g_2 - 2g_1 = 2$;
$^3D_2$—$^3P_1$, giving $4g_2 - 2g_1 = 5/3$.
Show that, given the error estimate of 20%, these cannot be distinguished reliably. However, for other cases, this might be the better route through the analysis.

When information is obtained more accurately and is combined from $4g_2 - 2g_1$ and $g_2 - g_1$, it is possible to obtain g_1 and g_2 separately, and even to permit an assessment of departures from pure LS coupling. But this carries us far beyond the intentions of the present chapter.

Problem 18.6 Overlapping σ components
Landé g-factors often work out to be ratios of simple numbers. Therefore two spectral components (say σ components of opposite handedness) may have exactly equal frequency, thereby giving misleading signals as to how many σ-components there are, and as to the relative intensities. In the event of such a complication, it is mandatory to use a circular polarizer to separate out the σ^+ components from σ^- (or vice versa). (Alternatively we may use a linear polarizer to isolate the π components if transverse-to-B observation is possible.)

To highlight the possibility of overlap, draw out the σ patterns (separating the two circular polarizations) and π patterns for the transitions:

- $^3D_3 \rightarrow {}^3F_2$ (the weakest σ^+ and σ^- components overlap exactly)
- $^3F_4 \rightarrow {}^3G_3$ (the weakest two σ^+ and σ^- overlap)
- $^3G_5 \rightarrow {}^3H_4$ (the weakest three σ^+ and σ^- overlap).

(This pattern continues indefinitely for transitions having $S = 1$.)

VIII Transitions

19 | The Einstein A and B coefficients

Intended readership: Soon after seeing a more conventional treatment.

19.1 Introduction

This chapter aims to do two things: to present a slightly unusual exegesis of the Einstein reasoning; and also to offer a finer-detail treatment that yields additional insights.

Atoms (or molecules or solids or nuclei) absorb and emit electromagnetic radiation (photons). The transitions may be electric-dipole, or electric-quadrupole, or magnetic-dipole, ... The *rates* of the reactions are made quantitative by means of the Einstein A and B coefficients.

For definiteness, we shall think of radiation interacting with atoms in a dilute gas, and we shall have in mind electric-dipole transitions. But most of the discussion is not limited to either of these special cases.[1]

The Einstein A and B coefficients are *properties of the atoms*. They can be calculated from quantum mechanics. But a brute-force calculation of A (in particular) requires advanced quantum mechanics: quantum electrodynamics (QED, Chapter 32). Fortunately, we can avoid the complications by exploiting a very clever trick invented by Einstein.[2]

19.2 Definitions

An atom has energy levels, two of which are labelled 1 and 2. Within these *levels*, there are individual *states*, probably characterized by their m_J-values (Fig. 19.1). (For the nomenclature, including *level* and *state*, see Chapter 16 sidenote 7.) We concentrate attention on one upper state b (energy E_b) and one lower state a (energy E_a). Transitions between b and a constitute a **spectral line** (in a Zeeman-effect context we'd call it a **component**), centred on angular frequency $\omega_{ba} = (E_b - E_a)/\hbar$. The spectral line has some width $\Delta\omega$ about which we need say nothing for now, save that it exists and $\Delta\omega \ll \omega_{ba}$.

There are N_a atoms in the lower state a and N_b in the upper state b (Fig. 19.2). Atoms in the upper state emit photons and fall to lower states; some go to state a. The number of atoms that emit photons in this decay from b to a in time δt is $N_b A_{ba}\, \delta t$, where this defines the coefficient A_{ba} describing the rate of **spontaneous emission** of photons

[1] In the case of a gas, we ignore frequency shifts caused by atomic motion. We might instead have magnetic-dipole transitions, undergone by ions embedded in a solid, for example, as is often the case with magnetic resonance.

[2] Though, as is discussed in § 19.9, Einstein himself used the same reasoning for a rather different purpose.

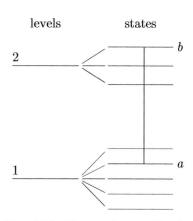

levels   states

2

1

Fig. 19.1 The notation used for describing levels (1 and 2) and states (such as a, b). The states are drawn separated so that we may see them individually; there may or may not be a magnetic field separating them.

We shall discuss a transition between states, exemplified by that drawn for $b \leftrightarrow a$.

For the special case shown, state b has quantum number $m_J = 1$, and has three possible decays, to $m_J = 0, 1, 2$, only one of which is drawn.

Essays in Physics: Thirty-two thoughtful essays on topics in undergraduate-level physics. Geoffrey Brooker Oxford University Press. © Geoffrey Brooker 2021. DOI: 10.1093/oso/9780198857242.003.0019

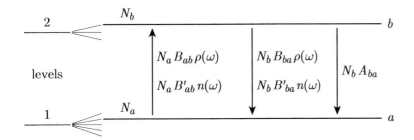

Fig. 19.2 An atom has quantum states a (a state within level 1) and b (a state within level 2), between which radiative transitions take place. These transitions are, in order: absorption, stimulated emission, spontaneous emission. Each reaction has its rate indicated alongside.

Suppose now that radiation, containing angular frequencies around ω_{ba}, is incident on the atoms. The radiation can cause atoms in state a to absorb energy, thereby raising them to state b (**absorption**). And the radiation can also cause atoms in state b to fall to state a, thereby emitting additional photons by **stimulated emission**. The stimulated reaction rates contain $n(\omega)$, defined by:

$$n(\omega) = \begin{pmatrix} \text{number of photons per quantum state} \\ \text{for angular frequencies around } \omega \end{pmatrix}. \qquad (19.1)$$

The numbers of atoms per second undergoing transitions are given by the three reaction rates (Fig. 19.2):[3]

$$\left.\begin{array}{l} \text{absorption rate for } a \to b \text{ is } N_a\, B'_{ab}\, n(\omega) \\[4pt] \text{stimulated-emission rate for } b \to a \text{ is } N_b\, B'_{ba}\, n(\omega) \\[4pt] \text{spontaneous-emission rate for } b \to a \text{ is } N_b\, A_{ba}. \end{array}\right\} \qquad (19.2)$$

We need a little more detail before the $n(\omega)$ in (19.1) is fully defined. At present (later we shall do things differently), we are imagining the occupied photon states to have both polarizations and to have their $\boldsymbol{k}$-vectors isotropically distributed over directions.[4]

The notation $n(\omega)$ indicates that n is allowed to depend on ω "in the large" (it need not be equally strong in the red and in the blue, say);[5] but over the linewidth $\Delta\omega$ surrounding ω_{ba} it is required to be uniform; we shall refer to such radiation as "white".

Digression

Enough has been said already for us to proceed at once to the reasoning of § 19.3 and to eqn (19.4). Nevertheless, we pause to define $\rho(\omega)$ such that $\rho(\omega)\,\mathrm{d}\omega$ is the energy density (i.e. per unit volume) of radiation occupying the angular-frequency range $\mathrm{d}\omega$. The radiation is, as above, isotropic, unpolarized and "white". Then from (6.14),

$$\rho(\omega)\,\mathrm{d}\omega = \underset{\substack{\text{number of} \\ \text{modes in} \\ \text{unit volume}}}{\frac{\omega^2\,\mathrm{d}\omega}{\pi^2 c^3}} \times \underset{\substack{\text{number } n(\omega) \\ \text{of photons} \\ \text{per mode}}}{n(\omega)} \times \underset{\substack{\text{energy} \\ \text{of each} \\ \text{photon}}}{\hbar\omega}. \qquad (19.3)$$

Alternative reaction rates, expressed in terms of $\rho(\omega)$, are entered on Fig. 19.2.[6] Textbook presentations of Einstein A and B usually work with $B\rho$ rather than $B'n$.

[3] It does not matter whether N_a and N_b refer to unit volume of the substance under discussion, or to the entire sample—so long, of course, as we are consistent in defining everything in the same way.

[4] That is, $n(\omega)$ is a function of ω only; for any given wave vector $\boldsymbol{k}$ it is the same for both polarizations $\boldsymbol{e_k}$.

However, experimental investigations most commonly use, or prepare, laser beams that are collimated and/or polarized. The reaction rates of (19.2) must be set up afresh if they are to handle such very different conditions. We deal with more realistic conditions in § 19.8.

[5] Any function of ω that is critically dependent on the distance of ω from line centre will be written as a function of $(\omega - \omega_{ba})$. Any quantity that depends in some more gentle way on ω will be written as a function of (ω). For the present, n is a "gentle" function of ω.

[6] It should be clear that a "mode" is a fully specified quantum state $|\boldsymbol{k}, \boldsymbol{e_k}\rangle$ available to the field, the same as the "quantum state" of (19.1).

It is my view that $n(\omega)$ is a more intuitive concept than $\rho(\omega)$: in (19.3) we need to understand $n(\omega)$ in order to construct $\rho(\omega)$. It is also the case that "photons per mode" is the quantity, and terminology, of preference in laser physics. And we show in § 19.8.1 that n links neatly to the ideas of quantum electrodynamics. By choosing to work with $B'n$, rather than $B\rho$, we are starting the way we mean to go on.

Although (19.3) is a digression for now, it will be needed for obtaining (19.15) in § 19.6.

19.3   The Einstein relations

As we have said, the "Einstein coefficients" B'_{ab}, B'_{ba}, A_{ba} are properties inherent in the atoms under discussion, and they can be calculated from scratch using advanced quantum theory (Chapter 32). But the reasoning in the present section gives a neat route to some of the same results.

We are about to play a very clever trick. The trick may be understood in any of the following ways:

- Even quantum electrodynamics is subject to restrictions as to what results it can conceivably yield. It *must not* assign to atoms any properties that would permit the building of a perpetual-motion machine.[7] So atoms interacting with radiation *must not* build a temperature difference from nothing. At the least, this is a consistency check that we must make. If our quantum mechanics were to fail this test it would be extremely embarrassing.

- If we are unable or unwilling to calculate transition rates by means of quantum electrodynamics, then the "consistency check" can be used on its own to tell us some restrictions that *must* apply to those transition rates. We have a student-accessible (non-QED) route to some answers: the ratios B'_{ab}/B'_{ba}, A_{ba}/B'_{ba}.

- It is even better. The B'-coefficients can be calculated (§ 19.6) without the apparatus of QED, because we may in imagination make $n(\omega)$ large. Then we have many photons per mode, and we are in the limit of large quantum numbers, where "one photon more or less makes little difference". We may treat the radiation classically (Maxwell's equations); the atoms are still treated by quantum mechanics, because they have discrete energy levels. Such a treatment is said to be **semi-classical**.[8] By contrast, we have no similar limiting case for finding A_{ba}. With spontaneous emission, we start with no photons and we end up with one. Neither number is large enough for us to take a classical limit. So for finding A_{ba} it's QED—or we do something devious. We choose to be devious: we obtain a B-coefficient by some route (perhaps the semi-classical calculation of § 19.6), and then the Einstein trick gives us the value of A_{ba}.

Our programme then is to put a large number of atoms,[9] in thermal equilibrium at temperature T, into black-body radiation also in thermal equilibrium and at *the same* temperature T. And we make sure that the reaction rates of Fig. 19.2 fit comfortably with this.

We now use eqns (19.2) to write the **rate equation**

$$\left(\frac{dN_b}{dt}\right)_{\text{reactions } a \,\leftrightarrow\, b} = N_a\, B'_{ab}\, n_{\text{bb}}(\omega_{ba}) - N_b\, B'_{ba}\, n_{\text{bb}}(\omega_{ba}) - N_b\, A_{ba}.$$

$$(19.4)$$

The angular frequency in $n_{\text{bb}}(\omega)$ has now been written as ω_{ba} because that is the frequency of interest for the transitions between states a and b.

In thermal equilibrium, N_a and N_b must be independent of time, so the expression just given must be zero. Setting it to zero and rearranging,

[7] Remember that any failure to conform to the Second Law of Thermodynamics would permit construction of a perpetual-motion machine. So we are about to require conformity with the Second Law.

[8] For this reason, the field is described by Maxwell's equations in § 19.6 where we calculate the B_{ab} coefficient for absorption. However, once this is done, we revert to using a quantum description of the field, as in the $n_{\text{bb}}(\omega_{ba})$ of eqn (19.4).

[9] We assume that the atoms (or molecules or nuclei) meet the conditions for having a Boltzmann distribution e.g. a dilute gas with atomic motion ignored, or "localized" atoms embedded in a solid. This is entirely for simplicity.

Since the transition is between single atomic *states* a and b, both upper and lower states have degeneracy 1; there are no extra factors for degeneracy in (19.6).

we obtain

$$n_{\mathrm{bb}}(\omega_{ba}) = \frac{N_b\,A_{ba}}{N_a\,B'_{ab} - N_b\,B'_{ba}} = \frac{A_{ba}/B'_{ba}}{\dfrac{B'_{ab}}{B'_{ba}}\dfrac{N_a}{N_b} - 1}. \tag{19.5}$$

In thermal equilibrium at temperature T the atoms have a Boltzmann distribution:

$$\frac{N_b}{N_a} = \exp\{-(E_b - E_a)/k_{\mathrm{B}}T\} = \exp(-\hbar\omega_{ba}/k_{\mathrm{B}}T). \tag{19.6}$$

Putting this into the expression for $n_{\mathrm{bb}}(\omega_{ba})$, we now have

$$n_{\mathrm{bb}}(\omega_{ba}) = \frac{A_{ba}/B'_{ba}}{\dfrac{B'_{ab}}{B'_{ba}}\,e^{\hbar\omega_{ba}/k_{\mathrm{B}}T} - 1}. \tag{19.7}$$

Given thermal equilibrium at a common T throughout, (19.7) must be identical with the Planck distribution for photons:

$$n_{\mathrm{bb}}(\omega_{ba}) = \frac{1}{e^{\hbar\omega_{ba}/k_{\mathrm{B}}T} - 1}. \tag{19.8}$$

Now we have said that B'_{ab}, B'_{ba} and A_{ba} are *atomic properties*: they can have nothing whatever to do with thermal equilibrium, or black-body radiation, or anything to do with the environment[10] within which the atoms are located. In particular, then, they can have nothing to do with temperature T. The only dependences[11] on T in eqns (19.7) and (19.8) are those explicitly on display (there is no T hiding inside some other symbol). The expressions must be identical for all T, so we may equate coefficients:[12]

$$B'_{ab} = B'_{ba} = A_{ba}. \tag{19.9}$$

These are the relations linking the Einstein A and B' coefficients.[13]

Let me say it once more. Relations (19.9) connect three *atomic properties* that hold under all conditions, even conditions very far from equilibrium. They even apply to the internal workings of a laser, which is certainly not a system in thermal equilibrium.[14]

19.4   The principle of detailed balance

The results in (19.9) are so important that we must check carefully what we have done. Is there any wriggle room?

One anxiety that can be raised concerns the use of only two quantum states a, b. Would the reasoning fail (or require modification) if more quantum states were included? And does state a have to be the atomic ground state?

Figure 19.3 illustrates the complications that lie in wait if we allow more levels and transitions, so that states a and b are not "isolated": there can be transitions into state a from other states, both above and below state a, and likewise transitions out. The same applies to state b. Fortunately, we can argue that none of this makes any difference. The **principle of detailed balance** says that thermal equilibrium is a dynamic equilibrium in which *equilibrium is never maintained by a cyclic process*.

[10] There can be exceptions if the atoms and radiation are confined within a small optical cavity.

[11] See problem 19.1 for a case where this may need a bit more thought.

[12] From (19.3) we may easily obtain the relations linking the unprimed Bs:

$$B_{ab} = B_{ba}; \qquad A_{ba} = \frac{\hbar\omega_{ba}^3}{\pi^2 c^3}\,B_{ba}.$$

At present we are concerned only with the ratios of coefficients A_{ba}, B'_{ab} (or B_{ab}), B'_{ba} (or B_{ba}). It needs the evaluation of only one of the coefficients, in due course, for everything to fall into place: eqn (19.16).

[13] It should be clear: the "information flow" in (19.9) runs from the B's to A_{ba}. The B's can be found as in § 19.6, that is, by fairly elementary physics. It is A_{ba} that requires (19.9) if we are to sidestep advanced methods.

[14] The misapprehension is common among students: that relations (19.9) are obtained via thermal equilibrium, so they can be true only when there is thermal equilibrium. Happily, I have been unable to find any textbook that encourages this error; the worst that can be said is that some could work harder to prevent so tempting a misunderstanding from taking root.

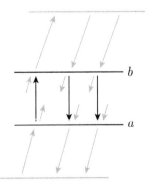

Fig. 19.3 An atom has multiple energy levels and, necessarily, multiple quantum states. Two states a, b are those discussed in § 19.2; they and their transitions are drawn in black. Other states and transitions are hinted at and are drawn in grey.

That is, we never have a "cyclic flow" (of particles or energy) such as $1 \to 2 \to 3 \to 1$ maintaining equilibrium. Every reaction "balances in detail": $1 \leftrightarrow 2 \leftrightarrow 3 \leftrightarrow 1$ and similarly. Incidentally, the principle of detailed balance is securely founded in both classical and quantum mechanics, so it may be applied with complete confidence.

The "balancing" of reactions $a \leftrightarrow b$ "in detail" means that the rate equation (19.4) (set to zero for the case of thermal equilibrium), holds quite independently of any other reactions that may be going on, even though those reactions may involve states a and b.

Therefore nothing endangers the generality of results (19.9). The atomic states do not need to be isolated from other states or levels; the lower state a does not need to be the atomic ground state.

There is an even more powerful consequence of detailed balance. In thermal equilibrium, we cannot have a cyclic flow of energy among field modes: $\mathbf{k}_1 \to \mathbf{k}_2 \to \mathbf{k}_3 \to \mathbf{k}_1$. Therefore a balancing of rates similar to that in (19.4) applies among field modes—when radiation and atoms are all in thermal equilibrium at temperature T. We pursue this idea in more detail in §19.8.

19.5   The transition's line profile

Consider the dependence of the atomic transition rates on the frequency of the radiation. The spectral line has a centre (angular) frequency ω_{ba} but radiation somewhat away from ω_{ba} still affects the atoms, and that effect must be taken into account.

Imagine that we illuminate the atoms by means of a tunable laser and that we tune the laser to scan through the profile of the spectral line centred on ω_{ba}. The absorption rises to a maximum at line centre then falls again. The same applies to stimulated emission. There is likewise a frequency dependence in the spontaneous emission, which may be observed by passing it through a high-resolution spectrometer: energy is emitted most strongly at line centre, with a fall-off on either side.[15]

These frequency dependences are accommodated by introducing a **line profile**, a **normalized lineshape function**, $g(\omega - \omega_{ba})$. Normalization means that[16]

$$\int_0^\infty g(\omega - \omega_{ba})\, \mathrm{d}\omega = 1. \tag{19.10}$$

Thus, for radiation absorbed from $\mathrm{d}\omega$ or emitted into $\mathrm{d}\omega$, we may "patch" the lineshape function $g(\omega - \omega_{ba})$ onto eqns (19.2):[17]

$$\left.\begin{aligned} \text{rate of absorption} &= N_a\, B'_{ab}\, n(\omega)\, g(\omega - \omega_{ba})\, \mathrm{d}\omega \\ \text{rate of stimulated emission} &= N_b\, B'_{ba}\, n(\omega)\, g(\omega - \omega_{ba})\, \mathrm{d}\omega \\ \text{rate of spontaneous emission} &= N_b\, A_{ba}\, g(\omega - \omega_{ba})\, \mathrm{d}\omega. \end{aligned}\right\} \tag{19.11}$$

The definitions have been made in such a way that, when we total over angular frequencies, we recover the reaction rates shown in Fig. 19.2.

The same lineshape function $g(\omega - \omega_{ba})$ applies to all three reactions in (19.11): see Exercise 19.1.

[15] We have mentioned a tunable laser, and a spectrometer, in order to give an operational definition of the lineshape function (line profile). Such experimental arrangements are unlikely to involve isotropic radiation. However, now that the explanation has been given, we revert to discussing an $n(\omega)$ that is isotropic.

A full description of a realistic absorption/emission process would specify the actual field mode from/into which the electromagnetic energy goes: the wave vector $\mathbf{k}$ and the polarization direction $\mathbf{e_k}$; see (19.17).

[16] Function g is critically dependent on the value of the angular frequency ω, relative to line centre ω_{ba}. We indicate this by writing g as a function of $(\omega - \omega_{ba})$.

In this chapter we make no attempt at calculating $g(\omega - \omega_{ba})$, even for "natural" (lifetime) broadening. The line profile is left as a shameless loose end. It is taken up in Chapter 32.

[17] We might think instead of defining a frequency dependent B'_{ab}, rather than introducing both B'_{ab} and $g(\omega - \omega_{ba})$. However, a "division of labour" is being set up in which B'_{ab} handles the overall strength of the absorption line while $g(\omega - \omega_{ba})$ handles the line profile. This division of labour is conventional, and convention is here very good indeed.

Exercise 19.1
Take again the special case where the atoms and radiation are in thermal equilibrium at a common temperature T. Use detailed balance to argue that the three reaction rates spelt out in (19.11) are governed by the same function $g(\omega - \omega_{ba})$.

19.6   The *B*-coefficient: semi-classical

This calculation is given in many textbooks, so we outline the steps only, leaving some details for problem 19.2.

Consider an absorption reaction from atomic state a to state b when the atom is subjected to an electric field[18] (classical) $\boldsymbol{E}_0 \cos(\omega t - \boldsymbol{k} \cdot \boldsymbol{r})$. The atomic states a, b are such that an electric-dipole transition between them is allowed. The Hamiltonian for interaction between the atom and the field is $\widehat{H}' = (e/m_e)\boldsymbol{A} \cdot \hat{\boldsymbol{p}}$, in which $\boldsymbol{A} = (\boldsymbol{E}_0/2i\omega)(e^{-i\omega t} - e^{i\omega t})$ is the vector potential and $\hat{\boldsymbol{p}}$ is the momentum operator.[19] The wave function for the atom is

$$\Psi = c_a \psi_a e^{-i\omega_a t} + c_b \psi_b e^{-i\omega_b t},$$

in which $\hbar\omega_a = E_a$, $\hbar\omega_b = E_b$, and $\omega_{ba} = \omega_b - \omega_a$. Initially $c_a = 1$ and $c_b = 0$.

The time-dependent Schrödinger equation tells us that

$$\dot{c}_b = \frac{-i}{\hbar} c_a \langle b|\widehat{H}'|a\rangle e^{i\omega_{ba}t}. \tag{19.12}$$

We look at the early-time part of the absorption reaction, where state a is undepleted,[20] so $c_a \approx 1$. Then (19.12) can be integrated over time to give

$$c_b = \frac{-e}{2im_e\hbar\omega} \boldsymbol{E}_0 \cdot \langle b|\hat{\boldsymbol{p}}|a\rangle \left(\frac{e^{i(\omega_{ba}-\omega)t} - 1}{(\omega_{ba} - \omega)} - \frac{e^{i(\omega_{ba}+\omega)t} - 1}{(\omega_{ba} + \omega)} \right). \tag{19.13}$$

The second term in large brackets can be neglected (*rotating-wave approximation*).[21] Also, there is a theorem in atomic physics[22] telling us that $\langle b|\hat{\boldsymbol{p}}|a\rangle = im_e\omega_{ba}\langle b|\boldsymbol{r}|a\rangle = im_e\omega_{ba}\boldsymbol{r}_{ba}$ so that

$$c_b = \frac{-e}{2\hbar} \frac{\omega_{ba}}{\omega} \boldsymbol{E}_0 \cdot \boldsymbol{r}_{ba} \frac{e^{i(\omega_{ba}-\omega)t} - 1}{\omega_{ba} - \omega},$$

in which we have written $\boldsymbol{r}_{ba}$ for $\langle b|\boldsymbol{r}|a\rangle$; we recognize $\boldsymbol{r}_{ba}$ as the electric-dipole matrix element ($-e$ omitted) linking state a with state b.

The early-time population of the upper atomic state is now given by

$$|c_b|^2 = \frac{e^2}{4\pi\epsilon_0} \frac{8\pi}{\hbar^2} \frac{\omega_{ba}^2}{\omega^2} \left(\frac{\epsilon_0}{2} |\boldsymbol{E}_0 \cdot \boldsymbol{r}_{ba}|^2 \right) \frac{\sin^2 \frac{1}{2}(\omega_{ba} - \omega)t}{(\omega_{ba} - \omega)^2}. \tag{19.14}$$

At this point we need to consider the vector dot product. Suppose (temporarily) that states a, b have the same value of m_J so that the transition between them has $\Delta m_J = 0$. Then the only non-zero matrix

[18] We approach a classical limit if the (field) quantum numbers are both large and "mixed", meaning that the radiation at angular frequency ω has a distribution of $n(\omega)$s around some large central value. If, by contrast, n is a precisely known integer, then the radiation cannot be classical, no matter how large n is—because we would have information about it on the scale of $\hbar$. We may take it that the conditions for the field to be in a classical limit are met here, without needing to go into detail.

[19] Some more details are being swept under the carpet here. In particular, we are setting the origin of $\boldsymbol{r}$ to be at the centre of the atom, and then approximating $e^{\pm i\boldsymbol{k} \cdot \boldsymbol{r}}$ to 1. These details are dealt with in problem 19.2, and more fully in Chapter 32.

[20] This step is a happy one. It means that state b does not have to be the only state receiving population from state a. Any other higher state that is being populated at the same time merely speeds up the depletion of state a, and has no effect within the approximation being made.

[21] Better reasoning starts from (32.14a), in which $e^{i(\omega_{ba}+\omega)t}$ does not appear in the first place.

[22] See e.g. Woodgate (1980) § 3.3.

[23] The definition of $\boldsymbol{E}_0$ makes it the peak value of the electric field, so $\epsilon_0 \boldsymbol{E}_0^2/2$ is the peak value of the electric energy density. The mean value is half of this. But there is also a magnetic-field energy density with the same mean. The total of the two field energies is $\epsilon_0 \boldsymbol{E}_0^2/2$. Expressing this in terms of the mean square field, instead of the peak field, we have field energy $\epsilon_0 \langle E_z^2 \rangle$.

[24] The addition of "inactive" field components E_x, E_y has a loose similarity to the addition of the magnetic field in sidenote 23.

The radiation was first made isotropic in eqn (19.3), in order to cater easily for black-body radiation. We continue with isotropic radiation for now; things will be done differently from § 19.7.

The factor $\frac{1}{3}$ is sometimes argued for by rotating the $\langle E_z^2 \rangle$ field round the atom and taking an average of $\cos^2 \theta$. This reasoning is acceptable, though I find it over-complicated.

What we must *not* do is change the rules in the middle of the game: randomizing the orientation of the atom over a continuous range of angles, because this violates our agreement (semiclassical) to treat the atom as a fully quantum system occupying specified states a, b only.

element linking these states is $\langle b|z|a \rangle$, and the field $\boldsymbol{E}$ that we have needed to supply lies in the z-direction also. The energy density of that field is $\epsilon_0 \boldsymbol{E}_0^2/2 = \epsilon_0 \langle E_z^2 \rangle$, where $\langle \, \rangle$ denotes a time average.[23]

The radiation imposed is to be "white" (not necessarily black-body), in line with the discussion following (19.2). Let there be many occupied field modes having non-zero E_z and angular frequencies close to ω. These modes have unrelated phases, so their effect on $|c_b|^2$ is to supply added terms to the squared field on the right of (19.14). The total (relevant) field energy is again represented by $\epsilon_0 \langle E_z^2 \rangle$, though $\langle E_z^2 \rangle$ represents an average taken over a stronger field than before.

$$\left(\frac{\epsilon_0}{2} \, |\boldsymbol{E}_0 \cdot \boldsymbol{r}_{ba}|^2 \right) \text{ is summed over field modes to yield } \epsilon_0 \langle E_z^2 \rangle \, |z_{ba}|^2 \, .$$

Next, make the radiation isotropic. Then $\langle E_x^2 \rangle$ and $\langle E_y^2 \rangle$ are present in addition to $\langle E_z^2 \rangle$. These field components do nothing to stimulate the $\Delta m_J = 0$ absorption, but they must be there and with equal strength.[24]

$$\left(\frac{\epsilon_0}{2} \, |\boldsymbol{E}_0 \cdot \boldsymbol{r}_{ba}|^2 \right) \text{ is summed to } \frac{1}{3} \times \epsilon_0 \left\{ \langle E_z^2 \rangle + \langle E_x^2 \rangle + \langle E_y^2 \rangle \right\} |z_{ba}|^2 \, .$$

We have said that the fields whose averages are being taken are all at or close to angular frequency ω. We can redescribe them as having an (isotropic now) energy density $\rho(\omega) \, d\omega$. Then the effect on $|c_b|^2$ is found by summing

$$\left(\frac{\epsilon_0}{2} \, |\boldsymbol{E}_0 \cdot \boldsymbol{r}_{ba}|^2 \right) \text{ to yield } \frac{1}{3} \rho(\omega) \, d\omega \, |z_{ba}|^2 \, .$$

Had we chosen to discuss a transition with $\Delta m_J = \pm 1$ the outcome would have been similar: *only one of three field components* (having appropriate polarization) *drives the absorption, but all three must be present*. Whichever Δm_J transition is involved, we have

$$|c_b|^2 = \frac{e^2}{4\pi\epsilon_0} \frac{8\pi}{3\hbar^2} \frac{\omega_{ba}^2}{\omega^2} \, |\boldsymbol{r}_{ba}|^2 \, \rho(\omega) \frac{\sin^2 \frac{1}{2}(\omega - \omega_{ba})t}{(\omega - \omega_{ba})^2} \, d\omega,$$

in which $\boldsymbol{r}_{ba}$ is whichever dipole-moment matrix element is non-zero for the transition under consideration.

As the final step we allow the radiation to have a $\rho(\omega)$ distributed over a range of frequencies ("white"). Then $\rho(\omega)$ is constant over the width of the spectral line, though we can make it fall off at low and high frequencies well away from ω_{ba} (calling upon $n(\omega)$ for the high frequencies). We can total over the frequencies by integrating over ω, while taking $(\omega_{ba}^2/\omega^2)\rho(\omega) \approx \rho(\omega_{ba})$ as constant.

$$\begin{aligned}
|c_b|^2 &= \frac{e^2}{4\pi\epsilon_0} \frac{8\pi}{3\hbar^2} \, |\boldsymbol{r}_{ba}|^2 \, \rho(\omega_{ba}) \int_0^\infty \frac{\sin^2 \frac{1}{2}(\omega - \omega_{ba})t}{(\omega - \omega_{ba})^2} \, d\omega \\
&= \frac{e^2}{4\pi\epsilon_0} \frac{8\pi}{3\hbar^2} \, |\boldsymbol{r}_{ba}|^2 \, \rho(\omega_{ba}) \frac{t}{2} \int_{-\omega_{ba}t/2}^\infty \frac{\sin^2 x}{x^2} \, dx \\
&\approx \frac{e^2}{4\pi\epsilon_0} \frac{8\pi}{3\hbar^2} \, |\boldsymbol{r}_{ba}|^2 \, \rho(\omega_{ba}) \frac{t}{2} \pi \\
&= \frac{4}{3} \frac{e^2}{4\pi\epsilon_0} \frac{\omega_{ba}^3}{\hbar c^3} \, |\boldsymbol{r}_{ba}|^2 \, n(\omega_{ba}) \, t.
\end{aligned} \tag{19.15}$$

In the integration, we have assumed that the time t after the start of the absorption is many periods of oscillation ($\omega_{ba}t \gg 1$) even though the reaction has as yet proceeded only a little way (the lower state remains undepleted).[25] We can now identify the Einstein coefficient B'_{ab} for absorption between states a and b, and with it the A coefficient:

$$A_{ba} = B'_{ab} \frac{|c_b|^2}{n(\omega_{ba})t} = \frac{4}{3} \frac{e^2}{4\pi\epsilon_0} \frac{\omega_{ba}^3}{\hbar c^3} |r_{ba}|^2 . \qquad (19.16)$$

In the above, we have made use of (19.3) and (19.9). Equation (19.16) gives the expected expression for the Einstein A coefficient.[26]

19.6.1 Aggregating atomic states into levels

Suppose we wish to find the rate A_{b1} at which state b decays into all states of level 1. Then $A_{b1} = \sum_a A_{ba}$. This decay rate is the same for all upper states b (otherwise a uniform distribution of atoms in states b would acquire a preferred orientation during decay) so we may name it more informatively as A_{21}.

It is not hard to find A_{21}, given one of the contributing A_{ba}, because there are intensity sum rules that relate the A_{ba} to each other.[27]

Treatments of the A and B coefficients often bundle together[28] the upper states (g_2 of them) into level 2, and the lower states (g_1 of them) into level 1. Such an aggregation, summing or averaging, destroys information needlessly. It may be possible to dis-aggregate, or un-average afterwards (example in the obtaining of (19.11)), but "if I was going there I wouldn't start from here". This is why our reasoning in §§ 19.2–6 has been set up to concentrate on individual atomic states a, b.

The next sections carry us even further in the direction of not-aggregating, this time not-aggregating field modes.

19.7 Individual field modes

Today's experiments most commonly illuminate a sample with a polarized laser beam, with the aim (or the effect) of inducing transitions between specified m_J-states in the upper and lower atomic levels.[29] And the laser's frequency width is narrow compared with the linewidth, so it can conveniently probe the shape of $g(\omega - \omega_{ba})$. For these experimental conditions, a more fine-grained discussion (not summing over wave vectors, polarizations or frequencies) would be more helpful.[30]

A single mode $|k, e_k\rangle$ of the radiation field is specified by its wave vector k and a unit vector e_k (one of two, both $\perp k$) giving the polarization direction. The mode will be occupied by $n = n(k, e_k)$ photons.

Figure 19.4 shows the reaction rates involving radiation in the selected field mode. Replacing (19.11) we have

$$\left.\begin{array}{r} \text{rate of absorption} = N_a \times \beta_{ab} \times n \\ \text{rate of simulated emission} = N_b \times \beta_{ba} \times n \\ \text{rate of spontaneous emission} = N_b \times \alpha_{ba}. \end{array}\right\} \qquad (19.17)$$

---

[25] In more detail: The occupation $|c_b|^2$ starts rising as t^2, but joins onto a linear rise when $\omega_{ba}t/2$ reaches about π. A similar glitch is displayed in Fig. 32.7.
The integral

$$\int_{-\infty}^{\infty} \frac{\sin^2 x}{x^2} \, dx = \pi.$$

[26] It is tempting to try to obtain the A-coefficient by applying Golden Rule Number Two to the spontaneous-emission process. This can be done, but it requires more care than is usually given. We need the full frequency dependence of (32.50) to show that a summation over all possible final states of the radiation (a) converges at all, and (b) focuses on frequencies so close to ω_{ba} as to permit the density of states and the matrix element (squared) to be taken outside the sum. This is investigated in problem 32.24. When all is done properly, it has to build on results obtained in Chapter 32, and there is no saving of labour.

[27] We have seen an application of sum rules, for a special case, in § 18.6. For a good discussion of this whole area, see Corney (1977) Chapters 4, 9.

[28] In this case, the ratio g_2/g_1 appears in a modified (19.6) and propagates from there to give $B'_{12}/B'_{21} = g_2/g_1$.
Confusingly, g_1, g_2 are usually called "degeneracies", suggesting that what is important is a common energy. Why is it acceptable for a level J to have its m_J-states bundled into $g_J = (2J+1)$? Why would it be quite wrong to bundle hydrogen 2s, 2p (they have nearly equal energies) into $g_2 = 8$?

[29] Even an "unpolarized" laser does not impose isotropic radiation on its target. If the laser beam travels in the z-direction, it irradiates the target with $\langle E_x^2 \rangle$ and $\langle E_y^2 \rangle$, but not with $\langle E_z^2 \rangle$.

[30] Equations (19.11) tell us the reaction rates from/into a given range $d\omega$ of angular frequency, but do not separate out individual wave vectors or polarizations.

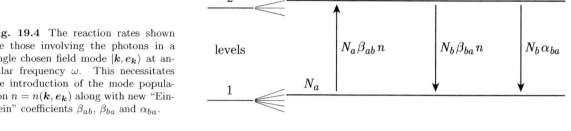

Fig. 19.4 The reaction rates shown are those involving the photons in a single chosen field mode $|\boldsymbol{k}, \boldsymbol{e_k}\rangle$ at angular frequency ω. This necessitates the introduction of the mode population $n = n(\boldsymbol{k}, \boldsymbol{e_k})$ along with new "Einstein" coefficients β_{ab}, β_{ba} and α_{ba}.

Here β_{ab} and β_{ba} are new kinds of B-coefficient and α_{ba} is a new A-coefficient.

Quantities $n(\boldsymbol{k}, \boldsymbol{e_k})$, $\beta_{a,b,\boldsymbol{k},\boldsymbol{e_k}}$, $\beta_{b,a,\boldsymbol{k},\boldsymbol{e_k}}$, $\alpha_{b,a,\boldsymbol{k},\boldsymbol{e_k}}$ are now all dependent upon $\boldsymbol{k}, \boldsymbol{e_k}$, but we shall suppress the $\boldsymbol{k}, \boldsymbol{e_k}$ to avoid clutter when the meaning is clear.[31]

[31] The dependence of $\alpha_{ba}, \beta_{ba}, \beta_{ab}$ on ω is now handled by their dependence on the magnitude of $\boldsymbol{k}$.

19.8   The Einstein reasoning—again

We obtain relations between the new "Einstein" coefficients by making an adaptation of the Einstein reasoning. We put the atoms into black-body radiation at temperature T: photon number $n = n(\boldsymbol{k}, \boldsymbol{e_k})$ is the number of photons in the selected field mode $|\boldsymbol{k}, \boldsymbol{e_k}\rangle$ whose angular frequency is ω. Detailed balance tells us that the reactions shown in Fig. 19.4 balance, regardless of what happens at other frequencies, other polarizations, or to other atomic states. Therefore we have in dynamic equilibrium (same $\boldsymbol{k}, \boldsymbol{e_k}$ throughout)

$$N_a\, \beta_{ab}\, n = N_b\, \beta_{ba}\, n + N_b\, \alpha_{ba}.$$

Rearranging results in

$$n = \frac{N_b\, \alpha_{ba}}{N_a\, \beta_{ab} - N_b\, \beta_{ba}} = \frac{\alpha_{ba}/\beta_{ba}}{\dfrac{\beta_{ab}}{\beta_{ba}}\dfrac{N_a}{N_b} - 1} = \frac{\alpha_{ba}/\beta_{ba}}{\dfrac{\beta_{ab}}{\beta_{ba}}\, e^{\hbar\omega_{ba}/k_{\mathrm{B}}T} - 1}. \tag{19.18}$$

The Planck distribution for photons is $n_{\mathrm{bb}} = (e^{\hbar\omega_{ba}/k_{\mathrm{B}}T} - 1)^{-1}$; equating coefficients we obtain

$$\beta_{ab} = \beta_{ba} = \alpha_{ba}, \tag{19.19}$$

[32] The expanded notation of (19.20) acts as a reminder that all three reaction coefficients in (19.17) and (19.19) involve precisely the same field mode $|\boldsymbol{k}, \boldsymbol{e_k}\rangle$: detailed balance again.

or, in full,[32]

$$\beta_{a,b,\boldsymbol{k},\boldsymbol{e_k}} = \beta_{b,a,\boldsymbol{k},\boldsymbol{e_k}} = \alpha_{b,a,\boldsymbol{k},\boldsymbol{e_k}}. \tag{19.20}$$

The coefficients $\beta_{ab}, \beta_{ba}, \alpha_{ba}$ must all depend upon angular frequency ω according to the lineshape function $g(\omega - \omega_{ba})$ of (19.11). Within (19.20), this is handled by the dependence upon $|\boldsymbol{k}|$. Problem 19.3 shows how $g(\omega - \omega_{ba})$ is accounted for by a sum over $\boldsymbol{k}, \boldsymbol{e_k}$:

$$A_{ba}\, g(\omega - \omega_{ba}) = \frac{V\,\omega^2}{2\pi^2 c^3} \int \frac{\mathrm{d}\Omega}{4\pi} \sum_{\boldsymbol{e_k}} \alpha_{b,a,\boldsymbol{k},\boldsymbol{e_k}}. \tag{19.21}$$

The full expression for $\alpha_{b,a,\boldsymbol{k},\boldsymbol{e_k}}$, not averaged over photon directions or polarizations, is given in (32.50).

19.8.1   New insights

Statements (19.20) are much "stronger" than those of (19.9) because they tell us about reactions involving individual field modes, and not just about sums/averages over those reactions.

We have achieved even more. We demonstrate this by writing

$$\frac{\text{rate of emission}}{\text{rate of absorption}} = \frac{N_b}{N_a}\frac{\beta_{ba}n + \alpha_{ba}}{\beta_{ab}n} = \frac{N_b}{N_a}\frac{(n+1)}{n}. \tag{19.22}$$

In the numerator, the n deals with stimulated emission, while the 1 deals with spontaneous emission.

To see why (19.22) is particularly insightful, we put together two statements from Chapter 31:

- The quantum operator $\widehat{\boldsymbol{E}}$ for the electric field, given in (31.24), contains operators $\hat{a}^+$ and $\hat{a}$ that add and remove photons to/from the radiation field.[33]

- The matrix elements for these operators are, in the present notation,

$$\langle n+1|\hat{a}^+|n\rangle = \sqrt{n+1}, \qquad \langle n-1|\hat{a}|n\rangle = \sqrt{n}.$$

A transition probability contains the matrix element squared, so it is wholly expected that the creation of a new photon should have $(n+1)$ in its reaction rate, while photon destruction should have n appearing in a similar way. These factors are on display in (19.22) and in Fig. 19.5.

At this point, we have aligned our notation and our results with those of quantum electrodynamics. Well, we did say that's what the Einstein trick is being used for .... The link just given may also make it easier for us to separate our findings (19.20) and (19.22) from any lingering association with thermal equilibrium or black-body radiation—or even detailed balance.

One motivation for setting up our presentation in §§ 19.2–4, using $B'n$ rather than $B\rho$, was to anticipate the shape of (19.22). The link with QED could not be made earlier because too many field modes were being summed over. But it did no harm to hint at results to come.

The question is often asked: "why do stimulated photons go the same way" as the stimulating photons? We may explain this in either of two ways—which are in fact equivalent. The first is to remember Fig. 19.4: the $\beta_{ba}n$ entered there gives the rate of filling of field mode $|\boldsymbol{k}, \boldsymbol{e_k}\rangle$ when the n stimulating photons arrive in the *same* mode $|\boldsymbol{k}, \boldsymbol{e_k}\rangle$. At the time this was argued for using detailed balance.

The second explanation is to note that the matrix element

$$\langle n+1|\hat{a}^+|n\rangle = \langle n+1, \boldsymbol{k}, \boldsymbol{e_k}|\hat{a}^+_{\boldsymbol{k},\boldsymbol{e_k}}|n, \boldsymbol{k}, \boldsymbol{e_k}\rangle$$

is non-zero only when the initial and final field states $|\boldsymbol{k}, \boldsymbol{e_k}\rangle$ are the same.

Equivalent? What we have just seen is really a demonstration from quantum mechanics that detailed balance holds for the reaction under discussion.

[33] The interaction between radiation and an atom involves operator $(e/m_e)\widehat{\boldsymbol{A}}\cdot\hat{\boldsymbol{p}}$, though we may be more used to seeing it given as $e\widehat{\boldsymbol{E}}\cdot\hat{\boldsymbol{r}}$. But $\widehat{\boldsymbol{A}}$, just as much as $\widehat{\boldsymbol{E}}$, contains operators $\hat{a}^+$ and $\hat{a}$. See (32.5) and (32.6).

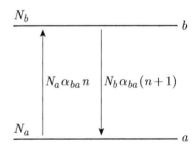

Fig. 19.5 The transition rates of Fig. 19.4 after making use of (19.19). The rates displayed are those to/from a single chosen mode of the radiation field—not necessarily at the resonance frequency; the lineshape function is hidden within $\alpha_{ba} = \alpha_{b,a,\boldsymbol{k},\boldsymbol{e_k}}$. The quantity n is the *initial* number of photons occupying the selected field mode $|\boldsymbol{k}, \boldsymbol{e_k}\rangle$—that is, before the absorption or emission takes place.

19.9 Summary and comments

Equation (19.19) shows that all three reactions (absorption, stimulated emission, spontaneous emission) involve the exact-same field modes $\boldsymbol{k}, \boldsymbol{e_k}$. This is, of course, why a laser works.

Equation (19.22), and the entries on Fig. 19.5, link the Einstein coefficients with results of quantum electrodynamics—a link that is rather thoroughly concealed in the conventional $B\rho$ presentation.

Elementary (semi-classical) calculations can give us the value of B_{ab} and hence A_{ba}. But finer details, α_{ba} and the spectral line profile $g(\omega - \omega_{ba})$, mostly require use of advanced methods (Chapter 32).[34]

The discussion of this chapter may have caused surprise by what was *not* said. The existence of the stimulated-emission process has been taken for granted, and not introduced as something new, something unexpected, something that Einstein was clever enough to invent. I take the view that, given what you already know of quantum mechanics, stimulated emission is no surprise, but something wholly to be expected.

Let me explain. Put our atom into electromagnetic radiation. If we start with the atom in state Ψ_a, interaction with the radiation (part of $\widehat{H}$) will cause Ψ to change with time[35] and to contain an increasing admixture of Ψ_b. This is the beginning of an absorption from state a to state b. Likewise, if we start with the atom in state Ψ_b, Ψ will acquire an admixture of Ψ_a: the beginning of a downward transition from state b to state a. The matrix elements appearing in the transition rates[36] are related by $\langle a|\widehat{H}|b\rangle = \langle b|\widehat{H}|a\rangle^*$. We cannot sensibly imagine absorption $(a \to b)$ without its being accompanied by stimulated emission $(b \to a)$.

Given this understanding of basic quantum mechanics, it was natural and inevitable that both B'_{ab} and B'_{ba} were included in the annotations on Fig. 19.2. Indeed, it is *spontaneous* emission that is hardest to understand, simply because it lies outside a semiclassical treatment. If anything was "patched" into Fig. 19.2, it was spontaneous emission.

What it meant to Einstein

It can be surprising to find out that the Einstein reasoning of §19.3 has been understood in very different ways at different times. Here we make a "for interest" foray into the history.

My description here is based on Pais (1982), in particular p. 407. In 1917 there were two big unsolved problems in physics: the spectral lines emitted/absorbed by atoms and molecules (why did they have the observed frequencies);[37] and the spectrum of black-body radiation (which was rather well described by Planck's law, but no one understood why). The Einstein reasoning "established a bridge between black-body radiation and Bohr's theory of spectra". In other words, these could no longer be thought of as two independent unsolved problems; they must be aspects of a bigger unsolved problem. Seeking a solution to either required the ambition to construct a "1917 theory of everything". We now call that solution quantum mechanics and quantum electrodynamics.

[34] The reader may object: A Lorentzian line shape (for "natural" broadening) follows directly from knowledge that the upper atomic state b decays exponentially with time, as $e^{-\Gamma t}$. Why should advanced methods be called for? Well, a detailed QED investigation (Chapter 32) shows that the decay of the upper state is not exactly exponential, and therefore $g(\omega - \omega_{ba})$ is not given exactly by a Lorentzian. The exponential/Lorentzian approximation is very good, but it shouldn't be assumed as though it was obvious.

[35] It is being assumed that $\widehat{H}$ has non-zero matrix elements linking state a with state b.

[36] Incidentally, it is the equality of things like $|\langle a|\widehat{H}|b\rangle|^2$ and $|\langle b|\widehat{H}|a\rangle|^2$ that is the quantum-mechanical reason for the principle of detailed balance.

[37] It was not even accepted that a "light-quantum"—if it existed—had energy $\hbar\omega$ related to the difference between two quantized atomic energies. And anyway, other than the Bohr theory for hydrogen (only four years old), there was no means of calculating the energy levels of atoms.

What it means to us now

We do not *need* the Einstein reasoning at all: coefficients B'_{ab}, B'_{ba} and A_{ba} (or $\beta_{ab} = \beta_{ba} = \alpha_{ba}$) come out from quantum electrodynamics without any need for us to mention thermal equilibrium, or black-body radiation, or detailed balance.

Nevertheless, a proper QED calculation is lengthy (Chapter 32). The opportunity to bypass it remains as helpful as it ever was.

Problems

Problem 19.1 Doppler broadening

In §§ 19.3 and 19.8 we said that we could equate coefficients because all dependence upon temperature T was on display. If the atoms are in a gas they move and their spectral line profile is broadened by Doppler broadening, giving a line profile dependent upon temperature. Does this upset the reasoning that leads to (19.9) and/or (19.19)?[38]

Problem 19.2 The semi-classical calculation for B'_{ab}

(1) The Hamiltonian $\widehat{H}'$ for interaction between radiation (vector potential $\boldsymbol{A}$) and an electron (momentum $\hat{\boldsymbol{p}}$) is[39]

$$\widehat{H}' = \frac{e}{2m_e}(\boldsymbol{A} \cdot \hat{\boldsymbol{p}} + \hat{\boldsymbol{p}} \cdot \boldsymbol{A}).$$

Take $\boldsymbol{A}$ as a sum over plane-wave modes, and show that then $\boldsymbol{A}$ commutes with $\hat{\boldsymbol{p}}$, so that $\widehat{H}' = (e/m_e)\boldsymbol{A} \cdot \hat{\boldsymbol{p}}$.

(2) Let a representative field mode have electric field $\boldsymbol{E}_0 \cos(\omega t - \boldsymbol{k} \cdot \boldsymbol{r})$. Show that[40]

$$\boldsymbol{A} = \frac{\boldsymbol{E}_0}{2\mathrm{i}\omega}\left(e^{-\mathrm{i}(\omega t - \boldsymbol{k} \cdot \boldsymbol{r})} - e^{\mathrm{i}(\omega t - \boldsymbol{k} \cdot \boldsymbol{r})}\right).$$

(3) Obtain eqn (19.12). You will need to find a reason why $\langle b|\widehat{H}'|b\rangle = 0$.

(4) Equation (19.12) for $\dot{c}_b$ includes

$$\langle b|\widehat{H}'|a\rangle \text{ containing } \int \psi_b^* \, e^{\pm \mathrm{i}\boldsymbol{k} \cdot \boldsymbol{r}} \, \hat{\boldsymbol{p}} \, \psi_a \mathrm{d}V.$$

Argue that $\psi_b^* \, \hat{\boldsymbol{p}} \, \psi_a$ extends over the volume of the atom, over distances of order a_0, the Bohr radius. Argue that $e^{\pm \mathrm{i}\boldsymbol{k} \cdot \boldsymbol{r}}$ approximates to 1 until $\boldsymbol{k}$ reaches values of order $1/a_0$, after which the exponential oscillates and cancels against itself in the integral. Show that when $|\boldsymbol{k}| = 1/a_0$ the angular frequency $\omega \sim c/a_0$, which is far above the resonance frequency for the transition of interest. Therefore we can replace $e^{\pm \mathrm{i}\boldsymbol{k} \cdot \boldsymbol{r}}$ with 1 as a good approximation.[41]

(5) Confirm the integration over time leading to eqn (19.13) for c_b. Investigate the term containing $e^{\mathrm{i}(\omega_{ba}+\omega)t}$, and confirm the validity of the rotating-wave approximation by which this term is discarded.

(6) Rehearse the atomic-physics theorem that $\langle b|\boldsymbol{p}|a\rangle = \mathrm{i}m_e \omega_{ba} \langle b|\boldsymbol{r}|a\rangle$.

(7) Check the reasoning in § 19.6 by which the field energy density involved in a transition is $\frac{1}{3}\rho(\omega)\,\mathrm{d}\omega$, with the $\frac{1}{3}$ coming from imposing an isotropic field on an oriented atom.

[38] *Hint:* For the reasoning of (19.5)–(19.9) to work, we need some realistic arrangement of atoms in equilibrium to obey a Boltzmann distribution. There is no need to be distracted by the fact that some atoms under some conditions behave otherwise. Think about this.

Can this be turned round to put a restriction on the temperature dependence of Doppler broadening?

[39] This Hamiltonian is non-relativistic so far as the electron is concerned.

[40] We are using the Coulomb gauge (radiation gauge) in which the scalar potential is zero, div $\boldsymbol{A} = 0$, and $\boldsymbol{E} = -\partial \boldsymbol{A}/\partial t$.

A different choice of phase is made in Chapter 32, eqn (32.5).

[41] This step is part of the **electric-dipole approximation**. The exponential must be treated more carefully if the $a \leftrightarrow b$ transition is "forbidden"; it must then be taken to a next approximation, perhaps accounting for an electric-quadrupole transition.

When $e^{\pm \mathrm{i}\boldsymbol{k} \cdot \boldsymbol{r}}$ is retained in the electric-dipole case, it has the effect of reducing the matrix elements for $\omega \gtrsim c/a_0$. This may seem unimportant, but it happens that some integrals over ω seem to diverge at large ω; some instances are encountered in Chapter 32. It is then necessary, even for electric-dipole transitions, to know that the matrix elements fall off at high frequencies, as convergence is rescued by taking account of that fall-off.

Problem 19.3 The link between α_{ba} and $g(\omega - \omega_{ba})$

Take a special case of eqns (19.17) in which $n(\boldsymbol{k}, \boldsymbol{e_k})$ is the same for all field modes $|\boldsymbol{k}, \boldsymbol{e_k}\rangle$ within the angular-frequency range $d\omega$. Then this $n(\boldsymbol{k}, \boldsymbol{e_k})$ is the same as the $n(\omega)$ of (19.11).

Let N_a atoms in atomic state a be illuminated by this radiation field. Show that the rate at which atoms fill an upper state b is

$$N_a \sum_{\text{modes } |\boldsymbol{k}, \boldsymbol{e_k}\rangle \text{ within } d\omega} \beta_{a,b,\boldsymbol{k},\boldsymbol{e_k}}\, n(\boldsymbol{k}, \boldsymbol{e_k}) = N_a\, n(\omega) \sum_{\boldsymbol{k},\boldsymbol{e_k}} \beta_{a,b,\boldsymbol{k},\boldsymbol{e_k}}.$$

Here β depends on the direction (ϑ, φ) of vector $\boldsymbol{k}$. Show that the rate of filling of atomic state b by photon absorption is[42]

$$N_a\, n(\omega) \frac{V \omega^2\, d\omega}{2\pi^2 c^3} \int \frac{d\Omega}{4\pi} \beta_{a,b,\boldsymbol{k},\boldsymbol{e_k}} = \text{also } N_a\, n(\omega)\, B'_{ab}\, g(\omega - \omega_{ba})\, d\omega.$$

Then

$$B'_{ba}\, g(\omega - \omega_{ba}) = \frac{V \omega^2}{2\pi^2 c^3} \int \frac{d\Omega}{4\pi} \beta_{b,a,\boldsymbol{k},\boldsymbol{e_k}}. \tag{19.23}$$

Summing over angular frequencies further gives

$$B'_{ba} = B'_{ba} \int g(\omega - \omega_{ba})\, d\omega = \int \frac{V \omega^2\, d\omega}{2\pi^2 c^3} \int \frac{d\Omega}{4\pi} \beta_{b,a,\boldsymbol{k},\boldsymbol{e_k}}.$$

We may recast this as a relation linking the A_{ba} coefficient to the αs

$$A_{ba} = \int \frac{V \omega^2\, d\omega}{2\pi^2 c^3} \int \frac{d\Omega}{4\pi} \alpha_{b,a,\boldsymbol{k},\boldsymbol{e_k}}. \tag{19.24}$$

Equation (19.24) is a consistency check, showing how A_{ba} is accounted for by summing over all radiative transitions (all photon destinations) that contribute to a spontaneous emission.[43]

Problem 19.4 Spontaneous emission of phonons

The Einstein A coefficient for spontaneous emission, in an electric-dipole transition, from state $|b\rangle$ to state $|a\rangle$ is given by (19.16):

$$A_{ba} = \frac{4}{3} \frac{e^2}{4\pi\epsilon_0} \frac{\omega_{ba}^3}{\hbar c^3} |\langle b|\hat{\boldsymbol{r}}|a\rangle|^2, \tag{19.16}$$

in which ω_{ba} is the angular frequency of radiation emitted (at line centre) in the transition from state b to state a.

(1) From this, estimate (by scaling) the radiative lifetime at $10\,\text{GHz}$ for (a) electric-dipole emission, (b) magnetic-dipole emission.[44]

(2) In a study of magnetic resonance in solids, we may put atoms into an excited state separated from a lower state by an energy equivalent to $10\,\text{GHz}$. The atoms descend by emitting phonons (relaxation time T_1), and the lifetime can be microseconds to seconds, not days. How can it be that phonon emission is so much more probable?
[Simple answer, based on what you can find in (19.16).]

(3) Given the weakness of magnetic-dipole transitions, we might expect that it would be difficult to prepare a radio-frequency $\boldsymbol{B}$-field strong enough to induce absorption or stimulated-emission reactions. In fact, it's quite easy. Why?

---

[42] Given that a, b are individual atomic states, there is only one polarization $\boldsymbol{e_k}$ associated with any $\boldsymbol{k}$ that can induce an absorption. We no longer sum over $\boldsymbol{e_k}$, and we do not introduce a factor 2 as was done in (19.3).

[43] Compare with (32.83).

[44] Assume that a typical visible-frequency transition has lifetime $\sim 10^{-8}\,\text{s}$. A magnetic-dipole transition is typically slower than an electric-dipole transition by a factor $\sim \alpha_{\text{fs}}^2$ where α_{fs} is the fine-structure constant.

IX Statistical mechanics

20 The Boltzmann distribution and molecular gases

Intended readership: towards the end of a first exposure to statistical mechanics.

20.1 Introduction: why Boltzmann?

Every atom or molecule is either a boson or a fermion. A gas of bosons obeys a Bose–Einstein distribution. A gas of fermions obeys a Fermi–Dirac distribution. So far, so clear. Yet "ordinary" gases are always described by giving them a Boltzmann distribution. Why?

We have no reason to doubt the correctness of the B–E or F–D distributions. Photons and phonons clearly obey a Planck distribution (a special case of B–E). Electrons in metals exhibit a F–D distribution. So what is it about ordinary atomic or molecular gases?

The answer is that an ordinary atomic or molecular gas always does obey B–E or F–D statistics, but the gas has such a low number density that the distribution is well approximated by Boltzmann. Of course, it isn't enough to say this: we must demonstrate that it's the case. The present chapter supplies that demonstration, which seems to be hard to find elsewhere.

Here is a paradox.[1] To show that Boltzmann always applies, the right thing to do first is to try as hard as we can to show that it does not.

If we give it our best shot, and we fail, then it will be demonstrated that Boltzmann indeed always does apply—as a good approximation, of course, not as an in-principle replacement for Fermi–Dirac or Bose–Einstein.

20.2 The Fermi–Dirac case: the gas ^{3}He

Let's start with the Fermi–Dirac case. We want to find a gas whose number density is so high that it's seriously Fermi–Dirac, rather than Boltzmann. To get the number density high, we must lower the temperature—without turning the gas into a liquid. So we look for the fermion gas with the lowest possible boiling temperature: ^{3}He.

[1] Actually it's not a paradox at all. On any occasion when we come up with a new idea, the first thing we should do is to look for evidence—not evidence in favour but evidence **against**. Every question has thousands of wrong answers and only one or two right answers. The overwhelming probability is that we've got it wrong, and in that case the sooner we find out the better. Moreover, the way in which we were wrong may offer clues as to what we should do instead.

This idea is often said to encapsulate "scientific method" but it's what ought to be done in any field of enquiry.

Essays in Physics: Thirty-two thoughtful essays on topics in undergraduate-level physics. Geoffrey Brooker, Oxford University Press. © Geoffrey Brooker 2021. DOI: 10.1093/oso/9780198857242.003.0020

At atmospheric pressure, ^{3}He boils at $3.19\,\mathrm{K}$. So we shall consider ^{3}He vapour at atmospheric pressure and at temperature $3.19\,\mathrm{K}$.

The Fermi–Dirac distribution tells us that the number of atoms per quantum state is[2]

$$n(\varepsilon_i) = \frac{N_i}{g_i} = \frac{1}{e^{(\varepsilon_i - \mu)/k_B T} + 1} = \frac{1}{A\,e^{\varepsilon_i/k_B T} + 1}, \qquad (20.1)$$

where ε_i is the energy of the state and A has been written as an abbreviation for $e^{-\mu/k_B T}$. If $A \gg 1$ then the $+1$ in the denominator is swamped by the first term,[3] and (20.1) reduces approximately to a Boltzmann distribution; and if not, not. So our interest will be focused on the numerical value of A.

Let there be N atoms in volume V at temperature T. Statistical mechanics gives us[4]

$$\frac{N}{V} = \int_0^\infty \frac{1}{A\,e^{\varepsilon/k_B T} + 1}\, 2\, \frac{p^2\,\mathrm{d}p}{2\pi^2 \hbar^3}. \qquad (20.2)$$

The factor 2 takes account of the facts that: the ^{3}He nucleus is a fermion with spin $\tfrac{1}{2}$; the atomic electrons (ground state) have zero total angular momentum; and so the whole atom has two spin orientations. The coefficient A is determined physically, as is the chemical potential μ (to which it is related), by the requirement that the gas contains the correct number N of atoms.[5] This is always the way in which the physics fixes the value of μ.

We change the variable of integration from p to $x = \varepsilon/k_B T$ and (20.2) reduces (problem 20.1) to

$$\int_0^\infty \frac{x^{1/2}}{A\,e^x + 1}\, \mathrm{d}x = \frac{N}{V}\, \frac{2\pi^2 \hbar^3}{(2mk_B T)^{3/2}}$$

$$= \frac{Nk_B T}{V}\, \frac{2\pi^2 \hbar^3}{(2m)^{3/2}\,(k_B T)^{5/2}}$$

$$= p_{atm}\, \frac{2\pi^2 \hbar^3}{(2m)^{3/2}\,(k_B\,T)^{5/2}}. \qquad (20.3)$$

Here m is the mass a ^{3}He atom and p_{atm} is atmospheric pressure.

Notice the $T^{5/2}$ in the denominator. It explains, if it was not obvious before, why in seeking to attain high number density we should choose a gas that permits the smallest possible T. (A small atomic mass m helps too, but not so strongly.)

In principle, (20.3) holds at any temperature at which the ^{3}He is gaseous, and it is an implicit equation giving A as a function of T. Nevertheless our interest is in applying it at the ^{3}He boiling temperature.

Put numbers in for yourself, and show that the right-hand side of (20.3) evaluates to 0.1818. Thus we have to solve for A in the implicit equation

$$\int_0^\infty \frac{\sqrt{x}}{A\,e^x + 1}\, \mathrm{d}x = 0.1818. \qquad (20.4)$$

It's possible to make an analytical reduction of (20.4) (problem 20.2), to the point where we can solve by successive approximation using an

[2] Here N_i is the number of atoms (on average) occupying a bundle of g_i quantum states with energies close to ε_i. The fraction N_i/g_i is supplied as a reminder of the kind of expression that the reader has probably worked out on the way to the Fermi–Dirac formula.

[3] We are using an origin of energy such that $\varepsilon_i \geqslant 0$, hence $A\,e^{\varepsilon_i/k_B T} \geqslant A$. The $+1$ is swamped even more when $\varepsilon_i > 0$.

[4] In (20.2), p is the momentum of an atom in the gas.

[5] We may say that μ (or A) is the only adjustable parameter on the right of (20.2), and we must have an adjustable parameter to make the integral evaluate to an N/V that is being imposed by us. Compare with the evaluation of the chemical potential for electrons in a solid; see Chapter 26.

ordinary pocket calculator. But there's no reason, with today's opportunities, not to bring out the big guns such as Mathematica.

In using powerful software, we can be more or less sophisticated. Here's a very simple program written for Mathematica:

```
a=3
NIntegrate[ Sqrt[x]/(a * Exp[x] + 1), {x, 0, Infinity} ]
```

In this we give a $= A$ a trial value, here 3, and ask for the numerical value of the integral (it comes out to be 0.2658, too big). We can easily replace this value of A by others until we have zeroed in on $A = 4.53$ which is the required result.[6]

You will have noticed (!) that the temperature of $3.19\,\mathrm{K}$ has been input to only three figures, and is raised to the $5/2$ power; so it is not justified to give more than three figures in the result for A.

Remember what we have come for. Return to (20.1), which now reads

$$n(\varepsilon) = \frac{1}{4.53\,\mathrm{e}^{\varepsilon/k_{\mathrm{B}}T} + 1}. \tag{20.5}$$

In this, the denominator takes its smallest value of 5.53 for $\varepsilon = 0$. If we discard the $+1$, we at worst replace 5.53 with 4.53, an error of 18%. Now this may not sound brilliant, but we have to remember that every physical quantity (internal energy, entropy, ...) is obtained by integrating after the fashion of (20.2). For all energies except zero, the error is smaller than 18%. We can guess that these integrals are good to around 10% or so.[7]

Remember also that the ^{3}He is on the point of liquefying at the temperature and pressure we have chosen, yet we have assumed it to be an ideal gas. There will be departures from ideality because of interatomic interactions, as well as those arising from the Fermi–Dirac statistics, and it wouldn't surprise us if those departures were also of order 10%. So we'd have some difficulty teasing out the "statistics" contribution from measured data. It seems best to say that 10% is the order of the unavoidable uncertainty in anything we calculate, for these conditions. And replacing Fermi–Dirac with Boltzmann really doesn't make anything worse.

20.3   The Bose–Einstein case: the gas ^{4}He

The densest boson gas we can think of is ^{4}He at its boiling temperature of $4.215\,\mathrm{K}$. So we perform an analysis similar to that used on ^{3}He.

The Bose–Einstein distribution is

$$n(\varepsilon_i) = \frac{N_i}{g_i} = \frac{1}{\mathrm{e}^{(\varepsilon_i-\mu)/k_{\mathrm{B}}T} - 1} = \frac{1}{B\,\mathrm{e}^{\varepsilon_i/k_{\mathrm{B}}T} - 1}, \tag{20.6}$$

where $B = \mathrm{e}^{-\mu/k_{\mathrm{B}}T}$.

The constant B is determined, as was A, by requiring that the correct number of atoms is present within the volume V. Thus

$$\frac{N}{V} = \int_0^\infty \frac{1}{B\,\mathrm{e}^{\varepsilon/k_{\mathrm{B}}T} - 1} \frac{p^2\,\mathrm{d}p}{2\pi^2\hbar^3}.$$

[6] The experienced programmer will see at once how to refine this program. We may plot a graph of the left-hand side of (20.4) as a function of A. We may use FindRoot[ ] to automate the search for the solution. And so on. But it doesn't take long to do a by-hand search for A, so refinement isn't worth it.

[7] Problem 20.1(2) shows that the error made in (20.2) by discarding the $+1$ is 7.5%.

There is no factor 2 now, because the ^{4}He nucleus has spin zero. Rearrange the integral for yourself to show that it can be shaped as:

$$\int_0^\infty \frac{\sqrt{x}}{B\,e^x - 1}\,\mathrm{d}x = p_{\mathrm{atm}} \frac{4\pi^2 \hbar^3}{(2m)^{3/2}\,(k_{\mathrm B}T)^{5/2}} = 0.1185. \qquad (20.7)$$

Here m is now the mass of a ^{4}He atom, and T is the boiling temperature of 4.215 K. Solution by Mathematica gives $B = 7.84$.

Return to (20.6), which has become (for this particular temperature)

$$n(\varepsilon) = \frac{1}{7.84\,e^{\varepsilon/k_{\mathrm B}T} - 1}. \qquad (20.8)$$

This would be a Boltzmann distribution if we could discard the -1 in the denominator. That would at worst change the denominator from 6.84 to 7.84, an error of 15%. This is a smaller error than the 18% that we found for ^{3}He. Thus ^{4}He is "more classical" than ^{3}He, in the sense that it is better approximated by a Boltzmann distribution.[8]

Problem 20.3 asks you to find out what happens for hydrogen vapour at its boiling temperature, and to show that hydrogen is even closer to having a Boltzmann distribution. This happens mainly because of the increased temperature, from 3 or 4 K to 20 K. Beside this, the smaller mass (2 instead of 3 or 4 daltons) is not very significant. And it should be obvious now that every other gas is even more closely Boltzmann.

We have now carried out the programme forecast at the beginning of this chapter. We have tried our hardest to find a gas that is significantly non-Boltzmann, and we've given it our best shot. Yet even ^{3}He does not show convincing departures from Boltzmann behaviour—unless you're a specialist who is prepared to put a lot of effort into analysing what's going on.

Conclusion: all "molecular" gases[9]—gases that a chemist would call a gas—can be described within statistical mechanics by means of a Boltzmann distribution, even though the underlying distribution must be Fermi–Dirac or Bose–Einstein. Everyone tells us this is the case, but the demonstration that it is so has required the "experimental arithmetic" given here.

[8] Problem 20.1(3) shows that the integral in (20.7) is changed by 4.6% when we discard the -1 in the denominator. By this test too, ^{4}He is "more classical" than ^{3}He.

[9] The point of "molecular" is not to insist on molecules—after all, the helium gases are monatomic. The idea is to exclude gases such as electrons in metals, or photons. Also excluded here are Bose–Einstein condensates, which require a completely different limiting case to be taken.

Problems

Problem 20.1 Obtaining (20.3); how wrong is Boltzmann?

(1) Do the manipulation that obtains (20.3) from (20.2). Evaluate the right-hand side of (20.3), and confirm the numerical value in (20.4).

(2) Using the solution $A = 4.53$, evaluate the left-hand side of (20.4) with the $+1$ in the denominator discarded. Show that the integral works out to $\sqrt{\pi}/(2\times 4.53) = 0.195$, differing from 0.1818 by 7.5%. A representative integral over the Fermi–Dirac distribution, with the $+1$ neglected, is thus shown to be in error by less than 10%.

(3) Similarly, obtain eqn (20.7) and show that the numerical value for its right-hand side is 0.1185. Using the solution $B = 7.84$, evaluate the

integral in (20.7) with the -1 discarded from the denominator, and show that the result is changed from 0.1185 by 4.6%.

Problem 20.2 Obtaining A for ^{3}He "by hand"
It was stated in the text that (20.4) can be solved with only an ordinary calculator, so there is no need to call upon the big guns. Let's show that this is possible.

Rewrite (20.4) in the form

$$\int_0^\infty \frac{\sqrt{x}}{A\,\mathrm{e}^x + 1}\,\mathrm{d}x = X, \tag{20.9}$$

where X stands for the right-hand side, known to be 0.1818.

(1) Ask the following: what would happen if A were so large that the $+1$ in the denominator could be safely discarded. Then we'd have

$$X = \int_0^\infty \frac{\sqrt{x}}{A\,\mathrm{e}^x}\,\mathrm{d}x = \frac{1}{A}\int_0^\infty \sqrt{x}\,\mathrm{e}^{-x}\,\mathrm{d}x.$$

To evaluate the integral, make the substitution $x = y^2$. Show that the outcome is $X = \sqrt{\pi}/(2A)$ or $A = \sqrt{\pi}/(2X) = 4.87$.

(2) Now the result of part (1) is not quite right, but it isn't too far off either. We can look for a way of improving it. If the $+1$ makes a small contribution to the denominator, we should be able to treat it as a small correction. Do a power-series expansion.[10]

[10] Don't be tempted to expand in powers of $A\,\mathrm{e}^x$, because this is greater than 1 and the series won't converge. You must expand in powers of something small—meaning smaller than what accompanies it.

$$X = \int_0^\infty \frac{\sqrt{x}}{A\,\mathrm{e}^x}\,\frac{1}{1 + A^{-1}\mathrm{e}^{-x}}\,\mathrm{d}x$$

$$= \frac{1}{A}\int_0^\infty \sqrt{x}\,\mathrm{e}^{-x}\left(1 - \frac{1}{A}\mathrm{e}^{-x} + \frac{1}{A^2}\mathrm{e}^{-2x} - \frac{1}{A^3}\mathrm{e}^{-3x} + \cdots\right)\mathrm{d}x.$$

(3) Look at the terms in the series. The first term we've already integrated by making the substitution $x = y^2$. The second integral can be evaluated in the same way by setting $2x = y^2$; and the third by setting $3x = y^2$. Then all the integrals become the same except for some tidying-up factors. Show that

$$X = \frac{1}{A}\frac{\sqrt{\pi}}{2}\left(1 - \frac{1}{2^{3/2}A} + \frac{1}{3^{3/2}A^2} - \frac{1}{4^{3/2}A^3} + \cdots\right).$$

[11] Remember the precept of §12.2.1: put on the right "the known and the smaller".

(4) Rearrange this into a form[11] suitable for finding A:

$$A = \frac{1}{X}\frac{\sqrt{\pi}}{2}\left(1 - \frac{1}{2^{3/2}A} + \frac{1}{3^{3/2}A^2} - \frac{1}{4^{3/2}A^3} + \cdots\right). \tag{20.10}$$

(5) Now remember exercise 12.1. As there, we have an almost-explicit equation for the unknown (there x, here A). We didn't make too bad an error at the beginning of this problem by keeping only the first term on the right of (20.10), thereby effectively setting $1/A = 0$ in the others. It can only improve things if we put a better $1/A$ into the bracket.

Our first attempt, keeping only the first term on the right of (20.10), gave $A = 4.87$. The next approximation is:

$$A = \frac{1}{X}\frac{\sqrt{\pi}}{2}\left(1 - \frac{1}{2^{3/2}4.87} + \cdots\right) = 4.52.$$

Continue this process until you have a result to a sufficient number of significant figures.

(6) Follow a similar route for the case of ^{4}He, and confirm that $B = 7.84$.

Problem 20.3 Hydrogen

The hydrogen molecule is a boson. The boiling temperature of hydrogen is about $20\,\mathrm{K}$. Find the right-hand side of the expression in (20.7) with the hydrogen values substituted in.[12] Show that hydrogen is "more Boltzmann" than ^{4}He.

[12] There is a complication with hydrogen because the combined nuclear-spin quantum number can be 0 or 1. At low temperatures you may have a low-temperature-equilibrium distribution (all molecules having spin zero) or a frozen-in high-temperature distribution (spin-1 three times as numerous as spin-0). Investigation should show that the required conclusion is not endangered by either possibility.

X Thermodynamics

Free energy

Intended readership: Towards the end of a course on thermodynamics.

This is a chapter that could (and perhaps should) appear in a textbook on thermodynamics. It provides the foundation needed for finding the energy of a electric or magnetic body in an applied field: Chapter 22.

21.1   Introduction: open systems

What is the energy of a body immersed in an externally applied magnetic field? This question will occupy us in Chapter 22. I raise the question here because it reminds us that there are many rather similar cases of "open systems": a tennis ball in a gravitational field; a quantum particle in a potential well; a body in contact with a heat bath . . . .

There is a good reason why it's possible to find our question quite teasing. Two bodies interact; who owns the interaction energy?

Here is a model to show that this is a serious question. Take two classical particles that interact: Fig. 21.1. Their kinetic energies are T_1, T_2 and their energy of interaction is Φ. The total energy of the combined system is $U = T_1 + T_2 + \Phi$. What is the energy of particle number 2? T_2? $T_2 + \Phi$? $T_2 + \frac{1}{2}\Phi$? Clearly, the only right answer to this question is "don't be silly".

And yet. Let the two bodies be the Earth and a tennis ball of mass m. If mass m is at height h, then $\Phi = mgh$. We have no hesitation in saying that this energy mgh is "the potential energy of the tennis ball". We don't agonize over the fact that the potential energy Φ is shared by both masses, because we treat the Earth and its surrounding gravitational field as "given". For this extreme case then, the common-sense energy of body 2, the tennis ball, is $(T_2 + \Phi)$.

We give a name to this common-sense energy, and call it the **free energy** of the body.

It may be clearer if we change the origin of energy. Take the Earth and its gravitational field as "given". Then bring up a body, perhaps a comet rather than a tennis ball, from infinity. When this process is finished, we shall have added to the Earth and its field:

- the energy of the body itself (its rest energy mc^2 which includes all thermal energies, and its kinetic energy if it's not stationary)
- the gravitational interaction energy $\Phi = -GMm/r$.

The free energy of the body is this added energy: the energy added to that of the Earth alone when the body is introduced.

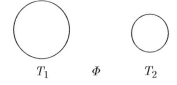

T_1 Φ T_2

Fig. 21.1 Two bodies with an interaction between them.

Essays in Physics: Thirty-two thoughtful essays on topics in undergraduate-level physics. Geoffrey Brooker, Oxford University Press. © Geoffrey Brooker 2021. DOI: 10.1093/oso/9780198857242.003.0021

[1] Well .... I've failed to find any textbook of thermodynamics that introduces the concept of free energy in just the way that I define it here (whatever name it might be given). Yet the idea is usually there, just below the surface.

One consequence is that I have had to devise my own terminology (*free energy, sample, system, controlled*) for use in this chapter and the next. The reader is warned that these terms are not standard. In particular, some authors (notably Landau–Lifshitz) use "free energy" as their name (old-fashioned now) for F.

[2] This may explain why F is sometimes known as the "Helmholtz free energy", and G as the "Gibbs free energy". These names are now frowned upon (see Royal Society (1975)), perhaps because F is correctly the free energy (my sense) only when the sample has V and T externally controlled; and G is correctly the free energy only when p and T are externally controlled. We should call them "Helmholtz energy" or "Helmholtz function" and similarly, not including the word "free".

The statements made here, as to when U, F, H or G is the free energy, are established below, in § 21.3 and examples 21.1–2.

[3] I refer to a "controlled-pressure" environment, rather than a "constant-pressure" environment, because we shall wish to impose changes on the pressure, thereby making it nonconstant. Likewise, an E-field or B-field will be "controlled" or "applied".

[4] Throughout this chapter and the next, the words *sample* and *system* are used with particular care.

We can now see how to answer the question posed at the beginning of this section. A magnetic body in a B-field has a free energy: the energy added when the body was introduced into the previously-empty B-field.

Reinventing wheels! Thermodynamics, especially as used by chemists and engineers, deals all the time with "open systems": cases where a body (which I'll henceforth call a "sample") interacts with an environment. The key to handling open systems is the concept of free energy. If a pVT sample is in contact with a heat bath (T controlled by our choice of bath), its free energy is $F = U - TS$. If it's in a controlled-pressure environment, the free energy is $H = U + pV$. If both, the free energy is G. The free energy that I introduced above, for a tennis ball or a comet, is exactly what thermodynamics means by free energy.[1,2]

Free energy is something we have been using for a long time: tennis balls. Physics might be easier to understand if the term "free energy" was used whenever we discuss a tennis-ball-type energy.

Another example: a quantum particle in a potential well. Something, not specified, supplies the potential well, and the potential energy $V(r)$ is really shared between well and particle, just like the Φ of Fig. 21.1. Even though this potential energy is in principle shared, the energy eigenvalue that we work out from the Schrödinger equation is the outcome of allocating all of the potential energy to the particle.

A surprise: When a sample interacts with a field (and I'm looking ahead to an electric or magnetic sample here), the energy you have to obtain first is the free energy. Only then will you have a securely defined function of state. Other energies, e.g. those defined in (22.5) and (22.8), are less fundamental, and—if needed at all—will be defined in some appropriate way starting from the free energy. This is the reverse of experience with pVT systems. See § 21.5.2 and Chapter 22.

21.2 Free energy defined

Use Fig. 21.2. A **sample** (meaning the body of interest) interacts with some **environment**. The whole thing is the **system**. The sample might be a tennis ball, and the environment the Earth and its gravitational field. Or the sample might be a mass of gas upon which the environment imposes a **controlled**[3] pressure and temperature. We shall always know what we wish to treat as sample and what as environment.

The environment has some "background" energy when the sample is completely absent (or removed). This is where we start. We introduce the sample, and with it some additional energy. We define[4]

$$\begin{pmatrix} \textbf{free energy} \\ \text{of the sample} \end{pmatrix} \equiv \begin{pmatrix} \text{all the energy present in the final system} \\ \text{because the sample is there.} \end{pmatrix}$$

Equivalently,

$$\begin{pmatrix} \textbf{free energy} \\ \text{of the sample} \end{pmatrix} \equiv \begin{pmatrix} \text{energy that the sample brought with it} \\ + \text{ energy we had to add in getting it in.} \end{pmatrix}$$

As suggested in Fig. 21.2, free energy may be located partly in the

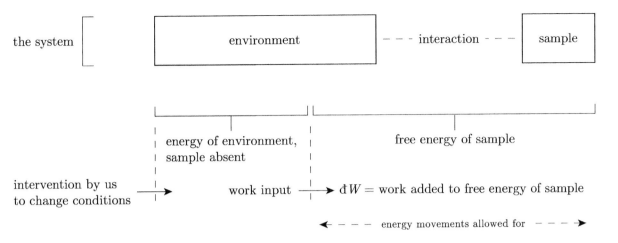

Fig. 21.2 A sample interacts with an environment. The free energy of the sample resides partly within the environment.

sample, partly in the environment, or even in no well-defined place. Where is the mgh energy of the tennis ball?[5,6]

It is likely that we shall wish to make some change to the system, including the sample. For example, we might increase the pressure of a controlled-pressure environment, or increase an applied magnetic field. Some energy is needed for changing the environment, and would be needed even in the absence of the sample; it may be large, but it's not interesting to us. Another part of the energy change, labelled đW on Fig. 21.2, adds to the free energy of the sample. We have

$$\delta(\text{free energy}) = đW. \qquad (21.1)$$

For calculation purposes, đW will always be input as mechanical work.

Equation (21.1) is more significant than it looks. During a change, there can be shifts of energy between sample and environment. The sample might expand, doing work against the atmosphere, or it might absorb heat from a heat bath. Such energy movements *do not affect the free energy*, because free energy counts all sample-related energy whether it has moved or not. It is this simplicity that motivates our defining free energy and using it as the energy quantity of choice when investigating something new. Free energy has had all kinds of distractions "designed out": the *only* thing that changes it is the đW in (21.1).

Contrast free energy with internal energy. For a pVT sample, the energy movements just discussed would, of course, entail changes of internal energy, because energy moves into and out of the sample. Working out all the gains and losses would be messy and distracting. Free energy is the user-friendly energy quantity for an open system.

These statements are brought to life in the discussions below.

- Section 21.3 gives us reassurance that the quantities U, F, H and G in the thermodynamics of pVT samples supply our free-energy needs according to the constraints imposed (on p, V, or T).

[5] The term *free* energy is a reference to the fact that the energy under consideration is not imprisoned within the boundary of the sample, but is free to spread itself round within some larger space. Free energy often has this property of being *delocalized*.

[6] A similar idea is used in a very different context. In nuclear physics, we may talk of the **separation energy** of a nucleon in a nucleus: the energy we have to give to the system to extract the chosen nucleon (leaving that nucleon at rest at infinity). The separation energy includes the kinetic energy that the nucleon had inside the nucleus, the potential energy that it had there (shared), and also the energy change that the remaining nucleons undergo as the daughter nucleus settles to its final quantum state. Our thermodynamic sample, the nucleon, in the given environment, has a separation energy that is minus the free energy.

In yet another context, it is separation energy that is being discussed when an atom's X-ray level is given energy $(Z - \sigma)^2/n^2$. This is the energy that is given to an atom to remove the specified electron. It is not the (negative of the) energy that the electron had when it was within the atom, because there is no such energy. If we were to remove two electrons, the resulting energy change would not be the sum of the two electrons' one-at-a-time separation energies.

- Examples 21.3 and 21.4 show how energy can move quite gymnastically between a sample and its environment, thereby illustrating how useful is an approach which concentrates on total energy and not on where energy is located. The examples also clarify what is meant by an "applied" field in cases where it's tempting to get it wrong.

- Chapter 22 deals in more detail with electric and magnetic systems.

21.3 Free-energy quantities for a pVT sample

Let's see how free energies are set up for a sample that's completely familiar.

21.3.1 The $+pV$ term for a sample at controlled pressure

In Fig. 21.3, a pVT sample is contained in the small cylinder. Initially, the sample is held at fixed ("controlled") volume V (the small piston is clamped to the small cylinder) and exerts pressure p on the small piston. Everything is held thermally isolated throughout, so the only inputs of energy will be in the form of mechanical work.

The large cylinder represents an environment that will in due course impose its pressure (also equal to p) on the sample; it could be used to model the atmosphere, for example. The volume of the large cylinder is fixed (for now, its piston is not allowed to move), and must be much greater than the volume V of the sample, so that inserting or removing the sample has negligible effect on the "environment" pressure.

Initially, the sample is located outside the "environment" cylinder (upper diagram). The sample (including its piston and cylinder, whose volume is negligible) is in mechanical equilibrium under balanced forces: the pressure from the environment acting downwards upon the small piston; and a force that we supply upwards from below. The space outside the large cylinder is a vacuum, so these are the only forces acting.

In the initial state, the energy possessed by the sample is its internal energy U. No energy resides anywhere else that can be anything to do with the sample.

Now push the sample slowly through the bottom wall of the "environment" cylinder. To bring the system to its final state, we input additional (sample-related) energy as mechanical work pV.

To complete the process, we unclamp the small piston, so that the sample is put into pressure equilibrium with the environment. This change in the constraints does not affect the energy of sample or environment.[7] The energy "there because the sample is there" is

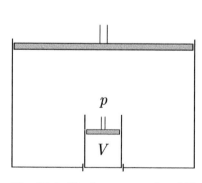

Fig. 21.3 The free energy of a pVT sample in a controlled-pressure environment is obtained by finding the work we have to do in introducing that sample into equilibrium with the environment.

[7] This is true macroscopically, but requires the more detailed examination given in § 21.3.2.

$$\begin{pmatrix} \text{free energy of sample (thermally isolated)} \\ \text{in controlled-pressure environment} \end{pmatrix} = U + pV = H. \quad (21.2)$$

Equation (21.2) tells us that the free energy consists of two pieces: the internal energy U that the sample "would have had anyway"; and $+pV$. The term $+pV$ is the "energy cost" of changing the environmental conditions round the sample, from controlled-V to controlled-p —even though the sample's p, V, T are not thereby changed.

What's important about free energy is how it changes when we make a small change to the environment. If we change the external pressure on the sample, allowing the volume V to adjust itself, we shall have[8]

$$\delta(\text{free energy}) = \delta H = \delta U + \delta(pV) = T\,\delta S - p\,\delta V + \delta(pV)$$
$$= T\,\delta S + V\,\delta p. \tag{21.3}$$

Expression (21.3) tells us the effect of imposing a change δp. Yet the algebra may look almost too glib. Can the effect of changing the externally applied pressure really be found so simply? We obtain reassurance in example 21.1, where the same result is obtained by a sequence of thermodynamic manoeuvres. We shall call manoeuvres of this type the **canonical procedure**: the clean and secure way for finding the change of a sample's free energy when environmental conditions are to be changed.

[8] The term $T\,\delta S$ is included here because it would look odd if it were omitted. However, we are holding the sample thermally isolated, so for the present discussion the $T\,\delta S$ term is zero.

Example 21.1 The $+V\,\delta p$ term in a change of free energy
Let's invent a direct proof[9] that the change of free energy, caused by changing pressure p by δp, is $+V\,\delta p$.

To change the external pressure, you need to do a large amount of work, compressing the material in the large cylinder of Fig. 21.3. Almost all of this is energy given to the environment, rather than to the sample. We need to tease out that part which is genuinely there because the sample is. Here's the recommended procedure.

- Reverse the process of §21.3.1: lock the small piston to its small cylinder, and then remove the sample completely from the "environment cylinder". Energy given to the system as work is $-pV$.

- Increase the pressure of the environment through δp by moving the large piston.[10] The work needed is clearly nothing to do with the sample; we do not trouble to evaluate it.

- Re-insert the sample, which still occupies volume V, into the environment that is now at pressure $p + \delta p$. Energy given to the system as work is $+(p + \delta p)V$.

- The small piston is subject to a small restraining force that stops it moving in pressure difference δp. Reduce this restraining force smoothly to zero until the sample is in equilibrium[11] with the new pressure $p + \delta p$. Work is input by the restraining force, but it is of second order in δp, so can be ignored.

The total sample-related work at the end of this agrees with eqn (21.3).

[9] It is not *necessary* to follow the canonical procedure; we didn't do so when obtaining (21.3). A *third* procedure is given in problem 21.1 at the end of this chapter. In problem 21.1, we change the environmental pressure with the sample still in place, moving the large piston in the lower diagram of Fig. 21.3. This is the worst of the three possibilities, and it is put on display to make just that point: to convince you that the canonical procedure is the right way to go.

[10] The small cylinder is held fixed, in the position drawn at the top of Fig. 21.3.

[11] This step may help us to understand why it was important to unclamp the small piston at the end of process of Fig. 21.3. The whole point is that the sample's relationship with its environment is being changed from controlled-V to controlled-p.

21.3.2   Thermal equilibrium and fluctuations

The reasoning of § 21.3.1 has described the sample "macroscopically", with p, V, T related via the equation of state. We must, however, look at finer detail: there are statistical fluctuations (here of p or V) about *mean* values that are set by the equation of state.

In the top diagram of Fig. 21.3, the small piston is clamped, so as to impose a fixed ("controlled") volume V on the sample. The sample exerts a pressure on the piston: the mean pressure is set by the equation of state, but the pressure is also subject to fluctuations about that mean. In the case of a gas, molecules collide with the piston, and exert pressure by exchanging momentum with it. Sometimes the number (or the violence) of the collisions is less than the average, sometimes more, so the pressure undergoes small fluctuations. For a sample of macroscopic size, these fluctuations are small in percentage terms, but it is important that they are there. In the case of a solid, there are rather similar pressure fluctuations arising from the random arrival of phonons.

Fluctuations are essential to thermal equilibrium: the dynamic equilibrium by which a sample cycles through its microstates. We cannot exert any control over pressure—either its mean value or its fluctuations—once we have decided to control V. And conversely.[12]

In the final state shown in the lower diagram of Fig. 21.3, it is pressure that is controlled externally. Volume V is free to adjust itself, to take up a mean value determined by the equation of state, and then to fluctuate slightly about that mean.[13] There has been a significant change to the sample's relationship with its surroundings, so it makes sense that there is an associated change in the sample-related energy.[14]

21.3.3   The $-TS$ term for a sample at controlled temperature

A sample starts off thermally isolated, and it is to be put into thermal equilibrium with a heat bath at temperature T_b. For the present discussion, the sample is held at constant volume, so the only inputs of energy *directly* to it are in the form of heat. At the same time, energy inputs to the *system* will be in the form of mechanical work, as always.

The first thing we did in § 21.3.1 was to invent a configuration in which the sample was completely outside (or removed from) the environment. This was the reference condition from which we started before inserting the sample, and to which we could return during example 21.1. We need to invent or discover a thermal equivalent here.

It will not do to start with the sample at the temperature T_b of the heat bath, because energy must have been moved around in getting to that condition. The reference condition must be independent of the current value of T_b. Let the sample start out thermally isolated, having temperature T_0, entropy $S(T_0)$ and internal energy $U(T_0)$.

The reference condition is shown in the upper diagram of Fig. 21.4: bath at temperature T_b and sample (isolated) at temperature T_0. We use

[12] It's obvious, but we spell it out for use later: there can never be an environment that controls both p and V.

Likewise, there can never be an environment that controls both T and S.

[13] We imagine, perhaps rather artificially, that fluctuations originating within the pressure environment are made negligible, and that the small piston is so light that its inertia does not impede changes of volume.

[14] Recapitulation. Look back at Fig. 21.3. In the initial state (top), the sample must be held at fixed volume: any fluctuation of volume would cause the small piston to invade the "environment" volume, and the sample would not be wholly decoupled from the environment. Conversely, in the final state (bottom), the sample must be allowed freely to change its volume, because that's what is meant by establishing a pressure equilibrium.

a reversible heat engine to get from here to a condition where the sample can be put into thermal equilibrium with the bath at temperature T_b. Our methods are modelled on those of § 21.3.1.

$$\begin{pmatrix}\text{Energy given as heat} \\ \text{to sample by heat engine}\end{pmatrix} = \int_{T_0}^{T_b} T\,\mathrm{d}S = U(T_b) - U(T_0).$$

A reversible heat engine conserves entropy, so

$$\begin{pmatrix}\text{entropy taken from bath} \\ \text{by heat engine}\end{pmatrix} = \text{entropy given to sample}$$
$$= S(T_b) - S(T_0);$$

$$\begin{pmatrix}\text{energy given as heat} \\ \text{to bath by heat engine}\end{pmatrix} = -T_b\{S(T_b) - S(T_0)\};$$

$$\begin{pmatrix}\text{energy } \mathrm{d}W \text{ given to} \\ \text{system as work}\end{pmatrix} = \text{net heat given out by engine}$$
$$= U(T_b) - U(T_0) - T_b\{S(T_b) - S(T_0)\}.$$

Our aim is to find all the energy that is present in the final configuration and is attributable to the sample. This consists of two pieces: the energy $U(T_0)$ which the sample had at the beginning; and the energy added as work in bringing the system to its final state. Adding these two we have:

$$\begin{pmatrix}\text{free energy of sample} \\ \text{at temperature } T_b\end{pmatrix} = \{U(T_b) - T_b\,S(T_b)\} + T_b\,S(T_0).$$

The last term here is untidy because it is dependent on the reference condition from which we started. To minimize the nuisance, choose T_0 close to absolute zero[15] so that $S(T_0) \approx 0$. Finally then,

free energy of sample at temperature T

$= $ energy in system because sample is present

$$= U - TS = F. \tag{21.4}$$

To complete the process, we put the sample at temperature T_b into thermal contact with the bath at temperature T_b. This change in the constraints leaves the energy F unaffected, but it is important in principle, just as unclamping the small piston was important as the final step in § 21.3.1.

The sample's free energy consists of two pieces: the internal energy U that the sample "would have had anyway"; and $-TS$. The term $-TS$ is the "energy cost" of changing the environmental conditions round the sample, from thermal isolation to controlled-T.

Now consider what happens if we change the temperature T of the heat bath,[16] allowing the sample to adjust by absorbing heat,

$$\delta(\text{free energy}) = \delta F = \delta U - \delta(TS) = T\,\delta S - p\,\delta V - \delta(TS)$$
$$= -S\,\delta T - p\,\delta V. \tag{21.5}$$

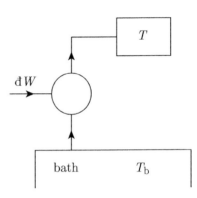

Fig. 21.4 The free energy of a sample in contact with a heat bath is obtained by finding the work $\mathrm{d}W$ that must be input to a reversible heat engine. The engine brings the sample from reference temperature T_0 to the bath's temperature T_b.

[15] The Third Law of Thermodynamics says that $S(0) = 0$: the entropy of any body in a true thermodynamic equilibrium tends to zero as the temperature approaches zero.

[16] The term $-p\,\delta V$ is included here because it would look odd if it were omitted. However, we are holding the sample at fixed volume, so for the present discussion the $-p\,\delta V$ term is zero.

Equation (21.5) tells us the change of free energy that results from changing the temperature of the environment by δT. As with the case of a pressure change δp, we must be sure that this is not misleadingly simple. Example 21.2 supplies that reassurance by pursuing a sequence of thermodynamic manoeuvres paralleling those in example 21.1.

Example 21.2 The $-S\,\delta T$ term in a change of free energy

To change the temperature of the heat bath, we need to input a large amount of heat energy, almost all of which has nothing to do with the sample. We need to tease out that part which is genuinely there because the sample is. Here is our (canonical) procedure.

- Use a reversible heat engine to bring the sample from temperature T_{b} to the reference temperature T_0.
- Raise the temperature of the heat bath from T_{b} to $T_{\mathrm{b}} + \delta T_{\mathrm{b}}$. The energy required has nothing to do with the sample, so we need not trouble to evaluate it.
- Use the reversible heat engine to bring the sample from temperature T_0 to temperature $T_{\mathrm{b}} + \delta T_{\mathrm{b}}$.
- Total the work input to the heat engine in the first and third steps this is the increase in the sample's free energy.
- The free-energy change comes out to be $-\big\{S(T_{\mathrm{b}}) - S(T_0)\big\}\delta T_{\mathrm{b}}$ (show this). Choosing $S(T_0) = 0$ removes the last term.

In example 21.1, the first step consisted in removing the sample completely from the environment: the sample had to "give back all of its volume" to the environment. The equivalent in example 21.2 is to make the sample "give back" entropy $S(T_{\mathrm{b}}) - S(T_0)$ to the environment. "Give back" may not sound like the obvious form of words, but our reasoning has shown it to be appropriate. Once this idea is articulated, it may seem less artificial[17] to set $S(T_0) = 0$.

Equilibrium and fluctuations may be discussed here in much the same way as in § 21.3.2. When the sample was in its reference condition, it was thermally isolated with fixed internal energy $U(T_0)$. In its final state, it is in thermal equilibrium with the heat bath at temperature T_{b}: small amounts of heat can pass, as fluctuations, between sample and bath. There has been a change in the sample's relationship with its environment, entailing a change in what the sample may adjust, and in how it may fluctuate. It is this change of external conditions that requires the sample-related energy to include the new term $-TS$.

21.4   Interpretation of thermodynamic potentials

It should be obvious that we can hybridize the arguments of §§ 21.3.1 and 21.3.3 to cover the case where a sample is both at controlled pressure

[17] The sample "gives back *all* of its entropy" to the environment, just as previously it "gave back *all* of its volume".

and controlled temperature. In that case, the free energy has both pV and $-TS$ added, making it

$$\left(\begin{array}{l}\text{free energy when both } T \text{ and } p \\ \text{are externally controlled}\end{array}\right) = U - TS + pV = G. \qquad (21.6)$$

21.4.1   "Controlled" variables

We develop ideas that have been hinted at in §§ 21.3.2 and 21.3.3.

Within the pair of variables (p, V), one—and only one—may be controlled by the sample's environment; the sample then adjusts the other (according to the equation of state) until equilibrium is achieved. Equilibrium permits and requires the "uncontrolled" variable to undergo small fluctuations around the equilibrium value.[18]

Let's see in a new way why δV must not appear in the change of controlled-pressure free energy. Imagine the sample to expand slightly through δV, perhaps just because of some small fluctuation. The sample (internal energy) receives work energy $-p\,\delta V$; the environment receives work energy $+p\,\delta V$. The total energy of sample and environment together does not change, to first order in δV. Of course: the *system* is in equilibrium when its total energy is a minimum, so its energy change (= the free-energy change of the *sample*) must be of second order in δV; there can be no term of first order in δV. And that's right: δH and δG contain no term in δV. Everything is looking beautifully consistent.

The pair of variables (S, T) may be discussed in similar terms to (p, V), though there are some differences of detail, discussed in the comment below.

The converse of the above. If the expression for an energy change contains δV, then that energy is a possible free energy for a sample at controlled V. The presence of δ in δV is a signal to us that V is not being allowed to fluctuate, so it must be subject to external control.

This "converse" is very powerful. If we have decided which two quantities to control, say T and p, then we may read off from (21.7) below which thermodynamic potential is the appropriate free energy: the one whose change contains δT and δp.

We catalogue in (21.7) what happens when four possible constraints are applied to a pVT sample:

$$\left.\begin{array}{ll} U & \delta U = \quad T\,\delta S - p\delta V \\ H = U + pV & \delta H = \quad T\,\delta S + V\,\delta p \\ F = U - TS & \delta F = -S\,\delta T - p\,\delta V \\ G = U - TS + pV & \delta G = -S\,\delta T + V\,\delta p \end{array}\right\} \begin{array}{l}\text{is the differential of the} \\ \text{free energy appropriate} \\ \text{to controlled}\end{array} \left.\begin{array}{ll} [S] & V \\ [S] & p \\ T & V \\ T & p \end{array}\right\}. \quad (21.7)$$

Comment on U and S

Internal energy U is the free energy of a sample held thermally isolated and at controlled volume. It meets the definition: "all the energy that is present because the sample is there". It is a free energy in a somewhat trivial sense, however, because the energy is wholly contained within the sample's volume V, and is not partly in the environment.[19]

[18] An anthropomorphic description might go: the sample fluctuates as it cycles through its microstates, always exploring small changes to make sure it still is at a free-energy minimum.

[19] This statement is particular to a pVT sample. We show in Chapter 22 that a sample in a field has a free energy (just how defined depends on what else is controlled) that is never confined within the sample's volume V.

[20] Remember the microcanonical ensemble in statistical mechanics.

[21] In Chapter 22, we find other quantities (electric dipole moment p and magnetic dipole moment m) that cannot be controlled directly by an externally applied constraint. So entropy is not exceptional in this regard.

[22] It will be clear why we have chosen to talk of "controlled" pressure, rather than "constant" pressure (for example). It makes little sense to say that pressure is held constant when discussing an equation containing δp.

[23] We may say that free energy is a **context-sensitive** quantity, meaning that its value depends upon the context—the parameters held under control by the environment.

[24] Refer to your favourite textbook on thermodynamics, and see how succinct are §§ 21.4.3 and 21.4.4.

A sample held thermally isolated does not quite have S controlled: the entropy S is permitted to fluctuate, while U is fixed.[20] Equilibrium is not a condition of minimum U, but of maximum S. Thus, when we consider fluctuations, and what is allowed to fluctuate, U (and H) does not quite fit the pattern of the other free energies.

However, to avoid always making carefully crafted statements, we let "controlled $[S]$" stand in (21.7) as proxy for "thermally isolated".[21]

21.4.2   Which quantity is the free energy?

You may feel a bit overwhelmed: a sample seems to have several different energies (U, H, F, G, ...), each with a claim to being free energy; how do you know which to work with? Hold onto this: there is only one free energy, and it is determined once you decide what in the sample's environment is to be fixed (held under control, rather than held always-constant).[22] You can't (for example) fix p and V simultaneously, so you can't be undecided between two possible free energies, one appropriate for controlled p and one for controlled V. As for having a menu of different energy functions, well, a tennis ball falling under gravity needs a different description from a tennis ball rolling on a slope. It really is as simple as that: different constraints need handling in different ways.[23]

21.4.3   Available work

A given sample is in equilibrium. Suppose it is to be brought to a new equilibrium, perhaps at a lower temperature. Then there is an amount of mechanical work that the sample could do, under ideal conditions, perhaps with the help of a reversible heat engine, as it approaches the new equilibrium. This is called the **available work**. Then

$$\text{available work} = -\text{d}W = -(\text{change of free energy}). \qquad (21.8)$$

The definition of free energy in § 21.2 makes this property trivial: the change of free energy is *defined* in terms of a possible input (or output) of reversible work. Moreover, (21.8) remains obvious whichever are the controlled variables, because free energy has a context-sensitive definition to make it so. We do not need to set up a separate argument about available work for each free-energy function (F or H or G ...) and its environmental constraints.[24]

21.4.4   Conditions for equilibrium

Thermodynamic potentials F and G (to take two cases out of four) have the following properties:

- for a sample subject to controlled V and T, $\qquad F = U - TS \qquad \rightarrow \text{minimum}$

- for a sample subject to controlled p and T, $\qquad G = U - TS + pV \rightarrow \text{minimum}$.

These statements have been obvious for a long time. Equilibrium happens when the free energy is a minimum because no work can then be extracted in getting the sample to any other state consistent with the external constraints. Just like a tennis ball rolling on a surface with a hollow.[25]

21.4.5 A short cut: Legendre transformations

We pick up an idea from § 21.4.1. Given a pair of variables that we wish to control, we may *recognize* the appropriate free energy in a list of possibilities such as (21.7). But we may do better: we may *construct* the free energy.

Imagine that we know so far of one function of state, say U whose differential is $\delta U = T \, \delta S - p \, \delta V$. Suppose we wish to find the free energy that applies to a sample in an environment that controls T and V. The free energy must be a function of state whose differential contains δT and δV. It is possible to "eyeball" that such a function is

$$U - TS, \quad \text{so that} \quad \delta(U - TS) = -S \, \delta T - p \, \delta V.$$

Even if we have never encountered $(U - TS)$ before, we deal with it now because it has the desired differentials.[26]

The operation of adding $-TS$ is known as a **Legendre transformation**. It has the effect of removing δS from the derivative and replacing it with a term containing δT.

All energy quantities in (21.7) are obtainable from each other by Legendre transformations.

I now state the general rule:[27]

> to obtain the free energy for controlled $X, Y, Z \ldots$, apply one or more Legendre transformations to a known energy function of state until you have a quantity whose differential contains δX, δY, δZ, ...

You might feel with me that this looks too glib to be true: surely it's too simple. I wouldn't be convinced until I had taken one or two examples and worked them out in full; we did, in § 21.3. The fact is that you *do* need the free energy to contain the right differentials, and that does actually fix things.[28,29]

21.4.6 Electric and magnetic dipoles

Let's look beyond the pVT samples considered so far. Suppose a sample possessing an electric dipole moment $\boldsymbol{p}$ is placed in a controlled ("applied") external electric field $\boldsymbol{E}_{\text{app}}$. The dipole moment $\boldsymbol{p}$ adjusts itself, and thereafter fluctuates slightly, as the sample comes to thermodynamic equilibrium in the given environment. The free-energy change contains $\delta \boldsymbol{E}_{\text{app}}$, the increment of the controlled quantity. Conversely, the free-energy change cannot contain $\delta \boldsymbol{p}$ because free energy must be a minimum with respect to variations of $\boldsymbol{p}$.

[25] Internal energy U is exceptional for the reasons given in the comment in § 21.4.1. Think about what happens to enthalpy H when we hold a sample thermally isolated at controlled pressure.

[26] A (change of) free energy cannot contain both δS and δT because we cannot impose external control of both S and T (compare with sidenote 12). Therefore in introducing δT we must remove completely the term containing δS. This fixes what we must add to U.

[27] Quantities X, Y, Z must be things that we can control by imposing a realistic environment on the sample. We know from the precedent of S that not all thermodynamic variables can be so controlled.

[28] The free energy for a sample under controlled T and V is $F = U - TS$. I used to wonder whether this expression should have an added constant, since the Legendre transformation yielding it is chosen by considering changes only. Fortunately, § 21.3 has shown that there is a direct physical meaning to the $+pV$ and $-TS$ terms, not just their changes, so no added constant is called for.

[29] Another example of the power of the Legendre transformation is given in Chapter 22 footnote 32.

Similar statements apply to a magnetic moment m in a controlled magnetic field B_{app}: the free-energy change may contain δB_{app}, but cannot contain δm.

In fact we have:

- controlled electric field E_{app}: $\delta(\text{free energy}) = -p \cdot \delta E_{app}$,
- controlled magnetic field B_{app}: $\delta(\text{free energy}) = -m \cdot \delta B_{app}$.

Detailed justifications are given in Chapter 22, but we can be sure of the zero coefficients of δp and δm now.

21.5 Energy and force in an electrostatic system

As we have just seen, free energy is not restricted to the handling of pVT samples. Above all, chemical reactions have their energetics determined by changes of G. Here we look at electromagnetism. We shall see how a non-pVT sample's free energy is handled, and how an "applied" field is understood. These ideas will be built upon further in Chapter 22.

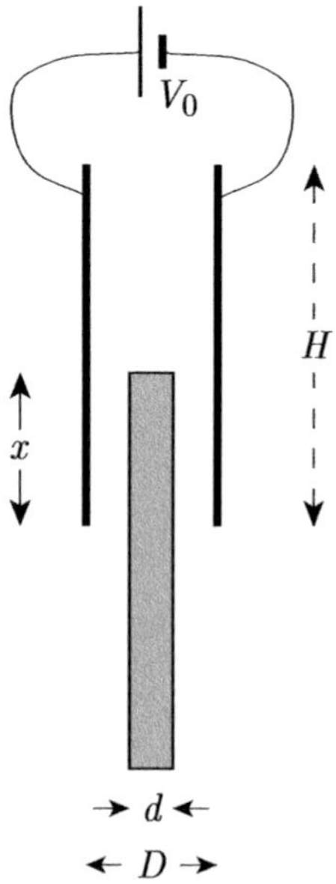

Fig. 21.5 A metal slab is supported against gravity by the electric field between capacitor plates.

Example 21.3 The force pulling a metal slab into a capacitor
A metal slab is suspended between the plates of a capacitor as shown in Fig. 21.5. A fixed potential difference V_0 is maintained across the capacitor plates by a voltage source.[30] To calculate: the upward force on the slab pulling it into the capacitor. The plates have spacing D, height H, and large width L ($L \gg D$) in the direction normal to the paper; the inserted slab has mass m, horizontal width L and thickness d. "Suspended" means that the slab is supported against gravity by the electric field itself.[31]

In the language used in this chapter, the slab is the "sample", and the whole arrangement is the "system".

Steps (1) and (2) below are provided to prepare our minds for the calculations that follow. They are not part of the logic. See § 21.5.1. In particular, fringing fields will introduce corrections into the expression for capacitance C, but they do not damage the results of steps (3) and (4).

(1) Show that the capacitance of the capacitor, with the slab in place, is (ignoring fringing fields)

$$C = \epsilon_0 L \left\{ \frac{H}{D} + x \left(\frac{1}{D-d} - \frac{1}{D} \right) \right\}.$$

(2) Remember that:

- the energy stored in the capacitor is $\frac{1}{2}CV_0^2$;
- the charge[32] on the capacitor's positive plate, which must have come from the voltage source, is CV_0;

[30] I use the neutral term "voltage source" so as not to hint at the nature of the source. I avoid calling it a "battery" because that suggests a chemical cell or cells, in which movement of charge is accompanied by absorption or emission of heat. To avoid distraction, we take it there is available a "non-thermodynamic" source of voltage that free from such complications.

[31] If this arrangement were implemented in the laboratory, we should need to prevent the slab from moving sideways into contact with the electrodes. We ignore such practicalities here, because our concern is entirely with vertical forces.

[32] There is, of course, a matching charge $-CV_0$ on the negative plate.

- this means that energy has been surrendered by the source, equal to $(CV_0)V_0 = CV_0^2$.

(3) Increase x by δx. The following changes of energy take place in the system (check this with care over signs):[33]

$$\left(\begin{array}{l}\text{gravitational potential energy}\\\text{of slab increases by}\end{array}\right) \quad mg\,\delta x$$

$$\left(\begin{array}{l}\text{energy stored in electric fields}\\\text{inside the capacitor increases by}\end{array}\right) \quad \tfrac{1}{2}\epsilon_0 L V_0^2 \left(\frac{1}{D-d}-\frac{1}{D}\right)\delta x$$

$$\left(\begin{array}{l}\text{energy stored in the voltage}\\\text{source increases by}\end{array}\right) \quad -\epsilon_0 L V_0^2 \left(\frac{1}{D-d}-\frac{1}{D}\right)\delta x.$$

(4) Check then that

$$\left(\begin{array}{l}\text{total change of energy}\\\text{consequent upon } \delta x\end{array}\right) = \delta x\left(mg - \frac{\epsilon_0 L V_0^2}{2}\frac{d}{D(D-d)}\right).$$

(5) Now for equilibrium the energy of the system must be a minimum, or at least stationary, so the coefficient of δx in the energy change must be zero; we echo the wording in § 21.4.1.[34] Then mg exactly balances the upward force pulling the slab into the capacitor. Therefore the force pulling the slab into the capacitor is $\epsilon_0 L V_0^2 d/\{2D(D-d)\}$.

---

[33] Notice that the last two quantities in part (3) differ by a sign and a factor 2. This very commonly happens in problems of this type, and it's something to look out for.

[34] It is safest if we avoid saying that x "can be adjusted by the system", because the equilibrium here is neutral, rather than stable; and in example 21.4 the equilibrium is unstable.

We may cross-reference the "principle of virtual work", which is another way of seeing that the coefficient of δx must be zero.

21.5.1   Discussion of example 21.3

The $\boldsymbol{E}$-field at the slab acts normal to the slab's (conducting) surface; most therefore acts "sideways", and does nothing to pull the slab into the capacitor. What pulls the slab in is the fringing field at the slab's top. Calculating the fringing field would be very messy, and that's why we don't attempt to find the force on the slab by summing $q\boldsymbol{E}$.

When we change x by δx, the fringing field moves up without change of shape and its energy remains unaltered;[35] what changes is the volumes occupied by the two regions of uniform field (beside the slab and above the slab), whose energy change we can calculate easily—as changes of $\int \tfrac{1}{2}\epsilon_0 E^2\,\mathrm{d}V$. The virtual-work method is more clever than perhaps we realised.

Now consider example 21.3 from the point of view of free energy. The "background" energy of the empty (slab-absent) system lies in the voltage source and in the electrostatic energy $\approx \tfrac{1}{2}(\epsilon_0 LH/D)V_0^2$ stored in the empty capacitor. This is an energy allocated by our accountancy to the environment alone. Additional energies are present after the slab—the sample—has been introduced. These additional energies are allocated to the sample and, by definition, they constitute the sample's free energy. Energy accountancy (virtual work) can consider a change of the total energy, or a change of the free energy; the results are the same since the "background" (sample absent) energy is unchanging.

Notice that in example 21.3 all algebraic terms relating to the slab contain x. More correctly, because fringing fields and their complications

[35] Note that the expressions in example 21.3 part (3) are *not* obtained by differentiating the expression for capacitance in part (1). They are obtained by thinking about the changed volumes occupied by regions of uniform field.

Incidentally, fringing fields located in front of and behind the paper do change their volume as x increases. Their energy is made negligible because it does not increase with L and we have made $L \gg D$.

are removed, *changes* of free energy contain δx. This is a useful signal that can help with the accountancy.

The division of physical quantities into sample-absent background and sample-associated change can—and should—be carried through for everything, not just energy. In particular, the "applied" electric field between the capacitor plates is V_0/D:

> the **applied field** $\boldsymbol{E}_{\text{app}}$ *or* $\boldsymbol{B}_{\text{app}}$ *is the field that is there when the sample is absent (not yet there or removed).*

All new things consequent on introducing the sample are now laid at the door of the sample. These include: the change of field in that part of the capacitor where the sample lies; the extra charge on that area of the capacitor plates; the extra charge delivered by the voltage source; and all the energy movements that these things entail.

Isn't something wrong? When it is in place, the slab "feels" a sideways E-field of $V_0/(D-d)$, plus a fringing field that pulls it into the capacitor; isn't it silly to say that the applied field experienced by the slab is a uniform V_0/D? As rebuttal, I'll say that the wrong thing has been identified as silly. In asking the question, our devil's advocate ignored charges induced onto the surfaces of the slab. These charges originate new fields and, together with additional surface charges induced on the capacitor plates, they increase the field from V_0/D to $V_0/(D-d)$. So the change of field from V_0/D to $V_0/(D - d)$ is due, directly and indirectly, to the slab's presence; it really does belong to the slab and not to the applied field.[36] This agrees with "dependent on δx" as a signal for identifying what's due to the slab.[37] What was silly was to muddle ourselves into thinking of all the field as "applied" when only part of it was. We're not discovering reasons for rejecting my definitions; we're finding out how to identify and cope with distractions.

Example 21.3 nicely illustrates how much goes on, even in a relatively simple system. In that part of the capacitor where the slab is, the electric field is strengthened (because V_0 is across an air gap of $D - d$ instead of D); the capacitor plates bear a redistributed charge; some charge has moved from voltage source to capacitor. Energy moves gymnastically all over the system. To do an accountancy job on the energy, you have to include all these energy movements. No energy changes can be neglected, no matter how far they are from the slab (don't forget the voltage source, no matter how long its connecting wires).[38] And some pieces of energy aren't in any particular place: where is the gravitational potential energy of the slab? The free energy here well merits the description: delocalized energy.

In the above discussion of example 21.3, we partitioned the total energy of the system into energy possessed by the empty environment, and the free energy of the sample. This separation has not simplified the reasoning at all: in particular, it does not give us a shorter route to the force. That couldn't have been expected, and wasn't the point. The intention was to elucidate by example the meaning of free energy, and the meaning of "applied field".

[36] If we say that charges on the slab feel an added field $V_0/(D-d)-V_0/D$, then we introduce a "self-force", in which the charges are counted as interacting with their own field.

[37] The increase of field strength beside the slab, from V_0/D to $V_0/(D-d)$, does not contain x or δx, though the volume of the affected region does. The algebraic signal—the presence of δx —is therefore helpful, but should not be relied upon exclusively. A quantity containing δx is definitely associated with the sample; a quantity not containing δx still might be.

[38] Compare with a pVT sample. If the sample is held at controlled pressure, energy $+pV$ resides in the pressure environment (perhaps the atmosphere). If the sample is in contact with a heat bath, energy $-TS$ resides in the heat bath. Energy is delocalized. And the whole purpose of "free energy" is to collect together all these energies that properly should be counted as "to do with the sample".

21.5.2 "First find the free energy (change)"

Before you can do any thermodynamics at all, you must obtain an authenticated function of state.[39] Only a function of state can have uniquely defined derivatives and yield Maxwell relations (for example). For a sample in a field, the only authenticated function of state is the free energy: find it first! We may be able to find the free energy explicitly, but we may have to make do with an expression for its small change, such as $-\boldsymbol{p} \cdot \delta \boldsymbol{E}_{\mathrm{app}}$.

If we want other thermodynamic potentials, they can be obtained safely from the free energy by one or more Legendre transformations.

In the case of the slab, the free energy has given us everything that we need. Perhaps other cases will be like this one, where there is no necessity for introducing other energies (one called U perhaps?) . . . .

21.6 Energy and force in a magnetic system

Example 21.4 An electromagnet attracting a piece of iron
A soft-iron ring [shown in Fig. 21.6] has a mean diameter of 0.3 m and a cross-sectional area of $0.01\,\mathrm{m}^2$. This ring, which is in a vertical plane, is divided horizontally into two equal parts separated by a small air gap of thickness x millimetres. The upper half is wound with a coil providing a constant 500 ampère turns. Show that the magnetic flux in the ring is given by

$$\Phi = \frac{3 \cdot 1 \times 10^{-3}}{x + 0 \cdot 47}\,\mathrm{Wb}.$$

For an infinitesimal change in x, calculate the work done by the external circuit against the back e.m.f. and find the change in the stored magnetic energy. Hence show that the upper half magnet will be able to lift the lower half provided x is less than about $1\,\mathrm{mm}$.
[Relative permeability of soft iron $= 1000$; density of iron $= 7800\,\mathrm{kg\,m}^{-3}$.]

[Shortened from Oxford Physics Finals 1983, paper 2a, question 5; solution in § 21.6.2.]

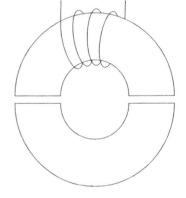

Fig. 21.6 The lower half of the soft-iron ring is supported against gravity by the magnetic field.

21.6.1 Discussion of example 21.4

There is a conceptual difficulty in example 21.4 that wasn't so clearly there in example 21.3. How is the coil fed with current? Obviously, from a battery. (?) But then the current is held constant only because there is some resistance in series with the coil. (With no resistance, $L\,\mathrm{d}I/\mathrm{d}t = \text{battery voltage} = \text{constant}$, and the current I would rise linearly with time.) Energy is being continually transferred out of the battery into heat generated in the resistance. Our system is not in thermodynamic equilibrium; what it's in is a *steady state*, where entropy is being continually generated.

How do we make arguments from thermodynamics on this irreversible system? If we move the lower piece of iron, it sets up an e.m.f. in the coil, and a changed amount of energy is delivered by the battery to the coil, and to the resistance. Can we disregard the "background" transfer of energy to the resistance and do accountancy on this tiny extra bit?[40] I wouldn't be convinced.

The difficulty is real, and it's usually swept under the carpet. To do thermodynamics, we need a proper reversible thermodynamic system which does not generate entropy when it's "doing nothing new".

We must invent some way of removing all irreversibility. First, we'll make the coil and wiring resistanceless, so a steady current can be maintained without energy dissipation; there is no steady voltage difference across the coil and none needed from the current generator. When the lower piece of iron moves,[41] it sets up an e.m.f. $-\mathrm{d}\phi/\mathrm{d}t$; the current generator must now maintain the current constant, against the e.m.f., and deliver whatever energy that requires. All changes in this redesigned system are thermodynamically reversible. But where can you buy a current generator like that? An in-principle reversible current generator is described in Robinson (1973*b*), p. 164; also in § 22.6 below. A reversible voltage generator, should we need one for an electrostatic problem, is also in Robinson (1973*b*), p. 163.

Comment: Everything that we found with electrostatics in example 21.3 applies here too. We can call the lower half-ring the sample and the rest of the system the environment. Energy associated with the lower half-ring's presence is its free energy. Changes in that free energy contain δx. Energy associated with δx is spread all round the system: in field energy, in the current generator, in gravitational potential energy.

We can set things up in this "free-energy" language if we wish—but we don't have to. In examples 21.3 and 21.4 there was no need to make a formal partitioning of the total energy into pieces possessed by the sample (free energy) and by the environment.[42] What was important was that we "think free energy", because the associated discipline reminds us to gather up *all* contributions to the energy *change*.

Looking back: example 21.3 has been given because it helps to clarify the meaning of "applied field". And example 21.4 demonstrates the need for a reversible current generator when dealing with a magnetic sample.

21.6.2   Solution to example 21.4

The coil has N turns and carries current I. The ring has circumference l (not including the air gap) and cross-sectional area A. The B- and H-fields are as indicated in Fig. 21.7.

We assume that all magnetic flux is confined to the iron and the air gap, with no "leakage".

The Maxwell equation div $\boldsymbol{B} = 0$ yields boundary condition $B_\mathrm{m} = B_\mathrm{a}$.

[40] I think this is what the examiners expected.

[41] The flux ϕ is given by (21.12). It threads the N turns of the coil, in contrast to the Φ of the question which threads one turn. Thus $\phi = N\Phi$.

[42] A complete separation would require us to establish the zero of free energy by removing the sample to infinity, an operation that would confer no benefit even if we could find a way of doing it.

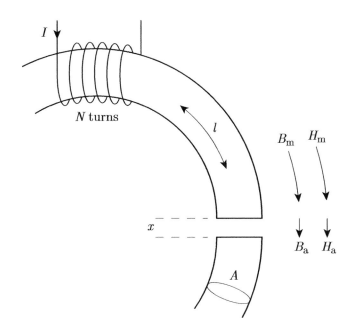

Fig. 21.7 The anchor ring of example 21.4. The fields in the ring and its air gap are shown.

The Maxwell equation curl $\boldsymbol{H} = \boldsymbol{J}$ gives

$$NI = \oint \boldsymbol{H} \cdot \mathrm{d}\boldsymbol{l} = H_\mathrm{m}l + H_\mathrm{a}2x \qquad (21.9)$$

$$= \frac{B_\mathrm{m}}{\mu_\mathrm{r}\mu_0}l + \frac{B_\mathrm{a}}{\mu_0}2x = \frac{2B_\mathrm{m}}{\mu_0}\left(\frac{l}{2\mu_\mathrm{r}} + x\right),$$

so

$$B_\mathrm{m} = \frac{\mu_0 NI}{2(l/2\mu_\mathrm{r} + x)}. \qquad (21.10)$$

$$\begin{pmatrix} \text{Magnetic flux } \varPhi \\ \text{threading area } A \text{ of ring} \end{pmatrix} = AB_\mathrm{m} = \frac{\mu_0 NIA}{2(l/2\mu_\mathrm{r} + x)}, \qquad (21.11)$$

$$\begin{pmatrix} \text{magnetic flux threading} \\ \text{the } N \text{ turns of the coil} \end{pmatrix} = \phi \quad = \frac{\mu_0 N^2 IA}{2(l/2\mu_\mathrm{r} + x)}, \qquad (21.12)$$

$$\begin{pmatrix} \text{energy stored in the} \\ \text{fields in ring and air} \end{pmatrix} \quad = Al \cdot \tfrac{1}{2}B_\mathrm{m}H_\mathrm{m} + A(2x) \cdot \tfrac{1}{2}B_\mathrm{a}H_\mathrm{a}$$

$$= \tfrac{1}{2}AB_\mathrm{m}(lH_\mathrm{m} + 2xH_\mathrm{a})$$

[using (21.9)]

$$= \tfrac{1}{2}AB_\mathrm{m}NI$$

[using (21.11)]

$$= \frac{\mu_0(NI)^2 A}{4(l/2\mu_\mathrm{r} + x)}. \qquad (21.13)$$

The flux $\varPhi$ asked for in the question is given by (21.11).

Now we consider the effect of increasing x by δx. We must list the energy changes that take place in the entire system, consequent upon δx. The coil is fed from a current generator,[43] both resistance-free, so the entire circuit encloses a constant magnetic flux ϕ_circ. A change of flux $\delta \phi$ through the coil is accompanied by an equal and opposite change $\delta \psi = -\delta \phi$ of flux through the current generator.[44] In turn, this changes

[43] This could conveniently be the Robinson current generator of § 22.6.

[44] When there is a change of flux, its rate of change is connected with an e.m.f.. We are not far from the back e.m.f. mentioned in the question. However, getting the sign right can be muddling. The Robinson current generator is a great help here.

the energy stored in the generator by $I \, \delta\psi$.

(Flux ϕ threading the coil, from (21.12), changes by)	$\delta\phi = -\dfrac{\mu_0 N^2 I A}{2(l/2\mu_\mathrm{r} + x)^2} \, \delta x,$
(flux ψ threading the current generator changes by)	$\delta\psi = +\dfrac{\mu_0 N^2 I A}{2(l/2\mu_\mathrm{r} + x)^2} \, \delta x,$
(energy stored in current generator, from (22.18), changes by)	$I \, \delta\psi = +\dfrac{\mu_0 N^2 I^2 A}{2(l/2\mu_\mathrm{r} + x)^2} \, \delta x,$

$$(21.14)$$

$$\left(\begin{array}{l}\text{energy stored in fields of ring and}\\ \text{air gap changes, from (21.13), by}\end{array}\right) \qquad -\frac{\mu_0 (NI)^2 A}{4(l/2\mu_\mathrm{r} + x)^2} \, \delta x,$$

$$(21.15)$$

$$\left(\begin{array}{l}\text{gravitational energy of half-ring}\\ \text{of mass } m \text{ changes by}\end{array}\right) \qquad -mg \, \delta x,$$

total change of energy is
$$\left(-mg + \frac{\mu_0 (NI)^2 A}{4(l/2\mu_\mathrm{r} + x)^2}\right) \delta x.$$

[45] The equilibrium here is unstable, so the free energy has a maximum, rather than a minimum. We must avoid saying (or thinking) that x is "adjusted by the system". The important point is that x is free to change, not controlled from outside. A mention of the "principle of virtual work" may again be helpful.

For equilibrium,[45] the coefficient of δx in the energy change must be zero. Then

$$x = \frac{NI}{2}\left(\frac{\mu_0 A}{mg}\right)^{1/2} - \frac{l}{2\mu_\mathrm{r}}, \qquad (21.16)$$

which works out to be just over $1 \, \mathrm{mm}$.

Comments

[46] The back-e.m.f. is $-\mathrm{d}\phi/\mathrm{d}t$, so work $-I(\mathrm{d}\phi/\mathrm{d}t)\mathrm{d}t$ is not far away. The remaining question requiring a clear head is the sign.

1. The only non-obvious step here is that leading to (21.14). It can be understood in terms of a back-e.m.f.,[46] but it is simplest to look ahead to (22.18) which explains it as a property of the Robinson generator.
2. The calculation here is not, in principle, more difficult than that of example 21.3. But people do tend to find this case more muddling. Notice how easy is the step leading to (21.14) when you can call upon a Robinson current generator; no muddle with the sign now.
3. Expressions (21.14) and (21.15) differ by a sign and a factor 2. This is no surprise given experience with example 21.3 (sidenote 33).

21.7   Key points in this chapter

A lot has happened in this chapter, some of it—to my surprise—a bit unconventional.

1. In a system consisting of interacting components, it's often helpful to identify one component of interest as the *sample*, and the rest as that sample's *environment*.[47]

[47] This step was helpful but not essential in example 21.4.

2. The essential first step in any thermodynamics is to find an energy that is a function of state—something proved to be a function of state.

   - The *free energy* $\equiv$ all the energy that is present because the sample is there.

If the free energy has been correctly identified, it will supply the required function of state, so the "essential first step" is

- *First find the (sample's) free energy.*

Other function-of-state energies can be built from here, probably by one or more Legendre transformations.

3. The value of the free energy depends upon what in the sample's environment is controlled: free energy is a context-sensitive concept. Within each pair (p, V), (S, T), $(\boldsymbol{p}, \boldsymbol{E})$, $(\boldsymbol{m}, \boldsymbol{B})$ only one quantity at a time can be controlled;[48] the other is free to fluctuate. The change of free energy contains the $\delta(\ )$ of each quantity being controlled.

4. For a sample in a field, the "applied" field is always the field present when the sample is absent (removed or not yet there).

5. When an electric or magnetic field is applied, it must be prepared using loss-free circuitry, so that a proper thermodynamic equilibrium is being discussed. In particular, when a magnetic field is prepared using a coil, feed a resistance-free coil using a Robinson current generator (Fig. 22.6).

6. If at all possible, design the layout so that the sample may be inserted/removed completely. The free energy is defined to be the work that is done by inserting the sample from infinity (plus the energy that the sample brought with it). If that's difficult to calculate, a free-energy change can be found using the "canonical procedure":[49]

   (a) remove sample completely from the environment
   (b) change the environment
   (c) replace sample into the changed environment.

   The sample's free-energy change is the total mechanical work done in steps (a) and (c).
   This procedure works well in the case of a small sample such as a dipole in an applied electric or magnetic field (§§ 22.3–4).[50]

7. There are usually good reasons why the environment should be imposed using equipment that is "large compared with the sample". The pressure source of Fig. 21.3 has a volume so large that its pressure is negligibly affected by inserting the sample; the heat bath of Fig. 21.4 has so large a thermal capacity that its temperature does not change when it receives heat from the heat engine. It will be no surprise that, in Chapter 22, applied fields $\boldsymbol{E}_{\mathrm{app}}, \boldsymbol{B}_{\mathrm{app}}$ are imposed using a large capacitor (Fig. 22.2) and a large coil (Fig. 22.4).

8. Sometimes a sample cannot be completely inserted/removed, either because of a physical obstacle or because the force it experiences is beyond simple evaluation. In such cases we may be able to impose a small change, perhaps a displacement δx, with calculable consequences. When a sample is so moved, as is the case in examples 21.3 and 21.4, then gather *all* terms containing δx before attempting to do accountancy on the energy changes.

[48] And of these, $\boldsymbol{p}$, $\boldsymbol{m}$ cannot, save exceptionally, be controlled by external means, while $[S]$ is a little out of pattern.

[49] Section 22.4.3 exhibits the high penalty that is exacted when this opportunity is available but not exploited.

[50] In Chapter 22 we are able to increase the fields $\boldsymbol{E}_{\mathrm{app}}$ and $\boldsymbol{B}_{\mathrm{app}}$ experienced by a dipole sample without a need to beef up the environment, and then things are even simpler.

Problem

Problem 21.1 The $+V\,\delta p$ term: exposing a poor route
Take the lower diagram of Fig. 21.3, with the sample's volume V_{sample} occupying space inside the much larger cylinder of volume V_{big}. Change the pressure being applied to the sample, the pressure in the big cylinder by moving the large piston, with the sample in place. Show that the work done on the system is

$$\text{work done} = -\int p\,dV = -p_{\text{env}}\,\delta V_{\text{big}} + V_{\text{sample}}\,\delta p. \qquad (21.17)$$

Interpret the two terms here as giving: the work that would have been done had the sample been absent; and the additional work $\text{d}W$ augmenting the free energy of the sample.

Solution to problem 21.1
The large piston enclosing the big cylinder's volume V_{big} was not allowed to move when the sample was pushed in (Fig. 21.3), so the volume available to the "environment" substance was reduced from V_{big} by V_{sample}. The pressure was thereby increased from p_{big} to[51]

$$p = p_{\text{big}} + \left(\frac{\partial p}{\partial V_{\text{big}}}\right)\delta V_{\text{big}} = p_{\text{big}} - \left(\frac{\partial p}{\partial V_{\text{big}}}\right)V_{\text{sample}}.$$

Now move the big piston, thereby doing mechanical $-p\,\delta V$ work on the system. The work input is

$$\text{work done} = -\left(p_{\text{big}} - \frac{\partial p}{\partial V_{\text{big}}}V_{\text{sample}}\right)\delta V_{\text{big}}.$$

Here the first term contains the pressure that the environmental substance would have if allowed to occupy the whole of volume V_{big}, so it is the work that would be done if the sample were absent. What remains is work supplied because the sample is present, so it is the change $\text{d}W$ of the sample's free energy:

$$\text{d}W = V_{\text{sample}}\frac{\partial p}{\partial V_{\text{big}}}\delta V_{\text{big}} = V_{\text{sample}}\,\delta p.$$

We have agreement with eqn (21.3).

Comment: In (21.17), pressure p_{big} is exactly like the "applied field" of § 21.5.1: it is the pressure in the large cylinder when the sample is absent.[52] This definition, and its use in $-p_{\text{env}}\,\delta V_{\text{big}}$, is necessary to ensure that sample-related energy is wholly excluded from the "work done on environment" term.

Look back at § 21.3.1: "The volume of the large cylinder ... must be much greater than the volume V_{sample} of the sample, so that inserting or removing the sample has negligible effect on the 'environment' pressure." Yet $\text{d}W$ has now come from that formerly-negligible change of 'environment' pressure. Surely a signal saying "inferior procedure".

Comment: All this is telling us that problem 21.1 started out from a strategy mistake: finding the total amount of work done on environment and sample together, and then having to extricate the $\text{d}W$ belonging to the sample. The "canonical procedure" is so much better.[53]

[51] This change of pressure was treated as negligible in § 21.3.1. Now we have to deal with relatively small quantities alongside the huge amount of work being used to change the environment. We make a Taylor expansion of the pressure, to first order in the small quantity $V_{\text{sample}}/V_{\text{big}}$.

[52] Having this demonstrated is a useful mind-clearing outcome from an otherwise disappointing calculation.

[53] Problems 22.10 and 22.11 find the free-energy change of a magnetic sample when the $\boldsymbol{B}$-field applied to it is changed. If the sample remains in place during the change (problem 22.11), the reasoning is far messier, just as we might expect on the precedent of problem 21.1.

Energy of a magnetic body: $-m \cdot \delta B$ or $+B \cdot \delta m$?

Intended readership: Towards the end of a course on thermodynamics, or within a course on magnetism in solid-state physics.

This essay describes a case study in getting un-muddled. It builds heavily on the discussion of free energy in Chapter 21.

22.1   Introduction

Reminiscences by Walter Marshall at the 1956 Varenna Summer School on Magnetism and Low Temperatures.[1]

> Within the first two days of the conference I acquired an inferiority complex which has remained with me throughout my life and which has remained secret and unconfessed until this moment.
>
> In the second lecture of the first day, Kittel told us that in magnetism, the work done was $-H \, dM$. In the first lecture of the second day, Gorter told us that the work done was $+M \, dH$. The English-speaking students were dismayed and the Italian students were perplexed. Neither lecturer was willing to acknowledge or understand the other's nomenclature, so Nicholas [Kurti] stepped to the blackboard and in a brief five minutes told us that in this matter, there had grown up an Oxford tradition and a Leiden tradition, that the two approaches were easily reconciled by a trivial piece of arithmetic which he quickly demonstrated. I wish he had not used the word "trivial". Some of the students decided that the laws of physics changed on crossing the English Channel. The remainder of us decided that we would avoid magnetic thermodynamics for the rest of our life. Despite several attempts subsequently, I have never understood this problem. I have now abandoned the attempt to understand it and I have learned to live with the psychological disadvantage of that word, "trivial". It is consistently my first thought each time I have met Nicholas for a quarter of a century.

The mystification is as alive and well now as it was in 1956. In principle we are dealing with an oddment topic in electromagnetism, but it turns up mainly in thermodynamic contexts, where it's an irritating and error-prone source of confusion.

[1] I now forget who gave me a typescript of this quotation, but was probably my late colleague D.T. Edmonds. I have been unable to verify the precise wording against a printed copy of the Summer School's proceedings.

269

Essays in Physics: Thirty-two thoughtful essays on topics in undergraduate-level physics. Geoffrey Brooker, Oxford University Press. © Geoffrey Brooker 2021. DOI: 10.1093/oso/9780198857242.003.0022

[2] Actually there are more than eight .... I do not identify the 20 books from which I have constructed this list (only 5 give $-\boldsymbol{m} \cdot \delta \boldsymbol{B}$), but the reader will have little difficulty in finding examples in at-least-several of the categories.

Several of the textbook discussions are in c.g.s. units and so omit the μ_0; I have translated into SI and not created even more variants from this cause alone.

[3] The two "it depends on how you do the accountancy" discussions that I have found are incompatible with each other.

[4] It was this muddle that led me to write Chapter 21, because I needed to know just what energy it is that is incremented by đW.

[5] Part? Equation (22.1) was introduced by referring to U as "internal energy", as is often done. This is an unhelpful description, since $-\boldsymbol{m} \cdot \delta \boldsymbol{B}$ is not "internal" to the sample. We have an "open system", so the appropriate energy quantity is the "free energy", as defined in Chapter 21. The U of (22.1) is the free energy if the sample is held under controlled [S], V and $\boldsymbol{B}$.

[6] A quick check on the sign in (22.3): Consider a simple paramagnet. Increase of m reduces the disorder of the atomic magnetic moments, so (statistical mechanics) it must reduce the entropy, and this property is unaffected by the holding constant—or not—of B. This makes $\partial m/\partial S$ negative and the right-hand side positive. Thus (22.3) as we give it predicts positive $(\partial T/\partial B)_S$; reduction of B reduces T. Expressions (5) and (6) in the catalogue predict the opposite sign, so that adiabatic demagnetization would result in heating.

[7] For anyone whose thermodynamics is rusty (s stands for "something"):

$$\delta(s) = T \,\delta S - m \,\delta B$$
$$= \left(\frac{\partial(s)}{\partial S}\right)_B \delta S + \left(\frac{\partial(s)}{\partial B}\right)_S \delta B.$$

Equating coefficients:

$$\left(\frac{\partial(s)}{\partial S}\right)_B = T, \quad \left(\frac{\partial(s)}{\partial B}\right)_S = -m.$$

Now form the double differential by two routes.

Consider a sample of magnetic material, whose total magnetic moment is $\boldsymbol{m}$, in an externally applied uniform field $\boldsymbol{B}$. Books on thermodynamics tell us that the sample has internal energy U with increment

$$(?) \qquad \delta U = T \,\delta S - p \,\delta V - \boldsymbol{m} \cdot \delta \boldsymbol{B} \qquad (22.1)$$

or

$$(?) \qquad \delta U = T \,\delta S - p \,\delta V + \boldsymbol{B} \cdot \delta \boldsymbol{m}, \qquad (22.2)$$

these two conflicting possibilities appearing about equally often. I take a less tolerant approach than Nicholas Kurti or Walter Marshall: (22.1) and (22.2) can't have equal claims to being right.

Things are even worse. I have found eight (!) different expressions for the "work" term đW, all given in respectable textbooks:[2]

(1) $-\boldsymbol{m} \cdot \delta \boldsymbol{B}$
(2) $-\boldsymbol{m} \cdot \delta(\mu_0 \boldsymbol{H})$
(3) $+\boldsymbol{B} \cdot \delta \boldsymbol{m}$
(4) $+\mu_0 \boldsymbol{H} \cdot \delta \boldsymbol{m}$
(5) $-\boldsymbol{B} \cdot \delta \boldsymbol{m}$
(6) $-\mu_0 \boldsymbol{H} \cdot \delta \boldsymbol{m}$
(7) $+\boldsymbol{B} \cdot \delta \boldsymbol{m}$ but offering $-\boldsymbol{m} \cdot \delta \boldsymbol{B}$ too, saying "it depends"
(8) $+\mu_0 \boldsymbol{H} \cdot \delta \boldsymbol{m}$ and $-\boldsymbol{m} \cdot \delta(\mu_0 \boldsymbol{H})$, saying "it depends".[3]

When things are this bad, we've got to rely on our own reasoning, not trusting even the most authoritative textbook. I record here my attempts at sorting it out, and you're invited to suspect and scrutinize my essay as critically as you would any other treatment.[4]

Satisfy yourself that the contentious quantity is the "work-input" đW, so it is the increment to the sample's *free energy* (which might be called U). I'll claim at once that (in my view) đ$W = -\boldsymbol{m} \cdot \delta \boldsymbol{B}$, thus part-endorsing (22.1).[5]

A surprise here is that it's—almost—possible to get away without deciding between the two possibilities (22.1) and (22.2)! Consider adiabatic demagnetization, a context in which this is often discussed. All that is needed for the thermodynamics is the Maxwell relation[6]

$$\left(\frac{\partial T}{\partial B}\right)_S = -\left(\frac{\partial m}{\partial S}\right)_B. \qquad (22.3)$$

To derive this,[7] we need a function of state whose differential is

$$\delta(\text{something}) = T \,\delta S - \boldsymbol{m} \cdot \delta \boldsymbol{B}.$$

(I ignore pV terms for simplicity here.) If (22.1) is true, the name for (something) is U; if (22.2) is true, then the name of (something) is $(U - \boldsymbol{m} \cdot \boldsymbol{B})$; but either way the Maxwell relation follows—if—if U or $(U - \boldsymbol{m} \cdot \boldsymbol{B})$ is first proved to be a function of state.

We can't leave things like this. The application is thermodynamics, but the root of the problem has to lie in electromagnetism. What is the energy of a magnetic sample placed in an externally applied magnetic field? There really must be a clear answer.

There is a similar difficulty when an electric dipole of moment p is placed in an electric field E. Some authorities give $\delta U = T\,\delta S - p \cdot \delta E$, others $\delta U = T\,\delta S + E \cdot \delta p$. Again, both statements can't be right.

The key to mastering all such problems is given in § 21.5.2: *first find the free energy.*

In this chapter we shall show that:

- the "work" increment $\mathrm{d}W$ to the free energy of an electric dipole p in field E_{app} is $-p \cdot \delta E_{\mathrm{app}}$
- the "work" increment $\mathrm{d}W$ to the free energy of a magnetic dipole m in field B_{app} is $-m \cdot \delta B_{\mathrm{app}}$.

Only when these assertions have been established shall we even have a authenticated function of state; and only then can any thermodynamics proceed at all.[8]

22.2   Electric dipole in an applied electric field I: scene setting

The electric and magnetic cases help with each other so we discuss both. We deal with the electric case first.

Exercise 22.1   Just to keep you awake

Think about an electric-dipole moment p in a non-uniform electric field E_{app}, and consider the following statements:

(?) the force on the dipole is simply the sum of qE terms, which add up to $(p \cdot \nabla)E_{\mathrm{app}}$

(?) the potential energy of the dipole in the field is $-p \cdot E_{\mathrm{app}}$

(?) the force on the dipole is minus the gradient of energy, so it's $-\nabla(-p \cdot E_{\mathrm{app}})$.

The two expressions for force are not the same: in the simple case where the dipole moment is proportional to the field, they differ by a factor 2. So at least one of the statements given is wrong. Which? And why?

We started off in § 22.1 by looking at two rival expressions for free-energy change, the electric versions of which are $-p \cdot \delta E$ and $+E \cdot \delta p$. We may be tempted to think that these expressions aren't very different: in a linear system for which $p \propto E$ they differ only by a sign. The difference is much deeper than that. To keep a clear head, it's good practice[9] to think always about a material for which p is *not* linearly related to E. In such a case, $-p \cdot \delta E$ and $+E \cdot \delta p$ are completely different. The same precaution will be taken when we deal with a magnetic dipole in a magnetic field.

[8] The First Law shows that in a thermodynamic cycle
$$\sum(\mathrm{d}Q + \mathrm{d}W) = 0,$$
thereby permitting the definition of a function of state U. Then a Carnot cycle shows that in a cycle of reversible changes
$$\sum \frac{\mathrm{d}Q}{T} = 0,$$
thereby permitting the definition of a function of state S. In "conventional" (some might say old-fashioned) treatments of thermodynamics, these arguments (reversible heat engines) are set up with admirable rigour—a rigour otherwise encountered only in the theorems of Euclidean geometry.

Yet the muddle exhibited in items (1)–(8) above shows that this rigour is not continued into many treatments of electric and magnetic samples. We have a right—and a duty—to demand better. Setting up a validated $\mathrm{d}W$, and hence a validated free energy, may seem tedious, but there's no avoiding it. The penalty for sloppiness is all too obvious. In fact, items (5) and (6) in the catalogue define quantities that are provably *not* functions of state (problem 22.1).

[9] A similar precaution is recommended in § 1.4.

22.2.1   The force on an electric dipole in an inhomogeneous field

[10] This is what we did in Chapter 21, examples 21.3 and 21.4.

Exercise 22.1 reminds us that we need a secure place to start from. In most other contexts, we'd build an incontrovertible expression for energy and we'd obtain the force by differentiating the energy.[10] But here the energy is in dispute, so we can't follow that route. To obtain a secure starting point, we have to find the force, and move from there to energy.

[11] Bookwork which the reader should easily supply; if not, problem 22.2. Henceforth, $\boldsymbol{E}_{\text{app}}$ and $\boldsymbol{B}_{\text{app}}$ represent "applied" fields so that unadorned letters $\boldsymbol{E}$ and $\boldsymbol{B}$ are available to describe an actual field.

The force $\boldsymbol{F}$ acting on electric dipole $\boldsymbol{p}$ in an inhomogeneous $\boldsymbol{E}_{\text{app}}$-field is $\boldsymbol{F} = (\boldsymbol{p} \cdot \nabla)\boldsymbol{E}_{\text{app}}$. This expression[11] is obtained directly from the Coulomb forces acting on the two charges of the dipole; we can't be in doubt that it's right.[12]

[12] The corresponding expression for the force on a magnetic dipole moment $\boldsymbol{m}$ is $(\boldsymbol{m} \cdot \nabla)\boldsymbol{B}_{\text{app}}$. Its derivation is again bookwork, but is distinctly less obvious: problem 22.4.

We shall find the (increment of) free energy of an electric dipole $\boldsymbol{p}$ by finding the work done by (or against) the known force $\boldsymbol{F}$ as the dipole is moved from a field-free region into the field $\boldsymbol{E}_{\text{app}}$.

22.2.2   Preparation of an applied electric field

The electric field $\boldsymbol{E}_{\text{app}}$ applied to our dipole will originate from two conducting capacitor plates, held at controlled potential difference by a voltage source (Fig. 22.1).[13] For now, the dipole sits in the middle between the plates. It is necessary for the plates to be "large", for two reasons. Let's explain.

[13] Alternatively, the plates may hold charges of controlled magnitude. We deal with the more complicated case of "controlled potential difference" because energy movements involving the voltage source occur and need to be allowed for.

Reason 1: image charges. The dipole's charges produce their own $\boldsymbol{E}$-field which adds to the applied field. The total $\boldsymbol{E}$-field is normal to the capacitor plates. This boundary condition is most easily dealt with if we think of image charges behind the plates. There are now three fields present, whose sum is the actual field:

To avoid distraction, the source is taken to be ideal, in the sense that it is reversible and has "no thermodynamics": no tendency to emit or absorb heat when charge moves between it and the capacitor plates. A "non-thermodynamic" voltage source is possible in principle (a chemical cell won't do, so we avoid words such as "battery"): see Robinson (1973*b*), p. 163.

- the field that the voltage source and plates produce when the dipole isn't there—the $\boldsymbol{E}_{\text{app}}$ defined in § 21.5.1
- the field from the dipole's own charges
- the field from the image charges.

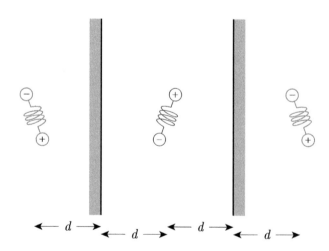

Fig. 22.1 An electric dipole placed between the plates of a capacitor. The first image dipole on each side is indicated in grey.

The dipole is free to adjust its orientation and magnitude as it is moved into the capacitor. We indicate this freedom to adjust by drawing a spring between the dipole's charges.

We don't want to be distracted by image charges, so we make a friendly choice of the plate spacing $2d$. The nearest image dipoles are distant $2d$ from the dipole (there is a whole chain of images of images but these are the nearest). The electric field from an image, at the original dipole, $\propto d^{-3}$, so it can be made negligibly small if we make d large (making the voltage large also so that $\boldsymbol{E}_{\mathrm{app}}$ takes its desired value).[14] What remains is our dipole in field $\boldsymbol{E}_{\mathrm{app}}$.

Reason 2: In due course, we shall evaluate a free-energy change by moving the dipole from infinity towards the capacitor—perhaps right in, but perhaps only into the fringing field. We shall need the field always to be near-uniform over the length of the dipole, so that the force exerted is $(\boldsymbol{p}\cdot\nabla)\boldsymbol{E}_{\mathrm{app}}$, not containing higher derivatives. This requires $2d$ to be much greater than the length of the dipole.[15]

22.3   Electric dipole in an applied electric field II: free energy

Only now have we put in place all the ideas needed for calculating the dipole's free energy. The calculation itself is almost trivial . . . .

To find the free energy of the dipole sample, we follow a procedure modelled on those in § 21.3. We pre-prepare the field $\boldsymbol{E}_{\mathrm{app}}$ by means of a large capacitor and a voltage source,[16] and with the sample completely outside that field. Then we introduce the sample quasi-statically, doing mechanical work đW on it. The sample's free energy is that đW, plus the energy that the sample brought with it.

It is open to us to make a change of $\boldsymbol{E}_{\mathrm{app}}$ along the lines in Chapter 21: remove sample, change field, reinsert sample (problem 22.3). But an even simpler recourse is available. We make the field in the middle of the capacitor greater than the desired $\boldsymbol{E}_{\mathrm{app}}$, so that the dipole reaches that desired $\boldsymbol{E}_{\mathrm{app}}$ when it is still in the fringing field. All we need do to increment the $\boldsymbol{E}_{\mathrm{app}}$ experienced by the dipole is to move the dipole to a place where the field is stronger.

During its introduction, the dipole moment $\boldsymbol{p}$ experiences a force pulling it into the field,[17] of $(\boldsymbol{p}\cdot\nabla)\boldsymbol{E}_{\mathrm{app}}$. To prevent the dipole from accelerating, we need to apply a restraining force $\boldsymbol{X} = -(\boldsymbol{p}\cdot\nabla)\boldsymbol{E}_{\mathrm{app}}$. Now allow the dipole to move through $\delta\boldsymbol{r}$. Work is added to the system by the restraining force, and it is

$$\delta(\text{free energy}) = \text{đ}W = \boldsymbol{X}\cdot\delta\boldsymbol{r} = -\delta\boldsymbol{r}\cdot(\boldsymbol{p}\cdot\nabla)\boldsymbol{E}_{\mathrm{app}}.$$

This expression can be tidied up quickly if we use tensor notation (repeated subscripts summed over), and make use of the fact that a static field has curl $\boldsymbol{E} = 0$ so that $\partial E_i/\partial x_j = \partial E_j/\partial x_i$:

$$\text{đ}W = -\delta x_i\, p_j\, \frac{\partial}{\partial x_j}E_i = -p_j\,\delta x_i\,\frac{\partial E_j}{\partial x_i} = -p_j\,\delta E_j = -\boldsymbol{p}\cdot\delta\boldsymbol{E}_{\mathrm{app}}. \quad (22.4)$$

In conformity with (21.1), this energy-as-work input is the increment in the dipole's free energy.

[14] If we don't let $d \to \infty$, the dipole is subject to a force from the image charges, and has a potential energy, even when no voltage difference is applied between the capacitor plates.

[15] The "dipole" need not be a point dipole, but could be a macroscopic body, uncharged but possessing an overall dipole moment, probably induced or modified by the applied field.

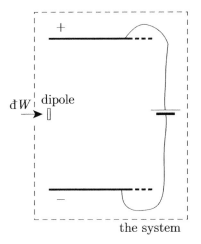

Fig. 22.2 An electric dipole $\boldsymbol{p}$ is located in the fringing field $\boldsymbol{E}_{\mathrm{app}}$ produced by a large capacitor connected to a voltage source. The dipole is moved through $\delta\boldsymbol{r}$ towards the capacitor. The dipole's free energy is incremented because a restraining force $\boldsymbol{X}$ does work đ$W = \boldsymbol{X}\cdot\delta\boldsymbol{r} = -\boldsymbol{p}\cdot\delta\boldsymbol{E}_{\mathrm{app}}$.

As the dipole is moved, there is a change in the charges induced on the capacitor plates, charges which must have moved from the voltage source. Clean accountancy requires that energy enter the *system* via only one route, so the system's boundaries should be drawn to enclose the voltage generator. Then energy moving between source and capacitor is an "internal transfer" whose magnitude need not be evaluated.

[16] Compare with the definition of "applied field" in § 21.5.1. Each case may help to explain the other: why this was the only reasonable way to set things up.

[17][From previous page] The electric dipole moment $\boldsymbol{p}$ is not fixed in magnitude or direction (except perhaps in an uninteresting special case), but is free to adapt to the field—to be always in equilibrium with the field—as the sample moves.

[18]It is represented by "energy movements allowed for" in Fig. 21.2.

We have chosen to connect the capacitor plates to a voltage source, rather than giving the plates fixed charges, in order to highlight the need for thought about the system boundary.

[19]Landau, Lifshitz and Pitaevskiĭ (1993) find the free energy by assembling charges onto the capacitor plates, once with the sample present and once with the sample absent. The sample's free energy (all the energy present because the sample is there) is the difference between these two energy inputs. To find a *change* of free energy, we have to undertake two more assemblies, this time with slightly increased charges. Both methods (theirs and ours) are valid. But our "work" route is more down-to-earth, and far simpler.

It should be no surprise that our method is simpler. Problem 21.1 has shown, rather forcefully, that changing conditions with a sample in place—just what Landau–Lifshitz do—is not a good idea. The point is further emphasized in problems 22.10 and 22.11.

Confusingly (for us) the translators of Landau–Lifshitz use "free energy" as their name for F (modified or not by $\boldsymbol{E}_{\mathrm{app}}$); their statements are not as close to ours as might appear.

[20]The relationship between $\boldsymbol{p}$ and $\boldsymbol{E}_{\mathrm{app}}$ may, of course, depend on temperature or pressure. So integration of the $\boldsymbol{p}(\boldsymbol{E})$ relation may well depend upon what other quantities are "controlled". It is the differential expression that is the same throughout.

Equation (22.4) gives the work done while moving the dipole a small distance into the field $\boldsymbol{E}_{\mathrm{app}}$. The total free energy is the work done in moving the dipole from infinity to its final position (plus the energy the dipole brought with it). However, $\boldsymbol{p}$ has an unspecified dependence upon $\boldsymbol{E}_{\mathrm{app}}$, and probably depends upon what else (pressure or volume for example) is being controlled in the sample's environment, so we cannot integrate at this stage and must be content with the differential expression (22.4)—just as we accept $-p\,\delta V$ or $T\,\delta S$ as useful differentials.

Comments

1. During introduction of the dipole (the "*sample*", terminology of Fig. 21.2), the only input of energy to the *system* is the work đW done by force $\boldsymbol{X}$. There is an associated movement of energy from voltage source to capacitor to field. The *system* must be defined so that it encloses *all* sample-related energy, so its boundaries must enclose the voltage source. Energy moving between source and capacitor is then an "internal transfer" and its movement does nothing to the sample's free energy—we must not think of including it in đW.[18]

2. The quantity we've been able to calculate, and prove to be a function of state (if hysteresis is absent), is the free energy (change). This is in line with experience in example 21.3 and statements in § 21.5.2. We see once again that it's the only sensible way to go.[19]

3. Free energy is located all over the place: some charge moved out of the voltage source when we introduced the dipole, so some of the free energy lies in the voltage source; the dipole's own self-field occupies a large volume round the dipole. Even if the sample hasn't absorbed heat, or changed volume, its free energy is delocalized, much as for the slab in example 21.3.

4. The absence of $\delta\boldsymbol{p}$ from đW agrees with expectations set out in § 21.4.6: $\boldsymbol{p}$ is adjusting itself to the applied field, until the free energy is a minimum with respect to variations of $\boldsymbol{p}$. Conversely (§ 21.4.1), we recognize in the presence in (22.4) of $\delta\boldsymbol{E}_{\mathrm{app}}$ a signal that $\boldsymbol{E}_{\mathrm{app}}$ is a "controlled" quantity—as of course it is.

5. Notice how simple and straightforward is our reasoning. By moving the sample within a slightly inhomogeneous field, we have been able to increment the $\boldsymbol{E}$-field it experiences without making any change to the capacitor or its voltage source. We could equally well set up the field by putting fixed-magnitude charges on the capacitor plates, or even make use of an electret (the electrical analogue of a magnet). There is very little here that could go wrong or generate the kind of confusion with which this chapter started.

22.3.1 A menu of free energies

The infinitesimal "work input" đ$W = -\boldsymbol{p}\cdot\delta\boldsymbol{E}_{\mathrm{app}}$ is the same whether the sample is held at controlled volume, or controlled pressure, or controlled temperature, or is thermally isolated. So this same quantity is the increment to all thermodynamic potentials.[20] We emphasize this by writing out the possible changes of free energy for different constraints.

$$
\begin{aligned}
U & & \delta U &= & T\,\delta S - p\delta V - \boldsymbol{p}\cdot\delta\boldsymbol{E}_{\text{app}} \\
H = U + pV & & \delta H &= & T\,\delta S + V\,\delta p - \boldsymbol{p}\cdot\delta\boldsymbol{E}_{\text{app}} \\
F = U - TS & & \delta F &= & -S\,\delta T - p\,\delta V - \boldsymbol{p}\cdot\delta\boldsymbol{E}_{\text{app}} \\
G = U - TS + pV & & \delta G &= & -S\,\delta T + V\,\delta p - \boldsymbol{p}\cdot\delta\boldsymbol{E}_{\text{app}}
\end{aligned}
$$

is the differential of the free energy appropriate to controlled

$$
\left\{
\begin{array}{ccc}
[S] & V & \boldsymbol{E}_{\text{app}} \\
[S] & p & \boldsymbol{E}_{\text{app}} \\
T & V & \boldsymbol{E}_{\text{app}} \\
T & p & \boldsymbol{E}_{\text{app}}
\end{array}
\right\}.
$$

$$(22.5)$$

The thermodynamic potentials U, F, G, H are natural extensions of the quantities represented by these symbols previously. With care (in particular care over "internal") we perpetuate the names also.

We repeat the observation in § 21.4.2. The free energy is a context-sensitive quantity; its value depends upon what in the sample's environment is "controlled". The appropriate free energy is determined uniquely, and without confusion, once we have decided upon the constraints to be applied.[21] In (22.5), $\boldsymbol{E}_{\text{app}}$ is added to the controllable quantities considered in Chapter 21, and it is handled in precisely the same way as p, V, T, S.

The thermodynamic potentials U, H, F, G are related to each other via Legendre transformations involving $+pV$ and $-TS$. In a similar way, we may apply the Legendre transformation of adding $+\boldsymbol{p}\cdot\boldsymbol{E}_{\text{app}}$; this yields four new functions of state (call them U', H', F', G') whose increments include $+\boldsymbol{E}_{\text{app}}\cdot\delta\boldsymbol{p}$. At this point, we have constructed eight certified functions of state. These in turn supply all the tools needed for performing thermodynamic analysis: for obtaining a full set of Maxwell relations appropriate to an electric-dipole sample.

We show below (§ 22.3.3) that the dipole moment $\boldsymbol{p}$ cannot be a "controlled" quantity. Hence U', H', F', G' cannot be free energies; but they remain functions of state and routes to Maxwell relations.

22.3.2   A model electric dipole: balls and spring

Sometimes it helps to have a simple model representing an electric dipole. I represent the dipole as two charges $\pm q$, held together by a spring and separated by distance a as in Fig. 22.3. In conformity with the precaution enunciated in § 22.2, the spring will be taken to have probably-non-linear elastic properties, so that the separation a of the charges is an unspecified function of the tension in the spring.[22]

Even when well away from an applied field, the dipole is surrounded by an $\boldsymbol{E}$-field, originating from its own charges $\pm q$. This field has energy $\int \frac{1}{2}\epsilon_0 \boldsymbol{E}^2 \, d\tau$. Whoever made the dipole had to supply this energy along with it.[23] If we pull the dipole longer, the field energy increases: work is done by the force we use to stretch the dipole. To that force, it feels just like pulling against a spring. So the dipole's own field is all "part of the spring". If we are worried about the Coulomb force which charge $+q$ exerts on charge $-q$ and vice versa, well, that's yet another way of looking at the field-energy part of the same spring.

Let's understand eqn (22.4) with the aid of the balls-and-spring model of an electric dipole shown in Fig. 22.3.

[21] The square brackets round $[S]$ perpetuate the convention introduced in § 21.4.1. Strictly, we have no means of applying external control on S, and what is meant is the near-equivalent of holding the sample thermally isolated.

Fig. 22.3 A balls-and-spring model for an electric dipole.

[22] The relationship between a and spring tension *is* assumed to be single-valued (though it might depend also on other variables such as temperature).

[23] In Chapter 1, such a field energy was said to contribute to the "self-energy" of the charges.

[24] Consider the dipole as one charge $-q$ at the origin where the electrostatic potential is V_0, and a second charge $+q$ at $\boldsymbol{a}$ where the potential is $V_0 + \boldsymbol{a}\cdot\nabla V$. The potential energy $\sum_i q_i V_i = q\boldsymbol{a}\cdot\nabla V = -\boldsymbol{p}\cdot\boldsymbol{E}_{\text{app}}$.

An alternative derivation prestretches the dipole to its final $|\boldsymbol{p}|$, and brings it from infinity to its final place, holding it always at right angles to $\boldsymbol{E}_{\text{app}}$ and therefore doing no work. The dipole is then rotated to its final orientation, with work done by a restraining torque $-\boldsymbol{p}\times\boldsymbol{E}_{\text{app}}$.

[25] It's possible that a real sample will have its dipole moment in a direction not parallel to $\boldsymbol{E}_{\text{app}}$, for reasons suggested in sidenotes 15 and 17.

[26] Suppose the free energy did consist of only the potential energy $-\boldsymbol{E}_{\text{app}}\cdot\boldsymbol{p}$. The system could minimize its energy by making $\boldsymbol{p}$ head for infinity. The charges would separate and fly in opposite directions. Clearly we need something like a spring to prevent this.

[27] This chapter is about the thermodynamics of macroscopic bodies. But the reasoning that justifies $-\boldsymbol{p}\cdot\delta\boldsymbol{E}_{\text{app}}$ can easily be adapted to a single atom or molecule. The balls-and-spring model makes a good place to start from.

[28] If this were done, there would be additions to the energy other than electrostatic .... But see sidenote 36 for an exception where $\boldsymbol{M}$ or $\boldsymbol{P}$ does have to be considered as the controlling—if not exactly controlled—quantity.

[29] During a fluctuation, there can be a non-zero $\delta\boldsymbol{p}$ even though no work is being input.

The dipole has electrostatic potential energy,[24] in the applied field, of $-\boldsymbol{p}\cdot\boldsymbol{E}_{\text{app}}$. This potential energy is really shared between the dipole and the sources of the applied field, but we allocate it all to the dipole (remember the tennis ball of § 21.1). That's what a free energy is.

In addition to its electrostatic potential energy, the dipole has elastic energy stored in its spring. The interesting energy is the sum of the two.

The dipole is sitting in the externally applied field and is in equilibrium. That is, each of the dipole's two charges experiences zero net force: the $q\boldsymbol{E}_{\text{app}}$ force is balanced by the spring. Now let the dipole elongate by δa (field unchanged). Energy added to the spring as the dipole elongates is (spring tension)$\delta a = (qE)\delta a = E(q\,\delta a) = E\,\delta|\boldsymbol{p}|$. Restoring vector notation[25] makes this energy change $+\boldsymbol{E}_{\text{app}}\cdot\delta\boldsymbol{p}$.

We can now do accountancy on the energy. We should remember experience in § 21.3, where we had to change pressure and temperature in order to give every kind of change a chance to show up. So here we'll have to change the applied $\boldsymbol{E}$-field.

$$\text{Increase of potential energy} = \delta(-\boldsymbol{p}\cdot\boldsymbol{E}_{\text{app}}) = -\boldsymbol{p}\cdot\delta\boldsymbol{E}_{\text{app}} - \boldsymbol{E}_{\text{app}}\cdot\delta\boldsymbol{p}$$
$$\text{increase of energy stored in spring} \qquad = \qquad\qquad +\boldsymbol{E}_{\text{app}}\cdot\delta\boldsymbol{p}$$
$$\text{total increase of free energy of dipole} \quad = -\boldsymbol{p}\cdot\delta\boldsymbol{E}_{\text{app}}. \qquad (22.6)$$

The result (22.6) agrees with (22.4), as it must.

This explains why the dipole's free energy has no first-order dependence on $\boldsymbol{p}$, no term in $\delta\boldsymbol{p}$ —something that might be counterintuitive at first. There is equilibrium when the free energy is a minimum. Around the minimum, a small increase of $\boldsymbol{p}$ reduces the potential energy in the field but increases the energy stored in the spring; these balance in first order, while in second order the total energy is increased.[26]

22.3.3 $\boldsymbol{E}$ can be "controlled", $\boldsymbol{p}$ not

Return to ideas discussed in § 21.3.2. A macroscopic[27] pVT sample confined to a controlled volume V has fluctuations in its pressure p; conversely, if it is held under controlled pressure there are fluctuations of volume V. Such fluctuations—explorations of microstates—are intrinsic to thermal equilibrium. In the case of our electric dipole, an environment that controls $\boldsymbol{E}_{\text{app}}$ sets the *mean* value of $\boldsymbol{p}$ via the dielectric equation of state, but it also permits the dipole moment $\boldsymbol{p}$ to fluctuate about that mean.

By contrast, there is no obvious way that we could hold $\boldsymbol{p}$ under control while allowing $\boldsymbol{E}_{\text{app}}$ to fluctuate in response. A ball-and-spring dipole might be physically stretched by grabbing and pulling,[28] but such a recourse is not available when the dipole moment is the total of many individual atomic dipole moments each reacting to its local environment.

It is for this reason that we stated in § 22.3.1 that $\boldsymbol{p}$ cannot be "controlled", and therefore U', H', F', G' cannot be free energies. It also follows that the energy $+\boldsymbol{E}_{\text{app}}\cdot\delta\boldsymbol{p}$ added to the spring does not represent the $\mathrm{d}W$ for an achievable work-input operation.[29]

22.4 Magnetic dipole in an applied magnetic field

We now consider the energy of a magnetic sample in an applied field, the fraught topic that motivates the whole of the present chapter.

For reasons explained in § 22.2, the sample's dipole moment m is assumed to be a not-necessarily-linear[30] function of field B_{app}.

We pre-prepare the field B_{app} by means of a large coil, and with the sample completely outside that field. Then we introduce the sample quasi-statically, doing mechanical work $\text{d}W$ on it. The sample's free energy is that work input, plus the energy the sample brought with it.

It is open to us to make a change of B_{app} along the lines in Chapter 21: remove sample, increase field, reinsert sample (problem 22.3 can be adapted easily). But an even simpler recourse is available. We make the field in the middle of the coil greater than the desired B_{app}, so that the dipole reaches the desired B_{app} when it is still in the fringing field. All we need do to increment the B_{app} experienced by the dipole is to move the dipole to a place where the field is stronger.

The coil is large compared with the sample so that the force on the dipole is $(m \cdot \nabla) B_{\text{app}}$ with negligible contributions from higher derivatives.

During introduction of the dipole m, we must prevent it from accelerating by applying a restraining force $X = -(m \cdot \nabla) B_{\text{app}}$. Now allow the dipole to move through δr. Work is added to the system by the restraining force and it is $\text{d}W = X \cdot \delta r = -\delta r \cdot (m \cdot \nabla) B_{\text{app}}$. This vector expression can be tidied up by writing it in components,[31] and it becomes (here B means B_{app}):

$$\delta(\text{free energy}) = \text{d}W = -\delta r \cdot (m \cdot \nabla) B_{\text{app}}$$

$$= -\delta x_i\, m_j\, \frac{\partial}{\partial x_j} B_i = -m_j\, \delta x_i\, \frac{\partial B_j}{\partial x_i} = -m_j\, \delta B_j = -m \cdot \delta B_{\text{app}}. \quad (22.7)$$

As was the case for the electric dipole, we cannot integrate the differential in (22.7): m is an unspecified function of B, and probably depends upon what other variables (pressure or volume, for example) are being controlled. But we can be content with the differential expression, alongside (perhaps) $T\,\delta S$ or $-p\,\delta V$.

Comments

1. The wording used above closely parallels that in § 22.3, because the arguments are in essence identical.

2. No use has been made of any illustrative model for the dipole, such as the balls-and-spring of § 22.3.2, so the proof is general.

3. The quantity we've calculated, and shown to be a function of state (in the absence of hysteresis), is the free energy (change). This is in line with experience in § 22.3, as well as in examples 21.3–21.4. Remember § 21.5.2: "first find the free energy".[32]

4. The "work" input $\text{d}W$ delivered by force X must be the only energy input to the *system* if we are to avoid muddle. The system must

[30] As with the electric-dipole case, we emphasize that the magnetic dipole moment is not fixed in magnitude or direction, but is free to adapt to the field—to be in equilibrium in the field—as the sample moves.

The magnetic moment m is allowed to depend on temperature, for example, as well as on B_{app}, so long as it is a single-valued function of the controlled conditions.

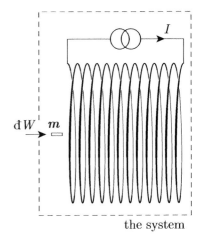

Fig. 22.4 A magnetic dipole is moved through δr into the fringing field B_{app} produced by a current-carrying coil.

As the dipole is moved, there is a change in the magnetic flux that it causes to link the turns of the coil, so energy moves from current generator to coil. Clean accountancy requires that energy enter the *system* via only one route, so the system's boundaries should be drawn to enclose the current generator. Compare with Fig. 22.2.

[31] The only "accessible" current $j_{\text{accessible}}$ is that flowing in the coil, and it is zero at the location of the sample. The field is time-independent, so $\partial D/\partial t = 0$. And the applied field is in vacuum, so $B = \mu_0 H$. Therefore curl $H = 0$ and curl $B = 0$, a property that is used for rearranging the partial derivatives. Compare with the derivation of (22.4).

[32] [From previous page] A different-looking derivation of (22.7) is given by Landau, Lifshitz and Pitaevskiĭ (1993) § 32. They build on the same ideas: a dipole moment not-necessarily-linearly related to field, and the applied field supplied by a large-distant coil driven by a reversible generator. They start from the $\int \boldsymbol{H} \cdot \delta \boldsymbol{B} \, \mathrm{d}\tau = \int \boldsymbol{j} \cdot \delta \boldsymbol{A} \, \mathrm{d}\tau$ energy change of the magnetic field, but they also add the energy change in the current generator.

The current generator is added by exploiting the idea in § 21.4.5. The current is to be controlled, so the free-energy change must contain $\delta \boldsymbol{j}$. This is achieved by the Legendre transformation of adding $-\int \boldsymbol{j} \cdot \boldsymbol{A} \, \mathrm{d}\tau$, bringing the free-energy change to contain $-\int \boldsymbol{A} \cdot \delta \boldsymbol{j} \, \mathrm{d}\tau = -\int \boldsymbol{B} \cdot \delta \boldsymbol{H} \, \mathrm{d}\tau$. This step is correct, though rather obscure.

There isn't a corresponding operation in the Landau–Lifshitz treatment of an electric dipole (their § 11) because it begins with fixed charges on the electrodes, rather than with a fixed potential difference. Had they used a voltage source to fix the potential difference, their electric case would also have needed to build in the energy of the source, and might have achieved this with a Legendre transformation. Compare with sidenotes 13 and 18: in our treatment of an electric dipole we included the energy in a voltage source, just in order to show that there can be such a contribution, easily handled.

have its boundaries drawn so as to make this the case. To clarify: During introduction of the sample, some energy moves from current generator to coil to field. This energy is sample-related, so it must be included in the sample's free energy throughout the process; equivalently, it must be within the *system* throughout. Compare with identical reasoning in the electric case of Fig. 22.2.

5. Free energy is located all over the place: energy moved out of the current generator when we introduced the dipole, so some of the free energy lies in the generator; the dipole's self-field occupies a large volume round the dipole. Even if the sample does not absorb heat from a heat bath, or change its volume in a controlled-pressure environment, the free energy is as "delocalized" as ever.

6. The absence of $\delta \boldsymbol{m}$ from $\mathrm{d}W$ agrees with expectations set out in § 21.4.6: $\boldsymbol{m}$ is adjusting itself to the applied field, until the free energy is a minimum with respect to variations of $\boldsymbol{m}$. Conversely (§ 21.4.1), we recognize in the presence of $\delta \boldsymbol{B}_{\mathrm{app}}$ a signal that $\boldsymbol{B}_{\mathrm{app}}$ is a "controlled" quantity—as of course it is.

7. Notice how simple and straightforward is our reasoning. By moving the sample within a slightly inhomogeneous field, we have been able to increment the $\boldsymbol{B}$-field it experiences without making any change to the coil or its current source. We could equally well set up the field by means of a permanent magnet. There is very little here that could go wrong or generate the kind of confusion with which this chapter started.

At this point we have secure reasoning in support of $-\boldsymbol{p} \cdot \delta \boldsymbol{E}_{\mathrm{app}}$ *and* $-\boldsymbol{m} \cdot \delta \boldsymbol{B}_{\mathrm{app}}$. *The main task of this chapter has been completed.* In what follows, we fill out consequences, and meet a textbook discussion head-on.

22.4.1   Magnetic sample: free energies

As in § 22.3.1, we give a menu of free-energy quantities.

$$
\left.
\begin{array}{ll}
U & \delta U = T \, \delta S - p \delta V - \boldsymbol{m} \cdot \delta \boldsymbol{B}_{\mathrm{app}} \\
H = U + pV & \delta H = T \, \delta S + V \, \delta p - \boldsymbol{m} \cdot \delta \boldsymbol{B}_{\mathrm{app}} \\
F = U - TS & \delta F = -S \, \delta T - p \, \delta V - \boldsymbol{m} \cdot \delta \boldsymbol{B}_{\mathrm{app}} \\
G = U - TS + pV & \delta G = -S \, \delta T + V \, \delta p - \boldsymbol{m} \cdot \delta \boldsymbol{B}_{\mathrm{app}}
\end{array}
\right\}
\text{is the differential of the free energy appropriate to controlled}
\left\{
\begin{array}{ccc}
[S] & V & \boldsymbol{B}_{\mathrm{app}} \\
[S] & p & \boldsymbol{B}_{\mathrm{app}} \\
T & V & \boldsymbol{B}_{\mathrm{app}} \\
T & p & \boldsymbol{B}_{\mathrm{app}}
\end{array}
\right\}.
$$

$$(22.8)$$

Comments about this menu exactly parallel those in § 22.3.1. The appropriate free energy is determined once we have decided what are the constraints applied to the magnetic sample. A further four functions of state (call them U', H', F', G') may be constructed by the Legendre transformation of adding $+\boldsymbol{m} \cdot \boldsymbol{B}_{\mathrm{app}}$. Such functions are not possible free energies, because there is no way[33] to impose external control on $\boldsymbol{m}$. But they are functions of state, and a source of Maxwell relations.

[33] Possible exception for a permanent magnet.

22.4.2 A magnetic dipole has a "spring"

Energy increment $+B_{\text{app}} \cdot \delta m$ can be understood as energy stored in a "spring" inside the magnetic sample. To substantiate this idea, we need to show that it works for a model dipole. I give two models that can make us comfortable with a magnetic "spring": one uses simple thermodynamics (problem 22.5), and the other uses the Zeeman effect in hyperfine structure (problem 22.6). For now, let us take it that there is such a thing as a spring with the energy increment given.[34]

The consistency of this idea may be confirmed as follows:[35]

$$
\begin{aligned}
\text{potential energy} &= -\, \boldsymbol{m} \cdot \boldsymbol{B}_{\text{app}} \\
\delta(\text{potential energy}) = \delta(-\boldsymbol{m} \cdot \boldsymbol{B}_{\text{app}}) &= -\, \boldsymbol{m} \cdot \delta\boldsymbol{B}_{\text{app}} \;-\; \boldsymbol{B}_{\text{app}} \cdot \delta\boldsymbol{m} \\
\delta(\text{energy stored in spring}) &= \hspace{3.5em} +\, \boldsymbol{B}_{\text{app}} \cdot \delta\boldsymbol{m} \\
\delta(\text{total free energy of dipole}) &= -\, \boldsymbol{m} \cdot \delta\boldsymbol{B}_{\text{app}}.
\end{aligned}
\tag{22.9}
$$

The pattern here exactly resembles that in (22.6), and helps to confirm that $+B_{\text{app}} \cdot \delta m$ is correctly identified as energy added to a "spring".

The applied field $\boldsymbol{B}_{\text{app}}$ is obviously a quantity that can be controlled by us: we control the current through the large coil that supplies $\boldsymbol{B}_{\text{app}}$. By contrast,[36] there is no way that we can exercise control of $\boldsymbol{m}$. The total magnetic dipole moment of the sample is the total of many individual atomic dipole moments each of which reacts to its local environment. Moreover, $\boldsymbol{m}$ can fluctuate through $\delta\boldsymbol{m}$ as the sample explores microstates, even though no work is being input. This is why thermodynamic potentials having differential $+B_{\text{app}} \cdot \delta m$ cannot be free energies.

22.4.3 Magnetic dipole: removing muddle

The question as to whether $\text{\dj}W$ should be $-\boldsymbol{m} \cdot \delta\boldsymbol{B}_{\text{app}}$ or $+\boldsymbol{B}_{\text{app}} \cdot \delta\boldsymbol{m}$ (or something else) is a source of muddle, as we saw at the beginning of this chapter. We must try to explain why it is so often found confusing. A good part of the answer lies in non-observance of points 2, 4, 7, 8 in § 21.7. These strategy mistakes do not make correct accountancy impossible, but they do make it difficult.

We examine one argument that is commonly presented, in order to see how it falls short. I am taking the view that even a correct analysis is not enough; we need also to understand why plausible alternatives will not do.

The coil of Fig. 22.5 carries current I and is filled with a sample whose magnetic-moment density is $\boldsymbol{M}$. Electromagnetic energy resides within the volume of the sample, in the surrounding field, and in the current generator. The coil is meant to be very long so that the field experienced by the sample is uniform, except for negligible regions of fringing field near the ends.

[34] *Of course* there's a spring. If $-\boldsymbol{m} \cdot \boldsymbol{B}_{\text{app}}$ were the only energy possessed by the magnetic dipole, the dipole would try to minimize its energy by making $\boldsymbol{m}$ head for infinity.

[35] To find the potential energy, pre-prepare the dipole to have its final magnitude, and clamp it to that value. Clamping means that energy cannot now enter or leave the spring. Start with the clamped dipole completely outside the field. Introduce the dipole into the field, inputting work $-\int \boldsymbol{m} \cdot \mathrm{d}\boldsymbol{B} = -\boldsymbol{m} \cdot \boldsymbol{B}_{\text{app}}$. Once this has been done the dipole moment can be un-clamped so that future changes are governed by (22.9).

[36] This section is concerned with the way in which a macroscopic magnetic sample responds to an externally applied (controlled) field $\boldsymbol{B}_{\text{app}}$. A very different case is that of a permanent magnet (or its electrical analogue the electret). The permanent-magnet regime requires a very different energetics from that of the present chapter, and is not pursued here.

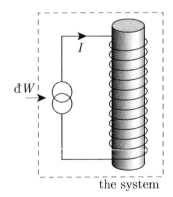

the system

Fig. 22.5 A magnetic sample fills a long coil. The coil is fed with current I from a current generator, details of which are not given. When we wish to increment the field, we increase I by δI, inputting the required energy "from the outside". A claimed energy accountancy is given in (22.10).

[37] Vectors $\boldsymbol{H}$ and $\boldsymbol{B}$ without subscripts are the actual fields in the volume of the sample.

[38] The field outside the coil should be thought about, since lines of $\boldsymbol{B}$ form closed loops. It is assumed to be negligible when the coil is long. I do not digress to investigate this point since the reasoning does not survive a more serious objection.

A variant argument considers a toroidal sample in a toroidal coil, which looks tidier but doesn't help.

[39] We have already asserted this for the electric and magnetic cases in § 22.3 comment 3 and in § 22.4 comment 5.

[40] Compare the treatment here with that of Landau, Lifshitz and Pitaevskiĭ, as described in sidenote 32; they are careful to include energy changes in the current generator.

[41] These include the two "it depends" accounts listed in § 22.1.

The wayward thought here seems to be "if the generator has got it then the sample hasn't". Equivalently: perhaps the *system* can optionally be redrawn so that the generator is outside? If that is the thinking, then it's backsliding from all we have done in Chapter 21 and here. Quite simply: the energy in contention contains $\boldsymbol{M}$, and that flag tells us it *must* be "to do with the sample"; remember § 21.7 item 8. No ambiguity. No choice. No "it depends".

[42] Remember point 7 in § 21.7. The use of a close fitting, rather than large, coil should have signalled that a non-optimum procedure was in prospect.

[43] But (22.7) gives a quicker and clearer route.

Increase the current by δI, and the energy contained in the fields changes, per unit volume, by

$$\boldsymbol{H}\cdot\delta\boldsymbol{B} = \boldsymbol{H}\cdot\delta(\mu_0\boldsymbol{H}) + \boldsymbol{H}\cdot\delta(\mu_0\boldsymbol{M})$$
$$= \boldsymbol{H}_{\text{app}}\cdot\delta\boldsymbol{B}_{\text{app}} + \mu_0\boldsymbol{H}_{\text{app}}\cdot\delta\boldsymbol{M}. \qquad (22.10)$$

Here, $\boldsymbol{H} = \boldsymbol{H}_{\text{app}}$ for a long coil because $\boldsymbol{H}$ relates to the current via curl $\boldsymbol{H} = \boldsymbol{j}$; it must be unaffected by the presence of the sample.[37,38] And $\boldsymbol{B}_{\text{app}} = \mu_0\boldsymbol{H}_{\text{app}}$ because these quantities refer to the empty coil in vacuum. The first term in (22.10) is identified as energy in the "empty environment"; the second term is then identified by difference as energy added to the sample. Therefore $đW = +\boldsymbol{B}_{\text{app}}\cdot\delta\boldsymbol{M}$ (integrated over the volume of the sample).

Critique of the above

- To increase current I, we have to supply additional energy to the current generator. Some of this energy addition depends upon $\boldsymbol{M}$ because the coil's inductance depends upon $\boldsymbol{M}$. After the change, some of the sample-related energy has moved into incrementing the fields, but some remains in the generator.[39,40] In the above, this in-generator energy was taken to be zero, since all added energy was assumed to be accounted for by $\boldsymbol{H}\cdot\delta\boldsymbol{B}$.

- Other discussions correctly identify the (change of) energy in the generator as $\delta(-\boldsymbol{M}\cdot\boldsymbol{B}_{\text{app}})$, but after that have difficulty deciding whether or not it is to do with the sample.[41]

- To sort this out, remember the precept of § 21.5.2: *For an open system, first find the free energy.* Only then can we have an authenticated function of state.

- Changing an environment with a sample in place is a recipe for difficult and untidy reasoning, as we have already seen in problem 21.1 and sidenote 19. We would do far better to follow the "canonical procedure" of Chapter 21: remove sample, change field, reintroduce sample. Yet the shapes of sample and coil of Fig. 22.5 have (needlessly) been set up to make such a calculation of the work impossible: a strategy error. It's even worse with a toroidal coil and sample.[42]

It is *possible* to succeed by changing the environment round an in-place sample. When all pieces of $đW$ are gathered together, we get $-\boldsymbol{M}\cdot\delta\boldsymbol{B}_{\text{app}}$, as we must.[43] But the reasoning is messy, as we've come to expect. I supply a fuller version in problem 22.11 because on the way it confirms statements made in this critique.

22.5 B or μ_0H?

In the case of a magnetic sample in an applied field, we've found that the change of free energy is $-\boldsymbol{m}\cdot\delta\boldsymbol{B}_{\text{app}}$. In this, I've written $\boldsymbol{B}_{\text{app}}$ for the applied field simply because we should write $\boldsymbol{B}$ unless we *really* mean $\mu_0\boldsymbol{H}$ —just as in electricity we write $\boldsymbol{E}$ rather than $\boldsymbol{D}/\epsilon_0$ unless we really mean $\boldsymbol{D}$. There is no difference between $\boldsymbol{B}_{\text{app}}$ and $\mu_0\boldsymbol{H}_{\text{app}}$

when the sample's surroundings are a vacuum. But not everything is a vacuum. What should we use in a non-vacuum?

Inevitably (!) both answers (and more) are to be found in print, so we've got to think. You and I might well have a prejudice in favour of treating $\boldsymbol{B}$ as the applied field, because $\boldsymbol{B}$ is the force vector in magnetism and $\boldsymbol{H}$ is a derived quantity (as with $\boldsymbol{E}$ and $\boldsymbol{D}$). Well, maybe. But go back to the time, about 60 years ago, when magnetism was taught in c.g.s. units: physicists said that $\boldsymbol{H}$ was the force vector acting on magnetic poles,[44] and $\boldsymbol{B}$ was the derived quantity ($\boldsymbol{H}$ and $\boldsymbol{B}$ corresponding to $\boldsymbol{E}$ and $\boldsymbol{D}$ and not the other way round). Both approaches can't be right, so we must mistrust both.

To find out, we need to make $\boldsymbol{B}_{\mathrm{app}}$ and $\mu_0 \boldsymbol{H}_{\mathrm{app}}$ different. So immerse the dipole $\boldsymbol{m}$ in a magnetic fluid within which $\boldsymbol{B} = \mu_{\mathrm{r}}\mu_0 \boldsymbol{H}$. Then the dipole's free-energy change might be $-\boldsymbol{m} \cdot \delta \boldsymbol{B}_{\mathrm{app}} = -\boldsymbol{m} \cdot \delta(\mu_{\mathrm{r}}\mu_0 \boldsymbol{H}_{\mathrm{app}})$ or $-\boldsymbol{m} \cdot \delta(\mu_0 \boldsymbol{H}_{\mathrm{app}})$ or something else. There are knock-on questions as to the force and torque experienced by $\boldsymbol{m}$.

Unfortunately, the answer is "something else". The free-energy change depends on the shape of the sample. Simple calculations can be done only for the case where the sample is ellipsoidal. If the ellipsoid is thinned to a needle the free-energy change is $-\boldsymbol{m} \cdot \delta(\mu_0 \boldsymbol{H}_{\mathrm{app}})$. If it is flattened to a disc, the free-energy change is $-\boldsymbol{m} \cdot \delta \boldsymbol{B}_{\mathrm{app}}$. And for an intermediate shape such as a sphere, it's in between.

In summary, there is no one-expression-fits-all-shapes for the free energy when a magnetic fluid surrounds a body having a magnetic moment $\boldsymbol{m}$. And neither $\boldsymbol{B}$ nor $\mu_0 \boldsymbol{H}$ is favoured within the range of possibilities.

I do not digress to prove the statements made in this section. How about a mini-project?[45]

You can take comfort from the fact that all this has caused muddle even to competent people for years. As late as 1971, it was worth doing an experiment, measuring the torque on a magnet immersed in a magnetic liquid, because people were genuinely uncertain what would happen (Whitworth and Stopes-Roe 1971).

22.6   The Robinson constant-current generator

Reference has been made, in §§ 21.6.1 and 22.4, to the need for an in-principle reversible current generator. Figure 22.6 shows such a device, originated by Robinson (1973b, Fig. 15.3).

Above the terminals ($x > 0$) is a current generator that feeds current I into a coil, its "load", which might be the coil of Fig. 22.4. All conductors are resistance-free, so I is a non-zero persistent current and the system is in a true thermodynamic equilibrium.

The two rails are vertical and are bridged electrically by the slider; the slider is free to move under gravity without friction. The slider's weight μg is supported by the $\boldsymbol{j} \times \boldsymbol{B}$ force. The $\boldsymbol{j} \times \boldsymbol{B}$ force is proportional to

[44] This piece of history, and habits arising from it, explains why $\boldsymbol{H}$ is still often written as the symbol of preference rather than the more correct $\boldsymbol{B}$.

The change of default symbol from $\boldsymbol{H}$ to $\boldsymbol{B}$ followed from a modernization of the fundamental definitions relating to magnetism in matter. Confusingly, this reform was implemented (in the UK at any rate) at the same time as the change from the c.g.s. to the SI system of definitions. Because of this historical coincidence, it can seem that the physics is changed by the change of units. It is only a bit like Walter Marshall's alarm that the laws of physics seem to change on crossing the Channel.

[45] It is easiest to look for textbook statements that deal with the analogous case of an electric dipole $\boldsymbol{p}$ embedded in a dielectric fluid having $\boldsymbol{D} = \epsilon_{\mathrm{r}}\epsilon_0 \boldsymbol{E}$. Then everything we need is to be found in Landau, Lifshitz and Pitaevskiĭ (1993) Chapter 2.

A body embedded in a polarizable fluid has a free-energy change of

$$-\int (\boldsymbol{D} - \epsilon_{\mathrm{r}}\epsilon_0 \boldsymbol{E}) \cdot \delta \boldsymbol{E}_{\mathrm{app}} \, \mathrm{d}V.$$

Here $\boldsymbol{E}$ and $\boldsymbol{D}$ are the actual field and displacement within the volume V of the sample. The bracket $(\boldsymbol{D} - \epsilon_{\mathrm{r}}\epsilon_0 \boldsymbol{E})$ is not equal to the dipole-moment density within the sample because of the presence of ϵ_{r}, but it is non-zero only within the volume of the sample.

The expression given above is a slight extension of what Landau–Lifshitz give (their § 11, problem), but the extension is easily made. Their statements hold only for a body whose $\boldsymbol{D}$ and $\boldsymbol{E}$ are linearly related to $\boldsymbol{E}_{\mathrm{app}}$. The extension I give does not make that assumption, needfully so. However, we require an expression that gives a change of free energy, which exists only for thermodynamic equilibrium. Therefore it is required that the relation between $\boldsymbol{D}, \boldsymbol{E}$ and $\boldsymbol{E}_{\mathrm{app}}$ must be single-valued.

To find $\boldsymbol{D}$ and $\boldsymbol{E}$ within the volume of the sample (which must be ellipsoidal), op. cit. (8.12).

The force and torque on the sample can be found from a free-energy change, for example the change that comes about if we move the sample into a stronger $\boldsymbol{E}_{\mathrm{app}}$. The force and torque may alternatively be found from the same reference, eqns (16.8) and (16.9).

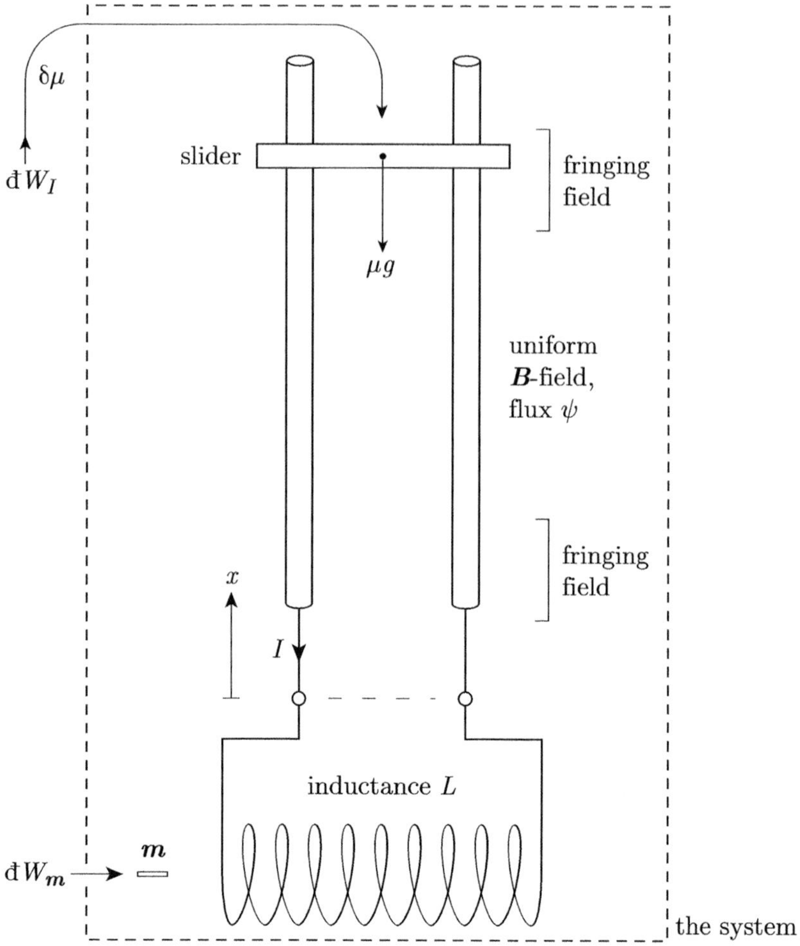

Fig. 22.6 The Robinson current generator.

The device supplies a "controlled" current I to the coil which is its load. Everything is resistance-free so that no energy is dissipated and the device is in thermodynamic equilibrium.

The rails are vertical and are bridged electrically by the slider which is free to move under gravity without friction.

In the region between the fringing fields, the $\boldsymbol{B}$-field (which faces out of the paper) is "uniform" in the sense that it is independent of x. It is, of course, dependent on distance from the conductors. The uniformity makes it possible to define an inductance λ per unit length of the rail structure (for values of x away from the fringing fields).

The slider's weight μg is supported by the $\boldsymbol{j}\times\boldsymbol{B}$ force. This force is shown in the text to be equal to $\frac{1}{2}\lambda I^2$, where I is the current flowing. For any given mass μ, the current is therefore set by the relation $\mu g = \frac{1}{2}\lambda I^2$.

A magnetic dipole $\boldsymbol{m}$ is shown, in the fringing field of the coil.

Work $đW$ can be input $(đW_{\boldsymbol{m}})$ by pushing the dipole further into the coil's fringing field. Alternatively $(đW_I)$ we may add mass $\delta\mu$ to the slider, thereby increasing the current I. Thus the circuit offers both of the possibilities of Figs. 22.4 and 22.5 for adding to the dipole's free energy. These possibilities are strictly either/or.

current I squared, so there is a definite current I for which this force is just able to support the slider. If the slider is in equilibrium, the current must have this value.

The entire circuit, meaning the generator and whatever it is connected to, encloses a constant magnetic flux ϕ. If, for any reason, the current decreases, the slider is not fully supported and starts to descend, the area available to the flux ϕ decreases, the $\boldsymbol{B}$-field increases to maintain the flux constant, the current rises for consistency with the increased $\boldsymbol{B}$, and equilibrium is restored when I is back up to its correct value. The system is therefore error-correcting in such a way as to do what it is intended for: to drive a constant current through the load.

Near the slider, and near the terminals at the bottom, there are "fringing fields" of complicated shape. But the rails are "long": the fringing fields are confined to regions near the ends, leaving plenty of space for a uniform-field region[46] between them. We are setting up for a "virtual work" analysis resembling that of example 21.3 (which had a long region of uniform field for the same reason).

[46] The field is far from uniform in directions across and normal to the diagram. It is uniform in the sense that it is independent of x.

22.6.1   Functioning of the current generator

We assemble the current generator by building up the current in it from zero by a sequence of manoeuvres:[47]

- Start with no current flowing, and raise the slider from the bottom, inputting work $\mu g x$ against gravity. Clamp the slider in place temporarily.
- Disconnect the generator from the load temporarily.
- Connect an external current source to the generator, and raise its current from 0 to I. When the current has reached the required value, attach a short circuit between the terminals and remove the current source; a persistent current I now flows through the rails, slider and short.[48] If we have set I correctly, the mass μ is now supported by the $\boldsymbol{j} \times \boldsymbol{B}$ force, and the clamp fixing the slider can be removed. Now use a similar procedure to build up a current I in the load circuit, and make it into a persistent current by means of a second short. Finally, the two circuits are joined together at the terminals; the two shorts carry a combined current of zero, so they can be removed without disruption. The circuit is now complete.

In the region of uniform $\boldsymbol{B}$-field, the rails have inductance λ per unit length, meaning that length δx is threaded by flux $(\lambda \, \delta x) I$. The inductance of the whole circuit is $\lambda x + L$, where L is the inductance of the circuit being driven; more precisely, LI includes the flux through the two regions of fringing field as well as through the load. The magnetic flux threading the whole circuit is

$$\text{flux threading entire circuit} = \phi = (\lambda x + L) I. \qquad (22.11)$$

In this, x and I have been set in independent operations,[49] and ϕ has ended up with whatever flux results.

The total energy that has been given to the circuit during assembly is $\mu g x + \frac{1}{2}(\lambda x + L) I^2$.

During circuit operation, ϕ is constant in the loss-free circuit; this fixes $(\lambda x + L) I$, meaning that x and I are now dependent on each other. Use (22.11) to eliminate I (or x) from the energy:

$$\text{energy in circuit} = \mu g x + \frac{1}{2}(\lambda x + L) I^2 = \mu g x + \frac{\phi^2}{2(\lambda x + L)}. \qquad (22.12)$$

If the slider is in equilibrium, the circuit's energy must be a minimum with respect to variation of x (or of I), otherwise some energy could be converted into kinetic energy of the slider. At a minimum there can be no change of energy to first order in δx. Imagine then that the slider is raised through δx. The fringing fields remain unaltered in shape and therefore in energy;[50] the uniform-field region lengthens by δx. These considerations permit us to find the energy change by differentiating (22.12). The reasoning resembles that in example 21.3.

$$0 = \frac{\partial(\text{energy})}{\partial x} = \mu g - \frac{\phi^2 \lambda}{2(\lambda x + L)^2},$$

[47] The magnetic sample is, of course, absent at this stage, as we are assembling the environment into which it will later be put.

[48] This procedure mimics a way in which a persistent current can be set up in a superconducting circuit, such as a superconducting magnet.

[49] During assembly, we can choose to position the slider at any reasonable height x without that choice affecting the final current I.

[50] The upper fringing field is simply translated upwards, while the lower fringing field does not move. We are saying that L is not changed by imposing δx, and so the change of $(\lambda x + L)$ is simply $\lambda \, \delta x$.

and the current I is then given by

$$I = \frac{\phi}{\lambda x + L} = \sqrt{\frac{2\mu g}{\lambda}}; \qquad \mu g = \tfrac{1}{2}\lambda I^2. \qquad (22.13)$$

Remembering example 21.3, we conclude

$$\text{upward } \boldsymbol{j} \times \boldsymbol{B} \text{ force acting on slider} = \tfrac{1}{2}\lambda I^2; \qquad (22.14)$$

this must be so whether or not this force is being balanced by weight μg or by some other force. All of the energy-minimization reasoning from eqn (22.12) to here has been a device for obtaining (22.14).

From now on, the generator will always be in equilibrium, changing at most quasi-statically. Unless we intervene to impose some additional force on the slider (which we do in problems 22.9 and 22.11), equation (22.13) will hold always as well. As forecast, the current I is determined by the weight μg of the slider and a quantity λ set by the geometry of the rails.

We backtrack to find the equilibrium value of the energy stored in the whole circuit, using (22.12), (22.13) and (22.11):[51]

$$\begin{aligned} \text{energy in circuit} &= \mu g x + \tfrac{1}{2}(\lambda x + L)I^2 = \tfrac{1}{2}\lambda I^2 x + \tfrac{1}{2}(\lambda x + L)I^2 \\ &= I(\lambda x I) + \tfrac{1}{2}LI^2 = I(\phi - LI) + \tfrac{1}{2}LI^2 \\ &= I\phi - \tfrac{1}{2}LI^2. \end{aligned} \qquad (22.15)$$

In the uniform-field part of the rail structure there is magnetic flux

$$\text{magnetic flux threading generator} = \psi = \lambda x I. \qquad (22.16)$$

Then:[52]

$$\text{energy stored in current generator} = \mu g x + \tfrac{1}{2}(\lambda x)I^2 = \lambda x I^2 = I\psi. \qquad (22.17)$$

Now in an application of the current generator we shall make changes to the "load" such that the flux ψ though the generator changes by $\delta\psi$. The current I remains constant, so

$$\text{energy change in current generator} = I\,\delta\psi. \qquad (22.18)$$

Statements (22.17) and (22.18) look innocuous, but they are all that we need for an application such as example 21.4.

22.7   Magnetic dipole moved into $\boldsymbol{B}$-field

Figure 22.6 shows a Robinson current generator feeding current into a large coil which might be coil of Fig. 22.4. A sample having magnetic dipole moment $\boldsymbol{m}$ is to be moved from infinity into the coil.[53] The coil and current generator constitute the sample's environment. The sample's magnetic moment $\boldsymbol{m}$ is some unspecified (but single-valued) function of the field $\boldsymbol{B}_{\text{app}}$ and possibly of other controlled variables.

The sample starts at a large distance from the coil, where the $\boldsymbol{B}$-field acting on it is zero. This is the "sample absent" condition. We move the sample into the coil (not necessarily right in) until it is at a location where the field acting on it (the field that was there before the sample

[51] This quantity is not of much interest here, but we use it in problem 22.8.

[52] We treat the "generator" as consisting of the slider and the region of uniform field. The fringing fields are treated as part of the load, just as their inductance is absorbed into L. Flux ϕ is different from ψ because it includes the flux threading the coil and the fringing fields.

[53] There is an associated work input $\mathrm{d}W_m$; $\mathrm{d}W_I$ is zero.

was introduced) is $\boldsymbol{B} = \boldsymbol{B}_{\mathrm{app}}$. We shall catalogue the changes and movements of energy that accompany the sample's introduction.

Work is done in pushing the sample into the field, after the manner of eqn (22.7). Move the sample through a distance $\delta\boldsymbol{r}$ towards the coil. The force pulling the dipole in is $(\boldsymbol{m} \cdot \nabla)\boldsymbol{B}_{\mathrm{app}}$, so the external force restraining the dipole is $-(\boldsymbol{m} \cdot \nabla)\boldsymbol{B}_{\mathrm{app}}$ and the work done by this force is (cf. eqn 22.7)

$$ đ(\text{work input to system}) = đW_{\boldsymbol{m}} = -\delta\boldsymbol{r} \cdot (\boldsymbol{m} \cdot \nabla)\boldsymbol{B}_{\mathrm{app}} = -\boldsymbol{m} \cdot \delta\boldsymbol{B}_{\mathrm{app}}. \tag{22.19} $$

When the sample is at a place where it experiences an applied field $\boldsymbol{B}_{\mathrm{app}}$ caused by current I, an additional flux threads the coil given by

$$ \begin{pmatrix} \text{flux threading coil caused by} \\ \text{magnetic moment } \boldsymbol{m} \text{ of sample} \end{pmatrix} = \boldsymbol{m} \cdot \boldsymbol{B}_{\mathrm{app}}/I. \tag{22.20} $$

See problem 22.7.

During movement of the sample through a small distance, the magnetic flux threading the coil increases by $\delta(\boldsymbol{m} \cdot \boldsymbol{B}_{\mathrm{app}})/I$. The flux threading the entire circuit cannot change, so an increase of flux through the coil must be accompanied by an equal and opposite change of flux ψ in the current generator: $\delta\psi = -\delta(\boldsymbol{m} \cdot \boldsymbol{B}_{\mathrm{app}})/I$. There is an increase in the energy stored in the generator, by (22.18), of

$$ \delta(\text{energy stored in generator}) = I\,\delta\psi = I \times \left\{ -\delta(\boldsymbol{m} \cdot \boldsymbol{B}_{\mathrm{app}})/I \right\} $$
$$ = \delta(-\boldsymbol{m} \cdot \boldsymbol{B}_{\mathrm{app}}). \tag{22.21} $$

These changes are accompanied (eqn (22.16)) by a rise of the slider through δx where

$$ \delta x = \frac{\delta\psi}{\lambda I} = \left(\frac{-\delta(\boldsymbol{m} \cdot \boldsymbol{B}_{\mathrm{app}})}{I} \right) \frac{1}{\lambda I} = -\left(\frac{1}{\lambda I^2} \right) \delta(\boldsymbol{m} \cdot \boldsymbol{B}_{\mathrm{app}}). \tag{22.22} $$

Comments

- When the sample is moved into the coil, the magnetic flux through the coil changes, the slider descends; there is a change of energy in the current generator, partly in the inductance of the rail structure, partly in the gravitational potential energy of the slider. Free energy is delocalized, as we have come to expect.
- The energy that moves between the current generator and the coil is wholly an "internal transfer" of energy within the "system" (definition of Fig. 21.2). It has no effect on the energy contained in the entire system, and therefore none on the sample's free energy.[54] The only thing that changes the free energy of the sample is the external work given by (22.19). This is of course in agreement with the đW of (22.7) calculated in § 22.4.
- Equation (22.21) can be integrated to show that energy contained within the generator includes $-\boldsymbol{m} \cdot \boldsymbol{B}_{\mathrm{app}}$, equal to the potential energy of the sample in the field.[55] We may see this as a departure from previous experience, where potential energy has often had no identifiable location. However, it is important not to over-interpret the exception. The potential energy is, as always, shared between

[54] Remember bullet points in § 22.4.3. We have no right to think in any other way about energy movements, at least until the free energy is securely identified.

[55] This is the energy that was omitted (at first) in the incorrect discussion of § 22.4.3.

[56] A permanent magnet could have been used to apply field $\boldsymbol{B}_{\text{app}}$ to the sample. The potential energy would then be given by $-\boldsymbol{m}_{\text{sample}}\cdot\boldsymbol{B}_{\text{from magnet}}$ or by $-\boldsymbol{m}_{\text{magnet}}\cdot\boldsymbol{B}_{\text{from sample}}$, these, of course, being equivalent expressions for the same shared energy. However, little more can be learned. We have chosen instead to use a current generator and coil because the detail revealed by doing so is much more informative.

[57] Taking this bullet point with sidenote 55, I make the wry observation that confusion has resulted from both paying too little attention to $-\boldsymbol{m}\cdot\boldsymbol{B}_{\text{app}}$ (forgetting that it is present in the current generator) and paying too much attention to it (including it but mis-interpreting the significance of its location).

[58] The reader is entitled to cry "overkill" here. In logic that's right. But when confusion is so widespread, there is a need to leave no possible wriggle room.

sample and environment, and is allocated to the sample in the definition of free energy. None of this is changed, or challenged. Indeed, the analysis of this section has shown that fragments of energy can be found all over the system, and it would surely be perverse to attach deep significance to the location of any one of these energy pieces. The only thing that matters is the total.[56]

- I think the last bullet point explains the two "it depends" accounts mentioned in § 22.1. These accounts apparently understand that potential energy $-\boldsymbol{m}\cdot\boldsymbol{B}_{\text{app}}$ "really is" in the generator, therefore "outside the sample", and so optionally "belonging to the sample" or not. This is not how a shared energy should be viewed.[57]

22.7.1   Changing the field round the sample

In § 22.4.3, we presented a calculation of the energy changes that take place when the field applied to a sample is changed with the sample in place; that calculation is faulty because the energy changes it finds are incomplete. We make a "friendly amendment" to that analysis below, finding *all* the energy movements, including those in the current generator. Problem 22.10 follows the "canonical procedure" of § 21.3.1: remove sample; change field; reinsert sample. Then problem 22.11 confronts § 22.4.3 directly, obtaining the same results by increasing the field with the sample remaining in place. By this somewhat painstaking route,[58] we show that the "work input" $\mathrm{d}W$ is always $-\boldsymbol{m}\cdot\delta\boldsymbol{B}_{\text{app}}$.

Problem 22.11 is a bit complicated, and we work up to it by first finding the $\mathrm{d}W$ needed for changing the current in an empty coil: problems 22.8 and 22.9.

Problems

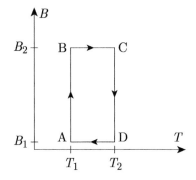

Fig. 22.7 A thermodynamic cycle. A sample is taken round cycle ABCD. It is shown in problem 22.1 that the sum of $T\,\mathrm{d}S+m\,\mathrm{d}B$, taken round the loop, is not zero, and therefore this expression is not the change of a function of state.

Problem 22.1 $+\boldsymbol{m}\cdot\delta\boldsymbol{B}_{\text{app}}$ yields a non-function of state

Define a thermodynamic function whose change is $T\,\delta S+m\,\delta B$, related (by a Legendre transformation) to item (5) in the list of § 22.1. Take a magnetic sample round the loop ABCD drawn in Fig. 22.7. The fields B_1, B_2 and the temperatures T_1, T_2 are so chosen that the sample is fully magnetized at B, C and fully unmagnetized at A, D (field B_1 is zero or close to zero). The magnetic entropy is then zero at B, C and $S_{m=0}=R\ln(2J+1)$ at A, D.

Show that

(a) On path AB, $\int T\,\mathrm{d}S=-T_1 R\ln(2J+1)$.

(b) On path BC, $\int T\,\mathrm{d}S=0$ and $\int m\,\mathrm{d}B=0$.

(c) On path CD, $\int T\,\mathrm{d}S=+T_2 R\ln(2J+1)$.

(d) On path DA, $\int T\,\mathrm{d}S=0$ and $\int m\,\mathrm{d}B=0$.

(e) Round the whole loop the change of the thermodynamic function is

$$(T_2-T_1)R\ln(2J+1)+\int_{B_1}^{B_2}\Big\{m(B,T_1)-m(B,T_2)\Big\}\mathrm{d}B.$$

It is known from experiment that magnetic materials have magnetization decreasing with temperature rise. Argue that both terms in the last expression are positive.[59] Argue that $T\,\mathrm{d}S + m\,\mathrm{d}B$ cannot represent the change of a function of state because it does not sum to zero round the loop. Argue that $T\,\mathrm{d}S - m\,\mathrm{d}B$ is not, by this argument, excluded from being the change of a function of state.

Argue that item (5) in the catalogue of § 22.1 is excluded from giving a function of state because what it defines is obtained from this problem's non-function by a Legendre transformation.

Problem 22.2 The force on an electric dipole moment
Show that the force on an electric dipole moment $\boldsymbol{p}$ in an inhomogeneous static electric field $\boldsymbol{E}_{\mathrm{app}}$ is $\boldsymbol{F} = (\boldsymbol{p}\cdot\nabla)\boldsymbol{E}_{\mathrm{app}}$. The reasoning can be modified from that in sidenote 24.

Problem 22.3 The free energy of an electric dipole, using the "canonical procedure"
We change the field $\boldsymbol{E}_{\mathrm{app}}$ experienced by an electric dipole $\boldsymbol{p}$ by first removing the dipole, then incrementing the field, then reinserting the dipole to its original location z_0. The dipole's path is along the z-axis with $z = 0$ outside the field.

The dipole experiences field $E_0(z)$ during its removal, changed to $E_0(z) + \delta E(z)$ during reinsertion.[60] Show that $\mathrm{d}W = -p\,\delta E$ where this is evaluated at the final location $z = z_0$.

Problem 22.4 The force on a magnetic dipole moment
Show that the force on a magnetic dipole moment $\boldsymbol{m}$ in an inhomogeneous static magnetic field $\boldsymbol{B}$ is $\boldsymbol{F} = (\boldsymbol{m}\cdot\nabla)\boldsymbol{B}$ (we write $\boldsymbol{B}$ for $\boldsymbol{B}_{\mathrm{app}}$ in this problem). The recommended route is as follows.[61,62]

A magnetic dipole moment $\boldsymbol{m}$ arises because charges move within some small region, while remaining within that region. The charges might constitute a current in a loop of conductor, but they might be charges orbiting within an atom. To cover all possibilities, we write the magnetic moment as[63]

$$\boldsymbol{m} = \tfrac{1}{2}\sum_a q_a\,\boldsymbol{r}_a \times \boldsymbol{v}_a, \qquad (22.23)$$

where the index a identifies the ath of the charges. Now that the notation has been explained, we shall drop the a, to make room for other subscripts.

Write the magnetic field, after the fashion of the electric field in problem 22.2, as $\boldsymbol{B} = \boldsymbol{B}_0 + (\boldsymbol{r}\cdot\nabla)\boldsymbol{B}$. Then the force on the moving charges is the Lorentz force:

$$\boldsymbol{F} = \sum q\,\boldsymbol{v} \times \boldsymbol{B} = \sum q\,\boldsymbol{v} \times \big\{\boldsymbol{B}_0 + (\boldsymbol{r}\cdot\nabla)\boldsymbol{B}\big\}$$
$$= \sum q\,\boldsymbol{v} \times \boldsymbol{B}_0 + \sum q\,\boldsymbol{v} \times (\boldsymbol{r}\cdot\nabla\boldsymbol{B}).$$

Now $\boldsymbol{v} = \mathrm{d}\boldsymbol{r}/\mathrm{d}t$, so process the last statement as follows:

$$\boldsymbol{F} = \sum q\,\boldsymbol{v} \times \boldsymbol{B}_0 + \sum q\frac{\mathrm{d}}{\mathrm{d}t}\big\{\boldsymbol{r} \times (\boldsymbol{r}\cdot\nabla\boldsymbol{B})\big\} - \sum q\,\boldsymbol{r} \times (\boldsymbol{v}\cdot\nabla\boldsymbol{B}).$$

[59] For a mole of atoms with angular momentum J, $S_{m=0} = R\ln(2J + 1)$, and this has been written for definiteness. But for (a) and (c) in the present argument all we need is the knowledge from statistical mechanics that $S_{\mathrm{saturated}} < S_{m=0}$.

[60] Even this application of the canonical procedure makes the calculation more complicated than is necessary, because you have to integrate the increased work over z. Find an even quicker way.

[61] In problem 22.4, it is assumed that the magnetic moment originates from loss-free currents circulating within the atoms of the sample. This is a reasonable (albeit classical) model if the atoms have magnetism arising wholly from orbital angular momentum, but that is rather unlikely. More realistic is the case where the magnetic moment arises, at least in part, from spin, most probably electron spin. In this case, each spin has potential energy $-\boldsymbol{\mu}\cdot\boldsymbol{B}_{\mathrm{app}}$ in field $\boldsymbol{B}_{\mathrm{app}}$, and this is the observational fact that leads us to say that it has magnetic moment $\boldsymbol{\mu}$. The force $\boldsymbol{f} = (\boldsymbol{\mu}\cdot\nabla)\boldsymbol{B}_{\mathrm{app}}$ can then be obtained by differentiating the energy with respect to position (there is no spring to worry about). Either way the force on the whole sample is $\boldsymbol{F} = \sum \boldsymbol{f} = (\boldsymbol{m}\cdot\nabla)\boldsymbol{B}_{\mathrm{app}}$.

[62] The thinking is taken from Landau and Lifshitz (1996) § 44, and the calculation is modified from Bleaney and Bleaney (2013) p. 106.

[63] In the first instance, we are here using a classical model to work out the magnetic moment of a single atom, with index a identifying the electrons within the atom. The value of expression (22.23) for $\boldsymbol{m}$ depends upon the origin chosen for $\boldsymbol{r}$. However, this dependence drops out at the stage when we take a time average. Keep an eye on this.

Combine the last two statements, writing half their sum:

$$\boldsymbol{F} = \sum q \frac{\mathrm{d}\boldsymbol{r}}{\mathrm{d}t} \times \boldsymbol{B}_0 + \frac{1}{2}\sum q \frac{\mathrm{d}}{\mathrm{d}t}\left\{ \boldsymbol{r} \times (\boldsymbol{r} \cdot \nabla \boldsymbol{B})\right\}$$
$$+ \tfrac{1}{2}\sum q\, \boldsymbol{v} \times (\boldsymbol{r} \cdot \nabla \boldsymbol{B}) - \tfrac{1}{2}\sum q\, \boldsymbol{r} \times (\boldsymbol{v} \cdot \nabla \boldsymbol{B}).$$

We shall now average over time, since the charges constituting the magnetic dipole moment may give a dipole moment that changes with time as the charges execute their finite motion.[64] Argue that the time average of any time derivative must be zero when the averaging time is made long. So the first two terms in the last equation drop out. We have

$$\overline{\boldsymbol{F}} = \tfrac{1}{2}\sum q\,\overline{\boldsymbol{v} \times (\boldsymbol{r} \cdot \nabla \boldsymbol{B})} - \tfrac{1}{2}\sum q\,\overline{\boldsymbol{r} \times (\boldsymbol{v} \cdot \nabla \boldsymbol{B})}.$$

Next, we need vector relation (3.24):

$$\boldsymbol{v} \times (\boldsymbol{r} \cdot \nabla \boldsymbol{B}) - \boldsymbol{r} \times (\boldsymbol{v} \cdot \nabla \boldsymbol{B})$$
$$= (\boldsymbol{r} \times \boldsymbol{v}) \times \operatorname{curl}\boldsymbol{B} - (\boldsymbol{r} \times \boldsymbol{v})\operatorname{div}\boldsymbol{B} + (\boldsymbol{r} \times \boldsymbol{v}) \cdot \nabla \boldsymbol{B}.$$

Argue that $\operatorname{curl}\boldsymbol{B}$ is zero at the location of the dipole, because $\boldsymbol{B}$ is the applied field, and the currents creating that field must be located elsewhere. The term containing $\operatorname{div}\boldsymbol{B}$ drops out as well, and we are left with

$$\overline{\boldsymbol{F}} = \tfrac{1}{2}\sum q\,\overline{(\boldsymbol{r} \times \boldsymbol{v})} \cdot \nabla \boldsymbol{B} = (\overline{\boldsymbol{m}} \cdot \nabla)\boldsymbol{B}. \qquad (22.24)$$

Note the re-use of (22.23) in this step.

Problem 22.5   An entropy-driven spring

Consider a sample that is thermally isolated and at controlled volume. It is in an externally applied magnetic field $\boldsymbol{B}_{\mathrm{app}}$. From (22.8) its free energy is U. The sample must be considered as an "open system" because U can be changed by $\mathrm{d}W = -\boldsymbol{m} \cdot \delta \boldsymbol{B}_{\mathrm{app}}$.

Let the sample be a solid[65] that is an assembly of N independent atoms,[66] each having angular momentum $\hbar \boldsymbol{J}$. It is to be placed in a uniform field $\boldsymbol{B}_{\mathrm{app}}$ in the z-direction, but to start with the field is zero. A representative atom in state m_J has a magnetic moment of fixed magnitude[67] $\mu_z = -g_J \mu_{\mathrm{B}} m_J$. We increase the field to its final value B_{app}, and the atom's contribution to the free energy changes by $\int -\mu_z\, \mathrm{d}B = -\mu_z B_{\mathrm{app}} = (g_J \mu_{\mathrm{B}}\, m_J)B_{\mathrm{app}}$.

Let there be N_{m_J} atoms present in the m_J state. Then[68]

$$U = \sum_{m_J} N_{m_J}(g_J \mu_{\mathrm{B}} m_J)B_{\mathrm{app}} = -\boldsymbol{m} \cdot \boldsymbol{B}_{\mathrm{app}}, \qquad (22.25)$$

the sum being taken over all m_J states (from $-J$ to J).[69]

The U of (22.25) is simply the potential energy of the sample in the field; this has come about because the atoms are unpolarizable. By contrast, the *sample* is polarizable because it can respond to a changed B_{app} by changing the N_{m_J}. Such changes of N_{m_J} change the "number of microstates" and thence the entropy; statistical mechanics is not far away.

[64] Suppose we have classical charges orbiting round each other. The magnetic moment is not then necessarily independent of time (for more than two charges), and the charges also have a time-dependent electric dipole moment. The averaging over time is done to remove such distractions.

[65] We specify a solid so that the atoms are localized, and thus obey Boltzmann statistics.

[66] It's assumed that the atoms don't interact with each other ("independent"), so there is no local-field correction: the $\boldsymbol{B}_{\mathrm{local}}$ acting on an atom is not made different from $\boldsymbol{B}_{\mathrm{app}}$ by the proximity of the other atoms.

The atoms do have interactions, e.g. with phonons in the lattice, just enough that they come to a proper thermodynamic equilibrium (progressing through microstates), but these interactions are weak enough not to affect the atoms' energies. This assumption, or an equivalent one, is usual in elementary statistical mechanics.

[67] Fixed magnitude because the field will always be weak enough that we have the weak-field Zeeman effect.

[68] To keep things simple, we assume that the temperature is low enough that energies from phonons and conduction electrons (if any) are negligible beside the magnetic energy of (22.25).

[69] The reader might well have written (22.25) at once, thinking it to be an obvious expression for "internal energy". However, there is no escaping the fact that we have an "open system" in which the sample interacts with the externally applied field and can exchange energy with it through a $\mathrm{d}W$ that has to be calculated.

Change the field and find the changes of energy that take place. (We ignore changes in the volume of the sample.) From (22.8) and (22.25)

$$\delta U = T \, \delta S - \boldsymbol{m} \cdot \delta \boldsymbol{B}_{\text{app}} = \delta(-\boldsymbol{m} \cdot \boldsymbol{B}_{\text{app}})$$

so that

$$-T \, \delta S = +\boldsymbol{B}_{\text{app}} \, \delta \boldsymbol{m}. \tag{22.26}$$

Referring to (22.9), we see that $-T \, \delta S$ supplies the change of energy that we there described as attributed to a "spring". We have a spring, and it is an entropy-driven spring.

Think about eqn (22.26). It should make good sense that any increase of magnetic moment $\boldsymbol{m}$ decreases the disorder in the atomic angular momenta, and therefore should entail a decrease in entropy. We have previously met this idea in (22.3).[70]

You may have encountered other instances of an entropy-driven spring, such as that in an idealized model for rubber.[71]

[70] There is little for the reader to do in this "problem" except follow the argument presented. My aim has been to supply a simple model for reasoning about an "open system".

[71] See, e.g., Riedi (1988), p. 145.

Problem 22.6 The energy of a single atom in a magnetic field
Here is another mind-clearing model, with about the same usefulness as that of problem 22.5.

It would be nice if we could discuss a single atom with magnetic dipole moment $\boldsymbol{\mu}$, in terms that resemble the charges-with-spring model of § 22.3.2 and eqns (22.6). At first sight, there's a difficulty, because in the Zeeman effect the magnetic moment of an atom is fixed in magnitude, for any given m_J-state. But we could think about using a very strong field and getting into the intermediate region between Zeeman and Paschen–Back. Then the atom's wave function becomes a linear combination with B-dependent coefficients; the magnetic moment is a function of field. Once we think of this, we realise that a more down-to-earth example can be found in hyperfine structure—down-to-earth because the B-field required to approach the Back–Goudsmit limit is modest, and if the

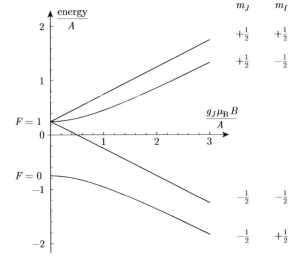

Fig. 22.8 The energy levels of a hydrogen (^{1}H) atom in a magnetic field. The levels shown are those for the 1s ground configuration.

atom is in its ground (electronic) state it can hang around there with infinite or near-infinite lifetime.

Think then about a hydrogen (^{1}H) atom. Its 1s ground configuration is split into four hyperfine states, which in zero external field have $F = 0$ and $F = 1$. The lines drawn on Fig. 22.8 show the variation of the four states' energies with $\boldsymbol{B}$-field; two of those lines are bent because the wave function is a $\boldsymbol{B}$-dependent mixture. On the bent lines, the total z-component of angular momentum is zero, so the atom's wave function ψ must be a linear combination of the building blocks:

$$\psi_0 = \left|m_J = -\tfrac{1}{2}, \, m_I = \tfrac{1}{2}\right\rangle, \qquad \psi_1 = \left|m_J = \tfrac{1}{2}, \, m_I = -\tfrac{1}{2}\right\rangle.$$

That is,

$$\psi = a\,\psi_0 + b\,\psi_1, \tag{22.27}$$

where a and b vary with the field (one set of values for each of the two bent lines).

The applied field $\boldsymbol{B} = \boldsymbol{B}_{\text{app}}$ will lie in the z-direction. Then the energy of the atom is determined by the Hamiltonian:

$$\widehat{H'} = A\widehat{\boldsymbol{I}} \cdot \widehat{\boldsymbol{J}} + \mu_{\text{B}} B \left(g_J \widehat{J_z} + g_I \widehat{I_z}\right). \tag{22.28}$$

That is, the atom's energy consists of two pieces:

- the hyperfine energy $A\langle \boldsymbol{I} \cdot \boldsymbol{J}\rangle$
- the potential energy in the field $\langle -\boldsymbol{m} \cdot \boldsymbol{B}\rangle = \mu_{\text{B}} B \langle g_J J_z + g_I I_z\rangle$.

[72] Solution at the end of this chapter.

Work out the quantum mechanics of this, and show that when a small change δB is made to the field we have energy change[72]

$$\delta E = -\langle \boldsymbol{m}\rangle \cdot \delta \boldsymbol{B}_{\text{app}} \tag{22.29}$$

and

$$A\,\delta\langle \boldsymbol{I} \cdot \boldsymbol{J}\rangle = +\boldsymbol{B}_{\text{app}} \cdot \delta\langle \boldsymbol{m}\rangle. \tag{22.30}$$

We are thus able to show that the hyperfine energy of the atom serves as the "spring" in this example, and the identification of the various pieces of energy proceeds exactly as it did in § 22.4.2 and eqn (22.9).

Problem 22.7 Flux through coil, caused by magnetic dipole

[73] *Suggestion:* Replace the sample by a small flat coil of n turns and area a carrying a current i that gives the small coil the same magnetic moment m as the sample; then $m = nai$. The magnetic flux threading the small coil is naB, so the mutual inductance of the two coils is naB/I. The mutual inductance is just that, mutual, so it also gives us the flux threading the main coil per unit of current in the small coil. Tidy up from here.

Prove eqn (22.20).[73]

Problem 22.8 Change current in an empty Robinson generator: easy
We have built a Robinson generator, as in § 22.6, so that it supplies current I_1. No magnetic sample is present. We wish to change the generator so that it supplies a new current I_2, with the same flux ϕ threading the entire circuit. It is simplest to imagine that we dismantle the system, and rebuild it afresh to generate the new current. This involves adding mass to the slider until the total slider mass is μ_2, with $\mu_2 g = \tfrac{1}{2}\lambda I_2^2$. Show from (22.15) that

$$\text{(change of energy in whole circuit)} = \phi(I_2 - I_1) - \tfrac{1}{2}L\big(I_2^2 - I_1^2\big), \tag{22.31}$$

and from (22.17) that

$$\text{(change of energy stored in current generator alone)} = I_2\psi_2 - I_1\psi_1. \tag{22.32}$$

Because we keep ϕ unchanged in this process, the energy changes must be the same as would be obtained had we caused the current to change by any other means. This is confirmed by brute force in problem 22.9.

Problem 22.9 Change current in an empty Robinson generator: brute force

Take a Robinson current generator, generating current I_1, again with no magnetic sample present. Push the slider (down if the current is to be increased) until the current is changed to I_2. Work is input by the force that pushes the slider.

(1) Show that the slider must be displaced from height x_1 to height x_2 with

$$\lambda x_1 + L = \frac{\phi}{I_1}, \qquad \lambda x_2 + L = \frac{\phi}{I_2}, \qquad x_2 - x_1 = -\frac{\phi}{\lambda}\frac{(I_2 - I_1)}{I_1 I_2}.$$

(2) Show that the work input in pushing the slider is

$$\int_{x_1}^{x_2} \left(\mu_1 g - \tfrac{1}{2}\lambda I^2\right)\mathrm{d}x = \tfrac{1}{2}\phi\frac{(I_2 - I_1)^2}{I_2}. \tag{22.33}$$

(3) For the system to function as a generator of the new current I_2, weight $(\mu_2 - \mu_1)g = \tfrac{1}{2}\lambda(I_2^2 - I_1^2)$ must be added to the slider,[74] after being raised from $x = 0$ to $x = x_2$. Show that the total work input for changing the current agrees with (22.31).

Problem 22.10 Change the B-field applied to a sample: sample moved

A Robinson current generator supplies current to a coil, with a sample present, as in Fig. 22.6. We start with current I_1 passing through the coil, and we change the current to I_2. The change need not be small. All quantities bear subscripts 1 and 2 similarly to indicate values before and after the change.

For this problem, we use the canonical procedure: remove the sample, change the field, then reinsert the sample.

(1) Argue that the work inputs from the outside are itemized as follows:

- Work done in removing the sample from the field B_1 is given by (22.19).[75] It is $-\boldsymbol{m}\cdot\mathrm{d}\boldsymbol{B}$, integrated from field B_1 to field zero.
- Work done in increasing the field with the sample absent is given by eqn (22.31).
- Work done in reinserting sample into field B_2 is $-\boldsymbol{m}\cdot\mathrm{d}\boldsymbol{B}$, integrated from field zero to B_2.

The work done directly on the sample is[76]

$$\int_{B_1}^{0} -m\,\mathrm{d}B + \int_{0}^{B_2} -m\,\mathrm{d}B = \int_{B_1}^{B_2} -m\,\mathrm{d}B,$$

The total work input is

$$\Delta W = \phi(I_2 - I_1) - \tfrac{1}{2}L(I_2^2 - I_1^2) \quad \text{work done on empty environment}$$

$$- \int_{B_1}^{B_2} m\,\mathrm{d}B \qquad\qquad \text{increase in free energy of sample.}$$

$$\tag{22.34}$$

[74] The raising of a mass such as $\mu_2 - \mu_1$ is hinted at in Fig. 22.6. However, what we have here is the mass that must be added when changing the current I with the sample absent. The $\delta\mu$ in Fig. 22.6 includes this but also includes the additional mass that must be raised when the sample is present.

[75] Throughout this problem and the next, B means B_{app}.

[76] It is assumed that there is a single-valued relationship between m and B: no hysteresis allowed.

If the changes of current and field are small, the change to the free energy of the sample reduces simply to $-m\,\delta B$. The energy added to the environment is of less interest (it's nothing to do with the sample) but then it didn't cost much effort to obtain it.

(2) Show that the above changes are accompanied by movements of the current generator's slider:

- $x_{1\,\text{full}}$ to $x_{1\,\text{empty}}$ as the sample is removed
- $x_{1\,\text{empty}}$ to $x_{2\,\text{empty}}$ as the current is increased
- $x_{2\,\text{empty}}$ to $x_{2\,\text{full}}$ as the sample is reinserted.

Find expressions for all three displacements, and show that the overall movement is $(x_{2\,\text{full}} - x_{1\,\text{full}})$ given by

$$\lambda(x_{2\,\text{full}} - x_{1\,\text{full}}) = -\phi \frac{I_2 - I_1}{I_1 I_2} - \frac{m_2 B_2}{I_2^2} + \frac{m_1 B_1}{I_1^2}. \qquad (22.35)$$

Problem 22.11 Change the $\boldsymbol{B}$-field with sample remaining in place
As a consistency check, we obtain the same results as in problem 22.10 but this time we increase the field while the sample remains in place.

(1) To increase the current from I_1 to I_2, we push down the slider in much the same way as we did in problem 22.9. But now the sample remains in place in the applied B-field, and its magnetic moment affects the magnetic flux movements and therefore the movement of the slider. It helps to write $\beta = B/I$, where β is a constant for the coil. Then the total flux threading coil and generator is, from (22.11) and (22.20),

$$\phi = (\lambda x + L)I + \frac{mB}{I} \quad \text{so} \quad \lambda x = \frac{\phi}{I} - \frac{mB}{I^2} - L = \frac{\phi}{I} - \beta \frac{m}{I} - L. \quad (22.36)$$

From this, show that $(x_{2\text{full}} - x_{1\text{full}})$ takes the same value as in (22.35).

(2) Find the work that is done in pushing the slider (down if the current is to be increased), using a procedure modelled on that of problem 22.9. The force needed is affected by the presence of the sample, so that

$$\lambda\,\mathrm{d}x = -\left\{ \frac{\phi}{I^2} + \beta \frac{\mathrm{d}}{\mathrm{d}I}\left(\frac{m}{I}\right) \right\}\mathrm{d}I.$$

Perform the integration[77] along the lines of (22.33) but with this changed $\mathrm{d}x$ and show that the work done on the slider is

$$\mathrm{d}W_I = \phi \frac{(I_2 - I_1)^2}{2I_2} + m_2 B_2 \frac{(I_2^2 - I_1^2)}{2I_2^2} - \int_{B_1}^{B_2} m\,\mathrm{d}B. \qquad (22.37)$$

(3) Energy must also be input to raise a mass $(\mu_2 - \mu_1)$ through height $x_{2\,\text{full}}$; height $x_{2\,\text{full}}$ can be obtained from (22.36). Add the work inputs and show that the total agrees with (22.34).

Comments

1. Problems 22.10 and 22.11 give the same results:[78] the increase of free energy is $-\boldsymbol{m} \cdot \delta \boldsymbol{B}_{\text{app}}$. This is no surprise because in both cases we evaluated the energy-as-work input $\mathrm{d}W$.

2. Problem 22.11 directly confronts the treatment in § 22.4.3. We have shown that, when work input to the current generator is included, the energy input is $-\boldsymbol{m} \cdot \delta \boldsymbol{B}_{\text{app}}$ and not the $+\boldsymbol{B}_{\text{app}} \cdot \delta \boldsymbol{m}$ of (22.10).

[77] Do not assume any specific relationship between m and $B = B_{\text{app}}$, save that it's single-valued. The integration can be done generally.

[78] In view of the fact that we are obtaining results in conflict with much that is in print, it was necessary to make searching consistency checks, giving any discrepancy a chance to show up.

3. There are two terms in the work-input expression (22.34). One does not contain m and is the energy gain of the empty system. The other contains m and is the increase in the sample's free energy. Had we not previously worked problem 22.10, we would have needed to use this (presence/absence of m) as an "algebraic signal" to separate the two terms and infer that $-m \cdot \delta B_{\text{app}}$ is the addition to the free energy of the sample.[79] By contrast, the canonical procedure (remove sample, change environment, re-insert sample) used in problem 22.10 exhibits the two contributions without need for any algebraic signal. There can be no doubt that problem 22.10 displays the superior reasoning.

4. The calculation that obtains (22.34) via (22.37) is considerably messier than that followed in problem 22.10. Yet another price is paid for changing the environment with the sample in place.

5. For all these reasons, even our "friendly amendment" to the treatment of § 22.4.3 is unable to make it look at all attractive.[80]

Answer to problem 22.6

Problem 22.6 The energy of a single atom in a magnetic field
The Hamiltonian is given by (22.28). In it, the nuclear g-factor g_I is of order $1/2000$ (on the definition I've used) and I'll ignore it henceforth; it clutters the algebra without adding any insight.

The wave functions and energies can be found in the usual way by working out matrix elements of the Hamiltonian, using (for the bent lines) wave functions ψ_0 and ψ_1 as basis set. I invite you to solve the problem that way. The four energies are given by a Breit–Rabi formula:

$$\left. \begin{aligned} E &= A(\tfrac{1}{4} \pm \tfrac{1}{2}x) \\ E &= A(-\tfrac{1}{4} \pm \tfrac{1}{2}\sqrt{1+x^2}) \end{aligned} \right\} \quad \text{where } x = g_J \mu_B B/A. \quad (22.38)$$

For the present solution, I'll adopt a different line of attack because it will make the required result seem inevitable.

For any wave function of the form (22.27), whether or not the coefficients a and b have the correct values to give an energy eigenstate, we can calculate the expectation value of the energy:

$$\langle E \rangle = A\langle \mathbf{I} \cdot \mathbf{J} \rangle + g_J \mu_B B \langle J_z \rangle. \quad (22.39)$$

We now consider the change of $\langle E \rangle$ that results from making small variations δa, δb in the coefficients of the wave function ψ:

$$\delta\langle E \rangle = A\,\delta\langle \mathbf{I} \cdot \mathbf{J} \rangle + g_J \mu_B B\,\delta\langle J_z \rangle;$$

the changes on the right can be written down (but need not be) in terms of δa and δb. (Variations δa and δb are not independent because the wave function has to remain normalized: $|a|^2 + |b|^2 = 1$.) If a and b have been randomly chosen, the change $\delta\langle E \rangle$ is of first order in δa. But there are values of a and b for which the change of $\langle E \rangle$ is only of second order. The **variational principle** states that when first-order changes δa, δb result in only second-order changes of $\langle E \rangle$, then we have an energy eigenstate and $\langle E \rangle$ is the energy E of that eigenstate.[81] Then a and b may be found by requiring

$$0 = \delta\langle E \rangle = A\,\delta\langle \mathbf{I} \cdot \mathbf{J} \rangle + g_J \mu_B B\,\delta\langle J_z \rangle \quad (22.40)$$

[79] We recall Chapter 21 sidenote 37: not every sample-dependent quantity contains m. So, although the algebraic signal does not mislead here, we ought to rely on it only if no better procedure is on offer—and then with great circumspection.

[80] We remember that problem 21.1 was a bad route to finding the $+V\,\delta p$ energy. We were warned.

[81] For the lowest energy, $\langle E \rangle$ is a minimum with respect to variations of a and b. But more generally, $\langle E \rangle$ is *stationary* with respect to such variations.

for values of δa and δb that are linked by $|a|^2 + |b|^2 = 1$. This constraint can be handled by the method of Lagrange multipliers; the outcome is the same as that of solving for E and ψ by the more conventional route. (If you're keen you can verify this.)

Next, we consider a small slow change of field from B to $B + \delta B$. An atom may experience such a change if it moves in and out of the fields of magnets, for example within an atomic-beam apparatus. The change of field entails changes of a and b as the wave function changes to a new linear combination—a new eigenstate for the changed Hamiltonian.[82] By differentiation of the eigenvalue E given by (22.39), we have

$$\delta E = A\,\delta\langle \boldsymbol{I} \cdot \boldsymbol{J}\rangle + g_J\mu_{\mathrm{B}}\,B\,\delta\langle J_z\rangle + g_J\mu_{\mathrm{B}}\,\langle J_z\rangle\,\delta B.$$

By virtue of the variational result (22.40), the first two terms here sum to zero:

$$\text{change of hyperfine energy } A\,\delta\langle \boldsymbol{I} \cdot \boldsymbol{J}\rangle = -Bg_J\mu_{\mathrm{B}}\,\delta\langle J_z\rangle = +\boldsymbol{B} \cdot \delta\boldsymbol{m}, \quad (22.41)$$

and we are left with only the final term which can be written as

$$\delta E = -\boldsymbol{m} \cdot \delta\boldsymbol{B}. \qquad (22.42)$$

It will now be clear from (22.41) that the hyperfine interaction $A\boldsymbol{I} \cdot \boldsymbol{J}$ is able to store energy within the atom in just the manner of a spring, taking up energy $+\boldsymbol{B} \cdot \delta\boldsymbol{m}$.

We can now repeat the statements of (22.9) in the present terms:

increase of potential energy $= \delta(-\boldsymbol{m} \cdot \boldsymbol{B})\quad = -\,\boldsymbol{m} \cdot \delta\boldsymbol{B}\, -\, \boldsymbol{B} \cdot \delta\boldsymbol{m}$

increase of energy stored in $A\langle \boldsymbol{I} \cdot \boldsymbol{J}\rangle$ spring $=\qquad\qquad\qquad +\,\boldsymbol{B} \cdot \delta\boldsymbol{m}$

total increase of atom's energy $\qquad\qquad = -\,\boldsymbol{m} \cdot \delta\boldsymbol{B}. \qquad (22.43)$

Comments

1. The accountancy in (22.43) exactly reproduces that of (22.9). It also resembles that of (22.6), even though the underlying model is very different and the spring is handled differently. This should give us confidence that a breakdown of the energy (electric or magnetic) along the lines of (22.43) should be general, independent of the model chosen to illustrate it.

2. You should be able to see that the variational result (22.41) is the magnetic equivalent of the statement in § 22.3.2 that an electric dipole's charges are in equilibrium under the combined forces from the applied field and the internal "spring".

[82] The point of "slow" is that the $\boldsymbol{B}$-field felt by the atom has no Fourier components that could induce a transition between the hyperfine states. The atom therefore moves along one of the lines drawn in Fig. 22.8, and does not jump between those lines.

XI Solid-state physics

The Debye theory of solid-state heat capacities

Intended readership: after a first look at the Debye theory, so as to get a better idea of where the cut-off comes from.

23.1 Introduction

The Debye theory is treated well in the standard textbooks on solid-state physics. There is, however, one detail that can be quite difficult to understand: the Debye cut-off, where the frequencies of vibrational modes are given a greatest value so as to accommodate the correct number of modes. This chapter introduces the theory in an unusual way, one aim being to make the cut-off seem reasonable.

23.2 The monatomic linear chain

The monatomic linear chain provides a simple model upon which we build a description of vibrational states, first for one dimension, and then by extension for a three-dimensional crystalline solid. It is assumed that the reader has already encountered the monatomic linear chain, so the description here will be brief.

The chain at rest consists of masses m, lying along the x-axis and spaced by a. The jth mass is displaced by y_j, which may be a longitudinal or a transverse displacement. The masses are connected by springs with spring constant β. Then the equation of motion for the jth mass is[1]

$$m\ddot{y}_j = \beta(y_{j+1} - y_j) - \beta(y_j - y_{j-1}) = \beta(y_{j+1} + y_{j-1} - 2y_j). \quad (23.1)$$

Look for a solution of the form $y_j = (\text{constant})\, e^{i(kx_j - \omega t)}$, with outcome

$$\omega = \sqrt{\frac{4\beta}{m}}\, \left|\sin(\tfrac{1}{2}ka)\right|. \quad (23.2)$$

This relation between ω and k is shown in Fig. 23.1.

We shall give the chain a finite length[2] Na, so that the e^{ikx} functions form a discrete set that can be used to expand any actual motion of the chain: a Fourier series. As discussed in § 6.3.1, the values of k can be made discrete in more than one way, according to how $e^{i(kx_j - \omega t)}$ is chosen to continue the y_j outside the basis range of length Na. We here

[1] Notice the mathematical resemblance between the expression on the right of (23.1) and a second derivative.

$$d^2y/dx^2 =$$
$$\lim_{a \to 0} \frac{y(x+a) + y(x-a) - 2y(x)}{a^2}.$$

If the limit is taken, as it can be when the wavelength greatly exceeds the interatomic spacing a, then (23.1) becomes the ordinary wave equation

$$\frac{\partial^2 y}{\partial t^2} = \frac{\beta a^2}{m} \frac{\partial^2 y}{\partial x^2}$$

with wave speed $\sqrt{\beta a^2/m}$. In the text, the limit is not taken, so (23.1) applies to waves whose wavelength is not necessarily large compared with a.

[2] To be precise, the chain has N gaps of width a, framed by $(N+1)$ atoms. If both ends are fixed, $(N-1)$ atoms are left free to move. If periodic boundary conditions are applied, atom N duplicates the motion of atom zero. Also, given the absence of external forces, one mode, having $k = 0$, representing a uniform motion of the entire chain, is excluded. Again, the chain supports $(N-1)$ independent motions.

The length Na may be carved out from a longer chain, just as in Chapter 6 we chose a large "quantization volume" V that might be smaller than the actual volume containing the waves under discussion.

Essays in Physics: Thirty-two thoughtful essays on topics in undergraduate-level physics. Geoffrey Brooker, Oxford University Press. © Geoffrey Brooker 2021. DOI: 10.1093/oso/9780198857242.003.0023

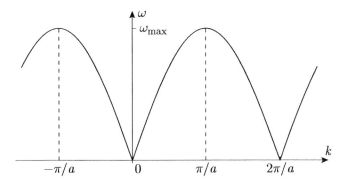

Fig. 23.1 The $\omega(k)$ relation for a monatomic linear chain whose masses have equilibrium separation a. All possible motions of the atoms making up the chain are described by e^{ikx} waves whose k-values lie within the first Brillouin zone: the region here chosen to be $-\pi/a < k \leqslant \pi/a$.

apply "periodic boundary conditions":[3]

$$y_{j+N} = y_j, \qquad \text{equivalently} \quad y(x + Na) = y(x), \qquad (23.3)$$

There are $(N-1)$ atoms whose initial displacements may be independently chosen: not N because a displacement having $k = 0$ is excluded—it could be made time dependent only by applying an external force to the chain as a whole. A full description in terms of the e^{ikx} must be achieved by using $(N - 1)$ such functions—$(N - 1)$ values of k—accompanied by $(N - 1)$ coefficients.

Condition (23.3) tells us that

$$\exp\{ik(x + Na)\} = \exp(ikx), \quad \text{so} \quad e^{ikNa} = 1,$$

giving

$$kNa = (\text{integer } r)2\pi, \quad \text{and} \quad k = k_r = \frac{2\pi}{a}\frac{r}{N}. \qquad (23.4)$$

Since we have the correct number of functions when r is given $(N - 1)$ different values, (23.4) tells us that the requisite values of k cover a range of $2\pi/a$. It is usually desirable for $-k$ to be treated on an equal footing with $+k$, and then we choose the first available k-range, the **first Brillouin zone**,[4] such that k lies between $-\pi/a$ and $+\pi/a$. This zone choice is delimited by broken lines in Fig. 23.1.

We can check directly that values of k outside the first zone duplicate those within it, meaning that they give rise to duplicate functions e^{ikx}:

$$\exp\{i(k + 2\pi/a)x_j\} = e^{ikx_j} e^{i2\pi(x_j/a)} = e^{ikx_j},$$

since all the x_j are integer multiples[5] of a. The reader may recognize $2\pi/a$ as a reciprocal lattice vector for the one-dimensional chain.

23.3 Vibrational modes in three dimensions

The ideas of the last section can be carried over to a three-dimensional crystalline solid composed of N atoms. The eigenfunctions for the vibration of atoms in the crystal have (by our choice) the form $e^{i(\boldsymbol{k}\cdot\boldsymbol{r}-\omega t)}$, in which the values of $\boldsymbol{k}$ are made discrete by a procedure similar to that for the one-dimensional case. The first Brillouin zone contains N values of $\boldsymbol{k}$. The N atoms can move in three dimensions, but this factor 3 is

[3] Actually, this step was anticipated when we looked for a solution in the form $e^{i(kx-\omega t)}$. Had the chain had both of its ends fixed, we would have started from $y_j \propto \sin(kx_j)e^{-i\omega t}$ with $0 \leqslant x_j \leqslant Na$ (problem 23.1(2)).

Compare with the discussion of waves in a continuous medium in §6.3.1. The choice of expansion-function set is unconnected with what happens at the ends of the chain, but is concerned with the usefulness of the functions chosen. Condition (23.3) is applied here because travelling waves are usually more suitable for the physics than are standing waves.

In §6.3.1, we found that all expansion functions fail at the boundaries of the quantization volume, different functions failing in different ways. Problem 23.1 asks you to consider whether a similar failure happens here.

[4] There is an infinite set of Brillouin zones. Each zone contains N values of k, and the ks within any one zone can be used to describe completely the motion of the chain. For more on Brillouin zones, see Chapters 24 and 28.

There are nit-picking differences in the precise k-values depending on whether N is even or odd. These are investigated in problem 23.1. Nevertheless, we are to use the monatomic chain as a model for a solid containing $\sim 10^{23}$ atoms, so it cannot matter whether N is even or odd; we ignore such details in the main text.

[5] We have not needed to specify where the origin of x lies, so we should say that $x_j = ja + x_0$. Show that the presence of x_0 does not upset the reasoning.

Pictures of these zones are given in books on solid-state physics. In particular, we mention Kittel (2005) Chapter 2, and Brillouin (1953) Chapters 6, 7. For the zones of a two-dimensional square lattice, an insightful case intermediate between one dimension and three, see Chapter 28.

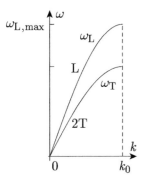

Fig. 23.2 The $\omega(k)$ relations for longitudinal (L) and transverse (2T) vibrations of an isotropic solid. The labelling indicates that each $\boldsymbol{k}$ has one longitudinal and two transverse waves associated with it. The total number of modes for N atoms is $3N$. The greatest value of k is k_0, given by (23.5).

[7] Since $k = |\boldsymbol{k}|$, we now plot positive values of k only.

[8] The word "average" appears repeatedly here, and is important.

[9] A formal demonstration of this should be given. See problem 23.2. For the extension of the reasoning to three dimensions, see your favourite book on solid-state physics.

[10] Statement (23.6) is often written as $C_{\mathrm{v}} = 3R$, for the special case where the quantity of solid material is one mole; $R = N_{\mathrm{Avogadro}} k_{\mathrm{B}}$ is the gas constant for one mole. We prefer to deal with a sample of general size, so N will not be set equal to Avogadro's number.

accounted for by the existence of three wave polarizations: each $\boldsymbol{k}$ is used three times over, with (in general) three values of ω.

There are complications that afflict real solids. The first Brillouin zone extends over a shape in $\boldsymbol{k}$-space that depends upon the crystal structure (and the same is of course true of higher zones).[6]

In order to keep things simple, we shall usually ignore the possible different crystal structures, and treat the $\omega(\boldsymbol{k})$ relation as isotropic: independent of the direction of $\boldsymbol{k}$. For an isotropic solid, waves are either longitudinal (sound) or transverse (shear waves). For each $k = |\boldsymbol{k}|$ there is one longitudinal wave and two transverse waves (Chapter 4). Thus Fig. 23.1 becomes changed into Fig. 23.2.[7] The angular frequency of the longitudinal wave is $\omega_{\mathrm{L}}(k)$, and similarly the angular frequency of the two transverse waves (degenerate by assumption) is $\omega_{\mathrm{T}}(k)$.

If we make the further simplifying assumption that the first Brillouin zone has a spherical boundary at $|\boldsymbol{k}| = k_0$, then for N atoms occupying volume V in real space

$$\frac{N}{V} = \frac{4\pi}{3}\frac{k_0^3}{(2\pi)^3}. \tag{23.5}$$

23.4 The internal energy and heat capacity

We consider a dielectric solid, so there is no contribution to the heat capacity from electrons. The excitations of interest are phonons only.

23.4.1 Classical theory

To understand the classical theory, we need the concept of a **degree of freedom**. This is a term in the energy that is quadratic in a coordinate or a momentum. Thus, for example, a harmonic oscillator having energy

$$E = \frac{p_y^2}{2m} + \frac{m\omega^2 y^2}{2}$$

accounts for two degrees of freedom, one from the p_y^2 term, one from the y^2 term. In this case, the degrees of freedom are the kinetic energy of the mass's motion, and the potential energy stored in the "spring" that supplies the restoring force. In thermal equilibrium, classical physics—the **equipartition of energy**—assigns average[8] energy $\frac{1}{2}k_{\mathrm{B}}T$ to each degree of freedom, so the harmonic oscillator has average thermal energy $k_{\mathrm{B}}T$.

A vibrational mode $y_j = A_k\,e^{\mathrm{i}(kx_j - \omega t)}$ represents one of the independent motions of a monatomic chain. It has kinetic energy of motion of the masses, and potential energy stored in the springs. So it too accounts for two degrees of freedom,[9] and possesses $k_{\mathrm{B}}T$ of thermal energy, on average. By extension, the same applies to each of the $e^{\mathrm{i}(\boldsymbol{k}\cdot\boldsymbol{r} - \omega t)}$ modes of a three-dimensional solid.

All motions of the atoms (three dimensions) are accounted for by the $3N$ modes in the first Brillouin zone. Therefore the average internal energy of the solid is $U = 3N \times k_{\mathrm{B}}T$. The heat capacity C_{v} is the derivative of this with respect to temperature, so the classical theory yields[10]

$$\text{classical:}\qquad C_\text{v} = 3Nk_\text{B}. \qquad (23.6)$$

We may contrast the above with a less satisfactory version, often presented, in which each atom is regarded as capable of oscillating (in three directions) about a mean position, with that oscillation unaffected by motion of the other atoms. Then there are $3N$ independent motions, each of which possesses two degrees of freedom and therefore, by equipartition, should have average thermal energy $k_\text{B}T$. The internal energy is $U = 3Nk_\text{B}T$ and the heat capacity is $C_\text{v} = 3Nk_\text{B}$. The answer is correct, but the reasoning is not, because the atomic motions described are *not* independent. It is the normal modes, within one Brillouin zone, that give the correct description of the independent motions.

23.5   Debye theory

We require a theory of heat capacities that is not restricted to the classical regime. Each vibrational mode in a solid is mathematically identical to a harmonic oscillator—as in the classical theory—but that mode's energy is now quantized, and has energy $(n + \tfrac{1}{2})\hbar\omega$, where n is an integer.[11]

We may say that each mode contains lumps of energy $\hbar\omega$, called **phonons**. The number of phonons in a mode is the n of the last paragraph, and its statistical-mechanics average obeys a Planck distribution[12]

$$\overline{n(\omega)} = \frac{1}{e^{\hbar\omega/k_\text{B}T} - 1},$$

while the mean energy in this mode is

$$\overline{\varepsilon(\omega)} = \frac{\hbar\omega}{e^{\hbar\omega/k_\text{B}T} - 1}. \qquad (23.7)$$

We must remember that longitudinal modes have angular frequency $\omega_\text{L}(\boldsymbol{k})$ while transverse modes have angular frequency $\omega_\text{T}(\boldsymbol{k})$, and there are two transverse modes for each $\boldsymbol{k}$.

The internal energy U of a three-dimensional solid can now be written down in general terms. An average energy $\overline{\varepsilon(\omega)}$ is assigned to each vibrational mode within the first Brillouin zone, and U is the sum of these energies.[13]

$$U = \sum_{\boldsymbol{k}\text{-modes in zone}} \left(\frac{\hbar\omega_\text{L}}{e^{\hbar\omega_\text{L}/k_\text{B}T} - 1} + 2\,\frac{\hbar\omega_\text{T}}{e^{\hbar\omega_\text{T}/k_\text{B}T} - 1} \right). \qquad (23.8)$$

Equation (23.8) is the starting point for approximations discussed below.

23.5.1   The Debye theory in the high-temperature limit

We know from Fig. 23.2 that there is an upper bound to the values of ω that phonons may have.[14] We consider here the limit where $k_\text{B}T \gg \hbar\omega$ is the case for all values of ω, even that at the maximum. Then in (23.8)

$$\frac{\hbar\omega}{e^{\hbar\omega/k_\text{B}T} - 1} \approx \frac{\hbar\omega}{\hbar\omega/k_\text{B}T} = k_\text{B}T$$

[11] For simplicity we ignore the zero-point energy $\tfrac{1}{2}\hbar\omega$ in the present chapter. The zero-point energy does not change if the solid is held at constant volume, though it is relevant to the energetics of, for example, thermal expansion (because the ωs change).

The quantization of the vibrations is not rehearsed here. For a monatomic chain, it follows very closely the quantization of vibrations of a continuous string, which is worked in detail in Chapter 31. The quantization for a monatomic chain is given by Kittel (1963) Chapter 2; Kittel (2005) Appendix C.

[12] There is a Planck distribution, not a Bose–Einstein, because a phonon–phonon reaction is possible in principle for which

$$\hbar\omega \rightleftharpoons \frac{1}{2}\hbar\omega + \frac{1}{2}\hbar\omega :$$

one phonon becomes two, or two become one. There is no conservation of phonon number. Then there is no chemical potential μ, or equivalently $\mu = 0$.

[13] The expression given here is *almost* general. We have simplified things by saying that the modes divide into longitudinal and transverse, and in giving the two transverse possibilities equal ω_T. For a general direction in a general crystal, waves have three polarizations, but those polarizations are not so simply described as longitudinal and transverse, and there is no twofold degeneracy. The bracket in eqn (23.8) is replaced by a sum over three terms with three different values of $\omega(\boldsymbol{k})$.

[14] We are not talking here about the Debye cutoff frequency, which is for later. We are referring to the fact that in Fig. 23.2 the ω_L curve has a ceiling $\omega_\text{L,max}$ at $k = k_0$.

[15]We may explain to ourselves why a classical limit holds. The number of phonons per mode approximates to $k_B T/\hbar\omega \gg 1$ for all ω. So one phonon more or less makes hardly any difference: we are in the limit of large quantum numbers.

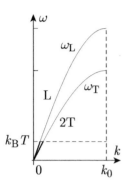

Fig. 23.3 When the temperature is low, the occupied vibrational modes are confined to the regions, marked heavy, close to the origin, where the $\omega(\mathbf{k})$ curves are straight.

The gradients ω/k of those curves give the speeds of low-frequency sound and shear waves, speeds that can be measured directly.

[16]It is not likely that the $\omega(\mathbf{k})$ relation for a three-dimensional crystal will have the detailed form shown in Fig. 23.2. Fortunately, this does not matter for the present subsection. The values of ω are being explored only near the origin, where the speed of sound (or of shear waves) is a measured quantity, and the linearity of the $\omega(\mathbf{k})$ relations can be trusted.

[17]Yes it is obvious. The exponential is the most complicated thing under the integral sign, so it's sensible to simplify that term as far as possible.

Incidentally, there is a point of good practice here. The substitution has had the effect of clearing messy symbols out from the integral until what remains is dimensionless and clearly evaluates to just some number. This does more for us than simplify the mathematics: the "messy symbols" contain all the physics of the problem, so separating them out has put the physics on display, even before the integral is dealt with.

for both terms in the big bracket. Then

$$U = \sum_{\text{zone}} (k_B T + 2k_B T) = 3k_B T \sum_{\substack{\mathbf{k}\text{-modes in zone}}} (1) = 3k_B T \times N. \quad (23.9)$$

In this way we reproduce the classical[15] finding for U, and therefore also that for C_v:

$$\text{high-temperature limit } k_B T \gg \hbar\omega_{\max}: \qquad C_v = 3Nk_B. \quad (23.10)$$

Comment: It is worth pausing here to see what has happened. Although eqn (23.8) was written down in a slightly simplified form, it should be obvious that the reasoning of (23.9) does not make use of that simplification, but is completely general.

We state the generality in another way. The finding (23.10) is *model independent*. It is independent of any assumption about the $\omega(\mathbf{k})$ relation for the phonons (save that ω has a ceiling); it has summed over the first Brillouin zone exactly, and not some approximation to it such as the spherical geometry of (23.5). This model independence is worth stressing, in view of the simplifying assumptions that have to be made in later parts of the present chapter.

23.5.2 The Debye theory in the low-temperature limit

We turn attention to the opposite limit from that of the last subsection. The interesting values of ω are those for which $\hbar\omega \lesssim k_B T$, and T is such that all these have $\omega \ll \omega_{\max}$. In Fig. 23.3 the "active" regions of the $\omega(\mathbf{k})$ curves are the straight-line sections close to the origin.

We shall introduce the assumption of isotropy (23.5) in order to make the mathematics manageable. Then longitudinal modes have $\omega_L = v_L k$ where $k = |\mathbf{k}|$ and v_L is the speed of longitudinal waves; v_L is, of course, the gradient of the upper curve of Fig. 23.3 taken at the origin.[16] Similarly $\omega_T = v_T k$ for transverse waves. Then (23.8) becomes

$$U = \int_{\text{zone}} \left(\frac{\hbar\omega_L}{e^{\hbar\omega_L/k_B T} - 1} + 2\frac{\hbar\omega_T}{e^{\hbar\omega_T/k_B T} - 1} \right) V \frac{4\pi k^2 \, dk}{(2\pi)^3}.$$

Given that the ks contributing significantly to the integral all lie close to the origin, the upper limit of integration can be extended to infinity without loss of accuracy. In addition, the variable of integration can now be changed from k to ω:

$$\frac{U}{V} = \int_0^\infty \left(\frac{\hbar\omega}{e^{\hbar\omega/k_B T} - 1} \right) \frac{4\pi}{(2\pi)^3} \left(\frac{\omega^2}{v_L^3} + \frac{2\omega^2}{v_T^3} \right) d\omega$$

$$= \frac{4\pi}{(2\pi)^3} \hbar \left(\frac{1}{v_L^3} + \frac{2}{v_T^3} \right) \left(\frac{k_B T}{\hbar} \right)^4 \int_0^\infty \frac{w^3 \, dw}{e^w - 1}.$$

Here we have made the obvious[17] substitution $w = \hbar\omega/k_B T$. All that remains is a standard integration that can be worked out (if you're ingenious) or looked up; it is $\pi^4/15$. Then, tidying, we have

$$\frac{C_v}{V} = \frac{2\pi^2}{15} \left(\frac{1}{v_L^3} + \frac{2}{v_T^3} \right) \frac{k_B^4}{\hbar^3} T^3. \quad (23.11)$$

Equation (23.11) displays the Debye T^3 law for the heat capacity of a solid at low temperatures.

Comment: The calculation of this subsection has incorporated approximations, but they have not been particularly drastic. The assumption of isotropy could be avoided by taking $\left(1/v_{\mathrm{L}}^3 + 2/v_{\mathrm{T}}^3\right)$ and averaging it over directions in the crystal. The extension of the integration range to infinity is valid for sufficiently low temperatures. There has been no need to introduce a Debye cut-off because the Planck distribution has imposed its own, much lower, cut-off. The T^3 in (23.11) is model-independent, with only the coefficient open to (slight) refinement.

23.5.3 The Debye theory at intermediate temperatures

The Debye theory gives results that are model-independent, or nearly so, in both the high-temperature and low-temperature limits. Only in the "middle" range of temperatures is there a need to give detailed consideration to the $\omega(\boldsymbol{k})$ relation for the phonons.

We return to (23.8), and re-state it as

$$U = \int \frac{\hbar\omega}{e^{\hbar\omega/k_{\mathrm{B}}T} - 1}\, \rho(\omega)\, \mathrm{d}\omega,$$

where $\rho(\omega)\,\mathrm{d}\omega$ is the number of modes (counting all three polarizations) within the range $\mathrm{d}\omega$ of angular frequency. Then

$$C_{\mathrm{v}} = \frac{\partial U}{\partial T} = k_{\mathrm{B}} \int \left(\frac{\hbar\omega}{k_{\mathrm{B}}T}\right)^2 \frac{e^{\hbar\omega/k_{\mathrm{B}}T}}{(e^{\hbar\omega/k_{\mathrm{B}}T} - 1)^2}\, \rho(\omega)\, \mathrm{d}\omega. \tag{23.12}$$

The integrand in (23.12) consists of a product: of a "density-of-states"[18] factor $\rho(\omega)$, and a statistical-weight factor

$$W(w) = \frac{w^2\, e^w}{(e^w - 1)^2}, \tag{23.13}$$

in which $w = \hbar\omega/k_{\mathrm{B}}T$. The dependence of C_{v} on temperature comes from changes with temperature in the multiplying factor linking ω to w.

Figure 23.4 shows $W(w)$ plotted against w. Because of the shape of the function $W(w)$, the integration of (23.12) takes an average over $\rho(\omega)$, covering a range that is roughly $0 \leqslant w \lesssim 5$.

Figure 23.5 shows how this averaging works in a practical case. The upper curve (Nilsson and Rolandson 1973) gives $\rho'(\nu)$, the density of phonon states in copper[19,20] (per interval of frequency ν and normalized so that $\int \rho'(\nu)\,\mathrm{d}\nu = 1$), while below are three plots of $W(w)$ corresponding to temperatures $T = \theta_{\mathrm{D}}/10$, $T = \theta_{\mathrm{D}}/3$ and $T = \theta_{\mathrm{D}}$, where θ_{D} is the Debye theta[21] defined below in (23.19); for copper $\theta_{\mathrm{D}} = 320\,\mathrm{K}$.

We start discussion of Fig. 23.5 by considering again the high- and low-temperature limits; those cases were previously treated in the most elementary way possible.

At high temperatures, the curve for $W(w)$ is "flat" at value 1, and the integral in (23.12) is simply $\int \rho(\omega)\,\mathrm{d}\omega = 3N$: the heat capacity takes the classical value $3Nk_{\mathrm{B}}$. We now see additionally from Fig. 23.5 that

[18] The term *density of states* should be reserved for the number of quantum states per unit of energy and per unit volume, hence the quotation marks. Here $\rho(\omega)$ is the number of states per $\mathrm{d}\omega$, not per increment of energy, and it applies to volume V, not to unit volume.

For each polarization, the "density of states" contributing to $\rho(\omega)$ can have a high-frequency cut-off as exemplified in Fig. 23.5. We are leaving these cut-offs for later consideration.

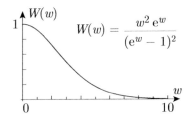

Fig. 23.4 The weight factor $W(w)$ of eqn (23.13).

[19] Copper is, of course, a metal, so it has an additional contribution to its heat capacity from conduction electrons. Here we are concerned only with the phonon contribution to the heat capacity. The electron contribution is, in any event, small, and can be accurately subtracted from experimental measurements to leave the phonon part.

[20] Since $\int \rho(\omega)\,\mathrm{d}\omega = 3N$, we have

$$\rho'(\nu) = \frac{2\pi}{3}\frac{\rho(\omega)}{N}.$$

[21] For now, we may take $k_{\mathrm{B}}\theta_{\mathrm{D}}$ as an estimate of the energy of phonons at the greatest value of ω.

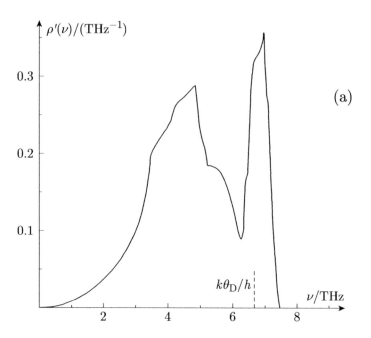

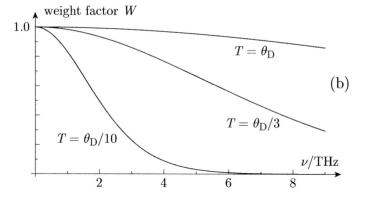

Fig. 23.5 (a) Upper curve: the normalized "density of states" $\rho'(\nu)$ is defined as a function of frequency ν such that $\rho'(\nu) \propto \rho(\omega)$ and $\int \rho'(\nu)\,d\nu = 1$; after Nilsson and Rolandson (1973).
(b) Lower curves: the weight factor $W(w)$, expressed as a function of frequency ν for three different temperatures: $\theta_D/10$, $\theta_D/3$ and θ_D, where θ_D is the Debye theta. The value of θ_D for copper is 320 K, a value also taken from Nilsson and Rolandson. Converted to $k\theta_D/h = 6.67\,\text{THz}$, this is marked on the frequency axis of the upper graph.

temperature $T = \theta_D$ is already high enough for those conditions to hold, very nearly. Consistent with this, θ_D appears high on the "shoulder" in the plot of the Debye C_v in Fig. 23.8.

In the low-temperature limit $T \ll \theta_D$, w runs over all interesting values while ω is still small. Thus we replace $\rho(\omega)$ by its low-ω limiting expression:

$$\rho(\omega)\,d\omega = V\,\frac{4\pi}{(2\pi)^3}\left(\frac{1}{v_L^3} + \frac{2}{v_T^3}\right)\omega^2\,d\omega \tag{23.14}$$

$$= V\,\frac{4\pi}{(2\pi)^3}\left(\frac{1}{v_L^3} + \frac{2}{v_T^3}\right)\left(\frac{k_B T}{\hbar}\right)^3 w^2\,dw.$$

Integration may now be taken over w, with w running to large values so that the upper limit may be extended to infinity. Then[22]

[22] Integrate by parts, and the integral becomes the same as that which led to (23.11).

$$\frac{C_v}{V} = k_B \frac{4\pi}{(2\pi)^3} \left(\frac{1}{v_L^3} + \frac{2}{v_T^3} \right) \left(\frac{k_B T}{\hbar} \right)^3 \int_0^\infty \frac{e^w}{(e^w - 1)^2} \, w^4 \, \mathrm{d}w$$

$$= k_B \frac{2\pi^2}{15} \left(\frac{1}{v_L^3} + \frac{2}{v_T^3} \right) \left(\frac{k_B T}{\hbar} \right)^3,$$

in agreement with (23.11).

Finally, we look at the implications for the "middle" range of temperatures. Even here, the integral in (23.12) is insensitive to the detailed form of $\rho(\omega)$. Imagine a substance with a different $\rho(\omega)$ for which the tall peak in Fig. 23.5(a) is truncated and some of its states are moved downwards in frequency to fill in the dip. For temperature $T = \theta_D/3$ these states change their weight W from 0.47 to 0.54, so the integral for C_v is little affected. At higher temperatures, the change of weight W would be less (the curve for W is shallower); and at lower temperatures the weight W is smaller, making these states give a less significant contribution to C_v overall. The value of the integral (23.12) for C_v is insensitive—remarkably so—to all details of the $\rho(\omega)$ in the integrand, and this is the case at all temperatures.

23.5.4   The Born–von Kármán theory

The graphical integrations of the last section explain the general behaviour of a solid's heat capacity, but they do not yield simple algebraic expressions, except in the two limiting cases; the "density of states" $\rho(\omega)$ is too complicated—and of course it is different for different solids.

A model for the three $\omega(\boldsymbol{k})$ relations is shown in Fig. 23.2. For each of the three wave polarizations there is a contribution to $\rho(\omega)$ of[23]

$$V \frac{4\pi k^2}{(2\pi)^3} \frac{\mathrm{d}k}{\mathrm{d}\omega}.$$

The resulting $\rho(\omega)$ could be calculated (we don't plot it), but even with this simple model an evaluation of C_v can be done only numerically. Given that the precise form of $\rho(\omega)$ has little effect on the calculated C_v, perhaps we could find a modified $\rho(\omega)$ that gives a simpler integration in (23.12), and yet does little harm to the heat capacity?

Two things are sacrosanct: the properties that gave rise to the high- and low-temperature limits calculated above. What fixed the high-temperature limit was the use of $\boldsymbol{k}$-values occupying a single Brillouin zone. What fixed the low-temperature limit was the gradients of the two $\omega(\boldsymbol{k})$ curves for small k. A reasonable simplification might therefore be to replace the curves of Fig. 23.2 by straight lines, tangent to the curves at the origin and continuing to the zone boundary: Fig. 23.6.[24]

We do not write out an expression for the heat capacity that results from making these changes to the $\omega(\boldsymbol{k})$ relation, and the consequent changes to $\rho(\omega)$. The point we would draw attention to is that our reasoning has led directly to the Born–von Kármán version of the Debye theory, originally constructed with the aim of improving the agreement between calculation and experiment.

[23] This expression assumes, for simplicity, that the $\omega(\boldsymbol{k})$ relation is isotropic. However, the important feature is the presence of the derivative $\mathrm{d}k/\mathrm{d}\omega$.

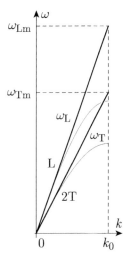

Fig. 23.6 A possible modification of the $\omega(k)$ relations, intended to simplify the integration (23.12) and yield an algebraic expression for the heat capacity of a solid. The extrapolated angular frequencies ω_{Lm}, ω_{Tm} are marked in preparation for Fig. 23.7.

[24] We remember that the quarter-sine curves have been carried over to three dimensions from the monatomic linear chain, so their shapes have no authority here. Rather, it is the gradients at the origin that are authoritative, because we are regarding these as the outcome of measurements of the speeds of sound and of shear waves.

23.5.5   The Debye theory proper

Given that the integration of (23.12) is highly insensitive to details of the $\rho(\omega)$ "density of states", it should be acceptable to make a further cruder, change to the $\omega(\boldsymbol{k})$ relation for vibrational waves in the solid.[25] That change is shown in Fig. 23.7. The two straight lines have unchanged gradients, but that for longitudinal waves is curtailed at ω_c, while that for transverse waves is extended beyond the Brillouin-zone boundary until it too reaches ω_c. The value of ω_c is chosen so that the total number of vibrational modes remains at $3N$.

The total number of vibrational modes has been left unchanged for the reason given in § 23.5.4: it is sacrosanct, otherwise the high-temperature (classical) limit will go wrong.

At this point, we have recovered the Debye theory in its original form. The Debye cut-off that terminates integration over ω at ω_c is explained by seeing its origin in the summing of modes that were formerly restricted to the first Brillouin zone.

We move towards obtaining an explicit expression for the Debye C_v. The expression for $\rho(\omega)$ can be taken from (23.14):

$$\rho(\omega) = V\,\frac{4\pi}{(2\pi)^3}\left(\frac{1}{v_L^3} + \frac{2}{v_T^3}\right)\omega^2 \quad \text{for } \omega \leqslant \omega_c, \tag{23.15}$$

with ω_c given by

$$\frac{3N}{V} = \frac{4\pi}{(2\pi)^3}\left(\frac{1}{v_L^3} + \frac{2}{v_T^3}\right)\int_0^{\omega_c}\omega^2\,\mathrm{d}\omega = \frac{1}{6\pi^2}\left(\frac{1}{v_L^3} + \frac{2}{v_T^3}\right)\omega_c^3. \tag{23.16}$$

To find the heat capacity, it is convenient to express things in terms of a characteristic temperature θ_D, the **Debye theta**, defined by

$$\hbar\omega_c = k_B\theta_D. \tag{23.17}$$

Then (we omit intermediate steps in the algebra)

$$C_v = 9Nk_B\,\frac{T^3}{\theta_D^3}\int_0^{\theta_D/T}\frac{e^w}{(e^w - 1)^2}\,w^4\,\mathrm{d}w$$

$$= 9Nk_B\,\frac{T^3}{\theta_D^3}\left\{4\int_0^{\theta_D/T}\frac{w^3}{e^w - 1}\,\mathrm{d}w - \frac{(\theta_D/T)^4}{e^{\theta_D/T} - 1}\right\}, \tag{23.18}$$

in which

$$\theta_D^3 = 18\pi^2\,\frac{N}{V}\,\frac{\hbar^3}{k_B^3}\left(\frac{1}{v_L^3} + \frac{2}{v_T^3}\right)^{-1}. \tag{23.19}$$

Expression (23.18) is plotted in Fig. 23.8.

The densities of states assumed by Born–von Kármán and by Debye are shown in Fig. 23.9. It may be seen that the Born–von Kármán version makes some attempt at reproducing the dip of Fig. 23.5(a), but the proportions around the dip are so poor that it's hardly worth the effort. The Debye version makes no attempt at reproducing the dip at all.

We should mention that the inadequacy of the Born–von Kármán theory, as it is given here, is a consequence of our simplifying assumptions introduced for pedagogical purposes. When a proper sum is taken over

[25] It is not hard to re-state the reasoning up to this point without imposing isotropy on the $\omega(\boldsymbol{k})$ relation of the phonons. Only now does isotropy (ω_c independent of the direction of $\boldsymbol{k}$) become unavoidable. We may regard the simplifying assumption of isotropy as part of "further, cruder".

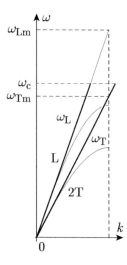

Fig. 23.7 The additional modification to the $\omega(\boldsymbol{k})$ relation assumed by Debye. The straight lines for longitudinal and transverse waves are made to terminate at a common frequency, at ω_c, rather than at a common k. However, ω_c is chosen so that the number of vibrational modes remains at $3N$.

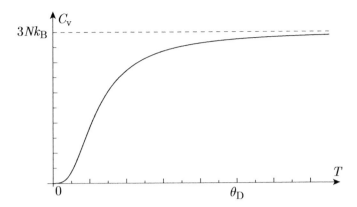

Fig. 23.8 The heat capacity of a dielectric solid, as predicted by the Debye theory. The parameter θ_D is defined in (23.19).

Note that θ_D lies high on the "shoulder" of the curve: a temperature of θ_D is high enough for the heat capacity to have nearly its classical value.

the k-values in the first Brillouin zone, with account taken of the variation of wave speed with k (in magnitude and direction), the density of states looks very like that of Fig. 23.5(a) (Nilsson and Rolandson 1973).

Note on the diagrams in this chapter
The measurements of Nilsson and Rolandson (1973), from which they calculated the density of states in Fig. 23.5(a), were undertaken on copper at a temperature of 80 K. For consistency with this, other quantities obtained from experiment have also been chosen to be appropriate to 80 K. The simplest waves in copper are those that propagate in the (001) direction, parallel to a cube edge. These waves have $\omega(k_z)$ relations that are close to quarter-sine curves, and have $v_L/v_T = 1.472$ at temperature 80 K (Fig. 24.3). Figure 23.2 and succeeding $\omega(k)$ curves have been drawn with these proportions. Likewise, the ω_c of Fig. 23.7 is drawn in correct relation to ω_{Lm} and ω_{Tm}. The one exception to the above is that we have used a Debye temperature θ_D of 320 K which is more nearly appropriate to room temperature.[26]

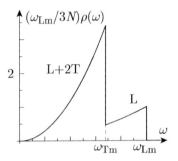

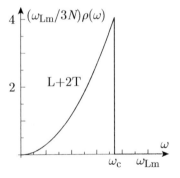

Fig. 23.9 The "density of states" $\rho(\omega)$ assumed by Born and von Kármán (upper curve) and by Debye (lower curve), both multiplied by a scaling factor $\omega_{Lm}/(3N)$. The solid is isotropic.

In the upper plot, the limiting angular frequencies are $\omega_{Lm} = v_L k_0$ and $\omega_{Tm} = v_T k_0$, where k_0 is $|k|$ at the Brillouin-zone boundary, taken as independent of direction.

23.6   Discussion

In this chapter, we have described the Debye theory in a way that shamelessly rewrites history. The Debye theory came first, and the Born–von Kármán presentation came later, as an attempt at improving on the numerical agreement between Debye and experiment. Yet we have reversed this sequence.

There is a good case for presenting the ideas in this non-historical order. It is usual for a study of solid-state physics to start with the monatomic linear chain and an understanding of Brillouin zones. It seems sensible for a presentation of the Debye theory to build upon this, rather than having Debye stand alone. Brillouin zones are implicit in the Debye theory, because a sum over modes in the first zone is the real reason for the existence of the cut-off.

The Debye theory has the fortunate feature that the calculated C_v is very insensitive to the precise $\omega(k)$ relation for the phonons: we can

[26] In the trade, departures from Debye predictions are often presented by forcing eqn (23.18) to apply, with θ_D made temperature dependent.

calculate C_V to good numerical accuracy without the need for knowledge of the input $\rho(\omega)$ to similar accuracy. Nature has been remarkably kind to us.

The converse is less happy. It would be nice if we could "undo" eqn (23.12) to find $\rho(\omega)$ from measurements of C_V. Such an operation is not prohibited in principle: we can regard (23.12) as an integral equation for $\rho(\omega)$ where the integral containing $\rho(\omega)$ evaluates to the known function $C_V(T)$. However, the insensitivity of C_V to the details of $\rho(\omega)$ means that we need to know C_V to many significant figures before we can gain any information at all, via this route, about $\rho(\omega)$. The practical limitations of heat-capacity measurement mean that there is little prospect of success at "doing phonon spectroscopy using a calorimeter".

Problems

Problem 23.1 Monatomic linear chain: details

(1) Look at the running-wave values of k_r defined by (23.4). Section 23.2 shows that $(N-1)$ atoms are free to move independently, so there must be $(N-1)$ coefficients, going with $(N-1)$ values of k_r, required to describe the motion. Those values of k_r must be "different", meaning that all k_r must lie in a single Brillouin zone (no duplicates). We choose the conventional first zone, which has $-\pi/a < k_r \leqslant \pi/a$. Show that r, k_r then run over the ranges

$$\left.\begin{array}{llll} \text{if } N \text{ is even,} & -\left(\dfrac{N}{2}-1\right) \leqslant r \leqslant \dfrac{N}{2}, & \dfrac{-\pi}{a}\left(1-\dfrac{2}{N}\right) \leqslant k_r \leqslant \dfrac{\pi}{a}, \\[3mm] \text{if } N \text{ is odd,} & -\left(\dfrac{N-1}{2}\right) \leqslant r \leqslant \left(\dfrac{N-1}{2}\right), & \dfrac{-\pi}{a}\left(1-\dfrac{1}{N}\right) \leqslant k_r \leqslant \dfrac{\pi}{a}\left(1-\dfrac{1}{N}\right). \end{array}\right\} \quad (23.20)$$

In conformity with § 23.2, $r \neq 0$, $k_r \neq 0$, because movement of the chain as a whole would require external forces not present.

(2) Take the monatomic chain again, but now hold both ends fixed, at $x=0$, $x=Na$. The number of atoms free to move is again $(N-1)$. Write down the eigenfunctions that we would choose to describe the chain's motion in such a case. Show that the permitted values of r and k are given by[27]

$$k = \dfrac{\pi}{a}\dfrac{r}{N} \quad \text{so that} \quad 1 \leqslant r \leqslant (N-1); \quad \dfrac{\pi}{a}\dfrac{1}{N} \leqslant k \leqslant \dfrac{\pi}{a}\left(1-\dfrac{1}{N}\right).$$

Investigate what happens if the chain has both ends free instead of fixed.

(3) We now have three eigenfunction sets available for describing the chain's motion: travelling waves, and two kinds of standing wave. Do their boundary conditions cause failure at the chain ends in a manner reminiscent of § 6.3.1?

Problem 23.2 A classical wave mode possesses two degrees of freedom
This property may sound obvious: a standing or travelling wave clearly has both kinetic and potential energy. Nevertheless, the definition of a

[27] The first Brillouin zone now covers a k-range of only π/a, but still accommodates $(N-1)$ modes, because the modes are twice as densely distributed along the k-axis in comparison with those in Fig. 23.1.

We do not need to write separate cases for odd and even N now.

degree of freedom is a formal one: a term in the energy proportional to either (displacement)^2 or (momentum)^2. For this purpose, the momentum must relate to the displacement according to the rules of Lagrangian mechanics. Nothing more earthy will do.[28]

We continue to consider a monatomic linear chain.[29] Set up a complex-exponential expansion for the displacement y_j of the jth mass m:

$$y_j = \sqrt{\frac{1}{Nm}} \sum_r q_r(t)\, e^{ik_r x_j}. \tag{23.21}$$

Here $x_j = ja$ is the location of the jth mass.[30]

The travelling-wave (periodic boundary) condition requires

$$e^{ik_r x_j} = e^{ik_r(x_j + Na)}, \quad \text{so} \quad k_r Na = (\text{integer } r) \times 2\pi,$$

giving

$$k_r = \frac{2\pi}{a}\frac{r}{N}, \tag{23.22}$$

in agreement with (23.4). The range of values permitted to r, and thence to k_r, is given by (23.20).

Conveniently, we make $j = 0$ identify the atom at one end of the chain; then $j = N$ identifies the other end. But the boundary condition makes displacement y_N duplicate y_0, so the independently moving atoms have

$$0 < j \leqslant N. \tag{23.23}$$

Displacement y_j must be real. We require $q_r(t)\,e^{ik_r x_j} + q_{-r}(t)\,e^{-ik_r x_j}$ to be real for each integer r, which implies[31]

$$q_{-r}(t) = q_r(t)^*. \tag{23.24}$$

Expansion (23.21) has been set up to look like a sum over travelling waves. But this Fourier expansion is, of course, general, capable of describing any physically permitted motion, travelling or standing. It all depends on the form of $q_r(t)$.

(1) Set up an orthogonality property. Consider e^{ikx} waves having $k = k_r$ and $k = k_s$, and sum over all the atoms that are free to move independently. Use (23.22) to show that

$$\sum_{j=0}^{N-1} \exp\left\{i(k_r - k_s)x_j\right\} = \sum_{j=0}^{N-1} \exp\left\{i\frac{2\pi}{a}\frac{(r-s)}{N}aj\right\}.$$

Recognize in this a geometric progression,[32] and show that the result is zero unless $r = s$, in which case the sum is $\sum_j 1 = N$. Thus

$$\sum_{j=0}^{N-1} \exp\left\{i(k_r - k_s)x_j\right\} = N\delta_{r,s}. \tag{23.25}$$

(2) Evaluate the total kinetic energy of all the masses of the chain using the following start.

$$(\text{KE}) = \frac{m}{2}\sum_{j=0}^{N-1} \dot{y}_j^2 = \frac{m}{2}\sum_{j=0}^{N-1}\left(\frac{1}{\sqrt{Nm}}\sum_r \dot{q}_r\, e^{ik_r x_j}\right)\left(\frac{1}{\sqrt{Nm}}\sum_s \dot{q}_s\, e^{ik_s x_j}\right).$$

[28] Consider a gas of molecules in a uniform gravitational field. A molecule at height z has potential energy mgz. We could make this contain the square of a coordinate y by defining $y = \sqrt{z}$. Explain why this trick cannot force the potential energy to have mean value $\frac{1}{2}k_B T$.

[29] Similar classical mechanics is set up in Chapter 31 for a continuous string—though there the aim is to progress to a quantum description.

[30] For simplicity, we take the x-origin to lie at one of the atoms. No such choice of origin was needed in § 23.2 or in problem 23.1.

The square root is included to forestall clutter in the expressions for kinetic and potential energy.

[31] The opportunity to combine r with $-r$ in this simple way arises from the symmetry of the first Brillouin zone, a symmetry built into the k-range in (23.20).

Show that when $r = N/2$, y_j is real for all j, so the absent partner having $r = -N/2$ is not needed to make y_j real. In turn, this makes $q_{N/2}(t)$ real without any need to impose a condition to make it so.

[32] There is a similarity to the mathematics that describes a diffraction grating.

Result (23.25) has a family resemblance to the Fourier-transform property

$$\int_{-\infty}^{\infty} e^{ikx}\, dx = 2\pi\,\delta(k).$$

Using (23.25), show that this reduces to

$$(\text{KE}) = \frac{1}{2} \sum_{(N-1) \text{ values of } r} \dot{q}_r \, \dot{q}_{-r}. \tag{23.26}$$

The sum is taken over the N values of r within a single Brillouin zone but excluding $r = 0$ since that represents a displacement ($k_r = 0$) of all atoms equally.

(3) By a similar procedure, show that the total potential energy held in all the springs of the chain is[33]

$$(\text{PE}) = \frac{1}{2} \sum_{j=0}^{N-1} \beta(y_{j+1} - y_j)^2 = \frac{1}{2} \sum_{(N-1) \text{ values of } r} (\omega_r)^2 \, q_r \, q_{-r}. \tag{23.27}$$

where $(\omega_r)^2 = (4\beta/m)\sin^2\left(\frac{1}{2}k_r a\right)$ as in (23.2).

(4) The Lagrangian for the chain is now[34]

$$\mathcal{L} = (\text{KE}) - (\text{PE}) = \frac{1}{2} \sum_{r=-(N-1)/2}^{(N-1)/2} \left\{ \dot{q}_r \, \dot{q}_{-r} - (\omega_r)^2 \, q_r \, q_{-r} \right\}. \tag{23.28}$$

Unfortunately we are not yet where we want to be.[35] The summand in (23.28) is not of the form $\dot{q}_r^2 - (\omega_r)^2 q_r^2$. Also, the rth terms duplicate the $-r$th terms, so there is a risk of double counting.

(5) Divide $q_r(t)$ into its real and imaginary parts:

$$q_r = \frac{c_r + is_r}{\sqrt{2}}, \quad q_{-r} = q_r^* = \frac{c_r - is_r}{\sqrt{2}}, \quad \text{so} \quad c_{-r} = c_r, \quad s_{-r} = -s_r.$$

Show that (23.21) can be reshaped into[36]

$$y_j = \sqrt{\frac{2}{Nm}} \sum_{r=1}^{(N-1)/2} \left\{ c_r \cos(k_r ja) - s_r \sin(k_r ja) \right\}. \tag{23.29}$$

This shows that c_r and s_r are coefficients for independent wave motions on the chain, and all are counted when r takes positive values only. Then

$$\dot{q}_r \, \dot{q}_{-r} = \frac{1}{2}(\dot{c}_r + i\dot{s}_r)(\dot{c}_r - i\dot{s}_r) = \frac{1}{2}(\dot{c}_r^2 + \dot{s}_r^2),$$

and this is not just a mathematical manipulation, but is a teasing out of the new independent motions, as they contribute to the kinetic energy.

(6) Find the kinetic energies and potential energies for the whole chain and show that the Lagrangian (23.28) becomes

$$\mathcal{L} = \frac{1}{2} \sum_{r=1}^{(N-1)/2} \left\{ \dot{c}_r^2 + \dot{s}_r^2 - (\omega_r)^2 c_r^2 - (\omega_r)^2 s_r^2 \right\}.$$

(7) Show that the momentum $\gamma_r = \partial\mathcal{L}/\partial\dot{c}_r = \dot{c}_r$ is conjugate to coordinate c_r and that the momentum conjugate to s_r is $\sigma_r = \partial\mathcal{L}/\partial\dot{s}_r = \dot{s}_r$. Show that the Hamiltonian becomes

$$\mathcal{H} = \sum_{r=1}^{(N-1)/2} \frac{1}{2}\left(\gamma_r^2 + \omega_r^2 c_r^2\right) + \sum_{r=1}^{(N-1)/2} \frac{1}{2}\left(\sigma_r^2 + \omega_r^2 s_r^2\right). \tag{23.30}$$

In (23.30) we can see squared coordinates and squared momenta representing a possible set of degrees of freedom for the monatomic chain

---

[33] *Hint:* Write a tidy expression for $(y_{j+1} - y_j)$ before squaring.

[34] From here on, we consider only the case where N is an odd integer. The case of even N is a little more complicated because $k_{N/2}$ is not paired with $k_{-N/2}$. The limits on r from (23.20) become $-(N-1)/2 \leqslant r \leqslant (N-1)/2$ with $r = 0$ omitted.

[35] I followed the procedure outlined above because it is what one does routinely when setting out to quantize a field, as is done in Chapter 31.

[36] Do not be tempted to see the cosine and sine in (23.29) as representing standing waves. We have said that any form of wave can be accommodated by a suitable choice of the coefficients. As a specific counterexample, let

$$c_r = \cos(\omega t), \quad s_r = -\sin(\omega t),$$

giving

$$\left\{ \cos(\omega t)\cos(k_r ja) + \sin(\omega t)\sin(k_r ja) \right\}$$
$$= \cos(\omega t - k_r ja),$$

clearly a travelling wave.

Assigning to each degree of freedom an average thermal energy $\frac{1}{2}k_\mathrm{B}T$, we find a total internal energy

$$U = (N-1)k_\mathrm{B}T. \qquad (23.31)$$

(8) Return to (23.21) and re-work the calculation taking the case where N is an even integer.

Whenever a one-dimensional chain is used to model the behaviour of atoms vibrating in a solid, N will be very large, so the distinction between $(N-1)$ and N (and between odd N and even) can safely be ignored.

Equation (23.31) holds for a one-dimensional chain. It should be obvious—or at least expected—that in three dimensions there will be a similar property with $U = 3Nk_\mathrm{B}T$.

Problem 23.3 Sound is an adiabatic motion?

When we study sound, we learn that sound is an adiabatic disturbance: there are temperature changes, coupled to the density changes, which cannot relax by thermal conduction (the distance through which heat would have to be conducted is too large). Yet the description of sound given in this chapter, as a purely mechanical vibration, seems to have nothing to do with adiabaticity. What is missing?

"Why do those cartwheels go backwards on the telly, Daddy?"
"Well son, they're in the wrong Brillouin zone . . . ."

Intended readership: To be read alongside other books' treatments of reactions where phonons or electrons undergo collisions, thereby obstructing the flow of heat or of electric current.

24.1   Introduction

In a dielectric crystalline solid, the thermal energy of the solid resides in the mechanical vibrations of the atoms. We have seen in Chapter 23 that those vibrations are described in terms of their normal modes. These in turn host a "gas" of quantized excitations called phonons. The thermal energy U of the solid is the energy of that gas. Consistently, the heat capacity $C_v = (\partial U/\partial T)_V$ is the energy needed to raise the temperature of the phonon gas. It must likewise be phonons that are the carriers of heat energy during thermal conduction.

In a metal, electrons are also available to possess thermal energy and to transport that energy from place to place. In the present chapter, we remove electrons from consideration by concentrating on a dielectric solid, in order not to have too many things to think about at once.

Phonons undergo collisions as they move about in the solid. That is why there is a thermal resistance.[1] Books on condensed-matter physics discuss what phonons can collide with: the boundaries of the crystal; boundaries of crystallites; other deviations from ideal crystallinity (dislocations, impurities); other phonons[2] ("normal processes"); other phonons ("Umklapp[3] processes"). Here we are concerned only with phonon–phonon collisions, normal and Umklapp: why does an Umklapp collision disrupt heat flow while a "normal" collision does not? what is so special about a reciprocal lattice vector that its presence switches a harmless collision into a disruptive one?

I devote a chapter to these topics because there are surprising subtleties, and there is at least one very tempting argument that needs discussion because it turns out to be wrong.

[1] It is also the reason why there can be a heat capacity. Without collisions—of some sort—there would be no mechanism by which added heat energy could be distributed among the phonons to yield a new thermal-equilibrium state at a raised temperature.

[2] One phonon "collides" with another because that other imposes on the atomic locations a departure from ideal periodicity; so a phonon–phonon collision is the result of yet another deviation from ideal crystallinity.

[3] *Umklapp* is usually capitalized because German nouns are.

Essays in Physics: Thirty-two thoughtful essays on topics in undergraduate-level physics. Geoffrey Brooker, Oxford University Press. © Geoffrey Brooker 2021. DOI: 10.1093/oso/9780198857242.003.0024

24.2   Phonon–phonon collisions and a non-linear stress–strain relation

Perhaps it's best to start by identifying conditions under which phonons *can not* collide: if the crystal's stress–strain relation is linear. Remember the monatomic chain of § 23.2. If the springs have a linear relation between force and extension, then the motions of the atoms separate out into normal modes that are exactly independent of each other.

Therefore any reaction that looks like a phonon–phonon collision can happen only because of some non-linearity in the interatomic forces: books speak of "anharmonicity", which is the same thing.

To deal mathematically with non-linearity is to invite messy calculation, so we handle things in the most elementary way possible. We start with a simple picture: Fig. 24.1.

Suppose we have a sound wave having wave vector k_1 and angular frequency ω_1; this wave might be thermally excited, or it might be injected from a transducer. A non-linearity means that the elastic properties of the solid are slightly changed by this wave, varying from peak to trough of the wave, that is, varying in space with a Fourier component having wave vector k_1.

Let a second wave k_2, ω_2 be incident in this region where the elastic properties are modified. It sees a periodic structure having wave vector k_1. That is, there are planes, perpendicular to k_1, on which the elastic "constants" are alternately high and low with period $2\pi/k_1$. Wave k_2 is Bragg-scattered from this structure to give a new wave k_3, ω_3. The Bragg condition is simply

$$k_3 = k_1 + k_2. \tag{24.1}$$

Moreover, the wavefronts of wave k_1 are moving (up-page in Fig. 24.1), so wave k_2 is reflected as from an array of moving mirrors. There is a Doppler shift such that

$$\omega_3 = \omega_1 + \omega_2. \tag{24.2}$$

Equations (24.1) and (24.2) have been given explanations entirely within classical physics. At the same time, there are no prizes awarded for multiplying both equations through by $\hbar$ and understanding the outcome as expressing the conservation of momentum and energy.[4]

Here is a second, more mathematical, presentation leading to the same results as above; it does little more than hint at what a full calculation would look like. Impose waves 1 and 2 as above, giving rise to strains[5]

$$\sigma_1 = A_1\, e^{i(k_1 \cdot r - \omega_1 t)} + \text{complex conjugate}$$

$$\sigma_2 = A_2\, e^{i(k_2 \cdot r - \omega_2 t)} + \text{complex conjugate}.$$

The stress contains terms linear in σ_1 and σ_2, but also a term involving the square of the strain $(\sigma_1 + \sigma_2)^2$, so it contains

$$2\sigma_1\sigma_2 = 2A_1 A_2 \exp\big\{i(k_1 + k_2) \cdot r - i(\omega_1 + \omega_2)t\big\} + \text{other terms.}$$

Thus the stress contains a term having angular frequency $(\omega_1 + \omega_2)$ and varying in space with wave vector $(k_1 + k_2)$. When this stress is fed as

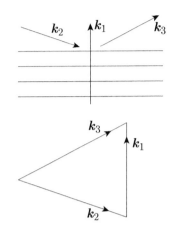

Fig. 24.1 A sound wave having wave vector k_1 travels up the page; its wave fronts are hinted at. Non-linearity in the underlying medium means that the elastic properties of that medium vary slightly from crest to trough of the wave, that is, in Fourier terms, with wave vector k_1. A second wave, k_2, is Bragg-scattered from these non-uniformities to yield an outgoing wave, k_3.

[4] This procedure, of multiplying a classical equation through by $\hbar$ and re-interpreting, is one of the clearest illustrations of the Correspondence Principle.

[5] Stress and strain are both second-rank tensors, so a stress related to the square of strain involves coefficients forming a tensor of rank 6. You can see why our presentation is little more than a "child's guide".

[6]More correctly: what is generated is an array of sources varying with time as $\exp(-i\omega_3 t) = \exp(-i\omega_1 t - i\omega_2 t)$ and with each source radiating into a wide range of directions. Consider a wave having angular frequency ω_3, starting out at one side of the crystal. After this wave has travelled to the other side of the crystal, it may or may not be in such a phase as to be augmented by the sources there (and everywhere in between). It is in the correct phase if $\boldsymbol{k}(\omega_3) = \boldsymbol{k}_1 + \boldsymbol{k}_2$, and this is the condition for a worthwhile amount of energy to be transferred into the new wave. This consideration is perhaps more familiar in other contexts where waves collide owing to a non-linearity, above all in non-linear optics, where the condition $\boldsymbol{k}_3 = \boldsymbol{k}_1 + \boldsymbol{k}_2$ is known as the condition for **phase matching**.

[7]Figure 24.2 exhibits the $\omega(k)$ relation for a monatomic chain, because that model system is likely to be familiar to the reader, and because it makes for a simple diagram. However, this restriction to one dimension makes the three k-vectors collinear, and makes the collision unrepresentative of any real collision in a three-dimensional solid. It has even been necessary to make a rather delicate choice for the proportions of the diagram in order for the collision to take place at all.

"source term" into an inhomogeneous wave equation it generates a new wave[6] having $\boldsymbol{k}_3 = \boldsymbol{k}_1 + \boldsymbol{k}_2$ and $\omega_3 = \omega_1 + \omega_2$. Of course, the energy in this new wave is obtained by depleting the energy in the original wave 1 and 2; that is what we learned when we multiplied through by $\hbar$.

A lattice wave, whether classical or quantum, has an ambiguity in its wave vector $\boldsymbol{k}$: its physics is unchanged if we add a reciprocal lattice vector $\boldsymbol{G}$. Corresponding to this is a Brillouin-zone structure to $\boldsymbol{k}$-space as mentioned in Chapter 23. All phonon physics can be described by means of $\boldsymbol{k}$-vectors lying within the first zone—though sometimes other possibilities are more convenient.

24.3 Normal and Umklapp collisions

Return to the monatomic linear chain of Fig. 23.1, but permitting waves to have both longitudinal and transverse polarizations. The dispersion relations for these two polarizations are sketched in Fig. 24.2. A transverse-wave mode having wave vector k_1 can be populated by phonons with energy $\hbar\omega_1$, and similarly for k_2 and (for longitudinal polarization) k_3.

A possible "collision" reaction (caused by a small non-linearity in the stress–strain relation for the springs) takes phonons at k_1 and k_2 and combines them into a single phonon at k_3. Energy is conserved, so $\omega_1 + \omega_2 = \omega_3$. As Fig. 24.2 is drawn,[7] k_3 lies to the right of $k = \pi/a$, so it is outside the first Brillouin zone—at least as that zone is conventionally carved out from the k-axis.

A standard textbook discussion (and it's correct) now goes as follows. All physics that is represented in Fig. 24.2 by wave vector k_3 can equally well be represented by $k_3' = k_3 - 2\pi/a$, where we have added a reciprocal

Fig. 24.2 The $\omega(k)$ relations for longitudinal (L) and transverse (T) waves on a monatomic linear chain (of atoms each connected to its nearest neighbours by springs). A small non-linearity in the springs causes phonons at k_1 and k_2 to be destroyed and to be replaced by a single phonon at k_3, with conservation of the total k and ω. For the case drawn, wave vector k_3 lies to the right of $k = \pi/a$, so it lies outside the first Brillouin zone, as conventionally drawn from $-\pi/a$ to $+\pi/a$. An alternative representation of the same physics as k_3 lies at $k_3' = k_3 - 2\pi/a$: a reciprocal lattice vector $G = -2\pi/a$ is added. Viewed this way, the reaction is an Umklapp collision. We observe that the group velocity $d\omega/dk$ is positive for k_1 and k_2, but is negative for k_3 and k_3': the direction of energy flow has been reversed.

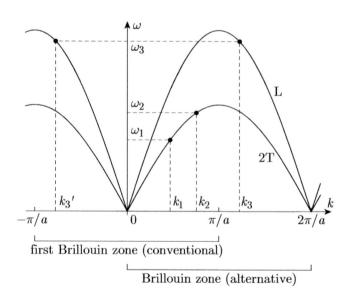

lattice vector $G = -2\pi/a$ to k_3. The conservation laws are respected, but (24.1) is changed into the Umklapp law[8]

$$k_3' = k_1 + k_2 + G. \qquad (24.3)$$

In fact, all phonon–phonon collisions (in three dimensions) fall into one of the following two types:[9]

$$\text{normal if} \qquad k_3 \rightleftharpoons k_1 + k_2, \qquad (24.4a)$$
$$\text{Umklapp if} \quad k_3' \rightleftharpoons k_1 + k_2 + G. \qquad (24.4b)$$

In (24.3–4), G is a reciprocal-lattice vector, for which the definition (full definition for a three-dimensional crystal) is given in books on solid-state physics.

Before we turn attention to the consequences for heat flow, we consider the implications of (24.4). Statements (24.4) seem to encapsulate a clear distinction, between collisions that do and do not require a G to balance the momentum-conservation equation. Yet I shall raise questions that will make the distinction murkier.

First: When we learn about reciprocal space and Brillouin zones, we are told that k-space can be partitioned into zones in more than one way, the only *necessary* requirement being that every physically distinct feature of the $\omega(k)$ dispersion relation must appear once and once only within each zone. In Fig. 24.2, this requirement is satisfied if we carve any piece of length $2\pi/a$ out of the k-axis.[10] Given this arbitrariness, we can consider the two possible choices of zone indicated in Fig. 24.2. The "conventional" choice is the obvious one to make, because in treating $+k$ and $-k$ on the same footing,[11] it treats the $\pm x$ directions symmetrically. But the "alternative" choice of zone is not wrong—at worst it may be somewhat perverse.[12] And in that "alternative" zone choice, there is no G to be inserted into the conservation equation for k: it seems the collision is not Umklapp when formerly it was. Perhaps the distinction between normal and Umklapp collisions is a little more slippery than we first thought . . . .

Second: We must not lose sight of the reason why we are considering phonon–phonon collisions: we are trying to find out which collisions obstruct the flow of heat down a temperature gradient. So, rather than getting hung up on words ("Umklapp" or "normal"), we should ask: is the collision of Fig. 24.2 one that obstructs heat flow, or not?

There is a feature of Fig. 24.2 that draws attention to itself: the two phonons (k_1 and k_2) we start with have group velocity $\mathrm{d}\omega/\mathrm{d}k$ positive, so they carry their energy from left to right, towards $+x$. The k_3 phonon they build has $\mathrm{d}\omega/\mathrm{d}k$ negative, so it carries its energy to the left, towards $-x$ (this independent of the zone choice or of any G). The collision reverses the transport of energy, so surely it should obstruct heat flow? There is a surprise here: a collision of this type (in one dimension or three) *does* obstruct heat flow, but the reversal of group velocity is not the reason,[13] however tempting the idea.

A harder look at the physics is inescapable.

[8] The quantity $\hbar k$ is often referred to as *crystal momentum* or *quasi-momentum*. We can see why it cannot be rigorously identified with a "real" momentum, since a G can be added without any change in the physics. Momentum conservation is saved by saying that $\hbar G$ is momentum "possessed by the crystal as a whole".

In Fig. 24.2, we can say that the addition of $G = -2\pi/a$ to k_3 amounts to shifting everything into the reduced-zone scheme.

[9] For simplicity, we discuss only "three-phonon" collisions in which two phonons combine to give one, or the inverse where one phonon splits into two. There can, of course, be more complicated reactions, such as a "four-phonon" reaction where there are two in and two out.

[10] Disconnected pieces of k-axis also suffice if appropriately chosen. Indeed, the conventional choice for the second zone makes it run from $-2\pi/a$ to $-\pi/a$ and from $+\pi/a$ to $+2\pi/a$. The gap is the location of the first zone.

[11] The undisplaced chain lies along the x-axis.

[12] The asymmetry makes it too inconvenient to be taken seriously for long.

[13] The present author confesses that he has been known to mislead students on this point.

24.4   Group velocity can be reversed even in the first zone

[14]Of course, copper is a metal, and electrons dominate over phonons in the transport of heat in a metal. I have presented these relations for copper because they are readily available, and copper has already been used as a standard example in Chapter 23. The idea is: if $\omega(\boldsymbol{k})$ relations can have this shape in copper, then something similar must happen in *some* of the dielectric crystals that the present chapter wishes to discuss.

We need a mind-clearing example on which to test ideas. Figure 24.3 shows the dispersion relations for phonon waves in copper, for two different directions in the crystal.[14] On the left we show the $\omega(\boldsymbol{k})$ relation for the [100] direction, to show that there can be three-dimensional waves whose behaviour closely resembles that of waves on a one-dimensional chain. More interesting is the $\omega(\boldsymbol{k})$ relation in the [110] direction shown on the right. For this direction, the two transversely polarized waves are no longer degenerate, so we have three different values of ω for each $\boldsymbol{k}$. In the [110] direction, the conventionally drawn Brillouin zone (the Wigner–Seitz cell) has a boundary at $k = \frac{3}{4}\sqrt{2}(2\pi/a)$ where a is the cube side for the face-centred-cubic lattice. Wave vector $\boldsymbol{k}_3$ lies within the conventional first Brillouin zone, and the associated group velocity faces leftwards on the diagram. We can have a "backwards" group velocity without the need to go into a higher-than-first zone.

Figure 24.3 marks two wave vectors $\boldsymbol{k}_1$ and $\boldsymbol{k}_2$ as well as $\boldsymbol{k}_3$. I have drawn these so that $\omega_1 + \omega_2 = \omega_3$: so far as energy is concerned, $\boldsymbol{k}_1$ and $\boldsymbol{k}_2$ might be collision partners yielding $\boldsymbol{k}_3$. Fortunately (measure the diagram!) $k_1 + k_2$ exceeds k_3. We can easily find (or imagine finding) $\boldsymbol{k}_1, \boldsymbol{k}_2$ above and below the plane of the paper such that $\boldsymbol{k}_1 + \boldsymbol{k}_2 = \boldsymbol{k}_3$ is satisfied (plenty of choices). The $\boldsymbol{k}_1, \boldsymbol{k}_2$ phonons must carry energy more or less to the right, while the $\boldsymbol{k}_3$ phonon carries energy to the left. This reaction $\boldsymbol{k}_1 + \boldsymbol{k}_2 \to \boldsymbol{k}_3$ involves a reversal of energy flow, even though all three wave vectors lie within the same Wigner–Seitz cell. Perhaps this reaction, with its reversal of energy flow, contributes to the obstruction of heat flow, in which case the adding of $\boldsymbol{G}$ is not essential to thermal resistance? Or perhaps not?

There is more to investigate.

Fig. 24.3 The $\omega(\boldsymbol{k})$ relations for phonons in copper (a face-centred crystal), for two different directions of the wave vector $\boldsymbol{k}$. After Svensson, Brockhouse and Rowe (1967). There is always one longitudinal polarization (L) and two transverse polarizations (T). For the [100] direction the two transverse polarizations are degenerate. Also in the [100] direction the Brillouin-zone boundary occurs at $k = 2\pi/a$ where a is the length of an edge of the face-centred cube. In the [110] direction, the zone boundary lies at $\frac{3}{4}\sqrt{2} \times (2\pi/a)$, but the plot has been extended into the next zone as far as $\sqrt{2}(2\pi/a)$. Wave vector $\boldsymbol{k}_3$ is a possible final state for a phonon–phonon collision $\boldsymbol{k}_1 + \boldsymbol{k}_2 \to \boldsymbol{k}_3$ as discussed in the text.

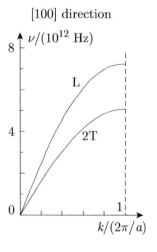

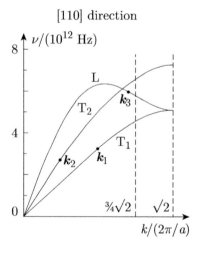

24.5 If **k** is conserved, heat flow does not decay

We shall show that a heat flow is not diminished by collisions if those collisions conserve **k** (i.e. if all collisions are "normal")—even if group velocity is reversed.[15] First, we need to think about **local thermodynamic equilibrium** (LTE).

We have a temperature gradient within a solid. That means the solid is not in precise thermodynamic equilibrium; it has no temperature that is uniform within it. How then can a temperature be defined at all? The answer lies in the concept of local thermodynamic equilibrium. This concept has application to a whole variety of non-equilibrium phenomena: in gases, in dielectric solids, in metals, and so on.

Consider a small volume within which the properties of our medium are nearly uniform. This volume has linear dimensions that are large compared with the mean free path for whatever collisions maintain near-equilibrium conditions. Yet it has dimensions small compared with the distance over which (macroscopic) conditions change significantly.[16] It is under these conditions that there is a "local" temperature within the small volume, and by extension "local" values of other thermodynamic variables. This does not prevent the temperature from varying (gradually) from place to place over large distances. LTE is the prerequisite for defining a temperature, a temperature gradient, and hence a thermal conductivity.

Local equilibrium for phonons permits the application of statistical mechanics to those phonons.

In near-equilibrium, there is a Planck distribution that determines how plentiful are phonons at $\boldsymbol{k}_1, \boldsymbol{k}_2, \boldsymbol{k}_3$, and that distribution controls[17] the rate of collision reactions $\boldsymbol{k}_1 + \boldsymbol{k}_2 \rightleftharpoons \boldsymbol{k}_3$. The reaction goes both ways, and (in near-equilibrium) at almost equal rates.

In dealing with the statistical mechanics, we need to introduce a refinement, equally applicable to (e.g.) a molecular gas, or to electrons in a metal. It is possible to have an "off-centre" distribution that meets all the conditions for equilibrium. All that is required is conservation of momentum. This idea is explored in problem 24.1.

In problem 24.1, it is shown that the particles of a boson gas can have a Bose–Einstein distribution of the form[18]

$$\frac{N_i}{g_i} = \frac{1}{\exp\left(\dfrac{\varepsilon_i - \mu - \boldsymbol{v} \cdot \boldsymbol{p}_i}{k_{\mathrm{B}}T}\right) - 1}, \tag{24.5}$$

in which N_i is the number of bosons in the ith bundle of quantum states, that bundle comprising g_i states close to energy ε_i and momentum $\boldsymbol{p}_i$. The quantity μ is the chemical potential, and $\boldsymbol{v}$ is a set of three Lagrange multipliers associated with conservation of the three components of $\sum_i N_i \boldsymbol{p}_i$. This kind of off-centre statistical distribution can always exist if collisions satisfy momentum conservation.

[15] To anticipate the key point: Reaction $\boldsymbol{k}_1 + \boldsymbol{k}_2 \rightarrow \boldsymbol{k}_3$ diminishes the energy flow, but the inverse reaction $\boldsymbol{k}_3 \rightarrow \boldsymbol{k}_1 + \boldsymbol{k}_2$ increases it. And in equilibrium both reactions proceed at the same rate.

[16] There has, of course, to be "headroom" between these two requirements, and that is what we shall understand by a "small" departure from a uniform thermodynamic equilibrium.

[17] Since phonons are bosons, the probability of a reaction that destroys phonons at $\boldsymbol{k}_1$ and $\boldsymbol{k}_2$ and creates one at $\boldsymbol{k}_3$ is proportional to

$$n(\boldsymbol{k}_1)\, n(\boldsymbol{k}_2) \left\{ n(\boldsymbol{k}_3) + 1 \right\}.$$

Problem 24.2 shows that this reaction rate fits nicely with the Planck distribution. Compare the $(n+1)$ here with the similar factor for photon emission in (19.22).

[18] There is a similar $-\boldsymbol{v} \cdot \boldsymbol{p}_i$ in the Fermi–Dirac distribution for fermions and in the Boltzmann distribution for near-classical molecules. See eqns (24.11).

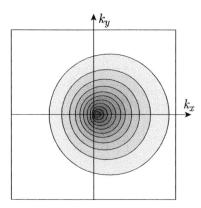

Fig. 24.4 An impression of the distribution for phonons (the darker the more) when the phonon distribution is in thermal equilibrium but is off-centred after the fashion of (24.6). The plot may describe a two-dimensional square lattice, or the $k_z = 0$ section for a cubic lattice. The square frame round the plot indicates the extent of the first Brillouin zone.

The off-centred distribution of phonons means that thermal energy is being transported towards positive x.

The $\omega(\boldsymbol{k})$ relations assumed are those of Fig. 23.6 (spherical symmetry assumed; transverse and longitudinal waves both included). For illustration, $\boldsymbol{v}$ has only an x-component, giving a displacement of the Planck distribution in the k_x-direction. We have given v/v_{T} the unrealistically large value of 0.2 (here v_{T} is the speed of low-frequency transverse waves, as in § 23.5.2).

Collisions mean that there are always fluctuations in the distribution N_i/g_i. Given momentum conservation, these fluctuations can do nothing to re-centre the distribution (24.5) or (24.6): on the contrary, they *maintain* it in its off-centred state, just as they maintain any more familiar equilibrium distribution.

The ideas just presented apply to phonons when those phonons undergo only "normal" collisions. One change: there is no conservation of phonon number, so μ does not appear in the statistical distribution. The distribution becomes the (modified) Planck expression

$$\frac{N_i}{g_i} = \frac{1}{\exp\left(\dfrac{\varepsilon_i - \boldsymbol{v}\cdot\boldsymbol{p}_i}{k_{\mathrm{B}}T}\right) - 1}. \tag{24.6}$$

Here $\varepsilon_i = \hbar\omega_i$ is the energy of a phonon having momentum $\boldsymbol{p}_i = \hbar\boldsymbol{k}_i$ and $\boldsymbol{v}$ is a measure of how far off-centre the distribution lies. With all collisions conserving $\sum_i N_i\boldsymbol{k}_i$ there is again no mechanism that can bring the distribution back to "centred". An impression of such an off-centred equilibrium Planck distribution is given in Fig. 24.4.

Distribution (24.6) has an important implication for heat transport. An off-centre distribution of phonons transports energy according to

$$\text{heat flow} = \boldsymbol{J} = \sum_i N_i\,\hbar\omega_i\,\frac{\partial\omega}{\partial\boldsymbol{k}},$$

a flow that is at least roughly in the direction of $\boldsymbol{v}$. That transport is not obstructed by the (normal only, remember) collisions that we have allowed to exist so far (they can't change the N_i). A non-zero heat flow, somehow set up, can continue without the need for any temperature gradient to drive it. The thermal resistance is zero.

This conclusion is surprisingly general. The statistical mechanics that leads to (24.6) makes no mention of the relation between ω and $\boldsymbol{k}$—nor even of Brillouin zones. The $\omega(\boldsymbol{k})$ relation can perfectly well have a region of negative group velocity, as in Fig. 24.3, and this does nothing to obstruct the transport of thermal energy by the phonons.

The reversed group velocity that we noticed when looking at Fig. 24.2 was so tempting that we had to investigate it, but it has turned out to be no more than a distraction, unconnected with the important physics. The reason is, of course, that a collision $\boldsymbol{k}_1 + \boldsymbol{k}_2 \to \boldsymbol{k}_3$ (as in Fig. 24.3) reduces the energy flow to the right, but $\boldsymbol{k}_3 \to \boldsymbol{k}_1 + \boldsymbol{k}_2$ increases it, and in equilibrium (LTE) both reactions take place with equal probability.

24.6 How Umklapp collisions obstruct heat flow

Start by thinking about low temperatures: the temperature regime represented in Fig. 23.3, where the only phonons thermally excited are those close to $\boldsymbol{k} = 0$. If we draw the first Brillouin zone in the obvious way, symmetrically about the origin, then all phonon momenta are well away from any zone boundary. Any phonon–phonon collision conserves

k, so all collisions are "normal". And resistance to heat flow is, from this cause, zero.

What about the arbitrariness in the drawing of Brillouin zones? Suppose we choose to draw the zone as we did in Fig. 24.2 (alternative choice), with one edge at the origin of k. With this choice, many of the possible collisions will involve an added G: any collision in which the three momenta are not all on the same side of the origin. But this is an entirely avoidable complication: all we need to do to make it go away is to draw the zone in the conventional way.[19] It follows, of course, that the collisions we have here forced to involve a G do not obstruct heat flow, but this property has been concealed by the bad zone choice.

Now raise the temperature, so that the populated vibrational modes extend out towards the half-way point across the conventional zone. We can keep a G out of the collision equations for as long as possible by centring the zone on the origin[20]—and the centring becomes more critical as the temperature is raised. Collisions remain both normal and non-obstructive.

A further rise of temperature causes some of the populated states to have k exceeding a half reciprocal lattice vector, and collisions necessarily put some k_3s outside the first zone, after the fashion of Fig. 24.2. An example of such a collision is shown in Fig. 24.5.

Figure 24.5 shows a portion of k-space according to the extended zone scheme. Everything drawn around $k = 0$ is repeated around each $k = G$; only one repetition is shown.

In Fig. 24.5, the reaction drawn shows k_1 and k_2 originating from regions where the phonon populations are enhanced by the off-centredness of the Planck distribution. This $k_1 + k_2 \to k_3$ reaction has a rate that is increased as compared with what happens with a "centred" distribution

[19] The chosen zone doesn't have to be exactly centred on $k = 0$. But it must have its boundaries well away from all the ks involved in a likely collision.

[20] In the language of the monatomic chain, the best we can do is to place the zone boundaries at $\pm\pi/a$. Such a symmetrical placing corresponds, in three dimensions, to applying the standard Wigner–Seitz construction in which the zone boundaries are given symmetrical locations in k-space.

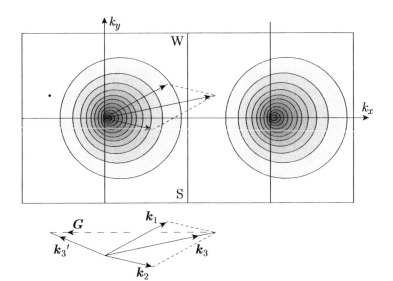

Fig. 24.5 A Brillouin zone for a square or cubic lattice is presented in the extended-zone scheme. Line WS is the zone boundary, drawn according to the usual Wigner–Seitz construction.

A phonon–phonon collision has $k_1 + k_2 \to k_3$, in which k_1, k_2 lie in the left-hand zone, but k_3 lies outside that zone. In the vector diagram at the foot of the figure, $k_3' = k_3 + G$ is the k-vector of the collision product, translated back into the first zone (reduced-zone scheme). The addition of G makes this an Umklapp collision.

The reaction drawn has its probability enhanced by the off-centredness of the phonon distribution; the inverse reaction is likewise partly suppressed. Phonons are taken from the surplus on the right and added to the deficit on the left: the distribution relaxes towards "centred". This happens only because k_3 lies to the right of boundary WS.

of phonons. Likewise, the reverse reaction rate is partly suppressed because k_3 lies in a region having diminished population. Overall, there is a movement of phonons in k-space from surplus to deficit. The heat flow diminishes as the phonon distribution relaxes towards "centred".

When the collision of Fig. 24.5 is re-described in the reduced-zone scheme, we have

$$k_3' \rightleftharpoons k_1 + k_2 + G,$$

which makes this an Umklapp collision according to (24.4b). Thus we associate Umklapp collisions with a relaxation of the phonon distribution towards "centred". Heat flow relaxes towards zero, unless replenished by an imposed temperature gradient—but that implies a non-zero thermal resistance. By contrast, "normal" collisions do not relax the distribution, and so they do not obstruct heat flow.

24.7   Summary

[21] We have seen that any other choice causes Gs to redistribute themselves in the k conservation equations in a way that can be extremely unhelpful.

For the purpose of this chapter's topic, k-space is best partitioned symmetrically into its Brillouin zones, according to the Wigner–Seitz rules.[21]

A phonon–phonon collision is described as *Umklapp* if the accountancy in k-vectors (reduced-zone scheme) includes a reciprocal lattice vector. In the case of a three-phonon collision, this means

$$k_3' \rightleftharpoons k_1 + k_2 + G.$$

A similar statement applies to a collision involving four phonon ks. Likewise, the same idea can be applied to electron–electron collisions or to electron–phonon collisions.

Collisions without a G conserve $\sum_i N_i k_i$, and thereby conserve any non-zero flow of heat carried by the phonons (given local thermodynamic equilibrium): these are *normal* collisions.

Collisions with a G—*Umklapp* collisions—cause any flow of heat energy to relax towards zero—unless, of course, there is a temperature gradient that replenishes the heat flow. But such a temperature gradient implies a non-zero thermal resistance.

A tempting idea, that obstruction of the heat flow is associated with a reversal of phonon group velocity, turns out to be incorrect.

Problems

[22] Remember § 21.3.2.

There will be some sort of container, but we can ignore collisions with the container provided the parcel of gas of interest is many mean free paths away from the container.

Problem 24.1   A moving gas
Think about a gas which, for simplicity, contains only one species of particle. I say "particle" so as to cover a more diverse range of possibilities than just molecules. Phonons are catered for, but as a special case later on. We are to imagine that the gas is moving relative to us as observers.

As always, the gas's particles collide with each other, thereby exploring microstates and maintaining local thermodynamic equilibrium (LTE).[22]

In any particle–particle collision, momentum is conserved; we exclude Umklapp collisions.

We now apply statistical mechanics. I'll do this in the most elementary way, the microcanonical ensemble, maximizing the entropy $S = k_{\mathrm{B}} \ln W$ subject to constraints.

Let the quantum states available to the particles be divided into bundles, where the ith bundle contains g_i quantum states, all close in energy to ε_i and in momentum to $\boldsymbol{p}_i$. The ith bundle is occupied by N_i particles. Then when the N_i undergo changes, perhaps because the gas is exploring its microstates,

$$\frac{\delta S}{k_{\mathrm{B}}} = \sum_i \frac{\partial \ln W}{\partial N_i} \, \delta N_i.$$

In the microcanonical way, we maximize S subject to the imposition of constraints on the total number N of particles, the total energy U, and now also on the total momentum $\boldsymbol{P}$ of the gas:

$$\delta N = \sum_i \delta N_i = 0, \qquad \delta U = \sum_i \varepsilon_i \, \delta N_i = 0, \qquad \delta \boldsymbol{P} = \sum_i \boldsymbol{p}_i \, \delta N_i = 0.$$

(1) Perform the maximization of $\ln W$ subject to the five constraints. You will have to introduce five Lagrange multipliers instead of the usual two. Show that the outcome is

$$\frac{\partial \ln W}{\partial N_i} = \alpha + \beta \varepsilon_i + \boldsymbol{\gamma} \cdot \boldsymbol{p}_i \tag{24.7}$$

for any state of the gas where that gas is in thermal equilibrium. Here α, β and the three components of $\boldsymbol{\gamma}$ are the five Lagrange multipliers.

(2) Argue that (24.7) holds for any equilibrium state of the gas, and that therefore it continues to hold during a move from one equilibrium state to another. Therefore, during such a change,

$$\frac{\delta S}{k_{\mathrm{B}}} = \sum_i \frac{\partial \ln W}{\partial N_i} \, \delta N_i = \sum_i (\alpha + \beta \varepsilon_i + \boldsymbol{\gamma} \cdot \boldsymbol{p}_i) \delta N_i = \alpha \, \delta N + \beta \, \delta U + \boldsymbol{\gamma} \cdot \delta \boldsymbol{P}.$$

(3) This shows that δU contains $\delta \boldsymbol{P}$. Turn things round to give

$$\delta U = T \, \delta S + \mu \, \delta N + \boldsymbol{v} \cdot \delta \boldsymbol{P} \tag{24.8}$$

so that, equating coefficients,

$$\alpha = -\frac{\mu}{k_{\mathrm{B}} T}, \qquad \beta = \frac{1}{k_{\mathrm{B}} T}, \qquad \boldsymbol{\gamma} = -\frac{\boldsymbol{v}}{k_{\mathrm{B}} T}. \tag{24.9}$$

Here μ is the chemical potential, and $\boldsymbol{v}$ is a quantity, so far uninterpreted, with the dimensions of velocity. Combine (24.7) and (24.9) to show that

$$\frac{\partial \ln W}{\partial N_i} = \frac{\varepsilon_i - \mu - \boldsymbol{v} \cdot \boldsymbol{p}_i}{k_{\mathrm{B}} T}, \tag{24.10}$$

a result that holds for all the different statistics.

(4) Now use the microcanonical expressions for W for bosons, fermions, and for the dilute limit of either of these to obtain from (24.10)

Boltzmann: $\qquad \dfrac{N_i}{g_i} = \exp\left(\dfrac{-(\varepsilon_i - \mu - \boldsymbol{v} \cdot \boldsymbol{p}_i)}{k_B T}\right),$ $\qquad$ (24.11a)

Bose–Einstein: $\qquad \dfrac{N_i}{g_i} = \dfrac{1}{\exp\left(\dfrac{\varepsilon_i - \mu - \boldsymbol{v} \cdot \boldsymbol{p}_i}{k_B T}\right) - 1},$ $\qquad$ (24.11b)

Fermi–Dirac: $\qquad \dfrac{N_i}{g_i} = \dfrac{1}{\exp\left(\dfrac{\varepsilon_i - \mu - \boldsymbol{v} \cdot \boldsymbol{p}_i}{k_B T}\right) + 1}.$ $\qquad$ (24.11c)

In no case has any assumption been made as to the relation between ε_i and $\boldsymbol{p}_i$, so these distributions are general extensions to the usual distributions.[23]

(5) Consider a parcel of a gas consisting of massive particles.[24] The parcel has total mass M and total momentum $\boldsymbol{P}$. Owing to its overall motion, the parcel has kinetic energy $\boldsymbol{P}^2/2M$. Use (24.8) to show that $\boldsymbol{v} = \boldsymbol{P}/M$, so that $\boldsymbol{v}$ is the velocity of the gas's centre of mass, a velocity possessed by the gas as a whole. This identification holds for any of the statistics of (24.11). But see part (7).

(6) Take a molecular gas obeying Boltzmann statistics. The ith molecule has kinetic energy $\varepsilon_i = p_i^2/2m$. In the rest frame Σ' of the gas's centre of mass, the Boltzmann distribution depends upon $\varepsilon_i' - \mu'$ and is independent of the direction of the molecule's momentum. Perform a Galilean transformation to a laboratory frame Σ relative to which the gas (and frame Σ') moves with velocity $\boldsymbol{v}$. Show that the distribution in the laboratory frame contains in the exponent[25]

$$\varepsilon_i' - \mu' = \varepsilon_i - \boldsymbol{v} \cdot \boldsymbol{p}_i - \mu + m\boldsymbol{v}^2.$$

Thus we have another way of seeing why the $\boldsymbol{v}$ in (24.11a) is the velocity of the gas's centre of mass.

(7) Show that (24.11b) holds even for the case where we have Bose–Einstein particles, such as phonons, whose number is not conserved: the removal of one constraint removes one Lagrange multiplier, and μ is absent as in (24.6). Notice that part (5) cannot be applied in this case.[26] Velocity $\boldsymbol{v}$ can no longer be identified as a centre-of-mass velocity. Nevertheless, $\boldsymbol{v}$ continues to be a measure of the "off-centredness" of the phonon distribution in $\boldsymbol{k}$-space.

Comment: Notice that, except in part (6), the statistical mechanics has nowhere assumed any particular relation between ε and $\boldsymbol{p}$; equivalently, in the case of phonons, any relation between ω and $\boldsymbol{k}$.

Problem 24.2 The Planck distribution and collision probabilities
In sidenote 17, we gave an expression claimed to be proportional to the rate at which phonons collide. That reaction rate must be compatible with the Planck distribution.[27] Here we give a rather crude derivation of the Planck distribution starting from the reaction rate.

[23] Some physical systems require special cases to be taken. For example, phonons and photons have $\mu = 0$; and photons have $\boldsymbol{v} = 0$.

[24] "Massive" particles might be bosons or fermions, or might constitute a gas of either dilute enough for Boltzmann statistics to apply. Excluded are phonons and other zero-mass quasi-particles; think about why.

[25] The chemical potential μ' is the energy we have to give to a particle at rest in Σ' in order to add it to the gas. The corresponding energy in frame Σ is μ. To bring the particle from rest in Σ to rest in Σ' we must give it $\frac{1}{2}m\boldsymbol{v}^2$, so $\mu = \mu' + \frac{1}{2}m\boldsymbol{v}^2$.
The $m\boldsymbol{v}^2$ in the displayed expression looks unexpected. Show that, when we normalize the Boltzmann distribution, $\exp(-m\boldsymbol{v}^2/k_B T)$ disappears into the normalization constant.

[26] Neither can part (6). There is no question of applying a Galilean transformation to phonons, since the underlying lattice remains at rest.

[27] More precisely, the equilibrium distribution must be maintained by collisions, so it is affected by the functional form of the collision rate. We must demand consistency.

For simplicity, we ignore the possibility of an "off-centre" Planck distribution of the kind discussed in problem 24.1.

Let $n = n(\omega)$ be the mean phonon occupation number of a vibrational mode in which the phonons have energy $\hbar\omega$. In equilibrium, $n(\omega)$ must be a function of ω only (that is, not dependent on the direction of travel of the phonon, and so not on $\boldsymbol{k}$), though it will depend also on temperature. Then[28]

$$\text{probability for } \omega_1 + \omega_2 \to \omega_3 \quad \propto \quad n_1\, n_2(n_3 + 1)$$
$$\text{probability for } \omega_3 \to \omega_1 + \omega_2 \quad \propto \quad n_3(n_1 + 1)(n_2 + 1).$$

We shall assume without proof that "proportional to" conceals the same geometrical factors for "forward" as for "backward" reactions.[29] Then in thermal equilibrium, the rates of these reactions must be equal:

$$n_1\, n_2(n_3 + 1) = n_3(n_1 + 1)(n_2 + 1),$$

which we may rearrange into

$$\frac{n_1}{n_1 + 1}\, \frac{n_2}{n_2 + 1} = \frac{n_3}{n_3 + 1}.$$

Phonons are not conserved in a three-phonon collision, but we have just discovered a replacement for the conservation law on number:[30]

$$\ln\left(\frac{n_1}{n_1 + 1}\right) + \ln\left(\frac{n_2}{n_2 + 1}\right) = \ln\left(\frac{n_3}{n_3 + 1}\right), \tag{24.12a}$$

together with

$$\omega_1 + \omega_2 = \omega_3. \tag{24.12b}$$

We have said that n is, on physical grounds, to be a function of ω, so the obvious conclusion from (24.12) is that

$$\ln\left(\frac{n(\omega)}{n(\omega) + 1}\right) = \text{constant} \times \omega. \tag{24.13}$$

Use (24.13) to show that

$$n(\omega) = \frac{1}{e^{(\text{constant})\omega} - 1},$$

and invent your own argument that the constant in this must be $\hbar/k_{\mathrm{B}}T$.

A similar analysis can be applied to show that a collision probability $\propto n_1 n_2(1 - n_3)(1 - n_4)$ applied to a four-fermion collision requires there to be a Fermi–Dirac distribution.

[28] We patch in factors $(n + 1)$ for the "recipient" quantum states following the photon precedent in (19.22) and Fig. 19.5.

[29] For any "normal" three-phonon collision $\boldsymbol{k}_1 + \boldsymbol{k}_2 = \boldsymbol{k}_3$. The point is that the resulting geometrical factors affect the forward and backward reactions equally. There is an argument from detailed balance not far away.

Think about the way in which the reaction rates relate to squared matrix elements for the reactions. Compare with Chapter 19 sidenote 36.

[30] You can try maximizing the entropy in the usual way, but putting in a Lagrange multiplier for the conservation law of (24.12a) in place of one on boson number. You'll find that the new Lagrange multiplier must be set to zero in order for the distribution to go over to Boltzmann in the limit of large ω.

25   Electrons and holes in semiconductors

Intended readership: Soon after seeing the standard textbook treatments that explain band structure and the existence of quasi-particles called electrons (conduction band) and holes (valence band).

25.1   Introduction

The main motivation for this chapter is to give a careful discussion of the concept of a hole in an almost-full valence band.

Electrons in crystalline solids have quantum states that are grouped, in energy, into **bands**.[1] A pure semiconductor, at absolute zero, has all quantum states within one band, the **valence band**,[2] filled with electrons, while the next higher band, the **conduction band**, is empty. At temperatures above zero, or if the semiconductor is doped with an impurity, there are some electrons (assumed few) in the states of the conduction band, and likewise there are a few empty states in the valence band. This chapter is concerned with both, but especially with empty states in the valence band, which are known as **holes**.

When a conduction current flows in a metal or semiconductor (not superconducting), electrons (in either or both of the conduction and valence bands) are accelerated by an applied electric field; electrical resistance arises because those electrons are scattered[3] by phonons or by (other) imperfections in the underlying crystal (impurity atoms or dislocations or crystallite boundaries).

Exercise 25.1 The free path for electrons in copper

Electrons in a periodic potential travel freely without being scattered by that potential: the Schrödinger equation has Bloch-function eigensolutions. Here we give a numerical confirmation. Any convenient crystalline material will serve as example, not necessarily a semiconductor.

Copper has one mobile electron originating from each atom. These mobile electrons occupy half the quantum states, and thus half the k-space volume, of the conduction band. Treat the conduction-band electrons as occupying a sphere[4] in k-space of radius k_F.

(1) Copper's crystalline structure is face-centred cubic. Work out the nearest-neighbour atom–atom spacing (data below).

---

[1] Figure 28.1 may give a reminder. Each *band* has quantum states whose k-values span a Brillouin *zone*. Conversely, each k, in the *reduced-zone scheme* (grey in Fig. 28.1), has several quantum states associated with it, one (doubled by spin) belonging to each band.

More jargon: The $\mu = \mu(T)$ in the electrons' Fermi–Dirac distribution is the **chemical potential** (often given the rather unfortunate name **Fermi level**). The zero-temperature value $\varepsilon_F = \mu(T{=}0)$ is the **Fermi energy**. An electron having energy $\mu(T)$ moves with velocity v_F (dependent on the direction of k), the **Fermi velocity**.

[2] More correctly: the valence band is the highest-in-energy of the filled bands. There are bands lying lower in energy, but these are usually thought of as filled atomic states, rather than as bands.

[3] There is no sense to a *mean* free path because electrons having energy ε well below μ cannot collide (there are no empty states for them to go into), and they contribute infinity to the average path. However, we can define a **relaxation time** τ that tells us how fast the electrons' distribution relaxes towards Fermi–Dirac after being perturbed. An interesting "free path" is $\lambda = v_F \tau$, a measure of the distance travelled (on average) by an electron whose energy is μ. For a description of the relaxation processes, see e.g. Singleton (2001) § 9.2.2; Ziman (1960) § 9.10.

Essays in Physics: Thirty-two thoughtful essays on topics in undergraduate-level physics. Geoffrey Brooker, Oxford University Press. © Geoffrey Brooker 2021. DOI: 10.1093/oso/9780198857242.003.0025

(2) Work out the number density N of electrons in the conduction band. Find a numerical value for k_F, the radius in k-space of the **Fermi surface**.

(3) Treat the effective mass of electrons on the Fermi surface as the same as the mass m_e of free electrons. Find the speed v_F (the Fermi speed) of electrons on the Fermi surface.

(4) Use Drude theory[5] to calculate the free path λ through which electrons travel between collisions (with whatever scatters them).

(5) Compare the free path with the interatomic spacing.

[Copper has the following properties: density = $8933\,\mathrm{kg\,m^{-3}}$; relative atomic mass = 63.54; resistivity at room temperature = $1.7 \times 10^{-8}\,\Omega\,\mathrm{m}$.]

Comments

1. My answers: interatomic spacing is $2.56 \times 10^{-10}\,\mathrm{m}$; the Fermi speed is $1.57 \times 10^6\,\mathrm{m\,s^{-1}}$; the free path at room temperature is $3.9 \times 10^{-8}\,\mathrm{m}$, which is 152 atomic spacings.

2. The Fermi speed is of order $c/200$, small enough that we don't need relativity; very much the same as the speed $\sim \alpha_{fs} c$ that electrons have inside atoms. (Well, they are confined in an average volume that's much the same ....)

3. If copper is cooled to liquid-helium temperatures, the conductivity rises, by as much as three orders of magnitude if the sample is very pure. Then the free path can be something like 10^5 atomic spacings.

4. Semiconductors are not very different. The physical properties of silicon are tabulated by Madelung (1996).[6] Charge is carried by conduction-band electrons and two types of valence-band hole, so there are several different relaxation times and several free paths. The best we can say is that at room temperature a free path is 100–200 atomic spacings, very much the same as in copper.[7] At low temperatures the free path rises by up to four orders of magnitude, so it can be as much as 10^6 atomic spacings.

[4] [From previous page] The conspicuous topological feature of copper's Fermi surface is the presence of "necks" where the Fermi surface reaches out to meet the Brillouin-zone boundary. Nevertheless, over most of the surface, the shape is not far from spherical.

[5] Electrical conductivity σ is given by

$$\sigma = \frac{Ne^2}{m_e}\tau,$$

in which τ is the relaxation time (sidenote 3) telling us how fast the electron distribution relaxes back to equilibrium after being perturbed. N is the number of mobile electrons per unit volume. See e.g. Bleaney and Bleaney (2013), eqn (3.12).

Do not accept a picture you may have been given when first meeting the Drude idea: that each collision brings an electron to rest. An electron undergoing a scattering event must find an empty quantum state to go into, and that means its initial and final states are close to the Fermi surface (within $\sim k_B T$); the initial and final speeds are both close to v_F. What should be said is that scattering randomizes the directions of the electron velocities, giving a near-zero after-scattering average velocity.

The Bleaney treatment is careful to get this right.

[6] I am indebted to Dr M.B. Johnston for help with the properties of silicon.

[7] We should not be surprised at this similarity. At room temperature the collisions are with phonons and the phonon spectrum is not so very different in metals and semiconductors.

The lengthened electron free path at low temperatures comes about, of course, because the phonons' Planck distribution "freezes out" the phonons.

Exercise 25.1 confirms that electrons in a solid "want" to travel forever. They fail only because occasional scatterings get in the way.

25.2   A model band structure

Figure 25.1 shows a section through k-space covering the full width of a Brillouin zone (reduced zone scheme). The lower curve shows a possible relation between energy ε and wave vector k for electrons in the valence band. The upper curve does the same for the conduction band.

The Brillouin zone is a volume in k-space, so our graph shows a section through k-space, taken through the origin $k = 0$. The k_x-width is a reciprocal lattice vector G. This somewhat crude picture suffices for the discussions of the present chapter.

If we were handling a one-dimensional model, electrons in a linear chain of atoms, the zone boundaries would lie at $k_x = \pm\pi/a$, where a is the lattice spacing. But three-dimensional crystals have other expressions for the locations of zone boundaries, so we leave these k_x-values unmarked.

For the discussion in the present chapter, it is irrelevant whether the semiconductor has a direct or an indirect band gap. A direct gap is shown for simplicity.

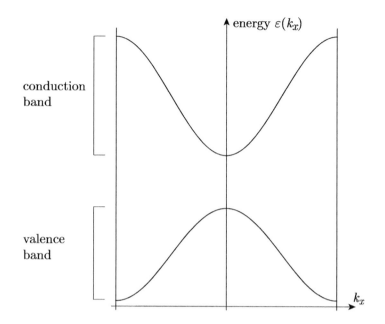

[8] A weak magnetic field can be dealt with by adding force $(-e)\boldsymbol{v} \times \boldsymbol{B}$ to (25.2). But this is a complication best avoided for now.

[9] The P_k in (25.1) shows that ψ_k is not an eigenstate of the usual momentum operator $-i\hbar\,\partial/\partial x$. In a periodic potential, the translation operator $\hat{T}$ of (25.15) replaces $\partial/\partial x$ and its eigenvalues e^{ika} account for "crystal momentum" $\hbar k$. We show this in problem 25.3.

A derivation of the equation of motion (25.2) is given in problem 25.5.

25.3   An electron in the conduction band

In a one-dimensional model, an electron experiences a periodic potential $V(x)$ having period a. The electron eigenfunctions have the Bloch form

$$\psi_{\boldsymbol{k}}(\boldsymbol{r}) \sim e^{i\boldsymbol{k}\cdot\boldsymbol{r}}\,P_{\boldsymbol{k}}(\boldsymbol{r}); \qquad \psi_k(x) \sim e^{ikx}\,P_k(x), \tag{25.1}$$

in which $P_k(x)$ is a periodic function of x with period a; $\psi_{\boldsymbol{k}}(\boldsymbol{r})$ and $P_{\boldsymbol{k}}(\boldsymbol{r})$ are the corresponding functions in three dimensions. The energy $\varepsilon(k_x)$ plotted in Fig. 25.1 is the associated energy. Each band has its own $P_{\boldsymbol{k}}(\boldsymbol{r})$ but, even within one band, $P_{\boldsymbol{k}}(\boldsymbol{r})$ depends upon $\boldsymbol{k}$.

Figure 25.2 shows an enlarged view of the $\varepsilon(k_x)$ relation for a single electron near the bottom of the conduction band of Fig. 25.1. In thermal equilibrium at absolute zero this electron occupies the lowest available quantum state, so it sits at the bottom of the band in a Bloch wave function having $k_x = 0$ (left panel).

Next imagine that an electric field E_x is applied to the semiconductor.[8] The equation of motion for the conduction-band electron is[9]

electron energy $\varepsilon(k_x)$

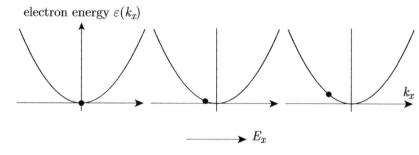

$$\frac{\mathrm{d}(\hbar \boldsymbol{k})}{\mathrm{d}t} = (-e)\boldsymbol{E}; \qquad \frac{\mathrm{d}(\hbar k_x)}{\mathrm{d}t} = (-e)E_x. \qquad (25.2)$$

The electron charge $-e$ is negative, so a field towards $+x$ causes the electron's ($\hbar k_x$) to become negative. The middle and right-hand panels of Fig. 25.2 show where the electron has got to (in $\boldsymbol{k}$-space) after a short time, and after double that time.

So far, we have treated the conduction-band electron as having a definite $\boldsymbol{k}$, meaning that its wave function has the Bloch form (25.1). A definite wave vector $\boldsymbol{k}$ means that the electron has a completely unknown location, being spread equally through the entire volume of the host crystal. This is too extreme: we wish to be able to discuss the electron's location, within limits set by the uncertainty principle. Therefore we now give the electron a fuzzy location by imagining that we build a wave packet, a wave packet that starts out centred on $k_x = 0$.

Let the wave packet span values of k_x within the range Δk_x. We want this to make only a small change to the interpretation of Fig. 25.2, so $\Delta k_x \ll G$, the k_x-width of the Brillouin zone. In turn, this means that the energy range $\Delta \varepsilon$ is small compared with the "height" of the conduction band. We can see the wave packet's Δk_x and $\Delta \varepsilon$ as represented by the size of the blob drawn in Fig. 25.2.

If the wave packet is sensibly constructed, it will occupy an x-range Δx given by $\Delta k_x \cdot \Delta x \sim 1$. This means that Δx must be many atomic spacings[10] —though it can still be small compared with the size of the host crystal. Problems 25.1 and 25.2 investigate the reasonableness of— indeed the necessity for—these conditions.

Consider next the motion of the conduction-band electron in real space. Its wave packet has *group velocity*[11]

$$\boldsymbol{v} = \frac{\mathrm{d}\varepsilon}{\mathrm{d}(\hbar \boldsymbol{k})}; \qquad v_x = \frac{\mathrm{d}\varepsilon}{\mathrm{d}(\hbar k_x)}, \qquad (25.3)$$

the gradient of the energy–wave-vector curve at the location of the blob. In the centre and right-hand panels of Fig. 25.2 this is negative so, under the influence of the applied electric field $E_x > 0$, the electron acquires a velocity v_x towards $-x$, getting bigger ($|v_x|$) with time.

The electron carries charge $-e$ at velocity v_x. Therefore it carries a current $j_x = (-e)v_x$, in the same direction as E_x. Exercise 25.2(3) asks you to think about this.

We may define an **effective mass** m_{e}^* for the conduction-band electron, according to[12]

$$\delta(\hbar k_x) = m_{\mathrm{e}}^* \,\delta v_x, \quad \text{so that} \quad \frac{1}{m_{\mathrm{e}}^*} = \frac{\delta v_x}{\delta(\hbar k_x)} = \frac{\mathrm{d}^2 \varepsilon}{\mathrm{d}(\hbar k_x)^2}. \qquad (25.4)$$

If the energy–wave-vector relation of Fig. 25.2 is parabolic within the k_x-range of interest, then the effective mass is constant, and we have[13]

$$(\hbar k_x) = m_{\mathrm{e}}^* v_x, \qquad \varepsilon = \tfrac{1}{2}m_{\mathrm{e}}^* v_x^2 = \frac{(\hbar k_x)^2}{2m_{\mathrm{e}}^*}, \qquad (25.5)$$

equations that have the same *form* as those governing an electron moving in vacuum. The difference is that the effective mass m_{e}^* replaces the free-particle mass m_{e}.

[10] $\Delta x \gg 1/G \sim$ (atomic spacing).

[11] Equation (25.3) requires a new derivation when the wave function contains $P_k(x)$ as in (25.1). An outline derivation is given in problem 25.4, and a consistency check is made in problem 25.5.

[12] For simplicity, eqn (25.4) assumes isotropy, so that m_{e}^* is a scalar (in general it is a second-rank tensor). A full definition of m_{e}^* should cater for dependence of $\varepsilon(\boldsymbol{k})$ on the direction of $\boldsymbol{k}$, but also on the magnitude $|\boldsymbol{k}|$ since the $\varepsilon(\boldsymbol{k})$ relation is not simple-parabolic except at small $\boldsymbol{k}$.

[13] The energy at the bottom of the conduction band is being taken as zero for this purpose.

25.4   Quasi-particles

In the discussion of § 25.3, we obtained almost all needful knowledge of the electron's behaviour by reading value, gradient and curvature from the $\varepsilon(\boldsymbol{k})$ relation. We must pause to understand the implications.

The band structure shown in Fig. 25.1 has come about because there is a periodic potential originating from the regular crystalline array of atomic cores. That potential has changed the free-electron wave function $e^{i\boldsymbol{k}\cdot\boldsymbol{r}}$ into a Bloch function $e^{i\boldsymbol{k}\cdot\boldsymbol{r}}P_{\text{periodic}}(\boldsymbol{r})$, still possessing a $\boldsymbol{k}$ but modified by the periodic-function part. In turn, that has affected the $\varepsilon(\boldsymbol{k})$ relation for the electron, as exemplified in Fig. 28.1; in particular, it is responsible for the energy gaps.

There is now no reason why there should be any resemblance between m_{e}^{*} and the mass m_{e} of a free electron.[14]

We handle this kind of thing by saying that the conduction-band electron is a **quasi-particle**, meaning that it has particle-like properties, but those properties are defined by the energy–wave-vector relation $\varepsilon(\boldsymbol{k})$, and this relation replaces the characteristics of the underlying electron.[15]

We have met quasi-particles before. In Chapters 23 and 24, we had a lattice of atoms connected to each other by "springs", and we described the motions of those atoms in terms of phonons possessing an $\varepsilon(\boldsymbol{k})$ relation such as that of Fig. 23.2. Everything that the atoms could do was accounted for by phonons, so we agreed to stop mentioning the atoms-and-springs and instead to talk exclusively about the phonons. In effect, all needful knowledge about the atoms was encapsulated in the $\varepsilon(\boldsymbol{k})$ relation, and thence was embedded into the dynamics of the phonons.

The same thing happens here. We agree to stop mentioning the crystal structure and its periodic potential, and instead we discuss everything in terms of the $\varepsilon(\boldsymbol{k})$ relation(s) and what the (quasi-) particles living on them do. We have found ourselves undertaking this journey for the case of a conduction-band electron.

It will turn out that an empty state at the top of the valence band has particle-like properties, so it constitutes another quasi-particle and is given the name "hole".

It is not to be taken for granted that a quasi-particle of some kind must exist. If one does exist, we are fortunate. If a conduction-band electron, or a valence-band unfilled state, is to be a valid quasi-particle, then rigorous tests must be applied—the more rigorous because we *want* the quasi-particle to be legitimate. Even for a conduction-band electron, we must look back with a critical eye: exercise 25.2.

[14] Figure 28.1 may emphasize this. At the bottom of the diagram, the lowest band's $\varepsilon(k_x)$ has similar curvature to the free-electron curve, and $m_{\text{e}}^{*} \approx m_{\text{e}}$. But a solid having four electrons per atom will have the bottom two bands full, and the conduction band is the next. This band has a sharply curved $\varepsilon(k_x)$ minimum that clearly has no resemblance to the free-electron curve.

Refer also back to Fig. 25.1. Away from the bottom of the conduction band, the energy–wave-vector curve is far from parabolic, so m_{e}^{*} is a function of k_x. Indeed, the curve has inflexions at which the effective mass becomes infinite, and at higher energies the effective mass is negative. Only at the bottom of a band is m_{e}^{*} constant—and even there it can be anisotropic, a second-rank tensor.

[15] It might be clearer if this quasi-particle were given a new name, to distinguish it from a free electron. But convention extends the name "electron" to cover the case of the conduction-band quasi-particle.

Exercise 25.2   A second look at a conduction-band electron

As forecast above, we must scrutinize the properties we have attached to a conduction-band electron. If the quasi-particle is to be a helpful concept, its properties must not be paradoxical.

(1) Start with the Bloch function $\psi_{\boldsymbol{k}}$ of (25.1). It has crystal momentum $\hbar\boldsymbol{k}$, meaning that $\psi_{\boldsymbol{k}}(\boldsymbol{r})$ is an eigenfunction of translation operators $\widehat{\boldsymbol{T}}$, three of them, each of which translates through one of the real-space lattice vectors.[16] A reciprocal-lattice vector $\boldsymbol{G}$ can be added to $\boldsymbol{k}$ without change to the physics. Revise the definition and understanding of the reduced-zone scheme, to be sure that the wave functions $\psi_{\boldsymbol{k}}$ having $\boldsymbol{k}$ within the reduced zone cover all our needs.

[16] Translation operators $\widehat{\boldsymbol{T}}$ are three-dimensional analogues of the $\widehat{T}$ introduced in sidenote 9.

(2) Work problems 25.1 and 25.2. Convince yourself that a wave packet is a legitimate concept, in that it can be built from a range of $\boldsymbol{k}$s that is small (in $\boldsymbol{k}$-space) compared with the size $\boldsymbol{G}$ of a Brillouin zone and yet is small (in real space) compared with the overall dimensions of a macroscopic crystal.

(3) Our crystal consists of ionic cores, accompanied by electrons filling the valence band, plus one "surplus" electron in the conduction band. The whole crystal is electrically neutral except for the charge of this one extra electron. Draw a Gaussian surface, perhaps a sphere, that is a little larger in extent than the electron's wave packet, and argue that that surface contains an electric charge $-e$. Look at a later time, after which the wave packet has moved, and draw a new surface enclosing the displaced wave packet. Argue that the charge $-e$ has moved to its new location with the group velocity, and that therefore there is an electric current

$$\boldsymbol{j} = (-e)\boldsymbol{v}_{\text{group}}. \tag{25.6}$$

(4) Scrutinize the reasoning[17] given by Kittel (2005) Chapter 8, pp 191–2, that derives the equation of motion (25.2).

[17] Kittel gives another derivation in his Appendix E. See problem 25.5 for a modified version of that derivation.

(5) Consider the change of the electron's energy under the influence of the electric field.

$$\frac{\mathrm{d}\varepsilon}{\mathrm{d}t} = \frac{\partial\varepsilon}{\partial(\hbar\boldsymbol{k})} \cdot \frac{\mathrm{d}(\hbar\boldsymbol{k})}{\mathrm{d}t} = \boldsymbol{v} \cdot (-e)\boldsymbol{E} = (-e)\boldsymbol{E} \cdot \frac{\mathrm{d}\boldsymbol{r}}{\mathrm{d}t} = (-e)\left(-\frac{\mathrm{d}\varphi}{\mathrm{d}t}\right),$$

where φ is the electrostatic potential given by $\boldsymbol{E} = -\nabla\varphi$. Then

$$\frac{\mathrm{d}}{\mathrm{d}t}\left(\varepsilon - e\varphi\right) = 0. \tag{25.7}$$

There is a similarity between this and the statement of energy conservation for an ordinary particle: ε is the kinetic energy and $-e\varphi$ is the potential energy. We see how completely the $\varepsilon(\boldsymbol{k})$ relation has "taken over" the properties of the electron. It is almost as if we had a real particle in vacuum, that particle possessing momentum $\boldsymbol{p} = \hbar\boldsymbol{k}$ and kinetic energy $\varepsilon(\boldsymbol{k})$.

Notice further that the conservation law of (25.7) holds wherever we are on the $\varepsilon(\boldsymbol{k})$ relation. It even holds near the top of a band where the effective mass is negative. We shall have occasion to remember this when thinking about holes.

25.5   Holes: a very wrong model: the cinema queue

Many books describe the motion of empty quantum states in a semiconductor by the cinema-queue model, or something equivalent whatever name it is given. This model is *seriously wrong.*[18]

We shall end up rejecting the cinema-queue model, Even so, an investigation of it is not a waste of time. It is often helpful to make a false start on a problem: by seeing what we have done wrong, we may get a clear idea of what should be done instead.

In one space dimension, let there be valence-band electrons[19] occupying locations on the x-axis with one location empty. This situation is modelled by a line of people queueing to enter a cinema, with each person representing an electron. One person has just entered the cinema leaving a space behind him.[20] The next person steps forward to fill the space, and a space appears behind that person. It is as if the original space had moved away from the cinema by one step. When a second person moves forward, the space appears to have moved another step further away, and so on. It is claimed that the empty space in the queue explains the behaviour of an empty state in the valence band.

It should be obvious at once that the model just outlined has taken a suspect limit: one in which an electron, or its absence, has a location specified (or specifiable) within $\Delta x \sim a$, where a is a lattice spacing. This is far from the condition $\Delta x \gg a$ established as necessary in § 25.3.[21]

More is wrong. When a person in the cinema queue steps forward to occupy a space, he must first accelerate using friction between his shoes and the ground; then when he has moved the requisite distance he must decelerate, again exchanging momentum with the ground. That's two exchanges of momentum within one lattice spacing. Yet exchanges of momentum by electrons are rare: the typical free path is hundreds to thousands of lattice spacings. The cinema-queue model builds in a wrong limit from the beginning.

Let's try to rescue the cinema-queue model by making momentum exchanges rare. Imagine that each person in the queue is on friction-free roller skates, and that the queue is made to move because the ground slopes slightly down towards the cinema, thereby providing an attractive potential-energy gradient. All people in the queue accelerate in the same way: the person behind the space cannot do anything to overtake the space, and the space remains between the same two people all the time, moving with them wherever they go. "Improving" the model has destroyed its essential feature. The queue space moves towards the cinema: the wrong way.

By contrast, we show in § 25.6 that an absent-electron wave packet has a natural description, free from paradox: it needs no cinema queue, or other homespun model, to explain it.

[18] You need to be able to recognize variants of the cinema-queue model, such as a bubble in a liquid, or a half-empty chemical bond, and identify them as similarly suspect.

[19] *Note on terminology*: Later on, when the concept of a hole is well established, we can be guided by the precedent of phonons, and replace all valence-band electrons by the set of empty states called holes. When that has been done, we can follow the practice of semiconductor physics and say "electrons" and "holes" with it understood that the "electrons" are in the conduction band while the "holes" are in the valence band.

At present, we must establish the validity of "holes" by discussing the valence-band electrons that they represent. During that discussion "electrons" will be electrons in the valence band. The usual jargon must be suspended until the reasoning is complete.

[20] Or her. The gender-neutral *they, their* is awkward, so it's easier to be unfashionable.

[21] Problem 25.2 explores what would go wrong if we did make an electron wave function (full or empty) localized to within about an atomic spacing.

25.6   Holes: the correct description of an almost-full valence band

Electrons in the valence or conduction band "try" to move as Newton's first law tells them, with constant (crystal) momentum: in a zeroth approximation their wave functions are Bloch eigenstates of (crystal) momentum $\hbar\boldsymbol{k}$. An electron can be in one such eigenstate for some considerable time before being scattered out, because collisions are infrequent.

Likewise, an $\hbar\boldsymbol{k}$ state that happens to be empty remains so until one of the infrequent collisions scatters an electron into it.

Our task now is to show that such an empty state, in the valence band, constitutes an entity having particle-like properties such that it can be usefully be called a quasi-particle.

The following discussion closely follows that given by Kittel (2005), Chapter 8. Kittel gives clear and correct statements of what is true. Alongside I supply my own preferred explanations.

The basic idea is simple. We have $\sim 10^{23}$ electrons occupying filled states in a semiconductor's valence band. There are far fewer empty states. An economical description should catalogue the empty states, rather than the filled states.[22]

In Fig. 25.3 we show just the top of the valence band from which one electron has been removed. The left panel notionally shows a thermal-equilibrium state for temperature zero, so the electrons have filled the available $\boldsymbol{k}$-states leaving one empty state at the top of the band.

In the middle and right-hand panels of Fig. 25.3, we have applied an electric field $E_x > 0$. All electrons have gained momentum $\hbar\,\delta k_x$ to the left according to (25.2). The two panels show "snapshots" of the electron distribution after a short time, and after double that time. Those electrons that frame the empty state have moved to states that are off-centre to the left.[23]

Comments
We have talked as though the electrons were distinguishable, so that we can follow the history of "named" electrons, in particular those framing

[22] We may be reminded of a figure–ground transformation. The most familiar of these is a white wine glass on a black background which may also be seen as two black faces on a white background. In that case, figure and ground are about equally significant. Here we can say that the "figure" is the electrons and the "ground" is the empty states. It is economical, and logically acceptable, to concentrate on the ground instead of the figure.

If you're ingenious, you may be able to see that Figs. 25.2 and 25.3 are figure–ground images of each other. This makes the validity of holes foreseeable.

There is also an obvious resemblance to the complementary behaviour of an electron and of a positron as a vacancy in the Dirac sea.

[23] You might like to colour red those electrons that are either side of the empty state, to emphasize that *everything* moves to the left, and not just the white circle.

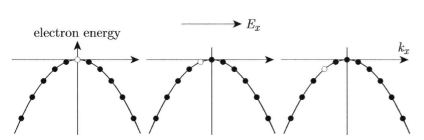

electron energy

E_x

k_x

Fig. 25.3 An enlarged view of the middle of the valence band of Fig. 25.1. In the left-hand diagram, there is an empty $\boldsymbol{k}$-state at the top of the band, at $k_x = 0$. This is the condition at time $t = 0$. An electric field E_x is applied, as a result of which all electrons gain momentum towards $-x$. The second and third diagrams show where the electrons have got to after a short time, and after double that time. The empty state is "framed" by the electrons on either side of it, so it too moves to the left in $\boldsymbol{k}$-space.

the empty state. This is, of course, not the way that identical particles should be described, but using correct words adds complication.

As we move from left to middle to right of Fig. 25.3, one electron disappears off the left-hand side of the diagram, and a "new" one appears at the right.

Out of sight, an electron reaches the left-hand edge of the Brillouin zone and reappears at the right-hand edge. This process causes no obstruction to the movement of the electrons in k-space. Sometimes it is said that an electron reaching a zone edge is "umklapped" or "Bragg-scattered" to the opposite side of the zone. Both of these descriptions come too close to suggesting that some scattering reaction takes place. Does an electron have to wait for a reaction to occur before it can find its way across the zone? It may help to remember that the location of a Brillouin zone in k-space is the subject of a choice. Figure 24.2 may act as a reminder, even though it applies to phonons. If we want to know what happens when an electron meets a zone edge, redraw the zone so that the edge is elsewhere, and it will be obvious that there is "nothing to happen" when the electron passes through this k.

We are going to describe the properties of the entire valence band (subscript h) in terms of properties of the one empty state. The sum of the ks of all the electrons in a filled band is zero. Then if one electron is removed from state k_e, the total momentum of the electrons in the valence band is

$$\hbar k_h = 0 - \hbar k_e. \tag{25.8}$$

Comment: Something we should *not* do appears now. A removed electron at k_e gives to the band the same momentum as is possessed by an actual electron[24] present at $-k_e$. Does that actual electron bear responsibility for all of the physics? Once the question is asked, the answer should be an obvious "no". But the question must be asked, just because this kind of muddle is so easy to get into.

Let's choose the origin of energy so that the energy of a filled valence band is zero. Then the removal of an electron having energy $\varepsilon_e(k_e)$ gives an energy[25] to the band of[26]

$$\varepsilon_h(k_h) = -\varepsilon_e(k_e). \tag{25.9}$$

There is no reason, as yet, to attach any particular physical significance to the outcomes in (25.8–9). It is the steps we take next that give substance to what we are doing.

25.6.1 Holes: group velocity and effective mass

To consider group velocity we must build a wave packet. Instead of fully-emptying a valence-electron quantum state at k_e, we'll part-empty quantum states around there, covering a range Δk_e. We can agree to see the open circles in Fig. 25.3 as representing such a part-emptying.[27] Where the wave packet for a single electron in the conduction band was

[24] Indeed, one way of doing the accountancy on momentum is to partner electrons at $\pm k$ to give zero total momentum except where the electron at $-k_e$ is not cancelled by one at $+k_e$. Such a procedure is mathematically impeccable, but has the demerit of drawing undue attention to the electron at $-k_e$.

[25] You may be able to think of a way in which the $\varepsilon(k)$ curve can be asymmetric, so that $\varepsilon(-k) \neq \varepsilon(k)$. Then $\varepsilon_h(k_h) = -\varepsilon_e(k_e) \neq -\varepsilon_e(-k_e)$, confirming that the electron at $-k_e$ cannot be of interest.

[26] You may need a clear head when confirming (25.9). The electron-removed valence band has momentum (the total for the remaining electrons) $\hbar k_h$ and energy $\varepsilon_h(k_h)$. These are definitions. What has been removed from the full band is an electron having momentum $\hbar k_e$ and energy $\varepsilon_e(k_e)$.

[27] The "width" Δk_e is small on the scale of the Brillouin-zone width. Then the wave packet has a "size" Δr that is large compared with an atomic spacing, but still small compared with macroscopic dimensions. For a discussion of the compatibility of these conditions, see problem 25.1 which applies as well here as to electrons in the conduction band.

a sort of lumpy wave function, here there is an equally valid wave packet that is a dip in the many-electron wave function.[28]

To understand the motion of the wave packet, we must think about what the electrons in the part-full states are doing. They "carry" the electron deficit (or part-absence) on their shoulders,[29] and do so with their group velocity $d\varepsilon_e/d(\hbar k_e)$.

The group velocity v_h for the electron-deficit wave packet is

$$v_h = \frac{d\varepsilon_e}{d(\hbar k_e)} = \frac{d(-\varepsilon_h)}{d(-\hbar k_h)} = \frac{d\varepsilon_h}{d(\hbar k_h)}. \tag{25.10}$$

In Fig. 25.3 this is a velocity towards $+x$.

Comment: You might wonder why the last expression wasn't written down at once: all we've done is write the group velocity for a particle whose momentum is $(\hbar k_h)$ and whose energy is ε_h? But that's the point. We're not allowed to say that there is any sort of "particle" having this momentum and energy, not yet anyway. We have *discovered* that a particle-like relation (25.10) holds, but such a relation could not legitimately have been assumed in advance.

Now think about the distribution of electric charge within the semiconductor. Construct reasoning exactly parallelling that in exercise 25.2(3) to show that there is a charge $+e$ located within the wave packet, and that this charge moves with the group velocity v_h. That is, there is a current flow

$$j = (+e)v_h. \tag{25.11}$$

Gathering results: The absence of an electron from the valence band is associated with a net positive charge $+e$, whose location can be represented by a wave packet, in just the same way as a conduction-band electron (charge $-e$) can be represented by a wave packet. The valence-band wave packet is accelerated[30] by an applied field E_x and carries its charge $+e$ with group velocity v_h, giving a current $j_x = (+e)(v_h)_x$. The current towards $+x$ is positive when E_x is positive. All this adds up to a cluster of properties that are particle-like. We now are at liberty to declare all this to be a quasi-particle that we call a **hole**.[31]

We may define an effective mass for the hole.[32]

$$\delta(\hbar k_h) = m_h^* \, \delta v_h,$$

so that

$$\frac{1}{m_h^*} = \frac{\delta v_h}{\delta(\hbar k_h)} = \frac{d^2\varepsilon_h}{d(\hbar k_h)^2} = \frac{d^2(-\varepsilon_e)}{d(-\hbar k_e)^2} = -\frac{d^2\varepsilon_e}{d(\hbar k_e)^2} = -\frac{1}{m_e^*}. \tag{25.12}$$

At the top of the valence band, the effective mass m_e^* for an electron is negative, so the effective mass m_h^* for the hole has come out to be positive. Again, the hole's properties are entirely particle-like.

25.7   Fermi–Dirac statistics for holes

We complete the present discussion of holes by showing that holes behave as though they were fermions having a Fermi–Dirac distribution.

[28] We usually discuss group velocity in the context of a moving lump on a wave amplitude. But all that's necessary is that the wave have some feature on its "envelope" that distinguishes one region from another. A dip is just as much described by a group velocity as is a lump.

[29] I have a mental picture of a line of soldiers carrying a long sausage on their shoulders. Most soldiers have the same height, but a few together are shorter than the rest. The sausage sags so as to have a dip where the soldiers are short, and that dip is carried along on the soldiers' shoulders. The velocity of the dip is the velocity of the soldiers who are carrying it. What else?

[30] Figure 25.3, centre and right panels, shows that the (group) velocity increases with time, so "accelerates" is appropriate.

[31] It is now even permissible to describe a hole using a wave function of Bloch form.

[32] As with eqn (25.4), the effective mass is a convenient way of describing quasi-particle properties, without usually contributing in an indispensable way to any reasoning.

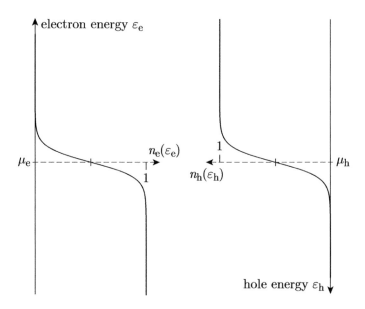

Fig. 25.4 Left: the Fermi–Dirac distribution for electrons having energy ε_e and chemical potential μ_e. Right: the Fermi–Dirac distribution for the states left empty by the electrons (i.e. the holes). The curve for holes can be fitted over that for the electrons, and differs only in a re-labelling of the axes.

If holes behave like particles, then they must be fermions, because there cannot be two in the same quantum state: a quantum state cannot be twice-emptied of electrons.

Figure 25.4 shows the Fermi–Dirac distribution $n_e(\varepsilon_e)$ for electrons (left panel) and $n_h(\varepsilon_h)$ for holes (right panel). Electron energy ε_e increases up-page in both panels. The hole energy ε_h increases down-page because of (25.9). In (25.9), the origin of energy was placed at the top of the valence band, a convention that has now served its purpose, so we permit a general energy origin:

$$\varepsilon_h = (\text{constant } C) - \varepsilon_e; \qquad \mu_h = C - \mu_e. \tag{25.13}$$

Given that electrons have a Fermi–Dirac distribution, we now show that holes have a Fermi–Dirac distribution as well. The probability for occupation of a quantum state by an electron of energy ε_e is

$$n_e(\varepsilon_e) = \frac{1}{\exp\{(\varepsilon_e - \mu_e)/k_BT\} + 1},$$

where μ_e is the chemical potential for electrons. The probability of occupation of that same state by a hole is the probability that the state is empty of an electron, so it is

$$n_h(\varepsilon_h) = 1 - n_e(\varepsilon_e) = 1 - n_e(C - \varepsilon_h)$$
$$= 1 - \frac{1}{\exp\{(C - \varepsilon_h - \mu_e)/k_BT\} + 1}$$
$$= \frac{1}{\exp\{(\varepsilon_h - (C - \mu_e))/k_BT\} + 1}$$
$$= \frac{1}{\exp\{(\varepsilon_h - \mu_h)/k_BT\} + 1}. \tag{25.14}$$

Therefore the distribution for holes has the Fermi–Dirac form. This result is used, in a slightly different form, in Chapter 26.

Problems

Problem 25.1 An electron wave packet

In § 25.3 we asserted that a wave packet could be constructed for which Δk is much less than the width of a Brillouin zone, yet Δr is much smaller than the dimension of the crystal (assumed macroscopic). We now investigate these conditions.

Use a one-dimensional model where the lattice spacing is a. Then the width of a Brillouin zone is $2\pi/a$. We are asserting that $\Delta k_x \ll 2\pi/a$ so that

$$\Delta x \sim \frac{1}{\Delta k_x} \gg \frac{1}{2\pi/a} = \frac{a}{2\pi}.$$

Thus a sensibly constructed wave packet much have a linear dimension much greater than a lattice spacing.

Take a numerical example. Let the host crystal have width $10\,\text{mm}$, which is roughly 10^8 atomic spacings. Argue that we could be comfortable with any wave packet whose Δk_x is between roughly $(2\pi/a) \times 10^{-2}$ and $(2\pi/a) \times 10^{-6}$. Find the associated values of Δx. The conditions specified at the beginning of this problem leave plenty of "headroom" between them.

Problem 25.2 An electron or hole localized within an atomic spacing

Following from problem 25.1, let us find out what goes wrong if we try to form an electron wave packet (conduction band) that is localized to within about an atomic spacing.

Refer to Fig. 25.1. We have $\Delta x \sim a$ and so $\Delta k_x \sim 1/a$. This is not far from $\Delta k_x = 2\pi/a = G$: the wave packet is constructed from Bloch states spread over the full k_x-width G of the Brillouin zone. Equivalently those states are taken from energies within the full height of the conduction band, which might well be several electron volts for a real crystal. The expectation value of the electron's energy is—let's say— $2\,\text{eV}$ above the bottom of the conduction band.

Remember that $k_B T \approx 25\,\text{meV}$ at room temperature. We are looking at energies of order $80\times$ this. Such an energetic state is not formed by ordinary thermal excitation. If it has been formed, then someone must have done something seriously violent to the crystal: perhaps the injection of a hard-X-ray photon.

Suppose that an electron wave packet *has* been formed with linear dimension of order a. Think of the ways in which the electron's energy is dissipated[33] into the kinetic energy of other electrons, or into phonons, or ... . Argue that the highly localized wave packet must have a very short lifetime. It is not the kind of excitation that moves from place to place as an unbroken entity.

The cinema-queue model takes it as given that a hole is the absence of an electron in the same way that a gap in a queue is the absence of

[33] Think again about electrons in a metal. The Fermi–Dirac distribution of electrons resembles that of fermions in a Fermi liquid (where there is no lattice to worry about, and the only thing one fermion can collide with is another fermion). The collision probability for a fermion at energy $\mu + s k_B T$ is proportional to

$$\frac{\pi^2 + s^2}{e^{-s} + 1}.$$

(Pines and Nozières 1966, p. 63). The most significant term here is usually the e^{-s} in the denominator, which suppresses collisions for fermions below μ owing to the shortage of empty states. This is consistent with a statement made in sidenote 3 of this chapter.

However, the important term for the present discussion is the s^2: a fermion well above the chemical potential (Fermi level) μ has many possible collision partners, and the higher it is in energy the shorter is its lifetime against collision. A similar property must hold in a solid where the electron has other possible collision partners.

a person. That is, the hole is an upside-down wave packet localized to within an atomic spacing. Argue, along the lines of the above, that such an excitation is possible, but involves part-emptying quantum states over the whole energy range of the valence band. The lifetime for collisional decay of this must be very short.

Problem 25.3 The Bloch wave function for an electron

Consider an electron in a one-dimensional periodic potential $V(x)$ having N periods of length a. Apply periodic boundary conditions so that $\psi(x)$ repeats when x is increased by Na (it's taken that N is arbitrary large). The Hamiltonian $\widehat{H}_0$ is periodic with period a, so

$$\widehat{H}_0(x + a) = \widehat{H}_0(x).$$

Because of the periodic boundary condition this does not stop at $x = Na$ (or multiples) but continues so as to hold for *all* x.

Define a translation operator $\widehat{T}$ that translates through a, so that for any function ψ and for all x

$$\widehat{T}\psi(x) = \psi(x + a). \tag{25.15}$$

With these definitions, $\widehat{T}$ commutes with $\widehat{H}_0$. Then these operators have a complete set of eigenfunctions in common. In finding the eigenfunctions of $\widehat{H}_0$, we may solve the simpler problem (§ 11.2.1) and instead find the eigenvalues and eigenfunctions of $\widehat{T}$.

Let $\widehat{T}$ have an eigenvalue λ, meaning that $\widehat{T}\psi = \lambda\psi$ when ψ is an eigenfunction. Then

$$\psi(x) = \psi(x + Na) = \widehat{T}^N \psi(x) = \lambda^N \psi(x), \quad \text{so} \quad \lambda^N = 1$$

and

$$\lambda = e^{ika} \quad \text{in which} \quad ka = r\,\frac{2\pi}{N}, \quad \text{or} \quad k = \frac{2\pi}{a}\,\frac{r}{N}.$$

Show that r is an integer, and takes N integer values before things repeat; we have found N distinct eigenvalues of $\widehat{T}$.

Now use (25.15) to show that the eigenfunction of $\widehat{T}$ accompanying eigenvalue e^{ika} has the Bloch form (25.1). Expressions of this form are eigenfunctions of $\widehat{H}_0$ as well as of $\widehat{T}$.

Problem 25.4 The group velocity for a Bloch wave function

Model a one-dimensional wave packet by adding two wave functions, having wave vectors k and $k + \Delta k$.

$$\Psi \sim e^{i(kx - \omega t)} P_k(x) + \exp\{i(k + \Delta k)x - i(\omega + \Delta\omega)t\} P_{k+\Delta k}(x)$$

$$= e^{i(kx - \omega t)} P_k \left(1 + \frac{P_{k+\Delta k}}{P_k} \exp\{i(\Delta k)x - i(\Delta\omega)t\} \right).$$

Let

$$\frac{P_{k+\Delta k}}{P_k} = 1 + \frac{1}{P_k}\frac{dP_k}{dk}\Delta k = 1 + (\alpha + i\beta)\Delta k \approx \exp\{(\alpha + i\beta)\Delta k\}$$

so that

$$\Psi = e^{i(kx - \omega t)} P_k \left\{ 1 + e^{\alpha\,\Delta k} \exp\{i(\Delta k)(x + \beta) - i(\Delta\omega)t\} \right\}.$$

In this, α and β are real and, because they derive from P_k, they are periodic functions of x with period a. As functions of x, they return to previous values when x is increased by a; in the large they are not "going anywhere". Then the wave packet Ψ has an interference maximum defined by $(\Delta k)(x + \beta) - (\Delta\omega)t = 0$ which propagates as

$$\frac{x + \beta}{t} = \frac{\Delta\omega}{\Delta k}.$$

Argue that the β in this is of no consequence, so the group velocity is given by the usual expression $\boldsymbol{v}_{\text{group}} = \mathrm{d}\omega/\mathrm{d}\boldsymbol{k}$.

Make all this formal by building a wave packet that contains a range of ks instead of just two.

Problem 25.5 The electron equation of motion[34]
Consider a single electron in a one-dimensional periodic potential having N periods of length a. Then (problem 25.3) the eigenfunctions of energy have the form

$$\psi_r = N^{-1/2}\,\mathrm{e}^{\mathrm{i}kx}\,P_r(x); \qquad k = \frac{2\pi}{a}\frac{r}{N}, \qquad (25.16)$$

in which r is an integer and takes N values.[35] In (25.16), $P_r(x)$ is a periodic function having period a; we must expect that its form may depend upon r.

(1) Prove the orthonormality of the Bloch functions.[36] Then

$$\int_0^{Na} \mathrm{d}x\,\psi_s^*\,\psi_r = \int_0^a \mathrm{d}x\,|P_r(x)|^2\,\delta_{r,s} = \delta_{r,s},$$

if we agree to choose the magnitudes of the periodic functions $P_r(x)$ so as to make the last integral equal to 1 for all integers r.

(2) Now consider an electron wave packet, built from several of the above eigenfunctions. We may write its wave function as[37]

$$\psi = \sum_r C_r \psi_r(x). \qquad (25.17)$$

Show that normalization of this ψ requires

$$\sum_r |C_r|^2 = 1. \qquad (25.18)$$

(3) Now consider what happens when we impose an electric field E in the $+x$-direction. The Hamiltonian is changed to $\widehat{H} = \widehat{H}_0 + eEx$. This must change the momentum of the electron; more precisely, the electron's wave packet must change its k-constituents.

We can think of the k in $\mathrm{e}^{\mathrm{i}kx}$ as changing with time, as is suggested in the drawings of Figs. 25.2, 3. However, in the present problem we are using time-independent eigenfunctions (25.16), and the motion of a wave packet is handled by the changes with time of the coefficients C_r in (25.17).

To proceed from here, we use again the translation operator $\widehat{T}$ introduced in problem 25.3.[38],[39] Show that

[34] This problem is intended as a "friendly amendment" to Appendix E of Kittel (2005).

[35] Elsewhere we have specified a wave function ψ and its periodic part $P(x)$ by wave vector k, because that was then appropriate. Now we are using fine detail in which k is identified by integer r, and the subscripts follow suit.

[36] The algebra required can be found in the derivation of (23.25).

[37] Increase of r by N repeats the same wave function with k in another Brillouin zone. I suggest you make r run from 0 to $(N - 1)$, which avoids having to deal separately with even N and odd N. If we wish to exclude $r = 0$ we can make $C_0 = 0$.

[38] This further use of $\widehat{T}$ is a neat trick to prevent $P_r(x)$ from complicating the algebra.

[39] Taking (25.18) and (25.19) together, we see that $|\langle T\rangle| \leqslant 1$. Equality is attained when ψ is built from a single Bloch function, meaning that a single C_r is present in the sum of (25.17).

$$\langle T \rangle = \int_0^{Na} \mathrm{d}x\, \psi^* \, \widehat{T} \psi = \int_0^{Na} \mathrm{d}x\, \psi^*(x)\, \psi(x+a)$$

$$= \sum_s C_s^* \sum_r C_r \int_0^{Na} \psi_s^*(x)\psi_r(x+a) = \sum_r |C_r|^2 \, \mathrm{e}^{\mathrm{i}2\pi r/N}. \quad (25.19)$$

(4) To find the time dependence of $\langle T \rangle$, show that

$$\langle [\widehat{T}, \widehat{H}] \rangle = \int_0^{Na} \mathrm{d}x\, \psi^*(x)\big(\widehat{T}(eEx) - (eEx)\widehat{T}\big)\psi(x)$$

$$= \int_0^{Na} \mathrm{d}x\, \psi^*(x)\Big(eE(x+a)\,\psi(x+a) - (eEx)\psi(x+a)\Big)$$

$$= \int_0^{Na} \mathrm{d}x\, \psi^*(x)\, eEa\, \psi(x+a) = eEa\langle T \rangle.$$

(5) The equation giving the time dependence of an expectation value is

$$\mathrm{i}\hbar \frac{\mathrm{d}}{\mathrm{d}t}\langle T \rangle = \langle [\widehat{T}, \widehat{H}] \rangle, \qquad \text{so that} \qquad \frac{\mathrm{d}}{\mathrm{d}t}\langle T \rangle = \frac{-\mathrm{i}eEa}{\hbar}\langle T \rangle,$$

giving[40]

$$\langle T \rangle = (\text{constant})\, \mathrm{e}^{-\mathrm{i}(eEa/\hbar)t}.$$

Backtracking to (25.19):

$$(\text{constant})\, \mathrm{e}^{-\mathrm{i}(eEa/\hbar)t} = \sum_r |C_r|^2 \, \mathrm{e}^{\mathrm{i}2\pi r/N} = \sum_r |C_r|^2 \, \mathrm{e}^{\mathrm{i}ka}. \quad (25.20)$$

As forecast, this shows that the coefficients C_r must depend upon time t.

(6) Now build a wave packet by letting the k-values in (25.20) occupy a small range $\Delta k \ll G$ around K. Then the range occupied by the ka in (25.20) is $(\Delta k)a \ll Ga \approx 1$. We may replace all exponents ka by a common Ka. Then (25.20) can be rearranged to[41]

$$\mathrm{e}^{\mathrm{i}Ka} = \frac{(\text{constant})}{\sum_r |C_r|^2}\, \mathrm{e}^{-\mathrm{i}(eEa/\hbar)t} = \mathrm{e}^{-\mathrm{i}(eEa/\hbar)(t-t_0)}.$$

(7) Think about the range of times for which the last equation can be trusted.[42] Allow time t to vary through only a small δt, and then

$$\frac{\mathrm{d}(\hbar K)}{\mathrm{d}t} = (-e)E. \quad (25.21)$$

This is the desired equation of motion. $\hbar K$ is the dominant momentum within the electron wave packet, and it changes in this way as the wave packet responds to the imposed force F.

(8) Look back to the manipulation that led to eqn (25.7). Trace that reasoning another way round, starting from (25.7) and (25.21), and use it to confirm that the group velocity $v_{\text{group}} = \partial \varepsilon / \partial(\hbar k)$.

[40] Sidenote 39 shows that the constant here has complex magnitude $\leqslant 1$.

[41] Incidentally, this makes $|\langle T \rangle| \approx 1$, though this fact has no particular significance for us.

[42] *Hint:* The value of K should not be made to change through anything approaching $2\pi/a$. If it were to do so, the wave packet could not remain compact because the electron's $\varepsilon(k)$ relation is dispersive.

The chemical potential for a semiconductor

Intended readership: At the point where doped semiconductors are first encountered.

26.1 Introduction

In a semiconductor, the electrons obey a Fermi–Dirac distribution

$$n(\varepsilon) = \frac{1}{\exp\{(\varepsilon - \mu)/(k_{\rm B} T)\} + 1}. \tag{26.1}$$

Here ε is energy. The quantity μ is variously known as the **chemical potential** or the **Fermi level**.[1] The value of μ is dependent on temperature T, but also has a very sensitive dependence on doping. At zero temperature μ is

$$\varepsilon_{\rm F} = \mu(T{=}0), \tag{26.2}$$

usually known as the **Fermi energy**. The Fermi energy $\varepsilon_{\rm F}$ inherits from μ a strong dependence on the doping of the semiconductor.

The Fermi distribution contains one adjustable parameter: μ. As is always the case in statistical mechanics, the value of μ is determined by arranging that the number of particles, here electrons, is correct.

To understand what is going on in a semiconductor, the first task is usually to get a good idea of where μ lies on an energy scale. This chapter shows how μ may be found, and exhibits the strong sensitivity of μ to doping, contrasted with a relatively weak dependence on temperature.

[1] The terminology introduced here is standard, but somewhat unfortunate. In particular, the term "Fermi level" strongly suggests that a quantum energy level must exist at energy μ, and that is not the case. In a semiconductor, μ often lies within the band gap where there is *no* quantum state—and for the avoidance of doubt: no quantum state is compelled to come into existence there. I therefore usually refer to μ as the "chemical potential"—which it is—even though the chemical potential (the Gibbs energy per electron) is perhaps not the most familiar of thermodynamic quantities.

26.2 A procedure for finding μ

Figure 26.1 shows schematically the valence band and conduction band of a representative semiconductor, with electron energy increasing up the page. Alongside is drawn a Fermi distribution for the electrons.

If μ is too high, an integration for the total number $N_{\rm electrons}$ present

$$N_{\rm electrons} = \int (\text{density of states}) \times n(\varepsilon)\, {\rm d}\varepsilon$$

will contain too large an $n(\varepsilon)$ (for all energies, though some more than others), and the integral will evaluate to too large a number. Conversely, if μ lies too low in energy, then the integral will come out too small. We present a trial-and-error method for finding μ in actual cases.

337

Essays in Physics: Thirty-two thoughtful essays on topics in undergraduate-level physics. Geoffrey Brooker, Oxford University Press. © Geoffrey Brooker 2021. DOI: 10.1093/oso/9780198857242.003.0026

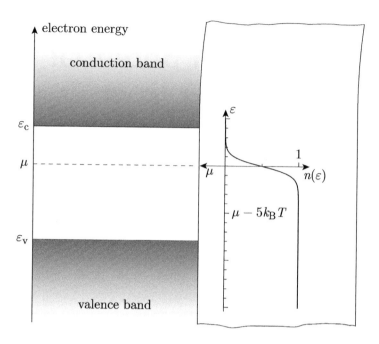

Fig. 26.1 The valence band and conduction band of a representative semiconductor. Alongside is drawn the Fermi distribution for the electrons, as if on a separate strip of paper that can be slid up and down. The temperature has been shown unrealistically high, in order for the shape of the Fermi distribution to be clearly visible. Each tick on the vertical axis represents an increment of $k_B T$.

[2] We repeat reasoning given in § 25.7.

[3] An "intrinsic" semiconductor is chemically pure, consisting of the semiconductor atoms only, with no added impurity. Impurity atoms would make their own, added, contribution to the density of states (§ 26.6), a contribution that is excluded here for simplicity.

[4] A further simplifying assumption is commonly made: that the energies of electrons in the two bands are both quadratic functions of wave vector k, parameterized by means of effective masses m_e^* (for the conduction band) and m_h^* (for the valence band):

$$\varepsilon_{cond} = \varepsilon_c + \frac{\hbar^2 k^2}{2m_e^*}, \quad \varepsilon_{val} = \varepsilon_v - \frac{\hbar^2 k^2}{2m_h^*}.$$

Here ε_c is the energy for electrons at the bottom of the conduction band, while ε_v is the energy for electrons at the top of the valence band. In these terms, the special case being taken in the present section is $m_e^* = m_h^*$ (together with isotropy). However, the only requirement for our reasoning here is that the two bands mirror each other.

Figure 26.1 shows the Fermi distribution as if drawn on a separate strip of paper that can be slid up and down relative to the valence and conduction bands. (In tutorial teaching I use just such a strip.) Then finding μ consists in finding where that movable piece of paper should be placed in order to represent correctly the location of the Fermi distribution in relation to the bands.

We need one mathematical property of the Fermi distribution: it has inversion symmetry about the point where $\varepsilon = \mu$ and $n = \frac{1}{2}$. Place the tip of your pencil on that point and, using it as pivot, rotate the strip through $180°$: the curve in its new orientation will lie exactly on top of the original curve. This shows that:

> *The probability for electron states in the valence band at energy $\mu - x k_B T$ to be empty is the same as the probability for electron states in the conduction band to be occupied at energy $\mu + x k_B T$.*

This may be shown formally as follows:[2]

$$1 - n(\varepsilon = \mu - xkT) = 1 - \frac{1}{e^{-x} + 1} = \frac{1}{e^x + 1} = n(\varepsilon = \mu + xkT).$$

The following sections make several applications of the ideas given here.

26.3   Intrinsic[3] semiconductor having equal densities of states in the two bands

For simplicity, we assume that the densities of states in the valence band (ρ_v) and conduction band (ρ_c) are mirror images of each other, as is suggested in Fig. 26.2.[4]

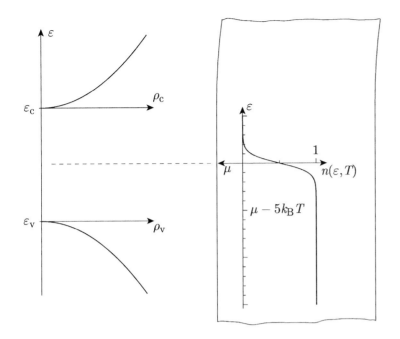

Fig. 26.2 The density of states ρ_c is plotted against the energy ε for electron quantum states in the conduction band. Similarly, the density of states ρ_v is plotted for electron states in the valence band. The two curves are drawn with a parabolic shape as in eqn (26.3), but the only property made use of is the mirror symmetry of the curves about the middle of the band gap (indicated by the broken line).

If the energy–wave-vector relation were isotropic—not usually the case in real semiconductors—the density of states ρ_c for electrons in the conduction band would be

$$\rho_c = \frac{(2m_e^*)^{3/2}\sqrt{\varepsilon - \varepsilon_c}}{2\pi^2\hbar^3}, \tag{26.3}$$

where ε_c is the energy at the bottom of the conduction band. A similar relation containing m_h^* and $\sqrt{\varepsilon_v - \varepsilon}$ would apply for electrons in the valence band.[5] This explains the quadratic shapes for the sketch graphs of ρ_c and ρ_v in Fig. 26.2. However, those shapes are intended to be illustrative only.

The number N_e of electrons in the conduction band and the number N_h of empty electron states in the valence band are given by integrals:

$$N_e = \int_{\varepsilon_c} \rho_c(\varepsilon)\, n(\varepsilon)\, d\varepsilon$$

$$= k_B T \int_0^\infty \rho_c(\varepsilon_c + x k_B T)\, \frac{1}{\exp\{(\varepsilon_c + x k_B T - \mu)/k_B T\} + 1}\, dx;$$

$$N_h = \int^{\varepsilon_v} \rho_v(\varepsilon)\{1 - n(\varepsilon)\}\, d\varepsilon$$

$$= k_B T \int_0^\infty \rho_v(\varepsilon_v - x k_B T)\, \frac{1}{\exp\{(\mu - \varepsilon_v + x k_B T)/k_B T\} + 1}\, dx.$$

Now for an intrinsic semiconductor, the number N_e of electrons in the conduction band must be the same as the number N_h of unoccupied states in the valence band. Also, $\rho_c(\varepsilon_c + x k_B T) = \rho_v(\varepsilon_v - x k_B T)$ since we are assuming that the valence-band density of states mirrors that of the conduction band. The two integrals are therefore made equal if we

[5] In this chapter, we deal with electrons and electron energies throughout. Occasionally we refer to empty electron states as "holes" as a convenient shorthand. The only explicit use of hole properties is in sidenote 4, where we replace the negative effective mass of top-of-valence-band electrons by the positive effective mass m_h^* associated with holes.

set $\varepsilon_c - \mu = \mu - \varepsilon_v$, so

$$\mu = \frac{\varepsilon_v + \varepsilon_c}{2}. \qquad (26.4$$

For an intrinsic semiconductor with symmetrical bands, the chemical potential lies half-way between valence and conduction bands.

Comment: Notice how weak have been the assumptions made in th above. We have not needed an explicit expression for the density c states in either band. The temperature has not been required to b small in relation to the band gap (though for a very high temperatur the energies of interest would extend so far above and below μ that th assumption of band symmetry would be improbable).

Comment: Look back to Fig. 26.1 and the slidable strip. Given th symmetry of the two bands, and the antisymmetry of the Fermi distri bution, it should be obvious by inspection that the right place to put is at the mid-point between the two band edges—as given by eqn (26.4) Whatever happens to filled states at an energy xk_BT above ε_c is the exactly mirrored by empty states at energy xk_BT below ε_v.

26.4   Intrinsic semiconductor having unequal densities of states

Take a more complicated case. Suppose that the densities of states i the valence and conduction bands are unequal. To make things simple suppose that ρ_v maintains the same shape, as a function of energy, a before, but is greater than before by a factor 5 throughout. We sha find how this affects the location of the chemical potential μ.

The problem we have set ourselves can be solved by writing out in tegrals, as was done in the last section. However, those integrals can ge a bit messy, so we adopt instead a trial-and-error approach using th slidable strip introduced in § 26.2.

The method can be seen from Fig. 26.3. We first make a guess (wron, but convenient) at the location of μ, then see how that guess should b corrected. Our first guess is that μ again lies at the mid-point betwee the bands. The number of electrons in the conduction band is unchange at $N_e' = N_e$. But the number of empty states in the valence band ha increased to $N_h' = 5N_h$. Electrons have been taken from the valenc band but not added to the conduction band: overall, the number o electrons present is too small. To put this right, we must make a improved estimate of μ, which clearly must be raised. Let μ be raise from the mid-point of the band gap by yk_BT.

At this point, we introduce an approximation that did not have to b made before. We assume that μ remains several k_BT below ε_c, so that al conduction-band states lie on the "Boltzmann tail" of the Fermi distribu tion. Likewise, all empty valence-band states lie on another Boltzman tail. Then the effect of raising μ by yk_BT is to increase the populatio of every conduction-band state by a factor e^y. The same factor necessar ily applies to the total of these populations, so $N_e^{new} = N_e' e^y = N_e e^y$

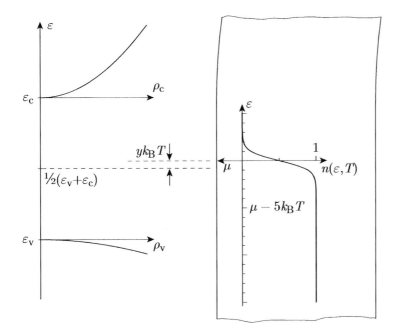

Fig. 26.3 The location of the chemical potential for an intrinsic semiconductor when the density of states in the valence band is 5× that in the conduction band.

Numerical values are calculated in the text for the case of silicon at room temperature. In relation to those values, the energies connected with $k_B T$ are drawn magnified by 3× compared with the band gap $\varepsilon_c - \varepsilon_v$, so that the detail can be seen clearly.

Similarly, the probability of every valence-band state being empty is reduced by a factor e^{-y}, giving $N_h^{\text{new}} = N_h' e^{-y} = 5 N_h e^{-y}$. To bring the ratio of these to 1 we need

$$1 = \frac{N_e^{\text{new}}}{N_h^{\text{new}}} = \frac{N_e}{5 N_h} e^{2y} = \frac{1}{5} e^{2y},$$

so that

$$e^{2y} = 5, \qquad y = \tfrac{1}{2} \ln 5 = 0.805.$$

To see the significance of this finding, we need to know some orders of magnitude. At a temperature of 293 K, $k_B T = 25.2$ meV, so the shift of μ is 20.3 meV. By contrast, the band gap of silicon is 1.14 eV. In proportion to the band gap, the shift of μ is quite tiny.[6]

If the disproportion between the densities of states were a factor 25, instead of a factor 5, the shift of μ would only be doubled, still small compared with the band gap.

If the temperature were reduced, the shift of μ, being proportional to T, would be reduced accordingly. In particular, the zero-temperature chemical potential $\varepsilon_F = \mu(T{=}0)$ lies at the mid-point of the band gap, irrespective of inequalities in the density of states.

We draw the following conclusion:

> *The location of μ is very insensitive to the densities of states in the valence and conduction bands. For intrinsic material, it lies at the mid-point of the band gap, within one or two $k_B T$, even for quite extreme asymmetries in the densities of states.*

We stress this because μ is extremely sensitive to doping, and the contrast between these two sensitivities is dramatic.

[6] In Fig. 26.3, $k_B T$ is represented on the right by 2.5 mm, and $y k_B T$ is drawn in proportion; if the band gap were drawn to the same scale it would occupy 113 mm vertically on the paper, a factor 3 greater than drawn.

26.5 The electron–hole product

For technological applications, semiconductors are usually doped so that the number of electrons available to occupy the valence and conduction bands is greater than that for intrinsic material (n-type doping) or less (p-type doping). Even the cleanest semiconductor samples contain impurities, even if at less than parts-per-million concentration, with observable consequences for the properties of the material.

Since we have seen that the densities of states in the valence and conduction bands have little effect on the location of μ, we shall think mostly about a model semiconductor for which the densities of states have the quadratic variations with energy similar to those shown in Fig. 26.2 and eqn (26.3).

Suppose that we have an n-type semiconductor, meaning that the number of electrons occupying the conduction band exceeds the number of empty states in the valence band (the difference accounted for by the partial ionization of "donor" impurities).[7] Then μ must lie above the middle of the band gap, say by $y k_B T$. Compared with intrinsic material, every conduction-band quantum state has its occupation increased by a factor e^y, and every "hole" state has its occupation changed by a factor e^{-y}. The product

$$(\text{number of electrons}) \times (\text{number of holes}) = \text{constant}, \tag{26.5}$$

independent of the concentration of dopant.[8] In obtaining this conclusion, we have again assumed that μ continues to lie several $k_B T$ away from either band edge, so that a Boltzmann-tail approximation holds.

The numbers of both electrons and holes depend exponentially upon temperature, roughly as $\exp\{-(\varepsilon_c - \mu)/k_B T\}$ in the case of conduction-band electrons. So the "constant" in (26.5) signifies a constant with respect to doping only.

To find the constant in (26.5), we need to assume a particular model for the densities of states in the two bands and to work out integrals accordingly. Problem 26.1 asks you to work this out explicitly, using the density of states in (26.3).

[7] I say "partial" ionization because the occupation by electrons of the donor states is controlled by the Fermi distribution, just as it is for other quantum states in the conduction band or the valence band. We must not assume, without checking, that donors (or acceptors) are somehow compelled to be completely ionized.

[8] Though the "constant" is strongly dependent upon temperature.

26.6 The chemical potential for a doped semiconductor

Consider a numerical example. Silicon is doped with 1 part in 10^9 of phosphorus, a donor impurity, giving rise to an n-type semiconductor. To find: the location of the chemical potential μ at a temperature of 300 K.

Density of silicon = $2329 \, \text{kg m}^{-3}$. Relative atomic mass of silicon = 28.09. Band gap of silicon = 1.14 eV. The electron–hole product for silicon at 300 K is $4.6 \times 10^{31} (\text{carrier m}^{-3})^2$. The ionization energy of phosphorus embedded in silicon is 45 meV.

For simplicity we'll take the densities of states in the two bands to be equal.

The number of donor atoms per unit volume can be worked out from the density of silicon and the relative atomic mass. When 1 atom in 10^9 of silicon is replaced by phosphorus, the concentration of donors works out to be $4.993 \times 10^{19} \, \mathrm{m}^{-3}$.

We find the chemical potential μ by a now-familiar trial-and-error method. We start by provisionally placing μ at the middle of the band gap. Since the densities of states in the conduction and valence bands are taken to be equal, the numbers of electrons and holes are at this stage equal, each given by the square root of the product: $6.782 \times 10^{15} \, \mathrm{m}^{-3}$. The donor levels are at energy $525 \, \mathrm{meV} = 20.31 k_B T$ above the assumed μ, so a Boltzmann-tail approximation[9] for the donor occupation is well justified. That occupation probability is $\exp(-20.31) = 1.52 \times 10^{-9}$, giving the number of electrons in donor states as $7.564 \times 10^{10} \, \mathrm{m}^{-3}$. These figures are recorded on Fig. 26.4.

Each donor atom supplies one electron, additional to those present in intrinsic silicon. Only 7.564×10^{10} "surplus" electrons are present, though there are 4.993×10^{19} donors. Clearly, μ has been set too low. As a new trial, we raise μ above the band centre by $y k_B T$. The consequences are shown on Fig. 26.4. The number of electrons in the conduction band, and the number in the donor states, are both raised by a factor e^y, while the number of empty states in the valence band is changed by e^{-y}. It is

[9] Aficionados will recognize a somewhat cavalier attitude here to details that they would regard as important. Each donor state resembles a quantum state of hydrogen, but with the "atom" embedded in a near-continuous medium with relative permittivity ϵ_r. The binding energy is reduced from that for hydrogen by a factor $(m_e^*/m_e)/\epsilon_r^2$, which explains why the ionization energy is only $45 \, \mathrm{meV}$. We ignore all of the hydrogen-like donor states except the ground state, treating higher states as belonging to the conduction band. The ground state is twofold degenerate because two spin orientations are possible, yet only one of those two states can be occupied. This requires an adjustment to the statistical mechanics which we ignore here.

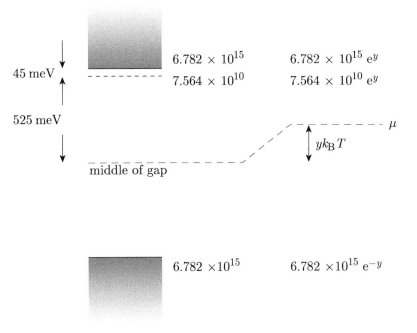

Fig. 26.4 The electron and hole concentrations in n-doped silicon. Donor levels lie $45 \, \mathrm{meV}$ below the conduction band, with (at 10^{-9} concentration) number density $4.993 \times 10^{19} \, \mathrm{atom \, m}^{-3}$.

As a first trial, μ is placed at the middle of the band gap, with the resulting number densities for electrons and holes shown alongside their energies. The excess of electrons over the intrinsic material works out at $7.564 \times 10^{10} \, \mathrm{m}^{-3}$, when there should be one electron originating from each donor.

To put this right, μ is raised by $y k_B T$, with consequences shown in the right-hand column. The calculation of y is performed in the text.

now easy to require the correct number of electrons to be present:

$$4.993 \times 10^{19} = (6.782 \times 10^{15} + 7.564 \times 10^{10})e^y - 6.782 \times 10^{15}\,e^{-y}$$
$$= 6.782 \times 10^{15}e^y + \text{negligible} - \text{negligible}$$

so

$$e^y = \frac{4.993 \times 10^{19}}{6.782 \times 10^{15}} = 7.362 \times 10^3, \qquad y = 8.904, \qquad yk_BT = 230\,\text{meV}.$$

This places μ at 230 meV above the middle of the band gap; it is shown in correct proportion in Fig. 26.4.

Compare the conductivity of this material with that of intrinsic silicon. The number of conduction-band electrons is greater by a factor $e^y = 7362$, while the valence-band holes have virtually disappeared. So the number of carriers is greater-than-intrinsic by a factor $\frac{1}{2}e^y$, and the conductivity is raised by the same factor of about 3700.

Comment: Notice how very small is the doping concentration that was assumed in this problem. I chose a doping of 1 in 10^9 to make just one point: the chemical potential μ is *extremely*—one might even say pathologically—sensitive to tiny quantities of impurity. If you want to prepare silicon with intrinsic conductivity, you have to recognize that you can't get that by achievable purification methods.[10]

Problem 26.2 asks you to find the location of the chemical potential for more realistic doping levels: 1 in 10^8, 1 in 10^7 and 1 in 10^6.

The location of the chemical potential is dependent on temperature, though not in so dramatic a way as the dependence on impurity concentration. You can see this at once: at absolute zero, all the donor states have re-acquired their electrons and are filled; the chemical potential must therefore lie *above* the donor levels. The approach to this may be seen in table 26.1.

[10] There are tricks by which intrinsic silicon and germanium can be simulated: causing lithium ions to drift through the semiconductor, propelled by an electric field, and letting Li$^+$ ions attach themselves to impurities, thereby neutralizing them. The resulting materials are known as "lithium-drifted silicon" and "lithium-drifted germanium".

Problems

Problem 26.1 electron–hole product
Use the density of states given in (26.3) to calculate the number of electrons in the conduction band, and show that it is

$$N_e = \frac{(2\pi m_e^* k_BT)^{3/2}}{4\pi^3\hbar^3}\exp\left(\frac{-(\varepsilon_c - \mu)}{k_BT}\right).$$

Find the corresponding expression for the number of holes. Show that the electron–hole product is

$$N_e\,N_h = \frac{(m_e^* m_h^*)^{3/2}(k_BT)^3}{2\pi^3\hbar^6}\exp\left(\frac{-(\text{band gap})}{k_BT}\right). \qquad (26.6)$$

Problem 26.2 Chemical potential of doped semiconductor
(1) In the example of Fig. 26.4, confirm the numerical values given in the text. Show that the donor states are about 10^{-5} occupied, i.e. the donors are almost entirely ionized with their electrons transferred to the

conduction band. It should be obvious that the occupation increases as
the temperature is lowered.

(2) Use the method given in the text to confirm the entries in the following table, which gives yk_BT, the location of μ above the middle of the
band gap, in milli-electron volts. You may assume that the electron–
hole product varies with temperature as in eqn (26.6). The values in
table 26.1 are obtained by using a Boltzmann-tail approximation. Show
that this approximation is justified for all entries in the table, though
it is least secure for the last item in the first row, where the calculated
μ lies only about $2k_BT$ below the donor level. (Even here, an iterative
solution reveals that the table entry is satisfactory. Try it.)

Table 26.1 The location of the chemical potential μ, in meV, above the middle of
the band gap in silicon.

temperature/K	dopant concentration			
	10^{-9}	10^{-8}	10^{-7}	10^{-6}
150	414	443	473	501
300	230	290	349	409

27 Heat capacity of electrons

Intended readership: When encountering the calculation of the heat capacity for electrons in a metal. The calculation presented here can substitute for other treatments.

27.1 Introduction

The heat capacity of a metallic solid contains contributions from conduction electrons, as well as (as with all solids) from lattice vibrations (phonons). The electron contribution is usually observable only at low temperatures; it is there proportional to temperature T and also to the density of states available to electrons at the Fermi surface.[1]

Many books on solid-state physics, or statistical mechanics, show that the electron contribution to the heat capacity is proportional to T, but fewer show convincingly that the coefficient contains the density of states—and does so whatever is the shape of the Fermi surface. We give an appropriate treatment here.[2]

27.2 The density of states

Let the number of quantum states available to electrons, in real-space volume V and in $\mathbf{k}$-space volume $\mathrm{d}\tau = \mathrm{d}^3\mathbf{k}$ be $V\mathrm{d}\tau$. From (6.9),

$$V\,\mathrm{d}\tau = \text{(number of states)}$$
$$= \frac{(\text{volume } V \text{ in real space}) \times (\text{volume } \mathrm{d}^3\mathbf{k} \text{ in wave-vector space})}{(2\pi)^3}$$
$$\times \text{(factor 2 for the electron spin)}. \qquad (27.1)$$

The density of states is defined as the number of quantum states (per unit volume) per unit of energy increment. We write it as

$$\text{density of states} = \rho(\varepsilon) \equiv \frac{\mathrm{d}\tau}{\mathrm{d}\varepsilon}. \qquad (27.2)$$

The quantum states within this $\mathrm{d}\tau$ are those[3] that lie within the given energy range $\mathrm{d}\varepsilon$.

The claim I have made in §27.1 is that the electron heat capacity is proportional to $\rho(\mu) = \left(\mathrm{d}\tau/\mathrm{d}\varepsilon\right)_{\varepsilon=\mu}$.

We are considering electrons in a metal. All energy bands are full except for the conduction band, so it is only the conduction-band electrons that can be excited thermally. Therefore we are considering the density of states for the conduction band only.[4]

---

[1] Since the phonon contribution varies as T^3 (Chapter 23), it falls faster with reduction of temperature, leaving the electron contribution dominant.

[2] Our calculation is also more-than-usually careful, in that it takes proper account of $\mathrm{d}\mu/\mathrm{d}T$, usually discarded.

[3] The energy–wave-vector relation $\varepsilon(\mathbf{k})$ in a real crystalline solid can be quite complicated, is different in different solids, and may not be well known. It is important, therefore, that we do not think of the volume element in $\mathbf{k}$-space as $4\pi k^2\,\mathrm{d}k$, because there is no spherical symmetry. Compare with (6.15) where isotropy *was* assumed.

In $\mathrm{d}\tau$ we implicitly total over all directions in $\mathbf{k}$-space so as to pick up all of the states within $\mathrm{d}\varepsilon$. If it helps, we may think that the states counted in $\mathrm{d}\tau$ are those that lie between a $\mathbf{k}$-space surface of energy ε and another of energy $\varepsilon + \mathrm{d}\varepsilon$.

[4] More correctly: there can be overlapping part-full bands, and then $\mathrm{d}\tau$ must, of course, include all quantum states within $\mathrm{d}\varepsilon$, whichever band they belong to.

Essays in Physics: Thirty-two thoughtful essays on topics in undergraduate-level physics. Geoffrey Brooker, Oxford University Press. © Geoffrey Brooker 2021. DOI: 10.1093/oso/9780198857242.003.0027

27.3   The number density and the internal energy U

The electrons in a solid have a Fermi–Dirac distribution

$$n(\varepsilon, T) = \frac{1}{\exp\{(\varepsilon - \mu)/(k_B T)\} + 1}, \qquad (27.3)$$

in which μ is the **chemical potential** or **Fermi level**.[5] The Fermi distribution falls from 1 to 0 over an energy range $\sim 5k_B T$ (see Figs. 26.1–3 and 27.1). In this chapter we take $k_B T$ to be "small", meaning small compared with the energy range over which $\rho(\varepsilon)$ changes.

The number N of electrons within a solid occupying real-space volume V is

N = sum of: (number $n(\varepsilon)$ of electrons in each quantum state)

$\times$ (number of quantum states $V\,d\tau$)

$$= V \int n(\varepsilon)\,d\tau = V \int_{\text{conduction band}} \frac{1}{e^{(\varepsilon - \mu)/k_B T} + 1} \frac{d\tau}{d\varepsilon}\,d\varepsilon. \quad (27.4)$$

In a similar way we can find the internal energy U:

U = sum of: (energy ε of each electron)

$\times$ (number $n(\varepsilon)$ of electrons in each quantum state)

$\times$ (number of quantum states $V\,d\tau$)

$$= V \int \varepsilon\, n(\varepsilon)\,d\tau = V \int_{\text{conduction band}} \varepsilon \frac{1}{e^{(\varepsilon - \mu)/k_B T} + 1} \frac{d\tau}{d\varepsilon}\,d\varepsilon. \quad (27.5)$$

We must understand the structure of eqns (27.4) and (27.5):

- The chemical potential μ is determined, as always, by fitting the correct number of electrons into the volume V; that is, it is determined by (27.4), an implicit equation for μ. This idea has been displayed in a dramatic way in Chapter 26 (albeit there for a semiconductor rather than a metal).

- The internal energy U can be obtained from (27.5). Within this, μ must be obtained from (27.4).

The heat capacity at constant volume is the derivative of U with respect to temperature:[6]

$$\frac{C_v}{V} = \frac{1}{V}\left(\frac{\partial U}{\partial T}\right)_V = \int_{\text{conduction band}} \varepsilon \frac{\partial n(\varepsilon, T)}{\partial T} \frac{d\tau}{d\varepsilon}\,d\varepsilon. \quad (27.6)$$

Likewise, since N/V is independent of temperature:

$$0 = \frac{d}{dT}\left(\frac{N}{V}\right) = \int_{\text{conduction band}} \frac{\partial n(\varepsilon, T)}{\partial T} \frac{d\tau}{d\varepsilon}\,d\varepsilon. \quad (27.7)$$

The programme to be followed (a slight change from that outlined in the bullets above but embodying the same principle) is to find $d\mu/dT$ from (27.7) and to use it in the evaluation of (27.6).

[5] For this terminology, see § 26.1.

Usually $n(\varepsilon, T)$ is abbreviated to $n(\varepsilon)$ or to n, and we keep it in mind that n depends upon T as well as on the electron energy ε.

This point requires more discussion. We are about to calculate C_v, the heat capacity that is measured when the metal is held at constant volume V. Constant volume means that the lattice spacing is not allowed to change with temperature, and so neither is the $\varepsilon(\boldsymbol{k})$ relation, and neither is the density of states $d\tau/d\varepsilon$. In turn, this means that the change of integration variable, as far back as (27.2), from $2d^3\boldsymbol{k}/(2\pi)^3 = d\tau$ to $d\varepsilon$, does not introduce any hidden dependence on temperature—which it would if the volume were allowed to change.

It also means that $n(\varepsilon, T)$ is a function of truly independent variables ε, T. The partial derivatives $\partial n/\partial \varepsilon$, $\partial n/\partial T$, needed in (27.6–8), can be worked out straightforwardly from (27.3).

[6] Equations (27.5) and (27.6) are usually written without explicit mention of V, and with a statement that U and C_v refer to unit volume. I have thought it an improvement of clarity to retain V as, for example, in sidenote 5. Here C_v and V are "extensive" quantities and their ratio is "intensive".

27.4   Calculation of the heat capacity

The derivative $\partial n/\partial T$ can be found by differentiating (27.3). However this contains a trap! There is an explicit T on display in (27.3), but there is also a temperature dependence hiding in μ. Therefore

$$\frac{\partial n}{\partial T} = \frac{-\mathrm{e}^{(\varepsilon-\mu)/k_\mathrm{B}T}}{\left(\mathrm{e}^{(\varepsilon-\mu)/k_\mathrm{B}T}+1\right)^2}\left(\frac{-(\varepsilon-\mu)}{k_\mathrm{B}T^2}-\frac{1}{k_\mathrm{B}T}\frac{\mathrm{d}\mu}{\mathrm{d}T}\right),$$

which can be tidied to

$$\frac{\partial n}{\partial T} = -\frac{\partial n}{\partial \varepsilon}\left(\frac{(\varepsilon-\mu)}{T}+\frac{\mathrm{d}\mu}{\mathrm{d}T}\right). \tag{27.8}$$

We shall see that the $\mathrm{d}\mu/\mathrm{d}T$ term makes a contribution to C_v of the same order as that from the other term; therefore we may not discard it as negligible (as might be tempting).[7]

Substitute (27.8) into (27.7) and (27.6).

$$0 = -\int_\mathrm{band}\left(\frac{\varepsilon-\mu}{T}+\frac{\mathrm{d}\mu}{\mathrm{d}T}\right)\rho(\varepsilon)\frac{\partial n}{\partial \varepsilon}\,\mathrm{d}\varepsilon; \tag{27.9a}$$

$$\frac{C_\mathrm{v}}{V} = -\int_\mathrm{band}\varepsilon\left(\frac{\varepsilon-\mu}{T}+\frac{\mathrm{d}\mu}{\mathrm{d}T}\right)\rho(\varepsilon)\frac{\partial n}{\partial \varepsilon}\,\mathrm{d}\varepsilon. \tag{27.9b}$$

Each of equations (27.9) has the form (27.15), evaluated in (27.18), so we may apply that theorem to them. The working is dealt with in problem 27.2. The results are:[8]

$$0 = \frac{\mathrm{d}\mu}{\mathrm{d}T}\rho(\mu)+\frac{\pi^2 k_\mathrm{B}^2 T}{3}\rho'(\mu)+\dots \tag{27.10a}$$

$$\frac{C_\mathrm{v}}{V} = \frac{\mathrm{d}\mu}{\mathrm{d}T}\mu\,\rho(\mu)+\frac{\pi^2 k_\mathrm{B}^2 T}{3}\left(\mu\,\rho'(\mu)+\rho(\mu)\right)+\dots. \tag{27.10b}$$

We are making expansions in powers of temperature T. Equation (27.10a) tells us that $\mathrm{d}\mu/\mathrm{d}T$ is of order T. Then (27.10b) contains three terms that are all of order T, making it clear that $\mathrm{d}\mu/\mathrm{d}T$ is not negligible:[9] the inclusion of $\mathrm{d}\mu/\mathrm{d}T$ and $\rho'(\mu)$ in our expressions, from (27.8) onwards, was necessary.

Equations (27.10) contain enough information for us to eliminate $\mathrm{d}\mu/\mathrm{d}T$ and hence find an expression for C_v containing only known (or knowable) quantities. Neat way: multiply (27.10a) through by μ and subtract. Two terms cancel, and we obtain

$$\frac{C_\mathrm{v}}{V} = \frac{\pi^2 k_\mathrm{B}^2 T}{3}\rho(\mu) = \frac{\pi^2 k_\mathrm{B}^2 T}{3}\left(\frac{\mathrm{d}\tau}{\mathrm{d}\varepsilon}\right)_{\varepsilon=\mu} \tag{27.11}$$

(plus terms containing higher powers of T). This is the expression that we have come for: the heat capacity contains the density of states $\rho(\mu)$ at the Fermi surface.[10]

Notice that, in obtaining (27.11), the density of states $\rho(\mu)$ has never been given an explicit expression: $\rho(\mu)$ can be whatever is required by the band structure of the crystal. Our demonstration that $C_\mathrm{v} \propto \rho(\mu)$ is general.

Turning this round: If we are trying to find out about the band structure of a crystal that is so far poorly characterized, then measurements

[7] Most textbook treatments either ignore $\mathrm{d}\mu/\mathrm{d}T$ or argue that it must be negligible. It is not negligible, but the algebra can be cleverly manipulated so as to give $\mathrm{d}\mu/\mathrm{d}T$ a small coefficient. That's not the same.

[8] Here $\rho'(\mu) = \left(\mathrm{d}\rho(\varepsilon)/\mathrm{d}\varepsilon\right)_{\varepsilon=\mu}$. Thus a prime denotes differentiation with respect to ε.

[9] Terms of order T^3 have been dropped from both equations (27.10); see problem 27.2.

[10] The density of states is evaluated at energy μ where μ is dependent upon temperature T. We may therefore wonder what value of μ should be used for evaluating the density of states: perhaps $\mu(T)$ or perhaps $\mu(T=0) = \varepsilon_\mathrm{F}$?

It does not matter.

We must remember that we are evaluating everything in a low-temperature limit, in which expansions are being made in powers of T. In $\left(\mathrm{d}\tau/\mathrm{d}\varepsilon\right)_{\varepsilon=\mu}$, any reasonable value of μ will yield C_v correct to the first power in T. Putting this another way: the effect on C_v of changing from $\mu(T)$ to $\mu(0)$ is to change the expression for C_v by a quantity of order T^3, which is of no interest (it belongs with the *next* term in the expansion of (27.18)). Refer to the final bulleted item in § 12.2.1: you can be brutal to the coefficient of a small quantity. Here the small quantity is T and the coefficient is the density of states.

of the electron heat capacity can help, because they tell us the density of states $\rho(\varepsilon=\mu)$ at the Fermi surface—on which more below.

Comment: The derivation of eqn (27.11) has been done with a minimum of trickery. I wanted to show that we can hammer out the algebra whether or not we see a slick way through. But we can now look back and see a trick. We are going to subtract $\mu \times$ one equation of (27.10) from the other. Using hindsight, go back to (27.9) and do the subtraction there:

$$\frac{C_{\mathrm{v}}}{V} = -\int_{\mathrm{band}} \left\{ (\varepsilon - \mu) \left(\frac{\varepsilon - \mu}{T} + \frac{\mathrm{d}\mu}{\mathrm{d}T} \right) \rho(\varepsilon) \right\} \frac{\partial n}{\partial \varepsilon} \, \mathrm{d}\varepsilon. \qquad (27.12)$$

This route[11] is explored in problem 27.3. When it is done, $\mathrm{d}\mu/\mathrm{d}T$ appears multiplied by T^2, and contributes a negligible $\sim T^3$ to C_{v}. This explains why $\mathrm{d}\mu/\mathrm{d}T$ is commonly omitted from the algebra in textbook calculations. However, the knowledge that the $\mathrm{d}\mu/\mathrm{d}T$ contribution to C_{v} is negligible can only be obtained by retaining $\mathrm{d}\mu/\mathrm{d}T$ and seeing what happens to it.

[11] This is the route followed by Kittel (2005), Chapter 6.
   See the comment after problem 27.3 for a reason why we might have thought (27.12) a judicious step, quite apart from its help with $\mathrm{d}\mu/\mathrm{d}T$.

27.5   Comparison with experiment

It is customary to compare experimental results with what (27.11) would give for an ideal Fermi gas. We define

$$r \equiv \frac{(\text{measured electron heat capacity})}{(\text{heat capacity for an ideal Fermi gas})}. \qquad (27.13)$$

From (27.11) this is also

$$r = \frac{\rho(\mu) \text{ for electrons in our metal}}{\text{calculated } \rho(\mu) \text{ for the ideal Fermi gas}}. \qquad (27.14)$$

In (27.13) and (27.14), the denominator is calculated by assuming there is a known number density for the electrons in the metal, a number density that can be carried over to the reference Fermi gas.[12]

A table of electron heat capacities is given by Kittel (2005) p. 146 Table 2. Only a selection of elements is assigned a Fermi-gas heat capacity, presumably because only these can be given a convincing electron number density.

In Kittel's table, the values of the ratio r range from 0.34 (Be) to 2.18 (Li). There is excellent order-of-magnitude agreement (never worse than a factor 3) between calculation and experiment. The ideas in the present chapter are convincingly confirmed. At the same time, the departures of r from 1 give information about the density of states at the Fermi surface, and hence about the metal's band structure.

[12] The number density will normally be obtained by counting the electrons in the conduction band. Electrons more tightly bound in the atoms won't be excited across the band gap by the small $k_{\mathrm{B}}T$ energies available. Things get untidy if there are overlapping bands.

Problems

Problem 27.1 The expansion in powers of T

We are to evaluate integrals having the form

$$I \equiv - \int_{\text{conduction band}} f(\varepsilon, T) \frac{\partial n}{\partial \varepsilon} \, d\varepsilon, \qquad (27.15)$$

in which $n = n(\varepsilon, T)$ is the Fermi distribution. The derivative $\partial n / \partial \varepsilon$ is a (negative) peaky function having peak width $\sim k_B T$, as shown in Fig. 27.1.

The temperature T is to be "low", meaning that[13]

(energy range within which $\partial n/\partial \varepsilon$ varies) $\sim k_B T$

$\ll$ (energy range within which $f(\varepsilon, T)$ varies significantly). (27.16)

More precisely, $f(\varepsilon, T) \, \partial n / \partial \varepsilon$ has a peak having energy width $\sim k_B T$ because $f(\varepsilon, T)$ cannot counter the exponential fall-off of $\partial n / \partial \varepsilon$ on either side of the peak. Turning this round, it is saying that $f(\varepsilon, T)$ must be smooth enough, as a function of ε, that condition (27.16) can be met—at all—and then that $k_B T$ is such that it *is* met.

Compare condition (27.16) with the similarly intentioned statement given following (27.3).

Each function $f(\varepsilon, T)$ (there is more than one) will be thought of as primarily a function of energy ε, as is suggested by its appearing in an integral over ε. Consistently, the derivatives are denoted by

$$f' = \frac{\partial f}{\partial \varepsilon}; \qquad f'' = \frac{\partial^2 f}{\partial \varepsilon^2}.$$

At the same time, each $f(\varepsilon, T)$ has a dependence on T as parameter (so for that matter does $n = n(\varepsilon, T)$, which is why $\partial n / \partial \varepsilon$ is written as a partial derivative), as well as on ε; hence the notation $f(\varepsilon, T)$, though we sometimes abbreviate it to $f(\varepsilon)$ or simply f.

It is convenient to express energies in units of $k_B T$, so we define x by

$$\varepsilon = \mu + k_B T x,$$

so that[14]

$$n = \frac{1}{e^x + 1}, \qquad \frac{dn}{dx} = \frac{-e^x}{(e^x + 1)^2}, \qquad \frac{\partial n}{\partial \varepsilon} \, d\varepsilon = \frac{dn}{dx} \, dx.$$

Given (27.16), function $f(\varepsilon)$ may be expanded in a Taylor series:

$$f(\varepsilon) = f(\mu) + (\varepsilon - \mu) f'(\mu) + \frac{(\varepsilon - \mu)^2}{2} f''(\mu) + \dots,$$

$$= f(\mu) + k_B T \, f'(\mu) \, x + \frac{(k_B T)^2}{2} f''(\mu) \, x^2 + \dots.$$

Then

$$I = -f(\mu) \int \frac{dn}{dx} \, dx - k_B T \, f'(\mu) \int x \frac{dn}{dx} \, dx$$

$$- \frac{(k_B T)^2}{2} f''(\mu) \int x^2 \frac{dn}{dx} \, dx + \dots. \qquad (27.17)$$

(1) Show that dn/dx is an even function of x, as claimed.

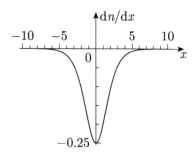

Fig. 27.1 The derivative dn/dx of the Fermi distribution function. Here the energy $\varepsilon = \mu + k_B T x$ defines x as the distance in energy from μ in units of $k_B T$. The function has a sharp dip centred on $x = 0$. It is an even function of x, consistent with oddness of $(n - \frac{1}{2})$ shown in Fig. 25.4.

(2) The integrations in eqn (27.17) are taken over energies within the metal's conduction band. The bottom of the band is many $k_B T$ below μ while the top is many $k_B T$ above. Show that at both limits the integrand is exponentially small, and hence the integration range for x can safely be extended to be from $-\infty$ to ∞, this for all three integrals.

(3) Show that the first term in (27.17) evaluates to $f(\mu)$, with corrections that are exponentially small.

(4) Show that the second term in (27.17) evaluates to zero because the integrand is an odd function of x integrated between symmetrical limits—again with corrections that are exponentially small.

(5) The third integral in (27.17) is just a number that we must try to find; it is $\pi^2/3$. Fill in the details in the following outline evaluation.

$$J \equiv -\int_{-\infty}^{\infty} x^2 \frac{dn}{dx}\, dx = 4 \int_0^{\infty} x\, n\, dx = -4 \int_0^{\infty} x \frac{(-e^{-x})}{1+e^{-x}}\, dx$$

$$= 4 \int_0^{\infty} \ln(1 + e^{-x})\, dx$$

$$= 4 \int_0^{\infty} \left(e^{-x} - \frac{e^{-2x}}{2} + \frac{e^{-3x}}{3} - \frac{e^{-4x}}{4} + \dots \right) dx$$

$$= 4 \int_0^{\infty} \left(e^{-x} - \frac{e^{-x}}{2^2} + \frac{e^{-x}}{3^2} - \frac{e^{-x}}{4^2} + \dots \right) dx$$

$$= 4 \left(1 - \frac{1}{2^2} + \frac{1}{3^2} - \frac{1}{4^2} + \dots \right).$$

In the first line, the integration range is halved because the integrand is an even function of x. In the first two lines there are two integrations by parts. The logarithm is expanded as an obvious power series, valid because $e^{-x} \leqslant 1$ everywhere within the integration range. Then there are substitutions that bring all exponents to $-x$ so that all integrations become the same.

It remains to evaluate the series. There is a standard way of handling such series: invent a function whose Fourier series has these terms as its expansion coefficients.

Consider the function[15]

$$\frac{\pi^2}{3} - x^2 = \sum_{n=0}^{\infty} A_n \cos(nx).$$

This expansion is to hold in the range $-\pi < x < \pi$. In the standard Fourier-series way, evaluate the coefficients A_n and show that

$$\frac{\pi^2}{3} - x^2 = 4 \sum_{n=1}^{\infty} \frac{(-1)^{(n+1)}}{n^2} \cos(nx).$$

Take the special case $x = 0$ and show that $J = \pi^2/3$.

(6) Assemble the contributions to the I of (27.15) and show that

$$I = -\int_0^{\infty} f(\varepsilon) \frac{\partial n}{\partial \varepsilon}\, d\varepsilon = f(\mu) + \frac{\pi^2 k_B^2 T^2}{6} f''(\mu) + \dots. \qquad (27.18)$$

Equation (27.18) is the result that we have come for.

[15] A more conventional route might be to Fourier-expand a sawtooth wave. The outcome is a series related to the one we want, and J can be found by playing tricks from there. However I decided to go directly to the required series.

How did I find the function to expand? I asked Mathematica to plot a graph of

$$\sum_{n=1}^{20} \frac{(-1)^{(n+1)}}{n^2} \cos(nx)$$

and found that $a - bx^2$ was a likely expression for the result. I could then find the Fourier series for $a - bx^2$ and equate coefficients. Everything fits, and the outcome is plotted in Fig. 27.2.

Incidentally, the fact that the coefficients A_n fall as n^{-2} shows that the function to be expanded (plotted over more than one period) has kinks (discontinuities of gradient) but no worse singularities. This puts a powerful constraint on the kind of function that would serve.

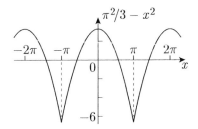

Fig. 27.2 The function that we have chosen to expand as a Fourier series. It is given by $\pi^2/3 - x^2$ within the range $-\pi < x < \pi$, and repeats outside that range with period 2π.

(7) Think about the μ appearing on the right of (27.18). Within the accuracy sought (order T^2), can this μ be replaced by ε_F: (a) in $f(\mu)$, (b) in $f''(\mu)$?

Problem 27.2 The derivation of eqns (27.10)
Deal first with (27.9a). The integral has the form (27.18) with

$$f(\varepsilon) = -\left(\frac{\varepsilon - \mu}{T} + \frac{d\mu}{dT}\right)\rho(\varepsilon).$$

For substitution into (27.18), we require $f(\mu)$ and $f''(\mu)$. To evaluate $f''(\mu)$ we differentiate $f(\varepsilon)$ with respect to ε using Leibniz's theorem. Show that

$$-f(\mu) = \frac{d\mu}{dT}\rho(\mu),$$

and

$$-f''(\mu) = \frac{2}{T}\rho'(\mu) + \frac{d\mu}{dT}\rho''(\mu).$$

Substitute these building blocks into (27.18) and show that

$$0 = \frac{d\mu}{dT}\left(\rho(\mu) + \frac{\pi^2 k_B^2 T^2}{6}\rho''(\mu)\right) + \frac{\pi^2 k_B^2 T}{3}\rho'(\mu) + \dots$$

$$\approx \frac{d\mu}{dT}\rho(\mu) + \frac{\pi^2 k_B^2 T}{3}\rho'(\mu) + \dots,$$

correct to order T, in agreement with (27.10a).
Perform a similar manipulation to obtain (27.10b).

Problem 27.3 A slicker route to (27.11)
Perform manipulation, after the fashion set out in the first part of problem 27.2, to evaluate the integral of (27.12). I think you will find $d\mu/dT$ appears with a coefficient of order T^2 and can therefore be dropped as giving a contribution to C_v of order T^3. Overall, the algebra should be slightly quicker than that used in the text. But, of course, you do have to invent the trick[16]. . . .

Comment: There is another, powerful, reason why eqn (27.12) is a smart move. Look back to eqn (27.9b), which contains ε in the integrand. We have not made explicit any choice for the origin of energy, so the integral in (27.9b) seems to take a value that changes if we move the origin of energy. That would be acceptable for U, but not for C_v. We are being told that we are missing something.
Of course it all comes out right in the end. In (27.10b),

$$\frac{C_v}{V} = \mu\left\{\frac{d\mu}{dT}\rho(\mu) + \frac{\pi^2 k_B^2 T}{3}\rho'(\mu)\right\} + \frac{\pi^2 k_B^2 T}{3}\rho(\mu).$$

The μ at the beginning of this depends on the energy origin, but is multiplied by $\{\} = 0$ according to (27.10a). In contrast to (27.9b), (27.12) has $(\varepsilon - \mu)$ in the integrand in place of ε, so the dependence on the energy origin is neatly removed, even before we integrate.

[16] And you still need (27.10a) to tell you that $d\mu/dT$ is of order T.

Electrons in a square lattice

Intended readership: On encountering Brillouin zones for electrons in one, two and three dimensions, in particular the discussion of a two-dimensional square lattice.

28.1 Introduction

When we first learn about the behaviour of electrons in a crystalline solid, we look at a one-dimensional chain of atoms, with electrons free to move in the one-dimensional periodic potential that the atoms impose (rather as a one-dimensional chain in Chapter 23 introduces phonons). This one-dimensional model suffices to introduce many of the important concepts: Bloch wave functions for the electrons; a modified energy–wave-vector relation for the electrons creating energy gaps; energy bands; Brillouin zones; the reduced zone scheme.

When we proceed to three dimensions, we encounter new phenomena that do not show up in a one-dimensional model: overlapping and non-overlapping bands; direct and indirect band gaps; anisotropic effective masses .... Inevitably, we were not prepared for these complications by a model confined to one dimension.

A two-dimensional model[1] provides a very useful intermediate case. It can introduce almost all of the complexities seen in three dimensions, and yet is much easier to think about. In particular, when wave vectors are confined to a plane we have the third dimension still available, along which to plot electron energy.

[1] The two-dimensional model is unashamedly taken from Kittel (2005) Chapter 9.

28.2 The empty lattice in one dimension

Before we discuss the main topic of this chapter, it will be helpful to revise the one-dimensional empty lattice.[2]

Electrons move freely along the x-axis. They have the free-electron wave function ψ_{free} and the free-electron energy $\varepsilon_{\text{free}}(k_x)$:

[2] Textbooks discuss the empty lattice, but do not always give it that name. I think the name helps.

$$\psi_{\text{free}} \propto e^{ik_x x}; \qquad \varepsilon_{\text{free}}(k_x) = \frac{(\hbar k_x)^2}{2m_e}. \qquad (28.1)$$

In due course we shall perturb the system by imposing a periodic potential whose space period in the x-direction is a. In preparation for

353

Essays in Physics: Thirty-two thoughtful essays on topics in undergraduate-level physics. Geoffrey Brooker, Oxford University Press. © Geoffrey Brooker 2021. DOI: 10.1093/oso/9780198857242.003.0028

[3]The lattice is not empty of electrons. It is the periodic potential that is "empty" (zero in magnitude).

that potential, we set up the Brillouin-zone structure in advance; the electrons remain free as before. This construct is called the **empty lattice**:[3] "lattice" because k-space is given the zone structure appropriate to a periodic potential having space-period a; "empty" because that potential is still zero. The procedure here should occasion no surprise: it is precisely what we do in Chapter 11, and above all in §11.5, where we anticipate application of a perturbation by setting up in advance the eigenfunctions (correct in first-order perturbation theory) that respect the symmetries of the perturbation to come: "prefabrication".

The energy–wave-vector relation for electrons in a one-dimensional lattice is shown in Fig. 28.1. A parabola (thickest line) $\varepsilon = (\hbar k_x)^2/2m_e$ is drawn centred on $k_x = 0$ to show what free electrons do. The

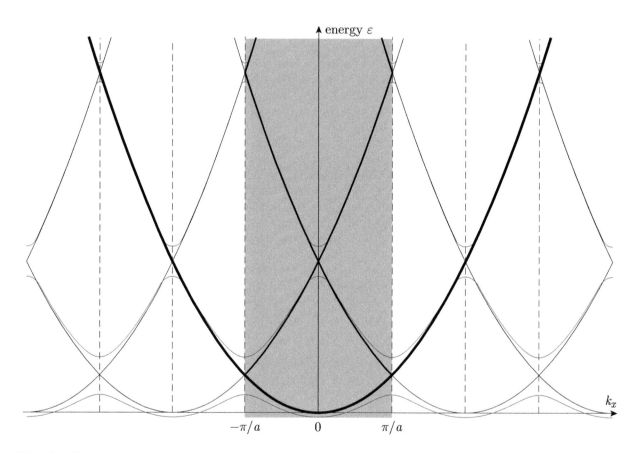

Fig. 28.1 Electrons are constrained to move in the x-direction only but are otherwise free. Their energy–wave-vector relation is parabolic: $\varepsilon(k_x) = (\hbar k_x)^2/2m_e$ (thickest line). In preparation for application of a periodic potential having real-space-period a, the k_x-axis is partitioned into Brillouin zones; at this stage the electrons are still free and we have the **empty lattice**. In the **periodic zone scheme**, the parabolic $\varepsilon(k_x)$ relation is repeated at k_x-intervals of a reciprocal-lattice vector $2\pi/a$ (thinnish lines but thickened in the central zone). Thinnest lines show an $\varepsilon(k_x)$ relation that might result from application of an actual periodic potential. The zone structure is already in place, and only the energy shifts modify the previous picture. The "avoided crossings" are exactly what we expect from experience elsewhere.

Every piece of the (repeated) $\varepsilon(k_x)$ curve appears within the first Brillouin zone: the range $-\pi/a \leqslant k_x \leqslant \pi/a$. We may use just those pieces, in which case we are using the **reduced zone scheme**, what's in the range highlighted in grey.

empty lattice is next catered for by imposing Brillouin zones in anticipation of a periodic potential having period a but still having zero strength. Everything now repeats at k_x-intervals of $2\pi/a$; in particular we draw additional $\varepsilon(k_x)$ parabolas at these intervals according to the **periodic zone scheme**. The diagram can also be understood in the **reduced zone scheme** (shaded region), by giving attention to the region $-\pi/a \leqslant k_x \leqslant \pi/a$ only.[4] Either way, the energy–wave-vector relations $\varepsilon(k_x)$ for the empty lattice are now completely specified.

The first two of the above parabolas cross at $k_x = \pi/a$; such crossings are most evident in the periodic zone scheme. While the lattice remains "empty" such crossings are permitted.

28.2.1   A one-dimensional non-empty lattice

The purpose of drawing the $\varepsilon(k_x)$ relation for the empty lattice is to make it as painless as possible to apply a non-zero periodic potential; the hard work has already been done. We now consider qualitatively what happens, the outcome being shown as the thinnest lines in Fig. 28.1.

Experience elsewhere tells us that energy-level crossings are "avoided" when a perturbation is applied (a perturbation such as to cause the wave functions associated with the two curves to form linear combinations), however weak the perturbation. Figure 28.1 indicates by thin lines the kind of thing that a periodic potential is likely to do.[5] Close to crossings of the original parabolas, the avoided crossings result in conspicuous changes to the shapes of the $\varepsilon(k_x)$ curves, with the imposition of band gaps. Elsewhere, the changes are less significant: quantitative, rather than qualitative.

The display of Fig. 28.1 should be revision for the reader. It has been given here so that there is no difficulty in seeing what is going on when we give similar discussions for electron behaviour in two dimensions.

28.3   The empty lattice in two dimensions

We illustrate by taking the simplest possible case: a two-dimensional square lattice whose period is a in both x- and y-directions. We need not draw the lattice in real space, as its structure is obvious. Electrons inhabit the lattice, just as they did in the one-dimensional case. To start with, the lattice is "empty", meaning that the periodic potential has zero magnitude.

In k-space, the empty lattice is represented by another square lattice, this one having period $2\pi/a$ in each direction. We partition k-space into Brillouin zones in preparation for the real-space periodic potential that is not yet there.

There is a standard way of partitioning k-space into Brillouin zones, the Wigner–Seitz construction. For details, see e.g. Brillouin (1953), Chapter 23.[6]

The first four Brillouin zones for a square lattice are shown in Fig. 28.2. The central square is the first zone. The four triangles adjoining it form

[4] To be strict, one or other of the end values should now be excluded, as in problem 23.1. If we follow pattern with eqns (23.20), then $-\pi/a < k_x \leqslant \pi/a$.

[5] For illustration, I modelled the periodic potential by a term $\propto \cos(2\pi x/a)$. The wave function for an electron is then a Bloch function, known to mathematicians as a Floquet solution to Mathieu's equation. However, the band gaps that result from this model decrease fast with increase of energy, and I have redrawn (by hand) the $\varepsilon(k_x)$ curves around the top two gaps to make their qualitative shapes more visible. See problem 28.1.

[6] Briefly: We plot the reciprocal-lattice points which lie at integer multiples of $2\pi/a$ in each of the k_x- and k_y-directions, forming a square lattice in k-space. A line is notionally drawn from the origin to each reciprocal-lattice point, and a plane is drawn bisecting that line orthogonally. These planes carve out volumes in k-space (strictly areas, but we are looking ahead to three dimensions). The smallest such volume is the first Brillouin zone. The next pieces, chosen symmetrically, form the second zone; the same happens with the third zone, and so on.

What is drawn in Fig. 28.2 is a small part of the Wigner–Seitz construction. Zones are neither repeated as they are in Fig. 28.1, nor are they translated into the first zone.

If the four second-zone fragments are translated through reciprocal-lattice vectors so as to overlap the volume/area of the first zone, they fill that volume without overlaps or gaps. The same happens with the third zone (eight pieces), and so on. We have then built the reduced zone scheme.

Fig. 28.2 The Brillouin-zone structure of **k**-space for a two-dimensional square lattice. The k_x and k_y-axes are drawn as broken lines, as they are not zone boundaries. Blobs indicate a few of the reciprocal-lattice points, lying at multiples of $2\pi/a$ in each direction from the origin. Each line denoting a zone boundary is drawn by bisecting a line from the origin to a reciprocal-lattice point. The outlines delimiting the first, second and third zones are shown by black lines of diminishing width. The outline enclosing the twelve fragments of the fourth zone is shown in grey.

The shaded circle indicates quantum states that are occupied if there are four electrons per atom in the empty lattice. The first zone is full, while the second, third and fourth zones are part-occupied.

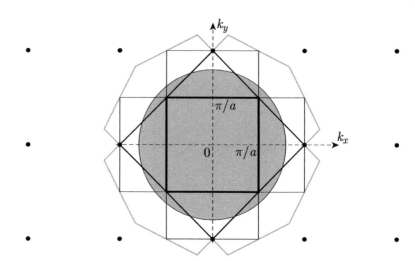

[7] Given the discussion of eigenstates in Chapter 6, we should note that we are using travelling-wave eigenfunctions, which we may say are the consequence of applying periodic boundary conditions.

We say that **k** fully identifies a quantum state because we are for now thinking in the manner of Fig. 28.2: the Wigner–Seitz step. If the Brillouin zones are translated into the reduced-zone scheme, each quantum state will need to be specified by saying which band it is in, as well as giving its **k**.

the second zone. The eight smaller triangles form the third zone; then twelve shapes give the fourth zone. Higher zones are not shown, to avoid clutter (they are in any case unoccupied by electrons).

In the square lattice, the energy eigenstates for an electron are Bloch functions, each specified by giving its location $\mathbf{k} = (k_x, k_y)$ in reciprocal space.[7] At this stage, the lattice is empty, the electrons are free, having energy $\varepsilon(\mathbf{k}) = \hbar^2(k_x^2 + k_y^2)/2m_e$, and the electron wave functions are simply $e^{i(\mathbf{k}\cdot\mathbf{r}-\omega t)}$.

The two-dimensional lattice has a great advantage for our understanding: we can use the third dimension to plot the energy $\varepsilon(\mathbf{k})$ for electrons inhabiting the real-space lattice. I have made a wooden model displaying this, shown in Fig. 28.3. The energy of the electron is indicated

Fig. 28.3 The first four Brillouin zones for an empty square lattice. In this wooden model, height represents the energy $\varepsilon(\mathbf{k})$ of electrons, plotted against **k**. Wave-vector space has been partitioned into Brillouin zones in preparation for a periodic potential (in real space) that has not yet been applied. Only four of the twelve fragments of the fourth zone are supplied; the model stands on a map of the extended-zone scheme that shows the footprints of some of the missing pieces. A circle represents the Fermi "surface" when four electrons per atom are present.

Small gaps have been left between the wooden fragments, partly to emphasize the location of the joins, partly to allow for material removed when the block was sawn into 17 pieces.

by height above the bottom of the bowl-shape. Had the bowl been cut accurately (which it wasn't) the bowl would have a parabolic section.

The model shown in Fig. 28.3 has all pieces present of the first, second and third zones. For the fourth zone, it has four of the twelve fragments; the underlying copy of the zone outlines, the same as in Fig. 28.2, shows the footprints of the remaining eight fragments (four visible).

28.3.1   The empty square lattice containing four mobile electrons per atom

Figure 28.2 has a shaded circular area, which indicates the region occupied by electrons (here four per atom). Those electrons are at near-zero temperature, so they occupy the lowest-energy states, which are those closest to the origin of k-space. The area of the circle is twice the area of the central square, the first Brillouin zone.[8]

The circular boundary, separating occupied states from empty states, is the Fermi surface. It may be seen marked on the wooden model (a little larger than it should be).

Our next step is to see what happens if we translate the pieces of each fragmented zone into the reduced-zone scheme. Figure 28.4 shows, in plan view, what happens when this is done. In (a), the four fragments of the second zone are shown translated[9] through reciprocal-lattice vectors until they cover the central square of Fig. 28.2. In (b), the same is done for the eight fragments of the third zone. In (c) is a modification in which the third-zone fragments are further translated in order to bring together the occupied electron states, though those states are now centred on a "corner" at $k = (\pi/a, \pi/a)$.

[8] The number of quantum states available to electrons, in area element $\mathrm{d}k_x\,\mathrm{d}k_y$, is proportional to $\mathrm{d}k_x\,\mathrm{d}k_y$ and independent of the location of k. Remember eqn (6.9). Therefore it is the *area* in Fig. 28.2 that is relevant. I stress this because it is possible to mislead oneself into thinking of electrons as resembling a liquid that is poured into the bowl of Fig. 28.3 up to the height of the Fermi level; if this were true there would be more electrons where the bowl is deeper . . . .

[9] *Translated.* Given the high symmetry of the physics, it is tempting to rotate triangular pieces when moving them, thinking that it makes no difference; this temptation must be resisted.

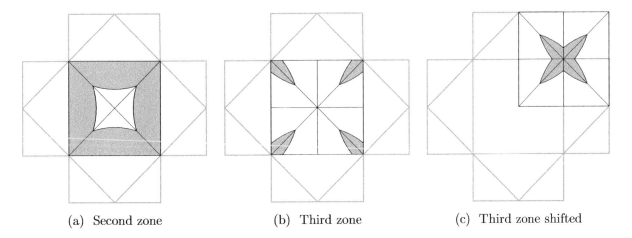

(a) Second zone   (b) Third zone   (c) Third zone shifted

Fig. 28.4 The (fragments of the) Brillouin zones of Fig. 28.2 have been translated through reciprocal-lattice vectors into the reduced zone scheme. The first zone is not shown, as it is just a filled square and un-translated. (a) On the left are the four fragments of the second zone. (b) In the middle are the eight fragments of the third zone. Because the third-zone states occupied by electrons form disjoint regions in the corners of the zone, it is usual to make further translations to yield the "shifted" third zone shown at the right (c). The fourth zone has a few filled states, also "in the corners", but is not shown.

Fig. 28.5 The second Brillouin zone for an empty square lattice. The four fragments of the second zone have been translated into the reduced zone scheme. Energy is represented by the height of the wooden surface; the Fermi surface is marked by the fragments of a circle near the top of the structure. The whole represents an energy band with a few unfilled states (we can call them holes) around the origin of k-space. The $\varepsilon(k)$ relation is far from having axial symmetry about $k = 0$, as follows from the non-circular Fermi surface, so the holes have an anisotropic effective mass.

The photographs of Figs. 28.3 and 28.5–7 were kindly taken by colleagues in the Oxford-Physics Photographic Department.

Although the constructions that led to Fig. 28.4 are straightforward, it is easy to lose sight of what is going on. What does the energy look like as a function of k? Why do the empty states in zone 2, and the filled states in zone 3, have the shapes that we have found? This is where the wooden model is useful.

Figure 28.5 shows the four pieces of the second Brillouin zone rearranged in the manner of Fig. 28.4(a). The $\varepsilon(k)$ relation has the shape of a tent with its apex at the centre. It is now rather easy to see why there are unoccupied-by-electrons quantum states at the zone centre. We have a not-quite-full band with a few empty states (holes if you wish) at the top.

Fig. 28.6 The third Brillouin zone for an empty square lattice. The eight fragments of the third zone have been translated into the reduced zone scheme. Energy is represented by the height of the wooden surface; the Fermi surface is marked by the fragments of a circle towards the corners of the structure. The whole represents an energy band with a few occupied states around the corners of the zone.

Fig. 28.7 The third Brillouin zone for an empty square lattice. Six of the eight fragments of the third zone have been translated through a new choice of reciprocal-lattice vectors so that the eight form a new block, centred at $(\pi/a, \pi/a)$. Energy is represented by the height of the wooden surface; the Fermi surface is marked by the fragments of a circle arranged around the middle of the structure. The whole represents an energy band with a few occupied states. The $\varepsilon(\mathbf{k})$ relation is far from having circular symmetry, so the electrons have a very anisotropic effective mass.

Figure 28.6 shows the eight pieces of the third Brillouin zone re-arranged in the manner of Fig. 28.4(b). The $\varepsilon(\mathbf{k})$ relation has great mountain ridges crossing at the middle of the zone, with a deep valley running down into each corner. Electrons occupying the lowest available energies will obviously settle in these corner regions.

It is inconvenient to handle the physics of these third-zone electrons when their quantum states are in disjoint pieces of $\mathbf{k}$-space. An electron, accelerated by an electric or magnetic field, may move between one region and another, and our description of its history will contain discontinuities (at zone boundaries) that the electron knows nothing about. It is clearly preferable to make the further translations through reciprocal-lattice vectors that lead to Fig. 28.4(c). The corresponding operation with the model results in Fig. 28.7.

28.3.2 A non-empty square lattice

We are now ready to consider what happens when a periodic potential (non-empty, in real space, having period a in each of the x- and y-directions) is applied. Reciprocal space ($\mathbf{k}$-space) has already been partitioned into Brillouin zones in preparation, so there is little disruption to the picture we have formed already.

The changes resemble those drawn in Fig. 28.1. Where levels would cross in the $\mathbf{k}$-space of the empty lattice, there are "avoided crossings". These crossings are easily identified: they are the sharp ridges where two pieces of wood meet in Figs. 28.5–7, and also the v-valleys in Figs. 28.6–7. It is easy to imagine how the periodic potential "rounds off" these sharp corners: eroding the ridges and silting up the valleys. We do not provide pictures of a model that has been subjected to these roundings; it is easily imagined.

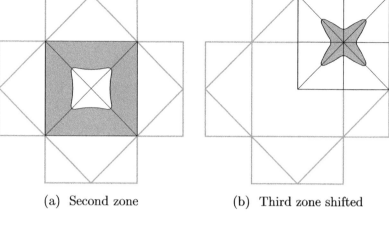

(a) Second zone   (b) Third zone shifted

Fig. 28.8 When a non-zero periodic potential is applied, quantum states whose energies were equal separate as "avoided crossings". Contours of constant energy on the k-plane become rounded, as shown here for the second zone and shifted third zone.

Erosion and silting affect the contours that you see on a geographical map, rounding off corners. The same must happen to contours of constant energy on our wooden model. In particular, the contour that is the Fermi surface must lose its sharp corners, both outward-facing and inward-facing. An impression of these changes is shown in Fig. 28.8.

The group velocity for electrons is $v_g = \partial\varepsilon/\partial k$, always normal to a surface of constant energy. In particular, at the Fermi level, v_g is always normal to the Fermi surface. The change in direction of the contours where they are smoothed profoundly affects the direction of the group velocity. In a magnetic field, electrons change momentum at the rate $(-e)v_g \times B$, so they move round the Fermi surface in k-space and follow closed orbits in real space (if not disrupted by collisions). The periodic potential completely changes the topology of these orbits, compared with those for free electrons. There are consequences for Landau levels, the de Haas–van Alphen effect, and so on.[10] Even though our model is two-dimensional, we can see from it how such dramatic changes can come about.

[10]The reader is referred to standard texts on solid-state physics, such as Kittel (ideally taking the best bits from the fourth edition (1971) and fifth edition (1976)).

28.4   Physics beyond the first dimension

Some of the phenomena that we have seen in two dimensions were already present in the one-dimensional model of Fig. 28.1. In one dimension, we had: a Brillouin-zone structure in k_x; a reduced zone scheme; avoided crossings in the $\varepsilon(k_x)$ relations; energy gaps between occupiable energy bands.[11] Here we concentrate on those phenomena that are newly demonstrated when we progress to two dimensions.

The first thing we would point out is the presence of overlapping bands, overlapping, that is, in energy. It should be easy to see that the energies of bands 1 and 2 overlap: compare the height of zone 1 at its corners with the lowest height of zone 2. Most conspicuously, the Fermi surface (Figs. 28.2, 3) extends over parts of zones 2, 3 and 4, so there is an overlap of three bands at this energy. If bands can overlap so easily

[11]The words "zone" and "band" are almost synonymous. We say "zone" when we are thinking about the quantum states available in k-space on one of the $\varepsilon(k)$ surfaces; we say "band" when we are thinking about the energies possessed by those same quantum states.

in a two-dimensional model, there should no difficulty in understanding how bands can overlap in a real three-dimensional crystal.

With some bands not full, and some not empty, the material represented by our model, having four electrons per atom in a square lattice, is a metallic conductor. Yet four electrons per atom are enough to fill two bands completely, if those bands supply the lowest energies available to electrons. A four-valent material could well be a semiconductor if the periodic potential is strong enough to remove the band overlap. And, of course, germanium and silicon are four-valent.

Refer to Fig. 28.5. The periodic potential smooths the ridges of the second zone, and in particular must lower the height of the apex. Similarly, silting must raise the bottom of the central hollow of Fig. 28.7. If the periodic potential is strong enough, it may lower the first and raise the second until there is an energy gap between them.[12] These are the conditions where the original metal is transformed into a semiconductor.

In the semiconductor, it will be possible to promote a few electrons from the top of zone 2 to the bottom of zone 3, but there will be an energy cost—the band gap—in so doing. Moreover, the top of band 2 lies at the origin of k-space, while the bottom of band 3 lies at $(\pi/a, \pi/a)$, at a zone corner.[13] The material under discussion has an **indirect band gap**, that is, one where an electron crossing the band gap, and having the least energy to cross, must be given momentum as well as energy.

Refer to Fig. 28.7. When the hollow in the middle has been smoothed by application of a periodic potential ("silting"), the energy–wave-vector relation becomes roughly parabolic. In the k_x- and k_y-directions (coordinates as in Fig. 28.2), the parabola is steep, so the electron energy is $(\hbar k)^2/2m^*$ with a rather small m^*. In the directions at 45°, the parabola is shallower, and the $\varepsilon(\boldsymbol{k})$ relation is described by a larger m^*: conduction-band electrons have an **anisotropic effective mass**, different for different directions. A similar discussion applies to the $\varepsilon(\boldsymbol{k})$ relation holding at the top of band 2. Thus valence-band holes likewise have an anisotropic effective mass.

[12] It is assumed that zone 4 will become empty as well.

[13] Actually $(\pm\pi/a, \pm\pi/a)$ and other locations, since the zone structure repeats (in the periodic zone scheme) at multiples of $2\pi/a$ in each direction.

28.5 Summary

We have discussed a two-dimensional lattice, and the energies available to electrons within it, as an intermediate step between one dimension and three. Phenomena that are commonplace in three-dimensional crystals appear naturally in our two-dimensional model, though they are absent from the introductory one-dimensional model of Fig. 28.1. We have found:

- Overlapping bands.
- The way in which bands may overlap, or not, depending on the strength of the periodic potential.
- The effect of a periodic potential on the $\varepsilon(\boldsymbol{k})$ relation for electrons in each band, with qualitative consequences for the topology of closed orbits round the Fermi surface.

- A demonstration that a complicated piece of physics (Figs. 28.5 and 28.7) can be accounted for as the result of applying simple rules: start from the empty lattice and apply a weak periodic potential together with "avoided crossings".
- A natural explanation for anisotropic effective masses possessed by conduction-band electrons and valence-band holes.
- An example of the way in which there can be an indirect band gap between valence band and conduction band.

[14] We again acknowledge Kittel for showing us the way.

Quite a good score for such a simple model.[14]

Problems

Problem 28.1 The periodic potential assumed in Fig. 28.1

Figure 28.1 shows energies appropriate to an electron in a one-dimensional periodic potential. In the units of the graph, where the parabolas cross at energies 1, 4, 9, 16, ..., the potential used in a Schrödinger equation was $1 \times \cos(2\pi x/a)$.

(1) Explain why application of the periodic potential lowers the energy for the lowest quantum state at $k = 0$.

(2) The second energy gap, correctly drawn, is only 0.12 high, and the third is 0.004 high, on the energy scale given. You may now see why I redrew those gaps in order to make them visible. Explain why a periodic potential of the magnitude given can be expected to have so little effect on quantum states at these energies.

Problem 28.2 A paradox

When we learn about band structure, we are told that sodium and copper are necessarily metals, because they have an odd number of electrons: one electron different from an inert gas. Chlorine is one electron different from an inert gas. Why is solid chlorine not a metal?

XII Electronics

29 | Negative feedback

Intended readership: After encountering enough analogue electronics to appreciate negative feedback and its advantages.

29.1 Introduction

Negative feedback is essential for almost all "control". The thing controlled might be: the output voltage of a stabilized power supply; the frequency of the oscillator in a cesium clock; the field of an electromagnet; the mirror spacing in a laser; the temperature of your house ....

When control is applied, the output (voltage, frequency, field, spacing temperature, ...) is measured and is compared with some desired value. The difference is used to apply a correction that brings the output closer to that intended value.

In this chapter, negative feedback is illustrated in the context of electronic "voltage" amplifiers.[1] It is to be understood that such circuits can be used as models, helpful for the design of all kinds of "control system".[2]

In the context of an electronic amplifier, negative feedback is applied with two aims in mind:

- to bring the ("closed-loop") gain to a required value, which includes giving it a desired dependence on frequency
- to reduce distortion (improve linearity).

One aim of this chapter is to give a more-than-usually thorough discussion: in particular of harmonic and intermodulation distortion, and of how negative feedback achieves their suppression. At the same time, we set the scene for Chapter 30, which shows how far we can go within limits set by the need for stability.

29.2 Basic definitions

A block diagram of a generic negative-feedback system is shown in Fig. 29.1. An input voltage v_{in} is to be amplified by a gain-factor G. The circuit contains an amplifying "gain block"[3] that gives voltage gain A. A fraction β of the output voltage is fed back to the beginning with such a sign as to reduce the overall gain (hence both "feedback" and "negative"). The labelling of the diagram shows that the gain block receives input $(v_{\text{in}} - \beta v_{\text{out}})$. Thus

$$v_{\text{out}} = G v_{\text{in}} = A(v_{\text{in}} - \beta v_{\text{out}}),$$

[1] That is: inputs and output are all voltages (rather than, say, currents).

[2] Remember how in Chapter 5 we found that electrical transmission lines furnish an illuminating model, and a toolbox, for all wave–boundary systems.

[3] We refer to the triangle symbol in Fig. 29.1 as the "gain block" to avoid using "amplifier" both for the amplifying component (gain A) and for the whole circuit (gain G).

Essays in Physics: Thirty-two thoughtful essays on topics in undergraduate-level physics. Geoffrey Brooker, Oxford University Press. © Geoffrey Brooker 2021. DOI: 10.1093/oso/9780198857242.003.0029

so that

$$G \equiv \frac{v_{\text{out}}}{v_{\text{in}}} = \frac{A}{1 + A\beta}. \tag{29.1}$$

Also[4]

$$v' = v_{\text{in}} - \beta v_{\text{out}} = \frac{v_{\text{out}}}{A} = \frac{G \, v_{\text{in}}}{A} = \frac{v_{\text{in}}}{1 + A\beta}. \tag{29.2}$$

A standard terminology attaches to eqns (29.1) and (29.2):

open-loop gain the voltage gain $A \equiv v_{\text{out}}/v'$ of the gain block alone
closed-loop gain the voltage gain $G \equiv v_{\text{out}}/v_{\text{in}}$ of the complete circuit
feedback fraction the fraction β of v_{out} that is fed back to the input
return difference the denominator $(1 + A\beta)$
loop gain the quantity $A\beta$ in the denominator.

To appreciate eqns (29.1–2), we need some background information.

Imperfections and uncertainties are located in the gain block, that is, in the open-loop gain A. Semiconducting components have a spread of characteristics, so that two circuits built to the same design will differ from each other. Semiconductor characteristics change quite fast with temperature, and most circuits heat up while they are running. Transistors (and FETs and vacuum valves) are inherently non-linear. And the frequency dependence of A may be quite different from that of the designed-for G.

Conversely, the feedback fraction β will be implemented with ordinary passive components (usually a combination of resistors and capacitors) that are highly linear and can be characterized accurately.

Consequently, it is desirable to minimize the effect of A on the circuit's performance.

If $|1 + A\beta|$, the (absolute value of the) return difference, is large then[5]

$$\frac{v_{\text{out}}}{v_{\text{in}}} = G = \frac{1}{\beta} \times \frac{A\beta}{1 + A\beta} \approx \frac{1}{\beta}, \tag{29.3}$$

almost independent of A; and

$$\frac{v_-}{v_{\text{in}}} = \frac{\beta v_{\text{out}}}{v_{\text{in}}} = \frac{A\beta}{1 + A\beta} \approx 1, \tag{29.4}$$

showing that v_- is made closely to "track" v_{in}.

Equation (29.3) explains why negative feedback achieves its success.[6] The circuit's overall performance, its closed-loop gain G, is now determined by β, and is hardly at all affected by A, or by uncertainties in A, or by the frequency dependence of A.

Equation (29.4) helps in later discussions, by showing what has to go on in the circuit's "innards".

29.2.1 Frequency dependence

Everything is dependent on frequency. Whatever is inside the gain block, its gain $A(\omega)$ must fall to nothing at very high frequencies. It is likely that we shall choose $\beta = \beta(\omega)$ to have a frequency dependence as well.

In this chapter and the next, voltages and currents vary with time as $e^{j\omega t}$. Therefore in circuit equations, ω always appears in the grouping

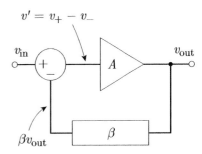

Fig. 29.1 A generic feedback system. The triangle symbol is an amplifying block ("gain block") giving voltage gain A. The rectangle takes a fraction β of the output voltage v_{out} and conveys it back to the "subtract" (or "inverting") input of the circle. The circle presents $v' = v_+ - v_- = v_{\text{in}} - \beta v_{\text{out}}$ to the input of the gain block.

[4] Some authors add βv_{out} to v_{in} at the amplifier's input instead of subtracting it, and then the denominator of G contains $(1 - A\beta)$ —which makes the name "return *difference*" seem more appropriate. This is purely a change of convention: either A or β is reversed in sign compared with ours. There is no strong reason to prefer either choice; you just have to be aware which convention is being used.

[5] When $|1 + A\beta|$ is large, the loop gain $A\beta$ must, of course, be large as well, and vice versa.

[6] Feedback systems other than voltage amplifiers have analogues of the return difference, with similar benefits resulting when that return difference is made large.

[7] This explains the modulus signs round $(1 + A\beta)$ in the setting-up of (29.3–4).

[8] Of course: an arrangement very similar to Fig. 29.1 is deliberately configured to be unstable when we intend to build an oscillator. At the present stage we simply point out that a possibly-zero denominator is, as always, a danger signal.

[9] There can be highly exceptional circuits (containing multiple feedback paths) for which different definitions of A, β are appropriate for different aspects of the circuit. One definition holds for the usual "return difference", but another gives the defect reduction of (29.8). The second of these is called "sensitivity" by Bode (1945), pp 47–8. I mention this complication for honesty only; it has no place in this chapter's exposition.

   Incidentally, the book by Bode is an intellectual tour de force. Although the book is ancient (vacuum valves are the only providers of gain), it contains almost everything you need to know about feedback, its merits, its analysis, its design rules, and the avoidance of instability.

[10] Strictly, this is the dependence upon angular frequency, rather than frequency. However, an insistence on "angular" would get in the way in the present chapter and the next. Whenever numerical values are stated, they are given in hertz, and are values of $f = \omega/2\pi$.

[11] "Error" in the sense of being wrong or undesired, not in its more usual physics sense of a quantified uncertainty (as "experimental error").

$(j\omega)$. Every quantity, such as A or β, that is frequency dependent is necessarily complex[7] (which implies phase shifting) as well.

A complex $A\beta$ may have a phase angle that is as much as $\pm\pi$ at some frequency or frequencies. Then $(1 + A\beta)$ may be zero at some frequency, raising the possibility that the feedback system may be unstable.[8] Unfortunately, the conditions for stability are not as obvious as one might hope, and we address them in Chapter 30.

29.3   Negative feedback: definition

Given that $A\beta$ depends upon frequency, and necessarily is a complex quantity, it is not obvious just what it means to have "negative" feedback. We need a formal definition.

I adopt the following:

Feedback is **negative** if $\begin{cases} \text{the circuit is stable} & \text{AND} & |G| < |A|\,; \\ \text{equivalently } |1 + A\beta| > 1. \end{cases}$

Feedback is **positive**   otherwise.

It is entirely possible for feedback to be negative at some frequencies, and positive at others, in the same circuit; see e.g. Fig. 30.6.

The definition above has been set up so that "negative" feedback is identical with "defect-reducing" feedback. The equivalence between "negative" and "defect-reducing" is established in § 29.6.[9]

I assert, in particular, that feedback can be negative even if $A\beta$ is pure imaginary, or negative. $A\beta = 999$, $A\beta = -1000j$ and $A\beta = -1001$ all give $|1 + A\beta| = 1000$, and all are equally effective at giving tight control—defect reduction—provided that the system is stable.

29.4   Effect of feedback on the frequency dependence of gain

Write $\gamma(\omega) = 1/\beta(\omega)$, where ω is angular frequency. Then $\gamma(\omega)$ represents the gain that we intend the circuit to provide when the return difference is large. Use $A_\omega = dA/d\omega$ as an abbreviation (in this section only) for the frequency dependence[10] of A, and similarly for other quantities. Then (problem 29.1)

$$\frac{G_\omega}{G} = \frac{\gamma_\omega}{\gamma} + \frac{A_\omega/A - \gamma_\omega/\gamma}{1 + A\beta}. \tag{29.5}$$

We can see the second term here as the error[11] in the frequency dependence of the amplifier, the extent to which A_ω/A affects the frequency dependence, and we can see this error being diminished by the factor $1/(1 + A\beta)$. Of course, the denominator is $1 + A(\omega)\beta(\omega)$, where A and β are evaluated at the frequency of interest.

29.4.1 An example of frequency dependence

Here is a standard tutorial example illustrating how negative feedback makes the closed-loop gain $G(\omega)$ follow $1/\beta(\omega)$, insensitive to the behaviour of $A(\omega)$, albeit over a range of frequencies ($\omega \lesssim \omega_c$) only. The amplifier in Fig. 29.2 has building blocks described by:[12]

$$A(\omega) = \frac{A_0}{1 + \mathrm{j}\omega/\omega_0}, \qquad \beta(\omega) = \text{constant}. \qquad (29.6)$$

The closed-loop gain (problem 29.2) is

$$G = \frac{A_0}{1 + A_0\beta + \mathrm{j}\omega/\omega_0} = \frac{G_0}{1 + \mathrm{j}\omega/\omega_c}, \qquad (29.7)$$

in which $\quad G_0 = \dfrac{A_0}{1 + A_0\beta} \quad$ and $\quad \omega_c = (1 + A_0\beta)\omega_0$.

In the example of Fig. 29.2, the open-loop gain of (29.6) is $A_0 = 10^4$ at zero frequency, changing smoothly to $A_0\omega_0/(\mathrm{j}\omega)$ from a **corner frequency** $\omega = \omega_0$. The feedback fraction $\beta = 1/100$. The closed-loop gain is $10^4/101 \approx 99$ at zero frequency, and remains controlled by $\beta(\omega)$ (which here is constant with frequency but it won't always be) up to the greatly raised corner frequency $\omega_c = 101\omega_0$, above which $|G(\omega)|$ falls along the same asymptote as the open-loop gain $|A(\omega)|$.

A simple rule emerges from Fig. 29.2. On a log–log graph draw the asymptotes for $|A|$ and for $|1/\beta|$. The closed-loop gain $|G|$ follows the lower of these asymptotes;[13] its curve can be sketched in by "rounding the corners", a procedure that may well be accurate enough for us.[14]

[12] As mentioned in § 29.2.1, the open-loop gain $A(\omega)$ must fall to nothing at high frequencies. Equation (29.6) gives one way in which this may happen, a model frequency dependence chosen for simplicity rather than for any realism. Nevertheless, most operational amplifiers are manufactured to have gain varying as (29.6). It is assumed that the user will apply feedback round the op-amp, and then (29.6) makes easy the achievement of stability.

[13] It should be near-obvious that the rule stated here holds for any frequency dependence of $\beta(\omega)$, even though it is the outcome of looking at a special tutorial example. Problem 29.2 asks you to confirm this.

[14] See Example 30.1 for an investigation of "accurate enough".

The display in Fig. 29.2 is of $|A|$ and $|G|$, that is, of each magnitude without regard to phase. Nevertheless, the frequency dependences of $|A|$ and $|G|$ necessarily imply phases possessed by A and G, as is indicated by the $\mathrm{j}\omega$ in (29.6–7). See also problem 29.2.

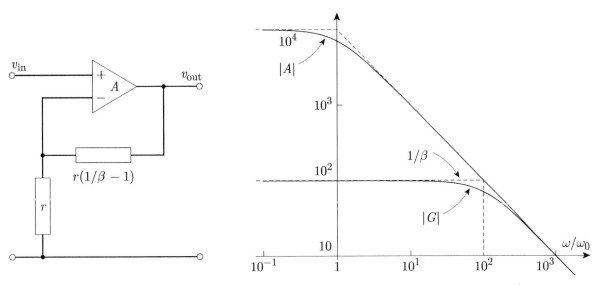

Fig. 29.2 A simple circuit, showing how feedback determines the frequency dependence of closed-loop gain $G(\omega)$. On the left is shown a gain block with "series feedback" round it. The feedback fraction $\beta = 10^{-2}$, frequency independent and therefore real. The gain block has open-loop gain $A(\omega)$ given by (29.6); the resulting closed-loop gain $G(\omega)$ is given by (29.7). On the right is a log–log plot of $|A(\omega)|$, $|1/\beta|$ and $|G(\omega)|$ versus angular frequency ω for a particular case where $A_0 = 10^4$, $1/\beta = 100$. Such a log-log plot is called a **Bode plot**. The properties of a Bode plot are discussed in more detail in § 30.3.

This rule shows how we may understand the frequency dependence for a newly-encountered circuit. And, given only a little experience, can guide us in the design of a feedback system.

29.5   Design procedure

We are to design a circuit to do a job. The circuit must give a specific gain $G(\omega)$ having a specified frequency dependence, and it has to deliver a specified quality (probably meaning a non-linearity subject to some upper limit). And, of course, it must be stable. As soon as we know the gain $G(\omega)$ required, we know what $\beta(\omega)$ must be. Whatever we do to adjust the circuit design to achieve stability or quality, there must be no tinkering with $\beta(\omega)$: meeting the specification is sacrosanct.[15]

- Fix $\beta(\omega)$, at least within the frequency range within which signals are to be handled, so as to give the required closed-loop gain $G(\omega)$.

After that,

- Choose $A(\omega)$ so as to give as large a return difference $(1 + A\beta)$ as brings circuit defects within tolerance. How large A can be made, and how it must depend on frequency, will be determined by the need for stability (Chapter 30).

It is easy to be misled by our own words. We may hear that quality is achieved by applying "lots of feedback". It is all too easy to misunderstand this as saying that we need a large β. It does not:[16] it means a *large return difference* $|1 + A\beta|$. And since β is fixed, the only way to enhance this is to increase the open-loop gain A. This is easy and cheap to do: add the stage giving gain a_1 in Fig. 29.3.

Take a numerical example. Suppose that we wish to build an audio-frequency amplifier to drive $100\,\mathrm{W}$ into an $8\,\Omega$ loudspeaker: the circuit must be capable of supplying a sinewave output voltage that ranges between $-40\,\mathrm{V}$ and $+40\,\mathrm{V}$ and a current between $-5\,\mathrm{A}$ and $+5\,\mathrm{A}$. This is the responsibility of the power-output stage, represented in Fig. 29.3 by the gain block whose voltage gain is a_2.

Power transistors have limited high-frequency response, and their gain is likely to be non-linear—just because the signal voltages are large. Suppose that the power stage has voltage gain a_2 (differential gain, a

[15]This isn't quite true. Suppose we intend to amplify signals at frequencies of $20\,\mathrm{kHz}$ and below. At the same time, stability requires that we understand and control the loop gain at frequencies up to $1\,\mathrm{MHz}$ or more. Then $\beta(\omega)$ is sacrosanct at frequencies below $20\,\mathrm{kHz}$, but we do have some opportunity to tinker with it at higher frequencies.

[16]When I was a student, it was common for books and lecturers to explain the merits of negative feedback by treating A as given (at a frequency of greatest interest), and then (notionally) adjusting β —"lots of feedback" literally. It was thereby shown that improved quality was achievable, though with a penalty of reduced G. Such a mindset, if it was ever helpful, is now long-obsolete.

It is possible that this unfortunate thinking was encouraged by introductory statements by Bode (1945) pp 32–3. When valves (vacuum tubes) were the only source of gain, the circuit designer tended to think of the number of valves in his circuit as fixed (at the least number possible), and tried to optimize within that limitation. Bode's thinking was as much influenced by that equipment limitation as everybody else's.

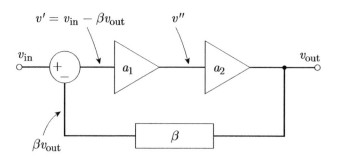

Fig. 29.3 An amplifier intended for supplying power into some load, perhaps a loudspeaker. Power is delivered to the output by the second gain block, whose voltage gain is a_2. Another gain block, having voltage gain a_1, may be placed before the power-output stage in order to provide an augmented return difference.

defined in the next section) somewhere between 100 and 400. At worst this stage requires an input voltage v'' of 0.4 V. A preceding stage, if we choose to supply one, needs to deliver only 0.4 V peak at its output. Such an added gain block, indicated in Fig. 29.3 as giving gain a_1, offers relatively few challenges to the designer, either in frequency response or linearity.

Adding gain a_1 in this way is almost penalty-free,[17] in cost, complexity, or additional non-linearity, and is entirely practical—up to the limit set by the need for stability.

29.6   Improvement of linearity

One reason why we may wish to implement negative feedback is to improve the linearity of an amplifying circuit. We next show that this can indeed happen.

29.6.1   Non-linearity: a tutorial example

Suppose that the gain block (thought of now as a single block after the fashion of Fig. 29.1) has an input–output relation such as that shown in Fig. 29.4. For simplicity we assume in this subsection that there is no frequency dependence in A, β or G; all three can therefore be taken to be real.

Consider the differential gain, meaning $\mathrm{d}v_\mathrm{out}/\mathrm{d}v'$, where v' is the voltage at the input of the gain block, as indicated in Fig. 29.1. We may call this A, understanding that it varies according to where we are along the curve of Fig. 29.4. The algebra of (29.1) applies, and the closed-loop differential gain $G \equiv \mathrm{d}v_\mathrm{out}/\mathrm{d}v_\mathrm{in} = A/(1 + A\beta)$. Then

$$\frac{1}{G} = \frac{1}{A} + \beta, \qquad \text{so} \qquad \frac{\delta G}{G^2} = \frac{\delta A}{A^2},$$

and

$$\frac{\delta G}{G} = \frac{\delta A}{A} \times \frac{G}{A} = \frac{\delta A}{A} \times \frac{1}{1 + A\beta}. \tag{29.8}$$

The fractional variations of G are smaller than those of A by the familiar factor $1/(1 + A\beta)$. The feedback circuit (gain G) can have a greatly improved linearity, as compared with the "naked" gain block (gain A).

The result just obtained is encouraging, but not as persuasive as we should like, because of the drastic simplifying assumption that A and β are both real. A practical circuit is very likely to be required to amplify at frequencies for which A or β or both are complex. This may be because our specification demands a frequency dependence in the closed-loop gain G, or because stability requires A to fall at frequencies well below the greatest "signal" frequency that we intend to amplify. Either way, the loop gain $A(\omega)\beta(\omega)$ is likely to be complex, even to be pure imaginary or to have a negative real part. What effect does this have on the linearity of the circuit? The reasoning given above doesn't tell us; we must do better.

[17] This is a statement made in the semiconductor age. It did not seem like that at the time when Bode's book was written. Then, augmenting gain A meant adding an extra valve: more chassis space, more heat, and more to go wrong. When an improved quality was demanded, those penalties had to be faced, but always with great reluctance.

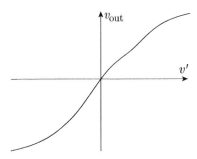

Fig. 29.4 A real amplifying block (the triangle in Fig. 29.1) is likely to have a non-linear relation between input voltage v' and output voltage v_out. The graph illustrates a possible non-linear relation.

29.6.2 Non-linearity: allowance for frequency dependence

Non-linearity and frequency dependence are now both in contemplation. There are phenomena that always turn up in such circumstances: the generation of harmonics and of sum and difference frequencies.

Suppose the amplifier is fed with an input signal v_{in} that is a sinewave at angular frequency ω. Non-linearity results in the generation of unwanted new frequencies that are harmonics of ω: (angular) frequencies 2ω, 3ω, 4ω, and so on. If negative feedback does what we hope, it will suppress the amplitudes of these new frequencies.

A further phenomenon shows up if we apply an input consisting of two (or more) frequencies, such as $v_{\text{in}} = v_1 \cos(\omega_1 t) + v_2 \cos(\omega_2 t)$. Non-linearity now generates angular frequencies $(\omega_1 + \omega_2)$, $|\omega_1 - \omega_2|$ and higher sums and differences, as well as $2\omega_1$, $2\omega_2$ and so on. As with harmonics, we wish feedback to suppress these unwanted frequencies, and we need to know how this is best achieved.

There are many possible non-linear functions that might describe v_{out} as a function of v_{in} and ω, and it is not feasible to handle such things with complete generality. Since the signature of non-linearity is the generation of new frequencies, I shall use the simplest model that achieves this: a term in A proportional to the square of the input voltage v'. Even with this simplest case, the algebra is somewhat heavy, and it is relegated to problems 29.5 (for harmonic generation) and 29.6 (for sum and difference frequencies).

A simple finding comes out from the (admittedly model) systems analysed in problems 29.5 and 29.6:

- Non-linearity generates frequencies that were not present in the input signal. The voltage amplitude of a new-and-unwanted angular frequency Ω is reduced by the feedback by the factor[18]

$$\frac{\text{reduction of}}{\text{voltage at } \Omega} = \frac{1}{|1 + A(\Omega)\,\beta(\Omega)|} = \frac{1}{|\text{return difference at } \Omega|}, \quad (29.9)$$

evaluated, that is, at the unwanted angular frequency Ω.

Finally, we should remark, and remark strongly, that negative feedback is not a panacea for distortion reduction. The best way to reduce distortion is to do it "at source" by paying careful attention to the performance of the gain block. Moreover, feedback-circuit design needs some care, because it can happen that for part of the time a transistor (often a small-signal transistor near the beginning of the gain block) is turned off altogether (because $v' = v_{\text{in}} - \beta v_{\text{out}}$ exceeds what it can handle);[19] feedback can do nothing to linearize the gain when $A = 0$.

29.7 Series and shunt feedback

There are two circuit topologies that can implement the feedback system that so far we have represented by Fig. 29.1. These are **series feedback** and **shunt feedback**. We must show that both implement (29.5, 8, 9).

[18] I have set out the mathematics in problems 29.5–6, even though the algebra is rather heavy, because of the importance of the finding bulleted here.

Negative feedback was a fashionable topic in the 1960s and '70s because amateurs enjoyed trying to build audio amplifiers with superb linearity. (This explains an unashamed emphasis in this chapter on the frequency range from zero to 20 kHz.) It was common for "experts" to make incorrect statements, and to mis-design accordingly. A common such mis-statement was "feedback isn't negative unless it's negative", the suggestion being that if $A\beta$ is imaginary then it does nothing to reduce distortion even if it's large. The words sound plausible enough to make us uneasy, and are the more in need of debunking for that.

Lest the reader gain an impression that harmonic generation and sum/frequency generation are the specialist concerns of an audiophile, I should mention that these same phenomena constitute the major part of non-linear optics. There, non-linearity is used constructively: the new frequencies are the desired outputs.

[19] A particular trap for the designer is "slew-rate limiting". A capacitor in the gain block controls the frequency dependence of the open-loop gain, and is put there as one of the measures needed for making the feedback loop stable. There is a greatest current that can be supplied for charging or discharging this capacitor, and this in turn sets an upper limit to $|\mathrm{d}v_{\text{capacitor}}/\mathrm{d}t|$, a quantity known as the **slew rate**. If the input signal is such as to demand a greater slew rate than can be supplied, then a large v' is generated as the circuit "tries" to correct an error that it is incapable of correcting.

29.7.1 Series feedback

Figure 29.5 represents a circuit in which there is a differential-amplifier gain block giving voltage gain $A(\omega)$ at angular frequency ω. The $+$ and $-$ inputs are "non-inverting" and "inverting", meaning that

$$v_{\text{out}} = A(v_+ - v_-);$$

voltages v_+, v_- receive the same gain apart from a sign. The impedances Z_1, Z_2 are connected so as to give "series feedback". It is understood that everything in this is frequency dependent.

Problem 29.3 shows that

$$G = \frac{v_{\text{out}}}{v_{\text{in}}} = \frac{A}{1 + A\,Z_1/(Z_1 + Z_2)}, \tag{29.10}$$

in which we may easily identify A with the open-loop gain and likewise identify $Z_1/(Z_1 + Z_2)$ with the feedback fraction β.

Notice also that

$$v_{\text{in}} - v_- = \frac{v_{\text{in}}}{1 + A\beta}, \tag{29.11}$$

in agreement with (29.2); when $|1 + A\beta| \gg 1$, the difference between v_- and v_{in} is much smaller than either. We may think of the circuit as forcing v_- to "track" the input voltage v_{in}. Once this is appreciated, it should be easy to "eyeball" that the circuit gives closed-loop gain $(Z_1 + Z_2)/Z_1$.

Negative "voltage" feedback reduces the circuit's output impedance, from that of the naked gain block, by factor $1/(1 + A\beta)$ (problem 29.3 for series feedback, problem 29.4 for shunt feedback): the output voltage is forced to be a faithfully enlarged replica of the input, hardly disrupted by the drawing of current into a load. The output voltage is being held under control—just what (voltage) negative feedback is for.[20]

Problem 29.3 also shows that series feedback can have the effect of increasing the input impedance of the circuit, above that of the naked gain block, by factor $(1 + A\beta)$; this property is specific to series feedback.

29.7.2 Shunt feedback

Problem 29.4 rehearses the analysis of the circuit in Fig. 29.6:

$$\text{closed-loop gain} = G \approx -\frac{Z_2}{Z_1}, \qquad \text{loop gain} = \alpha\,\frac{Z_1}{Z_1 + Z_2}. \tag{29.12}$$

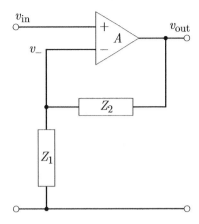

Fig. 29.5 An outline circuit in which the gain block gives open-loop voltage gain $A(\omega)$ and has "series feedback" $\beta = Z_1/(Z_1 + Z_2)$. The closed-loop gain

$$G \approx \frac{1}{\beta} = \frac{Z_1 + Z_2}{Z_1}.$$

[20] In this chapter, the circuits are arranged to have "voltage feedback", whether series or shunt, meaning that information about the output *voltage* is fed back, as βv_{out}, to the input. However, there are other possibilities. For example, one can feed back information about the *current* drawn into the load, and then we have "current feedback", resulting in an amplifier whose current output is held under control. A "current amplifier" has output impedance that is increased, rather than reduced, by its feedback.

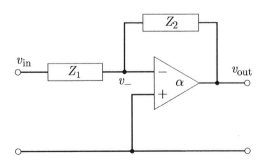

Fig. 29.6 An outline circuit in which an amplifier has "shunt feedback" provided by impedances Z_1 and Z_2. The closed-loop gain is

$$G \approx -Z_2/Z_1,$$

a property that should be obvious once it is realised that voltages "pivot" about a near-zero (often called a "virtual earth" or "virtual ground") at v_-. This is a special case of "tracking", where v_- tracks $v_+ = 0$.

However, there is more than one way of partitioning the loop gain into A and β, and two such decompositions are displayed in problem 29.4.

The voltage at the gain block's inverting input is

$$\frac{v_-}{v_{\text{in}}} = \frac{Z_2}{Z_1 + Z_2}\frac{1}{1 + A\beta}, \text{ which will usually be} \ll 1,$$

in line with the similar finding (29.11) for series feedback. This accounts for the common naming of the gain block's inverting input as a "virtual earth".

The circuit of Fig. 29.6 is sometimes called a "see-saw" because the voltages v_{in} and v_{out} "pivot" about $v_- \approx 0$. This makes it possible to see, without mathematics, how the circuit is intended to behave.

29.8   Summary

The definition given in § 29.3 makes "negative" feedback identical with "defect-reducing" feedback.

An imperfection, whether it's a departure of gain $G(\omega)$ from intention at angular frequency ω, or an unwanted angular frequency Ω generated by non-linearity, is suppressed by the factor $|1 + A(\Omega)\beta\Omega)|^{-1}$ — evaluated at that frequency Ω. The same factor appears in a reduction of the circuit's output impedance.

The closed-loop gain $G(\omega)$, and with it the feedback fraction $\beta(\omega)$, is set by the specification that the amplifier is required to meet. To suppress imperfections, we use the only remaining freedom by making $A(\omega)$, and with it $|1 + A(\omega)\beta(\omega)|$, as large as is practically possible. This point was made in § 29.5.

Series and shunt feedback configurations alike confer the advantages of negative feedback, because those advantages all follow from having a gain G of the form (29.1).

Problems

[21] *Hint:* It's simplest if you start from

$$\frac{1}{G} = \frac{1}{A} + \beta.$$

Problem 29.1 How negative feedback controls frequency response
Obtain (29.5) by differentiating (29.1).[21] Discuss G_ω/G in the two limiting cases $|1 + A\beta| \gg 1$ and $|A\beta| \ll 1$.

Problem 29.2 Negative feedback and frequency dependence
Confirm the algebra in § 29.4.1. Make sure you are comfortable with the interpretation given of (29.6–7) and Fig. 29.2.

In § 29.4.1 we give the rule "follow the lower asymptote". Show that this rule follows from $G = 1/(1/A + \beta)$ because G is dominated by the larger term in the denominator, whichever that is.

Discuss the phase of the output voltage $v_{\text{out}}(\omega)$ relative to that of a sinewave input; that is, look at $\phi_{\text{closed-loop}}$ in $G(\omega) = |G(\omega)|\,\mathrm{e}^{-\mathrm{j}\phi_{\text{closed-loop}}}$. Compare with $\phi_{\text{open-loop}}$ defined by $A(\omega) = |A(\omega)|\,\mathrm{e}^{-\mathrm{j}\phi_{\text{open-loop}}}$. Observe the way that the phases relate to the corner frequencies of Fig. 29.2. Figure 30.4 may help in showing what to look for.

Problem 29.3 Series feedback

Confirm the equations given in § 29.7.1. Assume for now that any current taken into the differential amplifier's two inputs is negligible beside that carried in the potential divider formed by Z_1 and Z_2.

In fact, the gain block of Fig. 29.5 is likely to draw small currents into its two inputs. Suppose that the important such current[22] can be attributed to an effective resistance r_{in} ($\gg Z_1, Z_2$, hiding inside the gain block and bridging the $+$ and $-$ inputs). Use (29.11) to show that series feedback raises the input impedance of the circuit from r_{in} to $r_{in}(1+A\beta)$.

Imagine that the gain block in Fig. 29.5 has output impedance z_{out}; think of z_{out} as a small resistor hiding inside the triangle just before the right-hand apex (after the amplification). Attach a load R_L between the actual output and the earth rail, and handle the circuit analysis by assuming that $R_L \ll (Z_1 + Z_2)$: current taken from the output into the voltage divider can be neglected beside that taken into the load. Show that the effect is to multiply the closed-loop voltage gain by

$$\frac{R_L}{R_L + z_{out}/(1 + A\beta)}.$$

Interpret this as showing that the output impedance has been reduced from z_{out} by the factor $1/(1 + A\beta)$.

[22] There may also be an impedance from v_+ to earth. This is not increased by the feedback arrangement, and may be the more important because it remains when r_{in} has been made negligible. A similar impedance from v_- to earth is shunted by Z_1 and is likely to be insignificant.

Problem 29.4 Shunt feedback

Use Fig. 29.6. The shunt-feedback amplifier is inherently inverting, so we build that fact into the mathematical description. Show that[23]

$$v_- = v_{in}\frac{Z_2}{Z_1 + Z_2} + v_{out}\frac{Z_1}{Z_1 + Z_2}.$$

Show that v_{out} can be expressed in the following forms:

$$G = \frac{v_{out}}{v_{in}} = \frac{Z_2}{Z_1 + Z_2} \times \frac{(-\alpha)}{1 + (-\alpha)\left(\frac{-Z_1}{Z_1+Z_2}\right)}, \quad (29.13)$$

$$G = \frac{v_{out}}{v_{in}} = \frac{\left(-\alpha \frac{Z_2}{Z_1+Z_2}\right)}{1 + \left(-\alpha \frac{Z_2}{Z_1+Z_2}\right)\left(\frac{-Z_1}{Z_2}\right)}. \quad (29.14)$$

Interpret eqn (29.13) as saying that:

- the signal is first attenuated by the factor $Z_2/(Z_1 + Z_2)$
- the signal is then amplified in an inverting[24] feedback amplifier whose open-loop gain is $A = -\alpha$ and whose feedback fraction is set by the potential divider to $\beta = -Z_1/(Z_1 + Z_2)$.

Interpret eqn (29.14) as saying that:

- the signal is amplified in an inverting amplifier whose open-loop gain is $A = -\alpha Z_2/(Z_1+Z_2)$ and whose feedback fraction is $\beta = -Z_1/Z_2$.

With either of the interpretations (29.13–14):[25]

$$\text{loop gain} = A\beta = \alpha\frac{Z_1}{Z_1 + Z_2}.$$

[23] *Suggestion*: Since the circuit equations are linear, you are allowed to obtain v_- by applying to the potential divider first v_{in} with $v_{out} = 0$, and then v_{out} with $v_{in} = 0$. This trick means that you can "eyeball" the expression for v_- without needing to do any manipulation.

[24] By calling the amplifier "inverting" we may give ourselves licence to drop the negative signs in A and β. I have not made use of this freedom.

[25] It is important for stability analysis that there is no ambiguity as to what is the loop gain.

Also, when the loop gain $A\beta$ is large, the closed-loop gain G approximates to $G \approx -Z_2/Z_1$. We confirm (29.12).

It may seem surprising that two interpretations are available here, but we have to remember that one expression for closed-loop gain G is being used to determine two quantities A and β, so their determination cannot be unique mathematically, whether or not it seems obvious physically.[26]

More importantly, the similarity of (29.13–14) to (29.1) tells us that the generic circuit configuration of Fig. 29.1 is applicable to shunt feedback, as well as to series, except for an inversion.

Show that the output impedance of the circuit of Fig. 29.6 is reduced from that of the naked gain block by a factor $1/(1 + A\beta)$.

Show that the input impedance of the circuit is (neglecting the input impedance of the gain block, that is, treating it as infinite)[27]

[27] Aficionados may recognize the factor $(1+\alpha)^{-1}$ here as akin to a similar factor in the Miller effect.

$$(Z_{\text{in}})_{\text{circuit}} = Z_1 + \frac{Z_2}{1+\alpha}.$$

Problem 29.5 The effect of feedback on harmonic distortion[28]
A block diagram of the amplifying circuit is given in Fig. 29.7.

The circuit of Fig. 29.7 is fed with a sinusoidal input voltage[29]

[28] This problem may look a bit intimidating. However, its findings are important and must be free from doubt.

$$v_{\text{in}} = \frac{v_0}{2}\, e^{j\omega t} + \frac{v_0^*}{2}\, e^{-j\omega t}. \tag{29.15}$$

We represent the imperfect gain block (the triangle in Fig. 29.1) as an ideal linear amplifier providing all the gain $A(\omega)$, followed by a non-linear element. It is realistic to locate the non-linearity at or near the output, since it is where the voltages are largest that it is most difficult to hold the circuit behaviour near-linear.

[29] We use a complex notation for all the usual reasons. At the same time, voltage v_{in} must be written as a real expression because we are going to square it.

The peak value of v_{in} is to be $|v_0|$, which explains why the coefficients are divided by 2, even though this slightly clutters the algebra. Similarly for the amplitude C in (29.16).

A voltage v appears at the place marked on the circuit diagram of Fig. 29.7. Let that voltage contain the sinusoidal component

$$v = \frac{C}{2}\, e^{j\omega t} + \frac{C^*}{2}\, e^{-j\omega t}, \tag{29.16}$$

together with distortion products to be found.[30]

[30] We shall keep $|C|$ roughly constant throughout the analysis, so that the non-linearity is "driven" by v to about the same extent. For this reason, other voltages in the circuit will usually be expressed in terms of C.

We model the nonlinearity[31] by the simplest possibility: a quadratic addition: $v_{\text{out}} = v + qv^2$. However, just as $A(\omega)$ and $\beta(\omega)$ are frequency dependent, so q may depend on the frequencies being combined. Thus

[31] This choice was introduced in § 29.6.2, as the simplest non-linearity that gives rise to new frequencies.

Fig. 29.7 The non-linear amplifier discussed in problems 29.5 and 29.6. The non-linearity is represented as located after a perfect amplifier that provides all of the required voltage gain A. The non-linearity makes a small addition to v_{out} proportional to the square of v; a graph of v_{out} versus v is therefore less wavy than that drawn in the rectangle.

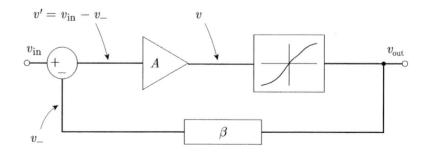

$$\text{``}qv^2\text{''} = q(\omega, \omega)\left(\frac{C}{2}\,\mathrm{e}^{\mathrm{j}\omega t}\right)\left(\frac{C}{2}\,\mathrm{e}^{\mathrm{j}\omega t}\right)$$

$$+ q(-\omega, -\omega)\left(\frac{C^*}{2}\,\mathrm{e}^{-\mathrm{j}\omega t}\right)\left(\frac{C^*}{2}\,\mathrm{e}^{-\mathrm{j}\omega t}\right)$$

$$+ 2q(\omega, -\omega)\left(\frac{C}{2}\,\mathrm{e}^{\mathrm{j}\omega t}\right)\left(\frac{C^*}{2}\,\mathrm{e}^{-\mathrm{j}\omega t}\right).$$

Here the two arguments of q are the two frequencies that are being added.[32] I shall save writing by omitting these arguments, and the reader is invited to restore them as he works this problem.

[32] There are requirements on function $q(\,)$ in order that the expression written should be real. There is no need to investigate those restrictions here.

The $\mathrm{e}^{\mathrm{j}2\omega t}$ term in v_{out} is affected by the feedback: it is multiplied by $B(2\omega)$ to give v_-, is subtracted from v_{in}, is multiplied by $A(2\omega)$, and is brought back to v. This explains how new frequencies appear in v. Other frequencies are affected similarly, so that (29.16) is modified to:

$$v = B\,\mathrm{e}^{\mathrm{j}0t} + \frac{C}{2}\,\mathrm{e}^{\mathrm{j}\omega t} + \frac{D}{2}\,\mathrm{e}^{\mathrm{j}2\omega t} + \frac{E}{2}\,\mathrm{e}^{\mathrm{j}3\omega t} \qquad (29.17)$$

+ higher-frequency terms (omitted) + complex conjugates of all terms written except that at zero frequency. We find the coefficients here, in terms of C, by following voltages round the feedback path, as outlined above, until they meet at v again.

We might expect that the procedure just outlined would set up for an iteration, gaining one order of accuracy for each passage round the feedback loop, but the calculation does not seem to lend itself to such a procedure. Instead, we require that v is recovered unchanged after voltages are "chased round the loop".

As the first step, we find v_{out}.

$$v_{\text{out}} = v + qv^2$$

$$= \left(B + qB^2 + q\frac{CC^*}{2} + q\frac{DD^*}{2} + q\frac{EE^*}{2}\right)\mathrm{e}^{\mathrm{j}0t}$$

$$+ \left(\frac{C}{2} + qBC + q\frac{DC^*}{2} + q\frac{ED^*}{2}\right)\mathrm{e}^{\mathrm{j}\omega t} \qquad + \text{c.c.}$$

$$+ \left(\frac{D}{2} + qBD + q\frac{C^2}{4} + q\frac{EC^*}{2}\right)\mathrm{e}^{\mathrm{j}2\omega t} \qquad + \text{c.c.}$$

$$+ \left(\frac{E}{2} + qBE + q\frac{CD}{2}\right)\mathrm{e}^{\mathrm{j}3\omega t} \qquad + \text{c.c.}$$

$$\approx \left(B + q\frac{CC^*}{2} + q\frac{DD^*}{2} + q\frac{EE^*}{2}\right)\mathrm{e}^{\mathrm{j}0t}$$

$$+ \frac{C}{2}\,\mathrm{e}^{\mathrm{j}\omega t} \qquad + \text{c.c.}$$

$$+ \left(\frac{D}{2} + q\frac{C^2}{4} + q\frac{EC^*}{2}\right)\mathrm{e}^{\mathrm{j}2\omega t} \qquad + \text{c.c.}$$

$$+ \left(\frac{E}{2} + q\frac{CD}{2}\right)\mathrm{e}^{\mathrm{j}3\omega t} \qquad + \text{c.c.} \quad (29.18)$$

Treat q as first-order small, and find reasons why the terms omitted in (29.18) are negligible. There are, in fact, three displayed terms that

can also be seen to be negligible, but it may be best to retain them an
let their smallness be revealed by the algebra.

It should be obvious what is going on in (29.18). You can see th
square of $e^{j\omega t}$ rectified to give a term at zero frequency, and also gene
ating its second harmonic. You can see $e^{j\omega t}$ multiplied by $e^{j2\omega t}$ to giv
the sum frequency. And you can interpret the other terms similarly.

Next, we chase voltages from v_{out} round the feedback loop back to
Show that outcome is:

$$
\begin{aligned}
v = &-\left(B + q\frac{CC^*}{2} + q\frac{DD^*}{2} + q\frac{EE^*}{2}\right) A(0)\beta(0)\, e^{j0t} \\
&+ \left(\frac{v_0}{2} - \frac{C}{2}\beta(\omega)\right) A(\omega)\, e^{j\omega t} && +\text{c.c.} \\
&- \left(\frac{D}{2} + q\frac{C^2}{4} + q\frac{EC^*}{2}\right) A(2\omega)\beta(2\omega)\, e^{j2\omega t} && +\text{c.c.} \\
&- \left(\frac{E}{2} + q\frac{CD}{2}\right) A(3\omega)\beta(3\omega)\, e^{j3\omega t} && +\text{c.c.} \quad (29.19
\end{aligned}
$$

For consistency, this expression for v must agree with (29.17). Equat
coefficients and obtain four equations relating v_0, B, D, E to C. Us
these equations to show that $E \sim qCD$ so that $E \ll C$ by at least on
factor q. Also, show that $EC^* \ll C^2$ so that $D \sim qC^2$. Then als
$DD^* \ll CC^*$ and $EE^* \ll CC^*$. These findings permit removal of thre
terms from your four equations.[33]

Substitute your findings into (29.18) and show that

$$
\text{linear gain} = \frac{C}{v_0} = \frac{A(\omega)}{1 + A(\omega)\beta(\omega)}. \quad (29.20
$$

This is what we know holds for a linear circuit, and we find that it
unaltered (to first order in q). Also show that

$$
\begin{aligned}
v_{\text{out}} = &\; q(\omega, -\omega)\frac{CC^*}{2}\frac{1}{1 + A(0)\beta(0)}\, e^{j0t} \\
&+ \frac{C}{2}\, e^{j\omega t} && +\text{c.c.} \\
&+ q(\omega, \omega)\frac{C^2}{4}\frac{1}{1 + A(2\omega)\beta(2\omega)}\, e^{j2\omega t} && +\text{c.c.} \\
&+ q(\omega, 2\omega)\frac{CD}{2}\frac{1}{1 + A(3\omega)\beta(3\omega)}\, e^{j3\omega t} + \text{c.c.} && (29.21
\end{aligned}
$$

Look at the terms in (29.21) having frequencies $0, 2\omega, 3\omega$. Argue tha
each is the product of two factors: what it would be in the presence c
the non-linearity but without feedback; and a reduction factor given b
the (return difference)$^{-1}$. The negative feedback does precisely what w
wish, in reducing the voltages at distortion frequencies.[34] At the sam
time, we confirm that the suppression factor for each such frequency
the return difference *at that "new" frequency.*

[33] These are the three terms forecast to be negligible. Try to find physics reasons why they should be negligible by looking at the way those terms combine frequencies.

[34] The return difference is likely to be complex, and therefore phase shifting, though this is likely to be of little or no interest. More important is the reduction of magnitude by the factor $|1 + A\beta|^{-1}$. We have a refutation of the thought in sidenote 18 that feedback might be ineffective at improving linearity unless $A\beta$ is real.

Comment on the third harmonic

In (29.21), the coefficient D is given by

$$D = -q(\omega, \omega)\frac{C^2}{2}\frac{A(2\omega)\beta(2\omega)}{1 + A(2\omega)\beta(2\omega)}$$

(show this). Then the third-harmonic term in v_{out} is more fully

$$-q(\omega, \omega)q(\omega, 2\omega)\frac{C^3}{4}\frac{A(2\omega)\beta(2\omega)}{1 + A(2\omega)\beta(2\omega)}\frac{1}{1 + A(3\omega)\beta(3\omega)}\,\mathrm{e}^{\mathrm{j}3\omega t}. \quad (29.22)$$

We have carried the newly generated frequencies as far as 3ω because this is the simplest of the second-order (q^2) terms, and in order to show how such higher-order terms can arise. Signals have to "pass twice" through the non-linear element before they can generate this term.

If there were no negative feedback, $\beta(2\omega) = 0$, there would be no term at angular frequency 3ω, so we have the unexpected finding that this frequency is added to the output because of the feedback: the circuit no longer has a purely quadratic distortion once feedback is applied. This is correct, but not very interesting. Few non-linearities are accurately quadratic, and if v_{out} contains a term in v^3 this too will generate 3ω by a simpler (first-order) mechanism.[35]

If the loop gain $A(3\omega)\beta(3\omega)$ is large enough to be interesting at frequency 3ω, it is likely that the loop gain $A(2\omega)\beta(2\omega)$ will be larger, large enough to make the penultimate factor in (29.22) close to 1. Then the 3ω term in v_{out} simplifies to

$$-q(\omega, \omega)\, q(\omega, 2\omega)\frac{C^3}{4}\frac{1}{1 + A(3\omega)\,\beta(3\omega)}\,\mathrm{e}^{\mathrm{j}3\omega t}.$$

This term may be reduced if we can increase the return difference. And, as we have come to expect, it is the return difference at frequency 3ω that is effective, whichever kind of non-linearity is responsible for the appearance of 3ω.

Problem 29.6 Sum and difference frequencies

Make an analysis similar to that in problem 29.5 when the input voltage to the circuit is

$$v_{\text{in}} = \frac{v_1}{2}\,\mathrm{e}^{\mathrm{j}\omega_1 t} + \frac{v_2}{2}\,\mathrm{e}^{\mathrm{j}\omega_2 t} + \text{complex conjugate}.$$

Concentrate attention[36,37] on the sum frequency $(\omega_1 + \omega_2)$ and the difference frequency $|\omega_1 - \omega_2|$, and show that these frequencies have amplitudes containing the expected denominator (return difference). This denominator is evaluated at $(\omega_1 + \omega_2)$ for the sum frequency, and at $|\omega_1 - \omega_2|$ for the difference frequency.

Once again, we find that negative feedback reduces the amplitude of each unwanted frequency, and the reduction factor is in line with the rule given in § 29.6.2.

[35] Therefore nothing said here about 3ω gives a basis for any thought that negative feedback might be undesirable.

A first-order outcome of any v^3 distortion is a term at 3ω, suppressed by a factor $|1 + A(3\omega)\beta(3\omega)|^{-1}$, according to reasoning that parallels that for the 2ω term.

[36] Since coefficients of complex exponentials are to be equated, we can, with care, discard all terms that we don't intend to process in this way. For example, since we're not now interested in frequency $2\omega_1$ we needn't include it in the algebra—unless we intend carrying things so far that $2\omega_1$ forms a sum frequency with $(\omega_2 - \omega_1)$, for example.

[37] In the audio world, the generation of sum and difference frequencies is called "intermodulation distortion".

30 Stability of negative feedback

Intended readership: When there is a need to bring to life the formal theorems concerning Nyquist plots, Bode plots ..., and to apply that knowledge in practice.

30.1 Introduction

Chapter 29 exhibits the advantages that accrue from the application of negative feedback in all kinds of control system.[1] The price of achieving those advantages is the requirement that the feedback loop be stable. We need to know how to ensure stability, how to recognize danger signals, how safety margins are to be assessed, and so on. The present chapter addresses these topics.

As with several other chapters in this book, I have unconventional things to say: here concerning "conditional" stability.

It is not our task to derive formal theorems that are well described elsewhere. So we quote those theorems here, and concentrate on displaying the meaning and the implications for design.

[1] As in Chapter 29, we shall describe feedback systems implemented as electronic voltage amplifiers, but it is to be understood that these are models that can be carried over to all kinds of control.

30.2 The Nyquist plot

The feedback system being described has the generic form shown in Fig. 29.1.[2] The open-loop gain is $A(\omega)$, the feedback fraction is $\beta(\omega)$,

[2] For the whole of this chapter we shall, for simplicity, assume that both A and β are non-zero, and probably at their greatest, at $\omega = 0$. In technical language, the gain block is a "baseband" amplifier, and its internal connections are "d.c.-coupled". If this is not the case, then the concepts and principles of this chapter must be applied to controlling the variation of $A(\omega)\beta(\omega)$ for $\omega \to 0$ as well as for $\omega \to \infty$.

In this semiconductor age, baseband amplification is the norm, as in commercial operational amplifiers. At the time when valves (vacuum tubes) were the only sources of gain, baseband amplification was hard to achieve, and consequently rare. Early treatments of stability, such as that by Bode (1945), must be understood accordingly.

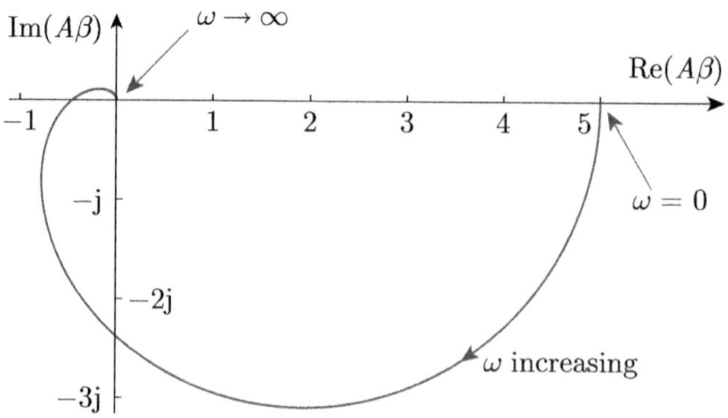

Fig. 30.1 A sample Nyquist plot. The loop gain $A(\omega)\beta(\omega)$ is plotted in the complex plane (positive j upwards). For any one ω the result is a point. All such points are joined to form a curve, plotted from $\omega = 0$ to $\omega \to \infty$. The Nyquist theorem says that the feedback system is stable provided that the curve does not "enclose" the point $-1 + 0\text{j}$. The system described by this curve is therefore stable.

The loop gain giving rise to this plot is given by (30.1).

Essays in Physics: Thirty-two thoughtful essays on topics in undergraduate-level physics. Geoffrey Brooker, Oxford University Press. © Geoffrey Brooker 2021. DOI: 10.1093/oso/9780198857242.003.0030

and the loop gain is $A(\omega)\beta(\omega)$. It is the loop gain that particularly concerns us now.

The loop gain $A(\omega)\beta(\omega)$ is a complex quantity and is plotted in the complex plane. As ω is varied from 0 to ∞, the loop gain traces out a path that is called the Nyquist plot. An example of such a plot is shown in Fig. 30.1.

The feedback system is stable provided that the Nyquist curve does not "enclose" the point $-1 + 0j$ on the negative real axis. By "enclose" we mean the following. Make a closed contour out of the path drawn, together with its mirror image in the real axis. The system is stable if the resulting closed curve does not encircle[3] the point -1. The system represented in Fig. 30.1 is stable.

The Nyquist plot provides the most fundamental and useful way of determining whether a feedback system is stable. But, of course, we need to understand what the plot means, how it may be calculated, how it might be measured, and how it may be given a desired shape.

Although the axes in Fig. 30.1 are labelled with the real and imaginary parts of $(A\beta)$, it is more useful to think in polar coordinates:

$$A\beta = |A\beta|\,e^{-j\phi}.$$

Thus we concentrate attention on the magnitude $|A\beta|$ and the phase lag ϕ, as indicated in Fig. 30.2. The phase is most usually a lag, so we define things so that ϕ is positive when it is a lag—measured downwards from the real axis as in Fig. 30.2.

A Nyquist plot can in principle be constructed by measuring $|A\beta|$ and ϕ for a circuit that we have built.[4] We may imagine opening the feedback loop, applying a sinewave signal $v_{\text{in}} = v_1\,e^{j\omega t}$ to the input of the gain block, and measuring what is transmitted through both gain block and feedback path (Fig. 30.3). The output is $v_{\text{in}} \times A\beta = v_{\text{in}}\,|A\beta| \times e^{-j\phi}$, in which it is $v_{\text{in}}\,|A\beta|$ and ϕ that are most easily measured. Values are obtained over a range of frequencies, and plotted as shown in Fig. 30.2.

Next, we have to understand what, in the design of the circuit, leads the Nyquist plot to have the shape that it has. This is all to do with the frequency dependence of $A\beta$, and it leads us to look at how that frequency dependence is best displayed: we return to the Bode plot, already seen in Fig. 29.2.

Aside: The Nyquist plots of Figs. 30.1 and 30.2 have been constructed from

$$A\beta = \frac{L_0}{1 + j\omega/\omega_0} \times \frac{1}{(1 + j\omega/\omega_1)^3}, \qquad (30.1)$$

with $L_0 = 5$, $\omega_1/\omega_0 = 10$. These values are very artificial, not representative of a practical circuit, and were chosen to give a curve whose main features can be seen without need for a distorted scale.[5] Even so, a detail is too small to see: the curve rotates through a full 2π and approaches the origin leftwards along the positive real axis.

[3] In imagination, lay a closed loop of string along this curve. Place the tip of your pencil on the point -1 and see if you can pull the string away without its getting wrapped round the pencil. If you can, the point -1 is not enclosed, and the feedback system is stable.

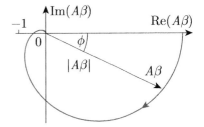

Fig. 30.2 A Nyquist plot may be thought of as the outcome of measuring $A\beta$ as in Fig. 30.3. The complex number $A\beta = |A\beta|\,e^{-j\phi}$ defining a point on the curve lies at radius $|A\beta|$ and at angle ϕ to the real axis.

[4] This procedure is somewhat notional, as measurements are not usually quite as easy as this description makes them sound. But the information required is available, by some route. For example, β is known from its circuit equations, which we shall have set up at the design stage.

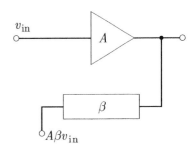

Fig. 30.3 An in-principle measurement of the loop gain of a feedback system. The feedback loop is opened, voltages $v_{\text{in}}(\omega)$ are applied to the input of the gain block, and the voltage after the feedback path is measured, in amplitude and phase, for each ω.

[5] Figure 30.13 and later Nyquist plots use a highly distorted radial scale in order to display huge values of $A\beta$, yet also show the vicinity of -1.

30.3   The Bode plot

[6] We have encountered a Bode plot in Fig. 29.2.

It may seem unduly complicated that we need now to consider a second kind of frequency-dependence graph.[6] However, it is the Bode plot that gives us the most intuitive view of a frequency dependence. Also, a knowledge of the shape of the Bode plot makes it easy to see the likely shape of the Nyquist plot; and conversely, the required shape of the Bode plot can be deduced from the desired shape of the Nyquist plot.

A Bode plot of $A\beta$ is a plot of $|A\beta|$ versus frequency f (or angular frequency ω), with logarithmic scales on both axes. We may plot the phase lag ϕ as well, as is done in Fig. 30.4, but a little experience permits us to sketch the graph for ϕ from knowledge of how $|A\beta|$ behaves.

[7] For a minimum-phase-lag network. See § 30.5 for explanation and the conditions for the circuit to be of the minimum-lag type.

The dependences of $|A\beta|$ and ϕ on frequency are intimately connected:[7] ω always appears in the circuit equations as $(j\omega)$.

The best way of seeing how the Bode and Nyquist plots link together is to look at a couple of tutorial examples, and this we now do.

Example 30.1 Loop gain falling as ω^{-1}
In our first example the loop gain is given by

$$A\beta = \frac{L_0}{1 + j\omega/\omega_0}, \qquad |A\beta| = \frac{L_0}{\sqrt{1 + (\omega/\omega_0)^2}}, \qquad \phi = \arctan(\omega/\omega_0).$$

$$(30.2)$$

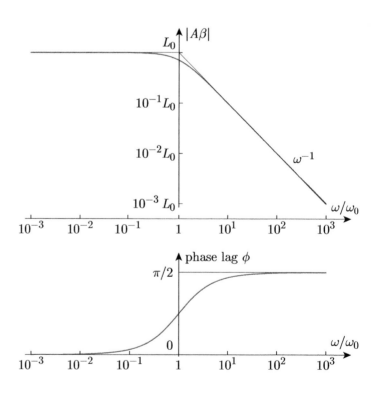

Fig. 30.4 An example of a Bode plot. The plot of $|A\beta|$ is logarithmic on both axes. The plot of phase lag is log–linear, meaning that the frequency is again plotted logarithmically, while the phase lag is plotted linearly.

For this illustrative plot, the loop gain $A\beta$ is given as a function of angular frequency ω by eqn (30.2). In the high-frequency limit $\omega/\omega_0 \gg 1$, the asymptote for $|A\beta|$ has slope -1 while the phase lag tends to $\pi/2$.

The frequency dependence shown here has been chosen primarily for its simplicity. Nevertheless, the gain of an operational amplifier very commonly depends on frequency in this way. The reason is given at the end of this example: the gain $A(\omega)$ is "rolled off" as ω^{-1} so as to make easy the construction of a stable negative-feedback circuit with this op-amp providing the open-loop gain.

The Bode plot for this is exhibited in Fig. 30.4.[8] The curve for $|A\beta|$ has two obvious asymptotes: at low frequencies $|A\beta|$ approximates to L_0, giving a horizontal asymptote; at high frequencies $|A\beta|$ approximates to $L_0(\omega_0/\omega)$, giving an asymptote falling with slope -1. The frequency $\omega = \omega_0$ where the asymptotes meet is referred to as a **corner frequency**.

It is worth appreciating the proportions of the Bode plot drawn in Fig. 30.4. First, notice how closely the curve for $|A\beta|$ hugs its two asymptotes: at $\omega = \omega_0$ the curve lies below the intersection of asymptotes by the unremarkable factor of $1/\sqrt{2}$. When ω is a factor 3 away from ω_0 in either direction, the curve lies below the appropriate asymptote by only about 5% (problem 30.1). Workers in this field always identify the corner frequencies of a Bode plot and draw in the straight-line asymptotes between them. They may then sketch in a likely actual curve by "rounding the corners" a bit, or they may not bother.[9]

By contrast, the phase is a much "slower" function of frequency. When ω is a factor 3 away from ω_0 (in either direction), the phase lag is not yet close to its asymptotic value of zero or $90°$: it differs from it by more than $18°$ (problem 30.1). A corner frequency extends its influence some distance away in frequency, so far as phase lag is concerned, even though the magnitude of $|A\beta|$ is relatively unaffected. This characteristic has a profound effect on the design of feedback systems (see § 30.4).

For a loop gain given by (30.2), the Nyquist plot is shown in Fig. 30.5. The curve is a semicircle, traced in the direction of the arrow as ω is increased. Problem 30.2 asks you to confirm the details.

The Nyquist curve of Fig. 30.5 does not "enclose" the point -1, so any feedback system having this loop gain is stable. Moreover, the distance from the point -1 to the curve always exceeds 1, so that $|1 + A\beta| > 1$. The feedback is therefore negative, and defect-reducing, at all frequencies, according to the definition of § 29.3 and the findings of Chapter 29.

Commercially supplied operational amplifiers often have an open-loop gain A that falls as $1/\omega$ above some rather low corner frequency ω_0 (often $\sim$10 Hz), over what may be five or more decades of frequency change, because this simplifies things for the user, who can add up to another $\pi/2$ of phase lag in his feedback fraction β before risking instability.

Figure 30.4 shows just one way in which the loop gain $A\beta$ of a feedback system may fall with frequency from its zero-frequency value of L_0. A steeper fall of $|A\beta|$ with frequency would necessarily be accompanied by a greater phase lag, as our next example shows.

Example 30.2 Loop gain falling as ω^{-2}
Consider

$$A\beta = \frac{L_0}{(1 + \mathrm{j}\omega/\omega_0)^2}, \qquad |A\beta| = \frac{L_0}{1 + (\omega/\omega_0)^2}, \qquad \phi = 2\arctan(\omega/\omega_0).$$
$$(30.3)$$

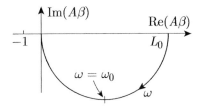

Fig. 30.5 The Nyquist plot associated with eqns (30.2) and Fig. 30.4. The curve is always distant from the point $-1 + \mathrm{j}0$ by more than 1, so the feedback that this represents is negative at all frequencies, according to the definition given in § 29.3.

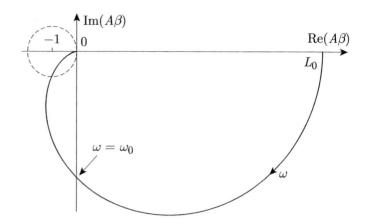

Fig. 30.6 A Nyquist plot for a system obeying eqn (30.3). The curve is drawn for $L_0 = 10$.

Because the loop gain falls at high frequencies like two powers of $(j\omega)^{-1}$, the phase lag there tends to π: the curve approaches the origin along the negative real axis. Where the curve passes inside the broken circle, the return difference $|1 + A\beta| < 1$: for frequencies in this range, the feedback is positive, and defect-increasing—though the circuit is stable.

The asymptotic (well above ω_0) behaviour is a fall of $|A\beta|$ as ω^{-2} accompanied by a phase lag of π; the Bode plot of $|A\beta|$ is given in Fig. 30.7. For behaviour near the corner frequency, see problem 30.1.

The Nyquist plot for (30.3) is shown in Fig. 30.6; the curve approaches the origin rightwards along the negative real axis. In order for detail to be visible, the curve has been plotted for an L_0 of only 10; had L_0 been given a more realistic value (say 1000), the curve would have passed dangerously close to the point -1.

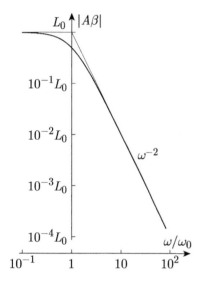

Fig. 30.7 The Bode plot representing the $|A\beta|$ given by (30.3). The corresponding Nyquist plot is Fig. 30.6, there drawn for the case $L_0 = 10$.

A general conclusion can be drawn, even from the two tutorial examples so far given. In order to keep the Nyquist curve away from the point -1, we must make $A\beta$ have a phase lag that is not too large: well below π where the curve passes near[10] the point -1. This in turn requires that the loop gain must be "rolled off" rather slowly with frequency: ω^{-1} is safe if unambitious; ω^{-2} is still a rather slow fall and yet is steep enough to be risky. And "risky" means we need to think next about safety margins . . . .

30.4 Safety margins

A good design must build in allowance for uncertainties: uncertainties in the open-loop gain A (because semiconductor devices are only loosely toleranced and change with temperature)[11] and uncertainties in the loop phase lag ϕ (because of parasitic effects from the circuit layout or from corner frequencies not well known). These allowances are conventionally expressed in terms of the **gain margin** and the **phase margin**.

---

[10] What "away from" and "near" mean is only hinted at here; they look ahead to eqn (30.6) and § 30.6.

[11] In a more general control system, there can be uncertainties of a more diverse nature. To give only one example: if we are controlling the spacing of the mirrors of a gas laser, then the laser frequency must be sensed so as to provide a correction signal. The sensor's response will depend upon the laser's output power, and that power is likely to change considerably as the laser ages. A loss of power will result in a reduced signal from the sensor, a reduction of loop gain for the entire system, and a consequent degradation of the control. Conversely, when this laser's discharge tube is replaced by a new one, the optical power may be so high as to precipitate instability in the control loop.

Figure 30.1 shows a Nyquist curve that crosses the negative real axis close to -0.5. If the loop gain $|A\beta|$ were (throughout) a factor just-over-2 greater than that plotted, the curve would pass through the point -1, or enclose that point, and the circuit would be unstable. We can say that the system has a **gain-increase margin**—an allowance for a possible increase in the loop gain—of about a factor 2. As a rough rule of thumb, a gain-increase margin of 2 should be regarded as barely adequate.

Figure 30.6 shows a Nyquist plot that passes fairly close to the point -1. Where $|A\beta| = 1$, the phase lag ϕ is 143°, leaving a **phase margin**[12] of 37°. An additional phase lag, from any unforeseen or poorly characterized cause, of 37° would bring the circuit to instability. As a rough rule of thumb, a phase margin of 30–45 degrees should be regarded as barely adequate.

This is where we need to remember the finding in § 30.3: the phase lag is a "slow" function of frequency. A corner frequency that looks far enough away to be harmless may add enough to the phase lag to bring the circuit to instability. This is illustrated in example 30.3.

[12] This phase lag is for the curve drawn in Fig. 30.6, which obeys (30.3) with $L_0 = 10$. If L_0 were larger, this arrangement, with $|A\beta|$ falling as ω^{-2}, would be unacceptable: too small a phase margin. In practice it could already be "near the edge", for reasons explored in example 30.3.

I define the phase margin as $(\pi - \phi)$ evaluated for the frequency at which $|A\beta| = 1$, leaving it permitted that $(\pi - \phi)$ might be smaller or even negative at other frequencies.

A more conservative design rule might set a "floor" value of $(\pi - \phi)$ that is to apply at this frequency *and at all lower frequencies*. Such a design rule is often adopted (I think needlessly), and when it is this $(\pi - \phi)$ may well be called *the* phase margin.

Example 30.3 An unsafe design
The loop gain we now consider is given by

$$A\beta = \frac{L_0}{(1 + j\omega/\omega_0)^2} \frac{1}{(1 + j\omega/\omega_1)^4}. \tag{30.4}$$

With the values $\omega_1/\omega_0 = 5\sqrt{10} \approx 16$ and $L_0 = 10$ this is displayed in the Bode plot of Fig. 30.9.

In the absence of the ω^{-6} section, the |loop gain| would reach 1 at $\omega/\omega_0 = 3$ with a phase margin of 37°, and this might be considered safe. However, the ω^{-6} section, in spite of its apparent remoteness from the region of interest, increases the phase lag of $A\beta$, so much so that at the new $|A\beta| = 1$ frequency the lag is 183°: instability. The Nyquist curve is plotted in Fig. 30.8; the curve encloses (passes above) the point -1, as is shown in the magnified fragment.

The transition from ω^{-2} to ω^{-6} may be thought somewhat extreme. It is not. It is quite likely that a modern feedback system contains one or more operational amplifiers. An operational amplifier contains a large number of transistors on the same chip, all of which are likely to deteriorate at about the same frequency.

The behaviour shown in Fig. 30.9 may easily escape notice because the offending corner frequency lies in a region where we are unlikely to conduct measurements. It is the more important to know (by measurement or otherwise) the loop phase lag at and around the frequency where we intend making $|A\beta| = 1$. Or, if we don't know it, then leave a generous safety margin.

Let me repeat that Figs. 30.8 and 30.9 have been printed by way of a horrible example, showing us what kind of design must be avoided. The rapid fall of loop gain above ω_1 is too steep or too near to the $|A\beta| = 1$ frequency—or both. Yet the zero-frequency loop gain is only 10. Our

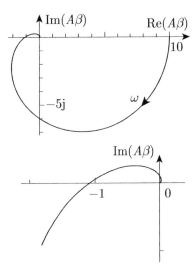

Fig. 30.8 The Nyquist plot for the system described by eqn (30.4) and Fig. 30.9. The additional phase lag, caused by the change to a steep fall of loop gain above ω_1, is enough to bring the curve to enclose the point -1.

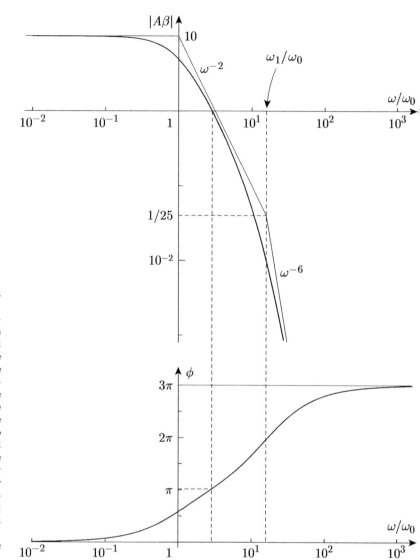

Fig. 30.9 The Bode plot for a circuit obeying eqn (30.4).

At first sight, the upper curve appears to be not much different from that described by eqn (30.3) and Fig. 30.7. However, the loop gain here has a transition from ω^{-2} to ω^{-6}. The steep (ω^{-6}) section starts at a frequency ω_1 that is 5× higher than the frequency at which the ω^{-2} asymptote crosses the line $|A\beta| = 1$. Even so, the ω^{-6} region exerts sufficient influence to bring the phase of the loop gain beyond π at the frequency where $|A\beta| = 1$; the circuit is unstable. Broken lines draw attention to the location of the corner frequency ω_1. Another broken line projects down from the point where $|A\beta|$ crosses 1 to highlight the phase lag ϕ at that frequency.

The Nyquist plot presenting the same information is given in Fig. 30.8.

instincts may well try to tell us that nothing can go wrong with a circuit design that is so unambitious, and so apparently similar to that of Figs. 30.6 and 30.7 (which are also drawn to have $L_0 = 10$).

Turning this round: a circuit that appears to conform to Fig. 30.6 may in fact have insufficient phase margin owing to a remote corner frequency, or a time delay, or to components lying at the ends of their tolerances. Be generous with phase margins!

30.5   Minimum phase-lag networks

In the model systems described in examples 30.1–3, the loop gain $A\beta$ has a fall with ω and a phase lag ϕ, the two being intimately related.

If there is any phase lag in $A\beta = |A\beta|\,\mathrm{e}^{-\mathrm{j}\phi}$, additional to that necessitated by the fall of $|A\beta|$, it is undesirable, because it moves the Nyquist curve in a clockwise direction, bringing it closer to, or beyond, the point -1. Such an excess phase lag can happen, and it is absent only if the circuit constitutes a **minimum phase-lag network**.

There are (at least) two ways in which a *non*-minimum phase lag can happen. The most obvious is a time delay τ, such as happens if a signal is sent along a transmission line. A signal at angular frequency ω acquires a phase factor $\mathrm{e}^{-\mathrm{j}\omega\tau}$, that is, a phase lag of $\omega\tau$. Such a phase lag is not accompanied by a fall of the transmitted signal with increase of ω —which would give a benefit to accompany the cost. A second way in which excess phase lag can be introduced is when a signal is fed forwards through circuitry by two routes and subtracted. An example of such a subtraction is shown in Fig. 30.10: the gain block amplifies $(v_+ - v_-)$ while the signals at v_+ and v_- have arrived by different routes. A circuit of the type shown in Fig. 30.10 is often called an "all-pass filter".

The voltage "gain" of the all-pass filter is (problem 30.3)

$$\frac{v_{\text{out}}}{v_{\text{in}}} = \frac{1 - \mathrm{j}\omega CR}{1 + \mathrm{j}\omega CR}; \tag{30.5}$$

it should be obvious that this complex number has magnitude 1 for all ω.

The signature of a "subtraction" non-minimum-phase circuit is a numerator like $(1 - \mathrm{j}\omega CR)$. Remember that any function of ω is not far away from a Fourier transform, in which ω is a complex variable. We can see the numerator of (30.5) as having a zero at $\omega = -\mathrm{j}/CR$, in the lower half of the complex j-plane. Conversely, the rule for avoiding excess phase lag is that there must be no such zero. In application to negative feedback, it is $A\beta$ that must lack any zero in the lower half of the ω-plane.

We do not present the mathematics that establishes the claims of this section. It suffices to give an idea of the danger signals, which can be followed up if a circuit design looks problematic. For all of this chapter, except the present section and part of the next, we assume without comment that $A\beta$ satisfies the condition for minimum phase lag.

A minimum phase-lag network has a phase lag ϕ that is predictable from the rate of fall of $|$gain$|$.[13] In a Bode presentation such as Fig. 30.4, the graph of ϕ versus frequency (lower curve) can be sketched from a knowledge of the shape of the $|A\beta|$ (upper) curve.[14] In particular:

- If the loop gain $|A\beta|$ falls as ω^{-1} over a lengthy frequency range, the loop phase lag ϕ in that range approximates to $\pi/2$.
- If the loop gain $|A\beta|$ falls as ω^{-n} over a lengthy frequency range, the loop phase lag ϕ in that range approximates to $n\pi/2$.

The point of "lengthy" may be seen from the finding in §§ 30.3–4: a "corner" between two different dependences on frequency affects the phase lag, even at some distance away.

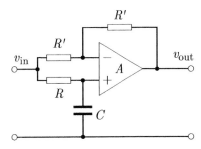

Fig. 30.10 An all-pass filter: a circuit which gives phase lag without a fall of gain with increasing ω. This is therefore an example of a *non*-minimum phase-lag circuit.

[13] The theorem—though not its proof—is quoted in eqn (30.6). There is a strong resemblance between that theorem and the Kramers–Kronig relations, and indeed the mathematics employed for derivation is very similar.

Any linear input–output relation, such as $v_{\text{out}} = G\,v_{\text{in}}$ or $\boldsymbol{D} = \epsilon\epsilon_0\boldsymbol{E}$, is subject to Kramers–Kronig relations linking the real and imaginary parts of $G(\omega)$ or of $\epsilon(\omega)$. There are two relations, one expressing $\mathrm{Im}\,G(\omega)$ as an integral of $\mathrm{Re}\,G(\omega')$ over ω', and the other expressing $\mathrm{Re}\,G(\omega)$ in terms of an integral of $\mathrm{Im}\,G(\omega')$ over ω'. The relations that we need here are quoted in § 30.5.1. These theorems and their derivations are spelt out in great detail in the classic work by Bode (1945) Chapter 14.

[14] The converse—obtaining $|A(\omega)\beta(\omega)|$ from knowledge of $\phi(\omega)$—is also possible: see sidenote 20.

[15] These rules were anticipated at the end of § 30.3.

[16] I consider "compensation" an infelicitous piece of jargon, as nothing is being compensated or cancelled out.

[17] Of course, $A\beta$ is determined by the designer's choice of $\beta(\omega)$ as well. This choice must yield stability. We can say that the "rolled-off" frequency dependence of $A(\omega)$ gives a head start in the right direction.

[18] The reader interested only in applications can skip this section. But I am assuming that a physicist will want to know mathematically what is going on.

[19] Bode (1945), eqn (14–11).

Table 30.1 The integral of the weight function $W(\omega'/\omega)$ over ω' from ω/y to ωy for selected values of y. This truncated integral is defined in (30.8).

limit y	integral, ω/y to ωy
1.694	0.5
2	0.582
3	0.726
10	0.919
∞	1

The bulleted points lead to three straightforward design principles:

- Within the range of frequencies that is of concern to stability, the circuit must have minimum phase lag.
- If we want a modest loop phase lag ϕ, say well below π, then we must accept a loop gain $|A\beta|$ that falls slowly (not as fast as ω^{-2}) with frequency.
- If we want a fast-falling loop gain, then we must somehow find it possible to accept a large loop phase lag.[15]

Designers of feedback systems talk of "rolling off" the loop gain $A\beta$, often as $1/\omega$, in order to achieve stability with good safety margins. Such a loop gain is shown in Fig. 30.4. Operational amplifiers commonly contain an internal capacitor, providing what is called "internal compensation",[16] to give to $A(\omega)$ this kind of frequency response.[17]

30.5.1 Interlude: the gain–phase relation

In § 30.5, we gave a rough idea of how the loop phase lag $\phi(\omega)$ links to the gradient of loop gain $|A\beta|$ on a Bode plot. Details:[18]

Let $A\beta = L(\omega)\,e^{-j\phi(\omega)}$ where $L(\omega)$ and $\phi(\omega)$ are real. "It can be shown"[19] that

$$\phi(\omega) = \frac{\pi}{2}\int_{\omega'=0}^{\omega'=\infty}\left\{-\frac{d\ln L(\omega')}{d\ln\omega'}\right\}\times W(\omega'/\omega)\,d(\ln\omega') \qquad (30.6)$$

+ any addition for a non-minimum-phase-lag circuit.

The "weight function" $W(\omega'/\omega)$ is given by

$$W(\omega'/\omega) = \frac{2}{\pi^2}\ln\left|\frac{\omega'/\omega+1}{\omega'/\omega-1}\right|; \qquad \int_{\omega'=0}^{\omega'=\infty}W(\omega'/\omega)\,d\ln\omega' = 1. \quad (30.7)$$

In (30.6) we can see the phase $\phi(\omega)$ obtained from $\{-d\ln L/d\ln\omega'\}$, the rate of fall of loop gain with frequency—in the log–log presentation of a Bode plot. If the "weight function" $W(\omega'/\omega)$ were $\delta(\ln\omega'-\ln\omega)$, the phase ϕ at ω would be determined by the rate of fall $\{-d\ln L/d\ln\omega\}$ at that same angular frequency ω. As it is, $\phi(\omega)$ is the outcome of taking a weighted average of $\{\}$ over angular frequencies ω' near to ω, the weighting function being W. The weight function $W(\omega'/\omega)$ is plotted in Fig. 30.11.

The constants in (30.6) and (30.7) have been so positioned that W is normalized. To give an idea of how "peaky" is W, we show in Table 30.1 the integral of it from ω/y to ωy for some choices of y. That is, the quantity tabulated is

$$\int_{\omega/y}^{\omega y}W(\omega'/\omega)\frac{d\omega'}{\omega'} = \int_{\ln(\omega/y)}^{\ln(\omega y)}W(\omega'/\omega)\,d\ln\omega'. \qquad (30.8)$$

The statements in § 30.5 may now be confirmed and explained: if $\{-d\ln L(\omega')/d\ln\omega'\} = n$ over a frequency range covering a factor 10 or more either side of ω, then it may be taken outside the integral in (30.6) and $\phi(\omega)$ approximates to $n\pi/2$. Conversely, the "width" of the weight

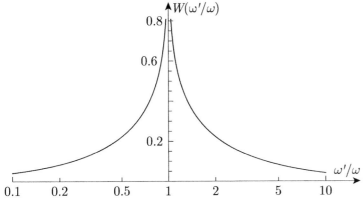

Fig. 30.11 The weight function $W(\omega'/\omega)$ in the relation between phase lag and gain fall. The horizontal scale is logarithmic, while the vertical scale is linear. When the horizontal scale is logarithmic, as it is here, the curve is symmetrical about $\omega'/\omega = 1$.

function $W(\omega'/\omega)$ makes quantitative the fact that even a fairly remote corner frequency has some effect on the phase lag.[20]

The integral in (30.6) gives the least phase lag that can accompany any given $\{-\mathrm{d}\ln L/\mathrm{d}\ln\omega'\}$: it's assumed we have a minimum-phase-lag system. Should there be a transmission-line-type time delay τ imposed on the signal as well as amplification, then the phase ϕ has $\omega\tau$ added to the expression in (30.6). Should there be a "subtraction–type" departure from minimum phase lag, then an added phase resembles that implicit in the formula of (30.5). The important fact is that these are all *additions* to the phase lag, and the best we can do is to make such additions zero.

[20] There exists a relation that gives $L(\omega)$ as an integral over $\phi(\omega')$. That relation is

$$\ln\frac{L(\omega)}{L(0)} = \frac{\pi\omega}{2}\int_{\omega'=0}^{\infty}\frac{\mathrm{d}}{\mathrm{d}\ln\omega'}\left(\frac{\phi(\omega')}{\omega'}\right)\times W(\omega'/\omega)\,\mathrm{d}\ln\omega'.$$

As with (30.6), there are additions if the circuit is not minimum-phase. There is no singularity in the integrand at $\omega' = 0$ because $\phi(0) = 0$.

We give the above relation merely to show that it exists; unlike (30.6) it is not particularly helpful.

30.6   Nyquist ("conditional") stability

We have seen that—even for a minimum-phase-lag network—a steep dependence of loop gain $|A\beta|$ on frequency implies a large phase lag ϕ. And a large phase lag is risky, most so at those frequencies where $|A\beta|$ is in the vicinity of 1. However, a steeper variation of $|A\beta|$, accompanied by a greater ϕ, is permissible at other frequencies, as we show next.

When the phase lag ϕ exceeds π over some range of frequencies (within which $|A\beta| > 1$ and yet the circuit is stable), we have **Nyquist stability**.

Figures 30.12 (b) and 30.13 show the Bode and Nyquist plots for a system whose loop gain is[21]

$$A\beta = \frac{L_0}{(1+\mathrm{j}f/f_0)^3}\times(1+\mathrm{j}f/f_1)^2. \tag{30.9}$$

The loop gain $|A\beta|$ varies as f^{-3} from $f = f_0$ to $f = f_1$. Above f_1 it changes to a shallower variation, as f^{-1}. The $A\beta$ of (30.9) is plotted for $L_0 = 10\,000$ and a ratio $f_1/f_0 = 2000^{1/3} = 12.6$ of corner frequencies.

As we must expect, the steep f^{-3} dependence of $A\beta$ implies a large loop phase lag ϕ. In fact, ϕ even exceeds π in the frequency range from $f_{\pi 1}$ to $f_{\pi 2}$ (marked in both figures).[22] Conversely, the f^{-1} region of (30.9) implies a more modest phase lag, which brings the Nyquist plot well below the point -1. In turn, this means the circuit is stable.

[21] As we move towards practical designs, with numerical values attached, it will be convenient to switch from using angular frequency ω to hertz-frequency f.

[22] If the f^{-3} region were lengthened, the phase lag ϕ within it would eventually reach $3\pi/2$. However, the f^{-3} region shown extends over only a factor 12.6 in frequency, so no frequency within that region is farther than $\sqrt{12.6}$ from a corner frequency whose influence is to "pull back" the phase from its limiting value of $3\pi/2$. This "pulling back" is in line with the breadth of the weight function displayed in Fig. 30.11.

Incidentally, Figs. 30.12 and 30.13 show rather well how key features of the Bode and Nyquist plots link together.

Fig. 30.12 Two possible ways of designing the loop gain for a feedback amplifier. In both cases, the drop-dead frequency of 1 MHz is a fixed point. Also a fixture is the f^{-1} asymptote there. Possibility (a) continues the asymptotic f^{-1} line downwards in frequency until it meets the "horizontal" asymptote at 10 000, which it does at 100 Hz. Possibility (b) follows f^{-1} down from 1 MHz for a factor 5, to 200 kHz, then rises as f^{-3} along a new asymptote which reaches 10 000 at 15.9 kHz. The loop gain is now level at 10 000 from zero frequency up to the corner at 15.9 kHz (rounding of the corner ignored).

Possibility (b) implements the $A\beta$ of eqn (30.9).

In the upper plot, the vertical distance between curves (b) and (a) illustrates how an enhanced $|A\beta|$ can be achieved. The price we pay for this is the large phase lag ϕ in the lower plot.

For possibility (b), the loop phase lag ϕ passes through π at frequencies $f_{\pi 1}, f_{\pi 2}$, between which ϕ exceeds π. The implication of this is displayed in a different way in Fig. 30.13, where it is emphasized that there is Nyquist stability.

We may see (a) as a consequence of conforming to the "more conservative" design rule of sidenote 12, in which $(\pi - \phi)_{\text{floor}} = \pi/2$.

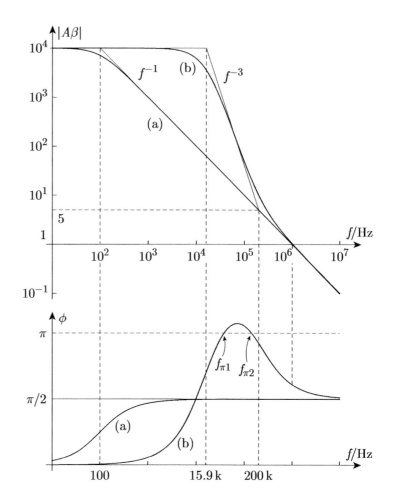

[23] For case (a) the Nyquist plot is shown in Fig. 30.5.

[24] A comprehensive account of the ingenious ways in which a Nyquist-stable circuit can become unstable (oscillating, or having an output that moves to a supply rail and stays there) is given by Bode (1945), pp 162–4. Bode's warnings have been much copied.

The potential advantage of a loop gain like (30.9) is shown by the "vertical" distance between curves (b) and (a) for $|A\beta|$ in Fig. 30.12: curve (b) offers an enhanced loop gain $|A\beta|$ at frequencies between 100 Hz and 200 kHz—if we can guarantee stability.[23]

Textbooks, and practising electronics engineers, usually warn that a feedback system having a Nyquist plot like Fig. 30.13 is a theoretical possibility, but one to be shunned. The system has the potential for instability if the loop gain falls, and "will find a way of oscillating".[24] Hence the usual name "conditional" stability. I disagree. There are pitfalls, but they can be understood, and they can be avoided. To signal this change of attitude, I refer to an $A\beta$ that crosses the negative real axis twice as giving "Nyquist stability", not "conditional stability".

The discussion above starts at the highest frequencies and works downwards from there. This generalizes. A feedback system is best "designed downwards" in frequency, and from this point of view it is the steep *rise* with falling f (below 200 kHz in Fig. 30.12 (b)) that is the benefit of a Nyquist-stable system. Read on.

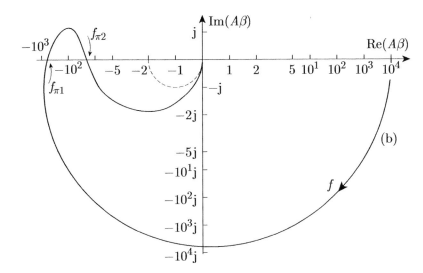

Fig. 30.13 The Nyquist plot accompanying possibility (b) of Fig. 30.12. The circuit is stable: the curve does not enclose the point −1. There is Nyquist stability, since there is a range of frequencies (from $f_{\pi 1}$ to $f_{\pi 2}$) within which the loop phase lag ϕ exceeds π.

The curve never enters the circle centred on the point −1, so the feedback is negative at all frequencies.

The system described here has a zero-frequency loop gain of 10 000, so the plot has had to be drawn on "rubber graph paper". The radius $|A\beta|$ is plotted linearly from 0 to 2, logarithmically above 10, and follows a smooth joining function (details not given) between 2 and 10.

30.7 A design procedure

Guided by § 30.6, I offer rules that govern the design of any feedback system (whether or not it exploits Nyquist stability).

(1) Work from the highest frequencies downwards.

(2) Locate the right-hand asymptote of Fig. 30.14: Working with a draft circuit, remove any and all components that restrict A or β at high frequencies. This may include inserting one or more capacitors (C_3 and C_4 of Fig. 30.19) so as to raise β to 1 at the highest frequencies. $|A\beta|$ will now follow an asymptote (probably varying as f^{-n} for some n) which has been pushed to the right (regardless of n) as far as is possible given the gain block's capabilities.

(3) Insert frequency dependent components into the circuit so that $|A\beta|$ now follows the f^{-1} asymptote of Fig. 30.14 (where it can, meaning to the left of the f^{-n} asymptote). Locate this asymptote as high as is prudent, recognizing that the higher it is the smaller is the circuit's phase margin.

(4) The f^{-1} asymptote crosses the line $|A\beta| = 1$ at frequency f_d, which we call the **drop-dead frequency**.[25] In step (3) we have made f_d as high as is practical. Make the f^{-1} asymptote continue downwards in frequency to about $f_d/5$, after which the loop gain may be made to depend more steeply on frequency. Ingenious choices start here.

In step (3), just how the f^{-1} is implemented is a matter for some opportunism, but a way can be found, perhaps by applying "local" negative feedback round a stage or stages within the gain block. There is also the capacitor(s) in the feedback path, which gives a $\beta \propto f$ asymptote within a limited range.

Figures 30.12 and 30.13 show these design rules in action. We have seen in step (4) that choices exist for frequencies below about $f_d/5$.

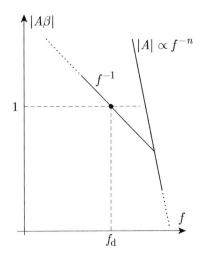

Fig. 30.14 Implementation of the design rules. The steep line at the right shows the f^{-n} fall of the open-loop gain $|A|$ possessed by the "naked" gain block, after all gain-reducing interventions have been removed and β has been set to 1. For the demonstration circuit of Fig. 30.19, $n = 2$ and the steep line passes through 1 at 3 MHz.

The line of slope −1 (this is a log–log plot) is a reduced loop gain $|A\beta|$, given this form by some appropriate intervention by us within A or β or both.

The upward-in-frequency continuation of the f^{-n} line is shown dotted as an expression of likely ignorance. The downward-in-frequency continuation of the f^{-1} line is shown dotted because it is where we have choices.

[25] [From previous page] I have adopted a name for f_d that is in use among my colleagues.

I define f_d to be the frequency where the f^{-1} asymptote passes through 1. By contrast, the unity-loop-gain frequency is where $|A\beta| = 1$. These frequencies are not very far apart, but the distinction between them is visible in Fig. 30.9.

[26] Here we imagine $|A\beta|$ to fall without change to its frequency dependence, so the Nyquist plot shrinks without change of shape—just what we shall prevent in later work. The factor under discussion here is, of course, really a factor $1/20.5$, but the meaning should be clear.

[27] Nothing here prevents the cautious designer from choosing a circuit intermediate between cases (a) and (b), such as using an f^{-2} dependence below 200 kHz. The design rules given at the beginning of this section continue to hold good.

[28] Moving the corner between f^{-3} and f^{-1} from $f_d/5$ to $f_d/3$ reduces the phase margin from $71°$ to $60°$, which should still be safe. But we dare not go much further. So the headroom between "unambitious" and "ambitious" is rather small.

Curve (a) of Fig. 30.12 represents a highly conservative design, in which $|A\beta|$ rises as f^{-1} all the way down in frequency until it stops at a corner frequency of 100 Hz; the loop gain at 20 kHz is only 50. For the other choice (b), the loop gain follows f^{-1} downwards from 1 MHz for a factor 5 only, after which it rises as f^{-3}, reaching its full value at corner frequency 15.9 kHz. At 20 kHz, the loop gain has the respectable value of 2427 (rounding of the corner allowed for, here only).

The Nyquist plot for Fig. 30.12 option (b) is shown in Fig. 30.13. The curve crosses the negative real axis twice, at frequencies $f_{\pi 1}$ and $f_{\pi 2}$. At $f_{\pi 2}$ the crossing is at $A\beta = -20.5$. We may say that there is a **gain-reduction margin** of 20.5, since $|A\beta|$ could withstand a misfortunate reduction [26] by a factor 20.5 before instability sets in.

The loop gain $|A\beta|$ is 1 at 1.04 MHz, where the loop phase lag ϕ is $109°$, giving a very safe phase margin of $71°$.

The greatest loop phase lag ϕ is $193°$, large enough to cause anxiety for a designer accustomed to conventional ideas of safety. Yet the gain and phase margins are good: our Nyquist-stable design example is not too scary.[27]

We may even think that case (b) in Fig. 30.12 has been designed somewhat unambitiously, in that the f^{-3} rise does not start until a frequency that is a factor 5 below f_d. This is the outcome of applying a rule of thumb suggested in design step (4). It is easy to reconsider this, locating the corner frequency according to one's choice of phase margin, but the rule of thumb is generally a good place to start from.[28]

Why does example (b) have a loop-gain variation as f^{-3}, and not something even steeper? There is no theoretical limit to how cunning you can be in getting the loop gain to vary steeply with frequency, but the more ambitious you are the harder you have to work. In a practical case it's usually pretty clear where diminishing returns set in.

We mentioned in § 30.6 a widespread belief that Nyquist stability is only for the foolhardy. Well, you do have to know what you're doing. If precautions are not taken, a circuit modelled on Fig. 30.12 (b) may well be a disaster. But measures are available to make the circuit safe, and they are described in the remainder of this chapter.

30.8   Non-linearity

In Chapter 29 we showed how negative feedback improves the linearity of an amplifying circuit. It may seem strange then that we are again to consider non-linearity here. The reason is that *some* non-linearity is unavoidable, and it is non-linearity that potentially endangers a system having Nyquist stability.

We are not talking here about small quadratic distortions. We are thinking of gross non-linearities, such as happen when we amplify a sinewave so large that the amplifier's output collides with a supply rail.

An amplifier necessarily exhibits non-linearity when we overload it. Of course, this is not a condition that we wish to encounter, but it is hard to prevent it from *ever* happening. So long as the amplifier comes out of overload in a reasonably graceful manner and continues to do its job, there is no problem. But it may not: it is up to us to understand what goes on here, and design in the required graceful behaviour.

Hitting the supply rails is not as improbable as it may sound. During switch-on, the supply-rail voltages grow from zero, and it is very likely that the amplifier will have its output at or close to a supply-rail voltage for a short while—with or without any signal input.

It is for reasons such as these that Nyquist stability has acquired a bad reputation.[29] An amplifier in overload has an effective loop gain that is reduced from its intended value. This reduction may shrink the Nyquist curve until it passes through −1; if so, the circuit goes unstable, probably oscillating. The oscillation can maintain the non-linear reduction of loop gain that makes instability possible.

The paragraphs above identify a danger signal. But they also reveal an escape route. The implicit assumption in the above was this: overload causes the Nyquist plot to shrink *without change of shape* until it makes contact with the point −1. There is no physical law that requires this. If "overload" non-linearity has unfriendly properties, then perhaps we can change it, or instead add a new non-linearity that is more to our liking. We show below that such a replacement/addition is possible, convenient and, if done properly, without disadvantage.

The idea is this. Whatever drives the circuit non-linear, whether it's an over-size signal we impose or an oscillation we are hoping to suppress, the new non-linearity is to reduce effective phase lag alongside reducing effective gain. The Nyquist plot is to *change shape at the same time as shrinking*—changing shape so as always to keep safely away from −1.

[29] Bode's "comprehensive" account of the risks associated with Nyquist stability, referred to in sidenote 24, includes those mentioned here.

30.9   Non-linear building blocks

30.9.1   A passive non-linear circuit

Figure 30.15 demonstrates a simple circuit that can implement a phase-reducing non-linearity.

As with the "overload" discussed in § 30.8, we from the start consider a "gross" non-linearity: diodes. If the phase lag is to undergo a major reduction on entry to the non-linear regime—and a major reduction is what we need—then nothing less than a "gross" non-linearity will do.

Consider first what happens when the voltage v_{in} applied to the circuit of Fig. 30.15 is too small to drive the diodes into conduction. Then the circuit is a low-pass filter having

$$\frac{v_{\text{out}}}{v_{\text{in}}} = \frac{1}{1 + j\omega CR};$$

the phase lag approaches 90° when $\omega CR \gg 1$. In the opposite limit, when large input voltages drive the diodes into heavy conduction, those diodes can be thought of—crudely—as a resistor r shunting the capacitor (Fig. 30.16); the capacitor has little effect now, and the output no longer

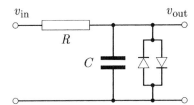

Fig. 30.15 A non-linearity, here implemented by diodes, can modify a circuit's frequency dependence, to an extent that depends on how hard the diodes are driven into conduction.

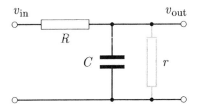

Fig. 30.16 The resistor r drawn grey shows qualitatively what the diodes of Fig. 30.15 are intended to achieve. The effective value of r depends on how heavily the diodes are driven into conduction.

suffers much phase lag. Diode non-linearity reduces the phase lag—also of course reducing the magnitude of the "gain" $|v_{\text{out}}/v_{\text{in}}|$.

The reader can supply his own sketched Bode and Nyquist plots for the circuit of Fig. 30.16, with and without resistor r. However, detailed investigation shows that a resistor r gives too crude a model of the diodes' effect. Plots (not given in this chapter) that take better account of the diodes show a less favourable reduction of phase lag.

We do not pursue Fig. 30.15 further here, because (additionally to the above) it is not a very helpful building block for implementing the design procedure of § 30.7. However, we do have an ideas-source that can be developed in better ways.

30.9.2   Active non-linear circuit modules

Figure 30.17 displays six modules, each of which can be used as an input or middle stage of a multi-stage amplifier.[30] It is a larger circuit, within which one (or more) of these modules is a part, whose feedback loop is to be made Nyquist-stable. The chosen module contributes part of the larger circuit's open-loop gain $A(\omega)$.

Each module incorporates a non-linear (diodes) "local" feedback loop which we take to be stable. Given careful design, the diodes never enter conduction while the amplifying module is handling a signal within its design range, but the diodes are there to wake up when needed for suppressing an incipient instability.

Figure 30.17 (a) shows an amplifying stage with local series feedback. If this is the first stage of a larger circuit, then overall shunt feedback (the feedback fraction β of the "outer loop") may be applied as suggested in grey at the top of the diagram.[31]

Figure 30.17 (b) shows an amplifying stage similar to (a), except that there is a faster variation of gain with frequency because of the two capacitors.

Figures 30.17 (c) and (d) show further implementations where the local feedback is applied in shunt. If one of these modules is used as the input stage of a multi-stage amplifier, then the outer feedback loop may be connected as shown in grey.

Circuits (e) and (f) are specific to the case where the gain block is an operational amplifier, probably "internally compensated" (Fig. 30.4). (Circuits (a) to (d) may have some more general gain block, subject of course to stability of the local feedback.) The gain of such an operational amplifier usually falls as f^{-1} over a wide frequency range, and this is hinted at in the Bode plot superimposed on the triangle symbol. It is the phase lag associated with this gain variation that is removed when the diodes are driven into conduction.[32]

There is, however, a restriction on the use of modules (c) to (f). We shall design the entire amplifying circuit so that the diodes do not conduct at signal frequencies (perhaps frequencies $\leqslant 20\,\text{kHz}$). Nevertheless there may be a small non-linear current in the diodes at such frequencies and this puts a non-linearity into the outer feedback β. If high linearity

---

[30] Each module of Fig. 30.17 gives an output voltage that is restricted—that is the point—to a diode offset of order 0.6 V. It is unlikely that the overall amplifier can be permitted such a restriction on its signal-output swing, so these modules are not usually appropriate for use as output stages. Hence "input or middle".

"Ordinary overload" of the output stage has therefore not been redesigned, but has been supplemented by an additional non-linearity elsewhere.

[31] Of course, the other stages must be overall-inverting so that this outer loop gives feedback that is negative.

[32] No capacitor is on show in the circuit diagrams (e) and (f). However there is a capacitor: it is internal to the op-amp and is there to control that op-amp's gain–frequency relation. In the jargon, the capacitor provides "internal compensation".

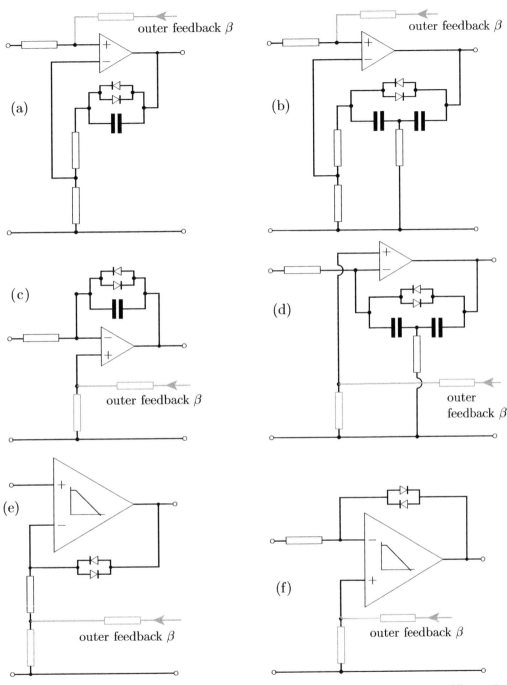

Fig. 30.17 Six "active" circuit modules that implement non-linearities inspired by that of Fig. 30.15. All are suitable for use as input or middle stages in a multi-stage amplifying circuit, though modules (c) to (f) may be best reserved for middle stages, for reasons discussed in the text and in the comments following problem 30.7. When one of these modules is used as the input stage, the outer feedback loop of the entire multi-stage amplifier may be connected as shown in grey. In (a) and (b) it is also possible to connect the outer feedback into a "split" vertical resistor after the fashion of (e).

The "split" vertical resistor in (a) and (b) makes it possible to adjust the overall gain of the module, independently of the frequency dependence, if the sum of the two component resistors is held constant.

[33] The discussion in the present chapter is as unsophisticated as I can make it. Local feedback loops within the outer loop are regarded simply as implementing building blocks whose own stability can be taken for granted. All discussion of stability and safety margins is focused on the outer loop.

If a more general discussion is called for, there are theorems governing systems having multiple feedback loops. See Bode (1945), § 8.8.

[34] See e.g. Baumwolspiner (1972).

[35] Vestiges of linear behaviour must not be unthinkingly assumed for the describing function. In particular, Kramers–Kronig relations do not link the real and imaginary parts of $g\,\mathrm{e}^{-\mathrm{j}\varphi}$; and likewise φ and $\ln g$ are not related by an analogue of (30.6). It will be no surprise that the actual φ tends to be greater (i.e. less friendly) than what (30.6) might suggest.

[36] In Fig. 30.18 we discuss a module that is a component within a larger feedback-amplifier circuit. A describing function gives the "local" gain of this module. We may plot the describing function of such a module in a Nyquist way. However, the quantity plotted is not a loop gain, so there is no significance to the curve's location relative to the point -1.

is what motivates the use of a Nyquist-stable loop gain, then it is safest to restrict circuits (c) to (f) to use as "middle" stages only. Problems 30.5–7 explore these statements.

Circuits 30.17 (a)–(f) are not thought of here as additional gain stages, to be slotted into "gaps" within a tentative circuit design—though they might be. Rather, their capacitors and diodes may (if convenient) be placed opportunistically round a gain stage or stages already present. We may see such opportunistic placing in the first stage of the demonstration circuit shown in Fig. 30.19.

Figures. 30.15 and 30.17 are developments of a single idea. Even so, they supply a useful variety of implementations that can be incorporated into a feedback amplifier or other control system.[33]

30.9.3 Describing function

The alert reader may have noticed a difficulty underlying the discussion of §§ 30.9.1–2. When a sinewave $V_0 \cos(\omega t)$ is input to a highly non-linear circuit, the output waveform must be seriously non-sinusoidal. How then is an effective gain, or phase lag, to be defined?

We take the non-sinusoidal output waveform and decompose it into a Fourier series: fundamental and harmonics. We discard the harmonics and treat the fundamental as a "best-fitting sinewave". If this best-fitting sinewave is $g V_0 \cos(\omega t + \varphi)$, then the circuit gives an effective gain $g(\omega)$ (real) accompanied by an effective phase lag $\varphi(\omega)$. An effective sinewave gain, known in the trade as the **describing function**,[34] is

$$\text{describing function} \equiv g(\omega)\,\mathrm{e}^{-\mathrm{j}\varphi(\omega)}. \tag{30.10}$$

All mention of "effective gain" and "effective phase lag" can now be understood as applying to $g(\omega)$ and to $\varphi(\omega)$: the describing function is set up, as far as is possible,[35] to mimic an ordinary linear gain.

Even at a single frequency ω, the values of $g(\omega), \varphi(\omega)$ depend also on how hard the non-linear circuit is driven: on the magnitude of V_0.

Nothing prevents us from displaying $g\,\mathrm{e}^{-\mathrm{j}\varphi}$ on a Nyquist-style plot: we plot g as radius and φ as polar angle.[36] However, because everything depends on V_0, there is now a family of Nyquist plots, each curve depending on how hard the circuit is driven. Figure 30.18 gives data from which a set of "effective" Nyquist plots can be sketched (problem 30.8).

We now have a respectable way of understanding how a Nyquist plot can "shrink without change of shape" or "'change shape at the same time as shrinking" when its circuit is driven into a non-linearity.

In the present chapter, it is assumed, and the assumption is usually acceptable, that the discarded Fourier harmonics are indeed negligible, so they have little effect in determining whether the overall feedback is stable or not. At any rate, generous safety margins are envisaged, so harmonics are unlikely to be harmful to stability even if their effect is not completely negligible.

30.9.4   Clipping amplifier having series feedback

We discuss the circuit of Fig. 30.18, a simplified version of Fig. 30.17(a).
It forms a part of what in Chapter 29 is called the "gain block" for an
entire feedback amplifier.

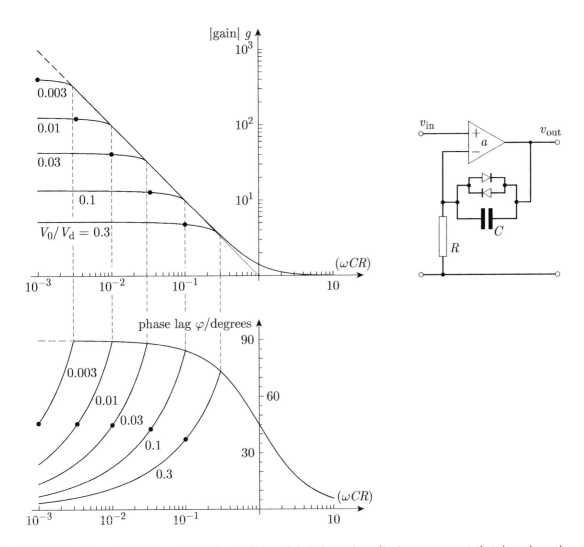

Fig. 30.18 A circuit module having series feedback. The module is driven into clipping to an extent that depends on the peak sinusoidal input voltage V_0.

In each plot, the uppermost curve shows the |gain| and phase lag ϕ for the linear regime where the diodes do not conduct. The upper graph also shows, as a thin line, the $1/(\omega CR)$ asymptote. Other curves show the effective gain g and phase lag φ (defined by a describing function) for selected values of V_0/V_d, where V_0 is the peak of the sinewave input voltage and V_d is the clipping voltage of the diodes.

We indicate by blobs what happens when the input voltage is 3× that which just brings the diodes to the threshold of conduction: we say that the circuit is then driven with a "clipping factor" of 3. In round numbers, a clipping factor 3 reduces the phase lag by a factor 2 and the gain by a factor between 2 and 2.5.

"Effective" Nyquist plots, perhaps one for each V_0/V_d, can be drawn by reading off values of g and of φ; their construction is left to the reader: problem 30.8.

The gain a of the triangle-symbol gain block in Fig. 30.18 is taken t be linear, which includes being free from slew-rate limiting, and to b high ($|a| \gg 1$) for all frequencies of interest.

Feedback makes v_- "track" the input voltage v_{in}; compare with eq (29.11). In the linear regime, where the diodes don't conduct,

$$\text{gain} = \frac{v_{\text{out}}}{v_{\text{in}}} = 1 + \frac{1}{\mathrm{j}\omega CR}; \qquad |\text{gain}| = \sqrt{1 + \frac{1}{(\omega CR)^2}};$$

$$\text{phase lag } \phi = \frac{\pi}{2} - \arctan(\omega CR).$$

The modulus of the voltage gain has asymptote $1/(\omega CR)$ for $\omega CR \ll 1$ and asymptote 1 for $\omega CR \gg 1$; the asymptotes[38] meet at a corner when $\omega CR = 1$.

The input to the module is $v_{\text{in}} = V_0 \cos(\omega t)$. Some input voltages V are large enough to drive the diodes into conduction during a part of each cycle. I have modelled the diodes in a simple way:

- during conduction, the forward voltage across a diode takes the fixe value V_{d} (we refer to this curtailment as "clipping");[39]
- when the forward voltage across a diode is less than V_{d}, the curren taken is zero.

At the threshold of diode conduction, V_0 takes the value

$$(V_0)_{\text{threshold}} = V_{\text{d}}\,(\omega CR),$$

and for larger inputs we define a

$$\text{clipping factor} = \frac{V_0}{(V_0)_{\text{threshold}}} = \frac{V_0}{V_{\text{d}}} \frac{1}{(\omega CR)}.$$

When the voltage output is non-sinusoidal owing to diode conduction it has its fundamental Fourier component extracted (describing func tion), and g, φ are calculated from it. Figure 30.18 shows how clippin reduces the gain g and phase lag φ for selected values of V_0/V_{d}.

Blobs on Fig. 30.18 show what happens to the circuit's gain and phas when the input voltage is such as to give a clipping factor of 3.

Figure 30.18 applies in the first instance to the circuit module (a) c Fig. 30.17. But it also applies to module (c) if that module is bein used as an input stage and we are considering how signals pass roun the outer feedback path.

30.9.5   Other clipping-amplifier modules

I have performed calculations, using the same simplifying assumption as yielded Fig. 30.18, on modules (c) and (d) of Fig. 30.17.[40] Results ar not displayed here as this chapter is more than technical enough. For tunately, the findings for all three modules can be summarized succinctl in a single rule of thumb:

- An input that gives a clipping factor 3 reduces the (describing function) gain g by a factor 2–2.5, and at the same time halve the (describing-function) phase lag φ.

[38] At the lowest frequencies, the gain does not rise to infinity, but reaches a ceiling at the zero-frequency gain of the gain block. This limit lies at frequencies far below the range of interest here, and is not taken account of in the graphs.

Likewise, not shown in Fig. 30.18 is a fall from the "level" asymptote at some frequency ($> 1/2\pi CR$) where the amplifying triangle gives up and $|a|$ falls through 1. Compare with the right-hand asymptote of Fig. 30.14, which is drawn dotted to indicate such a fall at a probably-uncertain frequency. (Remember that Fig. 30.18 is only a part of the whole gain block that gives the A of Chapter 29).

[39] I refer to "clipping" when diodes are driven into non-linear conduction. If the output stage of a multi-stage amplifier is driven non-linear by too large an input, or finds its output close to a supply rail for some other reason, this will be called "overload".

[40] The module of Fig. 30.17 (b) is more complicated to analyse than the others; I confess to its omission.

The circuit module of Fig. 30.18 (and by extension all the modules of Fig. 30.17) has the property envisaged at the end of § 30.8: its Nyquist plot changes shape in a helpful way when the module is driven into clipping. Moreover, that change is dramatic, in line with a "gross" non-linearity, even when the clipping factor is modest, only a factor 3 or so.

Given the availability of modules such as that of Fig. 30.18, for use as building-blocks within an "outer" feedback amplifier, Nyquist stability of that outer loop is a realistic prospect. We intend that any imagined incipient oscillation should drive the diodes in such a way as to pull the Nyquist curve away from the danger region near the point −1. Figure 30.19 shows that this possibility is achievable.

30.10   A demonstration circuit

Figure 30.19 exhibits a demonstration amplifier; its Nyquist plot is given in Fig. 30.20. The loop phase lag exceeds π for frequencies between $f_{\pi 1} = 651\,\text{Hz}$ and $f_{\pi 2} = 120\,\text{kHz}$, yet the diodes in the first stage make the whole circuit stable—Nyquist-stable.[41]

The circuit of Fig. 30.19 has component values listed in problem 30.9.

The amplifier is designed to have a closed-loop voltage gain $G = -10$ at frequencies from zero to $20\,\text{kHz}$; $|G|$ actually runs level to a corner frequency at $125\,\text{kHz}$. It is also to have an outer-loop gain $|A\beta|$ at $20\,\text{kHz}$ of at least 1000; it is actually 1074.[42]

The outer feedback loop has drop-dead frequency $628\,\text{kHz}$, where the f^{-1} asymptote of $|A\beta|$ passes through unity. The actual $|A\beta| = 1$ frequency is $622\,\text{kHz}$.[42]

The outer-loop phase lag ϕ exceeds π within the frequency range from $f_{\pi 1} = 651\,\text{Hz}$ to $f_{\pi 2} = 120\,\text{kHz}$. Nevertheless, at $622\,\text{kHz}$ the outer-loop

[41] I have constructed this circuit and subjected it to searching tests, trying hard to provoke it into instability or some other seriously unwelcome behaviour. And it seems to be as docile as you could wish. However, the signal source adds its output impedance to R_5, and the amplifier becomes unstable if the total exceeds about $20\,\text{k}\Omega$.

The crucial test, of course, is whether stability is dependent on the diodes: does the circuit go unstable if the diodes are disconnected? To my embarrassment, my design is so conservative that the circuit remains stable. However, it does become less tolerant of resistance in series with R_5.

Acknowledgements to several colleagues in the Oxford University Physics Department for their kind assistance in making available facilities for building and testing the circuit.

[42] This figure includes allowance for "rounding the corners" on a Bode plot.

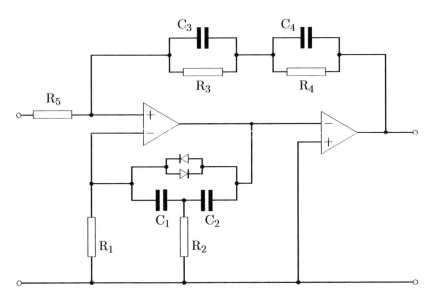

Fig. 30.19 A two-stage amplifying circuit, designed to implement Nyquist stability.

The gain blocks are taken to be opamps type TLO81, which are internally compensated (gain $\propto f^{-1}$ over a wide range) with a unity-gain frequency of $3\,\text{MHz}$. Component values are given in problem 30.9.

Fig. 30.20 The Nyquist plot for the outer-loop gain of the circuit of Fig. 30.19. The plot is calculated for linear small-signal conditions: the diodes are not conducting, and the output stage is not overloaded. The loop phase lag exceeds 180° between $f_{\pi 1} = 651\,\text{Hz}$ and $f_{\pi 2} = 120\,\text{kHz}$, and takes a greatest value of 223° at 25 kHz. At the frequency where $|A\beta| = 1$ the loop phase lag is 129°, leaving a phase margin of 51°.

As with Fig. 30.13, the plot is presented on "rubber graph paper", with $|A\beta|$ linear between 0 and 2, logarithmic above 10, and following an interpolation function in between.

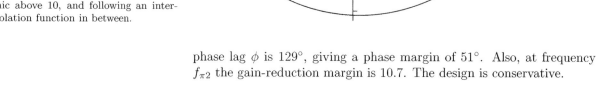

phase lag ϕ is 129°, giving a phase margin of 51°. Also, at frequency $f_{\pi 2}$ the gain-reduction margin is 10.7. The design is conservative.

30.10.1   How do we know that a Nyquist-stable circuit is in fact stable?

The amplifier of Fig. 30.19 can be used to exhibit the principle.

If an amplifier is unstable, it will be because it oscillates[43] at some frequency f. A sinewave of frequency f, traced round the circuit somewhat after the fashion of problem 29.5, undergoes non-linearity, the result of which is another sinewave (defined by a describing function). If there is oscillation at frequency f then this sinewave "joins onto itself":[44] $A\beta = -1$. This simple statement encapsulates two conditions:

- $|A\beta| = 1$: the loop gain has been brought down to 1 (from some higher value—otherwise oscillation cannot possibly start)
- $\varphi = \arg(A\beta) = \pi$.

Our demonstration circuit is so designed that there is no frequency f at which these conditions can be met together.[45] Let's explain.

In Figs. 30.12 and 13 we have introduced frequencies $f_{\pi 1}, f_{\pi 2}$ at which, in the linear regime, $A\beta$ is real and negative, equivalently the loop phase lag ϕ is π. For the circuit of Fig. 30.19, $f_{\pi 1} = 651\,\text{Hz}$, $f_{\pi 2} = 120\,\text{kHz}$. If the circuit is to oscillate, it will be in the range between these frequencies.

When diode non-linearity is added to this picture, it brings down the (describing-function) loop phase lag φ, enough that the second oscillation condition cannot be met, and is in fact precluded for all frequencies between $f_{\pi 1}$ and $f_{\pi 2}$, and therefore for all frequencies.

Consider the possibility of oscillation at frequency $f_{\pi 2}$. At this frequency, the linear $|A\beta| = 10.7$; the gain-reduction margin is 10.7. If

[43] A non-oscillating instability is also possible, if the output voltage moves towards a supply rail and stays there. This kind of instability would require a separate investigation, not given here.

[44] The negative sign comes from the subtraction in Fig. 29.1.

[45] When an oscillator is being designed, we attempt to make these two conditions compatible and orthogonal. Orthogonality means that changes of gain (non-linearity, temperature, ...) are designed to have minimal effect on the loop phase lag so as not to "pull" the frequency. In the present chapter, orthogonality is what we *don't* want.

the circuit is imagined to oscillate at $f_{\pi2}$, non-linearity must reduce the effective loop gain (describing function) by a factor $1/10.7$. For voltages large enough to do that, non-linearity is all-important because it reduces φ along with $|A\beta|$. Non-linearity occurs in the first-stage diodes, but also (a slight complication) in the output stage.

Still thinking about possible oscillation at frequency $f_{\pi2}$: overload in the second stage results in a peak output of about 12 V (TLO81 data). If this stage did not overload, it would give output lagging its input by 90°. We make the simplifying assumption that this phase lag remains close to 90° even during overload.[46]

A voltage of 12 V (peak) at the second-stage output is carried, via the outer feedback path, to the + input of the first op-amp, and from there to the diodes where, at frequency $f_{\pi2}$, it gives a "clipping factor" of about 8.6. According to the rule of thumb proposed in § 30.9.5, this should more-than-halve the phase lag that is contributed by that stage to the loop gain $A\beta$. The first-stage phase lag under linear conditions is 110°, so is reduced by the non-linearity by more than 55°. The circuit cannot be oscillating at $f_{\pi2} = 120$ kHz.

At frequencies below $f_{\pi2}$, there is more loop phase lag to be reduced by non-linearity, but there is also a larger loop gain available for its reduction. The clipping-reduced loop phase lag should be even smaller than at $f_{\pi2}$: there is *no* frequency at which the two conditions for oscillation can be met. The circuit is stable.

[46] The available output voltage is less than at lower frequencies, partly because of slew-rate limiting: a large sine-wave input to the second op-amp yields an output not too different from a triangle-wave whose fundamental Fourier component lags the input by 90°. The phase lag is therefore much the same under overload conditions as it is when the amplifier is operated in its linear regime.

Uncertainty here is well covered by the phase margin of 51°, together with the diode-induced reduction of the outer-loop phase lag ϕ.

Problems

Problem 30.1 The Bode plot of Fig. 30.4
Confirm the quantitative statements made in example 30.1 about the behaviour of $|A\beta|$ and ϕ in Fig. 30.4.

Make a similar analysis of the $A\beta$ for example 30.2. Show that $1/\sqrt{2}$ is changed to $\frac{1}{2}$, 5% to 11%, and 18° to 37°.

Investigate "dangerously close" at the end of example 30.2.

Problem 30.2 The Nyquist plot of Fig. 30.5
Show that the curve drawn in Fig. 30.5 is a semicircle, and that the annotations marked on it are correct.[47]

[47] *Hint:* Work out $(A\beta - L_0/2)$ and show that it is a complex number of magnitude $L_0/2$.

Problem 30.3 The all-pass filter
Work out the circuit equations for the circuit of Fig. 30.10, and show that

$$\frac{v_{out}}{v_{in}} = \frac{A}{A+2}\frac{1-j\omega CR}{1+j\omega CR}.$$

In a practical case, we shall ensure that $|A| \gg 1$ in the frequency range of interest, so this expression is dominated by its second factor, whose modulus is 1, independent of frequency.

Problem 30.4 The circuit of Fig. 30.16
Refer to Fig. 30.16. Show that without r the "gain" $|v_{\text{out}}/v_{\text{in}}|$ is 1 at low frequencies but falls as ω^{-1} from a corner at $\omega CR = 1$. Show that r changes the zero-frequency gain to $r/(R+r)$ and raises the corner frequency to be at $\omega CRr/(R+r) = 1$. Show that the phase-lag curve is shifted to the right by a factor $(R+r)/r$, and so r reduces the output's phase lag at frequencies between the two corners. This is the lag-reducing property sought by introducing the diodes in Fig. 30.15.

Problem 30.5 Gains for signal paths and feedback paths
In problems 30.5–7, we pick up statements made in § 30.9.2, some of which probably seemed obscure there.

Consider the circuit configuration shown in Fig. 30.21. The gain block has unequal gains for its two inputs, quite apart from the difference of sign: gain a for the upper input and gain $-\alpha$ for the lower input. The circuit is linear, so

$$v_{\text{out}} = av_{\text{upper}} + (-\alpha)v_{\text{lower}}.$$

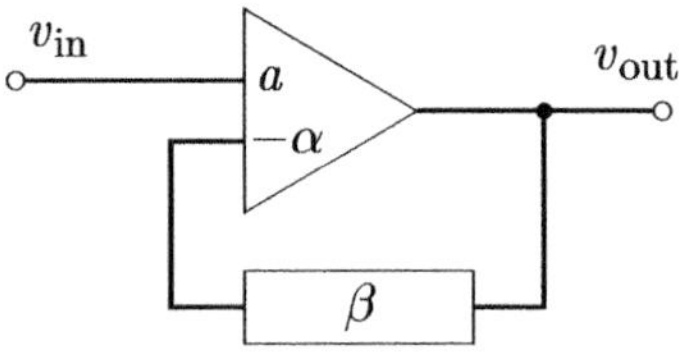

(1) Show that the closed-loop gain is

$$\frac{v_{\text{out}}}{v_{\text{in}}} = \frac{a}{1+\alpha\beta}, \qquad \text{which tends to} \qquad \frac{1}{\beta}\frac{a}{\alpha} \quad \text{as } |\alpha\beta| \to \infty. \qquad (30.11)$$

Thus, when the gains are unequal for the two "channels", the closed-loop gain contains their ratio a/α, as well as the usual $1/\beta$. This perhaps-surprising property has a nuisance value that we investigate in problems 30.6 and 30.7.

Problem 30.6 An amplifier having a badly chosen input stage
Consider the circuit of Fig. 30.22. We treat this amplifier using linear circuit theory: no diodes.
(1) Show that

$$v' = \frac{a_1(1+j\omega CR)\beta v_{\text{out}} - a_1 v_{\text{in}}}{1+j\omega CR(a_1+1)}.$$

(2) Argue that, in the absence of βv_{out}, the input stage gives gain

$$\frac{v'}{v_{\text{in}}} = \frac{-a_1}{1+j\omega CR(a_1+1)},$$

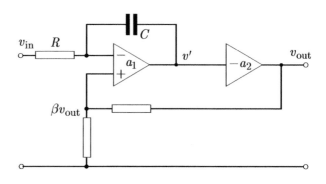

Fig. 30.21 In this feedback circuit, the gain block has gain a for voltages applied to the upper input, and $-\alpha$ for voltages applied to the lower input. A fraction β of the output voltage is fed back to the lower input.

Perhaps surprisingly, the closed-loop gain is not $1/\beta$ but is given by (30.11).

Fig. 30.22 An amplifier whose first stage is based on the module of Fig. 30.17(c). The diodes are omitted, as problem 30.6 deals with the circuit's linear behaviour only.

and that the gain corresponding to the a in (30.11) is $-a_2 \times$ this, so

$$\left(\frac{v_{out}}{v_{in}}\right)_{\beta=0} = a = \frac{a_1 a_2}{1 + j\omega CR(a_1 + 1)}. \tag{30.12}$$

Now set $v_{in} = 0$, disconnect the β feedback path, and apply a new input signal v_+ to the first amplifier's + input. Show that

$$\left(-\frac{v_{out}}{v_+}\right)_{\beta=0;\,v_{in}=0} = \alpha = \frac{a_1 a_2(1 + j\omega CR)}{1 + j\omega CR(a_1 + 1)}. \tag{30.13}$$

(3) Show that the entire circuit of Fig. 30.22 gives closed-loop gain

$$\frac{v_{out}}{v_{in}} = \frac{a_1 a_2}{1 + a_1 a_2 \beta(1 + j\omega CR) + (a_1 + 1)j\omega CR} = \frac{a}{1 + \alpha\beta}. \tag{30.14}$$

We confirm that the gains a and α are different, the fact that motivates this problem. This difference affects the frequency dependences of the gains, as is shown by eqns (30.12) and (30.13): $\alpha/a = 1 + j\omega CR$. Even in the limit $a_1 a_2 \to \infty$, v_{out}/v_{in} does not tend to $1/\beta$.

Problem 30.7 A better design
Consider the circuit of Fig. 30.23.
  Show that

$$\frac{v_{out}}{v_{in}} = \frac{-a_1 a_2(1 - \beta)}{a_1 a_2 \beta + 1 + \dfrac{a_1 j\omega CR}{1 + j\omega CR}}. \tag{30.15}$$

Show that when $a_1 a_2 \to \infty$, this tends to $-(1 - \beta)/\beta$ in conformity with (29.14).
  Note that the closed-loop gain is not quite independent of (ωCR), but the contribution of (ωCR) to v_{out}/v_{in} diminishes with increase of the local gain a_2. This was not the case with Fig. 30.22.

Comment: Comparison of eqns (30.14) and (30.15) shows why the module of Fig. 30.17 (a) is superior to that of (c) for use as the input stage of a feedback amplifier. Generalizing: series-feedback modules (a), (b), (e) are superior to shunt-feedback (c), (d), (f). This holds for "input" stages; for "middle" stages all modules are suitable.

Comment: A particular issue is linearity. The diodes in the circuits of Fig. 30.17 are there to short-circuit a capacitor at need, in order

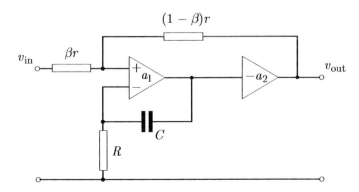

Fig. 30.23 An amplifier circuit whose first stage is based on the module of Fig. 30.17(a). Diodes are omitted as the analysis of problem 30.7 is concerned with linear behaviour only.

to engineer the stability of the outer feedback loop. Both expressions (30.14) and (30.15) contain C, and we can take that as an indicator that diode conduction will put *some* non-linearity into each amplifier's closed-loop gain, via an "effective" C.

In a carefully designed circuit, the diodes will not conduct when a signal v_{in} is applied that is within the design range of frequencies and voltages. (See problem 30.9(2) for confirmation of this in a specific case.) The diodes are there to wake up in the event of an incipient instability. Nevertheless, there may be a small non-linear current in the diodes even under "normal" conditions.[48] If this is of concern, Fig. 30.23 is preferable to Fig. 30.22, as is signalled by the less conspicuous location of C in the expression for the gain.

[48] We are repeating a point made in § 30.9.2.

Problem 30.8   "Effective" Nyquist plots for Fig. 30.18
Draw "effective" Nyquist plots using the curves in Fig. 30.18. That is, use the describing-function g for the radial coordinate and φ for the phase angle. You should be able to do this in a sketch-graph way without needing to read detailed numerical values from the plots.

You may decide to plot five curves, each for a chosen fixed V_0/V_d, but there are other possibilities.

Problem 30.9   The demonstration circuit of Fig. 30.19
I have chosen the following component values:

$$C_1 = 47\,\text{pF}, \quad C_2 = 470\,\text{pF}, \quad C_3 = 3.3\,\text{pF}, \quad C_4 = 47\,\text{pF},$$
$$R_1 = 15\,\text{k}\Omega, \quad R_2 = 2.2\,\text{k}\Omega, \quad R_3 = 15\,\text{k}\Omega, \quad R_4 = 27\,\text{k}\Omega, \quad R_5 = 4.3\,\text{k}\Omega.$$

The op-amps, type TLO81, each give voltage gain 2×10^5 from 0 to 15 Hz, then falling as f^{-1} to 1 at 3 MHz.

Show that the low-frequency closed-loop voltage gain

$$G = -\frac{R_3 + R_4}{R_5} = -\frac{15 + 27}{4.3} = -9.8.$$

[49] We are using the interpretation of (29.13). In the notation there, we ignore the initial attenuation $Z_2/(Z_1 + Z_2)$ as irrelevant to stability. Then the open-loop gain is $-\alpha$, here called A, and $\beta = -Z_1/(Z_1+Z_2)$, here the division factor that yields $-v_+/v_{out}$ when $v_{in} = 0$.

We further consider the circuit in two obvious blocks:[49] The open-loop gain $A = v_{out}/v_+$ is provided by the two op-amps together with the local-feedback components C_1, C_2, R_1, R_2; the diodes are treated as non-conducting when calculating the linear amplification. The feedback fraction β is provided by the voltage divider R_5 together with C_3, C_4, R_3, R_4. We deal with these blocks separately.

Show that the linear open-loop gain A has the following dependences on frequency:
$A = -4 \times 10^{10}$ at frequency $f = 0$
$|A| = 4 \times 10^{10}$ from $f = 0$ to 15 Hz
$|A|$ falls as f^{-2} to 11.6 kHz where it is 6.7×10^4
$|A|$ falls as f^{-3} to 126 kHz where it is 52.6
$|A|$ falls as f^{-2} to 277 kHz where it is 10.8
$|A|$ falls as f^{-1} to 3 MHz   where it is 1;
above 3 MHz the op-amps deteriorate in a manner not given.
The frequencies given are corner frequencies between different dependences on frequency f.

Next we turn attention to the feedback fraction β. The components form a voltage divider giving $v_+ = -\beta v_{\text{out}}$. The "lower end" of this divider is the impedance from v_+ to earth ("earth" meaning the zero-voltage rail). There are two impedances in parallel here

- R_5 in series with the output impedance of the signal source
- the input capacitance of the first op-amp, from its v_+ input to earth.

The input capacitance has two contributors: the capacitance from v_+ to earth; and the capacitance from v_+ to v_-. The first of these is constant. The second is affected by the feedback, and is reduced to the extent that v_+ tracks v_-. The tracking becomes less effective as the frequency is increased, and this capacitance then becomes more significant. For the preparation of Fig. 30.20 I've assumed that both capacitances sum to a constant 20 pF. Also, for Fig. 30.20, I've assumed that the signal source has zero output impedance[50] so the effective R_5 remains at $4.3\,\text{k}\Omega$.

(1) Show that, on these assumptions:

$\beta = -0.093$ at frequency $f = 0$

$|\beta| = 0.093$ from $f = 0$ to $125\,\text{kHz}$

$|\beta|$ rises as f to $283\,\text{kHz}$ where it is 0.21

$|\beta|$ runs level to $2.1\,\text{MHz}$ where it is 0.21

$|\beta|$ falls as f^{-1} to $3.2\,\text{MHz}$ where it levels at 0.13.

Draw Bode plots for $|A|$, for $|\beta|$, and for $|A\beta|$, showing how the corner frequencies for A and β work together. I have drawn the Nyquist plot for $A\beta$ in Fig. 30.20.[51]

(2) Consider conditions when a signal at 20 kHz is being handled. The output stage commences overload when its output voltage is 13.5 V. If the voltages present are just below those that overload the output stage, then the input to that stage is $13.5(20\,\text{kHz}/3\,\text{MHz}) = 90\,\text{mV}$. This is well below the voltage at which the diodes conduct. Confirm the statements in the text: that the diodes have no deleterious effect on the circuit at signal frequencies, even though they wake up if there is a need to suppress an instability.

[50] Experiments with a real circuit confirm the results of calculation: stability is conditional on the effective R_5 not being too big. There are signs of instability if the source's output impedance and R_5 add to more than about $20\,\text{k}\Omega$.

[51] The Nyquist plot of Fig. 30.20 exhibits a loop phase lag ϕ that takes a greatest value of $223°$, achieved at 25 kHz. It is this large phase lag that makes possible a large loop gain $|A\beta|$ — larger than would otherwise have been possible—at 20 kHz.

XIII Relatively advanced material

Quantization of waves: the stretched string

31

Intended readership: Third/fourth year, or within a theoretical-physics course that considers quantization of waves, above all of electromagnetic waves.

31.1 Introduction

We are familiar, from the beginning of our acquaintance with quantum physics, with the idea that waves have energy that comes in lumps. Above all, we learn that electromagnetic waves are quantized, and the lumps are called **photons**. Later, we may learn that sound waves in solids are similarly quantized, the lumps being called **phonons**. And there are other quantized waves too. Yet the demonstration that quantum mechanics yields these quantizations is not encountered until an advanced course—and even then it may seem to be remarkably complicated and obscure.

We attempt here to give a "soft introduction" by applying quantum rules to the quantization of transverse vibrations on a string.[1]

The simplest case, for calculation purposes, is that of a string having fixed ends, supporting standing waves. However, we know since Chapter 6 that running waves are more useful and realistic.[2] Running waves introduce the complication of degeneracy: waves having the same frequency may travel towards either $+x$ or $-x$. Degeneracy makes calculation somewhat untidy. Once the degeneracy is mastered (which is as far as this chapter goes) the extension to photons or phonons (three dimensions and more than one polarization) is relatively straightforward.

[1] Any other system supporting one-dimensional waves, such as an electrical transmission line, would do equally well.

[2] This shows up most clearly in Chapter 24, where the phonons involved in heat conduction are necessarily travelling so as to transport energy and (crystal) momentum.

31.2 Orientation: quantization procedure

In classical physics, the vibrations of a string are best described in terms of **normal modes** or **eigenfunctions**: those motions that can exist on their own, uncoupled to others,[3] and whose superposition can describe any overall motion that is physically allowed. The normal modes are standing waves (if the string has fixed ends) or running waves (if we prefer this eigenfunction set). With either choice of normal-mode set, the motion for any one mode is sinusoidal in time: there is an obvious similarity to a simple harmonic oscillator. It is the normal modes that

[3] We are working in a lowest approximation, where the equations of motion are linear, and therefore there can be a linear superposition of these independent motions.

Essays in Physics: Thirty-two thoughtful essays on topics in undergraduate-level physics. Geoffrey Brooker, Oxford University Press. © Geoffrey Brooker 2021. DOI: 10.1093/oso/9780198857242.003.0031

will be subjected to quantum rules, and it will be no surprise if we find ourselves setting up raising-and-lowering operators.

There is a standard procedure by which we quantize anything new. It consists of the following steps:

- choose "generalized coordinates" $q_r(t)$ appropriate to a description of the physical system in classical mechanics
- set up the system's Lagrangian $\mathcal{L}$ in terms of those coordinates
- use the Lagrangian to identify the momentum $p_r = \partial\mathcal{L}/\partial\dot{q}_r$ conjugate to each q_r
- "reinterpret" the observables $q_r(t)$, $p_r(t)$ as operators (time independent) whose commutation rule is $[\hat{q}_r, \hat{p}_r] = i\hbar$
- solve the quantum problem (Schrödinger representation) using these operators.

The above is the most elementary interface between classical and quantum physics; there are of course others. But our aim in this chapter is to use the most earthy route that we can.[4]

The classical generalized coordinates $q_r(t)$ may be anything appropriate to the problem in hand. Since the independent motions of our string are the normal modes, it is most natural to take $q_r(t)$ as the amplitude of the rth of these. We should not be alarmed that this seems to be stretching the idea of a "coordinate"; it is precisely the kind of thing that Lagrangian mechanics permits—and indeed encourages.

Given the procedure itemized above, we must start by setting up the physics of a stretched string using classical mechanics.

[4] It is assumed that the reader has already encountered Lagrangian mechanics, together with these rules interfacing classical mechanics with quantum mechanics. Lagrangian mechanics, applied to the rather similar case of a monatomic chain, is encountered in problem 23.2.

31.3 Classical mechanics of a stretched string

Let the stretched string lie, when undisturbed, along the x-axis from $x = 0$ to $x = L$, and let displacements of it, transverse, lie wholly in the y-direction. The string has mass per unit length ρ and tension τ. Then (problem 31.1)

$$\text{(kinetic energy)} = \frac{\rho}{2}\int_0^L \left(\frac{\partial y}{\partial t}\right)^2 \mathrm{d}x, \qquad (31.1\text{a})$$

$$\text{(potential energy)} = \frac{\tau}{2}\int_0^L \left(\frac{\partial y}{\partial x}\right)^2 \mathrm{d}x, \qquad (31.1\text{b})$$

$$\text{(mean energy flow towards }+x) = -\frac{\tau}{L}\int_0^L \frac{\partial y}{\partial x}\frac{\partial y}{\partial t}\,\mathrm{d}x. \qquad (31.1\text{c})$$

31.3.1 Choice of eigenfunctions: standing waves

Standing waves give the simplest case for us to quantize. The steps are spelt out in problem 31.4. The calculation is relegated to a problem so that the bulk of this chapter can concentrate on the more important (but messier) case of running waves.

31.3.2   Choice of eigenfunctions: travelling waves

The physical system we wish to think about is a very long string. We treat the string within a large "quantization length" L, a length unrelated to the (perhaps even larger) actual length of the string.

The string may be excited into a standing wave or a travelling wave, or a sum over such things. As forecast, we focus on the travelling-wave possibility; as in Chapter 6, periodic boundary conditions (a device for making the modes travelling, discrete, and forming a just-complete expansion set) are not far away.

Because travelling waves may travel along the string in either of two directions, each frequency of vibration is twofold degenerate. This complication is absent for standing waves (problem 31.4), but it must be faced now because of the greater usefulness of travelling waves.

31.3.3   The wave amplitudes: generalized coordinates

We anticipate the focusing on travelling waves by setting up an expansion of displacement y in terms of complex exponentials:[5,6]

$$ y = \frac{1}{\sqrt{\rho L}} \sum_{r=-\infty}^{\infty}{}' q_r(t)\, e^{ir2\pi x/L} = \frac{1}{\sqrt{\rho L}} \sum_{r=-\infty}^{\infty}{}' q_r(t)\, e^{irk_1 x}, \qquad (31.2) $$

where $k_1 = 2\pi/L$. The square root at the beginning is a convenient normalizing constant that prevents clutter in (31.4) below.

The sum in (31.2) must yield a real y, so $y = y^*$. For this, the integer r must run over negative as well as positive values. Given that there is a term $q_r\, e^{irk_1 x}$ in the sum of (31.2), there must be a companion containing $e^{(-i)rk_1 x} = e^{i(-r)k_1 x}$ if the two terms together are to sum to a real total for all x. And the entire "companion" must be the complex conjugate of the original. This requires that (cf. (23.24)):

$$ q_{-r}(t) = q_r(t)^* \quad \text{for all } r. \qquad (31.3) $$

It is to be understood that the $q_r(t)$ in (31.2), the coefficients of the $e^{irk_1 x}$ normal modes, are generalized coordinates, to be used in due course in Lagrangian mechanics, and thence in quantization.

We substitute amplitude y into expressions (31.1) for energy quantities. We must remember to use two different summation indices where there is a product of sums, but the orthogonality of the complex exponentials then simplifies things to a single sum.

$$ (KE) = \frac{\rho}{2} \frac{1}{\rho L} \int_0^L dx \sum_r{}' \dot{q}_r(t)\, e^{irk_1 x} \sum_s{}' \dot{q}_s(t)\, e^{isk_1 x} $$

$$ = \frac{1}{2} \sum_r{}' \sum_s{}' \dot{q}_r(t)\, \dot{q}_s(t) \int_0^L e^{irk_1 x}\, e^{isk_1 x}\, \frac{dx}{L} $$

$$ = \frac{1}{2} \sum_{r=-\infty}^{\infty}{}' \dot{q}_r\, \dot{q}_{-r}, \qquad (31.4) $$

[5] There are similarities to the discussion of vibrational motion of a monatomic linear chain of atoms, as rehearsed in problem 23.2. Each discussion may help with the understanding of the other.

Equations (31.1) show that we are to multiply together quantities obtained from y, so it is important that y is real. We are *not* using—we must not use—a "take-the-real-part-at-the-end" convention. Complex-exponential expressions remain available and convenient, but those expressions must be assembled in such a way as to give a y that is real overall.

We may also remember that quantum mechanics is set up so that its operators and wave functions are complex. We cannot introduce another complex convention without conflict.

[6] There is an irritating detail. A displacement $q_0(t)\, e^{i0k_1 x}$ represents a uniform displacement of the entire string. Given the absence of any externally applied forces, such a uniform displacement is time independent, and so takes no part in the physics under discussion. We shall understand that $\sum'$ excludes $r = 0$.

since the integral is $\delta_{r,-s}$. In a similar way, we evaluate the potential energy and the energy flow (problem 31.2):[7]

$$(\text{PE}) = \frac{1}{2} \sum_{r=-\infty}^{\infty}{}' (r\omega_1)^2\, q_r\, q_{-r}, \qquad (31.5)$$

$$(\text{mean energy flow to } +x) = \mathcal{F} = \frac{1}{L}\sqrt{\frac{\tau}{\rho}} \sum_{r=-\infty}^{\infty}{}' (-ir\omega_1)\, q_r\, \dot{q}_{-r}. \qquad (31.6)$$

Here

$$\omega_1 = \frac{2\pi}{L}\sqrt{\frac{\tau}{\rho}} = k_1\sqrt{\frac{\tau}{\rho}}; \qquad \omega_r = |r|\,\omega_1. \qquad (31.7)$$

Incidentally, condition (31.3) does more than guaranteeing that the wave amplitude y is real; rather obviously it also ensures that the three energy expressions are also real (problem 31.3).

At this stage, we have not calculated the form of $q_r(t)$, so we must not attach any meaning[8] (such as angular frequency of an oscillation) to ω_r or ω_1. To understand how the string behaves—in classical mechanics—we use the generalized coordinates $q_r(t)$ to set up a Lagrangian.

31.3.4   The Lagrangian and generalized momenta

We continue to use classical mechanics.

The Lagrangian $\mathcal{L}$ may be evaluated by making use of (31.4) and (31.5):

$$\mathcal{L} \equiv (\text{KE}) - (\text{PE}) = \frac{1}{2} \sum_{r=-\infty}^{\infty}{}' \left(\dot{q}_r\, \dot{q}_{-r} - \omega_r^2\, q_r\, q_{-r} \right). \qquad (31.8)$$

[9] Other choices of generalized coordinate are possible. Robinson (1974) § 6.5 quantizes waves on an electrical transmission line. In our language, he divides the string into limitingly small sections centred on $x_1, x_2, \dots x_i, \dots$, and takes the $y(x_i)$ as the generalized coordinates. The outcome is the same equations (31.15) and (31.16) as we obtain here, though after algebra that is slightly longer.

Also, eqn (23.29) contains generalized coordinates c_r, s_r as alternatives to q_r, q_{-r}; and these are used again in problem 31.5.

As forecast, the quantities q_r are to be used as the generalized coordinates for waves on the string.[9] Classical mechanics tells us that there is a momentum p_r conjugate to generalized coordinate q_r, and obtained by the rule $p_r = \partial\mathcal{L}/\partial\dot{q}_r$, so we have[10,11]

$$p_r = \dot{q}_{-r}. \qquad (31.9)$$

We make a consistency check by writing the Lagrangian equation of motion:

$$0 = \frac{\mathrm{d}}{\mathrm{d}t}\left(\frac{\partial\mathcal{L}}{\partial\dot{q}_{-r}} \right) - \frac{\partial\mathcal{L}}{\partial q_{-r}} = \frac{\mathrm{d}}{\mathrm{d}t}\dot{q}_r + \omega_r^2\, q_r = \ddot{q}_r + \omega_r^2\, q_r. \qquad (31.10)$$

Equation (31.10) shows that $q_r(t)$ executes a simple-harmonic oscillation $e^{\pm i\omega_r t}$ at angular frequency $\omega_r = |r|\,\omega_1$. And the terms in (31.2) have the form $e^{\pm i\omega_r t}\, e^{irk_1 x}$. This finding could have been obtained directly from (31.7), knowing that waves on a string travel with speed $\sqrt{\tau/\rho}$. So we have used advanced tools to obtain an elementary result. This is, of course, the point of a consistency check, and we have found that all is well.

Finally, we set up the Hamiltonian $\mathcal{H}$, using the standard recipe for obtaining it from the Lagrangian:

$$
\mathcal{H} = \sideset{}{'}\sum_{r=-\infty}^{\infty} p_r\,\dot{q}_r - \mathcal{L}
$$

$$
= \sideset{}{'}\sum_{r=-\infty}^{\infty} \dot{q}_{-r}\,\dot{q}_r - \frac{1}{2}\sideset{}{'}\sum_{r=-\infty}^{\infty} \dot{q}_r\,\dot{q}_{-r} + \frac{1}{2}\sideset{}{'}\sum_{r=-\infty}^{\infty} \omega_r^2\,q_r\,q_{-r}
$$

$$
= \frac{1}{2}\sideset{}{'}\sum_{r=-\infty}^{\infty} \left(p_r\,p_{-r} + \omega_r^2\,q_r\,q_{-r} \right). \tag{31.11}
$$

This expression is, of course, (kinetic energy) plus (potential energy).

A replacement of $\dot{q}_{-r}$ with p_r in (31.6) results in:

$$
\mathcal{F} = \frac{1}{L}\sqrt{\frac{\tau}{\rho}}\,\sideset{}{'}\sum_{r=-\infty}^{\infty} (-\mathrm{i}r\omega_1)q_r\,p_r. \tag{31.12}
$$

31.4   Quantization of waves on a string

In the last section, we obtained (classical) generalized coordinates q_r and their associated momenta p_r. These give us building blocks for making the transition from classical to quantum mechanics. We must "reinterpret" the coordinates and momenta as operators, and must impose a commutation rule on those operators.

The recipe given in § 31.2 is designed to yield operators $\hat{q}_r$, $\hat{p}_r$ in the **Schrödinger representation**. That is, the operators $\hat{q}_r$ and $\hat{p}_r$ are to be independent of time, because all time dependence of the physical system is handled by the wave function Ψ. This is even though the classical variables $q_r(t)$ and $p_r(t)$ from which we start do depend on time.[12]

Following the standard recipe, we replace q_r, p_r with operators $\hat{q}_r$, $\hat{p}_r$ having commutator $[\hat{q}_r, \hat{p}_r] = \mathrm{i}\hbar$. We now find out what these operators do for us.[13]

The best thing we can do with $\hat{q}_r$ and $\hat{p}_r$ is to move on at once to constructing raising and lowering operators. These operators will join onto the physics more easily than $\hat{q}_r$ and $\hat{p}_r$ themselves.

There are two pieces of background information that help us to construct the operators we need. The first is the case of the ordinary simple harmonic oscillator, for which the raising and lowering operators are:[14]

$$
\hat{a}^+ = \frac{1}{\sqrt{2\hbar\omega}}\,\frac{\hat{p} + \mathrm{i}\omega\hat{q}}{\mathrm{i}}, \qquad \hat{a} = \frac{1}{\sqrt{2\hbar\omega}}\,\frac{\hat{p} - \mathrm{i}\omega\hat{q}}{-\mathrm{i}}.
$$

The second is a finding in problem 31.3: In the classical mechanics of a stretched string, the groupings $(p_{-r} - \mathrm{i}\omega_r q_r)$, $(p_r + \mathrm{i}\omega_r q_{-r})$ pick out from the general y of (31.2) those parts that travel towards $\pm x$. Meaning: they travel towards $+x$ when $r > 0$, and travel towards $-x$ when $r < 0$. We seize this helpful property in making the quantum definitions of $\hat{a}_r^+, \hat{a}_r$ in (31.13):

[12] Since all time dependence is absent from the operators $\hat{q}_r$ and $\hat{p}_r$, the ω_1 and ω_r of (31.7) should now be regarded as just parameters, whose connection with angular frequency will be out of sight until it reappears in (31.23).

[13] There is something a bit unusual in the apparently innocuous statement of the commutation rule here. The classical quantities q_r, p_r are complex, while "observables" are usually real. In consequence, the operators $\hat{q}_r$, $\hat{p}_r$ are not Hermitian. This point is explored in problem 31.5. It is shown there that the commutation property $[\hat{q}_r, \hat{p}_r] = \mathrm{i}\hbar$ holds after all.

[14] The oscillator has mass m, displacement y and angular frequency ω. By defining $q = \sqrt{m}\,y$, we prevent m from cluttering the operators. This is similar to the uncluttering effected by $\sqrt{\rho L}$ in (31.2).

The $\pm\mathrm{i}$ could be omitted from the denominators of $\hat{a}^+$ and $\hat{a}$, and of the $\hat{a}_r^+, \hat{a}_r$ of (31.13). But we prefer to have no i in the q_r of (31.19).

$$\hat{a}_r = \frac{1}{\sqrt{2\hbar\omega_r}} \left(\frac{\hat{p}_{-r} - \mathrm{i}\omega_r\,\hat{q}_r}{-\mathrm{i}} \right), \qquad \hat{a}_r^+ = \frac{1}{\sqrt{2\hbar\omega_r}} \left(\frac{\hat{p}_r + \mathrm{i}\omega_r\,\hat{q}_{-r}}{\mathrm{i}} \right).$$

$$(31.13)$$

In these, $\omega_r = |r|\,\omega_1$ and ω_1 are given by (31.7).

The operators of (31.13) have the usual properties of raising and lowering operators (problem 31.5(6)):

$$[\hat{a}_r, \hat{a}_s] = 0 \quad \text{and} \quad [\hat{a}_r^+, \hat{a}_s^+] = 0 \text{ for all } r, s, \qquad (31.14\mathrm{a})$$

$$[\hat{a}_r, \hat{a}_r^+] = 1, \qquad [\hat{a}_r, \hat{a}_s^+] = 0 \text{ for } r \neq s. \qquad (31.14\mathrm{b})$$

We show in problem 31.6(3) that these operators $\hat{a}_r, \hat{a}_r^+$ remove/add one vibrational-energy quantum $\hbar\omega_r$ from/to the r-wave.

There is more. We have said that eqns (31.13) cleverly engineer the forecast "helpful property": the direction of wave travel is now given by $\mathrm{sign}(r)$; this has required a slight re-definition of the way that r counts, one that does not upset the energies of (31.8, 11, 12). See sidenote 18 and, for a classical equivalent, problem 31.3(2). We show this next.[15]

Operators for the total energy $\widehat{\mathcal{H}}$ and for the energy flow $\widehat{\mathcal{F}}$ can be expressed in terms of $\hat{a}_r, \hat{a}_r^+$. They are best displayed with positive and negative subscripts separated (no longer any prime on the sum). We find:[16]

$$\widehat{\mathcal{H}} = \sum_{r=1}^{\infty} \hbar\omega_r \left(\hat{a}_r^+\,\hat{a}_r + \tfrac{1}{2} \right) + \sum_{r=1}^{\infty} \hbar\omega_r \left(\hat{a}_{-r}^+\,\hat{a}_{-r} + \tfrac{1}{2} \right), \qquad (31.15)$$

$$\widehat{\mathcal{F}} = \sqrt{\frac{\tau}{\rho}} \sum_{r=1}^{\infty} \frac{\hbar\omega_r}{L} \left(\hat{a}_r^+\,\hat{a}_r - \hat{a}_{-r}^+\,\hat{a}_{-r} \right). \qquad (31.16)$$

Problem 31.6 rehearses standard harmonic-oscillator operator algebra to show that $\hat{a}_r^+\hat{a}_r$ is a "number operator" with eigenvalue N_r, an integer (positive or zero). Then (31.15) and (31.16) yield eigenvalues

$$\mathcal{E} = \sum_{r=1}^{\infty} \hbar\omega_r \left(N_r + \tfrac{1}{2} \right) + \sum_{r=1}^{\infty} \hbar\omega_r \left(N_{-r} + \tfrac{1}{2} \right), \qquad (31.17)$$

$$\mathcal{F} = \sqrt{\frac{\tau}{\rho}} \sum_{r=1}^{\infty} \frac{\hbar\omega_r}{L} \left(N_r - N_{-r} \right). \qquad (31.18)$$

In (31.18), we see N_r quanta[17] carrying energy $\hbar\omega_r$ towards $+x$. These quanta have energy $\hbar\omega_r/L$ per unit length of string and travel with velocity $\sqrt{\tau/\rho}$ towards $+x$. This is the case for all terms containing N_r with positive integer r. Terms involving N_{-r} have a similar interpretation but deal with energy travelling towards $-x$.

Comment on wave directions: There is a similarity to another piece of wave physics. In §9.2, we need to distinguish waves that travel outwards/inwards through a large hemisphere. Equation (9.3) picks out any wave that is travelling inwards (assumed or required to be absent there), while rejecting any contribution from an outgoing wave. The discrimination uses a combination $(G\,\nabla_r U - U\,\nabla_r G)$ of U and ∇U that closely parallels (31.28a).

---

[15] The $\mathrm{sign}(r)$ property holds for any expression containing $\hat{a}_r$ or $\hat{a}_r^+$, because these operators carry that property with them. In particular, it applies to the summand in (31.20).

[16] The expressions for energy and energy flow, obtained from classical mechanics in problem 31.3(4), closely resemble eqns (31.15) and (31.16).

[17] It is necessary to exclude the possibility that N_r quanta carry energy predominantly, but not wholly, in the $+x$-direction, and similarly for N_{-r}. The fact that the energy flow for the N_r quanta is

(energy density) × (velocity)

does what we need; any mixture would give a smaller energy flow—think about it.

31.5   The quantum expression for the vibration amplitude

We pick up the expression for operator $\hat{q}_r$ from problem 31.6(1);

$$\hat{q}_r = \sqrt{\frac{\hbar}{2\omega_r}}\left(\hat{a}_r + \hat{a}_{-r}^+\right), \tag{31.19}$$

and substitute it into (31.2) to find the string displacement $\hat{y}$:

$$\hat{y} = \frac{1}{\sqrt{\rho L}}\sum_{r=-\infty}^{\infty}{}' \hat{q}_r\, e^{irk_1 x} = \frac{1}{\sqrt{\rho L}}\sum_{r=-\infty}^{\infty}{}' \sqrt{\frac{\hbar}{2\omega_r}}\left(\hat{a}_r + \hat{a}_{-r}^+\right)e^{irk_1 x}$$

$$= \sqrt{\frac{\hbar}{2\rho L}}\sum_{r=-\infty}^{\infty}{}' \frac{1}{\sqrt{\omega_r}}\left(\hat{a}_r\, e^{irk_1 x} + \hat{a}_r^+\, e^{-irk_1 x}\right). \tag{31.20}$$

This has the standard "shape" for a wave amplitude in quantum mechanics; we see a similar shape in (31.24) below.[18]

31.6   The Heisenberg representation

It is sometimes insightful to reformulate quantum mechanics so that time dependence is shifted from the wave function to the operators. The physical outcomes are, of course, unaltered, but the meaning can sometimes be more clearly conveyed. When the shift of viewpoint has been made, we say we are using the **Heisenberg representation**. The recipe for making the transformation of an operator $\widehat{A}$ is

$$\widehat{A}_{\text{Heisenberg}} = e^{i\widehat{\mathcal{H}}t/\hbar}\,\widehat{A}_{\text{Schrödinger}}\,e^{-i\widehat{\mathcal{H}}t/\hbar}.$$

We apply this transformation to the raising and lowering operators. It will help the reasoning if we supply a state vector $|N_r\rangle$, whose energy is $(N_r + \tfrac{1}{2})\hbar\omega_r$, for the operators to work on. Then[19]

$$
\begin{aligned}
\left(\hat{a}_r\right)_{\text{Heis}}|N_r\rangle &= \exp\!\left(i\widehat{\mathcal{H}}t/\hbar\right)\hat{a}_r\,\exp\!\left(-i\widehat{\mathcal{H}}t/\hbar\right)|N_r\rangle \\
&= \exp\!\left(i\widehat{\mathcal{H}}t/\hbar\right)\hat{a}_r\,\exp\!\left(-i(N_r+\tfrac{1}{2})\omega_r t\right)|N_r\rangle \\
&= \exp\!\left(-i(N_r+\tfrac{1}{2})\omega_r t\right)\exp\!\left(i\widehat{\mathcal{H}}t/\hbar\right)\hat{a}_r\,|N_r\rangle \\
&= \exp\!\left(-i(N_r+\tfrac{1}{2})\omega_r t\right)\exp\!\left(i(N_r-1+\tfrac{1}{2})\omega_r t\right)\hat{a}_r\,|N_r\rangle \\
&= e^{-i\omega_r t}\,\hat{a}_r\,|N_r\rangle, \tag{31.21}
\end{aligned}
$$

$$
\begin{aligned}
\left(\hat{a}_r^+\right)_{\text{Heis}}|N_r\rangle &= \exp\!\left(i\widehat{\mathcal{H}}t/\hbar\right)\hat{a}_r^+\,\exp\!\left(-i\widehat{\mathcal{H}}t/\hbar\right)|N_r\rangle \\
&= \exp\!\left(i\widehat{\mathcal{H}}t/\hbar\right)\hat{a}_r^+\,\exp\!\left(-i(N_r+\tfrac{1}{2})\omega_r t\right)|N_r\rangle \\
&= \exp\!\left(-i(N_r+\tfrac{1}{2})\omega_r t\right)\exp\!\left(i\widehat{\mathcal{H}}t/\hbar\right)\hat{a}_r^+\,|N_r\rangle \\
&= \exp\!\left(-i(N_r+\tfrac{1}{2})\omega_r t\right)\exp\!\left(i(N_r+1+\tfrac{1}{2})\omega_r t\right)\hat{a}_r^+\,|N_r\rangle \\
&= e^{i\omega_r t}\,\hat{a}_r^+\,|N_r\rangle. \tag{31.22}
\end{aligned}
$$

[18] The summation index r in (31.2) does not have $+r$ associated with travel towards $+x$ (and vice versa). Yet that property is present in (31.20). It has been sneaked in in the last step, in which $\hat{a}_{-r}^+$ has had $-r$ replaced by r, changing the meaning of r but not affecting the total. This parallels a similar step in problem 31.3(2). If the reader feels this really is too sneaky, then look at (31.23): the right things have happened.

[19] In (31.21), we can detach the state vector again and write

$$(\hat{a}_r)_{\text{Heis}} = e^{-i\omega_r t}\,(\hat{a}_r)_{\text{Schr}},$$

since neither side depends on the value of N_r associated with the given integer r. Similarly, of course, for (31.22).

We substitute these into (31.20) to obtain the operator $\hat{y}$ in the Heisenberg representation:

$$\hat{y}_{\text{Heis}} = \sqrt{\frac{\hbar}{2\rho L}} \sum_{r=-\infty}^{\infty}{}' \frac{1}{\sqrt{\omega_r}}\left(\hat{a}_r\, e^{i(-\omega_r t + r k_1 x)} + \hat{a}_r^+\, e^{i(\omega_r t - r k_1 x)}\right). \quad (31.23)$$

Here the $\hat{a}_r, \hat{a}_r^+$ operators are still in the Schrödinger representation—not dependent upon time t—and continue to obey all of (31.13–18).

Expression (31.23) displays[20] in a new way the happy property that we have built into the definitions (31.13) of $\hat{a}_r$ and $\hat{a}_r^+$: when the subscript r is positive the wave travels towards $+x$, otherwise to $-x$.

Equation (31.23) reveals ω_r to be the angular frequency of the wave, an understanding that was concealed[21] somewhat while we were using the Schrödinger representation.

We can best bring (31.20) and (31.23) to life by making comparisons with other kinds of wave. The most important quantizations of waves concern electromagnetic waves (photons) and the vibrations of crystals (phonons). Their equations have shapes very similar to those given here. There is little difficulty in handling these more realistic systems if the case of the string has been worked first.

31.6.1   Electromagnetic waves

The electromagnetic field can be quantized using methods that exactly parallel those used above. The electric-field operator, in the Heisenberg representation, is (I quote):[22]

$$\widehat{\boldsymbol{E}}_{\text{Heis}} = i\sqrt{\frac{\hbar}{2\epsilon_0 V}} \sum_{\boldsymbol{k}, \boldsymbol{e_k}} \sqrt{\omega_k}\, \boldsymbol{e_k} \left\{\hat{a}_{\boldsymbol{k}}\, e^{i(\boldsymbol{k}\cdot\boldsymbol{r} - \omega t)} - \hat{a}_{\boldsymbol{k}}^+\, e^{-i(\boldsymbol{k}\cdot\boldsymbol{r} - \omega t)}\right\}. \quad (31.24)$$

Here: $V = L^3$ is the quantization volume; $\boldsymbol{k} = (2\pi/L)(l, m, n)$ where l, m, n are integers (positive, negative or zero, but not all zero); $\boldsymbol{e_k}$ is a vector identifying one[23] of the two polarization directions perpendicular to $\boldsymbol{k}$; $\omega_k = c\,|\boldsymbol{k}|$ is the angular frequency associated with a wave having wave vector $\boldsymbol{k}$. Operator $\hat{a}_{\boldsymbol{k}}^+$ adds one photon, of energy $\hbar\omega_k$ and momentum $\hbar\boldsymbol{k}$, to the field, while $\hat{a}_{\boldsymbol{k}}$ destroys such a photon.[24]

Expression (31.24) is a little more complicated than (31.23), as we must expect. But the similarities are more obvious than the differences. In particular, the raising and lowering operators, accompanying complex-exponential wave amplitudes, are conspicuous in both expressions, and indicate that the applications will be similar likewise.

An electron having charge $-e$ in an electromagnetic field experiences interaction $(e/m_e)\widehat{\boldsymbol{A}} \cdot \hat{\boldsymbol{p}}$, in which $\hat{\boldsymbol{p}}$ is the electron's momentum, and $\widehat{\boldsymbol{A}}$ is the field's vector potential.[25] When everything is quantized: $\hat{\boldsymbol{p}}$ (or $\hat{\boldsymbol{r}}$) is an operator that moves the electron from one state to another (I simplify slightly), whilst $\widehat{\boldsymbol{A}}$ (or $\widehat{\boldsymbol{E}}$) is an operator that must create or destroy a photon giving energy conservation overall. Operators $\hat{a}_{\boldsymbol{k}}^+$ and $\hat{a}_{\boldsymbol{k}}$, within $\widehat{\boldsymbol{A}}$, have work to do now: they create and destroy photons.[26]

By quantizing the vibrations of a string, we have shown that quantization of any wave results in a wave amplitude having the shape exhibited

---

[20] Remember that $\omega_r = |r|\,\omega_1$, so ω_r is always positive.

[21] It is to recover the understanding of ω_r as the angular frequency that we have taken the trouble to set up $\hat{y}$ in the Heisenberg representation.

[22] Equation (31.24) is sometimes presented in a slightly different form, which goes back to a choice we made in sidenote 14. If we do the equivalent of omitting $\pm i$ from the denominators of (31.13), the expression for $\widehat{\boldsymbol{E}}$ has no initial i and has a negative sign in front of $\hat{a}_{\boldsymbol{k}}$. Either form is acceptable provided, of course, one makes a choice and sticks to it. The choice depends on whether we prefer greater tidiness in the expression for $\widehat{\boldsymbol{E}}$ or in that for $\widehat{\boldsymbol{A}}$. In Chapter 32 I prefer greater tidiness in $\widehat{\boldsymbol{A}}$.

[23] The sum is taken over all modes of the field. This means a sum over all values of $\boldsymbol{k}$, and also a sum over the two polarization directions $\boldsymbol{e_k}$ accompanying each $\boldsymbol{k}$.

[24] In conformity with (31.23), the operators $\hat{a}_{\boldsymbol{k}}$ and $\hat{a}_{\boldsymbol{k}}^+$ remain in the Schrödinger representation, with no dependence on time.

Strictly, each operator should be given a label that identifies the polarization $\boldsymbol{e_k}$ of the photon that it creates or destroys, as well as the wave vector $\boldsymbol{k}$.

[25] We often write $e\widehat{\boldsymbol{E}} \cdot \hat{\boldsymbol{r}}$ in place of $(e/m_e)\widehat{\boldsymbol{A}} \cdot \hat{\boldsymbol{p}}$, especially when discussing radiative transitions in atoms. However, this replacement relies on making an approximation: the electric-dipole approximation. See Chapter 32.

[26] In this context, $\hat{a}_{\boldsymbol{k}}^+$ and $\hat{a}_{\boldsymbol{k}}$ are now usually called creation and annihilation operators, rather than raising and lowering.

in (31.20) or (31.23), with raising and lowering operators appearing as coefficients of complex exponentials. The shape we have constructed is to be seen from now on as "what we expect".

Problem 31.1 Energy and energy flow
Obtain the three equations (31.1). Work to the leading approximation only, so that the resulting equations of motion are linear. This means:

- Take the mass per unit length ρ as unchanged by the elongation of the string during the vibration.
- Take the tension τ of the string as unchanged by the stretching of the string during the vibration.
- In finding the elongation of the string, use small-angle approximations.

Confirm expression (31.1c) for the energy flow $\mathcal{F}$ by finding the rate at which string to the left of x does work on the string to the right of x. Then average this over the length of the string.[27]

[27] Remember that energy flow is usually "lumpy"; it is in the case of electromagnetic waves. So you remove a distraction by averaging.

Problem 31.2 Energy expressions as sums containing q_r
Confirm the working leading to (31.4). Supply the working leading to (31.5) and (31.6).

Problem 31.3 Waves travelling towards $\pm x$ in classical mechanics
The speed of a transverse wave travelling along a stretched string is $\sqrt{\tau/\rho}$, so the $\mathrm{e}^{\mathrm{i}r k_1 x}$ in the y of (31.2) is accompanied by

$$q_r = A_r \, \mathrm{e}^{-\mathrm{i}\omega_r t} + B_r \, \mathrm{e}^{\mathrm{i}\omega_r t},$$

in which

$$k_1 = \frac{2\pi}{L}, \qquad \omega_1 = \frac{2\pi}{L}\sqrt{\frac{\tau}{\rho}} \qquad \text{and} \qquad \omega_r = |r|\,\omega_1.$$

(1) Show, from $q_{-r} = q_r^*$, that $B_r = A_{-r}^*$, so that

$$q_r = A_r \, \mathrm{e}^{-\mathrm{i}\omega_r t} + A_{-r}^* \, \mathrm{e}^{\mathrm{i}\omega_r t}. \tag{31.25}$$

Substituting this into (31.2), we have

$$y = \frac{1}{\sqrt{\rho L}} \sum_{r=-\infty}^{\infty}{}' \left(A_r \, \mathrm{e}^{-\mathrm{i}\omega_r t} + A_{-r}^* \, \mathrm{e}^{\mathrm{i}\omega_r t} \right) \mathrm{e}^{\mathrm{i}r k_1 x}. \tag{31.26}$$

(2) In (31.26), the A_r term gives a wave travelling towards $\pm x$ when r is positive/negative, but the A_{-r}^* term does the opposite. So change r to $-r$ in the second term. Show that

$$y = \frac{1}{\sqrt{\rho L}} \sum_{r=-\infty}^{\infty}{}' \left(A_r \, \mathrm{e}^{\mathrm{i}(r k_1 x - \omega_r t)} + A_r^* \, \mathrm{e}^{\mathrm{i}(\omega_r t - r k_1 x)} \right). \tag{31.27}$$

Show from (31.25) that

$$(p_{-r} - \mathrm{i}\omega_r q_r) = \dot{q}_r - \mathrm{i}\omega_r q_r \quad = -2\mathrm{i}\omega_r \, A_r \, \mathrm{e}^{-\mathrm{i}\omega_r t} \tag{31.28a}$$

$$(p_r + \mathrm{i}\omega_r q_{-r}) = \dot{q}_{-r} + \mathrm{i}\omega_r q_{-r} = 2\mathrm{i}\omega_r \, A_r^* \, \mathrm{e}^{\mathrm{i}\omega_r t}. \tag{31.28b}$$

(3) Show that

$$y = \frac{1}{\sqrt{2\rho L}} \sum_{r=-\infty}^{\infty}{}' \frac{1}{\omega_r} \left\{ \frac{(p_{-r} - i\omega_r q_r)}{-i} e^{irk_1 x} + \frac{(p_r + i\omega_r q_{-r})}{i} e^{-irk_1 x} \right\}.$$

$$(31.29)$$

The inheritance from (31.27) shows that $(p_{-r} - i\omega_r q_r)$, $(p_r + i\omega_r q_{-r})$ associate the direction of travel (towards $\pm x$) with the sign of r.

Compare (31.29) with (31.20) and (31.13). Although (31.29) is, of course, entirely classical, we can see a "shape" that helps to explain where (31.13) came from.

(4) Confirm the wave directions by working out the energy from (31.11) and the energy flow from (31.12). Show that[28]

[28] Notice that the sums here are taken over positive r only, so that terms having positive and negative subscripts are displayed separately. Compare with eqns (31.15) and (31.16).

$$(\text{energy}) \; \mathcal{E} = \sum_{r=1}^{\infty} 2\omega_r^2 (A_r^* A_r + A_{-r}^* A_{-r}),$$

$$(\text{energy flux}) \; \mathcal{F} = \sqrt{\frac{\tau}{\rho}} \sum_{r=1}^{\infty} \frac{2\omega_r^2}{L} (A_r^* A_r - A_{-r}^* A_{-r}),$$

plus, in $\mathcal{F}$, terms (discarded) oscillating at angular frequency $2\omega_r$.

The signs in $\mathcal{F}$ confirm that waves travelling in the $+x$ and $-x$ directions have been cleverly associated with $\pm$ values of the r-subscripts.

All this, remember, has been done using classical mechanics.

Problem 31.4 Quantization of standing waves
In this problem we quantize the standing waves on a string that has ends fixed at $x = 0$ and $x = L$. The reasoning is somewhat simpler than that for the travelling-wave case, so it forms an easy introduction.

The standing-wave eigenfunctions are $\sin(r\pi x/L)$, in which r takes positive-integer values (> 0). We write a general displacement as[29]

[29] The coefficient $\sqrt{2/(\rho L)}$ is inserted into the definition of the rth amplitude $q_r(t)$ after the same fashion as in (31.2), to keep clutter out of the Lagrangian.

$$y = \sqrt{\frac{2}{\rho L}} \sum_{r=1}^{\infty} q_r(t) \sin(r\pi x/L).$$

(1) Using (31.1a) and (31.1b), show that:

$$(\text{KE}) = \frac{1}{2} \sum_{r=1}^{\infty} \dot{q}_r^2, \qquad (\text{PE}) = \frac{1}{2} \sum_{r=1}^{\infty} (r\omega_1)^2 q_r^2,$$

in which

$$\omega_1 = \frac{\pi}{L} \sqrt{\frac{\tau}{\rho}}$$

will turn out to be the angular frequency of the fundamental ($r = 1$) vibration mode of the string, whilst $\omega_r = r\omega_1$ is the angular frequency of the rth mode.[30]

[30] Note the factors 2 by which these expressions differ from those in § 31.3.3. Here each eigenfunction fits an integer number of *half* wavelengths into the length L of the string.

No modulus is taken in $\omega_r = r\omega_1$ because integer r is always positive in this problem.

(2) The Lagrangian $\mathcal{L}$ can be expressed neatly in terms of the generalized coordinates q_r:

$$\mathcal{L} \equiv (\text{KE}) - (\text{PE}) = \frac{1}{2} \sum_{r=1}^{\infty} \left(\dot{q}_r^2 - \omega_r^2 q_r^2 \right).$$

The momentum p_r conjugate to "coordinate" q_r is obtained by the rule $p_r = \partial \mathcal{L}/\partial \dot{q}_r$. Show that $p_r = \dot{q}_r$, and that the Hamiltonian becomes

$$\mathcal{H} = \sum_r p_r \dot{q}_r - \mathcal{L} = \frac{1}{2} \sum_{r=1}^{\infty} \left(p_r^2 + \omega_r^2 q_r^2 \right). \qquad (31.30)$$

The Hamiltonian is the sum of Hamiltonians for individual simple harmonic oscillators.[31]

Now we are ready to quantize the physical problem. Write the Hamiltonian in exactly the same form as (31.30), but with q_r and p_r (functions of time) replaced by their operators (independent of time), and with those operators given the commutation rule $[\hat{q}_r, \hat{p}_r] = i\hbar$.

We solve the quantum-mechanical problem by setting up raising and lowering operators, following a route familiar (!) from the quantum mechanics of a harmonic oscillator. Define[32]

$$\hat{a}_r = \frac{1}{\sqrt{2\hbar\omega_r}} \left(\frac{\hat{p}_r - i\omega_r \hat{q}_r}{-i} \right), \qquad \hat{a}_r^+ = \frac{1}{\sqrt{2\hbar\omega_r}} \left(\frac{\hat{p}_r + i\omega_r \hat{q}_r}{i} \right). \qquad (31.31)$$

Show that these operators have commutation rule

$$[\hat{a}_r, \hat{a}_s^+] = \delta_{rs},$$

meaning that the result is 1 if $r = s$ and zero if $r \neq s$.

(3) Show that the Hamiltonian takes the form[33]

$$\widehat{\mathcal{H}} = \sum_{r=1}^{\infty} \hbar\omega_r \left(\hat{a}_r^+ \hat{a}_r + \tfrac{1}{2} \right), \quad \text{with eigenvalues} \quad \mathcal{E} = \sum_{r=1}^{\infty} \left(N_r + \tfrac{1}{2} \right) \hbar\omega_r,$$

where N_r is an integer (positive or zero). Vibration mode $\sin(r\pi x/L)$ has its energy quantized in units of $\hbar\omega_r = \hbar(r\omega_1)$ (zero-point energy added).

Problem 31.5 Operators and commutation rules
This problem addresses anxieties raised in sidenotes 10 and 13.

(1) In the following, we again use running-wave eigenfunctions; c_r and s_r are real classical observables. Then we write

$$q_r = \frac{c_r + is_r}{\sqrt{2}}; \qquad q_{-r} = q_r^* \quad \text{so} \quad c_{-r} = c_r, \qquad s_{-r} = -s_r.$$

Show that the Lagrangian (31.8) becomes

$$\mathcal{L} = \frac{1}{4} \sum_{r=-\infty}^{\infty}{}' \left\{ \left(\dot{c}_r^2 - \omega_r^2 c_r^2 \right) + \left(\dot{s}_r^2 - \omega_r^2 s_r^2 \right) \right\}$$

$$= \frac{1}{2} \sum_{r=1}^{\infty} \left\{ \left(\dot{c}_r^2 - \omega_r^2 c_r^2 \right) + \left(\dot{s}_r^2 - \omega_r^2 s_r^2 \right) \right\}.$$

(2) Argue that c_r and s_r can be treated as independent generalized coordinates. Show that the momenta conjugate to coordinates c_r and s_r are γ_r and σ_r where

$$\gamma_r = \frac{\partial \mathcal{L}}{\partial \dot{c}_r} = \dot{c}_r, \qquad \sigma_r = \frac{\partial \mathcal{L}}{\partial \dot{s}_r} = \dot{s}_r; \qquad \gamma_{-r} = \gamma_r, \qquad \sigma_{-r} = -\sigma_r.$$

Then we also have $p_r = \dot{q}_{-r} = (\dot{c}_r - i\dot{s}_r)/\sqrt{2} = (\gamma_r - i\sigma_r)/\sqrt{2}$.

[31] Notice the simplification that has resulted from the use of standing-wave eigenfunctions. There is no degeneracy arising from two possible directions of travel, so there is no ugly mixing of q_r with q_{-r}. This is the reason for gaining experience with standing waves first.

There is little reason here for calculating the energy flux, since that flux oscillates about a mean value of zero.

[32] The basic shapes in (31.31) have to be found by trial or from precedents. Some choices remain; see sidenote 14.

[33] Showing that N_r is an integer requires some steps in the reasoning that can be "lifted" from a standard treatment of the harmonic oscillator. Those steps are rehearsed in problem 31.6 parts (3) and (4), though in a context where the eigenfunctions are travelling waves.

(3) Generalized coordinates c_r, s_r and their momenta γ_r, σ_r are all real in classical mechanics, so we can be sure that their quantization yields Hermitian operators with the commutation rules $[\hat{c}_r, \hat{\gamma}_r] = [\hat{s}_r, \hat{\sigma}_r] = i\hbar$.

(4) Show from the Hermiticity of $\hat{c}_r$ etc. that

$$(\hat{q}_r)^\dagger = \hat{q}_{-r}, \qquad (\hat{p}_r)^\dagger = \hat{p}_{-r},$$

so that $\hat{q}_r$ and $\hat{p}_r$ are not Hermitian.

(5) Show that nevertheless we have the "as usual" property

$$[\hat{q}_r, \hat{p}_r] = i\hbar.$$

Show that

$$\hat{a}_r^+ = (\hat{a}_r)^\dagger, \qquad (\hat{a}_r^+)^\dagger = \hat{a}_r,$$

so that $\hat{a}_r$ and $\hat{a}_r^+$ are Hermitian conjugates of each other.

(6) Obtain the commutation rules (31.14).

Problem 31.6 The number operators $\hat{a}_r^+ \hat{a}_r$

(1) Use eqns (31.13) to obtain four equations containing $\hat{p}_r$, $\hat{p}_{-r}$, $\omega_r \hat{q}_r$, $\omega_r \hat{q}_{-r}$ from which to obtain

$$\omega_r \hat{q}_r = \sqrt{\frac{\hbar \omega_r}{2}} \left(\hat{a}_{-r}^+ + \hat{a}_r \right), \qquad \hat{p}_r = i\sqrt{\frac{\hbar \omega_r}{2}} \left(\hat{a}_r^+ - \hat{a}_{-r} \right)$$

$$\omega_r \hat{q}_{-r} = \sqrt{\frac{\hbar \omega_r}{2}} \left(\hat{a}_r^+ + \hat{a}_{-r} \right) \qquad \hat{p}_{-r} = i\sqrt{\frac{\hbar \omega_r}{2}} \left(\hat{a}_{-r}^+ - \hat{a}_r \right).$$

Use these in (31.11) to obtain expression (31.15) for $\widehat{\mathcal{H}}$.

(2) Use the operators from item (1) to obtain from (31.12) the expression[34] (31.16) for $\widehat{\mathcal{F}}$.

(3) Use the commutation properties of the $\hat{a}_r^+$ and $\hat{a}_r$ operators to show that[35]

$$\widehat{\mathcal{H}}\, \hat{a}_s^+ = \hat{a}_s^+ (\widehat{\mathcal{H}} + \hbar \omega_s), \qquad \widehat{\mathcal{H}}\, \hat{a}_s = \hat{a}_s (\widehat{\mathcal{H}} - \hbar \omega_s). \tag{31.32}$$

Here s is a special-case value of r (integer r is still being summed over within $\widehat{\mathcal{H}}$). Use (31.32) to show that if ψ_s is an eigenfunction of $\widehat{\mathcal{H}}$ with energy $\mathcal{E}$, then $(\hat{a}_s^+ \psi_s)$ is an eigenfunction with energy $\mathcal{E} + \hbar \omega_s$. Similarly show that $(\hat{a}_s \psi_s)$ is an eigenfunction with energy $\mathcal{E} - \hbar \omega_s$.

(4) Using the fact that $\hat{a}_s$ cannot reduce the energy indefinitely, argue that there must be a least energy eigenvalue for the sth mode of excitation of the string, and show that this least energy is $\frac{1}{2}\hbar \omega_s$. Hence show that $\hat{a}_r^+ \hat{a}_r$ is the "number operator", whose eigenvalues are integers N_r (positive or zero). Use this finding to confirm eqns (31.17) and (31.18), giving the energy $\mathcal{E}$ and energy flow $\mathcal{F}$ in terms of the integers N_r and N_{-r}.

(5) Finally, obtain properties of the raising and lowering operators that are made use of in (19.22) and in Chapter 32:[36]

$$\langle N{+}1 | \hat{a}^+ | N \rangle = \sqrt{N+1}, \qquad \langle N{-}1 | \hat{a} | N \rangle = \sqrt{N}. \tag{31.33}$$

We have dropped the subscript r identifying the travelling-wave mode that is being excited, leaving it understood. Equivalently,

$$\hat{a}^+ | N \rangle = \sqrt{N+1}\, | N{+}1 \rangle, \qquad \hat{a} | N \rangle = \sqrt{N}\, | N{-}1 \rangle. \tag{31.34}$$

---

[34] *Hint*: Three terms in the sum disappear because they are odd in the summation index r whilst r is summed from $-\infty$ to ∞.

[35] The manipulations in parts (3) to (5) are standard bookwork for the harmonic oscillator. The similarity is expected, since the only property required of $\hat{a}_r, \hat{a}_r^+$ is their commutation rule (31.14b).

Note that $\hat{a}_s^+$ and $\hat{a}_s$ commute with all terms in the sum over r except the sth. So the presence of the sum merely makes the notation untidy, without obstructing the working.

[36] Strictly, the matrix elements in (31.33) and (31.34) are determined only within an arbitrary phase factor, propagating from a similar arbitrariness in the phase of each eigenfunction. It is conventional to define the phases of the quantum states so that the matrix elements are as given here.

Spontaneous emission of radiation

This chapter calculates two quantities introduced in Chapter 19 but not evaluated there: the modified Einstein A coefficient α_{bak}, and the spectral line profile $g(\omega - \omega_{ba})$. The reasoning uses QED but is not, on that account, particularly difficult.[1]

32.1 Introduction

In Chapter 19, we introduce a presentation of the Einstein coefficients that is "fine-gained" in that it gives the reaction rates to and from a specific mode $|\mathbf{k}, \mathbf{e_k}\rangle$ of the radiation ($\mathbf{k}$ is the wave vector while $\mathbf{e_k}$ is one of two unit polarization vectors, perpendicular to $\mathbf{k}$):[2]

$$\text{rate of absorption} = N_a\,\beta_{abk}\,n; \tag{32.1a}$$

$$\text{rate of stimulated emission} = N_b\,\beta_{bak}\,n; \tag{32.1b}$$

$$\text{rate of spontaneous emission} = N_b\,\alpha_{bak}. \tag{32.1c}$$

In absorption, N_a atoms initially in atomic quantum state a are raised to state b by absorbing a photon from radiation mode $|\mathbf{k}, \mathbf{e_k}\rangle$, there being $n = n(\mathbf{k}, \mathbf{e_k})$ photons initially in that mode. Similarly, N_b atoms initially in state b decay to state a with emission of a photon that enters field mode $|\mathbf{k}, \mathbf{e_k}\rangle$. It is shown in Chapter 19 that "Einstein" relations hold (eqn 19.19):

$$\beta_{abk} = \beta_{bak} = \alpha_{bak}. \tag{32.2}$$

A fully satisfactory treatment of radiation physics must obtain α_{bak}, which will also include[3] finding the spectral line profile $g(\omega_k - \omega_{ba})$.

A discussion of radiation physics usually starts with a semi-classical treatment of absorption (§ 19.6).[4] Intense "white" radiation irradiates N_a atoms. "Intense" means that the number of photons per mode is large, so that one photon more or less makes little difference. We are in the "limit of large quantum numbers" so that the field can be treated classically. At the same time, the atom is still treated by quantum mechanics. Hence "semi-classical".

"White", by contrast, is a two-edged sword: by totalling β_{abk} over $\mathbf{k}$ it totals over $g(\omega_k - \omega_{ba})$, thereby destroying information we need.

It is tempting to remove "white". Suppose we impose single-mode radiation, again near-classical, aiming to find β_{abk} instead of B_{ab}. The outcome is not absorption! it is undamped Rabi oscillation. To achieve

[1] It will be no surprise that the topic of this chapter has been thoroughly investigated, at a variety of different levels. My account, in spite of its length, is a "textbook" treatment, with no more sophistication than seems unavoidable. For an idea of what more there is, see e.g. Davidovich and Nussenzveig (1980). I am indebted to Professor C.J. Foot for drawing my attention to this reference.

[2] For brevity, we shall often (as here) use $\mathbf{k}$ as a label that identifies a mode, not just a wave vector. Thus α_{bak} is short for $\alpha_{b,a,\mathbf{k},\mathbf{e_k}}$ and n is short for $n(\mathbf{k}, \mathbf{e_k})$.

Equations (32.1) repeat (19.17) apart from a slight change of notation.

[3] For simplicity, this chapter is concerned with free-but-not-recoiling atoms, hence with "natural" broadening of the spectral line.

An assumed Lorentzian line profile $g_{\mathrm{Lor}}(\omega_k - \omega_{ba})$ is often "patched in" at this point. Equivalently, it is assumed that in spontaneous emission the upper-state population decays exponentially. However, such properties ought to be derived, not assumed. Moreover, we shall see that a Lorentzian line profile is not good enough, as it leads to an infinite rate of energy emission.

[4] In a conventional treatment, the "semi-classical" field has energy density $\rho(\omega)\,\mathrm{d}\omega$ in angular-frequency range $\mathrm{d}\omega$ (definition in (19.3)), and the reaction rates are $B_{ab}\rho$, $B_{ba}\rho$, and A_{ba}. Once the B_{ab} coefficient for absorption has been found by the semi-classical route, the A_{ba} coefficient is found via the Einstein trick.

Essays in Physics: Thirty-two thoughtful essays on topics in undergraduate-level physics. Geoffrey Brooker, Oxford University Press. © Geoffrey Brooker 2021. DOI: 10.1093/oso/9780198857242.003.0032

[5]Spontaneous emission turns up in a wider range of physics than one might at first imagine. It is the explanation for quantum noise in electrical circuits (black-body radiation at radio frequencies), observable at high frequencies and low temperatures where $\hbar\omega/k_B T \gtrsim 1$. See e.g. Robinson (1974); his equation (6.38) is clearly related to our (32.16) below. And spontaneous emission of phonons is discussed in problem 19.4.

In a wider context, Onley and Kumar (1992) treat a general two-level system that decays by emission of a particle, not necessarily a photon. Again, an equation reminiscent of (32.16) can be seen in their treatment.

Messiah (1965), his eqn (XXI.97), uses an operator formalism to show that an equation resembling (32.16) holds generally for emission of a particle, not necessarily a photon. The "shape" of (32.16) is obtained without specializing to a Laplace or Fourier transform.

The best treatment for photons I have found is in Barnett and Radmore (1997), Chapter 5.

[6]Thus we abandon both "intense" and "white".

[7]A classical electromagnetic field, upon which the Hamiltonian is based, is specified if we give both a vector ($\boldsymbol{A}$) and a scalar (φ) potential. Here we choose to express both $\widehat{H}_{\text{radiation}}$ and $\widehat{H}'$ in the "Coulomb gauge" in which div $\boldsymbol{A} = 0$. In the chosen gauge a plane wave has $\boldsymbol{A}$ parallel to the electric field $\boldsymbol{E}$, both transverse to the wave vector $\boldsymbol{k}$. A scalar potential φ exists, but it lives in $\widehat{H}_{\text{atom}}$ and is concerned solely with holding the atom together.

[8]Many books derive (32.5) and (32.6). If suffices to mention Messiah (1965), Chapter xxi, § 27 (albeit in cgs units).

In (32.5–6), V is the quantization volume to which periodic boundary conditions have been applied to make the field modes discrete and countable.

Comparison may be made with eqn (31.24) which gives the $\widehat{\boldsymbol{E}}$ operator in the Heisenberg representation. In the present chapter we handle everything in the Schrödinger representation in order to use the most earthy methods possible.

absorption, we need to introduce some physics that decoheres the Rabi oscillation. One way is to reintroduce a sum over frequencies—"white" radiation—but that totals over $\boldsymbol{k}$, just what we do not want to do.

The fundamental reason why Rabi oscillation is decohered is spontaneous emission from the upper state, a process excluded in any semiclassical treatment. Conclusion:

Only a fully quantum analysis will do.

Within a quantum treatment, the thing to calculate is the rate α_{bak} of spontaneous emission,[5] not absorption or stimulated emission.[6] To see why: If we try to analyse absorption, we shall apply radiation in mode $\boldsymbol{k}_0$: radiation will be re-emitted into mode $\boldsymbol{k}_0$ (stimulated emission) and will be spontaneous-emitted into many other modes. More is happening than would be the case if we had spontaneous emission on its own.

Our programme then is to supply an *ab initio* calculation for spontaneous emission: to find α_{bak} and $g(\omega_k - \omega_{ba})$. At the end of this we can find β_{abk} from α_{bak} via (32.2), a reversal of the customary procedure. And we can find (without tricks) the Einstein A_{ba} by totalling the rates for photon emission into all destinations $\boldsymbol{k}, \boldsymbol{e}_{\boldsymbol{k}}$: eqn (32.83).

32.2 The Hamiltonian

An atom interacts with radiation. The entire system has Hamiltonian

$$\widehat{H} = \widehat{H}_0 + \widehat{H}' = \widehat{H}_{\text{atom}} + \widehat{H}_{\text{radiation}} + \widehat{H}', \qquad (32.3)$$

where $\widehat{H}_{\text{atom}}$ is the Hamiltonian for the atom alone, $\widehat{H}_{\text{radiation}}$ is the Hamiltonian for the radiation alone, and $\widehat{H}'$ is the interaction between atom and radiation. For an atom with a single interesting electron (charge $-e$, mass m_e), it is usual to write

$$\widehat{H}' = \frac{e}{2m_e}(\hat{\boldsymbol{p}} \cdot \widehat{\boldsymbol{A}} + \widehat{\boldsymbol{A}} \cdot \hat{\boldsymbol{p}}) + \frac{e^2}{2m_e}\widehat{\boldsymbol{A}}^2, \qquad (32.4)$$

where $\hat{\boldsymbol{p}}$ is the operator for the electron's momentum and $\widehat{\boldsymbol{A}}$ is the operator for the field's vector potential.[7]

Expression (32.4) is a non-relativistic approximation so far as the electron is concerned: the kinetic energy is $\frac{1}{2}m_e v^2 = (\hat{\boldsymbol{p}} + e\widehat{\boldsymbol{A}})^2/2m_e$. Relativistic corrections exist and are smaller by factors of order α_{fs}^2, where α_{fs} is the fine-structure constant. The fact that small terms of this order have already been discarded here will cause us always to look mistrustfully at any calculated quantities of high order in α_{fs}.

In (32.4) the interesting terms are the first two. We shall discard the $e^2\widehat{\boldsymbol{A}}^2$ term, though our reason for doing so (§ 32.4.1) is not that usually given.

We expand the radiation field in terms of plane waves, a representative wave having wave vector $\boldsymbol{\kappa}$, polarization unit vector $\boldsymbol{e}_{\boldsymbol{\kappa}}$ (parallel to the field, normal to $\boldsymbol{\kappa}$), and angular frequency $\omega_{\boldsymbol{\kappa}}$. The field operators are:[8]

$$\widehat{A} = \sum_{\text{modes } \boldsymbol{\kappa}} \sqrt{\frac{\hbar}{2\epsilon_0 V \omega_{\boldsymbol{\kappa}}}} \, \boldsymbol{e}_{\boldsymbol{\kappa}} \left(\hat{a}_{\boldsymbol{\kappa}} \, \mathrm{e}^{\mathrm{i}\boldsymbol{\kappa}\cdot\boldsymbol{r}} + \hat{a}_{\boldsymbol{\kappa}}^{+} \, \mathrm{e}^{-\mathrm{i}\boldsymbol{\kappa}\cdot\boldsymbol{r}} \right); \qquad (32.5)$$

$$\widehat{E} = \mathrm{i} \sum_{\text{modes } \boldsymbol{\kappa}} \sqrt{\frac{\hbar\omega_{\boldsymbol{\kappa}}}{2\epsilon_0 V}} \, \boldsymbol{e}_{\boldsymbol{\kappa}} \left(\hat{a}_{\boldsymbol{\kappa}} \, \mathrm{e}^{\mathrm{i}\boldsymbol{\kappa}\cdot\boldsymbol{r}} - \hat{a}_{\boldsymbol{\kappa}}^{+} \, \mathrm{e}^{-\mathrm{i}\boldsymbol{\kappa}\cdot\boldsymbol{r}} \right). \qquad (32.6)$$

Here the sums are taken over all field modes, that is, over all discrete values of $\boldsymbol{\kappa}$ and over the two polarizations $\boldsymbol{e}_{\boldsymbol{\kappa}}$ associated with each $\boldsymbol{\kappa}$. The operators $\hat{a}_{\boldsymbol{\kappa}}^{+}$ and $\hat{a}_{\boldsymbol{\kappa}}$ are raising and lowering operators that add or remove a photon to/from field mode $|\boldsymbol{\kappa}, \boldsymbol{e}_{\boldsymbol{\kappa}}\rangle$, exactly as in (31.34).

A simplification can be made at once to the Hamiltonian $\widehat{H}'$. The term $\hat{\boldsymbol{p}} \cdot \widehat{\boldsymbol{A}}$, acting on a state $|\,\rangle$ of atom and field, contains

$$\hat{\boldsymbol{p}} \cdot \boldsymbol{e}_{\boldsymbol{\kappa}} \, \mathrm{e}^{-\mathrm{i}\boldsymbol{\kappa}\cdot\boldsymbol{r}} | \, \rangle = -\mathrm{i}\hbar \left(\boldsymbol{e}_{\boldsymbol{\kappa}} \cdot \nabla \right) \mathrm{e}^{-\mathrm{i}\boldsymbol{\kappa}\cdot\boldsymbol{r}} | \, \rangle.$$

The dot product $(\boldsymbol{e}_{\boldsymbol{\kappa}} \cdot \nabla)$ makes ∇ differentiate in a direction perpendicular to $\boldsymbol{\kappa}$, so it does not differentiate $\mathrm{e}^{-\mathrm{i}\boldsymbol{\kappa}\cdot\boldsymbol{r}}$ but passes through to operate on the electron part of $|\,\rangle$. This means that $\hat{\boldsymbol{p}}$ commutes with $\widehat{\boldsymbol{A}}$, and the two terms $\hat{\boldsymbol{p}} \cdot \widehat{\boldsymbol{A}} + \widehat{\boldsymbol{A}} \cdot \hat{\boldsymbol{p}}$ collapse to a single term $2\widehat{\boldsymbol{A}} \cdot \hat{\boldsymbol{p}}$.

The perturbing Hamiltonian now becomes

$$\widehat{H}' = \frac{e}{m_{\mathrm{e}}} \sum_{\text{modes } \boldsymbol{\kappa}} \sqrt{\frac{\hbar}{2\epsilon_0 V \omega_{\boldsymbol{\kappa}}}} \, \boldsymbol{e}_{\boldsymbol{\kappa}} \cdot \left(\hat{a}_{\boldsymbol{\kappa}} \, \mathrm{e}^{\mathrm{i}\boldsymbol{\kappa}\cdot\boldsymbol{r}} + \hat{a}_{\boldsymbol{\kappa}}^{+} \, \mathrm{e}^{-\mathrm{i}\boldsymbol{\kappa}\cdot\boldsymbol{r}} \right) \hat{\boldsymbol{p}}, \qquad (32.7)$$

in which the $\widehat{A}^2$ term has been dropped.

32.3 A model atom and transition

The present chapter concerns itself with calculating the rate of spontaneous emission, in order to show how such a calculation should proceed. For this pedagogical purpose, it is open to us to choose as simple an example as can be thought of. In particular, we choose an upper state b that decays to only one lower state a: we don't want the complication of a branching ratio.[9]

Figure 32.1 shows one way in which this may be arranged in hydrogen.

In the upper panel of Fig. 32.1, the $2\mathrm{p}_{3/2}$ states and 1s states of the hydrogen atom are drawn separated so that we may see them individually (there need be no magnetic field present). The two transitions drawn start and end at "pure states" (states that are eigenfunctions[10] of $\widehat{L}_z$, $\widehat{S}_z$ as well as of $\widehat{J}_z$) and so have only the decay paths drawn. The two $2\mathrm{p}_{1/2}$ states are omitted since their wave functions are both linear combinations of the $|m_L, m_S\rangle$.

In the lower panel of Fig. 32.1, things are further idealized by the removal of spin from the picture. The four states drawn all have the same value of m_S, either $+\frac{1}{2}$ or $-\frac{1}{2}$, and each m_S can be considered in isolation from the other m_S-possibility, since $\widehat{\boldsymbol{A}} \cdot \hat{\boldsymbol{p}}$ does not act to change spin. The labelling in Fig. 32.1 shows that it is physically acceptable to downplay spin in this way,[11] for the transitions having $\Delta m_L = \pm 1$.

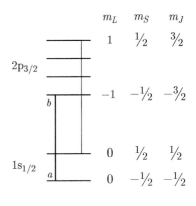

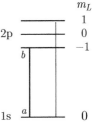

Fig. 32.1 Some electric-dipole allowed transitions in hydrogen. Quantum numbers m_L, m_S, m_J are indicated. The upper panel shows transitions that start and end on "pure states", eigenstates of $\widehat{L}_z$ and $\widehat{S}_z$ as well as of $\widehat{J}_z$; these transitions do not change m_S.

In the lower panel mention of spin is suppressed; it is almost as though spin did not exist. The atomic (space) wave functions for the states are the same as before, so the transition matrix elements are the same.

The transition $b \to a$ discussed in this chapter is indicated by heavy lines.

[9] Though this complication can't be wholly avoided; see § 32.11.

[10] We use capitals for angular-momentum quantum numbers L, S, J, even though only one electron is present, because symbols s, j have other uses later in the present chapter.

[11] States and transitions permitting such a simplification are special cases. For example, transitions not marked in Fig. 32.1 involve wave functions that are Clebsch–Gordan combinations of $|m_L, m_S\rangle$ wave functions. These states and their transitions preclude any spin-neglected description.

[12] The distinction between *level* and *state* is displayed in Fig. 19.1.

[13] But see problem 32.17 for terms $|\boldsymbol{k}\rangle\psi_b$ and $|0\rangle\psi_i$, ignored here.

[14] Of course, we expect the populations of higher-energy states to be negligibly small. But we should see it happening: problem 32.21.

[15] This progression can be seen in Fig. 32.8. Given that the energy range does not narrow beyond $\hbar A_{ba}$, there is no difficulty in understanding why many field modes are involved in the atomic transition $b \to a$.

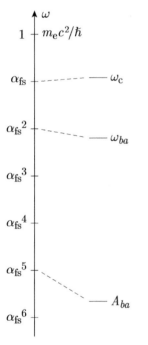

Fig. 32.2 An aide-memoire for the orders of magnitude of the (angular) frequencies appearing in this chapter. The number line is plotted on a logarithmic scale with each marked division representing a change by a factor α_{fs}. Actual values for hydrogen 2p–1s are drawn alongside, with broken lines joining them to the order that they are assigned by the algebra:

$$\omega_{\text{c}} = (m_e c^2/\hbar) \times (3/2)\alpha_{\text{fs}};$$
$$\omega_{ba} = (m_e c^2/\hbar) \times (3/8)\alpha_{\text{fs}}^2;$$
$$A_{ba} = (m_e c^2/\hbar) \times (2/3)^8 \alpha_{\text{fs}}^5.$$

Given these considerations, we choose to concentrate our attention on the single transition

$$\psi_b = \left|2\text{p}_{3/2},\, m_J = -\tfrac{3}{2}\right\rangle \quad \text{to} \quad \psi_a = \left|1\text{s}_{1/2},\, m_J = -\tfrac{1}{2}\right\rangle$$

otherwise known as

$$\psi_b = \left|2\text{p},\, m_L = -1\right\rangle \quad \text{to} \quad \psi_a = \left|1\text{s},\, m_L = 0\right\rangle.$$

32.3.1   Basic equations

The atom has initial *state*[12] $\psi_b\, e^{-i\omega_b t}$ together with an empty photon field $|0\rangle\, e^{-i0t}$. That atom-and-field state has coefficient $c_b(t)$ which takes initial value 1 but diminishes during photon emission.

In the first order of perturbation theory, operator $\widehat{H}'$ can add or remove just one photon, because it contains operators $\hat{a}_{\boldsymbol{\kappa}}^+, \hat{a}_{\boldsymbol{\kappa}}$ to the first power only. However, that one photon may part-occupy many field states $|\boldsymbol{k}, e_{\boldsymbol{k}}\rangle\, e^{-i\omega_{\boldsymbol{k}} t}$. Accompanying this is an atom, also part-occupying several states $\psi_i\, e^{-i\omega_i t}$, though mostly in the two states $\psi_b\, e^{-i\omega_b t}$ and $\psi_a\, e^{-i\omega_a t}$. The coefficient for atom and field together is $c_{\boldsymbol{k}i}(t)$. Putting this together, the state Ψ of the entire system is,[13] as a function of time t

$$\Psi = c_b(t)\, |0\rangle\, \psi_b\, e^{-i\omega_b t} + \sum_{\boldsymbol{k},\, i \neq b} c_{\boldsymbol{k}i}(t)\, |\boldsymbol{k}, e_{\boldsymbol{k}}\rangle\, e^{-i\omega_{\boldsymbol{k}} t}\, \psi_i\, e^{-i\omega_i t}. \quad (32.8)$$

Why the sum over atomic states i? Early in the transition, at times $t \ll 1/\omega_{ba}$, energy conservation has nothing to say, and probability (at this stage tiny but non-zero) can go into all atomic states ψ_i that are dipole-linked to ψ_b, even states *higher* in energy[14] than ψ_b. The sum over atomic states ψ_i in (32.8) must allow for this possibility, and not be restricted in advance to include only the dominant final state ψ_a.

Likewise, at early times $t \ll 1/\omega_{ba}$, probability can go into field modes $|\boldsymbol{k}, e_{\boldsymbol{k}}\rangle$ within a very wide energy range. That energy range narrows as the reaction proceeds, at first to $\sim\hbar/t$ and finally[15] to $\sim\hbar A_{ba}$. The sum over $\boldsymbol{k}, e_{\boldsymbol{k}}$ in (32.8) must allow for this too.

Our programme is first to calculate the time dependence $c_b(t)$ of the initial state $|0\rangle\psi_b$. Afterwards (§ 32.9) we can evaluate the filling $c_{\boldsymbol{k}i}(t)$ of each state $|\boldsymbol{k}, e_{\boldsymbol{k}}\rangle\psi_i$ which, in particular, will give the "Einstein" coefficient α_{bak} and the spectral line profile $g(\omega_{\boldsymbol{k}} - \omega_{ba})$.

We write the Schrödinger equation that describes the time evolution of the system:

$$i\hbar\, \frac{\partial\Psi}{\partial t} = \left(\widehat{H}_0 + \widehat{H}'\right)\Psi.$$

When we substitute into this the Ψ from (32.8), and cancel the "unperturbed" terms, we get (problem 32.1(1)):

$$i\hbar\left\{ \dot{c}_b\, |0\rangle\, \psi_b\, e^{-i\omega_b t} + \sum_{\boldsymbol{k},\, i \neq b} \dot{c}_{\boldsymbol{k}i}\, |\boldsymbol{k}\rangle\, \psi_i\, e^{-i(\omega_i + \omega_{\boldsymbol{k}})t} \right\}$$
$$= \widehat{H}'\left\{ c_b\, |0\rangle\, \psi_b\, e^{-i\omega_b t} + \sum_{\boldsymbol{k},\, i \neq b} c_{\boldsymbol{k}i}\, |\boldsymbol{k}\rangle\, \psi_i\, e^{-i(\omega_i + \omega_{\boldsymbol{k}})t} \right\}. \quad (32.9)$$

Remember that the ket $|\boldsymbol{k}\rangle$ represents a field state $\boldsymbol{k}, \boldsymbol{e_k}$ occupied by a single photon, while $|0\rangle$ represents the all-modes-empty field state.[16]

Equations governing the individual coefficients in (32.9) are extracted in a wholly conventional way. First, we multiply by $\psi_{i'}^* \langle \boldsymbol{k'}|$, where $i' \neq b$, and integrate over all space. In the sum on the right of (32.9) we have $\int \mathrm{d}V\, \psi_{i'}^* \langle \boldsymbol{k'} | \widehat{H}' | \boldsymbol{k} \rangle \psi_i$. This is zero for all $i', \boldsymbol{k'}$ since $\widehat{H}' | \boldsymbol{k} \rangle$ contains either two photons or none, and either way is orthogonal to $\langle \boldsymbol{k'} |$. Then,

$$\dot{c}_{\boldsymbol{k}i} = \frac{-\mathrm{i}}{\hbar}\, c_b \int \mathrm{d}V\, \psi_i^* \langle \boldsymbol{k} | \widehat{H}' | 0 \rangle\, \psi_b\, \mathrm{e}^{\mathrm{i}(\omega_{\boldsymbol{k}} - \omega_{bi})t} = C_{ib}(\boldsymbol{k})\, c_b\, \mathrm{e}^{\mathrm{i}(\omega_{\boldsymbol{k}} - \omega_{bi})t},$$

in which[17] $\omega_{bi} = \omega_b - \omega_i$. We tidy up by defining

$$C_{ib}(\boldsymbol{k}) = \frac{-\mathrm{i}}{\hbar} \int \mathrm{d}V\, \psi_i^* \langle \boldsymbol{k} | \widehat{H}' | 0 \rangle\, \psi_b \tag{32.10}$$

$$= \frac{-\mathrm{i}}{\hbar}\, \frac{e}{m_\mathrm{e}} \sum_{\boldsymbol{\kappa}} \sqrt{\frac{\hbar}{2\epsilon_0 V \omega_{\boldsymbol{\kappa}}}}\, \boldsymbol{e_\kappa} \cdot \int \mathrm{d}V\, \psi_i^* \Big\langle \boldsymbol{k} \Big| \big(\hat{a}_{\boldsymbol{\kappa}}\, \mathrm{e}^{\mathrm{i}\boldsymbol{\kappa}\cdot\boldsymbol{r}} + \hat{a}_{\boldsymbol{\kappa}}^+\, \mathrm{e}^{-\mathrm{i}\boldsymbol{\kappa}\cdot\boldsymbol{r}} \big) \hat{\boldsymbol{p}} \Big| 0 \Big\rangle \psi_b.$$

The operators $\hat{a}_{\boldsymbol{\kappa}}^+, \hat{a}_{\boldsymbol{\kappa}}$ have been carried over intact from the description of a harmonic oscillator or a vibrating string. Operator $\hat{a}_{\boldsymbol{\kappa}}$, operating on $|0\rangle$, tries to remove a photon from $|0\rangle$ which is empty already, so this operator gives zero. By (31.34), operator $\hat{a}_{\boldsymbol{\kappa}}^+$ has the effect $\hat{a}^+ |n\rangle = \sqrt{n+1}\, |n{+}1\rangle$ where n is the number of excitations initially present; and here n is that number for state $|0\rangle$, so $n = 0$. Moreover,[18] $\hat{a}_{\boldsymbol{\kappa}}^+ |0\rangle$ yields a state that is orthogonal to $|\boldsymbol{k}\rangle$ except in the one case where $\boldsymbol{\kappa} = \boldsymbol{k}$, so the sum over $\boldsymbol{\kappa}$ reduces to a single term having $\langle \boldsymbol{k} | \hat{a}_{\boldsymbol{k}}^+ | 0 \rangle = \sqrt{1} = 1$. Then (assistance in problem 32.1(2))

$$C_{ib}(\boldsymbol{k}) = \frac{-\mathrm{i}}{\hbar}\, \frac{e}{m_\mathrm{e}} \sqrt{\frac{\hbar}{2\epsilon_0 V \omega_{\boldsymbol{k}}}}\, \boldsymbol{e_k} \cdot \int \mathrm{d}V\, \psi_i^*\, \mathrm{e}^{-\mathrm{i}\boldsymbol{k}\cdot\boldsymbol{r}}\, \hat{\boldsymbol{p}}\, \psi_b \tag{32.11}$$

$$= \frac{-\mathrm{i}}{\hbar}\, \frac{e}{m_\mathrm{e}}\, \boldsymbol{\mathcal{A}} \cdot (\boldsymbol{J}_p)_{ib}, \tag{32.12}$$

in which[19]

$$\boldsymbol{\mathcal{A}} = \sqrt{\frac{\hbar}{2\epsilon_0 V \omega_{\boldsymbol{k}}}}\, \boldsymbol{e_k}; \qquad (\boldsymbol{J}_p)_{ib} = \int \mathrm{d}V\, \psi_i^*\, \mathrm{e}^{-\mathrm{i}\boldsymbol{k}\cdot\boldsymbol{r}}\, \hat{\boldsymbol{p}}\, \psi_b. \tag{32.13}$$

Next, we return to (32.9) and pick out the coefficient $\dot{c}_b$ associated with $|0\rangle \psi_b$. The manipulation closely parallels that spelt out above (problem 32.1(2) and (3)). Summarizing:[20]

$$\dot{c}_b(t) = -\sum_{\boldsymbol{k},\, i \neq b} C_{ib}(\boldsymbol{k})^*\, c_{\boldsymbol{k}i}(t)\, \mathrm{e}^{\mathrm{i}(\omega_{bi} - \omega_{\boldsymbol{k}})t}; \tag{32.14a}$$

$$\dot{c}_{\boldsymbol{k}i}(t) = C_{ib}(\boldsymbol{k})\, c_b(t)\, \mathrm{e}^{\mathrm{i}(\omega_{\boldsymbol{k}} - \omega_{bi})t}. \tag{32.14b}$$

In writing eqns (32.14), we have emphasized that c_b and $c_{\boldsymbol{k}i}$ depend on time by writing them as $c_b(t)$ and similarly. Likewise we emphasize that C_{ib} is not a universal constant[21] but depends on $\boldsymbol{k}$, both via the $\omega_{\boldsymbol{k}}$ under the square root in (32.11) and via $\mathrm{e}^{-\mathrm{i}\boldsymbol{k}\cdot\boldsymbol{r}}$.

[16] There is no question that field state $\boldsymbol{k}, \boldsymbol{e_k}$ might become occupied by two or more photons because the $\widehat{H}'$ of (32.7) is being applied in first order only. Thus the notation $|\boldsymbol{k}, \boldsymbol{e_k}\rangle$ is unambiguous: one photon in that field state.

[17] Note the sign: ω_{bi} is negative if state i lies above b in energy.

[18] This step is the only one in the present chapter that makes any explicit use of quantum electrodynamics (QED)!

[19] The unit vector $\boldsymbol{e_k}$, and with it $\boldsymbol{\mathcal{A}}$, is dependent on the angular-momentum change when state b decays to state a—in our case $\Delta m_\mathrm{atom} = +1$ but the algebra so far is general. Also, $\boldsymbol{e_k}$ depends on the direction ϑ, φ to which the radiation travels, while $\boldsymbol{\mathcal{A}}$ depends on $|\boldsymbol{k}|$ via $\omega_{\boldsymbol{k}}$. We have omitted subscripts that might have indicated these dependences because we need other subscripts to indicate the components $\mathcal{A}_X, \mathcal{A}_Y$ of $\boldsymbol{\mathcal{A}}$.

[20] Some observations should be made about eqns (32.14).

First: all angular frequencies are abbreviations for energies or energy differences. There is no time dependence treated via a Fourier "$\mathrm{e}^{-\mathrm{i}\omega t}$ and take the real part at the end", a convention disallowed given the different complex convention within quantum mechanics.

Second, we observe that the angular frequency (meaning energy) of the photons appears in a grouping $\mathrm{e}^{\pm \mathrm{i}(\omega_{bi} - \omega_{\boldsymbol{k}})t}$. This looks like the outcome of a rotating-wave approximation, but it is not, because no partner time dependence $\mathrm{e}^{\pm \mathrm{i}(\omega_{bi} + \omega_{\boldsymbol{k}})t}$ has been discarded.

[21] It will be helpful later for us to remember that in C_{ib} the first subscript refers to the final state and the second to the initial state. We may see this in (32.10) where ψ_i is associated with $|\boldsymbol{k}\rangle$ while ψ_b is associated with $|0\rangle$. This is regardless of what letters appear in the subscripts.

32.4 Solution of eqns (32.14)

Equations (32.14) constitute a set of simultaneous equations for the time dependence of a phenomenon for which initial values are given. The mathematical method of choice for handling such initial-value problems is the Laplace transform. We therefore define transformed quantities after the fashion

$$\tilde{c}(s) = \int_0^\infty c(t)\, e^{-st}\, dt. \tag{32.15}$$

Standard bookwork properties of the Laplace transform are:[22]

$$\text{transform of } \dot{c}(t) \quad \text{is} \quad s\tilde{c}(s) - c(t{=}0{+});$$

$$\text{transform of } c(t)\, e^{i\omega t} \quad \text{is} \quad \tilde{c}(s - i\omega).$$

Taking the transforms of eqns (32.14), we have (problem 32.2)

$$s\,\widetilde{c_{\boldsymbol{k}i}}(s) = C_{ib}(\boldsymbol{k})\,\tilde{c}_b\big(s - i\omega_{\boldsymbol{k}} + i\omega_{bi}\big),$$

$$\therefore \quad \big(s + i\omega_{\boldsymbol{k}} - i\omega_{bi}\big)\widetilde{c_{\boldsymbol{k}i}}\big(s + i\omega_{\boldsymbol{k}} - i\omega_{bi}\big) = C_{ib}(\boldsymbol{k})\,\tilde{c}_b(s);$$

$$s\,\tilde{c}_b(s) - 1 = -\sum_{\boldsymbol{k},\, i\neq b} C_{ib}(\boldsymbol{k})^*\,\widetilde{c_{\boldsymbol{k}i}}\big(s + i\omega_{\boldsymbol{k}} - i\omega_{bi}\big)$$

$$= -\sum_{\boldsymbol{k},\, i\neq b} C_{ib}(\boldsymbol{k})^*\,\frac{C_{ib}(\boldsymbol{k})\,\tilde{c}_b(s)}{s + i\omega_{\boldsymbol{k}} - i\omega_{bi}},$$

so that

$$\tilde{c}_b(s) = \frac{1}{\displaystyle s + \sum_{\boldsymbol{k},\, i\neq b} \frac{|C_{ib}(\boldsymbol{k})|^2}{s + i(\omega_{\boldsymbol{k}} - \omega_{bi})}}. \tag{32.16}$$

Equation (32.16) supplies a formal solution to the problem for finding $\widetilde{c}_b(s)$. All that remains is to evaluate the sum and then invert the Laplace transform to yield $c_b(t)$. (!) However, we need a good deal more information before the inversion can be done. And the physics insights we are aiming for are not yet on display. In particular, it is not until (32.37) that we see hints of a Lorentzian spectrum, while an exponential decay does not appear until (32.46).

32.4.1 Looking backwards and forwards

Looking back: has anything been brushed under the carpet? Looking forwards: what things—even if commonly advocated—should we *not* do next?

(a) The $\widehat{A}^2$ term in the Hamiltonian

I have followed almost every textbook treatment in saying that the kinetic energy of an electron of charge $-e$ and mass m_{e} is given by the "non-relativistic" expression (32.4)

$$\text{KE} = \frac{(m_{\mathrm{e}}v)^2}{2m_{\mathrm{e}}} = \frac{(\boldsymbol{p} + e\boldsymbol{A})^2}{2m_{\mathrm{e}}} = \frac{p^2}{2m_{\mathrm{e}}} + e\,\frac{\boldsymbol{p}\cdot\boldsymbol{A} + \boldsymbol{A}\cdot\boldsymbol{p}}{2m_{\mathrm{e}}} + \frac{e^2}{2m_{\mathrm{e}}}\,\boldsymbol{A}^2,$$

[22] The reader may find the Laplace transform less familiar than the Fourier transform. The Laplace transform has it built into the definitions that $c_b(t) = 0$ for $t < 0$, so it is ideal for our initial-value problem. If a Fourier transform is taken instead, and $c_b(t)$ is again set to zero for $t < 0$, then similar algebra results with $-i\omega$ replacing s. But the insertion of initial values requires work that is made easier with the Laplace transform.

By setting $c_b = 0$ for times $t < 0$, we are making a mathematical idealization, in which there is a time $t = 0$ at which the atom is step-function excited to the single excited state b and the starting gun is fired for the decay to proceed for $t > 0$. A more general treatment might include details of the excitation process as a function of time before $t = 0$. But each such excitation route would require a separate analysis covering the time from $t = -\infty$ to 0 as well as from $t = 0$ to ∞. We are hoping that our model will convey the essentials of all such reactions. This simplifying assumption is conventional: see e.g. Heitler (1954) § 16.

which is then "reinterpreted" as an operator forming part of the Hamiltonian.

When we take a sum over all normal modes of the electromagnetic field (quantized) the term in $e^2\boldsymbol{A}^2$ sums to infinity—even if only the zero-point energy is included. Physicists can become blasé about infinities, and treat them as minor nuisances to be argued away by some process such as "mass renormalization". However, the infinity in $e^2\boldsymbol{A}^2$ is best looked at in a different way.

Our discussion started by taking a low-speed (non-relativistic) limit for the kinetic energy of the electron, but it found itself having to deal with a term $e^2\boldsymbol{A}^2$ in which the photons contributing to $\boldsymbol{A}$ travel at the speed of light. At best, this seems an odd mixture of low-speed and high-speed physics that surely ought to be treated with caution.

The safe procedure must be to take cognizance of relativity from the beginning, for both electron and field: we use the Dirac equation.

This is not the place to give a detailed discussion of the Dirac equation, but we can give the briefest of outlines. We have the Hamiltonian[23]

$$\mathsf{H} = \beta m_e c^2 - e\varphi + c\boldsymbol{\alpha} \cdot (\hat{\boldsymbol{p}} + e\widehat{\boldsymbol{A}}).$$

In this, $\mathsf{H}, \beta, \alpha_x, \alpha_y, \alpha_z$ are five $4{\times}4$ matrices that act on a wave function that is a 4-element column matrix.[24] The important point for now is that there is no sign of an $\widehat{\boldsymbol{A}}^2$ term.[25]

Fortunately, the transition matrix elements calculated from $ce\boldsymbol{\alpha} \cdot \widehat{\boldsymbol{A}}$ are the same (to order α_{fs}^2 and with $\mathrm{e}^{-\mathrm{i}\boldsymbol{k}\cdot\boldsymbol{r}} \approx 1$) as those calculated from $(e/m_e)\widehat{\boldsymbol{A}} \cdot \hat{\boldsymbol{p}}$, so Dirac imposes no nasty surprises on the usual non-relativistic discussion of transitions. We are permitted to continue using the non-relativistic $(e/m_e)\widehat{\boldsymbol{A}} \cdot \hat{\boldsymbol{p}}$ as the interaction Hamiltonian.[26]

We should turn round what has just been said. Matrix elements set up in (32.10) and (32.12), using Schrödinger wave functions and a Schrödinger $\widehat{H}' = (e/m_e)\widehat{\boldsymbol{A}} \cdot \hat{\boldsymbol{p}}$, undergo corrections (to both) of order α_{fs}^2 when Dirac replaces Schrödinger.[27] Incidentally, eqn (32.16) retains its shape unaltered when $C_{ib}(\boldsymbol{k})$ is given its relativistic value.

(b) Evaluation of matrix elements must take account of $\mathrm{e}^{-\mathrm{i}\boldsymbol{k}\cdot\boldsymbol{r}}$

The form of the expression in (32.16) determines the direction we shall need to take. The sum Σ in the denominator sums over all atomic states i, and all field modes $|\boldsymbol{k}, \boldsymbol{e}_{\boldsymbol{k}}\rangle$, that are connected to $|0\rangle\psi_b$ by a non-zero $C_{ib}(\boldsymbol{k})$. In particular, the sum includes all field modes, even those that are far higher in energy than $\hbar\omega_b$. Indeed, the sum over field modes pays so much attention to high-energy modes that it is in danger of diverging! If we set $\mathrm{e}^{-\mathrm{i}\boldsymbol{k}\cdot\boldsymbol{r}} \approx 1$, even for large $\omega_{\boldsymbol{k}}$, we have $C_{ib}(\boldsymbol{k}) \propto \omega_{\boldsymbol{k}}^{-1/2}$ and

$$\sum_{\boldsymbol{k}} \sim \int^\infty \frac{|C_{ib}(\boldsymbol{k})|^2}{\mathrm{i}\omega_{\boldsymbol{k}}}\, \omega_{\boldsymbol{k}}^2\, \mathrm{d}\omega_{\boldsymbol{k}} \sim \int^\infty \frac{(\omega^{-1/2})^2}{\omega}\, \omega^2\, \mathrm{d}\omega = \int^\infty \mathrm{d}\omega = \infty,$$

diverging at the upper limit.[28] Clearly, this ultraviolet catastrophe is self-inflicted: it will not do to set $\mathrm{e}^{-\mathrm{i}\boldsymbol{k}\cdot\boldsymbol{r}} \approx 1$. When $\boldsymbol{k} \cdot \boldsymbol{r} \gtrsim 1$, the integration in (32.11) cancels against itself.[29] This must first happen when $\omega \sim c/a_0$, which is close to the ω_c of Fig. 32.2 and of (32.17).

[23] *Reminder:* e is the proton charge; the electron charge is $-e$. And the Dirac equation is being used to describe an electron.

[24] The term $-e\varphi$ in the Hamiltonian is multiplied by a $4{\times}4$ unit matrix so that every term in H is a 4×4 matrix operator. In the context of hydrogen, φ is the electrostatic potential (positive) surrounding the nucleus. Then $-e\varphi$ is the potential-energy function for the electron in that field; it holds the atom together and is not connected with the radiation.

[25] A favourite saying of my late colleague D.T. Edmonds: "The second law of thermodynamics says you can't get out of a piece of mathematics anything that didn't go in at the beginning."

The $e^2\boldsymbol{A}^2$ term is made to appear in a textbook "non-relativistic expansion" that assumes $(e^2/2m_e)\boldsymbol{A}^2 \ll m_e c^2$. But—in the present chapter— quantization makes the $e^2\boldsymbol{A}^2$ term infinite, much *greater* than $m_e c^2$. The expansion is, at the least, suspect.

[26] But see sidenote 67 for a more careful statement.

[27] As already said in § 32.2, this knowledge will determine what small quantities may or may not be retained later in the calculation.

[28] Look back to (32.5) and (32.6). Had we replaced $\widehat{\boldsymbol{A}} \cdot \hat{\boldsymbol{p}}$ with $\widehat{\boldsymbol{E}} \cdot \hat{\boldsymbol{r}}$, the sum would have diverged as $\int^\infty \omega\, \mathrm{d}\omega$.

[29] The integral in (32.11) has the form of a three-dimensional Fourier transform. Think first of a one-dimensional function $f(x)$ that has a discontinuity in its nth derivative; its Fourier transform falls as $k^{-(n+1)}$ for large k. A similar result in three dimensions tells us that, if $\psi_i^* \hat{\boldsymbol{p}} \psi_b$ exits from the origin as r^n, then its transform falls as $k^{-(n+3)}$ for large k (equivalently as $\omega^{-(n+3)}$ for large ω). For the case 2p–1s, $n = 1$ and we may see the ω^{-4} in the $(1 + \kappa^2)^{-2}$ of (32.23).

[30] In a similar way, a factor $e^{-i\mathbf{k}\cdot\mathbf{r}}$, usually ignored, is necessary to rescue us from an ultraviolet catastrophe in the derivation of Golden Rule Number Two. See problem 32.24.

[31] Many books (e.g. Schiff (1965), p. 409) point out that matrix elements $\langle a|e^{-i\mathbf{k}\cdot\mathbf{r}}\,\hat{\mathbf{p}}|b\rangle$ and $\langle a|e^{-i\mathbf{k}\cdot\mathbf{r}}\,\hat{\mathbf{r}}|b\rangle$ require the $e^{-i\mathbf{k}\cdot\mathbf{r}}$ if they are to be exact. But usually they proceed thereafter to the electric-dipole approximation, applying it to cases where it suffices. It happens that our calculation is less forgiving.

[32] This range means that the sum in (32.16) is not "focused" onto values of ω close to ω_{ba}. We may be expecting a resonance denominator governing the behaviour of $\tilde{c}_b(s)$, and one will appear eventually, but then it arises from the *use* of the sum in (32.16), not from its *preparation*.

Some authors obtain (32.16) and recognize that the sum over $\mathbf{k}$ is in danger of diverging. They note that "new physics" must enter when photons have enough energy $\hbar\omega \sim m_e c^2$ to cause pair production. The terms in the sum then become incomplete, and the least-worst thing to do is to truncate the sum at this energy. However, (32.24) shows that $e^{-i\mathbf{k}\cdot\mathbf{r}}$ causes a fall-off when $\kappa \gtrsim 1$, at an energy $\hbar\omega_c \sim m_e c^2 \times \alpha_{fs}$, well below $m_e c^2$.

Here's what really happens at energy $\hbar\omega \sim m_e c^2$. Exponential $e^{-i\mathbf{k}\cdot\mathbf{r}}$ oscillates so fast over the dimensions of the atom that each (matrix element)^2 is reduced by the factor $(1+\kappa^2)^{-4} \sim \alpha_{fs}^8$ in (32.24). There is negligible contribution to the sum in (32.16) from energies $\sim m_e c^2$, no need to truncate the sum, and no need to feel that inconvenient physics has been waved away.

Of course, pair production is still going to cut in at $\hbar\omega \sim m_e c^2$. But it's a higher-order process, requiring perturbation theory to be taken beyond first order.

This explains why the exponentials $e^{\pm i\mathbf{\kappa}\cdot\mathbf{r}}$ have been retained, as far back as the expressions for $\widehat{A}$ and $\widehat{E}$ in (32.5) and (32.6).[30]

The effect of $e^{-i\mathbf{k}\cdot\mathbf{r}}$ is often said to describe "retardation", as in "retarded potential". The electromagnetic field at a field point $\mathbf{R}$ receives instructions from the movement of charge in the radiating atom. What arrives first comes from the near side of the atom, followed later by what comes from the far side. The factor $e^{-i\mathbf{k}\cdot\mathbf{r}}$ handles this range of time delays[31] as it affects an oscillation having angular frequency $\omega = c\,|\mathbf{k}|$.

(c) Operator $\widehat{A}\cdot\hat{\mathbf{p}}$ may not be replaced with $\widehat{E}\cdot\hat{\mathbf{r}}$

Two new quantities require definition now:

$$\omega_c \equiv \frac{3c}{2a_0} = \frac{3}{2}\frac{m_e c^2}{\hbar}\alpha_{fs} \quad \text{and} \quad \kappa \equiv \frac{\omega_k}{\omega_c} = \frac{2a_0}{3c}\omega_k, \tag{32.17}$$

in which a_0 is the Bohr radius.

It is shown in problem 32.7(2) that for hydrogen 2p–1s

$$\left\langle a \left| \langle \mathbf{k}|\widehat{E}\cdot\hat{\mathbf{r}}|0\rangle \right| b \right\rangle = \frac{\omega_k}{\omega_{ba}}\frac{1}{(1+\kappa^2)}\left\langle a \left| \left\langle \mathbf{k}\left| \frac{\widehat{A}\cdot\hat{\mathbf{p}}}{m_e}\right|0\right\rangle \right| b \right\rangle. \tag{32.18}$$

In the sum of (32.16), the contributing values of ω_k range[32] from zero to roughly $\omega_c \sim \omega_{ba}/\alpha_{fs}$. Clearly the $(\omega_k/\omega_{ba})(1+\kappa^2)^{-1}$ in (32.18) cannot be approximated to 1 across that range: $\widehat{E}\cdot\hat{\mathbf{r}}$ is not an acceptable replacement for $\widehat{A}\cdot\hat{\mathbf{p}}/m_e$.

(d) Electric dipole transitions

There is a more familiar way of handling $e^{-i\mathbf{k}\cdot\mathbf{r}}$ which expands

$$e^{-i\mathbf{k}\cdot\mathbf{r}} = 1 - i\mathbf{k}\cdot\mathbf{r} + \dots. \tag{32.19}$$

This expansion is used to account for higher-order transitions, such as electric quadrupole. That same reasoning can be applied here to show that higher-order transitions are weaker than electric-dipole by at least two powers of α_{fs}. Our attention can therefore be given exclusively to electric-dipole transitions, meaning those having $\langle i|\mathbf{r}|b\rangle \neq 0$. However the customary step of replacing $e^{-i\mathbf{k}\cdot\mathbf{r}}$ with 1 must *not* be taken, as the exponential is needed to control the fall-off of the matrix elements at large k, and hence the convergence of $\sum_{\mathbf{k}}$ in (32.16).

We should avoid difficulties with words. The 2p–1s transition we are concerned with has $\langle a|\hat{\mathbf{r}}|b\rangle \neq 0$: it is an *electric-dipole transition*. At the same time, the strictures of this subsection show that we shall *not* be making use of the family of mathematical steps collectively known as the *electric-dipole approximation*.

32.5   Evaluation of the sum in (32.16)

The sum over i in (32.16) contains a term having $i = a$, the state in which the atom ends up (mostly). We separate out that term:

$$\tilde{d}_b(s) \equiv \tilde{c}_b(s + i\omega_{ba}) = \left\{ \text{transform of } d_b(t) = c_b(t)\, e^{-i\omega_{ba}t} \right\}$$

$$= \cfrac{1}{s + i\omega_{ba} + \displaystyle\sum_{\boldsymbol{k},\, i \neq b} \cfrac{|C_{ib}(\boldsymbol{k})|^2}{(s + i\omega_{ia}) + i\omega_{\boldsymbol{k}}}} \qquad (32.20)$$

$$= \cfrac{1}{s + i\omega_{ba} + \displaystyle\sum_{\boldsymbol{k}} \cfrac{|C_{ab}(\boldsymbol{k})|^2}{s + i\omega_{\boldsymbol{k}}} + \displaystyle\sum_{\boldsymbol{k},\, j \neq a,b} \cfrac{|C_{jb}(\boldsymbol{k})|^2}{(s + i\omega_{ja}) + i\omega_{\boldsymbol{k}}}}. \qquad (32.21)$$

Our chosen 2p–1s transition has only one (relevant) atomic state below the 2p initial state, so the states $j \neq a$ all lie above 2p in energy. We expect their effect on $\tilde{d}_b(s)$ to be negligible, so we ignore them for now. We return to (32.21) in problem 32.22, and there investigate whether and how much the physics is affected when the sum over states j is restored.

32.5.1   Elliptical polarization

In Fig. 32.3, the atom sits at the origin, and has its wave functions defined relative to the xyz axes. For the transition identified in Fig. 32.1, the atom's electric-dipole moment traces a circle in the xy plane, as indicated by the heavy line. It rotates in the direction of the arrow (problem 32.3).

Radiation can travel away in all directions. We concentrate on that whose $\boldsymbol{k}$-vector faces in the (ϑ, φ) direction, the Z-direction in Fig. 32.3. The radiation's $\boldsymbol{A}$-field (or $\boldsymbol{E}$-field) lies in the XY plane ($\perp Z$). Looking back at the atom, along $-\boldsymbol{k}$, we see that A_X is driven by dipole $r_{x'}$, and A_Y is driven by $r_{y'}$. Dipole $r_{x'}$ appears foreshortened by $\cos\vartheta$, while there is no foreshortening of $r_{y'}$. This makes the radiation elliptically polarized (problem 32.8). We may guess[33] that it has a complex $\boldsymbol{A}$ whose vector direction $\boldsymbol{e_k}$ looks something like $\boldsymbol{e}_X \cos\vartheta + i\boldsymbol{e}_Y$, where $\boldsymbol{e}_X$ and $\boldsymbol{e}_Y$ are unit vectors in the X and Y directions. Thus

$$\boldsymbol{A} = |\boldsymbol{A}|\, \boldsymbol{e_k} = |\boldsymbol{A}| \left(\frac{\cos\vartheta\, \boldsymbol{e}_X + i\boldsymbol{e}_Y}{\sqrt{1 + \cos^2\vartheta}} \right). \qquad (32.22)$$

This form for $\boldsymbol{A}$ is confirmed in problem 32.8.

32.5.2   Evaluation of $|C_{ab}(\boldsymbol{k})|^2$

I have broken the evaluation into several small pieces which are rehearsed in problems 32.3 and 32.6–9. The outcome is:

$$C_{ab}(\boldsymbol{k}) = \frac{-e\,\omega_{ba}}{\sqrt{2\epsilon_0 V \hbar \omega_{\boldsymbol{k}}}}\, |\boldsymbol{r}_{ab}| \sqrt{\frac{1 + \cos^2\vartheta}{2}}\, e^{-i\varphi}\, \frac{1}{(1 + \kappa^2)^2}. \qquad (32.23)$$

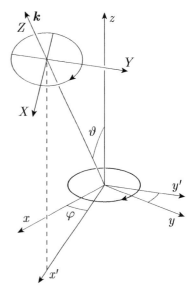

Fig. 32.3 The atom has its quantum states referred to the x, y, z axes, and its electric dipole moment rotates in the xy plane: eqn (32.58). We look at outgoing radiation that enters a plane-wave mode having wave vector $\boldsymbol{k} = (k, \vartheta, \varphi)$ and elliptical polarization. The unit polarization vector $\boldsymbol{e_k}$ is perpendicular to $\boldsymbol{k}$, so it is composed of unit vectors $\boldsymbol{e}_X, \boldsymbol{e}_Y$ in the X and Y directions.

[33] A complex representation, pairing $\boldsymbol{e}_X$ with $i\boldsymbol{e}_Y$, is given by Born and Wolf (1999) § 1.4.3; Landau and Lifshitz (1996) § 48. The relative signs accompanying $\boldsymbol{e}_X$ and $i\boldsymbol{e}_Y$ depend on the sense of the field's rotation.

This squares to give

$$|C_{ab}(\boldsymbol{k})|^2 = \frac{\pi c^3}{V} \frac{A_{ba}}{\omega_{ba}\,\omega_{\boldsymbol{k}}} \left(\frac{3}{4}(1 + \cos^2 \vartheta)\right) \frac{1}{(1 + \kappa^2)^4}, \tag{32.24}$$

in which

$$A_{ba} = \frac{4}{3} \frac{e^2}{4\pi\epsilon_0} \frac{\omega_{ba}^3 \,|\boldsymbol{r}_{ab}|^2}{\hbar c^3}; \qquad \boldsymbol{r}_{ab} \equiv \int \psi_a^* \,\boldsymbol{r}\, \psi_b \,\mathrm{d}V. \tag{32.25}$$

[34] For a transition linking state b to state i, the A-coefficient is A_{bi} and contains ω_{bi}^3. If state i lies higher in energy than state b, then, in $|C_{ib}(\boldsymbol{k})|^2$, A_{bi} and its ω_{bi} must be replaced by A_{ib} and ω_{ib}.

For the moment, A_{ba} is a convenient abbreviation for the expression[34] in (32.25); it is shown in (32.46) that it is the Einstein A-coefficient for the transition $b \to a$.

A factor $p(\vartheta, \varphi)$ in (32.24) gives the angular dependence[35] of the radiation emitted into directions (ϑ, φ). For $\Delta m_{\mathrm{atom}} = +1$ it is

[35] The angular distribution $p(\vartheta, \varphi)$ of (32.26) is just what we expect for emission from an atom undergoing a σ transition, as shown, for example, in Fig. 18.2.

$$p(\vartheta, \varphi) \equiv \left(\frac{3}{4}(1 + \cos^2 \vartheta)\right); \qquad \int p(\vartheta, \varphi)\, \frac{\mathrm{d}\Omega}{4\pi} = 1. \tag{32.26}$$

The matrix element $\boldsymbol{r}_{ab}$ in (32.25) is the "usual" electric-dipole matrix element linking states a, b; it does not involve $\mathrm{e}^{-\mathrm{i}\boldsymbol{k}\cdot\boldsymbol{r}}$ which in (32.24) is dealt with by $(1 + \kappa^2)^{-4}$.

A factor $f_{ab}(\boldsymbol{k})$ in (32.24), given for 2p–1s by

$$f_{ab}(\boldsymbol{k}) \equiv \frac{1}{(1 + \kappa^2)^4} = \frac{1}{\{1 + (\omega_{\boldsymbol{k}}/\omega_{\mathrm{c}})^2\}^4}, \tag{32.27}$$

is 1 at zero frequency, but falls away at high frequencies, a fall-off caused by $\mathrm{e}^{-\mathrm{i}\boldsymbol{k}\cdot\boldsymbol{r}}$ and just what we need to prevent the ultra-violet catastrophe discussed in § 32.4.1(b).

Equation (32.27) shows that ω_{c}, defined in (32.17), is the cut-off frequency above which the matrix element $\int \psi_a^* \,\mathrm{e}^{-\mathrm{i}\boldsymbol{k}\cdot\boldsymbol{r}} \,\hat{\boldsymbol{p}}\, \psi_b \,\mathrm{d}V$ falls at high frequencies. As a check: $\kappa \sim 1$ happens when $1/k \sim 2a_0/3$ which is about the size of the atom.

Conveniently, $f_{ab}(\boldsymbol{k})$ is independent of ϑ, φ for 2p–1s, so $p(\vartheta, \varphi)$ and $f_{ab}(\boldsymbol{k})$ do their jobs independently in (32.24). We take advantage of this by normalizing $p(\vartheta, \varphi)$, meaning that $p(\vartheta, \varphi)$ gives 1 when averaged over all directions ϑ, φ.

[36] The adventurous reader could examine the transition from $|3d, -2\rangle$ to $|2p, -1\rangle$, in which the m_l values (-2 and -1) are again chosen to make spin suppressible. The cut-off function $f_{ab}(\boldsymbol{k})$ depends on both κ and ϑ, and there are other complications.

It is easy to see why the cut-off function can in general depend on the direction of $\boldsymbol{k}$. If $\psi_i^* \,\hat{\boldsymbol{p}}\, \psi_b$ is long and thin, then the cut-off function $f_{ib}(\boldsymbol{k})$ must fall soonest when $\boldsymbol{k}$ lies "lengthways".

Every transition $b \to i$ necessarily has a cut-off function f_{ib} because its matrix element is given by an integral containing $\mathrm{e}^{-\mathrm{i}\boldsymbol{k}\cdot\boldsymbol{r}}$. If we forget this cut-off (replacing $\mathrm{e}^{-\mathrm{i}\boldsymbol{k}\cdot\boldsymbol{r}}$ by 1), we may be warned when an integral or sum, such as that in the denominator of (32.16), appears to be divergent.

Transitions other than 2p–1s have cut-off functions that are different from that in (32.27). Fortunately, the precise form of a cut-off does not matter (as long as there *is* a cut-off at $\omega \sim \omega_{\mathrm{c}}$) for evaluation of sums such as those in (32.21); see problem 32.15.[36]

32.6 The first sum in the $\widetilde{d_b}(s)$ of (32.21)

The first sum in the denominator of (32.21) is a sum over the $\boldsymbol{k}$-states of the outgoing radiation, that is a sum over the magnitude of $\boldsymbol{k}$, over its directions of travel, and over its associated polarizations $\boldsymbol{e}_{\boldsymbol{k}}$—though here we have only one polarization.

The number of $\boldsymbol{k}$-states, of a single polarization, in $\boldsymbol{k}$-range dk and solid-angle range $d\Omega$ is $Vk^2\,dk\,d\Omega/(2\pi)^3$, so that

$$\sum_{\text{modes }\boldsymbol{k}} = \int \frac{V}{2\pi^2 c^3}\,\omega_{\boldsymbol{k}}^2\,d\omega_{\boldsymbol{k}}\int\frac{d\Omega}{4\pi}.$$

Then

$$\sum_{\boldsymbol{k}}\frac{|C_{ab}(\boldsymbol{k})|^2}{s+\mathrm{i}\omega_{\boldsymbol{k}}} = \frac{V}{2\pi^2 c^3}\int_0^\infty d\omega_{\boldsymbol{k}}\,\frac{\omega_{\boldsymbol{k}}^2}{s+\mathrm{i}\omega_{\boldsymbol{k}}}\int\frac{d\Omega}{4\pi}$$

$$\times\,\frac{\pi c^3}{V}\,\frac{A_{ba}}{\omega_{ba}\,\omega_{\boldsymbol{k}}}\,p(\vartheta,\varphi)\,f_{ab}(\boldsymbol{k})$$

$$= \frac{A_{ba}}{2\pi\omega_{ba}}\int_0^\infty d\omega_{\boldsymbol{k}}\,\frac{\omega_{\boldsymbol{k}}}{s+\mathrm{i}\omega_{\boldsymbol{k}}}\int\frac{d\Omega}{4\pi}\,p(\vartheta,\varphi)\,f_{ab}(\boldsymbol{k})$$

$$= \frac{A_{ba}}{2\pi\omega_{ba}}\int_0^\infty d\omega_{\boldsymbol{k}}\,\frac{\omega_{\boldsymbol{k}}}{s+\mathrm{i}\omega_{\boldsymbol{k}}}\,\frac{1}{(1+\kappa^2)^4}, \tag{32.28}$$

since $f_{ab}(\boldsymbol{k})$ is independent of ϑ,φ and $p(\vartheta,\varphi)$ is normalized.

To evaluate the integral in (32.28) set $s=-\mathrm{i}\omega_c\sigma$ so that

$$\sum_{\boldsymbol{k}}\frac{|C_{ab}(\boldsymbol{k})|^2}{s+\mathrm{i}\omega_{\boldsymbol{k}}} = \frac{\omega_c}{2\pi\mathrm{i}}\frac{A_{ba}}{\omega_{ba}}\int_0^\infty\frac{\kappa}{\kappa-\sigma}\,\frac{1}{(1+\kappa^2)^4}\,d\kappa$$

$$= \frac{\omega_c}{2\pi}\frac{A_{ba}}{\omega_{ba}}\Big\{-\mathrm{i}\delta(\sigma)+\gamma(\sigma)\Big\}, \tag{32.29}$$

where the integral in this is evaluated in problem 32.10; the algebra is tedious but not difficult.[37,38] The results for δ and γ are:

$$\delta(\sigma) = \frac{1}{(1+\sigma^2)^4}\left\{\begin{array}{l}-\sigma\ln|\sigma|+\dfrac{5\pi}{32}-\dfrac{11}{12}\sigma-\dfrac{15\pi}{32}\sigma^2\\[2mm]-\dfrac{3}{2}\sigma^3-\dfrac{5\pi}{32}\sigma^4-\dfrac{3}{4}\sigma^5-\dfrac{\pi}{32}\sigma^6-\dfrac{1}{6}\sigma^7\end{array}\right\};$$

$$\tag{32.30a}$$

$$\gamma(\sigma) = \frac{1}{(1+\sigma^2)^4}\,(-\sigma)\arg(-\sigma). \tag{32.30b}$$

These expressions hold for all complex values of σ, therefore of s.

The first sum in the denominator of $\widetilde{d_b}(s)$ has yielded an expression containing $\ln(-\mathrm{i}s/\omega_c)=\ln(-\sigma)=\ln|\sigma|+\mathrm{i}\arg(-\sigma)$. This has a branch cut on the negative imaginary s-axis as is indicated by a heavy line in Fig. 32.4.[39] The imaginary part $\mathrm{i}\arg(-\sigma)$ of $\ln(-\sigma)$ has been allocated to $\gamma(\sigma)$.

The next section shows that, for evaluating $d_b(t)$, the only interesting values of $\widetilde{d_b}(s)$ are those on either side of the branch cut where, by Fig. 32.4,[40] $(-\sigma)$ has argument $\pm\pi$:

$$\gamma(\sigma) = \begin{cases}\pm\dfrac{\pi\sigma}{(1+\sigma^2)^4} & \text{for } \sigma>0 \text{ on } \begin{smallmatrix}\text{right}\\\text{left}\end{smallmatrix}\text{ side of cut}\\[2mm]0 & \text{for } \sigma<0.\end{cases} \tag{32.31}$$

Given that $\arg(-\sigma)$ has been dealt with, we can take the σ in (32.31) to be real. Define, for real positive σ,

$$\Gamma(\sigma) = \frac{\omega_c}{2\pi}\frac{A_{ba}}{\omega_{ba}}\Big\{\gamma_{\text{right}}(\sigma)-\gamma_{\text{left}}(\sigma)\Big\} = \omega_c\frac{A_{ba}}{\omega_{ba}}\frac{\sigma}{(1+\sigma^2)^4}. \tag{32.32}$$

[37] It should be obvious from (32.28) that the cut-off $f_{ab}=(1+\kappa^2)^{-4}$ cannot be approximated to 1, because then the integral will diverge. Because we can, we handle the integral of (32.28) without approximating, even though what follows is a bit heavy.

[38] In (32.29), the separation into δ and $\mathrm{i}\sigma$ is forward-looking, knowing the uses that are going to be made of δ and σ; it is not (yet) a separation into real and imaginary parts.

[39] We may see the logarithm and its branch cut coming by looking back to the $(s+\mathrm{i}\omega_{\boldsymbol{k}})$ in the denominator of (32.28).

[40] An alternative route to the argument $\pm\pi$ is given in sidenote 91.

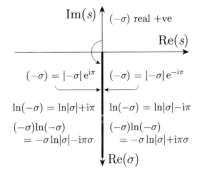

Fig. 32.4 In (32.29), $\delta(\sigma)+\mathrm{i}\gamma(\sigma)$ is a function of $s=-\mathrm{i}\omega_c\sigma$, in which s is complex, and so in the first instance is σ. Along the negative imaginary s-axis, there is a branch cut and $\gamma(\sigma)$ takes different values on either side of the cut, as given by (32.31).

Then (32.29) yields

$$\left(\sum_{k} \frac{|C_{ab}(\boldsymbol{k})|^2}{s + \mathrm{i}\omega_{\boldsymbol{k}}}\right)_{\text{beside cut}} = -\mathrm{i}\omega_{ba}\Delta(\sigma) \pm \frac{\Gamma(\sigma)}{2} \text{ on } \begin{smallmatrix}\text{right}\\\text{left}\end{smallmatrix} \text{ side, } \quad (32.33)$$

in which

$$\Delta(\sigma) = \frac{\omega_c}{2\pi\omega_{ba}} \frac{A_{ba}}{\omega_{ba}} \delta(\sigma) \qquad (32.34)$$

$$= \frac{\omega_c}{2\pi\omega_{ba}} \frac{A_{ba}}{\omega_{ba}} \frac{1}{(1+\sigma^2)^4} \left\{ -\sigma \ln|\sigma| + \frac{5\pi}{32} - \frac{11}{12}\sigma \right.$$

$$\left. - \frac{15\pi}{32}\sigma^2 - \frac{3}{2}\sigma^3 - \frac{5\pi}{32}\sigma^4 - \frac{3}{4}\sigma^5 - \frac{\pi}{32}\sigma^6 - \frac{1}{6}\sigma^7 \right\}, \quad (32.35)$$

and $\Gamma(\sigma)$ is given by (32.32).

The expression for $\Delta(\sigma)$ in (32.35) looks formidable, but fortunately it turns out to be negligible (Fig. 32.6).

32.7   The complex-integration path for finding $d_b(t)$

Section 32.6 has shown that the first sum in the denominator of $\widetilde{d}_b(s)$ (eqn 32.21) yields a function of s that includes $\ln(-\mathrm{i}s/\omega_c) = \ln(-\sigma)$ which has a branch cut along the negative imaginary s-axis, starting at the origin. In order to see the significance of this, we consider the inverse Laplace transform by which $d_b(t)$ is obtained from $\widetilde{d}_b(s)$.

The inverse transform is

$$d_b(t) = \frac{1}{2\pi\mathrm{i}} \int_{\beta-\mathrm{i}\infty}^{\beta+\mathrm{i}\infty} \widetilde{d}_b(s)\,\mathrm{e}^{st}\,\mathrm{d}s, \qquad (32.36)$$

in which β must be chosen to make the integration path lie to the right of all singularities of the integrand. A possible integration path is shown as the vertical line at the right of Fig. 32.5.

The branch cut is the only singularity possessed by $\widetilde{d}_b(s)$; there are no singularities or poles anywhere else in the complex plane. The justification for this statement is obtained in problem 32.11.

This property permits us to extend the integration path to form the closed contour shown in Fig. 32.5. The contour encloses no singularity of $\widetilde{d}_b(s)$, so the integral round the closed path is zero. The contour must avoid the branch cut, as is indicated.

On the large semicircle, e^{st} is negligible—for $t > 0$ —because s has a negative real part. The integrand has no infinity at $s = 0$, so there is no contribution from the small circle when that circle is shrunk to limiting smallness. What remains is the integration up and down the sides of the branch cut. Therefore[41]

$$d_b(t) = \frac{1}{2\pi\mathrm{i}} \int_{-\mathrm{i}\infty+\varepsilon}^{\mathrm{i}0+\varepsilon} \widetilde{d}_b(s)\,\mathrm{e}^{st}\,\mathrm{d}s + \frac{1}{2\pi\mathrm{i}} \int_{\mathrm{i}0-\varepsilon}^{-\mathrm{i}\infty-\varepsilon} \widetilde{d}_b(s)\,\mathrm{e}^{st}\,\mathrm{d}s.$$

This explains why we have said that $\widetilde{d}_b(s)$ is needed at locations alongside the cut only. For performing the integrations we express s as[42]

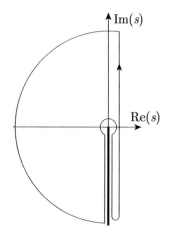

Fig. 32.5 The contour used for evaluating the inverse Laplace transform giving $d_b(t)$ from $\widetilde{d}_b(s)$. The thick line shows the branch cut which the contour must avoid.

[41] Here ε is a positive infinitesimal that identifies on which side of the branch cut we are integrating s.

[42] The use of symbol ω (without subscript $\boldsymbol{k}$) will be convenient in future discussion. It must be distinguished from the $\omega_{\boldsymbol{k}}$ in (32.28) which has now been summed over.

$$s = -\mathrm{i}\omega \pm \varepsilon = -\mathrm{i}\omega_c \sigma \pm \varepsilon.$$

The integrations giving $d_b(t)$ now read

$$d_b(t) = \frac{1}{2\pi\mathrm{i}} \int_\infty^0 \frac{1}{\omega - \omega_{ba} + \omega_{ba}\Delta(\sigma) + \mathrm{i}\Gamma(\sigma)/2} \, \mathrm{e}^{-\mathrm{i}\omega t} \, \mathrm{d}\omega$$

$$+ \frac{1}{2\pi\mathrm{i}} \int_0^\infty \frac{1}{\omega - \omega_{ba} + \omega_{ba}\Delta(\sigma) - \mathrm{i}\Gamma(\sigma)/2} \, \mathrm{e}^{-\mathrm{i}\omega t} \, \mathrm{d}\omega$$

$$= \int_0^\infty \frac{\Gamma(\sigma)/2\pi}{\left\{\omega - \omega_{ba} + \omega_{ba}\Delta(\sigma)\right\}^2 + \Gamma(\sigma)^2/4} \, \mathrm{e}^{-\mathrm{i}\omega t} \, \mathrm{d}\omega \qquad (32.37)$$

or rather, remembering the starting point of the Laplace transform,[43]

$$d_b(t) = \begin{cases} \displaystyle\int_0^\infty \frac{\Gamma/2\pi}{\left\{\omega - \omega_{ba}(1 - \Delta)\right\}^2 + \Gamma^2/4} \, \mathrm{e}^{-\mathrm{i}\omega t} \, \mathrm{d}\omega & \text{for } t > 0 \\[4mm] 0 & \text{for } t < 0. \end{cases} \qquad (32.38)$$

The presence of $\mathrm{e}^{-\mathrm{i}\omega t}$ in (32.38) tells us that ω can represent angular frequency. Indeed, $d_b(t)$ is the inverse Fourier transform[44] of

$$\overline{d_b}(\omega) = \begin{cases} \displaystyle\frac{\Gamma(\sigma)/2\pi}{\left\{\omega - \omega_{ba}[1 - \Delta(\sigma)]\right\}^2 + \Gamma(\sigma)^2/4} & \text{for } \omega > 0 \\[4mm] 0 & \text{for } \omega < 0. \end{cases} \qquad (32.39)$$

Within this, we may express $\Gamma(\sigma)$ in terms of $\omega = \omega_c\sigma$:

$$\Gamma(\sigma) = \frac{A_{ba}}{\omega_{ba}} \frac{\omega}{\left\{1 + (\omega/\omega_c)^2\right\}^4} \qquad \text{for } \omega > 0, \qquad (32.40)$$

in which $\omega > 0$ is necessitated by (32.31) or (32.39).

Expression (32.39) may be seen as signalling that there is a Lorentzian or near-Lorentzian resonance, and it does not mislead. However, Δ and Γ are not constants but are dependent on σ, equivalently on $\omega = \omega_c\sigma$. More importantly, a spectral line profile $g(\omega - \omega_{ba})$ does not come from here: it must be found from the frequency spectrum of $|c_{ak}|^2$, and (32.39) is far from telling us that.[45] There is more investigation to be done.

32.7.1   Reversibility and dissipation

Look back to eqn (32.14a). If there were no radiative transition, we would have $C_{ib}(\boldsymbol{k})^* = 0$ for all $i, \boldsymbol{k}$, and then, for $t > 0$,

$$\dot{c}_b = 0; \qquad c_b(t) = 1; \qquad d_b(t) = \mathrm{e}^{-\mathrm{i}\omega_{ba}t}.$$

Were we not aiming for an *ab initio* solution, we might reasonably think of patching[46] onto this a slow (in relation to ω_{ba}) decay as $\mathrm{e}^{-t/\tau}$. This would yield a Laplace transform

$$\widetilde{d_b}(s) \propto 1/(s + \mathrm{i}\omega_{ba} + 1/\tau), \quad \text{having a pole at } s = -\mathrm{i}\omega_{ba} - 1/\tau.$$

Such a single pole yields an exponential decay when the inverse Laplace transform is taken, and the decay rate is given by $1/\tau$, the negative real part of the pole's location.

Even though this $\widetilde{d_b}(s)$ is not what we have in the calculation of the present chapter, it supplies helpful insights. The pole's location, and

[43] The zero value for $t < 0$ can be confirmed by redoing the contour integration. The contour must now be completed by a large semicircle round the right-hand half of the complex s-plane (e^{st} blows up for $\mathrm{Re}(s) < 0$ when $t < 0$), and the integrand has no singularities within this new contour.

Notice that the zero of $d_b(t)$ for $t < 0$ does not come from evaluation of the integral in (32.38) extended to $t < 0$. It comes from a discontinuous change in the rule for evaluating $d_b(t)$.

[44] The inverse Fourier transform is more often presented with $(\mathrm{d}\omega/2\pi)$ at the end, compensated by the removal of 2π from under $\Gamma(\sigma)$. Whichever choice is made, there is a slight untidiness somewhere.

[45] Indeed, the algebra never needs to identify the ω of (32.39) with an angular frequency of the emitted radiation!

In partial recognition of this, I retain different symbols for the $\kappa = \omega_k/\omega_c$ of (32.17) and $\sigma = \omega/\omega_c$ in (32.40).

[46] Just such a patching is what one does in setting up the optical Bloch equations. See e.g. Foot (2005), § 7.5.2.

[47] At root, quantum mechanics is time-reversible, a fact consistent with the non-decay noted here.

[48] The density of states is hiding in Γ.

[49] One departure from exponential behaviour is investigated in problem 32.19, and another in problem 32.20.

[50] The Poincaré cycle is a concept in statistical mechanics. A physical system in equilibrium passes rapidly through its microstates (owing to collisions or other small perturbations). The number of microstates is finite, albeit extremely large, so eventually the system must revisit microstates it has been in before. This revisiting, and the time that elapses before it happens, is called the Poincaré cycle.

The Poincaré cycle explains how some far-from-equilibrium initial state (say a gas all concentrated at one end of a box) can relax to near-uniformity while undergoing only time-reversible interactions. What results is "near-uniform" because it has random fluctuations, mostly small, but some large, of gas density. The original state can be revisited, as a giant fluctuation, but only after the Poincaré cycle; and this timescale works out to be greater than the age of the Universe (for samples of reasonable size). The original relaxation can therefore safely be thought of as irreversible and entropy-increasing.

The radiation process under discussion here does not exactly correspond to the picture just given, but we can argue that it possesses a Poincaré cycle in much the same way. Problem 32.12 asks you to think about this.

[51] The investigation of this chapter, leading to the near-exponential decay of (32.46), can provide a model to make us comfortable with other exponential decays, such as those described by T_1 and T_2 in magnetic resonance.

[52] We repeat the warning from Chapter 12. If we keep only some terms of a given order, there is a risk that the neglected terms are larger and of opposite sign. Therefore the terms we contemplate retaining cannot even be trusted to carry us in the right direction.

the associated decay rate $1/\tau$, tell us how to "read" singularities in the complex s-plane. We may think of the branch cut of Fig. 32.5 as a dense distribution of poles (problem 32.11). These poles all lie *on* the imaginary s-axis, and what they contribute to $d_b(t)$ individually does not decay[47] with time t.

A full picture emerges only when we look at what the poles supply collectively. Each pole contributes to one of the ωs in (32.39), and the integration over ω is a sum over all of the field modes into which energy becomes distributed.[48] The probability/occupation, originally concentrated in the atom's upper state, becomes distributed among a large number of field modes. This looks very much like dissipation: $d_b(t)$ diminishes while order changes to disorder. Whether this leads to a near-exponential decay of $d_b(t)$ remains to be determined.[49]

We are reminded of randomizations in statistical physics. Molecules of a gas may start off at one end of a box, then spread out to a near-uniform distribution with an increase of entropy. Yet the gas molecules are subject to quantum rules that are time-reversible, so nothing prohibits resurgence into the original now-improbable configuration. The paradox is resolved by introducing the idea of the Poincaré cycle.[50]

The field modes used in (32.5) are travelling waves, often associated with periodic boundary conditions (§ 6.3). When such a wave crosses a bounding surface of the quantization volume V, a replica of it enters through the opposite face. Field energy never leaves volume V, and is therefore always available for a Poincaré resurgence.

The larger the number of field modes that become populated/occupied (the larger we make the quantization volume V), the longer is the time before a Poincaré resurgence happens. If the number of modes is made infinite, which in effect is what we do when we coalesce the many poles into a branch cut, then the resurgence is pushed out to infinite time.

We can understand how an apparently dissipative, near-exponential decay can happen for the atom's upper state.[51]

32.7.2   The "frequency shift" Δ in (32.39)

Up to this point, the mathematical solution of the Schrödinger equation has been exact within our assumptions: a *first-order* perturbation applied to the decay of an atom's excited state, such as a hydrogen 2p state decaying to 1s. Now, as far back as eqn (32.4) we have said that this treatment of spontaneous emission cannot be really exact, however careful we have been, but is the beginning of an expansion in powers of the fine-structure constant α_{fs}. A quantity of high power in α_{fs} can be significant only if we have retained *all* terms of that order.[52]

Return to (32.35) and refer to Fig. 32.2.

$$\Delta(\sigma) = \frac{\omega_c}{2\pi\omega_{ba}} \frac{A_{ba}}{\omega_{ba}} \delta(\sigma) \sim (\alpha_{\text{fs}})^{-1} \times (\alpha_{\text{fs}})^3 \times (\text{order unity}) \sim (\alpha_{\text{fs}})^2. \tag{32.41}$$

The fact that $\delta(\sigma)$ is of order unity, regardless of the value of σ, is shown in Fig. 32.6.

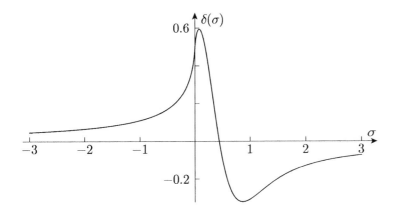

Fig. 32.6 The function $\delta(\sigma)$ defined by (32.30a). Function $\delta(\sigma)$ never goes outside the range ± 0.6, so it is of order unity or less for all values of σ.

Variable $\sigma = is/\omega_c$, real on the imaginary s-axis. At the peak of the "resonance" in (32.38), $\omega \approx \omega_{ba}$, so δ is evaluated at $\sigma \approx \omega_{ba}/\omega_c \sim \alpha_{fs}$, close to $\sigma = 0$.

In the denominators of (32.38) and (32.39), we have $\omega_{ba}(1 - \Delta)$, so Δ appears beside 1: we are looking at a correction that is α_{fs}^2 smaller than the leading term. Terms of this order have not been retained consistently—indeed some were explicitly discarded in the discussion of the Dirac equation in §32.4.1—so the only right thing to do with the Δ in (32.39) is to discard it: it has nothing to tell us about any departure of the "resonance frequency" from ω_{ba}, not even the sign.[53]

32.7.3 The Γ in the denominator of (32.39)

We next turn attention to the function $\Gamma(\sigma)$ defined by (32.32) or (32.40). For hydrogen 2p–1s we have

$$\frac{A_{ba}}{\omega_{ba}} = \frac{2^{11}}{3^9}\,\alpha_{fs}^3 = 0.104\,\alpha_{fs}^3, \tag{32.42}$$

that is, A_{ba}/ω_{ba} is of order α_{fs}^3. This is of a higher order in α_{fs} than the Δ that we have just rejected. Nevertheless we must retain Γ in both numerator and denominator of (32.39): in the numerator because without it nothing is left to describe the behaviour of $d_b(t)$; and in the denominator because omitting it would cause the integral in (32.38) to have convergence problems.

We here consider the $\Gamma(\sigma)$ in the denominator of (32.39).

When $\omega = \omega_{ba}$, (32.40) tells us that $\Gamma = A_{ba}(1 + \text{order } \alpha_{fs}^2) \sim \omega_{ba}\alpha_{fs}^3$. Therefore the "resonance peak" of $\overline{d_b}(\omega)$ is narrow; within its width Γ hardly changes so we can replace Γ by A_{ba}. Away from resonance, $\Gamma^2/4$ is negligible beside $(\omega - \omega_{ba})^2$, and can safely be replaced by $A_{ba}^2/4$ because this is negligible also. Fuller reasoning is pursued in problem 32.13. The outcome is that (32.39) can be replaced by[54]

$$\overline{d_b}(\omega) \approx \frac{\Gamma(\sigma)/2\pi}{(\omega - \omega_{ba})^2 + A_{ba}^2/4} \quad \text{for } \omega > 0, \tag{32.43}$$

where the error introduced in this step is of order α_{fs}^2.

It is worth stressing: the approximations between (32.38) and (32.43) involve the discarding of quantities that *must* be neglected because they are of an order in α_{fs} that is beyond the accuracy of our analysis.

[53] Some books that I have consulted say that Δ is infinite. Or that it needs rescuing from infinity by a space smearing of the wave function, perhaps associated with a Foldy–Wouthuysen representation of the Dirac equation. Such an infinity seems to be an artefact of setting $e^{-i\mathbf{k}\cdot\mathbf{r}} \approx 1$ when working out matrix elements. Since there is no infinity when $e^{-i\mathbf{k}\cdot\mathbf{r}}$ is properly handled, no rescue is called for. We may see this clearly in (32.28): the final factor $(1 + \kappa^2)^{-4}$ originates from $e^{-i\mathbf{k}\cdot\mathbf{r}}$. If $e^{-i\mathbf{k}\cdot\mathbf{r}}$ is omitted (replaced by 1) the integral diverges; with it there is no problem.

[54] Note that $\Gamma(\sigma)$ remains unapproximated in the numerator.

32.8   The initial-state coefficient $c_b(t)$

We first evaluate $d_b(t)$, then find $c_b(t)$ from there.

Equations (32.38) and (32.43) contain $\Gamma(\sigma)$ in the numerator. This $\Gamma(\sigma)$ is not swamped by the addition of something much larger so we retain it in full. Rewriting (32.38) with the latest $\overline{d_b}(\omega)$ from (32.43) and $\Gamma(\sigma)$ from (32.40):

$$d_b(t) = \int_0^\infty \frac{A_{ba}/2\pi}{(\omega - \omega_{ba})^2 + A_{ba}^2/4} \, \frac{\omega}{\omega_{ba}} \, \frac{1}{\{1 + (\omega/\omega_c)^2\}^4} \, \mathrm{e}^{-\mathrm{i}\omega t} \, \mathrm{d}\omega. \quad (32.44)$$

Evaluation of the integral in (32.44) is a little untidy, so we deal with it in problem 32.14.[55] The outcome is, for $t > 0$,

$$d_b(t) = \mathrm{e}^{-\mathrm{i}\omega_{ba}t} \, \mathrm{e}^{-A_{ba}t/2}(1 + \text{order } \alpha_{\mathrm{fs}}^2) + \text{order } \alpha_{\mathrm{fs}}^3 \ln \alpha_{\mathrm{fs}}.$$

Then, finally,[56]

$$c_b(t) = \begin{cases} \mathrm{e}^{-A_{ba}t/2} + \text{order } \alpha_{\mathrm{fs}}^2 \text{ and } \alpha_{\mathrm{fs}}^3 \ln \alpha_{\mathrm{fs}} & \text{for } t > 0 \\ 0 & \text{for } t < 0. \end{cases} \quad (32.45)$$

In particular, we draw attention to the finding that

$$|c_b(t)|^2 = \mathrm{e}^{-A_{ba}t} \quad (32.46)$$

for $t > 0$ within claimed accuracy, so it is now established that the A_{ba} defined in (32.25) is the Einstein A_{ba}-coefficient for the decay $b \to a$.

At the same time there have been warnings. In §32.7 we integrated round a branch cut in the complex s-plane, and in §32.7.1 we noted that such an integration cannot result in an exact exponential decay.

We can identify two regimes in which the actual $c_b(t)$ behaves in a way different from the simple exponential in the leading term of (32.45):

- Problem 32.20(2) shows that, at the beginning of the transition $c_b(t)$ departs from 1 with a fall that starts out as t^2. By contrast, the exponential of (32.45) predicts a change linear with t.
- At the end of the transition, for very large times, $d_b(t)$ diminishes as t^{-2}, so its decay is slower than exponential (problem 32.19).[57]

The departures from exponential behaviour itemized here are not alarming: they are small and fall comfortably within the stated error of α_{fs}^2 or $\alpha_{\mathrm{fs}}^3 \ln \alpha_{\mathrm{fs}}$; see also §32.10.

32.9   The occupation of final states $a, \boldsymbol{k}$

This chapter started by saying that two quantities were to be evaluated via quantum electrodynamics: the state–state Einstein coefficient α_{bak} and the spectral line profile $g(\omega - \omega_{ba})$. So far we have discussed neither, because the decay behaviour of the upper state $b = 2\mathrm{p}$ had to be found first. Completion of our task is now quite straightforward.

We substitute (32.45) back into (32.14b) and solve directly[58] for the time dependence of $c_{\boldsymbol{k}a}$. We have, for $t > 0$,

$$\dot{c}_{\boldsymbol{k}a}(t) = C_{ab}(\boldsymbol{k}) \, \mathrm{e}^{\mathrm{i}(\omega_{\boldsymbol{k}} - \omega_{ba})t} \, \mathrm{e}^{-A_{ba}t/2}.$$

[55] We do the job again in problem 32.15, this time replacing $\{1 + (\omega/\omega_c)^2\}^{-4}$ with a step cut-off at ω_c. The result is the same, confirming that the form of the cut-off is not critical.

[56] Of course, at time $t = 0$, $c_b(t=0+)$ is exactly equal to 1. But expression (32.45) has been approximated by omission of terms of order α_{fs}^2: in the omission of $\Delta(\sigma)$ and in handling the integration in (32.44). The finding in (32.45) is as close as we can expect to get.

[57] We note that not-quite-exponential decays are already known. See e.g., Kelkar and Nowakowski (2010) and references therein. I am indebted to Dr C.V. Sukumar for making me aware of this reference.

[58] By "directly" we mean that there is no need to obtain $\widetilde{c_{\boldsymbol{k}a}}(s)$ and then invert the Laplace transform. The Laplace-transform route is available, of course, and yields (32.48) as it must. It is, however, lengthy compared with what is done here.

Integration of the last equation is easy, and we find

$$c_{\boldsymbol{k}a}(t) = C_{ab}(\boldsymbol{k}) \frac{1 - e^{\{i(\omega_{\boldsymbol{k}} - \omega_{ba}) - A_{ba}/2\}t}}{i(\omega_{ba} - \omega_{\boldsymbol{k}}) + A_{ba}/2}; \qquad (32.47)$$

$$|c_{\boldsymbol{k}a}(t)|^2 = \frac{|C_{ab}(\boldsymbol{k})|^2}{(\omega_{\boldsymbol{k}} - \omega_{ba})^2 + A_{ba}^2/4}$$
$$\times \left\{ (1 - e^{-A_{ba}t/2})^2 + 4\, e^{-A_{ba}t/2} \sin^2 \tfrac{1}{2}(\omega_{\boldsymbol{k}} - \omega_{ba})t \right\}. \quad (32.48)$$

For the next two subsections we shall be concerned with $|c_{\boldsymbol{k}a}(\infty)|^2$, the population of state $|a, \boldsymbol{k}\rangle$ when the reaction giving rise to this state is complete. Thus[59]

$$|c_{\boldsymbol{k}a}(\infty)|^2 = \frac{|C_{ab}(\boldsymbol{k})|^2}{(\omega_{\boldsymbol{k}} - \omega_{ba})^2 + A_{ba}^2/4}. \qquad (32.49)$$

We return to considering the time dependence in § 32.10.

[59] There are clear signs of a Lorentzian line shape in the denominator of (32.49). But there is more frequency dependence hiding in $|C_{ab}(\boldsymbol{k})|^2$, hence the need for the investigations in the next two subsections.

32.9.1   The modified Einstein coefficient $\alpha_{ba\boldsymbol{k}}$

From early in this chapter, in (32.8), we have been considering a single atom, initially excited to state b, and then undergoing radiative decay, a decay that we have been following as a function of time t. Let us change viewpoint now, looking at many atoms excited at random times, and looking at average rates of decay.[60] The language is that of "rate equations" beloved of laser physicists.

Let there be many atoms present, and let upper state b be filled (by some process unspecified) at a steady rate W_b atoms per second. There will be N_b atoms populating state b, each decaying at rate A_{ba}. Then in a steady state, $W_b = N_b A_{ba}$. Each atom has probability $|c_{\boldsymbol{k}a}(\infty)|^2$ of decaying[61] to final state $|a, \boldsymbol{k}\rangle$, so the rate at which this final state is filled is

$$W_b\, |c_{\boldsymbol{k}a}(\infty)|^2 = N_b \alpha_{ba\boldsymbol{k}},$$

by definition (32.1c) of the modified Einstein coefficient $\alpha_{ba\boldsymbol{k}}$. Thus

$$\alpha_{ba\boldsymbol{k}} = \frac{W_b}{N_b} |c_{\boldsymbol{k}a}(\infty)|^2 = A_{ba}\, |c_{\boldsymbol{k}a}(\infty)|^2 = A_{ba} \frac{|C_{ab}(\boldsymbol{k})|^2}{(\omega_{\boldsymbol{k}} - \omega_{ba})^2 + A_{ba}^2/4}$$
$$= \frac{2\pi^2 c^3}{V} \frac{A_{ba}}{\omega_{ba}} \frac{A_{ba}/2\pi}{(\omega_{\boldsymbol{k}} - \omega_{ba})^2 + A_{ba}^2/4} \times \frac{1}{\omega_{\boldsymbol{k}}}\, p(\vartheta, \varphi)\, f_{ab}(\boldsymbol{k}), \quad (32.50)$$

[60] This is a return to the language of eqns (32.1).

[61] The decay must go (almost wholly) to atomic state a, but it can go to any of many possible field modes $|\boldsymbol{k}, \boldsymbol{e_k}\rangle$. The probability $|c_{\boldsymbol{k}a}(\infty)|^2$ caters for this "branching" and is far smaller than 1.

using (32.24). In this, $p(\vartheta, \varphi)$ is the projection factor from eqn (32.26) showing how the atom's rotating dipole moment projects onto the (elliptical) polarization state $\boldsymbol{e_k}$ of the outgoing radiation;[62] while $f_{ab}(\boldsymbol{k})$ is the cut-off function of (32.27) that accounts for the diminution of the transition probability at high frequencies owing to $e^{-i\boldsymbol{k}\cdot\boldsymbol{r}}$.

The fact that we have been able to obtain expression (32.50)—at all—confirms the respectability of the "fine-grained" picture introduced in Chapter 19.

When $\alpha_{ba\boldsymbol{k}}$ is summed over all final states $|\boldsymbol{k}, \boldsymbol{e_k}\rangle$ of the radiation, it yields A_{ba}, as it must; see problem 32.18.

[62] Strictly, we should write $\alpha_{b,a\boldsymbol{k},\boldsymbol{e_k}}$ for the transition to the defined field mode $|\boldsymbol{k}, \boldsymbol{e_k}\rangle$; there is, of course, another possible polarization orthogonal to $\boldsymbol{e_k}$, and this has its own α, equal to zero.

[63] In § 19.2–5 we followed convention by aggregating field modes into an energy density $\rho(\omega)$ on the way to obtaining the Einstein B'_{ab}-coefficient.

However, it was then necessary to "dis-aggregate", to disentangle the summing over frequencies by patching in $g(\omega - \omega_{ba})$. This procedure was justifiable, but awkward. That is one reason why we are now using a "bottom-up" approach. There is no objection to aggregating transitions: it is a confidence-making procedure, as dis-aggregation, or un-averaging, is not.

[64] Since modes $\boldsymbol{k}, \boldsymbol{e}_{\boldsymbol{k}}$ within $d\omega$ have been aggregated, it is appropriate to write ω, rather than $\omega_{\boldsymbol{k}}$, as ω is no longer being thought of as dependent on $\boldsymbol{k}$.

Up to this point, we have maintained a notational distinction between $\omega_{\boldsymbol{k}}$ (angular frequency contained in the radiated field as in (32.14b)) and ω (a Fourier component of the atom's $\overline{d_b}(\omega)$ as in (32.39)). It will do no harm for this distinction to become a little blurred, since our equations are linear: an $e^{-i\omega t}$ accompanying $\overline{d_b}(\omega)$ must be involved in the radiation of the $e^{-i\omega t}$ component of $c_{\boldsymbol{k}a}(t)$.

[65] We make use of several quantities with slightly different definitions. To clarify:

$|c_{\omega a}(t)|^2 \, d\omega$
$= \dfrac{V \omega^2 \, d\omega}{2\pi^2 c^3} \int \dfrac{d\Omega}{4\pi} \, |c_{\boldsymbol{k}a}(t)|^2;$

$|c_a(t)|^2 = \int |c_{\omega a}(t)|^2 \, d\omega$
$= \dfrac{V}{2\pi^2 c^3} \int d\omega \, \omega^2 \int \dfrac{d\Omega}{4\pi} \, |c_{\boldsymbol{k}a}(t)|^2.$

We should perhaps point out once again that $c_{\boldsymbol{k}a}$ gives the occupation of atomic state a and field mode $|\boldsymbol{k}, \boldsymbol{e}_{\boldsymbol{k}}\rangle$ having a single polarization state $\boldsymbol{e}_{\boldsymbol{k}}$. There is therefore no factor 2 inserted to encompass two polarizations, as is commonly done in other contexts.

[66] We see yet again that the $f_{ab}(\boldsymbol{k})$ in (32.51) cannot be neglected (replaced by 1); if that were done the integral of (32.52) would diverge logarithmically.

32.9.2   The spectral line profile

It will help us to see the significance of the expressions in (32.49) and (32.50) if we aggregate transitions to some extent.[63] We sum over directions of photon travel and (trivially) over polarizations, but retain the dependence on frequency. Thus we seek the **spectral line profile**:[64,65]

probability $g(\omega - \omega_{ba}) \, d\omega = |c_{\omega a}(\infty)|^2 \, d\omega$ that the atom ends up in its lower state a and that radiation has been emitted into modes within $d\omega$, regardless of that radiation's direction of travel or polarization

$$= V \frac{\omega^2 \, d\omega}{2\pi^2 c^3} \int \frac{d\Omega}{4\pi} \, |c_{\boldsymbol{k}a}(\infty)|^2 = \frac{V \omega^2 \, d\omega}{2\pi^2 c^3} \int \frac{d\Omega}{4\pi} \frac{|C_{ab}(\boldsymbol{k})|^2}{(\omega - \omega_{ba})^2 + A_{ba}^2/4}$$

$$= \frac{A_{ba}/2\pi}{(\omega - \omega_{ba})^2 + A_{ba}^2/4} \frac{\omega \, d\omega}{\omega_{ba}} \int \frac{d\Omega}{4\pi} \, p(\vartheta, \varphi) \, f_{ab}(\boldsymbol{k}).$$

For hydrogen 2p–1s, $f_{ab}(\boldsymbol{k})$ is independent of the direction of wave vector $\boldsymbol{k}$, so it passes through the integration over Ω. The integration over $p(\vartheta, \varphi)$ then gives 1. The desired probability gives

$$g(\omega - \omega_{ba}) = \frac{A_{ba}/2\pi}{(\omega - \omega_{ba})^2 + A_{ba}^2/4} \frac{\omega}{\omega_{ba}} \frac{1}{\left\{1 + (\omega/\omega_c)^2\right\}^4}. \tag{32.51}$$

This is dominated by its first factor: a Lorentzian line profile.

Expressions containing elements of (32.51) have been on display since (32.39) and (32.44). However, only now is it clear that this mathematical shape has anything to do with a spectral line profile.

As a consistency check, we find:

Probability $|c_a(\infty)|^2$ that the atom ends up in its lower state a, regardless of which radiation modes become populated

$$= \int_0^\infty |c_{\omega a}(\infty)|^2 \, d\omega = \int_0^\infty g(\omega - \omega_{ba}) \, d\omega$$

$$= \int_0^\infty \frac{A_{ba}/2\pi}{(\omega - \omega_{ba})^2 + A_{ba}^2/4} \frac{\omega}{\omega_{ba}} \frac{1}{\left\{1 + (\omega/\omega_c)^2\right\}^4} \, d\omega = 1, \tag{32.52}$$

by application of (32.74). As a byproduct we have just shown that $g(\omega - \omega_{ba})$ is the *normalized* line profile.[66]

Equation (32.52) is not exact because it contains an uncertainty of order α_{fs}^2 propagating from (32.74). Thus $|c_a(\infty)|^2 = 1 + O(\alpha_{\text{fs}}^2)$. State b decays mostly to state a, but there is room for a little of the probability to go elsewhere, which can only be to higher states j. These higher-state occupations can be at most of order α_{fs}^2, as is confirmed in § 32.11.

Problem 32.16 asks you to refine consistency check (32.52) to give

$$|c_a(t)|^2 = 1 - e^{-A_{ba}t} = 1 - |c_b(t)|^2, \tag{32.53}$$

showing that the atomic population in states a and b together is 1 at all times—but again within α_{fs}^2.

32.10   Time dependence of final states a, k

The main aim of the present chapter has now been achieved: the obtaining of α_{bak} in eqn (32.50) and of $g(\omega - \omega_{ba})$ in eqn (32.51). Nevertheless, there are some details that we choose to present next because they give us a fuller idea of what is going on physically.

The filling of the lower atomic state a

Return to the picture in which a single atom is excited to state b at time $t = 0$, and we follow the dependence of the system on time t thereafter. Two conflicting behaviours can be found here: $|c_a(t)|^2$ starting out quadratically with t (problem 32.20), and (problem 32.16) $|c_a(t)|^2$ rising as $1 - e^{-A_{ba}t}$, which implies a start linear with time. These behaviours can be reconciled, within α_{fs}^2, but the matter needs investigation.

Combine (32.48) with (32.49) and aggregate the states $\boldsymbol{k}$ within $\mathrm{d}\omega$.

$$|c_{\boldsymbol{k}a}(t)|^2 = |c_{\boldsymbol{k}a}(\infty)|^2 \left\{ \ \right\};$$

$$|c_{\omega a}(t)|^2 = |c_{\omega a}(\infty)|^2 \left\{ \ \right\}$$

$$= g(\omega - \omega_{ba}) \left\{ \left(1 - e^{-A_{ba}t/2}\right)^2 + 4 e^{-A_{ba}t/2} \sin^2 \tfrac{1}{2}(\omega - \omega_{ba})t \right\}.$$
$$(32.54)$$

The probability, after time t, that the atom is in state a and a photon has been emitted, the photon going into any frequency or direction, is

$$|c_a(t)|^2 = \int |c_{\omega a}(t)|^2 \, \mathrm{d}\omega = \int_0^\infty \mathrm{d}\omega \, g(\omega - \omega_{ba})$$

$$\times \left\{ \left(1 - e^{-A_{ba}t/2}\right)^2 + 4 e^{-A_{ba}t/2} \sin^2 \tfrac{1}{2}(\omega - \omega_{ba})t \right\}. \quad (32.55)$$

Figure 32.7 displays plots of $|c_a(t)|^2$, evaluated from (32.55). The initial rise (eqn 32.56) of $|c_a(t)|^2$ is as t^2. But this quadratic regime lasts for a time $\sim 1/\omega_c$ only, and the curve joins the $(1 - e^{-A_{ba}t})$ of

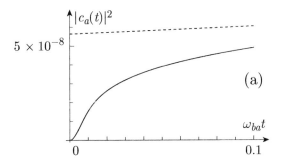
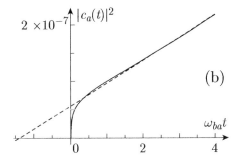

Fig. 32.7 The occupation probability $|c_a(t)|^2$ of the lower atomic state a as a function of time t, as given by (32.55).

(a) The initial increase of $|c_a(t)|^2$ is as t^2, but this lasts for a time of only about $1/\omega_c$, and there is a steep rise thereafter. Note that the time range displayed is a small fraction of the period of the transition frequency ω_{ba}. It is more usefully thought of as about 9 periods of the cutoff frequency ω_c.

(b) When the time range displayed is increased by a factor 40, we see a linear rise, the beginning of the exponential form $(1 - e^{-A_{ba}t})$ expected from (32.53). The broken line shows $A_{ba}t$ but with a time shift of $-1.395/\omega_{ba} = -0.222(2\pi/\omega_{ba})$, less than a period at the resonance frequency.

(32.53) (with a slight shift) after a time that is a small fraction of the resonance period, around $\omega_{ba}t \approx 0.3$. Thus, except for a brief glitch at the beginning of the decay, $|c_a(t)|^2$ follows $1 - e^{-A_{ba}t}$ very closely.

The broken line in Fig. 32.7 is $(1 - e^{-A_{ba}t})$ plus an added vertical offset of 5.64×10^{-8}. This offset is far smaller than the $\sim \alpha_{\text{fs}}^2$ that we have understood to be the possible calculation error. Perhaps we were just lucky. At any rate, the behaviour displayed in Fig. 32.7 comfortably reconciles the apparent contradiction flagged up at the beginning of this section. We just have a slight refinement of (32.53); using (32.86):

$$|c_a(t)|^2 \text{ starts as } \frac{A_{ba}}{12\pi\omega_{ba}}(\omega_c t)^2 \text{ changing soon to } 1 - e^{-A_{ba}t}. \quad (32.56)$$

The filling of photon states k, e_k

It was stated in § 32.3.1 that radiation initially enters field modes $|k, e_k\rangle$ within a wide energy range, and that that energy range narrows as the reaction proceeds. We are now in a position to investigate.

Graphs are given in Fig. 32.8 of $|c_{\omega a}(t)|^2$ for several values of ω in the vicinity of resonance ω_{ba}. The graphs in (a) and (b) show $|c_{\omega a}(t)|^2$ (solid curves) for $|\omega - \omega_{ba}|$ equal to zero and to A_{ba}. On the same plots we show the $g(\omega - \omega_{ba})(1 - e^{-A_{ba}t})$ that one might have expected on the precedent of Fig. 32.7. The resemblance is not close. We therefore

Fig. 32.8 The quantity $|c_{\omega a}(t)|^2\, d\omega$ is the probability that atomic state a is occupied at time t, accompanied by a photon within frequency range $d\omega$, regardless of the direction of travel of that photon.

The behaviour of $|c_{\omega a}(t)|^2$ is shown for selected values of $|\omega - \omega_{ba}|$ (solid lines). The time dependence is given by the { } of (32.54), and the final value by $g(\omega - \omega_{ba})$.

In (a) there is exact resonance: $\omega = \omega_{ba}$. In (b), $|\omega - \omega_{ba}| = A_{ba}$, away from resonance by A_{ba}, the FWHM of the resonance.

Had we not seen (32.54), we might have imagined that all frequency components $|c_{\omega a}(t)|^2$ would behave as $g(\omega - \omega_{ba})(1 - e^{-A_{ba}t})$, sharing the exponential of (32.53). The broken curves in (a) and (b) plot this expression to show that it is far from correct.

(c) shows the behaviour of $|c_{\omega a}(t)|^2$ for several angular frequencies ω in the vicinity of ω_{ba}. All $|c_{\omega a}(t)|^2$ rise initially as t^2 with the same coefficient, but then break away towards their final values. The number attached to each curve is the value of $|\omega - \omega_{ba}|/A_{ba}$.

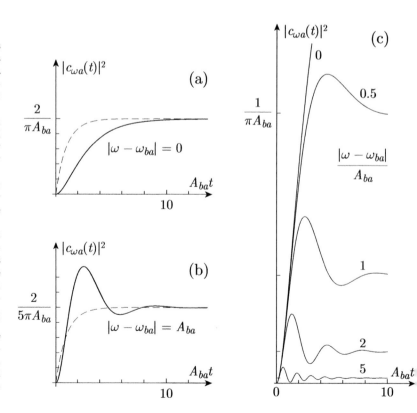

have to see the $(1 - e^{-A_{ba}t})$ of Fig. 32.7 as an "emergent property" that appears only when all radiation frequencies are summed.

All curves in Fig. 32.8(c) follow a similar initial dependence upon time: problem 32.20(1). This is in line with the statement in § 32.3.1 that, initially, while energy conservation has nothing to say, all field states fill in much the same way.

These curves also explain how the frequency distribution narrows with increase of time: frequencies far from resonance do not weaken—they plateau—but they are overtaken by frequencies closer to resonance which continue to strengthen over a longer time.

32.11   The occupation of higher-energy atomic states

By "higher-energy" we mean atomic states j higher in energy than the initial state b.

Surely these higher-energy atomic states cannot have any significant effect on the physics? Well, higher-energy *field* states have had to be considered, even requiring care to prevent sums over them from diverging. If for field states, then perhaps for atomic states also. Investigation is called for.

- The final occupation $|c_j(\infty)|^2$ of higher-energy state j is of order $\alpha_{\mathrm{fs}}^3 \ln(1/\alpha_{\mathrm{fs}})$ and therefore negligible (problem 32.21).
- When higher states are allowed for, they add to the radiation-induced shift $-\omega_{ba} \Delta$ of the resonance frequency, but only slightly, and Δ remains of order α_{fs}^2 (problem 32.22).
- The occupation $|c_b(t)|^2$ of the upper state, and its decay $\propto e^{-A_{ba}t}$, are unaffected, to order $\alpha_{\mathrm{fs}}^3 \ln(1/\alpha_{\mathrm{fs}})$, by inclusion of upper states j in the evaluation (problem 32.23).

32.12   Conclusions

We have evaluated a quantity introduced in Chapter 19: the Einstein coefficient $\alpha_{ba\boldsymbol{k}}$ for spontaneous emission from atomic state b, into atomic final state a and into a specific mode $|\boldsymbol{k}, \boldsymbol{e_k}\rangle$ of the outgoing radiation. We also have the normalized spectral line profile $g(\omega_{\boldsymbol{k}} - \omega_{ba})$.

$$\alpha_{ba\boldsymbol{k}} = \frac{2\pi^2 c^3}{V} \frac{A_{ba}}{\omega_{ba}\, \omega_{\boldsymbol{k}}} \frac{A_{ba}/2\pi}{(\omega_{\boldsymbol{k}} - \omega_{ba})^2 + A_{ba}^2/4}\, p(\vartheta, \varphi)\, f_{ab}(\boldsymbol{k}); \qquad (32.50)$$

$$g(\omega_{\boldsymbol{k}} - \omega_{ba}) = \frac{A_{ba}/2\pi}{(\omega_{\boldsymbol{k}} - \omega_{ba})^2 + A_{ba}^2/4} \frac{\omega_{\boldsymbol{k}}}{\omega_{ba}}\, f_{ab}(\boldsymbol{k}), \qquad (32.51)$$

both for 2p → 1s; the cut-off function is

$$f_{ab}(\boldsymbol{k}) = \frac{1}{\{1 + (\omega_{\boldsymbol{k}}/\omega_{\mathrm{c}})^2\}^4}. \qquad (32.27)$$

[67] In § 32.4.1, I said that matrix elements calculated from Schrödinger wave functions agree (to order α_{fs}^2 and with $e^{-i\boldsymbol{k}\cdot\boldsymbol{r}}$ set to 1) with those found using Dirac wave functions. In fact, Dirac gives slightly different results for frequencies $\omega \gtrsim \omega_c$, that is, when $e^{-i\boldsymbol{k}\cdot\boldsymbol{r}}$ is allowed for. Since there is a cut-off at $\omega \sim \omega_c$, these differences are of little consequence.

[68] The imagined excitation raises the atom step-function-wise from state a to state b at time $t = 0$. Equation (32.14a) shows that the achievement of this requires an intervention that is a δ-function of time. However this is done, it is required, on our assumptions, to populate state b only, leaving all higher states j initially empty. Even so, coefficient c_b is necessarily given Fourier components that extend up to ω_{jb} and beyond. It is therefore not surprising that our calculations have revealed weak excitation to higher states j. See eqn (32.14b).

Results (32.50) and (32.51) are the outcome of first-order perturbation theory, and are subject to corrections of order α_{fs}^2.

Both α_{bak} and $g(\omega - \omega_{ba})$ have been evaluated for the special-case 2p–1s decay identified in Fig. 32.1, but it should be possible to carry over the calculation method to other cases.

In both (32.50) and (32.51), we see a Lorentzian spectrum. For most purposes, this dominates the frequency dependence. However, in both expressions the final factor $f_{ab}(\boldsymbol{k})$ reduces things at frequencies of order $\omega_c \sim c/a_0$, a cut-off that is required if we need to integrate over frequencies, otherwise such an integration is in danger of diverging.[67]

Equation (32.16) contains a sum over all possible final atomic states, even states that are higher in energy than the initial state b. A small probability of a transition to such a state is not ruled out, since the atom is imagined to be shock-excited[68] to state b at time $t = 0$. But investigation (§ 32.11) has shown that these higher-energy states make negligible contributions to the physics, as we should expect.

There are implications of our findings for wider areas of physics. In particular, we mention Golden Rule No. 2. The reasoning that leads to this Rule requires the introduction of a high-energy cut-off, similar to the $f_{ab}(\boldsymbol{k})$ of our radiation emission. Without that cut-off, which is rarely introduced, the sum over final states of the outgoing particle(s) can be divergent. See problem 32.24.

Finally, we remark that the calculation we have presented is lengthy, but not excessively difficult if taken a piece at a time, and it leads to results we can be comfortable with. It seems strange that it is hard to find in the literature.

Problems

Problem 32.1 The basic equations
(1) Obtain (32.9).
(2) Show from (32.7) that

$$\langle \boldsymbol{k}|\widehat{H}'|0\rangle = \frac{e}{m_e}\sqrt{\frac{\hbar}{2\epsilon_0 V \omega_{\boldsymbol{k}}}}\, e^{-i\boldsymbol{k}\cdot\boldsymbol{r}}\, \boldsymbol{e_k}\cdot\hat{\boldsymbol{p}};$$

$$\langle 0|\widehat{H}'|\boldsymbol{k}\rangle = \frac{e}{m_e}\sqrt{\frac{\hbar}{2\epsilon_0 V \omega_{\boldsymbol{k}}}}\, e^{i\boldsymbol{k}\cdot\boldsymbol{r}}\, \boldsymbol{e_k}\cdot\hat{\boldsymbol{p}}.$$

[69] This step may look improbable because of the $e^{\pm i\boldsymbol{k}\cdot\boldsymbol{r}}$ in $\widehat{H}'$. Depending on how you handle things, you may have to remember that $\boldsymbol{e_k}\cdot\hat{\boldsymbol{p}}$ does not differentiate $e^{\pm i\boldsymbol{k}\cdot\boldsymbol{r}}$.

(3) Show from the Hermitian property of $\widehat{H}'$ that[69]

$$C_{ij}(-\boldsymbol{k}) = -C_{ji}(\boldsymbol{k})^*.$$

Obtain (32.14a).

Problem 32.2 Obtaining (32.16)
Perform the elimination that leads from (32.14) to (32.16) and thence also to (32.21).

[70] See e.g. Woodgate (1980) Tables 2.1, 2.2; Foot (2005) Tables 2.1, 2.2.

Problem 32.3 The sense of rotation of the atom's dipole moment
The wave functions for our chosen states of hydrogen are:[70]

$$\psi_a = \psi_{1s} = \frac{1}{a_0^{3/2}} 2\,e^{-r/a_0} \times \sqrt{\frac{1}{4\pi}}, \qquad (32.57a)$$

$$\psi_b = \psi_{2p} = \frac{1}{(2a_0)^{3/2}} \frac{2}{\sqrt{3}} \frac{r}{2a_0}\,e^{-r/2a_0} \times \sqrt{\frac{3}{8\pi}}\,\sin\theta\,e^{-i\phi}, \qquad (32.57b)$$

in which a_0 is the Bohr radius.

(1) Define the electric-dipole moment (charge $-e$ omitted)

$$\boldsymbol{r}_{ab} = \int \psi_a^* \,\boldsymbol{r}\, \psi_b\, dV = \int \psi_a^* (x\boldsymbol{e}_x + y\boldsymbol{e}_y + z\boldsymbol{e}_z)\psi_b\, dV$$
$$= x_{ab}\boldsymbol{e}_x + y_{ab}\boldsymbol{e}_y + z_{ab}\boldsymbol{e}_z,$$

in which (for this case) x_{ab} is real, $y_{ab} = -ix_{ab}$ and $z_{ab} = 0$.
  We define the dot product of complex vectors such that

$$\left|\boldsymbol{r}_{ab}\right|^2 = \boldsymbol{r}_{ab}^* \cdot \boldsymbol{r}_{ab} = (x_{ab}^*\boldsymbol{e}_x + y_{ab}^*\boldsymbol{e}_y)\cdot(x_{ab}\boldsymbol{e}_x + y_{ab}\boldsymbol{e}_y) = \left|x_{ab}\right|^2 + \left|y_{ab}\right|^2.$$

Then

$$x_{ab} = \frac{\left|\boldsymbol{r}_{ab}\right|}{\sqrt{2}}, \qquad y_{ab} = \frac{-i\left|\boldsymbol{r}_{ab}\right|}{\sqrt{2}}, \qquad z_{ab} = 0. \qquad (32.58)$$

(2) During the transition between states b and a, the radiating atom has a wave function that is a linear combination[71] of Ψ_b and Ψ_a:

$$\Psi = c_a\,\psi_a\,e^{-i\omega_a t} + c_b\,\psi_b\,e^{-i\omega_b t}.$$

Show that, with α defined by $c_a^* c_b = \left|c_a c_b\right| e^{i\alpha}$,

$$\langle x\rangle = \left|c_a c_b\right|\left|\boldsymbol{r}_{ab}\right|\sqrt{2}\cos(\omega_{ba}t - \alpha);$$
$$\langle y\rangle = -\left|c_a c_b\right|\left|\boldsymbol{r}_{ab}\right|\sqrt{2}\sin(\omega_{ba}t - \alpha),$$

Show that the atom's dipole moment $\langle \boldsymbol{r}\rangle$ rotates[72] from x towards $-y$, i.e. negatively, towards $-\phi$, as is indicated in Fig. 32.3.

Problem 32.4 The coordinates for calculating matrix elements
We are looking ahead to the evaluation of matrix elements in (32.60). For that purpose we must do integrations in the XYZ coordinates, because only then does $e^{-i\boldsymbol{k}\cdot\boldsymbol{r}}$ take a (fairly) simple form.
  In order to differentiate the atom's 2p wave function[73] in the X,Y,Z directions, we must resolve $r\sin\theta\,e^{-i\phi}$ onto X,Y,Z. The geometry we need is shown in Fig. 32.9.

(1) Show that

$$x = x'\cos\varphi - y'\sin\varphi, \qquad y = x'\sin\varphi + y'\cos\varphi,$$
$$x' = X\cos\vartheta + Z\sin\vartheta, \qquad y' = Y,$$
$$r\sin\theta\,e^{-i\phi} = x - iy = e^{-i\varphi}(x' - iy') = e^{-i\varphi}(X\cos\vartheta - iY + Z\sin\vartheta).$$

(2) Show that[74]

$$\hat{p}_X\,\psi_{2p} = \frac{-i\hbar\,e^{-i\varphi}}{8\sqrt{\pi}\,a_0^{5/2}}\,e^{-r/2a_0}\left\{\frac{-1}{2a_0}\frac{X}{r}\left(X\cos\vartheta - iY + Z\sin\vartheta\right) + \cos\vartheta\right\},$$

$$\hat{p}_Y\,\psi_{2p} = \frac{-i\hbar\,e^{-i\varphi}}{8\sqrt{\pi}\,a_0^{5/2}}\,e^{-r/2a_0}\left\{\frac{-1}{2a_0}\frac{Y}{r}\left(X\cos\vartheta - iY + Z\sin\vartheta\right) - i\right\}.$$

[72] The sense of rotation is determined by the angular momenta (z-components) of the two wave functions ψ_a, ψ_b that are being linearly combined. It does not depend on whether a reaction is taking place, or on whether that reaction is going from a to b or from b to a.
  Consistency check: $\Delta m_{\mathrm{atom}} = +1$ so $\Delta m_{\mathrm{radiation}} = -1$. The negative rotation of the dipole moment is such as to deliver negative z-angular momentum into the radiation field.

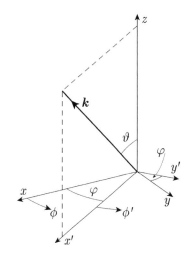

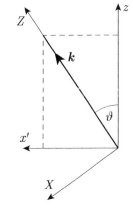

Fig. 32.9 The geometry set up in Fig. 32.3 and used in problem 32.4.

[73] [From previous page] In problem 32.4(2) we may project xyz onto XYZ before differentiating, as is suggested here. Or we may differentiate first and then project $\hat{p}_x \psi_{2p}$, $\hat{p}_y \psi_{2p}$, $\hat{p}_z \psi_{2p}$ onto the XYZ directions. The important thing is that the integrations of (32.60) must be done in XYZ coordinates in order to cater for $\mathrm{e}^{-i\mathbf{k}\cdot\mathbf{r}} = \mathrm{e}^{-ikZ}$.

[74] [From previous page] The algebra of problem 32.4(2) might seem a little simpler if we transposed the desired matrix element, because ψ_{1s} is spherically symmetrical and so can be differentiated with equal ease in any direction. However, at the end one then has to take the Hermitian conjugate, which needs thought since $\mathrm{e}^{-i\mathbf{k}\cdot\mathbf{r}}$ is not Hermitian.

[75] The indefinite integrals may readily be verified by differentiating their right-hand sides. Note that the two indefinite integrals obey the mnemonic for integration of $\mathrm{e}^{ax} \genfrac{}{}{0pt}{}{\cos}{\sin}(bx)$: differentiate; change the sign of the thing that wasn't there; divide by $(a^2 + b^2)$.

Show that, in $(X, Y, Z) = (r, \Theta, \Phi)$ coordinates,

$$X = r \sin\Theta \cos\Phi, \qquad Y = r \sin\Theta \sin\Phi, \qquad Z = r \cos\Theta,$$

and that

$$\mathrm{e}^{-i\mathbf{k}\cdot\mathbf{r}} = \mathrm{e}^{-ikZ} = \mathrm{e}^{-ikr \cos\Theta}.$$

Problem 32.5 Some integrals needed in problem 32.6
Obtain the following integrals over Θ:

$$\int_0^\pi \mathrm{d}\Theta \sin\Theta \, \mathrm{e}^{-i\kappa\rho\cos\Theta} = \int_{-1}^1 \mathrm{d}c \, \mathrm{e}^{-i\kappa\rho c} = \frac{2\sin(\kappa\rho)}{(\kappa\rho)};$$

$$\int_0^\pi \mathrm{d}\Theta \sin\Theta \cos^2\Theta \, \mathrm{e}^{-i\kappa\rho\cos\Theta} = \frac{2\sin(\kappa\rho)}{(\kappa\rho)} + \frac{4\cos(\kappa\rho)}{(\kappa\rho)^2} - \frac{4\sin(\kappa\rho)}{(\kappa\rho)^3};$$

$$\int_0^\pi \mathrm{d}\Theta \sin^3\Theta \, \mathrm{e}^{-i\kappa\rho\cos\Theta} = 4\left(\frac{-\cos(\kappa\rho)}{(\kappa\rho)^2} + \frac{\sin(\kappa\rho)}{(\kappa\rho)^3}\right).$$

The results of integration here are all real, so it does not matter whether the exponentials $\mathrm{e}^{-i\kappa\rho\cos\Theta}$ contain a negative or positive sign in the exponent. The same applies to eqns (32.59).
Obtain the following integrals[75] over ρ:

$$\int \mathrm{e}^{-\rho} \cos(\kappa\rho) \, \mathrm{d}\rho = \frac{-\mathrm{e}^{-\rho}\cos(\kappa\rho) + \kappa\,\mathrm{e}^{-\rho}\sin(\kappa\rho)}{1 + \kappa^2};$$

$$\int \mathrm{e}^{-\rho} \sin(\kappa\rho) \, \mathrm{d}\rho = \frac{-\mathrm{e}^{-\rho}\sin(\kappa\rho) - \kappa\,\mathrm{e}^{-\rho}\cos(\kappa\rho)}{1 + \kappa^2};$$

$$\int_0^\infty \mathrm{e}^{-\rho} \cos(\kappa\rho) \, \mathrm{d}\rho = \frac{1}{1 + \kappa^2};$$

$$\int_0^\infty \mathrm{e}^{-\rho} \sin(\kappa\rho) \, \mathrm{d}\rho = \frac{\kappa}{1 + \kappa^2}.$$

Find definite integrals over $\rho\,\mathrm{e}^{-\rho}\cos(\kappa\rho)$, $\rho\,\mathrm{e}^{-\rho}\sin(\kappa\rho)$, $\rho^2\,\mathrm{e}^{-\rho}\cos(\kappa\rho)$, $\rho^2\,\mathrm{e}^{-\rho}\sin(\kappa\rho)$. Use these to obtain:

$$\int_0^\infty \mathrm{d}\rho\,\rho^2\,\mathrm{e}^{-\rho} \int_0^\pi \mathrm{d}\Theta \sin\Theta \, \mathrm{e}^{-i\kappa\rho\cos\Theta} = \frac{4}{(1 + \kappa^2)^2}; \qquad (32.59\mathrm{a})$$

$$\int_0^\infty \mathrm{d}\rho\,\rho^3\,\mathrm{e}^{-\rho} \int_0^\pi \mathrm{d}\Theta \sin^3\Theta \, \mathrm{e}^{-i\kappa\rho\cos\Theta} = \frac{8}{(1 + \kappa^2)^2}; \qquad (32.59\mathrm{b})$$

$$\int_0^\infty \mathrm{d}\rho\,\rho^3\,\mathrm{e}^{-\rho} \int_0^\pi \mathrm{d}\Theta \sin\Theta \cos^2\Theta \, \mathrm{e}^{-i\kappa\rho\cos\Theta} = \frac{4(1 - 3\kappa^2)}{(1 + \kappa^2)^3}; \qquad (32.59\mathrm{c})$$

$$\int_0^\infty \mathrm{d}\rho\,\rho^4\,\mathrm{e}^{-\rho} \int_0^\pi \mathrm{d}\Theta \sin^3\Theta \, \mathrm{e}^{-i\kappa\rho\cos\Theta} = \frac{32}{(1 + \kappa^2)^3}. \qquad (32.59\mathrm{d})$$

[76] Integrals containing Z and p_Z are non-zero, but are of little interest given that $\mathcal{A}_Z = 0$ (remember div $\mathbf{A} = 0$).

Problem 32.6 The 2p–1s matrix elements for hydrogen
(1) Problems 32.4 and 32.5 have provided the tools we need for evaluating matrix elements, and from them the $C_{ab}(\mathbf{k})$ of (32.12). Show that[76]

$$J_X \equiv \int \psi_{1s}^* \, e^{-i\mathbf{k}\cdot\mathbf{r}} \, X \, \psi_{2p} \, dV = \frac{2^7}{3^5} \, a_0 \, \frac{e^{-i\varphi} \cos\vartheta}{(1+\kappa^2)^3} \qquad (32.60a)$$

$$J_Y \equiv \int \psi_{1s}^* \, e^{-i\mathbf{k}\cdot\mathbf{r}} \, Y \, \psi_{2p} \, dV = (-i) \frac{2^7}{3^5} \, a_0 \, \frac{e^{-i\varphi}}{(1+\kappa^2)^3} \qquad (32.60b)$$

$$J_{pX} \equiv \int \psi_{1s}^* \, e^{-i\mathbf{k}\cdot\mathbf{r}} \, \hat{p}_X \, \psi_{2p} \, dV = i \frac{(-16)}{81} \, \frac{\hbar}{a_0} \, \frac{e^{-i\varphi} \cos\vartheta}{(1+\kappa^2)^2} \qquad (32.60c)$$

$$J_{pY} \equiv \int \psi_{1s}^* \, e^{-i\mathbf{k}\cdot\mathbf{r}} \, \hat{p}_Y \, \psi_{2p} \, dV = \frac{(-16)}{81} \, \frac{\hbar}{a_0} \, \frac{e^{-i\varphi}}{(1+\kappa^2)^2}. \qquad (32.60d)$$

(2) Use the known energies of the 1s, 2p quantum states of hydrogen to show that[77]

$$J_{pX} = (-i\omega_{ba})m_e J_X (1+\kappa^2), \qquad J_{pY} = (-i\omega_{ba})m_e J_Y (1+\kappa^2). \quad (32.61)$$

Problem 32.7 Matrix elements for momentum and position
(1) There is a theorem in atomic physics that, for any atomic states a, b,

$$\langle a|\hat{\mathbf{p}}|b\rangle = -i\omega_{ba} m_e \langle a|\hat{\mathbf{r}}|b\rangle. \qquad (32.62)$$

Rehearse the derivation of this.[78]

Notice the absence of $e^{-i\mathbf{k}\cdot\mathbf{r}}$ from both matrix elements in (32.62). There is no generalization of (32.62) to matrix elements containing $e^{-i\mathbf{k}\cdot\mathbf{r}}$. Confirm this by writing (32.61) in a changed notation as[79]

$$\langle a|e^{-i\mathbf{k}\cdot\mathbf{r}} \, \hat{\mathbf{p}}|b\rangle = -i\omega_{ba} m_e \langle a|e^{-i\mathbf{k}\cdot\mathbf{r}} \, \hat{\mathbf{r}}|b\rangle \times (1+\kappa^2). \qquad (32.63)$$

Argue that the factor $(1+\kappa^2)$ in (32.63) must be special to the 2p–1s transition.[80]
(2) Show from (32.5) and (32.6) that

$$\langle \mathbf{k}|\widehat{\mathbf{E}}|0\rangle = -i\omega_{\mathbf{k}} \langle \mathbf{k}|\widehat{\mathbf{A}}|0\rangle.$$

Combine this with (32.63) to obtain

$$\left\langle a \middle| \left\langle \mathbf{k}|\widehat{\mathbf{E}}\cdot\hat{\mathbf{r}}|0\right\rangle \middle| b \right\rangle = \frac{\omega_{\mathbf{k}}}{\omega_{ba}} \frac{1}{(1+\kappa^2)} \left\langle a \middle| \left\langle \mathbf{k}\middle|\frac{\widehat{\mathbf{A}}\cdot\hat{\mathbf{p}}}{m_e}\middle|0\right\rangle \middle| b \right\rangle. \qquad (32.64)$$

Equation (32.64) is used in § 32.4.1(c) to show that $\widehat{\mathbf{E}}\cdot\hat{\mathbf{r}}$ is not equivalent to $\widehat{\mathbf{A}}\cdot\hat{\mathbf{p}}/m_e$.

Problem 32.8 Elliptical polarization
Show from (32.60c) and (32.60d) that

$$J_{pX} = i J_{pY} \cos\vartheta.$$

The $(\mathbf{J}_p)_{ab}$ of (32.13) is multiplied by $\mathbf{e}_{\mathbf{k}}$ in $C_{ab}(\mathbf{k})$, so we need only the result of resolving it onto the XY plane. Thus

$$(\mathbf{J}_p)_{ab} = J_{pX} \, \mathbf{e}_X + J_{pY} \, \mathbf{e}_Y + (\text{don't care}) \, \mathbf{e}_Z$$
$$= i J_{pY} \sqrt{1+\cos^2\vartheta} \, \frac{\cos\vartheta \, \mathbf{e}_X - i\mathbf{e}_Y}{\sqrt{1+\cos^2\vartheta}} = i J_{pY} \sqrt{1+\cos^2\vartheta} \, \mathbf{e}_{\mathbf{J}}, \quad (32.65)$$

shaped so that the final factor is a unit vector[81] $\mathbf{e}_{\mathbf{J}}$.

---

[77] *Suggestion:* Show first that for hydrogen 2p–1s
$$\frac{\hbar}{a_0} = \frac{8}{3} \omega_{ba} \, m_e \, a_0.$$

[78] Assistance in Woodgate (1980) § 3.3; Bransden and Joachain (2003) § 4.3. We can confirm this, in the special case 2p–1s, from eqns (32.60) or (32.63), but property (32.62) is general.

[79] We discard J_{pZ} and J_Z.

[80] Evidence confirming that this is a special-case factor can be obtained by investigating some other transition, such as that suggested in sidenote 36.

[81] Check that $\mathbf{e}_{\mathbf{J}}^* \cdot \mathbf{e}_{\mathbf{J}} = 1$.

Combine (32.14b) with (32.11–13) and show that

$$\dot{c}_{ka} \sim (J_p)_{ab}\, \mathrm{e}^{-\mathrm{i}\omega_{ba}t} \cdot \boldsymbol{\mathcal{A}}\,\mathrm{e}^{\mathrm{i}\omega_k t} \sim \boldsymbol{e_J}\,\mathrm{e}^{-\mathrm{i}\omega_{ba}t} \cdot \boldsymbol{e_k}\,\mathrm{e}^{\mathrm{i}\omega_k t}. \qquad (32.66)$$

Since $\boldsymbol{e_J}$ and $\boldsymbol{e_k}$ are unit vectors, their dot product $|\boldsymbol{e_J}\cdot\boldsymbol{e_k}| \leqslant 1$; show that this holds even though $\boldsymbol{e_J}$ and $\boldsymbol{e_k}$ are complex.[82]

[82] Suggestion: Let

$$\boldsymbol{e_J} = \frac{a\,\boldsymbol{e_X} + b\,\mathrm{e}^{\mathrm{i}\varepsilon}\,\boldsymbol{e_Y}}{\sqrt{a^2+b^2}},$$

$$\boldsymbol{e_k} = \frac{c\,\boldsymbol{e_X} + d\,\mathrm{e}^{\mathrm{i}\delta}\,\boldsymbol{e_Y}}{\sqrt{c^2+d^2}}$$

in which $a,b,c,d,\varepsilon,\delta$ are real. Show that these are normalized. Show that

$$1 - |\boldsymbol{e_J}\cdot\boldsymbol{e_k}|^2 \geqslant 0,$$

for all choices of $a,b,c,d,\varepsilon,\delta$.

The radiation emitted has field vector $\boldsymbol{\mathcal{A}}$, and is "driven" by $(J_p)_{ab}$. Argue that the radiation will be polarized such that $|\boldsymbol{e_J}\cdot\boldsymbol{e_k}|$ takes its greatest possible value of 1. Show that, within a phase factor,

$$\boldsymbol{e_k} = \boldsymbol{e_J}^* = \frac{\cos\vartheta\,\boldsymbol{e_X} + \mathrm{i}\boldsymbol{e_Y}}{\sqrt{1+\cos^2\vartheta}}; \qquad \boldsymbol{\mathcal{A}} = |\boldsymbol{\mathcal{A}}|\left(\frac{\cos\vartheta\,\boldsymbol{e_X} + \mathrm{i}\boldsymbol{e_Y}}{\sqrt{1+\cos^2\vartheta}}\right). \qquad (32.67)$$

This confirms the polarization vector $\boldsymbol{e_k}$ suggested in (32.22).

Refer to (32.66). Show that, in the xy plane ($\vartheta = 0$),

$$\boldsymbol{e_J}\,\mathrm{e}^{-\mathrm{i}\omega_{ba}t} \sim (\boldsymbol{e}_x - \mathrm{i}\boldsymbol{e}_y)\,\mathrm{e}^{-\mathrm{i}\omega_{ba}t} \quad \text{and} \quad \boldsymbol{e_k}\,\mathrm{e}^{\mathrm{i}\omega_k t} \sim (\boldsymbol{e}_x + \mathrm{i}\boldsymbol{e}_y)\,\mathrm{e}^{\mathrm{i}\omega_k t}$$

both rotate in the same sense as is drawn in Fig. 32.3.

Problem 32.9 Assembling $C_{ab}(\boldsymbol{k})$

Show from (32.60d) that

$$J_{pY} = \langle a|\mathrm{e}^{-\mathrm{i}\boldsymbol{k}\cdot\boldsymbol{r}}\,\hat{p}_Y|b\rangle = \langle a|\hat{p}_Y|b\rangle \frac{1}{(1+\kappa^2)^2}$$

$$= -\mathrm{i}\omega_{ba}m_e\left(\frac{-\sin\varphi\,\langle a|x|b\rangle + \cos\varphi\,\langle a|y|b\rangle}{(1+\kappa^2)^2}\right)$$

$$= -\omega_{ba}m_e\,\frac{|\boldsymbol{r}_{ab}|}{\sqrt{2}}\,\frac{\mathrm{e}^{-\mathrm{i}\varphi}}{(1+\kappa^2)^2},$$

using (32.62) and (32.58). From (32.65) and (32.12) show that

$$C_{ab}(\boldsymbol{k}) = \frac{-\mathrm{i}}{\hbar}\frac{e}{m_e}\,\boldsymbol{\mathcal{A}}\cdot(J_p)_{ab} = \frac{-\mathrm{i}}{\hbar}\frac{e}{m_e}\,|\boldsymbol{\mathcal{A}}|\,\mathrm{i}J_{pY}\sqrt{1+\cos^2\vartheta}\,\boldsymbol{e_k}\cdot\boldsymbol{e_J},$$

and we have argued in problem 32.8 that $\boldsymbol{e_k}\cdot\boldsymbol{e_J} = 1$. Tidy up to show from (32.13) that we have (32.23), repeated here:

$$C_{ab}(\boldsymbol{k}) = \frac{-e\,\omega_{ba}}{\sqrt{2\epsilon_0 V\hbar\omega_k}}\,|\boldsymbol{r}_{ab}|\sqrt{\frac{1+\cos^2\vartheta}{2}}\,\mathrm{e}^{-\mathrm{i}\varphi}\,\frac{1}{(1+\kappa^2)^2} \qquad (32.68)$$

Problem 32.10 The integral in the denominator of $\tilde{d}_b(s)$

The required integral is that in expression (32.29)

$$\delta(\sigma) + \mathrm{i}\gamma(\sigma) = \int_0^{\infty} \frac{\kappa}{\kappa - \sigma}\,\frac{1}{(1+\kappa^2)^4}\,\mathrm{d}\kappa.$$

[83] The big { } factorizes but the factorization is not helpful.

Put the integrand of this into partial fractions and show that[83]

$$\frac{\kappa}{\kappa-\sigma}\,\frac{1}{(1+\kappa^2)^4} = \frac{1}{(1+\sigma^2)^4}\,\frac{\kappa}{\kappa-\sigma}$$

$$+ \frac{1}{(1+\sigma^2)^4}\,\frac{1}{(1+\kappa^2)^4}\left\{\begin{array}{l} -\kappa^8 - \sigma\,\kappa^7 - (\sigma^2+4)\kappa^6 - \sigma(\sigma^2+4)\kappa^5 \\ -(\sigma^4 + 4\sigma^2 + 6)\kappa^4 \\ -\sigma(\sigma^4 + 4\sigma^2 + 6)\kappa^3 \\ -(\sigma^6 + 4\sigma^4 + 6\sigma^2 + 4)\kappa^2 \\ -\sigma(\sigma^6 + 4\sigma^4 + 6\sigma^2 + 4)\kappa \end{array}\right\}.$$

Perform the integrations over κ (all elementary),[84] and show that we end up with

$$\delta(\sigma) + i\gamma(\sigma)$$

$$= \frac{1}{(1+\sigma^2)^4} \left\{ \begin{array}{l} -\sigma \ln(-\sigma) + \dfrac{5\pi}{32} - \dfrac{11}{12}\sigma - \dfrac{15\pi}{32}\sigma^2 \\[2mm] -\dfrac{3}{2}\sigma^3 - \dfrac{5\pi}{32}\sigma^4 - \dfrac{3}{4}\sigma^5 - \dfrac{\pi}{32}\sigma^6 - \dfrac{1}{6}\sigma^7 \end{array} \right\}. \quad (32.69)$$

This expression has a branch cut along the real-positive σ-axis, coming from the logarithm. Write the logarithm as $\ln(-\sigma) = \ln|\sigma| + i\arg(-\sigma)$ and confirm that a possible sharing of these pieces between δ and γ is that given in (32.30).

The reasoning of § 32.7 shows that the integration needed for evaluating $d_b(t)$ is carried out on a path round the branch cut in the s-plane. Confirm the statements made in the annotations on the two sides of Fig. 32.4, and hence obtain (32.31).

Problem 32.11 All singularities of $\widetilde{d}_b(s)$ lie on the negative imaginary s-axis

Consider the denominator of (32.21), where only the first sum over electromagnetic-field modes is included, and that sum is still thought of as a sum over discrete modes, rather than as an integral.

Let $s = -iz$ where z may be complex.[85] Show that the denominator of (32.21) can be put into the form

$$-i\left(z - \omega_{ba} - \sum_{\boldsymbol{k}} \frac{|C_{ab}(\boldsymbol{k})|^2}{z - \omega_{\boldsymbol{k}}} \right). \quad (32.70)$$

Here ( ) is the ratio of two large polynomials in z, and everything in it except possibly z is real. This denominator of $\widetilde{d}_b(s)$ goes to infinity whenever z passes through one of the values of $\omega_{\boldsymbol{k}}$ (all are positive), and in between these infinities it must pass through zero. The zeros are the roots of the equation

$$z - \omega_{ba} = \sum_{\boldsymbol{k}} \frac{|C_{ab}(\boldsymbol{k})|^2}{z - \omega_{\boldsymbol{k}}}. \quad (32.71)$$

The behaviour of the functions on the two sides of (32.71) is illustrated graphically in Fig. 32.10. The curve labelled Σ represents the sum on the right of (32.71), while the straight line is $z - \omega_{ba}$; the solutions of (32.71) are the intersections of the straight line with the curve.

The model of (32.71) used for plotting Fig. 32.10 includes five discrete values of $\omega_{\boldsymbol{k}}$ so it has the form

$$\frac{\text{(polynomial in } z \text{ of order 6)}}{\text{(polynomial in } z \text{ of order 5)}};$$

argue that it must have six zeros. There are six intersections on the graph, so all zeros are accounted for—all six at real values of z.

Extend this reasoning to an arbitrary number of values of $\boldsymbol{k}$, showing that all roots for z are again real.[86]

[84] The integral that we started with is manifestly convergent for large κ. But the first three terms in the partial-fractions expression result in integrals that are divergent at the upper limit. Therefore we must integrate to a finite-but-large upper limit K and afterwards take the limit $K \to \infty$.

[85] In the problems, x, y, z are sometimes, as here, used as "scratchpad" variables. There should be no confusion with the atom's coordinates of Fig. 32.3.

[86] This finding is physically necessary. Given that all coefficients in (32.71) are real, the roots for z must be either real or complex-conjugate pairs. If there is a complex-conjugate pair, then one member of the pair will lie to the right of the imaginary s-axis, and a pole there in $\widetilde{d}_b(s)$ will give exponential growth of $d_b(t)$.

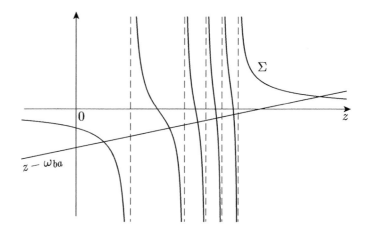

Fig. 32.10 A specimen graphical solution of eqn (32.71). For this illustration, there are five values of ω_k, so the polynomial to be solved has order 6. The sum goes to infinity, as a function of z, at each of the ω_k-values, so five times. It is easy to see that there must necessarily be six real roots, and all roots (except possibly the left-most) for z are positive.

All values of z that solve the plot in Fig. 32.10 are positive except possibly the left-most. To investigate this, we can find which curve lies higher at $z = 0$. The straight line has height $-\omega_{ba}$ where it crosses the vertical axis. In the sum, set $z = 0$ and approximate the sum to an integral, as was done in § 32.6. Show that the height of the Σ curve where it crosses the vertical axis is (compare with (32.29))

$$\sum_k \frac{|C_{ab}(\mathbf{k})|^2}{-\omega_k} = \frac{-\omega_c}{2\pi} \frac{A_{ba}}{\omega_{ba}} \delta(0) = -\omega_{ba} \left(\frac{\omega_c}{\omega_{ba}} \frac{A_{ba}}{\omega_{ba}} \frac{5\pi/32}{2\pi} \right)$$

$$= -\omega_{ba} \times (\text{order } \alpha_{\text{fs}}^2).$$

Thus the Σ curve lies higher at $z = 0$. The left-most solution of (32.71) does indeed lie to the right of the z-origin, at a z-value that is positive.

Expression (32.70) is the denominator (approximated) of $\tilde{d}_b(s)$ (eqn 32.16); wherever the denominator is zero, $\tilde{d}_b(s)$ has a pole. All of these poles lie at real-positive values of z, which means values of s on its negative imaginary axis. And, with a polynomial in the denominator, nothing worse than a pole can happen.

Given that the number of field modes is extremely large, there is a very dense concentration of poles of $\tilde{d}_b(s)$ on the negative imaginary s-axis. If we replace the sum over modes by an integral these poles coalesce into a line singularity. Of course, this line singularity is just the same as the branch cut in the logarithm of (32.69).

Argue that it is safe to deform the integration path of Fig. 32.5 as is shown there.

One detail in the above needs a closer look. In the sum over modes $|\mathbf{k}, e_{\mathbf{k}}\rangle$, there often will be several modes having exactly the same ω_k. Argue that such degeneracy can be accommodated in our reasoning without compromising the conclusion.

Problem 32.12 The Poincaré cycle and photon emission
Think about the points raised in § 32.7.1.

We have said that the energy, originally present in the excited atom, gets itself distributed among a large number of field modes. We have also said that, with "periodic boundary conditions", that energy remains for all time within the quantization volume V. A Poincaré resurgence is therefore possible. However, it now requires that all the field modes reacquire the (relative) phases that they had at the time of emission.[87] This is somewhat different from the resurgence usually contemplated in statistical mechanics, where it is occupation numbers that reacquire their former values—or is that the case? Think about it.

Problem 32.13 $\Gamma(\omega) \approx A_{ba}$ in denominators
Consider the denominator of (32.39), simplified by omission of Δ to

$$(\omega - \omega_{ba})^2 + \Gamma(\sigma)^2/4 + \text{order } \alpha_{\text{fs}}^2.$$

Write

$$\Gamma(\sigma) = \frac{A_{ba}}{\omega_{ba}} \frac{\omega}{(1 + \sigma^2)^4} = 2\eta\,\omega$$

in which $\eta \sim \alpha_{\text{fs}}^3$. Complete the square and show that

$$(\omega - \omega_{ba})^2 + \Gamma(\sigma)^2/4 = (\omega - \omega_{ba})^2 + A_{ba}^2/4$$

plus an additional error no larger than α_{fs}^2. This validates the denominator displayed in (32.43).

Problem 32.14 The exponential decay of (32.45)
The integral giving $d_b(t)$ is that in (32.44).
 We play the trick of extending the integration range of (32.44) back to $-\infty$, in order to apply a contour integration.

$$d_b(t) = \int_{-\infty}^{\infty} \frac{A_{ba}/2\pi}{(\omega - \omega_{ba})^2 + A_{ba}^2/4} \frac{\omega}{\omega_{ba}} \frac{1}{\{1 + (\omega/\omega_c)^2\}^4} e^{-i\omega t}\,d\omega - \int_{-\infty}^{0}.$$

The first integral can be considered as taken along the real ω-axis, and that path can be continued into a closed contour by adding a large semicircle enclosing the lower half of the ω-plane as shown in Fig. 32.11; on this semicircle $e^{-i\omega t}$ is exponentially small. Within the contour, the integrand has two poles: a simple pole at $\omega = \omega_{ba} - iA_{ba}/2$; and a fourth-order pole at $\omega = -i\omega_c$.
 Evaluate the integrand close to the pole at $\omega = \omega_{ba} - iA_{ba}/2$, and show that it behaves as

$$\frac{1}{(\omega - \omega_{ba} + iA_{ba}/2)} \times \frac{i}{2\pi} \frac{(1 + \text{order } \alpha_{\text{fs}}^3)}{(1 + \text{order } \alpha_{\text{fs}}^2)^4} e^{(-i\omega_{ba} - A_{ba}/2)t}.$$

The integral is $-2\pi i \times$ the residue, negative because the contour is traced in the negative sense. Show that this pole's contribution to $d_b(t)$ is

$$d_b(t)_{\text{part}} = e^{-i\omega_{ba}t} e^{-A_{ba}t/2} \times (1 + \text{order } \alpha_{\text{fs}}^2).$$

 Next consider the fourth-order pole at $\omega = -i\omega_c$. To deal with it, set $y = \omega/\omega_c + i$ so that the pole lies at $y = 0$. Show that the integrand, as a function of y is

$$\frac{A_{ba}}{2\pi\omega_{ba}} e^{-\omega_c t} \frac{1}{(y - i - \omega_{ba}/\omega_c)^2 + A_{ba}^2/4\omega_c^2} \frac{(y - i)}{(y - 2i)^4} e^{-i\omega_c t y} \times \frac{1}{y^4}.$$

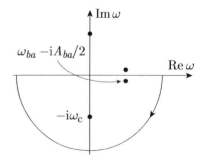

Fig. 32.11 The contour used for evaluating part of $d_b(t)$. There are simple poles at $\omega = \omega_{ba} \pm iA_{ba}/2$ and fourth-order poles at $\omega = \pm i\omega_c$. Two of these lie within the contour. The pole locations are drawn far out of scale since $\omega_c \sim \omega_{ba}/\alpha_{\text{fs}}$ and $A_{ba} \sim \omega_{ba}\alpha_{\text{fs}}^3$.

To find the residue at the pole we must construct a Laurent series (a power series in y but starting at y^{-4}), and the residue at the pole contains the third derivative of what multiplies y^{-4}. We can get an idea of this by neglecting the small quantities $\omega_{ba}/\omega_c \sim \alpha_{fs}$ and $A_{ba}^2/\omega_c^2 \sim \alpha_{fs}^8$. Then the residue at the pole contains the third derivative with respect to y of

$$\frac{A_{ba}}{2\pi\omega_{ba}} e^{-\omega_c t} \frac{e^{-i\omega_c ty}}{(y-i)(y-2i)^4}.$$

To keep the algebra under control, set $t=0$ in the last fraction. We need not work out the third derivative of this fraction, as it is obviously of order unity. The residue at the fourth-order pole is then of order $(A_{ba}/\omega_{ba})e^{-\omega_c t} \sim \alpha_{fs}^3$, and can/must be neglected.[88]

Finally, we must look at the integral from $-\infty$ to 0. This turns out to be small, so we are allowed to be brutal to it.[89] Simplify the integral by taking the special case $t=0$, and neglecting $A_{ba}^2/4$ beside $(\omega - \omega_{ba})^2$ (both approximations amount to finding an upper bound). Also, replace $\{1 + (\omega/\omega_c)^2\}^{-4}$ by a step cut-off at $\omega = -\omega_c$. The integral becomes[90]

$$-\frac{A_{ba}}{2\pi\omega_{ba}} \int_{-\omega_c}^{0} \frac{\omega}{(\omega - \omega_{ba})^2} \, d\omega = \frac{A_{ba}}{2\pi\omega_{ba}} \left\{ \ln\left(\frac{\omega_c}{\omega_{ba}} + 1\right) - \frac{\omega_c}{\omega_c + \omega_{ba}} \right\}$$

$$= \text{order } \alpha_{fs}^3 \ln(1/\alpha_{fs}) + \text{order } \alpha_{fs}^3.$$

Gathering threads:

$$d_b(t) = e^{-i\omega_{ba}t} e^{-A_{ba}t/2} + \text{order } \alpha_{fs}^2, \tag{32.72}$$

in agreement with (32.45). Also, we have the values of integrals needed later:

$$\int_0^\infty \frac{A_{ba}/2\pi}{(\omega - \omega_{ba})^2 + A_{ba}^2/4} \frac{\omega}{\omega_{ba}} \frac{1}{\{1 + (\omega/\omega_c)^2\}^4} e^{-i\omega t} \, d\omega = e^{-i\omega_{ba}t} e^{-A_{ba}t/2} \tag{32.73}$$

for $t > 0$, with the special case for $t = 0+$

$$\int_0^\infty \frac{A_{ba}/2\pi}{(\omega - \omega_{ba})^2 + A_{ba}^2/4} \frac{\omega}{\omega_{ba}} \frac{1}{\{1 + (\omega/\omega_c)^2\}^4} \, d\omega = 1, \tag{32.74}$$

both results having an uncertainty of order α_{fs}^2.

Examine the way in which ω_c has passed through the calculation. You should see that ω_c can be given any sensible value without significant effect on $d_b(t)$. There is a signal here that the form of the cut-off function, and the frequency at which cut-off happens, are both highly non-critical. The only necessity is that there *be* a cut-off to ensure convergence of integrals taken over ω.

The next problem shows in a different way that the form of the cut-off is not critical.

Problem 32.15 A step-function high-frequency cut-off
In the sum of (32.28), replace the cut-off function $(1 + \kappa^2)^{-4}$ by a step cut at $\kappa = 1$, equivalently at $\omega = \omega_c$. Show that the sum evaluates to[91]

---

[88] Before we leave the fourth-order pole altogether, we may notice that it gives a decay as $e^{-\omega_c t}$. We are therefore being told that $d_b(t)$ may exhibit a small glitch at the beginning of the radiation process, the glitch lasting for a time of order $1/\omega_c$, that is, occupying a tiny fraction of an optical period $2\pi/\omega_{ba}$. This fits neatly with the glitch discussed in § 32.10.

[89] Final bulleted point in § 12.2.1.

[90] Note that $\ln(1/\alpha_{fs})$ is only about 5, so we do not neglect 1 beside it.

[91] Here is an alternative to the reasoning of Fig. 32.4. Show that
$$\int_0^1 \frac{d\kappa}{\kappa - \sigma} = \ln\left|\frac{1-\sigma}{\sigma}\right|$$
when $\sigma < 0$ and when $\sigma > 1$. When $0 < \sigma < 1$ there is a pole in the integrand, and σ must be made (at least slightly) complex in order to go round it. If $\mathrm{Im}\,\sigma \gtrless 0$, the κ-integration path passes below/above σ, and can be indented with a small semicircle below/above the real κ-axis. Show that
$$\int_0^1 \frac{d\kappa}{\kappa - \sigma} = \ln\left|\frac{1-\sigma}{\sigma}\right| \pm i\pi$$
when $\mathrm{Im}\,\sigma \gtrless 0$, that is, on the two sides of the branch cut, the same finding as in Fig. 32.4.

$$\sum_k \frac{|C_{ab}(\boldsymbol{k})|^2}{s+i\omega_k} = \frac{A_{ba}}{2\pi\omega_{ba}}\int_0^{\omega_c} \frac{\omega_k}{s+i\omega_k}\,d\omega_k = \frac{\omega_c}{2\pi i}\frac{A_{ba}}{\omega_{ba}}\int_0^1 \frac{\kappa}{\kappa-\sigma}\,d\kappa$$

$$= \frac{\omega_c}{2\pi i}\frac{A_{ba}}{\omega_{ba}}\left\{1+\sigma\ln\left|\frac{1-\sigma}{-\sigma}\right| + i\sigma\arg\left(\frac{1-\sigma}{-\sigma}\right)\right\},$$

in which $\kappa = \omega/\omega_c$, $\sigma = is/\omega_c$. Make contact with the notation of (32.29) by setting

$$\delta(\sigma) = 1 + \sigma\ln\left|\frac{1-\sigma}{-\sigma}\right|; \qquad \gamma(\sigma) = \sigma\arg\left(\frac{1-\sigma}{-\sigma}\right).$$

There is again a branch cut in the s-plane, much as drawn in Fig. 32.4. Show that the cut runs from $\sigma = 0$ to $\sigma = 1$ (not to ∞) and just beside it $\gamma_{\text{right}} = \pi\sigma$, $\gamma_{\text{left}} = -\pi\sigma$.

Show that $d_b(t)$ is given by an expression very similar to (32.38):

$$d_b(t) = \int_0^{\omega_c} \frac{\Gamma(\sigma)/2\pi}{\left\{\omega-\omega_{ba}(1-\Delta)\right\}^2 + \Gamma(\sigma)^2/4}\,e^{-i\omega t}\,d\omega, \qquad (32.75)$$

in which $s = -i\omega$, $\sigma = \omega/\omega_c$ (real now). In (32.75) we have[92] (cf. (32.34)).

$$\Delta(\sigma) = \frac{\omega_c}{2\pi\omega_{ba}}\frac{A_{ba}}{\omega_{ba}}\delta(\sigma) = \frac{\omega_c}{2\pi\omega_{ba}}\frac{A_{ba}}{\omega_{ba}}\left\{1+\sigma\ln\left(\frac{1-\sigma}{\sigma}\right)\right\}.$$

This is of order α_{fs}^2, so shifts the resonance frequency ω_{ba} by a negligible amount, as expected.

Also in (32.75) (cf. (32.32)),

$$\Gamma(\sigma) = \frac{\omega_c}{2\pi}\frac{A_{ba}}{\omega_{ba}}(\gamma_{\text{right}}-\gamma_{\text{left}}) = \frac{A_{ba}}{\omega_{ba}}\omega_c\sigma = \frac{A_{ba}}{\omega_{ba}}\omega \quad \text{for } 0<\omega<\omega_c.$$

Adapt the reasoning of problem 32.13 to show that $\Gamma(\sigma)$ can be replaced by A_{ba} in the denominator of (32.75), giving

$$d_b(t) = \frac{A_{ba}}{2\pi\omega_{ba}}\int_0^{\omega_c} \frac{\omega}{(\omega-\omega_{ba})^2 + A_{ba}^2/4}\,e^{-i\omega t}\,d\omega.$$

Find a way of evaluating this integral, approximately,[93] and show that it yields

$$d_b(t) = e^{-i\omega_{ba}t}e^{-A_{ba}t/2}$$

plus errors of order α_{fs}^3 and $\alpha_{\text{fs}}^3\ln\alpha_{\text{fs}}$.

This finding is the same as (32.45), so we make no error (in excess of that already made) by replacing the "soft" cut-off $\left\{1+(\omega/\omega_c)^2\right\}^{-4}$ with a step-function cut-off at $\omega = \omega_c$.

Problem 32.16 The occupation $|c_a(t)|^2$ of the lower atomic state Include the function of time t in (32.48). Sum over the field modes to show that the occupation of the lower atomic state is[94]

$$|c_a(t)|^2 = 1 - e^{-A_{ba}t}, \qquad (32.76)$$

thereby making the total atomic population $|c_a(t)|^2 + |c_b(t)|^2 = 1$, as it should be.

Trace where errors involving α_{fs} affect the steps of this calculation, giving the $+O(\alpha_{\text{fs}}^2)$ of (32.53).

[92] It is no longer the case that $\Delta(\sigma)$ is guaranteed to be of order α_{fs}^2 for all σ, because $\ln(1-\sigma)$ goes logarithmically to $-\infty$ when σ approaches 1. Estimate the range of frequencies within which Δ is significant, and show that it is so narrow that Δ can be neglected after all. *Suggestion:* I suggest that Δ is "significant" if

$$\sigma\ln\left(\frac{1-\sigma}{\sigma}\right) \sim -\alpha_{\text{fs}}^{-2}.$$

See if you agree.

[93] No clues given, this time.

[94] *Suggestion:* Make use of equations (32.73) and (32.74).

Problem 32.17 A more general Ψ

The Ψ of (32.8) is not fully general. During the reaction, some probability may remain in ψ_b at the same time as some is in $|\boldsymbol{k}\rangle$. That requires Ψ to contain a new $c_{\boldsymbol{k}b}(t)\,|\boldsymbol{k}\rangle\,\psi_b(t)\,\mathrm{e}^{-\mathrm{i}(\omega_{\boldsymbol{k}}+\omega_b)t}$ and similarly a new $c_{0a}\,|0\rangle\,\psi_a(t)\,\mathrm{e}^{-\mathrm{i}\omega_a t}$. We show now that the omitted terms are small—small enough that their omission was permissible.

Write a fuller Ψ, replacing that of (32.8), as

$$\Psi = c_{0b}(t)\,|0\rangle\,\psi_b\,\mathrm{e}^{-\mathrm{i}\omega_b t} + \sum_{\boldsymbol{k},\,i\neq b} c_{\boldsymbol{k}i}(t)\,|\boldsymbol{k},\boldsymbol{e}_{\boldsymbol{k}}\rangle\,\psi_i\,\mathrm{e}^{-\mathrm{i}(\omega_{\boldsymbol{k}}+\omega_i)t}$$

$$+ \sum_{i\neq b} c_{0i}(t)\,|0\rangle\,\psi_i\,\mathrm{e}^{-\mathrm{i}\omega_i t} + \sum_{\boldsymbol{k}} c_{\boldsymbol{k}b}(t)\,|\boldsymbol{k},\boldsymbol{e}_{\boldsymbol{k}}\rangle\,\psi_b\,\mathrm{e}^{-\mathrm{i}(\omega_{\boldsymbol{k}}+\omega_b)t}, \quad (32.77)$$

in which the sums on the second line are new.

Write the Schrödinger equation that replaces (32.9) and extract from it the four equations that replace (32.14):

$$\dot{c}_{0b} = -\sum_{\boldsymbol{k},\,i\neq b} C_{ib}(\boldsymbol{k})^*\,c_{\boldsymbol{k}i}\,\mathrm{e}^{\mathrm{i}(-\omega_{\boldsymbol{k}}+\omega_{bi})t} - \sum_{\boldsymbol{k}} C_{bb}(\boldsymbol{k})^*\,c_{\boldsymbol{k}b}\,\mathrm{e}^{-\mathrm{i}\omega_{\boldsymbol{k}}t}; \quad (32.78)$$

$$\dot{c}_{\boldsymbol{k}i} = C_{ib}(\boldsymbol{k})\,c_{0b}\,\mathrm{e}^{\mathrm{i}(\omega_{\boldsymbol{k}}-\omega_{bi})t} + \sum_{j\neq b} C_{ij}(\boldsymbol{k})\,c_{0j}\,\mathrm{e}^{\mathrm{i}(\omega_{\boldsymbol{k}}+\omega_{ij})t}; \quad (32.79)$$

$$\dot{c}_{0j} = -\sum_{\boldsymbol{k},\,i\neq b} C_{ij}(\boldsymbol{k})^*\,c_{\boldsymbol{k}i}\,\mathrm{e}^{\mathrm{i}(-\omega_{\boldsymbol{k}}+\omega_{ji})t} - \sum_{\boldsymbol{k}} C_{bj}(\boldsymbol{k})^*\,c_{\boldsymbol{k}b}\,\mathrm{e}^{\mathrm{i}(-\omega_{\boldsymbol{k}}+\omega_{jb})t};$$

$$\qquad (32.80)$$

$$\dot{c}_{\boldsymbol{k}b} = C_{bb}(\boldsymbol{k})\,c_{0b}\,\mathrm{e}^{\mathrm{i}\omega_{\boldsymbol{k}}t} + \sum_{i\neq b} C_{bi}(\boldsymbol{k})\,c_{0i}\,\mathrm{e}^{\mathrm{i}(\omega_{\boldsymbol{k}}+\omega_{bi})t}. \quad (32.81)$$

Here $i\neq b$ and $j\neq b$ refer to any atomic states other than ψ_b; in particular, j is not now restricted to mean a state higher in energy than ψ_b.

Take Laplace transforms to get rid of the time derivatives, and solve between the transforms of eqns (32.78–79) to obtain

$$\widetilde{c_{0b}}(s)\left\{ s + \sum_{\boldsymbol{k},\,i} \frac{|C_{ib}(\boldsymbol{k})|^2}{s+\mathrm{i}\omega_{\boldsymbol{k}}-\mathrm{i}\omega_{bi}} \right\}$$

$$= 1 - \sum_{\boldsymbol{k},i,j} \frac{C_{ib}(\boldsymbol{k})^*\,C_{ij}(\boldsymbol{k})\,\widetilde{c_{0j}}(s+\mathrm{i}\omega_{jb})}{s+\mathrm{i}\omega_{\boldsymbol{k}}-\mathrm{i}\omega_{bi}} - \sum_{\boldsymbol{k}} C_{bb}(\boldsymbol{k})^*\,\widetilde{c_{\boldsymbol{k}b}}(s+\mathrm{i}\omega_{\boldsymbol{k}}).$$

$$\qquad (32.82)$$

Equation (32.82) replaces (32.16), introducing the two additional sums on the right, sums that contain the new coefficients $\widetilde{c_{0j}}$ and $\widetilde{c_{\boldsymbol{k}b}}$. The contributions from these new terms can be shown to be negligible ($\sim \alpha_{\mathrm{fs}}^4$ for the first and at most $\sim \alpha_{\mathrm{fs}}^2$ for the second). Investigate.[95]

[95] The information needed comes from processing eqns (32.80–81) in a manner similar to that which gave (32.82); the manipulation can be somewhat heavy. The clue as to orders of magnitude is in (32.29).

Problem 32.18 Another consistency check

The A_{ba} for the decay of atomic state b with radiation entering all possible radiation modes should be given by the sum over modes

$$A_{ba} = \int \frac{V\,\omega^2\,\mathrm{d}\omega}{2\pi^2 c^3} \int \frac{\mathrm{d}\Omega}{4\pi}\,\alpha_{ba\boldsymbol{k}}, \quad (32.83)$$

where $\alpha_{ba\boldsymbol{k}}$ is given by (32.50). Show from (32.74) that this is so.

Problem 32.19 The behaviour of $d_b(t)$ for long times
Equation (32.44) can be written in the form

$$d_b(t) = \int_{-\infty}^{\infty} \overline{d_b}(\omega)\,\mathrm{e}^{-\mathrm{i}\omega t}\,\mathrm{d}\omega,$$

in which

$$\overline{d_b}(\omega) = \begin{cases} \dfrac{A_{ba}/2\pi}{(\omega - \omega_{ba})^2 + A_{ba}^2/4}\,\dfrac{\omega}{\omega_{ba}}\,\dfrac{1}{\{1 + (\omega/\omega_c)^2\}^4} & \text{for } \omega > 0 \\[2ex] 0 & \text{for } \omega < 0. \end{cases}$$

Rehearse the Fourier-transform theorem which says that, if $\overline{d_b}(\omega)$ has a discontinuity at the origin[96] of height h_0, then $d_b(t)$ must fall for large $|t|$ as $h_0/(\mathrm{i}t)$. Likewise, if $\overline{d_b}(\omega)$ is a continuous function of ω but has a discontinuity h_1 in its first derivative, then $d_b(t)$ falls for large $|t|$ as $-h_1/t^2$.

The expression above for $\overline{d_b}(\omega)$ has a discontinuity h_1 in its first derivative at $\omega = 0$, given by

$$h_1 = \frac{A_{ba}}{2\pi\omega_{ba}}\,\frac{1}{\omega_{ba}^2 + A_{ba}^2/4} \approx \frac{A_{ba}}{2\pi\omega_{ba}^3}.$$

Therefore $d_b(t)$ must fall as $-h_1/t^2$ for large t.

Assume that $d_b(t)$ varies according to (32.45) as $\mathrm{e}^{-\mathrm{i}\omega_{ba}t}\mathrm{e}^{-A_{ba}t/2}$ until the t^{-2} regime takes over. We can estimate that this will happen when

$$\mathrm{e}^{-A_{ba}t/2} \approx \frac{A_{ba}}{2\pi\omega_{ba}^3}\frac{1}{t^2}.$$

Rearrange this into an equation for $x = A_{ba}t/2$ that can be solved iteratively. Using numerical values for A_{ba}/ω_{ba} for hydrogen 2p–1s from (32.42), I get $x = 62.6$, so that the changeover takes place when $A_{ba}t/2$ is of order 63 and $\mathrm{e}^{-A_{ba}t/2} \approx \mathrm{e}^{-63} \approx 7 \times 10^{-28}$. So late a departure from exponential decay is unobservable because there is by then nothing left of the upper-state population, and therefore no remaining emission of radiation.[97] Do your own solution and check whether you agree with my figures.

Problem 32.20 What happens at very early times
We ignore all atomic final states except for the dominant state $a = 1s$.

The occupation of final field states, per $\mathrm{d}\omega$, with photon propagation directions summed over, is given by the $|c_{\omega a}(t)|^2$ of (32.54).

(1) Expand the function of time $\{\ \}$ in (32.54) as far as t^2, and show that what multiplies t^2 cancels against the resonance denominator in $g(\omega - \omega_{ba})$, leaving

$$|c_{\omega a}(t)|^2 = t^2\,\frac{A_{ba}}{2\pi\omega_{ba}}\,\frac{\omega}{\{1 + (\omega/\omega_c)^2\}^4}. \tag{32.84}$$

For the frequencies displayed in Fig. 32.8, $\omega \approx \omega_{ba}$, so all curves there start with the same $|c_{\omega a}(t)|^2 \approx (A_{ba}/2\pi)t^2$, before each breaks away when its $\sin^2\frac{1}{2}(\omega - \omega_{ba})t$ ceases to approximate to $(\omega - \omega_{ba})^2 t^2/4$.

For $|c_a(t)|^2$, a t^2 rise holds when all contributing frequencies start as t^2. Equation (32.84) shows that these frequencies extend up to $\sim\omega_c$, so the

[96] The function of time contains a complex exponential if the singularity occurs at a value of ω away from $\omega = 0$.

[97] The changeover between exponential and t^{-2} takes place at a very late stage in the decay because $A_{ba}/\omega_{ba} \sim \alpha_{\mathrm{fs}}^3$.
The order α_{fs}^3 applies to electric-dipole transitions in atomic physics. A higher order applies to higher-order transitions: electric quadrupole, magnetic dipole, ... However, only one of the factors α_{fs} (in the electric-dipole case) comes from the radiation process, while the other two come from the fact that the atom is held together by electrical forces. In nuclear physics, where the nucleus is held together by the strong force, the order is only α_{fs} for an electric-dipole transition, and then a more careful discussion may be called for. It is no accident that attempted observations of such a changeover have been undertaken in a nuclear/particle-physics context.
I am indebted to Dr C.V. Sukumar for drawing my attention to these differences.
The t^{-2} behaviour of this problem, and the t^2 behaviour of problem 32.20, are very similar to the findings of Onley and Kumar (1992), though, for the reasons given above, those authors are interested in transitions involving nuclei.

quadratic rise lasts only as long as $\omega_c t \lesssim 1$. This very-short-duration quadratic start can be seen in Fig. 32.7.

(2) Take eqns (32.14) and set $\omega_c t \lesssim 1$ so that the complex exponentials $e^{\pm i(\omega_k - \omega_{ba})t} \approx 1$ for all contributing ω_k. Show that in this limit

$$c_{ka} \approx C_{ab}(k)t; \qquad c_b \approx 1 - \frac{t^2}{2} \sum_k |C_{ab}(k)|^2 . \tag{32.85}$$

From here or by integrating (32.84) over ω, show that

$$|c_a(t)|^2 = \frac{A_{ba}}{12\pi\omega_{ba}} (\omega_c t)^2 \quad \text{when } \omega_c t \lesssim 1. \tag{32.86}$$

(3) Discuss how the reasoning here is affected by errors propagating from the $\sim \alpha_{\rm fs}^2$ in (32.45) and shown as shifts in Fig. 32.7. Since everything on display in Fig. 32.7 is of order $\alpha_{\rm fs}^3$, can we trust those plots at all?

Problem 32.21 The occupation of higher-energy atomic states
States higher in energy than $b = 2p$ are labelled by index j. Return to eqn (32.14b):

$$\dot{c}_{kj} = c_b(t) C_{jb}(k) e^{i(\omega_k + \omega_{jb})t} .$$

Follow the route that led to (32.49), and show that the final occupation of state j, k, e_k is

$$|c_{kj}(\infty)|^2 = \frac{|C_{jb}(k)|^2}{(\omega_k + \omega_{jb})^2 + A_{ba}^2/4}. \tag{32.87}$$

Note that the denominator is off-resonance for all ω_k, giving us a good reason for expecting that the $|c_{kj}(\infty)|^2$ will turn out to be small.

The final atomic states j may belong to nd or ns—everything dipole-connected to $b = 2p$. This gives a range of possibilities[98] for $C_{jb}(k)$. Since we seek only the order of magnitude of a small quantity, we ignore all complications and copy (32.24)[99] with j replacing a:

$$|C_{jb}(k)|^2 = \frac{\pi c^3}{V} \frac{A_{jb}}{\omega_{jb}} \frac{1}{\omega_k} p(\vartheta, \varphi) f_{jb}(k). \tag{32.88}$$

Show that on this assumption[100]

$$|c_j(\infty)|^2 = \frac{A_{jb}}{2\pi\omega_{jb}} \int d\omega \frac{\omega}{(\omega_k + \omega_{jb})^2 + A_{jb}^2/4} f_{jb}(k).$$

Argue that $f_{jb}(k)$ cannot be very different from $f_{ab}(k)$, so we can reuse $f_{ab} = (1 + \kappa^2)^{-4}$ or replace f_{jb} by a step cut-off at ω_c, the same ω_c as before.[101]

Show from here that $|c_j(\infty)|^2 \sim \alpha_{\rm fs}^3 \ln(1/\alpha_{\rm fs})$ or smaller, and this holds whichever higher state j is considered.

Problem 32.22 Effect of higher states on the frequency shift Δ
(1) Return to (32.21) in which higher-energy atomic states (above b) are labelled by index j. Let the second sum in the denominator be called Σ_2:

$$\Sigma_2 = \sum_{k, \, j \neq a, b} \frac{|C_{jb}(k)|^2}{(s + i\omega_{ja}) + i\omega_k}.$$

---

[98] If the final level is nd then states dipole-connected to $|2p, m_J = -\frac{3}{2}\rangle$ may have $m_J = -\frac{1}{2}, -\frac{3}{2}, -\frac{5}{2}$. Transitions may have $\Delta m_J = +1, 0, -1$, all happening at once with a branching ratio. Only $\Delta m_J = +1$ yields radiation having the polarization e_k of (32.67).

[99] Here A_{jb} is the Einstein A coefficient for the spontaneous-emission transition $j \to b$. There is no such spontaneous emission here, but A_{jb} is a convenient measure of the strength of the transition $b \to j$ containing, as it does, the squared matrix element $|r_{jb}|^2$.

[100] As part of the simplification, I have used the same angular distribution $p(\vartheta, \varphi)$ for all transitions $b \to j$ and integrated over it according to (32.26).

[101] Hint: Compare the spatial extent of $\psi_j^* \, \hat{p} \, \psi_b$ with that of $\psi_a^* \, \hat{p} \, \psi_b$, and find a reason why these are not greatly different.

As in problem 32.21, the variety of final atomic states makes $|C_{jb}(\mathbf{k})|^2$ complicated. Fortunately, we expect Σ_2 to be small, so we may be brutal to it and follow the steps that led to (32.88). Use (32.88) in the sum for Σ_2.

$$\Sigma_2 \approx \frac{\omega_c}{2\pi\mathrm{i}} \sum_j \frac{A_{jb}}{\omega_{jb}} \int_0^\infty \frac{\kappa}{\kappa - (\sigma - \omega_{ja}/\omega_c)} f_{jb}(\mathbf{k}) \,\mathrm{d}\kappa.$$

In the last integral, all $f_{jb}(\mathbf{k})$ can be taken to be about the same, as argued in problem 32.21. We here continue use of $f_{jb}(\mathbf{k}) = (1 + \kappa^2)^{-4}$, so as to apply existing algebra. Then from (32.29)

$$\Sigma_2 = \frac{\omega_c}{2\pi} \sum_j \frac{A_{jb}}{\omega_{jb}} \left\{ -\mathrm{i}\delta(\sigma - \omega_{ja}/\omega_c) + \gamma(\sigma - \omega_{ja}/\omega_c) \right\}. \qquad (32.89)$$

(2) Refer back to (32.33). Sum Σ_2 is now added to the sum on the left, causing Δ and Γ to be augmented.[102] Show that

addition to $\Delta = \dfrac{\omega_c}{2\pi\omega_{ba}} \displaystyle\sum_j \frac{A_{jb}}{\omega_{jb}} \delta(\sigma - \omega_{ja}/\omega_c);$ \hfill (32.90)

addition to $\Gamma = \omega_c \displaystyle\sum_j \frac{A_{jb}}{\omega_{jb}} \frac{\sigma - \omega_{ja}/\omega_c}{\left\{1 + (\sigma - \omega_{ja}/\omega_c)^2\right\}^4}.$ \hfill (32.91)

Combining (32.90) with (32.35), we have

$$\Delta = \frac{\omega_c}{2\pi\omega_{ba}} \left(\frac{A_{ba}}{\omega_{ba}} \delta(\sigma) + \sum_{j \neq a} \frac{A_{jb}}{\omega_{jb}} \delta(\sigma - \omega_{ja}/\omega_c) \right).$$

Figure 32.6 shows that $\delta(\sigma)$ may have either sign, depending on the value of σ, but in any event $|\delta| < 1$, so we may find an upper bound for $|\Delta|$. Show that for 2p–1s

$$|\Delta| < \frac{\omega_c}{2\pi\omega_{ba}} \left(\frac{A_{ba}}{\omega_{ba}} + \sum_{j \neq a} \frac{A_{jb}}{\omega_{jb}} \right) \qquad (32.92)$$

$$= \frac{8}{3\pi c^2} \left(\omega_{ba}^2 \,|\mathbf{r}_{ab}|^2 + \sum_{j \neq a} \omega_{jb}^2 \,|\mathbf{r}_{jb}|^2 \right).$$

There is an infinite number of higher-energy atomic states j dipole-connected to the initial state $\psi_b = \psi_{2\mathrm{p}}$, and the sum here is taken over all such states, including unbound states above the energy of atomic ionization. We even need to check whether the sum over j is convergent. Here's my way through, for you to check or improve. We have the sum rule:[103]

$$\sum_i \omega_{ib} \,|\mathbf{r}_{ib}|^2 = \frac{\hbar}{2m_e}. \qquad (32.93)$$

In the sum of (32.93) we must separate out state a from all the others because ω_{ab} is negative while all other ω_{ib} are positive. Show that[104]

[102] Equation (32.31) shows that $\gamma(\sigma - \omega_{ja}/\omega_c)$ takes different values on each side of a branch cut, but starting at $\sigma = \omega_{ja}/\omega_c$. It follows that the addition to Γ in (32.91) applies for $\sigma - \omega_{ja}/\omega_c > 0$ only.

The branch cut does not need to be modified, since the original γ still has a cut from $\sigma = 0$ to $\sigma = \infty$.

[103] See e.g. Woodgate (1980) § 3.3 and problem 3.5; Corney (1977) §§ 4.8 and 4.8.1. The rule given is a version of the f-sum rule.

The sum over i in (32.93) encompasses all states, including $a = 1\mathrm{s}$, that are dipole connected to state $b = 2\mathrm{p}$.

The fact that the sum in (32.93) is convergent goes a long way to reassuring us that the sum in $|\Delta|$ may be convergent also.

[104] All bound states have $\omega_{jb} < \omega_{\infty b}$ but this is not the case for unbound states (above the atomic ionization level). Therefore the first step introduces an approximation, not an upper bound.

$$|\Delta| \approx \frac{8}{3\pi c^2} \left(\omega_{ba}^2 |\mathbf{r}_{ab}|^2 + \omega_{\infty b} \sum_{j \ne a} \omega_{jb} |\mathbf{r}_{jb}|^2 \right)$$

$$= \frac{8}{3\pi c^2} \left\{ \omega_{ba}^2 |\mathbf{r}_{ab}|^2 + \omega_{\infty b} \left(\frac{\hbar}{2m_e} + \omega_{ba} |\mathbf{r}_{ab}|^2 \right) \right\}$$

$$= \frac{8}{3\pi c^2} \omega_{ba}^2 |\mathbf{r}_{ab}|^2 \left\{ 1 + \left(\frac{3^8}{2^{13}} + \frac{1}{3} \right) \right\} = \frac{8}{3\pi c^2} \omega_{ba}^2 |\mathbf{r}_{ab}|^2 (1 + 1.13)$$

$$= \frac{2^{12}}{3^9 \pi} (2.13) \alpha_{\text{fs}}^2 = 0.14 \alpha_{\text{fs}}^2.$$

Backtrack to show that (compare eqn 32.41)

$$|\Delta| \approx \frac{\omega_c}{2\pi \omega_{ba}} \frac{A_{ba}}{\omega_{ba}} \times 2.13.$$

The radiation-induced shift $-\omega_{ba}\Delta$ of the resonance frequency remains negligible, only doubled compared with that found in (32.41).

(3) Backtrack to (32.92) and show that

$$\sum_{j \ne a} \frac{A_{jb}}{\omega_{jb}} \sim \frac{2\pi \omega_{ba}}{\omega_c} |\Delta| - \frac{A_{ba}}{\omega_{ba}} \sim \alpha_{\text{fs}}^3.$$

Argue that, since every term in this sum is positive, A_{jb}/ω_{jb} must be of order α_{fs}^3 or smaller for all states j.

Problem 32.23 Effect of higher states on $d_b(t)$
Show from (32.91) that the contribution of higher states j to Γ is

$$\left. \begin{array}{ll} \displaystyle \sum_j \frac{A_{jb}}{\omega_{jb}} \frac{\omega - \omega_{ja}}{\left\{ 1 + (\omega - \omega_{ja})^2/\omega_c^2 \right\}^4} & \text{for } \omega - \omega_{ja} > 0 \\[12pt] 0 & \text{otherwise.} \end{array} \right\} \qquad (32.94)$$

This addition to Γ carries through to an addition to $d_b(t)$ in (32.38) Show that the addition to $d_b(t)$ is

$$\int_{\omega_{ja}}^{\infty} \frac{d\omega}{2\pi} \frac{(\text{addition to } \Gamma)}{(\omega - \omega_{ba})^2 + A_{ba}^2/4} e^{-i\omega t}.$$

[105] *Suggestion:* Find a reason why $A_{ba}^2/4$ can be neglected. Be prepared to apply a step cut-off to the integral at $\omega = \omega_c + \omega_{ja}$. Be prepared then to find an upper bound, such as the value at time $t = 0$.

Show[105] that this is of order $\alpha_{\text{fs}}^2 \ln(1/\alpha_{\text{fs}})$.

The small occupation that enters atomic states higher in energy than $E_b = \hbar\omega_b$ is seen to have negligible effect on the occupation of the upper state $b = 2\text{p}$, and therefore on its decay with time.

Problem 32.24 Golden Rule Number Two
To calculate: the early occupation probability $|c_a(t)|^2$ of the lower atomic state, during spontaneous emission of electromagnetic radiation.

By "early" we now mean times during which the upper-atomic-state is not significantly depleted: $A_{ba}t \ll 1$, but $\omega_{ba}t \gg 1$ so not the very-early times of problem 32.20. Then (32.45) gives $c_b = e^{-A_{ba}t/2} \approx 1$.

[106] We drop the subscript $\mathbf{k}$ in $\omega_{\mathbf{k}}$ to avoid clutter.

Show that, within this approximation, (32.48) becomes[106]

$$|c_{\mathbf{k}a}(t)|^2 = |C_{ab}(\mathbf{k})|^2 \times \frac{4\sin^2 \frac{1}{2}(\omega - \omega_{ba})t}{(\omega - \omega_{ba})^2 + A_{ba}^2/4}. \qquad (32.95)$$

The quantity of interest is $|c_a(t)|^2$, the occupation of the final atomic state a, with no interest taken in where the photon goes, in frequency, direction or polarization. Therefore we require[107]

$$|c_a(t)|^2 = \sum_{\text{modes}} |c_{\boldsymbol{k}a}(t)|^2 = V \int_0^\infty \frac{\mathrm{d}\omega\,\omega^2}{2\pi^2 c^3} \int \frac{\mathrm{d}\Omega}{4\pi}\,|c_{\boldsymbol{k}a}(t)|^2.$$

We shall pay particular attention to the powers of ω appearing in the integrand.

The "density of states" $\rho(\omega)$ for photons is defined by[108]

$$\rho(\omega)\,\mathrm{d}(\hbar\omega) = \frac{\omega^2\,\mathrm{d}\omega}{2\pi^2 c^3},$$

so

$$\frac{\omega^2}{2\pi^2 c^3} = \hbar\rho(\omega) = \hbar\rho(\omega_{ba})\,\frac{\omega^2}{\omega_{ba}^2}.$$

We make a slight change to the notation in (32.10) by writing $H'_{ab}(\boldsymbol{k})$ for the integral $\int \psi_a^* \langle \boldsymbol{k}|\widehat{H}'|0\rangle\,\psi_b\,\mathrm{d}V$ so that

$$C_{ab}(\boldsymbol{k}) = \frac{-\mathrm{i}}{\hbar}\,H'_{ab}(\boldsymbol{k}).$$

We also define H^2 by

$$H^2 = \left(V \int \frac{\mathrm{d}\Omega}{4\pi}\,|H'_{ab}(\boldsymbol{k})|^2 \right)_{\omega=\omega_{ba}}.$$

From (32.24),

$$V \int \frac{\mathrm{d}\Omega}{4\pi}\,|H'_{ab}(\boldsymbol{k})|^2 = \hbar^2 \int \frac{\mathrm{d}\Omega}{4\pi} V\,|C_{ab}(\boldsymbol{k})|^2 = \pi\hbar^2 c^3\,\frac{A_{ba}}{\omega_{ba}}\,\frac{f_{ab}(\boldsymbol{k})}{\omega}$$

$$= H^2 \times \frac{\omega_{ba}}{\omega}\,\frac{f_{ab}(\boldsymbol{k})}{\{f_{ab}(\boldsymbol{k})\}_{\omega=\omega_{ba}}} \approx H^2\,\frac{\omega_{ba}}{\omega}\,f_{ab}(\boldsymbol{k}).$$

Assemble pieces and show that

$$|c_a(t)|^2 = H^2\,\frac{\rho(\omega_{ba})}{\hbar\omega_{ba}} \int_0^\infty \mathrm{d}\omega\,\omega\,\frac{4\sin^2\frac{1}{2}(\omega-\omega_{ba})t}{(\omega-\omega_{ba})^2 + A_{ba}^2/4}\,f_{ab}(\boldsymbol{k}). \quad (32.96)$$

Trace the powers of ω that have contributed to (32.96), in particular at high frequencies: ω^2 from the density of states, ω^{-1} from the vector potential in $|H'_{ab}|^2$, ω^{-2} in the resonance denominator, and finally the frequency dependence in $f_{ab}(\boldsymbol{k})$. If we set aside f_{ab}, for large ω these give $\omega^2 \times \omega^{-1} \times \omega^{-2} \propto \omega^{-1}$, and the integral in (32.96) is divergent—not a surprise given what we have seen before, but a warning that care is needed.

First, we show in Fig. 32.12 that the $A_{ba}^2/4$ in the denominator of (32.96) can be neglected. Its effect is to impose a "notch" in the integrand, but that notch is narrow enough to be ignored when $A_{ba}t$ is small, a condition we set at the beginning of this problem.

It should suffice to replace $f_{ab}(\boldsymbol{k})$ with a step cut-off at $\omega = \omega_{\mathrm{c}}$, so

$$|c_a(t)|^2 = \frac{4}{\hbar}\,H^2\,\frac{\rho(\omega_{ba})}{\omega_{ba}} \int_0^{\omega_{\mathrm{c}}} \mathrm{d}\omega\,\omega\,\frac{\sin^2\frac{1}{2}(\omega-\omega_{ba})t}{(\omega-\omega_{ba})^2}. \quad (32.97)$$

Write $\omega = \omega_{ba} + (\omega-\omega_{ba})$. Then the integral in (32.97) is

$$\omega_{ba} \int_0^{\omega_{\mathrm{c}}} \mathrm{d}\omega\,\frac{\sin^2\frac{1}{2}(\omega-\omega_{ba})t}{(\omega-\omega_{ba})^2} + \int_0^{\omega_{\mathrm{c}}} \mathrm{d}\omega\,\frac{\sin^2\frac{1}{2}(\omega-\omega_{ba})t}{(\omega-\omega_{ba})}. \quad (32.98)$$

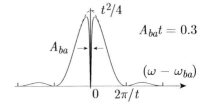

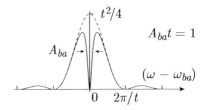

Fig. 32.12 The function
$$\frac{\sin^2\frac{1}{2}(\omega-\omega_{ba})t}{(\omega-\omega_{ba})^2 + A_{ba}^2/4}$$
drawn for two values of A_{ba}. The broken line shows the same function with A_{ba} set to zero.

When $A_{ba}t \ll 1$ the notch is so narrow that the curve approximates to the broken line. and the area under it is

$$\frac{t}{2} \int_{-\omega_{ba}/2}^{(\omega_{\mathrm{c}}-\omega_{ba})t/2} \mathrm{d}x\,\frac{\sin^2 x}{x^2}$$
$$\approx \frac{t}{2} \int_{-\infty}^{\infty} \mathrm{d}x\,\frac{\sin^2 x}{x^2} = \frac{\pi t}{2}.$$

Our processing of (32.97) differs from that in every book I have consulted. Usually two errors are made that happen to compensate for each other. First, it is said (explicitly or implicitly) that

$$\frac{4\sin^2\frac{1}{2}(\omega-\omega_{ba})t}{(\omega-\omega_{ba})^2} \approx \frac{\pi}{2}\,t\,\delta(\omega-\omega_{ba}).$$

Yet the $(\omega-\omega_{ba})^{-2}$ does not fall fast enough at large ω to overcome the ω^2 in $\rho(\omega)$; it "fails to focus" attention on frequencies close to ω_{ba}, making the δ-function an inadmissible replacement. Second, $f_{ab}(\mathbf{k})$ is omitted (replaced by 1); without it the δ-function is even more inadmissible (if that's possible) since the integration of (32.96) is then divergent.

Show that when $\omega_{ba}t \gg 1$ the first integral in (32.98) gives $(\pi/2)\omega_{ba}t$, so that it yields

$$\frac{|c_a(t)|^2}{t} = \frac{2\pi}{\hbar}\left(V\int\frac{d\Omega}{4\pi}\,|H'_{ab}|^2\right)_{\omega=\omega_{ba}}\rho(\omega_{ba}), \qquad (32.99)$$

which is our statement of GR2.

Refer back to (32.53) The right-hand side of (32.99) is nothing new: just A_{ba} separated into building blocks.[109]

Can you see the second term in (32.98) as contributing to the glitch of Fig. 32.7?

Comments on result (32.99)

The "squared matrix element" H^2 has had to be defined as an average over the emission directions of the outgoing photon (and with a V if that hasn't been catered for elsewhere). Once that is recognized, H^2 is defined just as we would wish when understanding GR2.

Remember that the density of states $\rho(\omega_{ba})$ has been defined to count only half of the possible field states, only those whose polarization entitles them to be excited by the atom's dipole moment.

With these understandings, we may say that GR2 correctly describes the emission of a photon by a radiating atom.

However, some things have happened that we might not have expected before reading the present chapter.

The integration in (32.96) threatens to diverge because the density of states rises with ω as ω^2. The integral is prevented from diverging by a fall-off $f_{ab}(\mathbf{k})$ in the matrix element at large ω, a fall-off that traces back to the $e^{-i\mathbf{k}\cdot\mathbf{r}}$ in (32.11). Physically, the source $\psi_a^*\,\hat{\mathbf{p}}\,\psi_b$ driving the radiation has a finite spatial extent, which makes $e^{-i\mathbf{k}\cdot\mathbf{r}}$ cancel against itself for large ω, when integrated over the volume of the source.

Investigations of particle-emission reactions usually claim that the shape of the integrand in (32.96) means that the integration over ω is "focused" on a small range of frequencies around ω_{ba}. This has turned out to be the case, though in an unusual way given that $f_{ab}(\mathbf{k})$ is needed to ensure convergence—via the ω_c in the second integral of (32.98).

We should ask whether a high-frequency fall-off of the matrix element is a save-the-day special to photon emission, or whether it is needed by all reactions in which a particle or particles is emitted. I claim that all such reactions are equally needy. Whether or not a sum over final states is convergent depends upon the behaviour of the summand at high energies. At energies high enough to endanger convergence, *all* emitted particles are relativistic, and their density of states varies as ω^2, as here. Convergence then relies upon a fall-off of the matrix element with ω, a fall-off that we trace back to $e^{-i\mathbf{k}\cdot\mathbf{r}}$ integrated over a finite-sized source. Given relativistic behaviour of the emitted particle(s), something similar always applies. Turning this round: if we've killed off the infinity for photon emission, we've done it for everything.

Bibliography

Abramowitz, M. and Stegun, I. A. (1965). *Handbook of mathematical functions*. Dover, New York. Previously published by National Bureau of Standards in 1964.

Accad, Y., Pekeris, C. L. and Schiff, B. (1971). S and P states of the helium isoelectronic sequence up to $Z = 10$. *Physical Review A*, **4**, 516–536.

Adkins, C. J. (1983). *Equilibrium thermodynamics*, third edn. Cambridge University Press, Cambridge.

Ashcroft, N. W. and Mermin, N. D. (1976). *Solid state physics*. Saunders, Philadelphia.

Bardeen, J., Cooper, L. N. and Schrieffer, J. R. (1957). Theory of Superconductivity. *Physical Review*, **108**, 1175–1204.

Barnett, S. M. and Radmore, P. M. (1997). *Methods in theoretical quantum optics*. Oxford University Press, Oxford.

Baumwolspiner, M. (1972). Stability considerations in nonlinear feedback structures as applied to active networks. *Bell System Technical Journal*, **51**, 2029–2063.

Blakemore, J. S. (1974). *Solid state physics*, second edn. Saunders, Philadelphia.

Bleaney, B. I. and Bleaney, B. (2013). *Electricity and magnetism*, third edn. Oxford University Press, Oxford. Reissued, with corrections, as an Oxford Classic Text, 2013, from the original third ed., 1976.

Bode, H. W. (1945). *Network analysis and feedback amplifier design*. Van Nostrand, NY. subsequently reprinted by R. E. Krieger, Huntington, NY, 1975.

Born, M. and Wolf, E. (1999). *Principles of optics*, seventh edn. Cambridge University Press, Cambridge.

Bouwkamp, C. J. (1954). Diffraction theory. *Reports on progress in physics*, **17**, 35–100.

Bransden, B. H. and Joachain, C. J. (2003). *Physics of atoms and molecules*, second edn. Prentice Hall (Pearson Education), Harlow, Essex.

Brillouin, L. (1953). *Wave propagation in periodic structures: electric filters and crystal lattices*, second edn. Dover, New York.

Brooker, G. A. (2006). *Modern classical optics*. Oxford University Press, Oxford. Reprint, with corrections, of original 2003.

Brooker, G. A. (2008). Diffraction at an ideally conducting slit. *Journal of Modern Optics*, **55**, 423–445.

Budker, D., Kimball, D. F. and DeMille, D. P. (2008). *Atomic physics: an exploration through problems and solutions*, second edn. Oxford University Press, Oxford.

Cagnac, B. and Pebay-Peyroula, J.-C. (1975). *Modern atomic physics II: quantum theory and its applications*. Macmillan, London.

Cohen-Tannoudji, C., Diu, B. and Laloë, F. (1977). *Quantum mechanics* Vol. 1. Wiley, New York.

Cohen-Tannoudji, C., Dupont-roc, J. and Grynberg, G. (1992). *Atom photon interactions: basic processes and applications*. John Wiley & Sons, New York.

Condon, E. U. and Shortley, G. H. (1951). *The theory of atomic spectra* Cambridge University Press, Cambridge. Reprint with corrections of 1935 printing.

Corney, A. (1977). *Atomic and laser spectroscopy*. Oxford University Press, Oxford.

Cottingham, W. N. and Greenwood, D. A. (1991). *Electricity and magnetism*. Cambridge University Press, Cambridge.

Davidovich, L. and Nussenzveig, H. M. (1980). Theory of natural line shape. In *Foundations of radiation theory and quantum electroynamics* (ed. A. O. Barut), pp. 83–108. Plenum Press, New York.

Davis, C. C. (1996). *Lasers and electro-optics*. Cambridge University Press, Cambridge.

Dekker, A. J. (1958). *Solid state physics*. Macmillan, London.

Drake, G. W. F. (1996). High precision calculations for helium. In *Atomic, molecular & optical physics handbook* (ed. G. W. F. Drake) pp. 154–171. American Institute of Physics, Woodbury, NY.

Dugdale, J. S. (1996). *Entropy and its physical meaning*. Taylor and Francis, London.

Eastham, D. A. (1986). *Atomic physics of lasers*. Taylor and Francis London.

Eisberg, R. M. (1961). *Fundamentals of modern physics*. Wiley, New York.

Eisberg, R. M. and Resnick, R. (1985). *Quantum physics of atoms molecules, solids, nuclei and particles*, second edn. Wiley, New York.

Enge, H. A. (1966). *Introduction to nuclear physics*. Addison-Wesley Reading, Massachusetts.

Feynman, R. P., Leighton, R. B. and Sands, M. (1964). *The Feynman lectures on physics*. Addison-Wesley, Reading, Massachusetts.

Finn, C. B. P. (1993). *Thermal physics*, second edn. Chapman and Hall London.

Fock, V. A. (1964). *The theory of space time and gravitation*, second edn. Pergamon, Oxford.

Foot, C. J. (2005). *Atomic physics*. Oxford University Press, Oxford.

Fox, M. (2006). *Quantum optics*. Oxford University Press, Oxford.

French, A. P. (1968). *Special relativity*. Nelson, London.

Guénault, A. M. (Tony) (1995). *Statistical physics*, second edn. Chapman and Hall, London.

Haken, H. and Wolf, H. C. (2005). *The physics of atoms and quanta: introduction to experiments and theory*, seventh edn. Springer-Verlag, Berlin.

Heitler, W. (1954). *The quantum theory of radiation*, third edn. Oxford University Press, Oxford.

Herzberg, G. (1944). *Atomic spectra and atomic structure*, second edn. Dover, New York.

Hook, J. R. and Hall, H. E. (1991). *Solid state physics*, second edn. Wiley, Chichester.

Jackson, J. D. (1999). *Classical electrodynamics*, third edn. Wiley, New York.

Joos, G. (1934). *Theoretical physics*. Blackie, London.

Keen, B. E., Matthews, P. W. and Wilks, J. (1965). The acoustic impedance of liquid helium-3. *Proceedings of the Royal Society*, **A284**, 125–136.

Kelkar, N. G. and Nowakowski, M. (2010). No classical limit of quantum decay for broad states. *J. Phys. A: Math. Theor.*, **43**, 385308 (1–9).

Kittel, C. (1963). *Quantum theory of solids*. Wiley, New York.

Kittel, C. (2005). *Introduction to solid state physics*, eighth edn. Wiley, Hoboken, N.J.

Kleppner, D., Littman, M. G. and Zimmerman, M. L. (1981). Highly excited atoms. *Scientific American*, **244** (May), 108–122.

Koppelmann, G. (1969). Multiple-beam interference and natural modes in open resonators. In *Progress in optics* (ed. E. Wolf), vol. VII, pp. 3–66. North Holland, Amsterdam.

Kramida, A., Ralchenko, Yu., Reader, J. and NIST ASD Team. (2021). NIST Atomic Spectra Database (ver. 5.8) [online]. Available: `https://physics.nist.gov/PhysRefData/ASD/levels_form.html` [2021, May 4]. National Institute of Standards and Technology, Gaithersburg, MD.

Kuhn, H. G. (1969). *Atomic spectra*, second edn. Longman, London.

Landau, L. D. and Lifshitz, E. M. (1986). *Theory of elasticity*, third edn. Course of theoretical physics, vol. 7. Butterworth-Heinemann, Oxford. Formerly published by Pergamon, Oxford.

Landau, L. D. and Lifshitz, E. M. (1991). *Quantum mechanics*, third edn. Course of theoretical physics, vol. 3. Butterworth Heinemann, Oxford. Formerly published by Pergamon, Oxford.

Landau, L. D. and Lifshitz, E. M. (1996). *The classical theory of fields*, fourth edn. Course of theoretical physics, vol. 2. Butterworth Heinemann, Oxford. Formerly published by Pergamon, Oxford.

Landau, L. D., Lifshitz, E. M. and Pitaevskiĭ, L. P. (1993). *Electrodynamics of continuous media*, second edn. Course of theoretical physics, vol. 8. Butterworth-Heinemann, Oxford. Formerly published by Pergamon, Oxford.

Landsberg, P. T. (1978). *Thermodynamics and statistical mechanics*. Oxford University Press, Oxford.

Leighton, R. B. (1959). *Principles of modern physics*. McGraw Hill, New York.

Lifshitz, E. M. and Pitaevskiĭ, L. P. (1995). *Physical kinetics*. Landau–Lifshitz course of theoretical physics, vol. 10. Butterworth-Heinemann, Oxford. Formerly published by Pergamon, Oxford.

Liverts, E. Z. and Barnea, N. (2011). S-states of helium-like ions. *Computer Physics Communications*, **182**, 1790–1795.

Lorrain, P. and Corson, D. R. (1970). *Electromagnetic fields and waves*, second edn. W. H. Freeman, San Francisco.

Madelung, O. (ed) (1996). *Semiconductors—basic data*, second revised edn. Springer, Berlin.

Mandl, F. (1988). *Statistical physics*, second edn. Manchester Physics Series. Wiley, Chichester, UK.

Messiah, A. (1965). *Quantum mechanics*. Vol. 2. North-Holland, Amsterdam.

Moore, C. E. (1949). *Atomic energy levels: Circular of the NBS 467*. National Bureau of Standards, Washington.

Newman, F. G. and Searle, V. H. L. (1957). *The general properties of matter*, fifth edn. Arnold, London.

Nilsson, G. and Rolandson, S. (1973). Lattice dynamics of copper at 80 K. *Physical Review B*, **7B**, 2393–2400.

NIST. (2021). `https://physics.nist.gov/PhysRefData/ASD/levels_form.html`. See also Kramida *et al.*

O'Brien, Flann. [O'Nolan, Brian.] (1974). *The third policeman*. Picador edition by Pan Books, London.

Onley, D. S. and Kumar, A. (1992). Time dependence in quantum mechanics—Study of a simple decaying system. *American journal of physics*, **60**, 432–439.

Pais, A. (1982). *Subtle is the Lord*. Oxford University Press, Oxford.

Pekeris, C. L. (1959). $1\,^1$S and $2\,^3$S states of helium. *Physical Review*, **115**, 1216–1221.

Pershan, P. S. (1963). Nonlinear optical properties of solids: energy considerations. *Physical Review*, **130**, 919–929.

Pines, D. and Nozières, P. (1966). *The theory of quantum liquids*. Vol. I, normal Fermi liquids. W. A. Benjamin Inc., New York.

Pippard, A. B. (1964). *The elements of classical thermodynamics*. Cambridge University Press, Cambridge.

Pointon, A. J. (1967). *Introduction to statistical physics*. Longman, London.

Rae, A. I. M. and Napolitano, J. (2016). *Quantum mechanics*, 6th edn. CRC Press, part of the Taylor & Francis Group, Boca Raton.

Reif, F. (1965). *Fundamentals of statistical and thermal physics*. McGraw Hill, Auckland.

Richtmyer, F. K., Kennard, E. H. and Cooper, J. N. (1969). *Introduction to modern physics*, sixth edn. McGraw Hill, New York.

Riedi, P. C. (1988). *Thermal physics*, second edn. Oxford University Press, Oxford.

Rindler, W. (2001). *Relativity: special, general, and cosmological*. Oxford University Press, Oxford.

Roberts, J. K. and Miller, A. R. (1954). *Heat and thermodynamics*, fourth edn. Blackie, London.

Robinson, F. N. H. (1973*a*). *Electromagnetism*. Oxford University Press, Oxford.

Robinson, F. N. H. (1973*b*). *Macroscopic electromagnetism*. Pergamon, Oxford.

Robinson, F. N. H. (1974). *Noise and fluctuations in electronic devices and circuits*. Oxford University Press, Oxford.

Robinson, F. N. H. (1995). *An introduction to special relativity and its applications*. World Scientific, Singapore.

Rosenberg, H. M. (1988). *The solid state*, third edn. Oxford University Press, Oxford.

Royal Society (1975). *Quantities, units, and symbols*, second edn. The Royal Society, London.

Saha, M. N. and Srivastava, B. N. (1950). *A treatise on heat*, third edn. Indian Press, Allahabad.

Schiff, B., Lifson, H., Pekeris, C. L. and Rabinowitz, P. (1965). $2\,^{1,3}$P, $3\,^{1,3}$P, and $4\,^{1,3}$P states of He and the $2\,^1$P state of Li$^+$. *Physical Review A*, **140**, A1104–A1121.

Schiff, L. I. (1968). *Quantum mechanics*, third edn. McGraw Hill, New York.

Silfvast, W. T. (1996). *Laser fundamentals*. Cambridge University Press, Cambridge.

Singleton, J. (2001). *Band theory and electronic properties of solids*. Oxford University Press, Oxford.

Slater, J. C. (1928). Central fields and Rydberg formulas in wave mechanics. *Physical Review*, **31**, 333–343.

Slater, J. C. (1929). The theory of complex spectra. *Physical Review*, **34**, 1293–1322.

Slepian, J. (1942). Energy and energy flow in the electromagnetic field. *Journal of Applied Physics*, **13**, 512–518.

Smith, G. and Tomkins, F. S. (1976). The absorption spectrum of europium. *Philosophical transactions of the Royal Society of London. Series A, Mathematical and physical sciences*, **283**, 345–365.

Sonntag, R. E. and Van Wylen, G. J. (1971). *Introduction to thermodynamics: classical and statistical*. Wiley, New York.

Svelto, O. (1998). *Principles of lasers*, fourth edn. Plenum Press, New York.

Svensson, E. C., Brockhouse, B. N. and Rowe, J. M. (1967). Crystal dynamics of copper. *Physical Review*, **155**, 619–32.

Thorne, A. P., Litzén, U. and Johansson, S. (1999). *Spectrophysics: principles and applications*. Springer, Berlin.

Turner, R. E. and Betts, D. S. (1974). *Introductory statistical mechanics*. Sussex University Press.

van der Straten, P. and Metcalf, H. J. (2016). *Atoms and molecules interacting with light*. Cambridge University Presss, Cambridge.

Whitworth, R. W. and Stopes-Roe, H. V. (1971). Experimental demonstration that the couple on a bar magnet depends on H, not B *Nature*, **234**, 31–33.

Wilks, J. (1961). *The third law of thermodynamics*. Oxford University Press, Oxford.

Willmott, J. C. (1995). *Atomic physics*. John Wiley, London.

Woodgate, G. K. (1980). *Elementary atomic structure*, second edn Oxford University Press, Oxford.

Yariv, A. (1997). *Optical electronics in modern communications*, fifth edn. Oxford University Press, Oxford.

Zemansky, M. W. and Dittman, R. H. (1981). *Heat and thermodynamics*, sixth edn. McGraw Hill, New York.

Ziman, J. M. (1960). *Electrons and phonons*. Oxford University Press Oxford.

Index

461